THE LIQUID CRYSTALS BOOK SERIES

NEMATIC AND CHOLESTERIC LIQUID CRYSTALS

CONCEPTS AND PHYSICAL PROPERTIES ILLUSTRATED BY EXPERIMENTS

THE LIQUID CRYSTALS BOOK SERIES

Edited by
G.W. GRAY, J.W. GOODBY & A. FUKUDA

The Liquid Crystals book series publishes authoritative accounts of all aspects of the field, ranging from the basic fundamentals to the forefront of research; from the physics of liquid crystals to their chemical and biological properties; and, from their self-assembling structures to their applications in devices. The series will provide readers new to liquid crystals with a firm grounding in the subject, while experienced scientists and liquid crystallographers will find that the series is an indispensable resource.

PUBLISHED TITLES

Introduction to Liquid Crystals: Chemistry and Physics
By Peter J. Collings and Michael Hird

The Static and Dynamic Continuum Theory of Liquid Crystals:
A Mathematical Introduction
By Iain W. Stewart

Crystals that Flow: Classic Papers from the History of Liquid Crystals
Compiled with translation and commentary by Timothy J. Sluckin, David A. Dunmur, Horst Stegemeyer

THE LIQUID CRYSTALS BOOK SERIES

NEMATIC AND CHOLESTERIC LIQUID CRYSTALS

CONCEPTS AND PHYSICAL PROPERTIES ILLUSTRATED BY EXPERIMENTS

Patrick Oswald
Pawel Pieranski

TRANSLATED BY Doru Constantin

With the support of Merck KgaA (LC division), Rolic Research Ltd, and the Laboratoire de Physique des Solides d'Orsay

Taylor & Francis
Taylor & Francis Group

Boca Raton London New York Singapore

A CRC title, part of the Taylor & Francis imprint, a member of the Taylor & Francis Group, the academic division of T&F Informa plc.

Cover Illustration: "Cholesterics Dance." A cholesteric liquid crystal can become partially unwound, forming isolated fingers (solitons) when placed between two surfaces treated to ensure homeotropic anchoring (molecules perpendicular to the surfaces). Under the action of an AC electric field, some fingers (those endowed with a suitable symmetry) start to drift and assume the shape of an Archimedes spiral. This photograph, taken by Slavomir Pirkl, shows a unique arrangement composed of three such spirals centered on the vertices of an equilateral triangle. This photograph (shown here with color added) was taken under the microscope between crossed polarizers.

Published in 2005 by
CRC Press
Taylor & Francis Group
6000 Broken Sound Parkway NW
Boca Raton, FL 33487-2742

Printed in the United States of America on acid-free paper
10 9 8 7 6 5 4 3 2 1

International Standard Book Number-10: 0-415-32140-9 (Hardcover)
International Standard Book Number-13: 978-0-415-32140-2 (Hardcover)

Library of Congress Cataloging-in-Publication Data

Catalog record is available from the Library of Congress

T&F informa

Taylor & Francis Group
is the Academic Division of T&F Informa plc.

Visit the Taylor & Francis Web site at
http://www.taylorandfrancis.com

and the CRC Press Web site at
http://www.crcpress.com

CONTENTS

Nematic and Cholesteric Liquid Crystals

PART B: MESOPHASES WITH AN ORIENTATIONAL ORDER

Smectic and Columnar Liquid Crystals (in preparation)

Foreword

Having witnessed, thirty or forty years ago in Orsay, the revival of liquid crystal research, Patrick Oswald and Pawel Pieranski's book shows me that this area is still very open. Of course, the impact of new applications probably has shifted from the industrial manufacturing of displays to the understanding of biological processes, and justly so, one may say, since the discoverer of the field, F. Reinitzer, saw in the textures of mesophases the hallmark of life!

This richness of behavior was then thoroughly studied by O. Lehmann, but the main merit of F. Grandjean and my grandfather G. Friedel was no doubt that they obtained, by careful observation of the simplest cases, a clear comprehension of the structure of these bodies at the molecular level. By distinguishing the most ordered state possible and the textures produced by the appearance of defects, they opened the way for a systematic study of the equilibrium phases exhibited by these anisotropic liquids; they also started the study of the defects leading to these deformations, especially the rotation dislocations or disclinations.

If G. Friedel's 1922 paper paved the way, it was P.-G. de Gennes' first book on liquid crystals, in 1974, that rekindled this research activity, still very active to this day. P. Oswald and P. Pieranski participated, among many others, in this revival. This book is a result of thorough personal research and of the graduate courses taught by the two authors, one at the École Normale Supérieure de Lyon and the other at the University of Paris VII (Denis Diderot).

One may wonder about the value of this new work, in an area already covered by a significant number of recent publications, also cited by the authors.

The first immediate answer is found just by reading the table of contents. If the two volumes are split according to the classical separation between what one might call three-dimensional anisotropic liquids (nematics and cholesterics) and those that are liquid only along two directions or even just (smectic and columnar phases, respectively), we notice the important weight given to topics that have only received limited coverage elsewhere in the literature, among them surface phenomena (anchoring, faceting) and the dynamics of growth or deformation. From this point of view, we can only regret the absence of any discussion on mesomorphic polymers.

To me, however, what fully justifies this work is its style, which strives to be as clear and as close to the experimental side as possible. A genuine pedagogical effort was deployed, no doubt a result of the authors' continuous contact with students.

There are only three minor criticisms I would like to offer to the authors:

1. The use of arrows to denote the molecular orientation is only justified in the magnetic case, when the elementary magnets exhibit dipolar order; it is however not adapted to the quadrupolar molecular order observed in liquid crystals. The by now classical use of nail-shaped symbols reminds one of this property and, by the length of the nail, represents the orientation of the molecules with respect to the plane of the image, going from the dash (parallel) to the dot (perpendicular to the plane). Notwithstanding a cautionary note of the authors, I fear that the use of both notations might be confusing to the reader.

2. To simplify their discussion of disclinations, the authors replace the Volterra process of general rotation by de Gennes' process, where the rotation is distributed over the surface of the slit. This is perfectly justified and easier to understand in the case of nematics, where all translations are relaxed by liquid movements. However, the equivalence of the two processes does not hold in the other cases, where liquid relaxation is more complex: in cholesterics, for instance, this presentation poses difficulties in accounting for the distribution of translation dislocations involved in the curvature of disclinations. Similarly, one can also regret the too brief discussion of the topological analysis of defects, which shows its full strength in complex smectics. Finally, concerning the description of Blue Phases, the seminal role of J.-F. Sadoc and M. Kléman in the concepts of projection from a curved space and in the analysis of twist is not sufficiently emphasized.

3. The growth instabilities as such merit a short appendix, rather than an entire chapter. The role of B. Chalmers as a forerunner might have been mentioned, without necessarily proceeding all the way to the distinction made by G. Friedel, in his crystallography lessons (1926), between equilibrium facets, characterized by the law of P. Curie and the Wulff construction, and the development of slow growth facets in dendritic growth. I am nevertheless certain that it was by studying the shapes of mesomorphic "rods" that G. Friedel was able to make this distinction, in the early 1920s.

Overall, this book leaves both a solid and stimulating impression. Everything relating to molecular configuration, to optics (however complex, as shown by Ch. Maugin), to dielectric properties and to (anti) ferroelectrics, to elasticity, the laminar flows and the buckling towards the first instabilities, is described in an authoritative and detailed manner, as are the properties of textures and plastic deformations, involving the appearance of the defects typical for these phases. But, if the authors were directly involved in the study of these aspects, it might have been noted that the topic itself is more complex and this study is far from complete.

Of course, I appreciate the emphasis given, at the beginning of the book, to the contribution of G. Friedel to the field. The 1922 paper in *Annales de Physique*, which established his reputation, was written upon his arrival at Strasbourg University in 1919 to head the Institute of Mineralogy. Coming from the École des Mines in Saint-Étienne, where he worked and which he directed, he could not become a university professor, according to the rules of the time, for lack of both B.Sc. and Ph.D degrees. He most certainly wrote this paper, mainly describing his own activities, in order to prove that a thesis posed him no problems. This paper struck a note among the specialists and, notably, defined the main terms of the domain, inspired by his penchant for the humanities and a daughter who was good at Greek. This success undoubtedly overshadowed in the minds of readers the original contributions of Friedel's friend F. Grandjean, as Y. Bouligand likes to point out. On the other hand, of course, this paper cannot mention the subsequent work of G. Friedel. It is for instance certain that he saw the Grandjean walls in cholesterics as singular lines, bordering the additional half-pitch on one side of the wall, initially sketched by Grandjean as in Figure B.VII.18 of the present work. His joint paper with my father E. Friedel states this clearly at the liquid crystal conference in 1930, published in *Zeitschrift für Kristallographie* (1931, pp. 1 and 297). After having situated this line on one of the surfaces of the mica wedge (used by Grandjean long before Cano), it appears, according to my father, that G. Friedel realized it was more obvious to place it in the middle of the sample, as proposed much later by researchers in Orsay (cf. figure B.VII.26). At any rate, it is certain that he likened the Grandjean wall to what later would be called a translation dislocation line of the edge type, that he observed its thermal displacement, its pinning on impurities as well as the possibility, in the wedge geometry, of producing a weakly disoriented sub-grain boundary between two slightly mismatched ordered phases. It was only in 1939 that J.M. Burgers predicted such sub-grain boundaries in crystals, and only as late as 1947 did P. Lacombe find them in aluminum. This 1931 conference also discusses the direct X-ray measurements of the layered structure of smectics, performed by E. Friedel in M. de Broglie's laboratory (*C. R. Acad. Sci.* **176**, 738, 1922 and **180**, 209, 1925, cf. *Zeitschrift für Kristallographie*, 1931, pp. 134 and 325).

More generally, it is noteworthy that such a rich and varied area of research went unnoticed by the mainstream scientists until the end of the 1960s. And when recognition finally came, the surge of interest was prompted by the applications. At first glance, this is very surprising, especially since the experiments, though not always easy to perform, are not complex, and since in the 1940s and 1950s the physicists were absolutely ready for the mesoscopic scale arguments, constantly employed ever since. A paper by F.C. Frank in the

1940s had even emphasized the existence of rotation dislocations, but no one, least of all myself, understood the essential role of liquid relaxation in the creation and movement of these lines in liquid crystals. P.-G. de Gennes would certainly point out that the applications relaunched the scientific investigation. The lack of interest in liquid crystals may also have resulted from their apparent lack of applications, at a time of intense research on crystal dislocations and their role in plasticity and growth. At any rate, this example should inspire a certain modesty and warn us against too early abandoning a research field considered to be exhausted.

In conclusion, I would like to cite again G. Friedel at the 1930 Conference (p. 320), saying that none of the three approaches – the naturalist, the physicist, and the mathematician – should be neglected and that a healthy balance must be preserved amongst them! It is no doubt in this spirit that the present work was written.

Jacques FRIEDEL

To our wives, Jocelyne and Héléna

Preface to the English edition

Liquid crystals were discovered at the end of the 19th century. With properties intermediate between those of solids and liquids, they are known primarily for their widespread use in displays. However, even more important is the fact that they allow us to perform (sometimes very simple) experiments which provide insight into fundamental problems of modern physics, such as phase transitions, frustration, hydrodynamics, defects, nonlinear optics, or surface instabilities. It is in this spirit that we conceived this book, without attempting to exhaustively describe the physical properties of liquid crystals. From this point of view, the book is far from complete; it only reflects a very small part of the enormous work done on this subject over the last thirty years. On the other hand, it provides a thorough treatment of topics reflecting the research interests of the authors, such as growth phenomena, flow and thermal instabilities, or anchoring transitions, which are not described in detail in the fundamental (and already well-known) treatises by P.-G. de Gennes and J. Prost [1], S. Chandrasekhar [2], or P. J. Collings and M. Hird [3]. We also recommend M. Kléman's book on the physics of defects [4] as well as P.M. Chaikin and T.C. Lubensky's book [5], which discusses the general principles of condensed matter physics, often using liquid crystals as an illustration.

We would also like to warn the reader that we use the tensor conventions usually employed in mechanics, and not those of Ericksen, which are employed in references [1] and [2]. Thus, if $\underline{\sigma}$ is a second-rank tensor and $\mathbf{n}$ a vector, $\underline{\sigma}\,\mathbf{n}$ is a vector with components $\sigma_{ij}\, n_j$ (and not $\sigma_{ij}\, n_i$). This difference is essential for all materials exhibiting orientational order, such as nematics, where the stress tensor $\underline{\sigma}$ is not symmetric. Similarly, div $\underline{\sigma}$ is a vector of components $\sigma_{ij,j}$ (and not $\sigma_{ij,i}$) and the torque associated with the antisymmetric part of the tensor is given by $-e_{ijk}\sigma_{jk}$ (instead of $e_{ijk}\sigma_{jk}$). Let us also point out that we sometimes use arrows ↑ to represent the director field $\mathbf{n}$ in uniaxial nematics, while the classical representation, which takes into account the $\mathbf{n} \Leftrightarrow -\mathbf{n}$ symmetry is by "nails" (⊥). In our case, these two representations are strictly equivalent, the head of an arrow corresponding to the head of a nail. Finally, we added at the end of this volume a chapter on growth instabilities. Although it goes beyond the domain of liquid crystals, this chapter is very useful for understanding the growth experiments described in chapter B.VI, as well as two other chapters in the second volume, entitled *Smectic and Columnar Liquid Crystals: Concepts and Physical Properties Illustrated by Experiments*, dedicated to the growth of columnar and smectic phases.

In conclusion, we would like to thank all those who contributed to the French and English versions of this book. We thank first of all Professor

J. Friedel for his thorough reading of the French manuscript and his observations and informed comments on the history of liquid crystals. We are also grateful to Mr. J. Bechhoefer who suggested many very useful corrections and to Mrs. J. Vidal who corrected the numerous errors which had found their way into the chemical formulas and the associated nomenclature; we also owe her thanks for the calculation and the graphical representation of several molecular configurations.

The translation of the first volume into English is the masterpiece of Mr. Doru Constantin to whom we address especially warm thanks for a very enthusiastic, careful, and punctual accomplishment of this task. Our thanks also go to the financial sponsors of this translation: Mr. B. Bigot from École Normale Supérieure de Lyon, Mr. J. Charvolin from Laboratoire de Physique des Solides in Orsay, Dr. M. Schadt from Rolic Ltd, and Drs. J.-P. Caquet, W. Becker, and J. Gehlhaus from Merck. We also thank Mr. V. Bergeron who was the first critical reader of the English version. The final proof was made by Prof. G.W. Gray, the editor of the Taylor & Francis series on Liquid Crystals. We are grateful to him for numerous corrections and informed comments on the historical and chemical aspects of liquid crystals.

We also thank warmly Ms. C. Andreasen who helped us prepare the PDF printer-ready final files of this book.

Last, but not least, we thank all those with whom we collaborated on the topics presented in this book. In particular, one of the authors (P.P.) wishes to acknowledge his friendly and fruitful collaboration with R. Hornreich on Blue Phases.

[1] De Gennes P.-G., Prost J., *The Physics of Liquid Crystals*, Oxford University Press, Oxford, 1993.

[2] Chandrasekhar S., *Liquid Crystals*, Cambridge University Press, Cambridge, 1992.

[3] Collings P.J., Hird M., *Introduction to Liquid Crystals*, Taylor & Francis, London, 1997.

[4] Kléman M., *Points, Lines and Walls in Liquid Crystals, Magnetic Systems, and Various Ordered Media*, John Wiley & Sons, Chichester, 1983.

[5] Chaikin P.M., Lubensky T.C., *Principles of Condensed Matter Physics*, Cambridge University Press, Cambridge, 1995.

Part A

OVERVIEW

Chapter A.I

Some history

In this introductory chapter we briefly recall the pioneers in liquid crystals. The first experimental observations date from the 19th century, but liquid crystal physics really began in the early 1920s with G. Friedel. This period was followed by almost half a century of inactivity. It was only in the early 1970s, particularly under the influence of P.-G. de Gennes, that liquid crystal physics entered a stage of expansion that continues to the present day.

Useful complementary information on liquid crystal history can be found in the two papers by H. Kelker published in 1973 and 1988 in *Molecular Crystals and Liquid Crystals* (vol. 21, pp. 1–48 and vol. 165, pp. 1–43) and in a special issue of *Liquid Crystals* (vol. 5, 1989).

I.1 Georges Friedel and liquid crystals

Georges Friedel (Fig. A.I.1), the famous French crystallographer of the early 20th century, was the first to suggest that liquid crystals are states of matter in their own right, intermediates between isotropic liquids and crystals. Before Friedel, several scientists of the 19th and 20th centuries had observed liquid crystals without understanding that they were dealing with new states of matter, separated from the liquid state and the crystalline solid state by sharp boundaries (phase transitions). Friedel took this decisive step and described his discovery in the now famous treatise entitled "Mesomorphic States of Matter," published in *Annales de Physique* in 1922 [1]. It is well worth citing some of the crucial passages from this fascinating work. From the very beginning, Friedel declares without false modesty that he intends to bring some order into the state of confusion created by his famous but ill-inspired forerunner, Otto Lehmann:

*"By this name ("**mesomorphic states of matter**") I will designate the uncommon states exhibited by the bodies that Lehmann reported on since 1889 under the name of **liquid crystals** or **crystalline fluids**. Upon these denominations, ill-chosen but incessantly repeated for thirty years, many people imagine that the peculiar bodies to which Lehmann had the great merit of drawing attention, but that he was wrong to misname, are nothing more than*

crystallized substances, differing from the previously known ones simply by their higher or lower degree of fluidity. In fact, all this is about something else completely and infinitely more interesting than merely more or less fluid crystals."

Fig. A.I.1 Georges Friedel, author of "Mesomorphic States of Matter" [1].

Friedel further details his way of understanding the nature of the bodies Lehmann called **fliessende Kristalle** or **flüssige Kristalle**:

"The specificity of Lehmann's substances lies not in their more or less fluid state, but rather in their extremely particular structures, always the same and whose number is very small. I hope that the rest of this work will demonstrate that the bodies reported by Lehmann represent two ***entirely new forms of matter****, invariably separated from the crystalline form and the amorphous form (isotropic liquid) by discontinuities, as the crystalline form and the amorphous form are always separated by a discontinuity."*

Finally, Friedel names the new states of matter (nematic, cholesteric, and smectic) drawing his inspiration from the Greek language:

"I shall call ***smectics*** *(σμεγμα, soap) the forms, bodies, phases, etc. of the first kind (part of the fliessende Kr., Schleimig flüssige Kr. of Lehmann; liquids with conics), because soaps, in usual temperature conditions, belong to this group and because potassium oleates, in particular, are the first bodies of this group to have been signaled. I shall call* ***nematics*** *(νημα, thread) the forms, phases, etc. of the second type (flüssige Kr., Tropfbar flüssige Kr. of Lehmann; liquids with filaments) because of the linear discontinuities, winding like threads, that are their salient feature."*

It is also important to specify that Friedel put forward the distinction between two types of nematic mesophases:

"One should distinguish in nematic bodies two different types, between which one can observe ***continuous transitions****, as we shall see. I will call the first type:* ***nematics proper*** *(liquids with filaments or* noyaux*, positive nematic liquids); and the second:* ***the cholesteric type*** *(liquids with Grandjean planes, negative nematic liquids)."*

These four excerpts from Friedel's treatise merit a few historical remarks, especially concerning the forerunners of Friedel.

I.2 The discovery of birefringence in fluid biological substances by Buffon, Virchow, and Mettenheimer: lyotropic liquid crystals

The seeds of the idea that condensed matter can exist in states different from the isotropic liquid and the crystalline solid appeared about the mid-19th century.

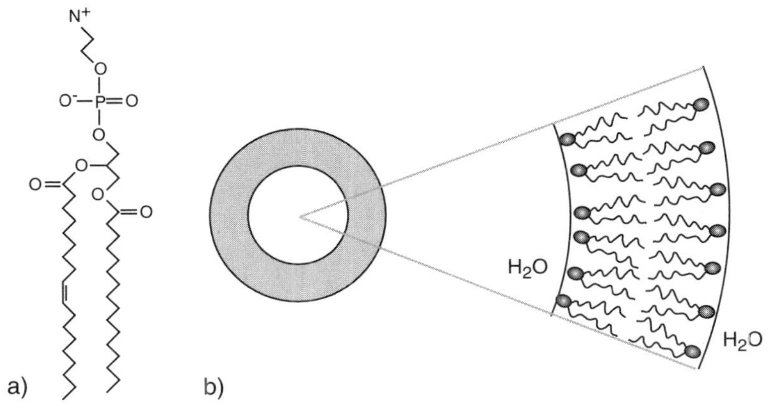

Fig. A.I.2 Phospholipids are the main ingredients of the cell membrane. a) Chemical formula of phosphatidylcholine (lecithin); b) in aqueous solution, phospholipids gather in bilayers.

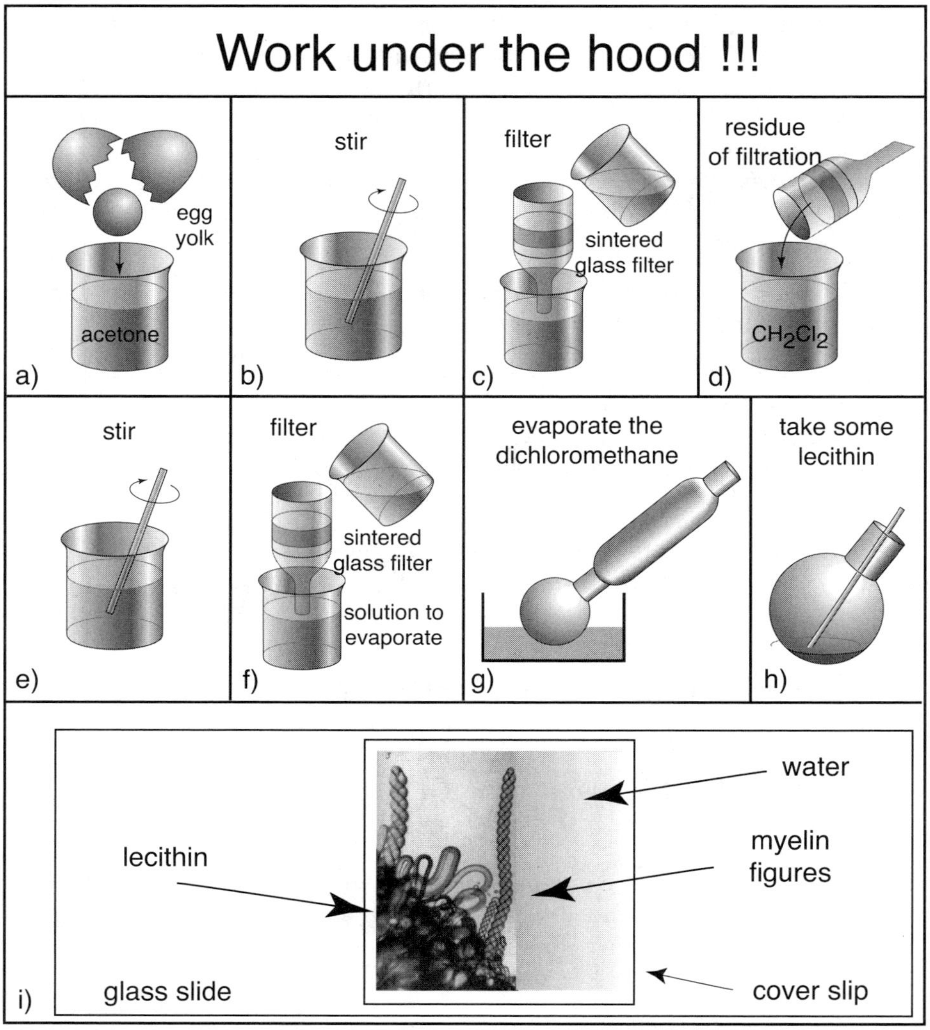

Fig. A.I.3 Recipe for extracting lecithin from egg yolk and producing myelin figures: a-b) mix an egg yolk with 40 ml of acetone; c) filter; d-e) mix the residue of filtration with 40 ml of dichloromethane (CH_2Cl_2); f) filter; g) evaporate the dichloromethane using an appropriate device (rotating evaporator); h) take some lecithin, place it on a microscope slide, and squeeze it under a cover slip; i) add water to the preparation and observe the myelin figures under the polarizing microscope.

At that time, biologists started using the polarizing microscope for studying vegetable and animal tissues, as well as the substances extracted from them (the publication in 1861 of a manual entitled "Die Untersuchung der Pflanzen und Tiergewebe im polarisierten Licht" by G. Valentin shows that the polarizing microscope was among the biologists' instruments at that time).

Observed in 1857 under the polarizing microscope by the ophthalmologist Mettenheimer, myelin, a soft substance extracted from the nerves, and identified by Virchow in 1850, turned out to be particularly interesting. Indeed, by placing myelin in contact with water, Mettenheimer observed the growth of tubular objects at the myelin-water interface. However, Mettenheimer was not the first (as it is often maintained) to have observed these tubes, now called myelin figures [2]. Similar observations can be traced back to Buffon's complete works, published in 1840, where he speaks of "writhing eels" obtained by dilacerating wheat or rye ergot in water. Obviously, it was not fish, but myelin figures formed by the ergosterol and lecithins contained in these substances.

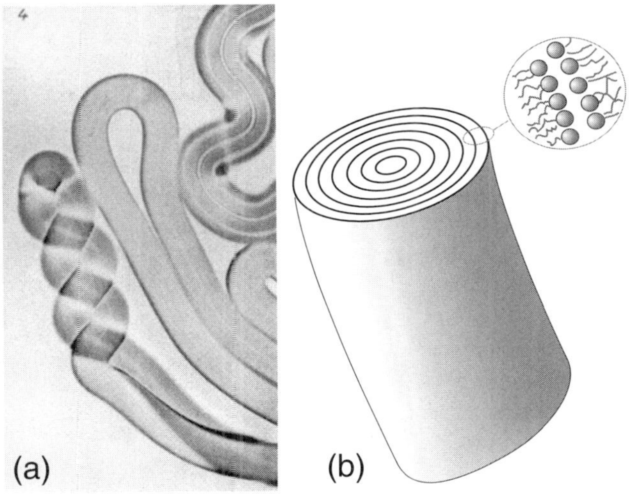

Fig. A.I.4 Myelin figures: a) observed under a microscope (by J. Nageotte); b) concentric cylinder structure made of phospholipid bilayers

We know today that myelin consists mostly of phospholipids, molecules that are the main ingredient of cell membranes [3]. The extent of these membranes in nerves is enormous, because of their ramified shape, explaining their high phospholipid content. Egg yolk is another source for phospholipids, and has an important lecithin concentration (Fig. A.I.2). The recipe for extracting lecithin from egg yolk is very simple (Fig. A.I.3), and allows one to reproduce the Virchow-Mettenheimer experiment very easily.

At high concentrations, phospholipids in water are arranged in bilayers. We shall see later (in chapter C.III of Vol. 2 on defects in smectics) that bilayers can take various forms and organize in different manners. For instance, they can fit into each other under the form of concentric cylinders. These cylinders are birefringent and constitute the main elements of the myelin textures visible in figure A.I.4.

The arrangment of phospholipids in bilayers is a consequence of their hybrid (**amphiphilic**) molecular anatomy. Indeed, one end of the molecule is hydrophilic, while the other is hydrophobic. The same amphiphilic character is found in soaps which, chemically speaking, are nothing other than fatty acid salts, and therefore derivatives of biological substances. In his treatise (last phrase in section I.1), Friedel alludes to potassium oleate (Fig. A.I.5) whose aqueous solutions, studied at that time by Perrin, also presented, in certain concentrations, a structure of stacked-up bilayers.

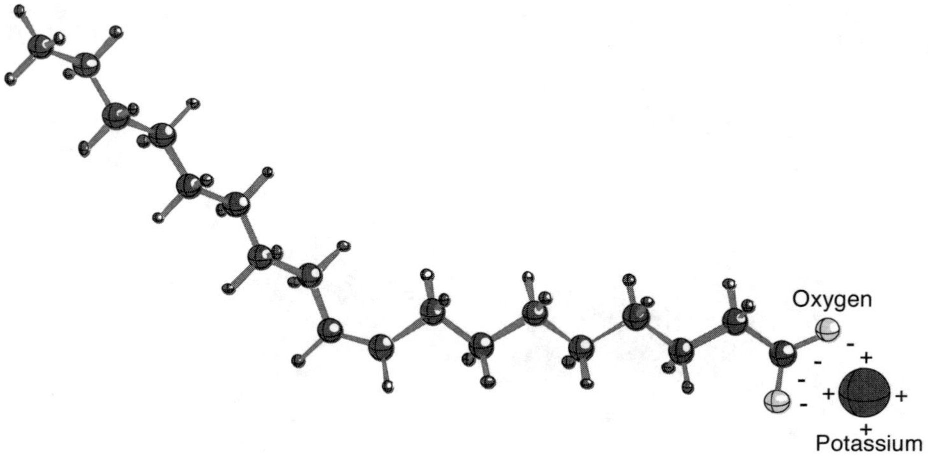

Fig. A.I.5 Molecular structure of potassium oleate. The chain is unsaturated, with a double bond at C_9.

In conclusion, the first mesophases were **discovered in aqueous solutions of amphiphilic molecules**. Today, they are termed **lyotropic liquid crystals**, in contrast to **thermotropic** phases that do not contain water.

I.3 Observation of the surprising behavior of cholesteryl esters by Planer and Reinitzer: thermotropic liquid crystals

Friedel knew that liquid birefringent phases existed not only in solutions, but also in pure bodies. Once more, it is the biologists who set the physicists on the right track. The story begins with cholesterol [3], a substance extracted from plants. In the 19th century, its chemical structure was still unknown;

however, it was classified among alcohols, since one could prepare cholesteryl esters (Fig. A.I.6) by reaction with fatty acids. The biologists Planer in 1861 [4] and Reinitzer in 1888 [5] noticed its opaqueness and the iridescent colors exhibited by these esters upon melting from the crystalline phase or upon cooling from the isotropic liquid. Reinitzer also knew that the optical properties of materials interested some physicists of his time.

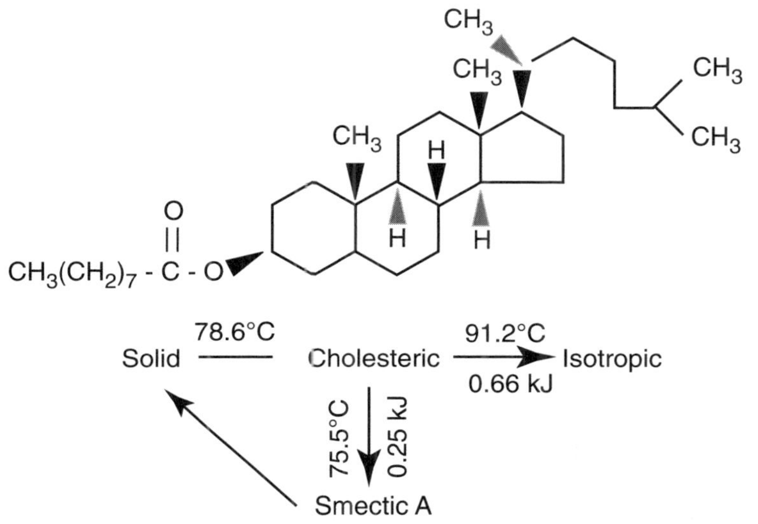

Fig. A.I.6 Molecular structure of cholesteryl nonanoate and its phase transitions. The structures of the cholesteric and smectic phases will be described in the following chapter.

In particular, Reinitzer knew of Otto Lehmann for his work on crystal growth and the changes of crystalline structure induced, for instance, by temperature (crystalline polymorphism). That is why, in 1888, he sent Lehmann a 16-page letter, presenting in detail his surprising observations on the properties of cholesteryl esters. Here is an excerpt:

"The two substances (cholesteryl esters) hold such beautiful and peculiar phenomena that I hope they will be of the greatest interest to you.(...) ***The substance (cholesteryl benzoate) has two melting points, if one may say so****. At about 145.5°C it melts, forming a turbid, but completely fluid liquid, that suddenly becomes completely clear at about 178.5°C. On cooling it, blue and violet colors appear that rapidly get dim, leaving the substance opaque but still fluid. If one cools it further, one sees the blue and violet colors reappear and, immediately afterwards, the substance solidifies forming a white crystalline mass (...) The turbidity that appears upon heating is not due to crystals, but rather to the liquid forming oily streaks."*

I.4 Fliessende Kristalle or "the flowing crystals" of Otto Lehmann

The physicist and the biologist started corresponding. After confirming Reinitzer's findings with his own observations, Lehmann published in 1889 in the *Zeitschrift für Physikalische Chemie* a paper entitled "Über fliessende Kristalle," thus coining the name "liquid crystals" [6]. What surprised Lehmann the most was that Reinitzer's substance, although fluid, was birefringent, one of the features of solid crystals. The way he put it was:

"If one trusts the interpretation of the experimental observations, this is so far the only case when a crystalline substance, showing considerable birefringence, has such a weak mechanical strength that it can hardly resist the action of its own weight."

The publication of Lehmann's paper was immediately followed by the observation of a similar combination of birefringence and fluidity in other substances, such as *para*-azoxyanisole (PAA) synthesized and studied in Heidelberg by Gattermann, who communicated his results to Lehmann right away. Emphasizing that the viscosity of this new substance was much lower than in cholesteryl esters, Lehmann introduced in 1890 another designation: "Kristalline Flüssigkeiten" or crystalline liquids.

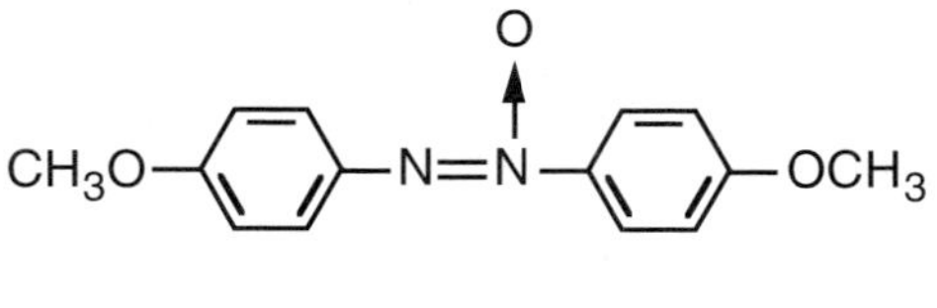

Di-methoxyazoxybenzene (*p*-azoxyanisole)

$$\text{Solid} \xrightarrow[29.57\ \text{kJ}]{118.2°\text{C}} \text{Nematic} \xrightarrow[0.57\ \text{kJ}]{135.3°\text{C}} \text{Isotropic}$$

Fig. A.I.7 Molecular structure and sequence of phase transitions for PAA (para-azoxyanisole).

The years following the seminal works of Lehmann saw an increasing volume of similar observations, and the question of the real molecular structure of "liquid crystals" or "crystalline liquids" became more and more important. Heated debates opposed those that pleaded for chemically and physically homogeneous phases to the ones holding that the optical turbidity was a result of either demixing impure substances or incomplete melting of the crystalline phase. And even for those considering the phases homogeneous, the question was if they were to be classified among (flowing) crystals or, on the contrary, in a category of their own.

It is in this state of conceptual and terminological confusion that Georges Friedel came up with the idea that these phases were **distinct states of matter whose molecular structures were intermediary (*meso-morphic*) between the crystals and ordinary liquids**. At the same time, he proposed a classification of mesomorphic states which is still in use today and that will be described (in its modern form) in the next chapter.

BIBLIOGRAPHY

[1] Friedel G., "Mesomorphic States of Matter," *Annales de Physique*, **18** (1922) 273.

[2] Bouligand Y., private communication.

[3] Stryer L., chapter 10, "Introduction to biological membranes," *Biochemistry*, W.H. Freeman, New York, 1975.

[4] Planer P., *Liebigs Ann.*, **118** (1861) 25.

[5] Reinitzer F., "Contributions to the knowledge of Cholesterol," *Liq. Cryst.*, **5** (1989) 7. English translation of the original paper published in 1888 in *Monatshefte für Chemie*, **9**, pp. 421–441.

[6] Lehmann O., "Über fliessende Kristalle," *Zs. Phys. Ch.*, **4** (1889) 462.

Chapter A.II

Modern classification of liquid crystals

We begin this chapter with a brief summary of the principles used by Georges Friedel to classify liquid crystals (section II.1). As will be seen, this classification is not complete and it can be extended by employing the very general notion of broken symmetry (section II.2). We then return to the classification of smectic and columnar phases (sections II.3 and II.4, respectively).

II.1 The terminology introduced by Georges Friedel

II.1.a) Polymorphism and mesomorphic states

As stated in the previous chapter, Friedel was the first to speak of "mesomorphic states." By this designation of liquid crystalline phases, Friedel intended, on the one hand, to avoid the semantic controversies triggered by the discrepancy of the terms "crystalline" and "liquid" and, on the other hand, to point out that they were genuine states of matter whose molecular properties are intermediary (*meso*-morphic) between those of crystals and those of ordinary liquids. By reading certain passages of his 1922 treatise (see the previous chapter), one notices that, by speaking of mesomorphic *states*, in the plural, Friedel already knew that at least two structures could combine crystalline anisotropy and fluidity. Perhaps his idea of mesomorphic polymorphism was an analogy with the crystalline polymorphism, a notion well established by that time. As a matter of fact, Friedel knew very well that the same substance could display different crystalline structures depending on the temperature or pressure. In his other work, "Lessons in crystallography," Friedel actually gave a complete overview of the 230 space groups allowing for the classification of all possible crystalline structures in three dimensions [1a]. However, since he only knew two mesomorphic structures, Friedel's idea of mesomorphic polymorphism was a jump ahead. The future proved him right, as the number of mesophases with different symmetries known today is large, with the count still ongoing.

II.1.b) Nematic and cholesteric phases

If the term "mesomorphic" and the concept of mesomorphic polymorphism stood the test of time, one can wonder if Friedel's other terminological proposals were equally correct. The last passage from Friedel's treatise cited in the previous chapter shows that the choice of the name "nematic" was based on microscope observations of "linear discontinuities" (thread-like texture shown in figure A.II.1). Polarized light observations allowed Friedel to deduce the configuration of the optical axes around these lines and to characterize them as orientational singularities of the optical axis (similar observations had also been made by Lehmann and, later, by Grandjean).

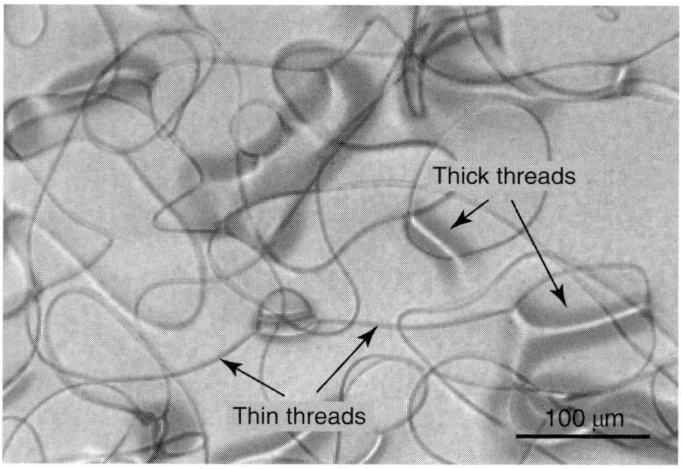

Fig. A.II.1 Thread-like texture observed in a free drop of uniaxial nematic liquid crystal. Thin and thick threads correspond to two different kinds of line defects.

Was Friedel right in speaking of line defects? One might doubt it, since today condensed phases are classified by the type of symmetry breaking they present with respect to the liquid isotropic phase. We shall see that, according to these criteria, a "nematic" is a phase whose only broken symmetry is orientational and whose order parameter is quadrupolar. Yet we shall also find out that the line defects observed by Friedel are nothing other than topological defects (nowadays termed disclinations) of the nematic order parameter. These defects are a distinctive mark of uniaxial quadrupolar orientational long-range order. Once again, Friedel's choice of terms was accurate, although one must note that the quadrupolar moment can be uniaxial or biaxial. A corresponding distinction should therefore be made between uniaxial and biaxial nematics. In most of the cases, the nematic phase is uniaxial. It took half a century to find the first examples of a biaxial nematic phase that imposed a revision of Friedel's nomenclature. We will return to this topic in the chapter on nematics.

Friedel was also right to consider that the cholesteric phase (see figure A.II.5) was only a particular (chiral) type of nematic phase. This is how he describes it in his treatise:

"Two important conclusions seem to emerge from this series of observations: 1. Cholesteric bodies are nothing more than a very special type of nematic bodies. Upon the gradual disappearance of rotatory power and of the related structural properties: (e.g., Grandjean planes, commas, etc.) what remains is a nematic proper. There is a continuous passage between the cholesteric and the nematic types..."

Finally, he explains that:

"The transition from the right-handed to the left-handed body or vice versa is done by changing the sign of structure twist. This twist takes place about the normal to the preparation plane when the layered structure is uniform; it is undergone by a positive uniaxial nematic body whose optical axis is parallel to the preparation plane and hence it takes place about the normal to its optical axis, thereby giving rise to a negative uniaxial arrangement. It is this twist that, together with the negative uniaxiality, determines the rotatory power."

Friedel was once again right, as we will show in chapter B.VII, dedicated to the cholesteric phase.

II.1.c) Smectic phases

Polarized light observations led Friedel to the conclusion that in the second type of mesophases molecules are arranged in layers. Although their thickness is fixed, these layers can bend and give rise to textures that exhibit singular lines in the shape of "focal conics" (Fig. A.II.2).

These geometrical objects were invented by Dupin, a mathematician from the early 19th century (see the chapter on defects in smectic phases). Instead of trying to coin a term inspired by focal conics, Friedel put forward the term "smectics," as this layered molecular structure is very frequent in aqueous solutions of soaps. In order to confirm his conclusions on the structure of the smectic mesophase, Friedel considered the use of X-rays [1b]:

Fig. A.II.2 Focal conics texture observed in a free smectic A drop placed on a glass plate.

"Let us point out that the mesurements of Perrin and Wells on soaps are in principle sufficient to validate the structural hypothesis suggested by so many facts. There is however another way of revealing the equidistant planes in all smectic bodies and to measure their distances in pure bodies, better defined than those used by Perrin, namely the use of X-rays. (...) The repeat distance is probably of the order of several tens of Angstroms so that, in order to have a satisfying reflexion angle, long wavelength X-rays should be employed."

According to broken symmetry criteria, nowadays one calls smectic a phase where molecules are organized in layers, the translational symmetry being broken in **at least** one direction in space.

This definition does not specify the way molecules are set up inside the layers, or define the type of stacking, or the shape of the layers, which need not be flat. Hence, there is a very large number of distinct smectic phases. Their names are formed by adding one or several letters to the name proposed by Friedel. We provide some examples later.

II.2 Modern definition of mesophases; broken symmetries; short- and long-distance order

Boundaries between different liquid crystalline phases are not always easily determined. Sometimes it is even hard to say if one is dealing with a mesophase or not. One of the possible criteria involves the concept of broken symmetry. As explained by Anderson [2a], this idea, stemming from the work of P. Curie on magnetism [2b] and from that of L. Landau on second-order phase transitions [2c], is based upon the fact that the symmetry of a system of particles can be lower than that of the interaction hamiltonian.

In a homogeneous and isotropic space, the hamiltonian for a system of interacting particles must be invariant with respect to the elements of the group G [3]:

$$G = t \times SO_3^{(L)} \qquad \text{(A.II.1)}$$

comprising the following symmetries:

t – the group of 3D translations; it consists of all the translations $\mathbf{t} = x\mathbf{i} + y\mathbf{j} + z\mathbf{k}$ with x,y,z real;

$SO_3^{(L)}$ – the group of space rotations; its elements are the matrices $\underline{R}$ fulfilling $\det(\underline{R}) = \pm 1$.

Put another way:

$$\text{for all } g \in G \text{ one must have } gH = H \qquad \text{(A.II.2)}$$

with H being the hamiltonian. Remarkably, in spite of this interaction symmetry G, all thermodynamic equilibrium states, except for the gas and the isotropic liquid, have lower symmetries G' such that:

$$G' \in G \qquad \text{(A.II.3)}$$

where G' is a sub-group of G. This lowering of hamiltonian symmetry is termed symmetry breaking. Here, "symmetry of the system states" refers, for instance, to the symmetry of density distribution, molecular orientation, or dipolar magnetic or multipolar electric moments, etc.

In the gas or the isotropic liquid phases, there is no symmetry breaking. In periodic 3D crystals, both translational and orientational symmetries are broken, as the mass (or charge) density

$$<\rho(r)> = \rho_0 + \Sigma\, \rho(q)\exp(iqr) \qquad \text{(A.II.4)}$$

is only invariant with respect to the elements of one of the 230 space groups, rather than with respect to all translations t and all rotations $\underline{R}$. In particular, simple translations belonging to these space groups are discrete and define the Bravais lattices well known to crystallographers.

By definition, symmetry breaking is *less important* in mesophases than in 3D crystals; these phases can therefore be seen as *intermediaries* between 3D crystals and the liquid isotropic phase.

From now on, whenever it is called for, we shall analyze the symmetry breaking characteristic to the phase studied. To provide an idea of liquid crystalline diversity, figure A.II.5 presents a simplified diagram of mesophase classification. This diagram merits some comments:

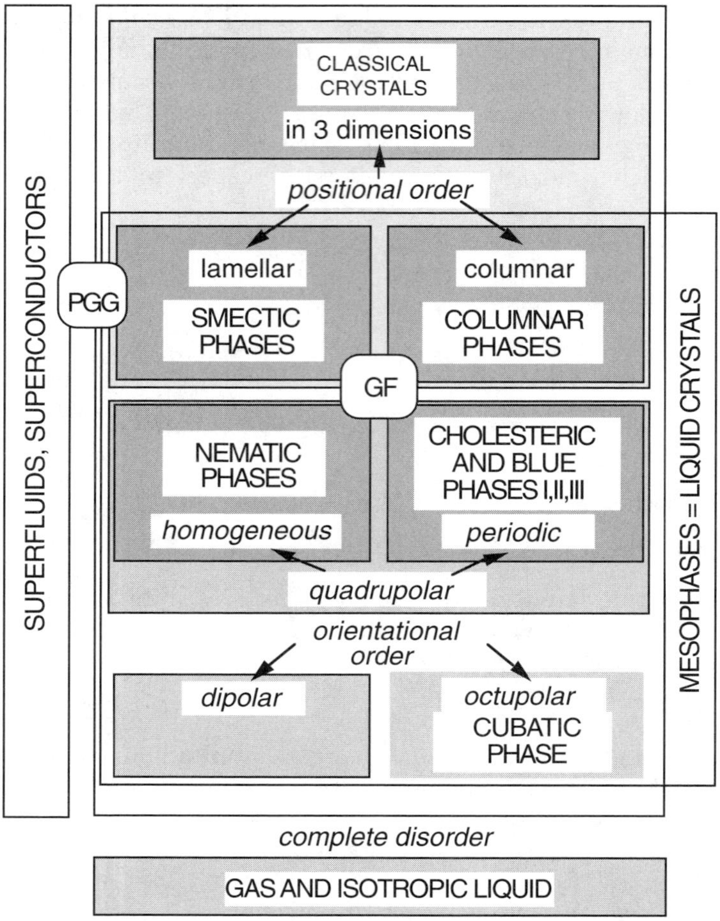

Fig. A.II.3 Mesophase classification diagram. PGG indicates the analogy established by P.-G. de Gennes between quantum liquids and smectics (see chapter C.I). GF denotes the terminology proposed by G. Friedel.

– All classification of condensed matter phases only in terms of symmetries is incomplete, as a system can exhibit two distinct phases of the same symmetry, separated by a discontinuous phase transition. The classic example is that of the liquid-gas transition. To understand the difference between these two phases one must look at the nature of the local order and calculate the correlation functions of the corresponding order parameter. Similar transitions exist in liquid crystals, such as those between smectic phases with the same symmetry, but with locally different molecular arrangements (for instance, the smectic A_1-smectic A_2 transition [4]).

– In this diagram, nematic phases belong to the family of mesophases presenting only orientational long-range order. This order is quadrupolar (the concept will be rigorously defined in chapter B.I) and spatially homogeneous. When the mesogenic material is composed of chiral molecules, quadrupolar order is spatially modulated, which does not imply the appearance of long-distance positional order. Cholesteryl esters present such phases, one of which was termed "cholesteric" by Friedel. In this particular case, the modulation appears along one space direction. We know now that, between the isotropic and the cholesteric phases, these materials also exhibit the so-called "Blue Phases" (I, II, and III), where quadrupolar order varies periodically in the three space directions. It is even probable that they are responsible for the blue and violet colors mentioned by Reinitzer in his letter to Lehmann (see the previous chapter). However, as these phases only exist in a very narrow temperature range (≈ 3°C), they were only identified in the late 1970s.

– The diagram also contains a fourth phase, not present in the Friedel classification, called columnar or canonic (the last denomination proposed by Sir Charles Frank). In this type of mesophase, the molecules (usually discoidal or cone shaped) are stacked in one-dimensional columns that can present various spatial arrangements.

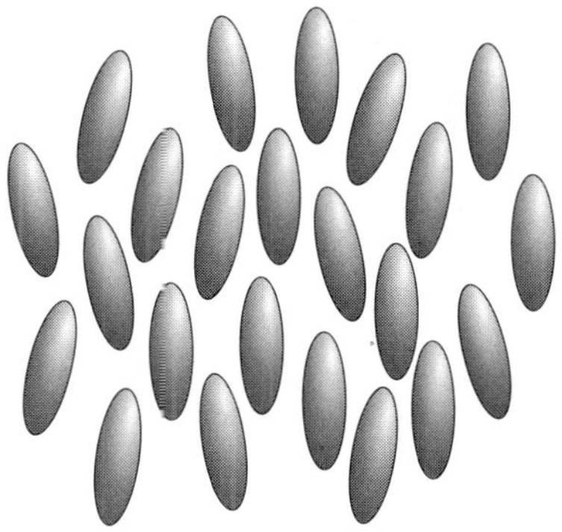

Fig. A.II.4 Schematic representation of the nematic uniaxial phase.

– Note that order is sometimes intermediate between short and long range. This is the case for two-dimensional systems where correlation functions can algebraically decay with the distance r (say $1/r^n$). The resulting behavior is in between exponential decay (defining short-range order) and convergence towards a finite value for $r \rightarrow \infty$ (characteristic of long-range order). As an example one can consider hexatic phases, predicted by Kosterlitz and Thouless [5], and subsequently discovered in smectics (chapter C.V).

– Finally, phases of dipolar or octupolar orientational order have never been experimentally observed.

Figure A.II.4 depicts the nematic uniaxial phase. Every molecule is represented by a prolate spheroid. This rendering is suggestive of the fact that molecules can rotate freely about their major axes. One should notice that the three-dimensional orientational order is long range.

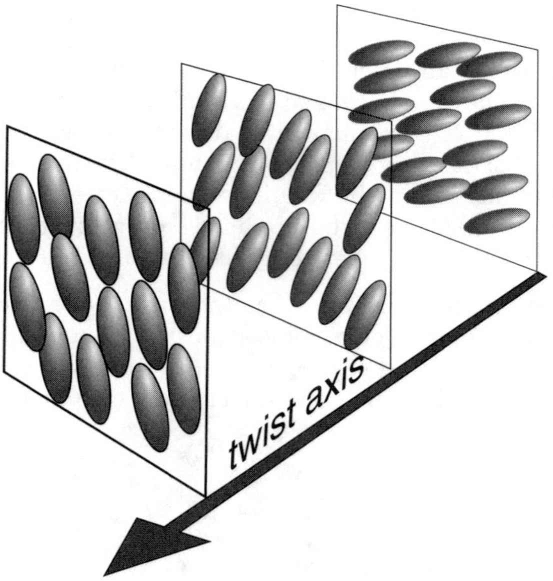

Fig. A.II.5 Schematic representation of the cholesteric phase. The planes have no real existence, the centers of mass of the molecules being randomly distributed.

In the cholesteric phase, the average molecular orientation turns about a space direction called the helicoidal axis. This twisted chiral phase is shown in figure A.II.5.

These two phases are fluid, the centers of mass of the molecules being randomly distributed. The molecular arrangement in Blue Phases is more complicated and will be described in chapter B.VIII. In the two following sections we shall thoroughly explain the classification of smectic and columnar phases.

II.3 Classification of smectic phases

These phases are distinguished by the different molecular positioning in the layers and in the transverse direction. In particular, rod-like molecules (see the following chapter on molecular architecture) can be parallel or tilted with respect to the layer's normal axis [6]. The figure A.II.6 table summarizes the primary smectic phases.

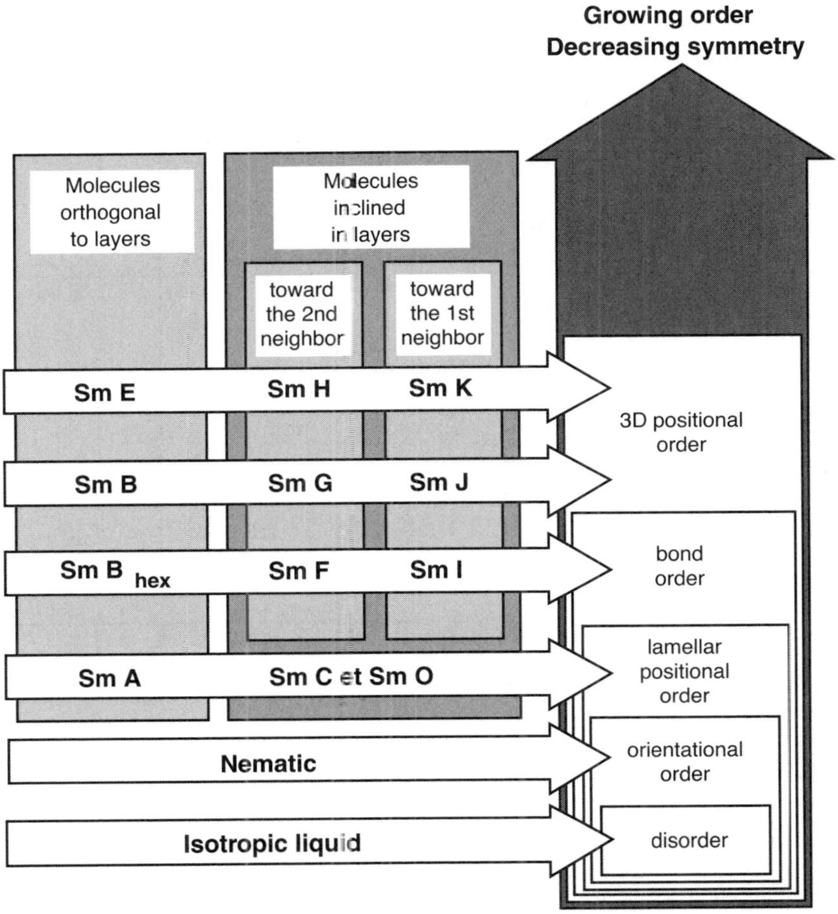

Fig. A.II.6 Classification of the main smectic phases.

To begin, a distinction is made in order between:

- 3D orientational order of the molecular major axis,

- orientational order of bonds between nearest neighbor molecules (in the plane of the layers) and, finally,

- positional order of the molecular center of mass in the layer plane and in the transverse direction.

In each case, one should distinguish between short-range order (S.R.O.), long-range order (L.R.O.) or intermediate (Q.L.R.O. for "quasi-long-range order") as in hexatics. These data are gathered in the figure A.II.7 table for the primary smectic phases, which will now be described in detail.

Phases	Orientational order of the molecular major axis	Orientational order of bonds in nearest neighbor molecules	Positional order across the layers	Positional order in the layers
Smectic A (SmA)	L.R.O.	S.R.O.	L.R.O.*	S.R.O.
Smectics C and O (SmC, SmO)	L.R.O.	S.R.O.	L.R.O.*	S.R.O.
Smectic B hexatic (SmB_{hex})	L.R.O.	Q.L.R.O.	L.R.O.*	S.R.O.
Smectic F (SmF)	L.R.O.	Q.L.R.O.	L.R.O.*	S.R.O.
Smectic I (SmI)	L.R.O.	Q.L.R.O.	L.R.O.*	S.R.O.
Smectic B (SmB)	L.R.O.	L.R.O.	L.R.O.	L.R.O.
Smectic G (SmG)	L.R.O.	L.R.O.	L.R.O.	L.R.O.
Smectic J (SmJ)	L.R.O.	L.R.O.	L.R.O.	L.R.O.
Smectic E (SmE)	L.R.O.	L.R.O.	L.R.O.	L.R.O.
Smectic H (SmH)	L.R.O.	L.R.O.	L.R.O.	L.R.O.
Smectic K (SmK)	L.R.O.	L.R.O.	L.R.O.	L.R.O.

Fig. A.II.7 Types of order present in three-dimensional smectic phases. The asterisk * indicates that, strictly speaking, lamellar order in fluid layer smectics is Q.L.R. because of the Landau-Peierls instability (unbounded 1D crystals cannot exist).

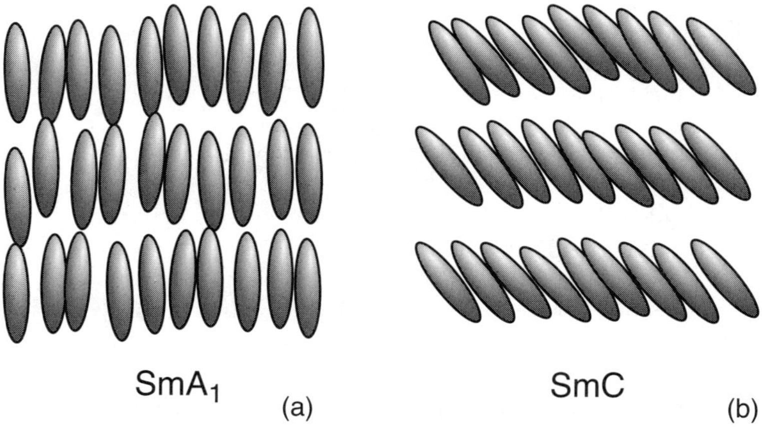

Fig. A.II.8 Schematic representation of the smectic A and smectic C phases.

In the smectic A phase (Fig. A.II.8a), molecules are arranged in fluid layers that can slide one with respect to the other. On average, molecules are perpendicular to the layer plane. This phase is therefore optically uniaxial, with an optical axis normal to the layers. The layer thickness d can vary with temperature and the molecular nature. It can be equal to the molecular length l ($d = l$, SmA_1), twice the molecular length ($d = 2l$, SmA_2) or have an intermediate value ($d = xl$ with $1 < x < 2$, SmA_d). These different types of smectic A phases are separated by discontinuous phase transitions.

In the smectic C phase (Fig. A.II.8b), molecules are tilted with respect to the layer normal. As in the A phase, fluid layers can slide over each other. This phase is optically biaxial.

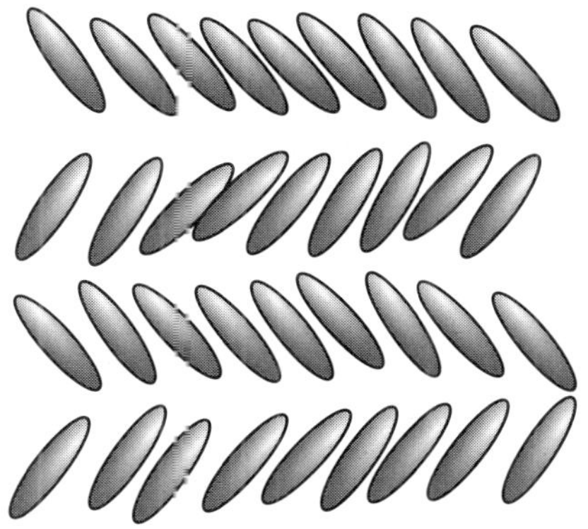

Fig. A.II.9 Schematic representation of the smectic O phase.

The smectic O phase (Fig. A.II.9) is another "tilted version" of the smectic A [7]; this time, molecules are placed in a "herringbone" pattern, alternately tilted to the left and to the right in successive layers. This phase is also optically biaxial.

Hexatic phases are more ordered than the previously discussed phases, from which they are distinguished by the existence of quasi-long-range orientational order of geometrical bonds between nearest neighbor molecules. Thus, in the hexatic B phase, the molecules are normal to the layers and locally disposed on a hexagonal lattice (Fig. A.II.10). The lattice orientation is preserved over quasi-long distances, but the positional order of the molecular center of mass is only short range due to the presence of a high density of free (i.e., very mobile) dislocations breaking the translational symmetries. One therefore speaks of hexatic order, a rather subtle concept to be discussed in chapter C.VI. The layers are fluid, as they cannot sustain shear stress without flowing, so they too slide over each other. This phase is optically uniaxial, the optical axis being normal to the layers. Smectic F and I phases are "tilted versions" of the hexatic B phase (Fig. A.II.11). However, they are optically biaxial.

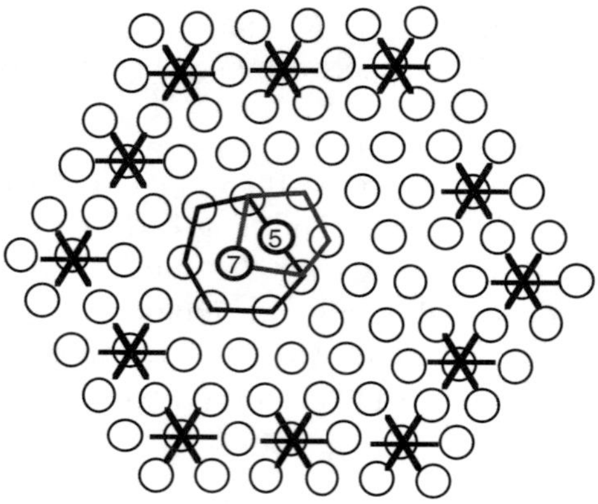

Fig. A.II.10 Locally hexagonal lattice of a hexatic phase. Although the geometrical bonds between nearest neighbor molecules exhibit quasi-long-range order, the positional order is short range due to an important density of "free dislocations" like that shown in the center of the drawing.

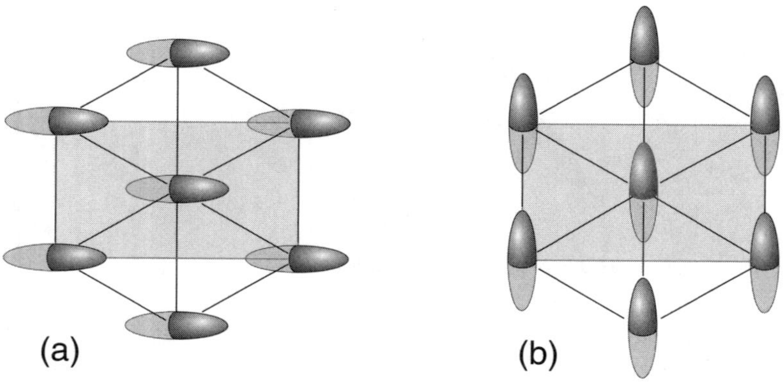

Fig. A.II.11 Local arrangement of molecules in the plane of the layers in smectic F (a) and I (b) phases. Ellipses represent molecules tilted with respect to the plane of the layers. In both phases, the hexagonal lattice is slightly deformed (in fact, it becomes rectangular).

Finally, crystalline smectics are, by definition, more ordered than hexatic smectics. In these systems, molecules are placed on the sites of a 3D lattice with positional long-range order. However, as we will see, they preserve some degrees of freedom.

Let us begin with smectics B and E. In both phases, molecules are on average normal to the layers, but their rotation about their major axes is different. Thus, in the B phase molecules are free to turn and they are placed on the sites of a hexagonal lattice. The phase is therefore optically uniaxial, the optical axis being normal to the layers. In the E phase, on the contrary, this rotation is partially hindered and the molecules exhibit herringbone stacking in the plane of the layers (Fig. A.II.12) [8]. This particular stacking induces a slight distortion of the hexagonal lattice that becomes rectangular ($a \neq b$). The phase is therefore optically biaxial. Smectic G and J and smectic H and K phases are tilted versions of the B and E phases, respectively.

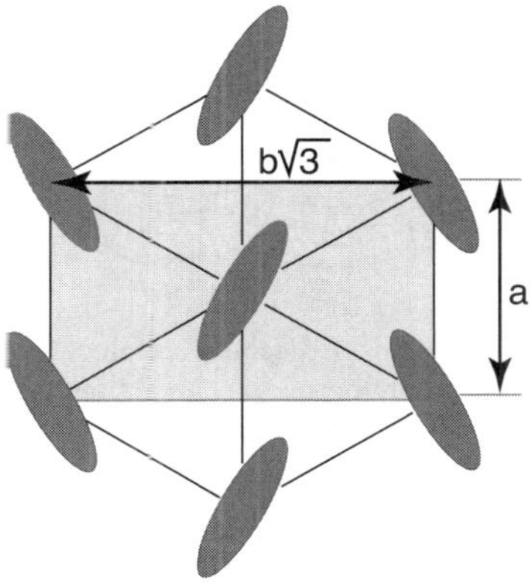

Fig. A.II.12 Herringbone molecular stacking in the smectic E phase (here the segments represent sections normal to the molecular major axis and not the tilted molecules as in Fig. A.II.11). The lattice has rectangular symmetry ($a \neq b$) and the phase is optically biaxial.

Let us emphasize that crystalline smectic phases are more properly classified among plastic crystals, as the molecules, although placed on the sites of a 3D lattice, retain a certain rotational freedom, contrary to true crystals where all rotations are frozen. Some authors drop the term smectic (Sm) for such phases, referring to them simply as B, G, K, etc. phases, or cryst. B, cryst. G, cryst. K, etc. phases.

II.4 Classification of columnar phases

In these phases, discovered in 1977 by the Chandrasekhar group in India [9], discoidal [10] or cone-like [11] molecules stack up to form "infinitely long" columns, free to slide with respect to each other (no positional correlation between columns). The molecules can be normal to the column axis or have a certain tilt angle. Strictly speaking, molecular positional order along the columns is short range, although in some cases the stacking is very regular: one speaks then of "ordered" phases, as opposed to the so-called "disordered" phases, where the distance between molecules along a column has strong fluctuations. There is however no clear-cut distinction between these two kinds of phases that must all be considered as one-dimensional fluids. On the other hand, the columns themselves are parallel and form a two-dimensional lattice that can be hexagonal (Fig. A.II.13), rectangular, or oblique. From this point of view, columnar phases are two-dimensional crystals. Figure A.II.14 shows some of the experimentally observed structures, but the list is by no means exhaustive.

Fig. A.II.13 Hexagonal D_{hd} phase formed of discoidal molecules (sometimes misnamed discotic). The letter "D" stands for the disk-like shape of the molecules. The index "h" indicates the nature of the lattice, hexagonal in this case. The second index is optional and specifies the nature of the stacking in a column: "d" for disordered and "o" for ordered.

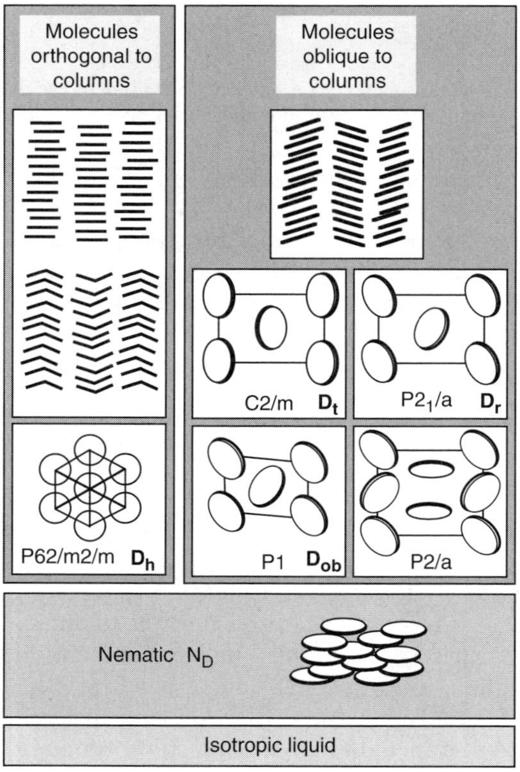

Fig. A.II.14 Some examples of columnar mesophases. In certain materials a nematic N_D phase appears between the columnar phase and the isotropic liquid. For every columnar phase the symmetry group is indicated. D_h and D_r phases are the most common. Note that the letter D stands for the shape of the molecule, while the index specifies the nature of the lattice ("h" for hexagonal, "r" for rectangular, and "ob" for oblique). A singular example of a triangular lattice with four columns per unit cell will be given in chapter C.IX [12].

Note that in the hexagonal phase the molecules are not always normal to the column axis. Indeed, hexagonal symmetry is preserved if the molecules are free to rotate about this axis [12]. Finally, certain rod-like molecules, called phasmids because of their six aliphatic chains [13], can yield columnar phases of hexagonal (Φ_h) or oblique (Φ_{ob}) symmetry. Here, the letter Φ refers to the "phasmidic" nature of the molecule.

II.5 Chiral smectic phases

The preceding classification of smectic phases only applies to non-chiral substances. What happens then when the molecules are chiral? Several structures can appear. One of them, called SmC*, is a twisted version of the SmC phase [14] and is characterized by the fact that the molecules are tilted and turn from one layer to the next with a constant angle about the layer normal. This phase can be ferroelectric after unwinding and can therefore have interesting applications. This topic will be presented in the next volume.

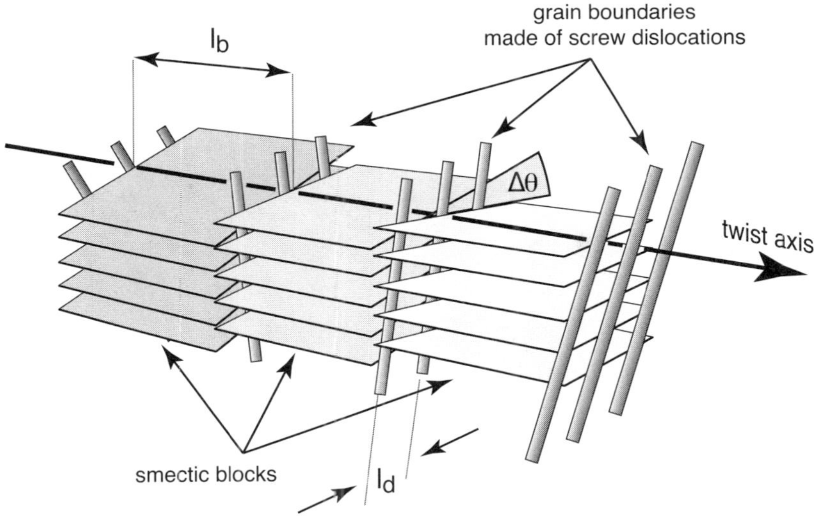

Fig. A.II.15 Renn and Lubensky model of the TGBA phase (from "twist-grain-boundary"). In this twisted smectic phase, the smectic A layers form blocks of width l_b separated by screw dislocation walls. Inside each wall, dislocations are parallel and equidistant, but their orientation turns with an angle $\Delta\theta$ between two successive walls, so they are on the average normal to the smectic layers and therefore retain the "screw" character. Note that $\Delta\theta$ is also the angle of the layers between two successive blocks.

Another solution, theoretically predicted by de Gennes in 1973 by analogy with superconductors [15] and experimentally evidenced 25 years later

[16], is obtained by slicing the smectic A (or smectic C) phase in blocks, subsequently assembled in a helical structure. Between two neighboring blocks appears a wall of screw dislocations (Fig. A.II.15). The phase is therefore called TGB, from "twist-grain-boundary." An entire chapter of the second volume is devoted to this phase.

Later, we shall also discuss the recent discovery of so-called Smectic Blue Phases [17,18], still under study (chapter C.V of the second volume). The next chapter discusses the materials that can exhibit mesomorphic phases.

BIBLIOGRAPHY

[1] a) Friedel G., *Leçons de Cristallographie*, Librairie Scientifique Albert Blanchard, Paris, 1964. In passing, let us mention that the classification of crystal symmetries had been completed before X-ray diffraction enabled "seeing" the position of atoms in crystals. Relying on their intuition, mineralogists such as Haüy and Bravais inferred the periodical arrangement of atoms from the crystal faceting. Mathematicians like Schönflies described all the possible symmetries of periodical three-dimensional atomic lattices.
b) We have learned that, as early as 1919, G. Friedel had asked his son, E. Friedel, to confirm by X-ray scattering (in the De Broglie laboratory) the layered structure of smectics (J. Friedel, private communication).

[2] a) Anderson P.W., chapter 2, "Basic Principles I: Broken Symmetry," in *Basic Notions of Condensed Matter Physics*, Addison-Wesley Publishing Company, New York, 1984.
b) Curie P., *Ann. Chimie Phys.*, **5** (1895) 289.
c) Landau L., Lifchitz E., *Statistical Physics*, Mir, Moscow, 1967.

[3] Salomaa M.M., Volovik G. E., *Rev. Mod. Phys.*, **59** (1987) 533.

[4] The first smectic A-smectic A transition was discovered in the DB5-TBBA mixture by Sigaud G., Hardouin F., Achard M.F., *Phys. Lett.*, **72A** (1979) 24. For a detailed overview of this kind of transition see the book by de Gennes P.-G. and Prost J., *The Physics of Liquid Crystals*, Oxford University Press, Oxford, 1995.

[5] Thouless D., "Condensed matter physics in less than three dimensions," in *The New Physics*, Ed. Davies P., Cambridge University Press, Cambridge, 1992.

[6] Pershan P.S., in *Structure of Liquid Crystal Phases*, World Scientific, Singapore, 1988.

[7] a) Levelut A.M., Germain C., Keller P., Liebert L., *J. Physique (France)*, **44** (1983) 623.
b) Galerne Y., Liebert L., *Phys. Rev. Lett.*, **64** (1990) 906.
c) Hamelin P., Ph.D. Thesis, University of Paris XI, Orsay, 1991.

[8] a) Levelut A.M., Doucet J., Lambert M., *J. Physique (France)*, **35** (1974) 773.
b) Levelut A.M., *J. Physique Colloq. (Paris)*, **37** (1976) C3.51.
c) Doucet J. in *Molecular Physics of Liquid Crystals*, Eds. Gray G.W. and Luckhurst G.R., Academic Press, New York, 1979.

[9] Chandrasekhar S., Sadashiva B.K., Suresh K.A., *Pramana*, **9** (1977) 471.

[10] Destrade C., Nguyen H.T., Gasparoux H., Malthête J., Levelut A.M., *Mol. Cryst. Liq. Cryst.*, **71** (1981) 111.

[11] a) Malthête J., Collet A., *Nouv. J. Chim.*, **9** (1985) 151.
b) Levelut A.M., Malthête J., Collet A., *J. Physique (France)*, **47** (1986) 351.
c) Zimmermann H., Poupko R., Luz Z., Billard J., *Z. Naturforsch. A: Phys., Phys. Chem., Kosmophys.*, **40A** (1985) 149; ibid., **41A** (1986) 1137.
d) Malthête J., Collet A., *J. Am. Chem. Soc.*, **109** (1987) 7545.
e) For a review see Collet A., Dutasta J.P., Lozach B., Canceill J., "Cyclotriveratrylenes and cryptophanes: their synthesis and applications to host-guest chemistry and to the design of new materials," in *Topics in Current Chemistry*, Vol. 165, Springer-Verlag, Heidelberg, 1993, p. 104.

[12] Levelut A.M., Oswald P., Ghanem A., Malthête J., *J. Physique (France)*, **45** (1984) 745.

[13] Malthête J., Levelut A.M., Nguyen H.T., *J. Physique Lett.*, **46** (1985) L-880.

[14] Meyer R.B., Liébert L., Strzelecki L., Keller P., "Ferroelectric liquid crystals," *J. Physique (France)*, **36** (1975) L-69. This paper reports the discovery of the first ferroelectric mesophase.

[15] De Gennes P.-G., "On analogy between superconductors and smectics A," *Sol. State Comm.*, **10** (1972) 753.

[16] Goodby J.W., Waugh M.A., Stein S.M., Chin E., Pindak R., Patel J.S., "Characterization of a new helical smectic liquid crystal," *Nature*, **337** (1989) 449.

[17] a) Li M.H., Lause V., Nguyen H.T., Sigaud G., Barois Ph., Isaert N., *Liq. Cryst.*, **23** (1997) 389.
b) Pansu B., Li M.H., Nguyen H.T., *J. Phys. II (France)*, **7** (1997) 751.
c) Pansu B., Grelet E., Li M.H., Nguyen H.T., *Phys. Rev. E*, **62** (2000) 658.
d) Grelet E., Pansu B., Li M.H., Nguyen H.T., *Phys. Rev. Lett.*, **86** (2001) 3791.
e) Grelet E., Pansu B., Nguyen H.T., *Phys. Rev. E*, **64** (2001) 10703.
f) Grelet E., "Etude structurale des phases bleues smectiques," Ph.D. Thesis, University of Paris VII, 2001.

[18] Kamien R., *J. Phys. II (France)*, **7** (1997) 743.

Chapter A.III

Mesogenic anatomy

Mesophases are found in all kinds of materials. To specify the nature of the mesogenic material one distinguishes between:

1. Thermotropic LCs (low molecular mass species)
2. Lyotropic LCs (aqueous solutions of amphiphiles)
3. Polymeric LCs (reticulated or not)
4. Colloidal LCs (suspensions of colloidal particles, the best known being the tobacco mosaic virus)

This distinction is important because, even though a mesophase always has, by definition, the same symmetries and the same laws of macroscopic behavior, its physical constants strongly depend on the chosen material. Thus, thermotropic nematics flow very easily (with viscosities ranging from that of water to that of glycerine) while polymeric (unreticulated) nematics are extremely viscous.

The rest of this chapter is devoted to describing, using concrete examples, the general architecture of mesogenic molecules. The examples are deliberately chosen from thermotropic (section III.1) and lyotropic (section III.2) liquid crystals, which constitute the main topic of this book. Special attention will be paid to the concept of molecular frustration, which is the key to mesomorphic polymorphism. At the end of the chapter, diblock copolymers (section III.3) and colloidal liquid crystals (section III.4) will be discussed briefly.

III.1 Thermotropic liquid crystals

Practically all mesogenic substances known today were obtained by synthesis [1, 2]. Their number is impressive and they are thought to account for about 5%

of the synthesized organic products. The work achieved by chemists in this regard is admirable, for they continuously design new mesogenic molecules that engender new phases.

Nowadays, new synthesis is no longer performed at random; on the contrary, it is the result of reflection based upon a considerable body of knowledge, acquired over almost a century. The essential question, which has been asked by Vorländer, Lehmann, and Friedel is the following:

"For what reason does a given substance go through mesomorphic states instead of passing directly from the crystalline solid to the isotropic liquid?"

III.1.a) Hybrid molecular form and molecular frustration

To find the first clues, let us consider the phase sequences of figure A.III.1.

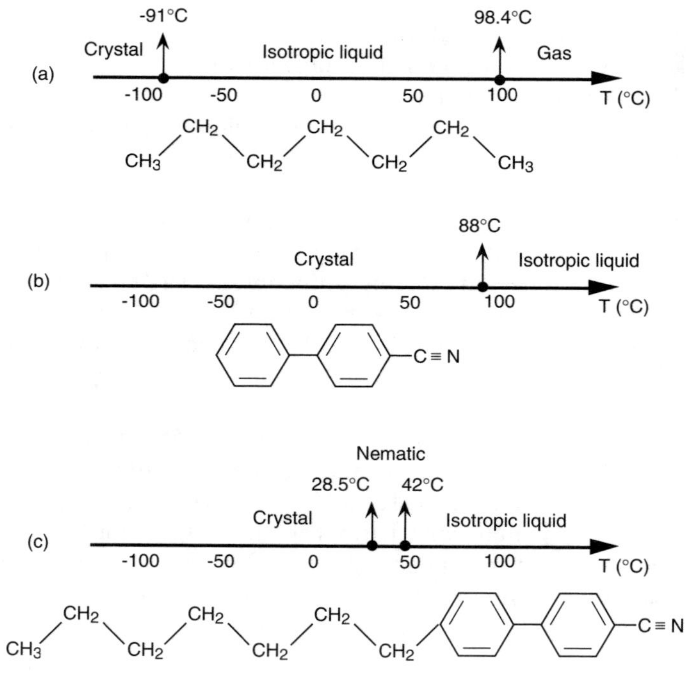

Fig. A.III.1 Phase sequence of three organic substances; the third one, known as 7CB, is mesogenic. It is obtained by covalently bonding two very dissimilar and non-mesogenic molecules, heptane (A) and 4-cyanobiphenyl (B).

The first two sequences belong to very dissimilar and non-mesogenic molecules, heptane (A) and 4-cyanobiphenyl (B). The third sequence is that of the substance C (A-B) obtained by chemically bonding the first two molecules.

These species were not randomly chosen; indeed, mesogenic A-B, known as 7CB, is a component of many nematic mixtures used in liquid crystal displays [3, 4].

Thermodynamically, at fixed temperature and pressure a substance chooses between two possible phases α or β the one that has the lowest chemical potential $\mu_{\alpha,\beta}$ (T,p). For a given pressure, μ is a function of temperature and depends on the phase α or β of the system. For heptane and 4-cyanobiphenyl, the solid crystalline phase S is in competition with the liquid phase L. The S-L transition takes place at the temperature at which the curves $\mu_L(T)$ and $\mu_S(T)$ cross.

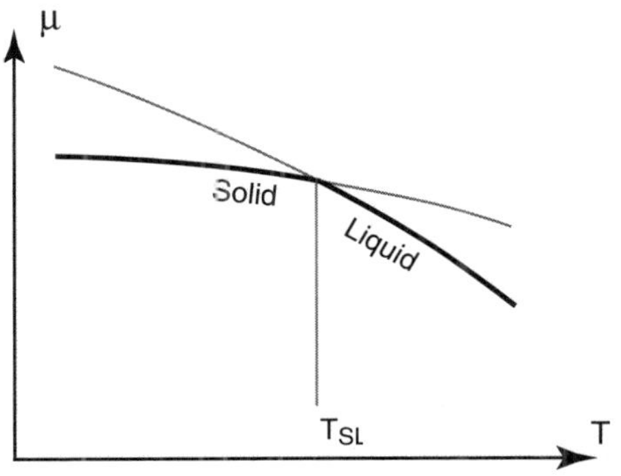

Fig. A.III.2 Chemical potential variation close to a first-order transition between the solid crystalline phase (S) and the liquid phase (L).

In order to understand better the microscopic nature of this transition, one should recall that the chemical potential μ is the Gibbs free energy G per molecule:

$$\mu = G/N \quad \text{with} \quad G = U - TS + pV \qquad \text{(A.III.1)}$$

where G is the thermodynamic potential to be minimized at fixed T and p.

In a three-dimensional crystal, molecules are placed on the sites of a crystalline lattice. When T = 0, they are optimally stacked in order to minimize the potential energy U of intermolecular attraction. At constant pressure p, let the temperature increase up to $T \geq T_{SL}$. At the melting point T_{SL}, the entropy $\Delta S = S_L - S_S$, gained as the molecules quit their fixed positions, exactly balances the increase $\Delta U = U_L - U_S$ in internal energy due to this transformation and the work $p\Delta V = p(V_L - V_S)$ necessary for dilating the system. Consequently, melting takes place when:

$$\Delta G_{SL} = \Delta U - T\Delta S + p\Delta V = 0 \qquad \text{(A.III.2)}$$

Hence, when the attraction between the molecules is stronger, it is more difficult for the substance to melt, so the melting temperature is higher.

The attraction between two organic molecules can be due to:
1. Direct interaction between their permanent electric dipoles – this interaction is attractive for certain positions of the molecules;
2. Permanent dipole-induced dipole interaction – this interaction depends on the molecular polarizability;
3. Induced dipole-induced dipole interaction named after van der Waals – this interaction also depends on the molecular polarizability.

In alkanes such as heptane, the first two kinds of interaction do not exist, because the molecule bears no permanent dipole; moreover, dispersion forces are weak due to the low polarizability of these molecules which have no double bond. Therefore, it is not surprising that at ambient pressure, heptane melts at a temperature as low as $T_{SLA} = -91°C$ [5] (and boils at 98.4°C).

On the other hand, 4-cyanobiphenyl melts at a much higher temperature, around $T_{SLB} = 88°C$. The height of the melting point is due to the presence of the cyano group –C≡N, which has a strong dipolar moment, and to the high polarizability of the aromatic rings. In conclusion, the 7CB molecule consists of two parts, each of them having a classical phase sequence but with very different melting temperatures.

Let us now see how mesophases are created. As shown in figure A.III.3, merely mixing the two substances is not enough. Indeed, the phase diagram of the binary mixture (A+B) is altogether classical, comprising a liquid isotropic phase and two crystalline phases of almost pure A or B (the phase diagram for the case of a diluted binary mixture will be discussed in detail in chapter B.IX).

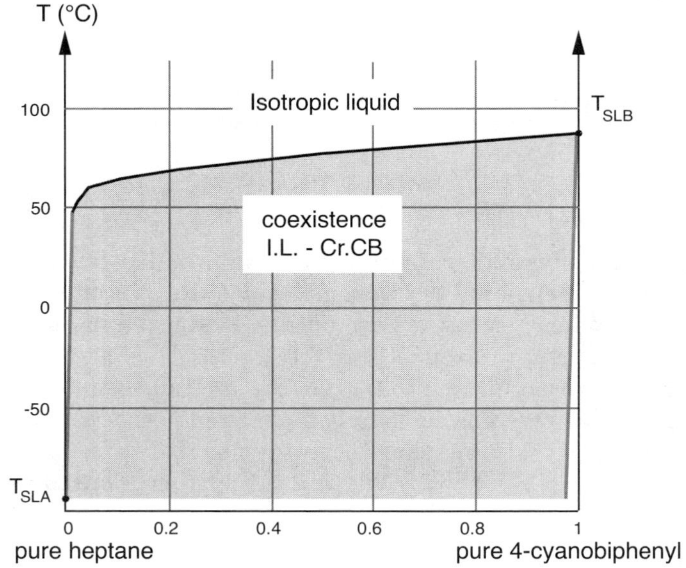

Fig. A.III.3 Phase diagram of the binary mixture heptane / 4-cyanobiphenyl. The thick line corresponds to measurements performed on mixtures of 1.9%, 4.9%, and 50% molar concentration. Thin lines are hypothetical and have been included for clarity.

When we consider an equimolar mixture of the two components A and B (one mole of A for one mole of B), the diagram shows that this combination can be homogeneous only at high enough temperature in the liquid isotropic phase. Below about 77°C, phase separation occurs between a heptane-rich isotropic liquid and a crystal consisting of almost pure 4-cyanobiphenyl. This separation is the sign of chemical incompatibility between the two coexisting species.

The situation is completely different when the molecules are chemically associated by a covalent bond:

In this case the phase separation is impossible and the system must find a different solution to the conflict opposing the two parts of the molecule. Forming mesophases is one solution to this *molecular frustration.*

In the case of the nCB family, polymorphism (the phase sequence as a function of temperature) depends on the length n of the aliphatic chain C_nH_{2n+1}. As shown by the diagram in figure A.III.4, a nematic phase appears between the crystalline and liquid isotropic phases for $n \geq 4$.

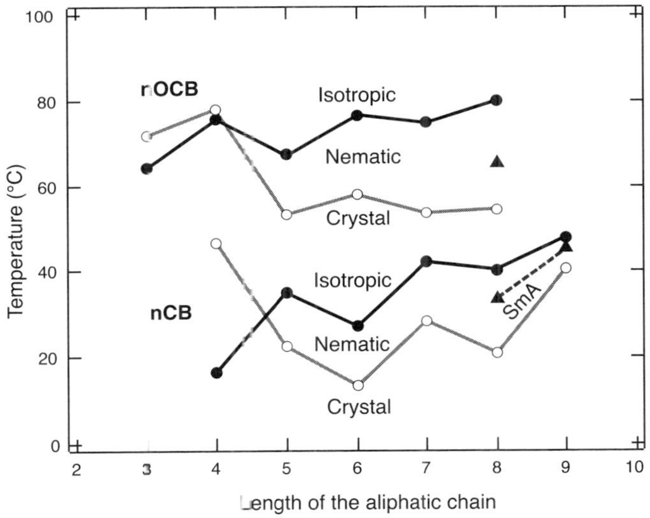

Fig. A.III.4 Polymorphism of the corresponding series nCB and nOCB [3].

For 8CB and 9CB, a smectic A phase appears between the crystal and the nematic phase. In the alkoxy analogues 3OCB and 4OCB and in 4CB, the nematic phase only appears upon cooling (in this case, the N-I transition is said to be "monotropic"). Notice that for 7CB, the nematic to isotropic transition temperature is about 40°C, while the 50% cyanobiphenyl-heptane mixture only becomes homogeneous above 77°C (Fig. A.III.3).

III.1.b) Importance of small structural details

The diagram in figure A.III.4 also contains data on another homologous series, that of the nOCBs. Molecules in this family differ from the previous ones only by the presence of an oxygen atom between the biphenyl component and the aliphatic chain (Fig. A.III.5).

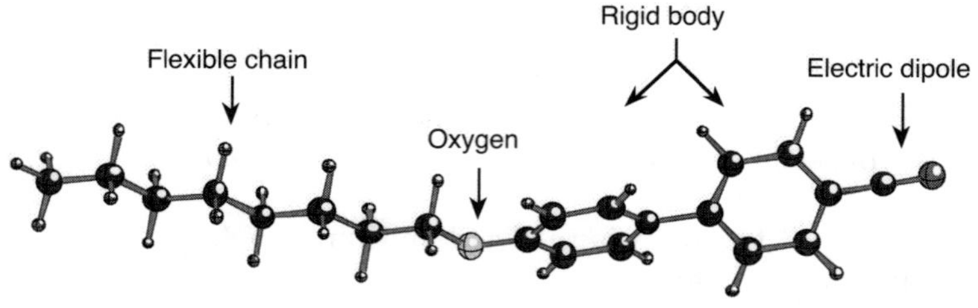

Fig. A.III.5 Molecular structure of 8OCB. The molecule has a cyano group at its extremity.

This apparently minor detail has however a very strong influence on the polymorphism. Indeed, the diagram in figure A.III.4 shows that the crystal-nematic or nematic-isotropic liquid transition temperatures are 20°C higher in nOCB than in nCB for the same length of the aliphatic chain. The diagram also shows that the transition temperatures have a "sawtooth" dependence on the length of the aliphatic chain. This feature, known as the "even-odd effect," is also present at the solid-liquid transition in alkanes. It is related to the parity effect detected by NMR measurements in the evolution of the orientational order parameter along the aliphatic chain as a function of the number n of segments [6].

III.1.c) Eutectic mixtures

It is evident that the pure substances of type nCB or nOCB cannot be directly used in liquid crystal displays since the temperature range of the nematic phase $\Delta T_N = T_{N/I} - T_{Cr/N}$ is either too narrow or at too high temperature. Hence the question:

How can one extend the temperature range ΔT_N of the nematic phase without reducing the nematic-isotropic transition temperature $T_{N/I}$?

The simplest method consists of blending the different products in proportions corresponding to the eutectic point. This idea comes to mind when investigating the phase diagrams of binary metallic alloys. For certain metals mixed in suitable proportions (called eutectic), the transition to the crystalline phases is sharp, without a coexistence range with the isotropic liquid. This direct crystallization takes place at a temperature lower than the melting points of the two pure metals. A good example is that of the soldering alloy: a eutectic mixture of lead and tin.

Applying the principle of eutectic mixtures to mesophases relies upon the fact that the latent heat $\Delta H_{N,I}$ at the nematic-isotropic transition is about 20 to 50 times lower than the latent heat $\Delta H_{Cr/N}$ at the crystal-nematic transition. Therefore, almost all the increase in entropy due to the loss of crystalline order takes place at the crystal-nematic transition.

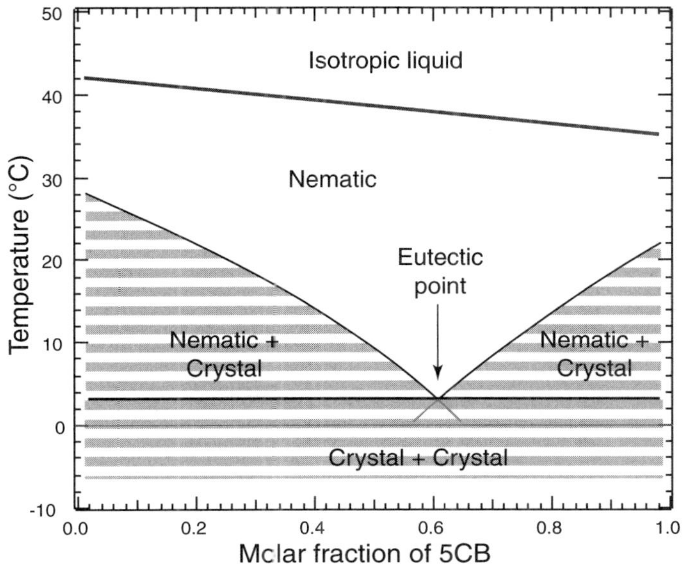

Fig. A.III.6 Theoretical phase diagram fcr the 5CB/7CB blend computed from the Schröder-van Laar law. Solidus lines are not shown.

The proportions of eutectic mixtures can be very easily determined in the following way. Let ΔH_i be the latent heat at the crystal-nematic transition for component i and T_i the transition temperature. The Schröder-van Laar law gives the molar ratio x_i of component i dissolved in the nematic phase at $T < T_i$ (assuming ideal behavior):

$$\ln x_i = \frac{\Delta H_i}{R}\left(\frac{1}{T_i} - \frac{1}{T}\right) \qquad \text{(A.III.3)}$$

The eutectic composition is defined by the relation:

$$\sum_{i=1}^{N} x_i = 1 \qquad \text{(A.III.4)}$$

since at the eutectic point the liquidus lines of the crystal/nematic coexistence ranges converge to one point. These lines generalize to hypersurfaces in a mixture with more than two components. Using the 5CB/7CB blend demonstrates this approach. The latent heat for the crystal-nematic phase transition in these materials is [3] $\Delta H_{5CB} = 4.1$ kcal/mole and $\Delta H_{7CB} = 6.2$ kcal/mole, respectively. Knowing the transition temperatures (Fig. A.III.4) one can compute the molar ratios $x_{5CB}(T)$ and $x_{7CB}(T)$ from eq. A.III.3: $x_{5CB} \approx 60\%$ and $x_{7CB} \approx 40\%$. Graphical determination (Fig. A.III.6) of the eutectic point satisfying eq. A.III.4 yields $T_{eutectic} \approx 3°C$.

For this mixture, the solidification temperature is decreased to 3°C. The goal of reducing the lower limit of the nematic domain has therefore been achieved.

The nematic-isotropic transition temperature of the mixture still needs to be determined. In principle, this transition is not sharp, as for a pure product, but rather takes place over a temperature interval $\Delta T_{N/I}$ where the two phases coexist. At thermal equilibrium, the composition of the nematic and isotropic phases is given by the van Laar equations:

$$\ln\left(\frac{x_i^N}{x_i^I}\right) = \frac{\Delta H_i}{R}\left(\frac{1}{T_i^{NI}} - \frac{1}{T}\right) \qquad i = \text{5CB and 7CB} \qquad \text{(A.III.5a)}$$

with

$$x_{5CB}^N = 1 - x_{7CB}^N \quad ; \quad x_{5CB}^I = 1 - x_{7CB}^I \qquad \text{(A.III.5b)}$$

This law applies when the products are miscible in all proportions in the two phases, which is true in the case under consideration. Solving these equations with respect to x_{5CB}^N and x_{5CB}^I yields:

$$x_{5CB}^N = \frac{1 - \exp(A_{7CB})}{\exp(A_{5CB}) - \exp(A_{7CB})} \quad \text{and} \quad x_{5CB}^I = x_{5CB}^N \exp(A_{5CB}) \qquad \text{(A.III.6a)}$$

with

$$A_{5CB} = \frac{\Delta H_{5CB}^{NI}}{R}\left(\frac{1}{T_{5CB}^{NI}} - \frac{1}{T}\right) \quad \text{and} \quad A_{7CB} = \frac{\Delta H_{7CB}^{NI}}{R}\left(\frac{1}{T_{7CB}^{NI}} - \frac{1}{T}\right) \qquad \text{(A.III.6b)}$$

Knowing that $\Delta H_{5CB}^{NI} \approx \Delta H_{7CB}^{NI} \approx 100R$, the concentrations x_{5CB}^N and x_{5CB}^I determined as a function of temperature give a very narrow coexistence range (almost invisible at the scale of figure A.III.6). Consequently, for the eutectic composition, the nematic-isotropic transition temperature is around 38°C.

To further extend the temperature range of the nematic phase, more than two products have to be mixed. For example, in a blend containing 36% 7CB, 18% 3CB, 15% 5OCB, 12% 7OCB, and 19% 8OCB, the nematic phase exists between $T_{CrN} = 0°C$ and $T_{NI} = 61°C$.

III.1.d) Precursors and "relatives" of cyanobiphenyls

With respect to other longer-known families of mesomorphic molecules, cyanobiphenyls stand out because of their chemical stability and high dielectric anisotropy.

Among the precursors, one should first of all mention PAA or *para*-azoxyanisole, a mesogenic synthetic substance already known in Friedel's time. PAA belongs to the di-alkyloxyazoxybenzenes, of chemical structure described in figure A.III.7a. Like biphenyls, molecules of this group consist of a rigid polarizable body, bearing a permanent dipolar moment. They do however include two aliphatic chains C_nH_{2n+1} instead of one, linked by oxygen atoms to the extremities of the central body. Mesomorphic polymorphism of these molecules depends on the length n of the aliphatic chains, as shown in figure A.III.8. For n < 7, only the nematic phase is present. When n ≥ 7, the smectic C phase appears between the crystalline and nematic phases. Finally, for n ≥ 10 the nematic phase is replaced by a smectic C phase.

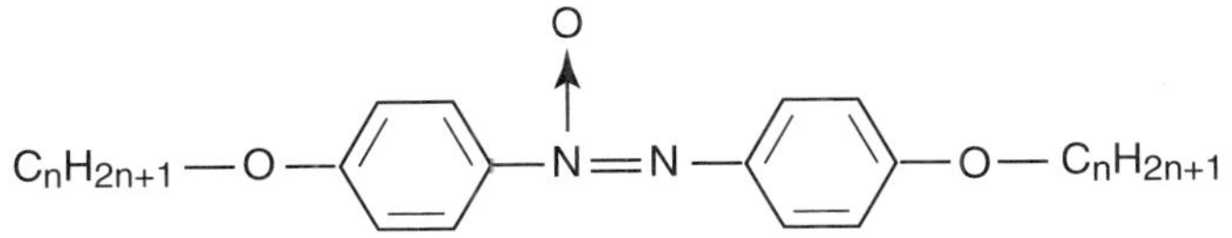

(a)

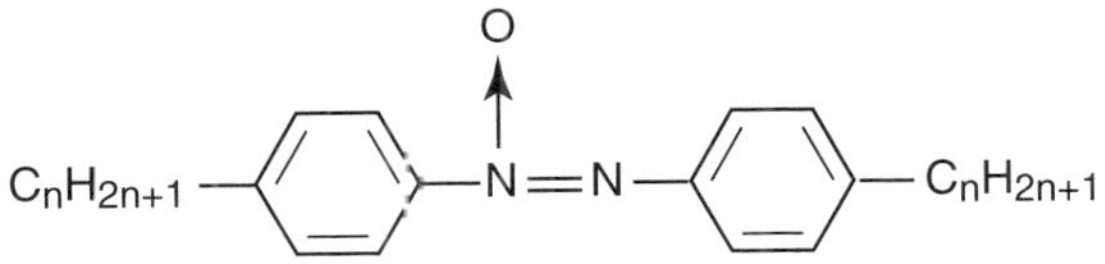

(b)

Fig. A.III.7 Family portrait of: a) di-alkyloxyazoxybenzenes (n = 1 corresponds to PAA), and b) di-alkylazoxybenzenes.

Figure A.III.7b displays the chemical formula of the di-alkylazoxybenzenes, a group closely related to that of PAA. Aliphatic chains are directly attached (without the oxygen atom) to the same central body. The effect of this small structural change resembles the one already mentioned when discussing the difference between nOCB and nCB: as shown by the diagram in figure A.III.8, the melting temperature of the crystalline phase and the temperature range of the mesomorphic phases are reduced by removing the oxygen atom. This gradually brings us to the following conclusions:

1. Lengthening the aliphatic chains favors smectic order, the molecules being situated in well-defined layers;

2. Attaching the aliphatic chains to the central body by an oxygen atom increases the transition temperatures.

Another, less expected, effect of inserting an oxygen atom is the replacement of the smectic C phase by smectic A.

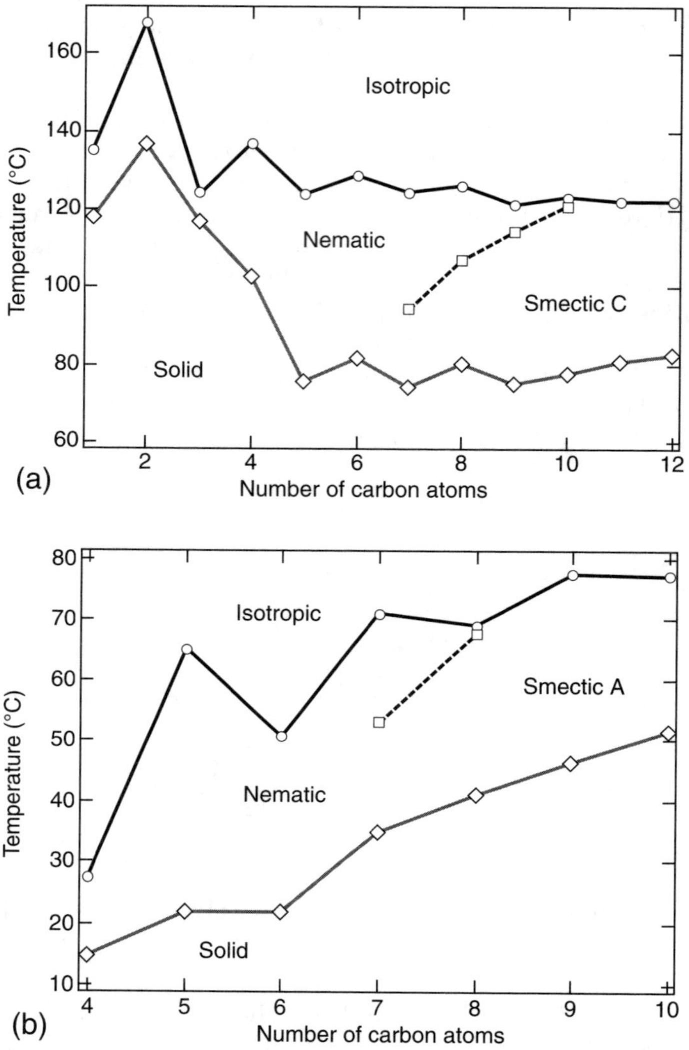

Fig. A.III.8 Mesomorphic polymorphism of the di-alkyloxyazoxybenzenes (n = 1; PAA) (a) and di-alkylazoxybenzenes (b).

Once involved in the game of molecular design, one can try linking one of the chains directly to the rigid body, while the other is attached by means of an oxygen atom. A sort of crossbreed is then obtained.

The family of *p*-alkyloxybenzilidene-*p*-n-alkylanilines provides an example of such molecular hybrids (Fig. A.III.9).

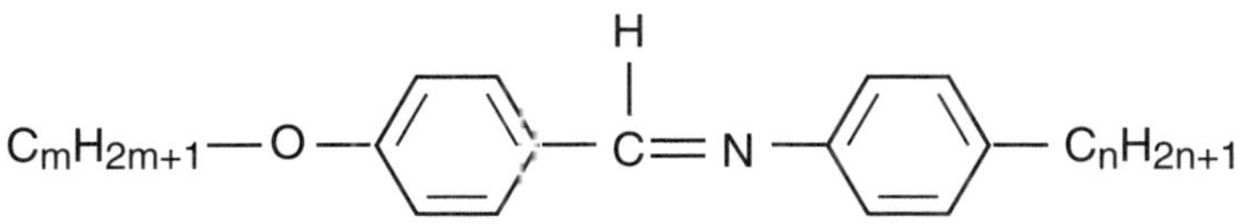

Fig. A.III.9 Family portrait of p-alkyloxybenzilidene-*p*-n-alkylanilines, denoted by mO.n. The most famous member is MBBA, the first nematic at room temperature. MBBA corresponds to m = 1 and n = 4 (1O.4).

The best-known member of this family is MBBA, the first nematic at room temperature. Indeed, as the diagram in figure A.III.10 shows, its crystal-nematic melting temperature is only 21°C (thus, MBBA is nematic in the summertime, but not during the winter). The other members of the group with larger n and/or m are known for exhibiting very rich mesomorphic polymorphism. For instance, the phase sequence of 5O.6 is as follows:

$$\mathrm{Cr} \xrightarrow{32.5°\mathrm{C}} \mathrm{SmG} \xrightarrow{40°\mathrm{C}} \mathrm{SmF} \xrightarrow{43°\mathrm{C}} \mathrm{SmB} \xrightarrow{51°\mathrm{C}} \mathrm{SmC} \xrightarrow{53°\mathrm{C}} \mathrm{SmA} \xrightarrow{61°\mathrm{C}} \mathrm{N} \xrightarrow{73°\mathrm{C}} \mathrm{Iso}$$

Unfortunately, this family is notorious for low chemical stability. The bond between two aromatic cycles, known as a Schiff base linkage, is indeed easily broken in the presence of water. Therefore, the product gradually decomposes and its content of impurities (decomposition by-products) increases.

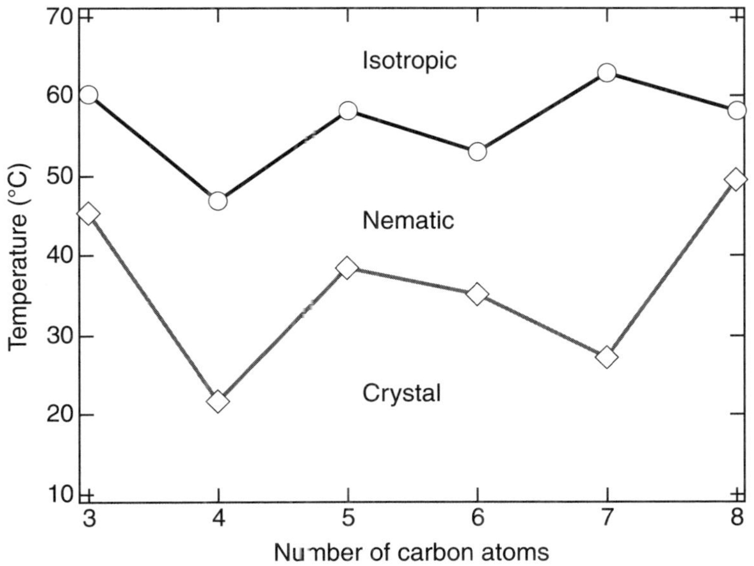

Fig. A.III.10 Mesomorphic polymorphism of the 1O.n. family; MBBA corresponds to n = 4 (1O.4).

III.1.e) General architecture of smectogenic and nematogenic products: calamitic molecules

In all the previous examples, the mesomorphic molecule was formed by a central rigid body, including at least two aromatic groups, to which are attached one or more aliphatic chains. These rod-like (calamitic) molecules often present nematic and/or smectic phases. Their general formulae are given in figure A.III.11. As the table indicates, there is a multitude of possible combinations. However, this list is still far from complete, as shown in the review article by Demus [3].

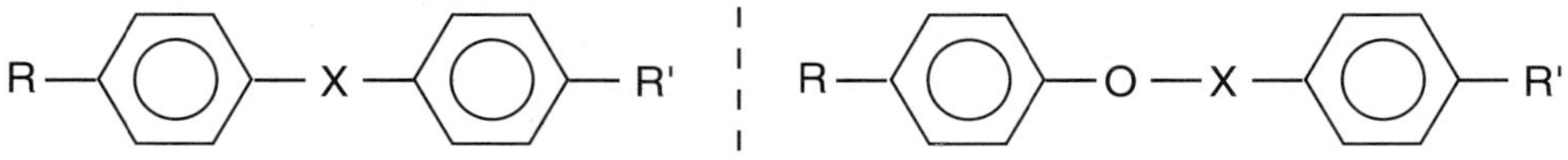

R	X	R'
C_nH_{2n+1} –	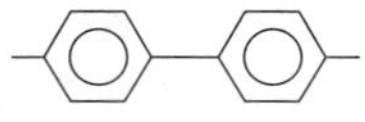	R
C_nH_{2n+1}-O –	– CH = N –	– C ≡ N –
C_nH_{2n+1}-COO –	– COO –	– Cl
C_nH_{2n+1}-OCOO –	– C ≡ C –	– Br
		– F
	– COO  COO –	– NO_2
	– COO  OOC –	
	– OOC  OOC –	
	– C ≡ C –	

Fig. A.III.11 Some examples of calamitic molecules (according to de Gennes and Prost).

III.1.f) Other examples of mesomorphic molecules: discoidal molecules

Columnar phases, presented in the previous chapter, consist of disk-like molecules piled up in parallel columns that give a two-dimensional lattice. The first discoidal molecule to exhibit a columnar phase was a hexasubstituted derivate of benzene. This compound, consisting of a rigid core (here, a benzene group) to which are attached six flexible chains, belongs to the homologous series of benzene-hexa-n-alkanoates (BHn for short) where n is the number of carbon atoms on the aliphatic chain (Fig. A.III.12). Only compounds with n = 7 to 9 give mesomorphic phases. As one can see from the phase sequence of BH7 (Fig. A.III.12) the columnar phase spans only a few degrees in temperature. As this is a general feature of the series, chemists are still looking for molecules that would exhibit richer polymorphism over a wider temperature range.

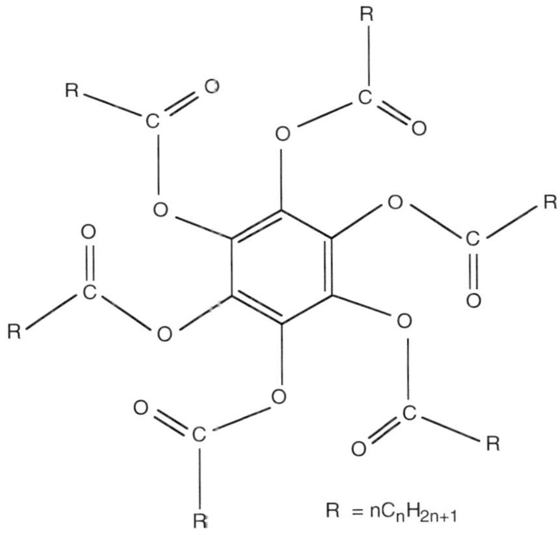

$$\text{Crystal} \xrightarrow{81.2°C} \text{Columnar hexagonal phase} \xrightarrow{87°C} \text{Isotropic liquid}$$

Fig. A.III.12 Homologous series of BHn (benzene-hexa-n-alkanoates). It is to this series that the first columnar phase belongs, discovered in 1977. The phase sequence presented is that of BH7 (from ref. [7]).

The synthesis of such molecules has progressed in recent years. Among the most studied are the series of triphenylenes (Fig. A.III.13a) [8] and truxenes (Fig. A.III.13b) [9]. While triphenylene ethers and esters (CnHET and CnHAT, respectively) only give columnar phases (Fig. A.III.14), truxene esters

(CnHATX) also present a discotic nematic phase N_D (Fig. A.III.15a).

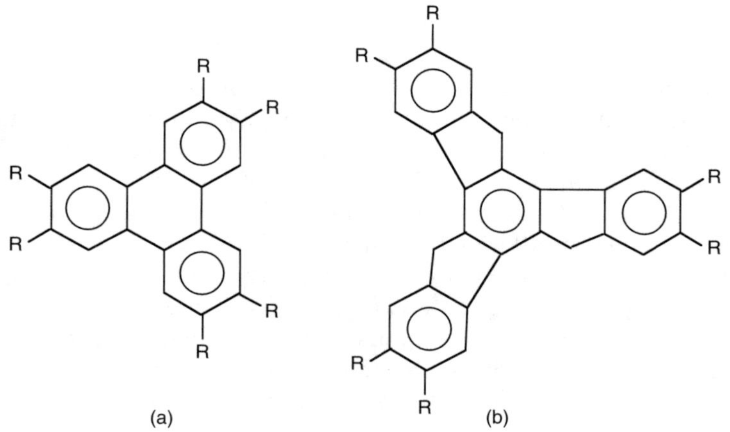

R n = number of C atoms	$-OC_nH_{2n+1}$ (ether)	$-OOC-C_nH_{2n+1}$ (ester)	$-OOC-\varphi-OC_nH_{2n+1}$ (benzoate)
Triphenylene (T)	CnHET	CnHAT	CnHBT
Truxene (TX)	CnHETX	CnHATX	CnHBTX

Fig. A.III.13 Homologous series of triphenylenes (a) [8] and truxenes (b) [9]. Compounds with three different types of R groups have been synthesized, the short names of which are given in the table. The letter φ represents a benzene ring. C_n chains are linear.

This last series shows that, as for smectic phases, columnar phases are more stable than the nematic phase when the aliphatic chains are lengthened. It would seem natural, and it is the case in our example, that the nematic phase, when present, appears at higher temperature than the columnar phase, but this is not a systematic effect. Indeed, one can have the N_D phase at lower temperature than the columnar phase(s), as in the surprising series of truxene esters (CnHATX) (Fig. A.III.15b). Moreover, some compounds even exhibit "re-entrant" behavior. This is the case with C11HBTX, for phase sequence [11]:

$$\text{Crystal} \xrightarrow{90^\circ C} D_r \xrightarrow{137^\circ C} N_D \xrightarrow{171^\circ C} D_r \xrightarrow{284^\circ C} N_D \xrightarrow{297^\circ C} \text{Isotropic liquid}$$

We notice the existence of two "reentrant" phases (N_D and D_r). This "re-entrance" phenomenon also appears in certain calamitic compounds [12].

Finally, imaginative chemists, whose creativity seems to have no limits, have synthesized cone-shaped molecules that stack together, giving rise to columns [13]. Such mesophases are called "pyramidic" (P) [14]. One example of such a molecule is depicted in figure A.III.16.

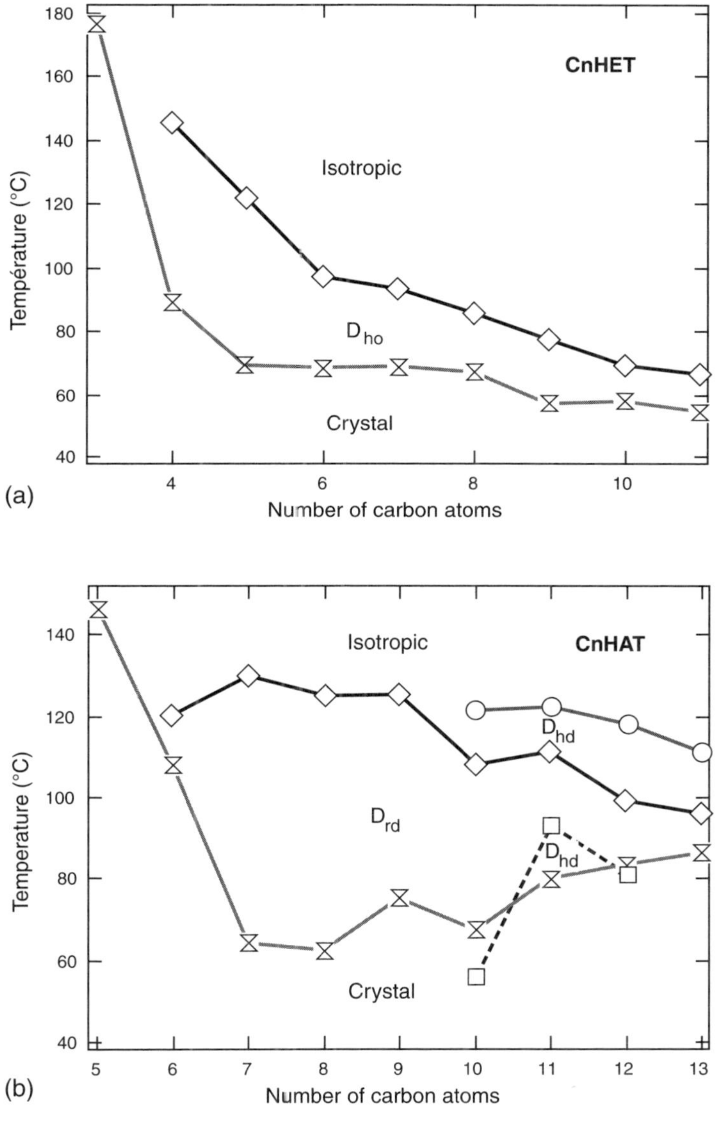

Fig. A.III.14 Phase transitions in the homologous series of CnHET (a) and CnHAT (b) (from ref. [10]).

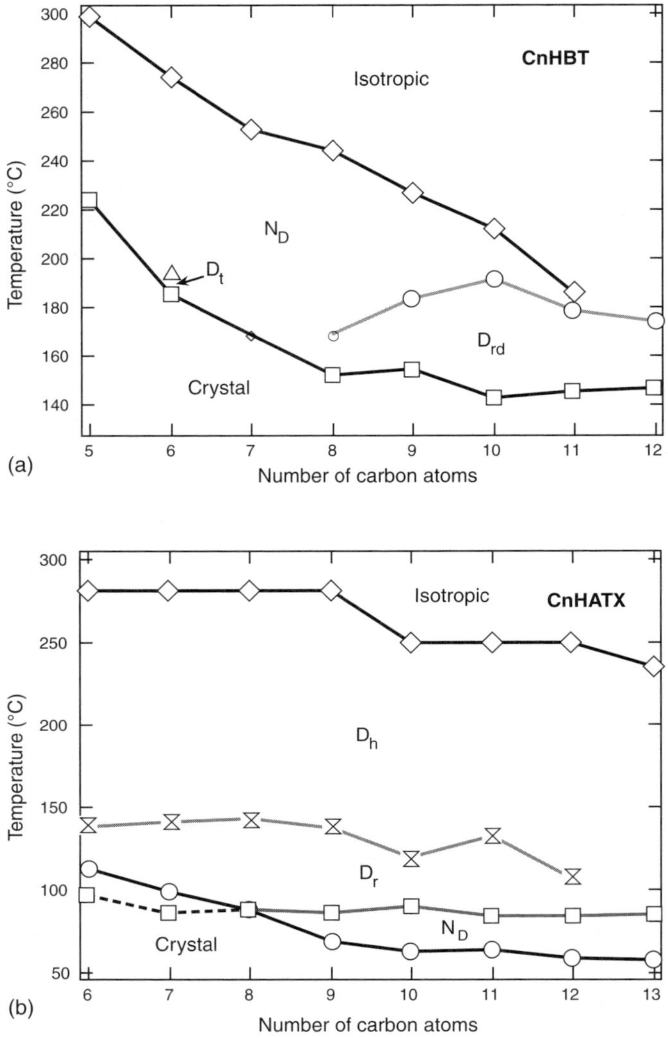

Fig. A.III.15 Phase transitions in the homologous series of CnHBT (a) and CnHATX (b). In series (b), the nematic phase is sandwiched between the solid and a columnar phase (from ref. [10]).

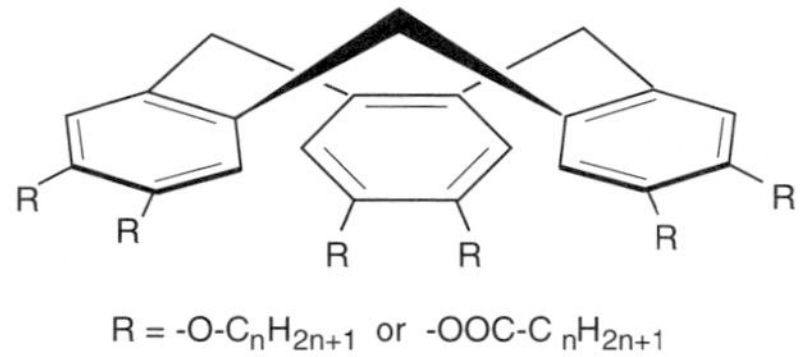

Fig. A.III.16 Example of a cone-shaped molecule (cyclotricatechylene hexaether or hexaester) producing columnar mesophases (called "pyramidic") (from ref. [13a]).

III.2 Lyotropic liquid crystals

As we pointed out in chapter A.I, the first mesophases were discovered in mixtures of myelin and water (myelinic figures). These biological molecules belong to the family of amphiphilic molecules which, by definition, possess two parts with very different affinities.

For instance, in phospholipids, which are the main component of the cell membrane, part of the molecule is hydrophobic (and avoids water at all costs), while the other part is hydrophilic (and looks for an aqueous environment) (Fig. A.III.17a). Surface-active molecules (surfactants) such as sodium dodecyl sulphate (SDS, Fig. A.III.17b) also belong to this category of molecules. Indeed, this surfactant molecule contains a hydrophobic linear alkyl chain (dodecyl, which is insoluble in water) and an ionic group $SO_4^-Na^+$, very hydrophilic because it is ionizable.

A last and more subtle example of amphiphilic molecules is that of non-ionic surfactants like $C_{12}EO_6$, represented in figure A.III.17d. In this case, the hydrophobic sequence (also called lipophilic) is a linear alkyl chain, C_{12}, as in SDS, while the hydrophilic sequence is a six-unit oligomer of ethylene oxide, soluble in water because of the hydrogen bonds it can form with it.

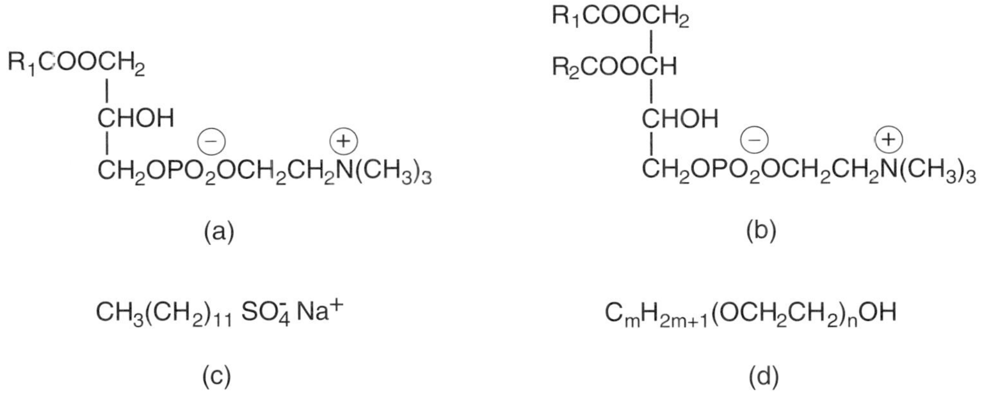

Fig. A.III.17 Examples of amphiphilic molecules: a) phospholipid with one saturated chain, 10 to 16 carbon atoms long (lysolecithin); b) double-chained phospholipid (phosphatidylcholine or lecithin). The R_1 chain is saturated, while R_2 contains a double bond C=C. The chains contain 12 carbon atoms; c) sodium dodecyl sulphate (SDS), an anionic detergent; d) non-ionic surfactant of the C_mEO_n series. $C_{12}EO_6$ is the hexa-ethylene glycol mono-n-dodecyl ether.

Let us now examine what happens when a surfactant is dissolved in water (Fig. A.III.18).

At very low concentrations, an equilibrium is established between molecules adsorbed in a film at the air-water interface with those dissolved in the bulk solution. The driving force for adsorption is the reduction of the surface tension (hence their name of surface-active molecules). Increasing the surfactant concentration leads to a saturated film at the interface and the formation of aggregates in the bulk water phase.

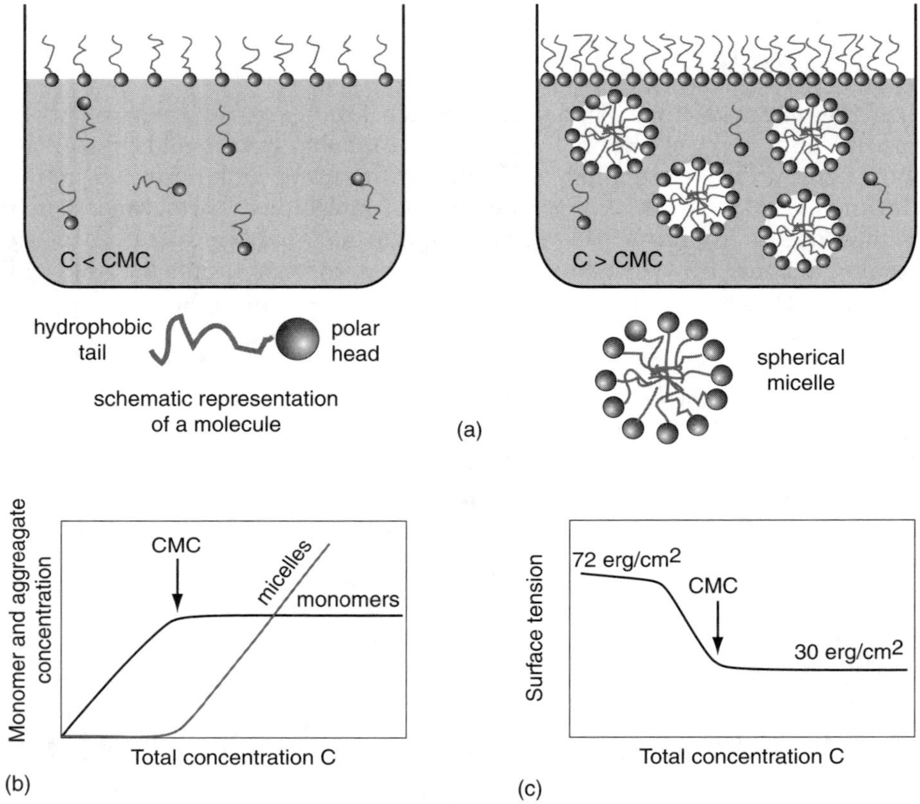

Fig. A.III.18 a) Surfactant molecules form a monolayer at the water surface. The film is saturated when the total concentration reaches the CMC. Above this concentration, micelles are formed; b) typical behavior of the monomer and aggregate (micelle) concentration as a function of the total surfactant concentration; c) evolution of the surface tension.

These changes are reflected in the surface tension versus concentration isotherm. As the total concentration increases, the film at the air-water interface becomes richer in surfactant, with a corresponding decrease in the surface tension. Once the surface film reaches saturation, no further decreases in the surface tension are observed and additional molecules form aggregates in the bulk solution. For simple single chain surfactants, the most common aggregate structure is that of spheres which are called micelles. As such, the surfactant concentration at which they form is referred to as the critical micelle concentration (CMC). This concentration is experimentally determined by the point where the surface tension isotherm develops a plateau (Fig. A.III.18). To provide an idea, the CMC is typically 8×10^{-3} M for SDS. Its value is however much smaller for lysolecithin, which has a C_{16} chain (CMC $\approx 7\times10^{-6}$ M). Thus, the CMC varies strongly from one molecule to another, and is very sensitive to the nature of the polar head and its counterion (if any). It also depends very strongly on the length of the aliphatic chain. For instance, the CMC of lysolecithin increases by a factor of 10 when its aliphatic chain is shortened by two carbon atoms.

One should notice that the aggregates formed at the so-called CMC are not always spherical. For example, double-chained lipids (like lecithin) form vesicles, while lysolecithin (which only has one tail) produces elongated micelles of cylindrical conformation. Thus, the shape of the aggregate crucially depends on the molecular structure [15]. More specifically, it depends on a dimensionless quantity named the "stacking parameter P." Of purely geometric origin, this quantity combines the optimal head-group area a_0 of the molecule (at the water-oil interface), its volume v, and the length of the aliphatic chain l_c (Fig. A.III.19):

$$P = \frac{v}{a_0 l_c} \qquad \text{(A.III.7)}$$

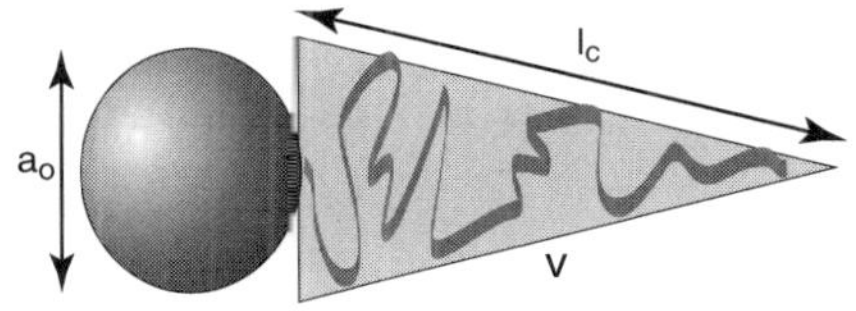

Fig. A.III.19 Each surfactant molecule is characterized by its area per polar head a_0, as well as by its length l_c and the volume v of its aliphatic chain.

This parameter characterizes a specific amphiphilic molecule, and its value will determine the shape of the aggregates formed.

Treating the chains as an incompressible liquid, the volume for a saturated chain containing n carbon atoms, this volume can be computed using the Tanford formula [16]:

$$v(\text{Å}^3) = 27.4 + 26.9\,n \qquad \text{(A.III.8)}$$

Experience shows that in many cases the area per polar head can be considered constant for a given system. However, the chain length can vary up to a maximal stretching length l_{max} given by the second Tanford formula [16]:

$$l_{max}(\text{Å}) = 1.5 + 1.265\,n \qquad \text{(A.III.9)}$$

The condition for the existence of an aggregate of a certain shape is therefore given by:

$$l_c < l_{max} \qquad \text{(A.III.10)}$$

Several shapes can simultaneously fulfill this condition: the system chooses the most symmetric form that increases its mixing entropy (e.g., sphere, cylinder, or plane (bilayer) (Fig. A.III.20)).

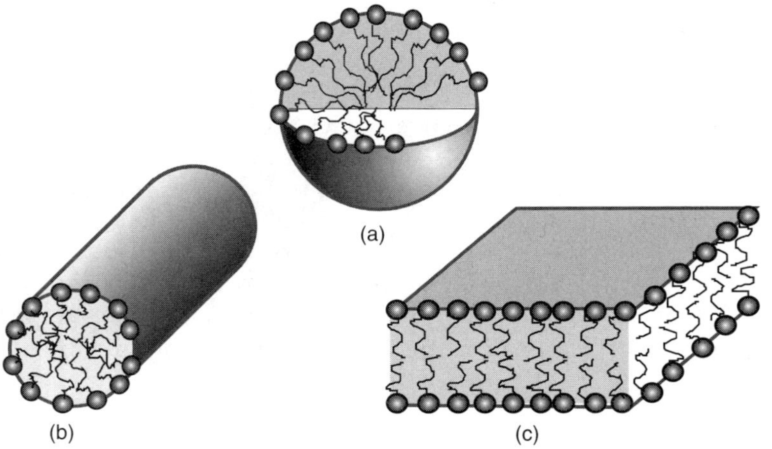

Fig. A.III.20 In order to keep their aliphatic tails from getting wet while exposing the polar heads to the water, surfactant molecules gather in a monolayer that can form a sphere (a), a cylinder (b), or a plane (c). In this last case, two monolayers bring their aliphatic sides together forming a bilayer (or lamella).

Simple geometrical considerations show that surfactant molecules form:
– spherical micelles when

$$P = \frac{v}{a_o l_c} < 1/3 \qquad \text{(A.III.11a)}$$

– elongated and locally cylindrical micelles when

$$1/3 < P = \frac{v}{a_o l_c} < 1/2 \qquad \text{(A.III.11b)}$$

– and vesicles (locally lamellar, Fig. A.III.21) when

$$1/2 < P = \frac{v}{a_o l_c} < 1 \qquad \text{(A.III.11c)}$$

For $P \approx 1$, the lamellae prefer being flat and try to minimize their total curvature. This configuration is achieved in swollen lamellar phases (very high water concentration) or in the L_3 (sponge) phase, which is an isotropic phase of fluctuating membranes. Locally, the lamellae have zero total curvature and negative gaussian curvature (see section B.II.2 for the definition of the curvatures).

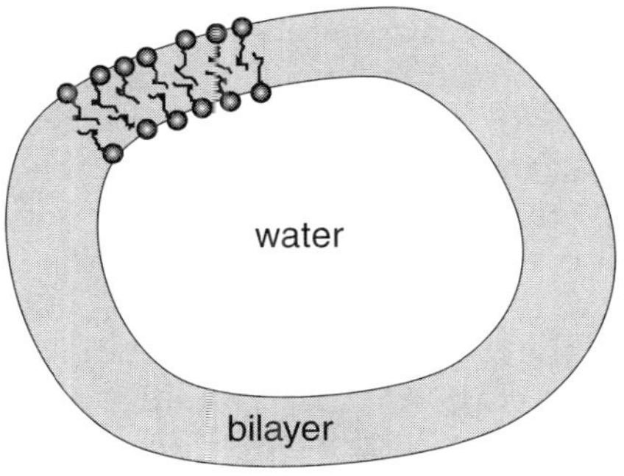

Fig. A.III.21 Water-filled vesicle.

Finally, one can have P > 1. In this case, the system forms inverted phases, consisting of water-filled aggregates with the aliphatic chains extending out of the aggregates. Citing examples, P = 0.37 for SDS and P = 0.84 for lecithin. Hence, SDS will form slightly elongated micelles (the value being close to the limit between the sphere and the cylinder) while lecithin gives vesicles.

In very dilute systems, the only important quantities are the mixing entropy and the binding energy α for two molecules in an aggregate. In particular, one can prove that [15]:

$$\mathrm{CMC} \approx \exp(-\alpha) \qquad \text{(A.III.12)}$$

However, with increasing surfactant concentration, the various interactions between aggregates (steric, electrostatic, etc.) become dominant and can lead to the formation of mesomorphic phases. For instance, a solution of elongated micelles can acquire nematic order if the concentration is high enough. The fundamental unit is no longer a single rod-like molecule, but rather a molecular structure (micelle) containing hundreds of amphiphilic molecules. In spite of this essential difference, thermotropic and lyotropic nematics have identical macroscopic behavior (the same defects, the same optical and rheological properties, etc.), so long as they present the same symmetries. On the other hand, their physical constants are sometimes very different. For example, the birefringence is much weaker in lyotropics than in thermotropics. Let us equally mention that micelles are not necessarily axisymmetric; this can be due to the presence of a cosurfactant (e.g., an alcohol) that renders them dissymmetric (ordinary ellipsoid). The nematic phase can then become biaxial (see chapter B.1), a feature that has never been observed in thermotropic phases of small molecules.

The nematic phase is by no means the only possible mesophase. In fact, it is rather seldom observed in lyotropic systems. On the contrary, the lamellar L_α and hexagonal H_α phases (Fig. A.III.22) are much more frequent; they often exist in large concentration and temperature ranges. The phase diagram of the

$C_{12}EO_6$/water binary mixture represented in figure A.III.23 provides a good example [17]. Let us recall that in the L_α phase the molecules form stacked lamellae, while the H_α phase consists of cylinders placed on a hexagonal lattice. Notice that in the $C_{12}EO_6$/water system a narrow ribbon of phase V_1 separates the lamellar and hexagonal phases. This phase is optically isotropic but very viscous. X-ray diffraction measurements have revealed the existence of cubic symmetry. In this phase, of space group Ia3d [18], surfactant molecules form two infinite aggregates that are interwoven but not connected. The aqueous medium is also continuous (Fig. A.III.24). Finally, the micellar phases denoted by the letters L_1 and L_2 in figure A.III.23 are direct and inverted, respectively.

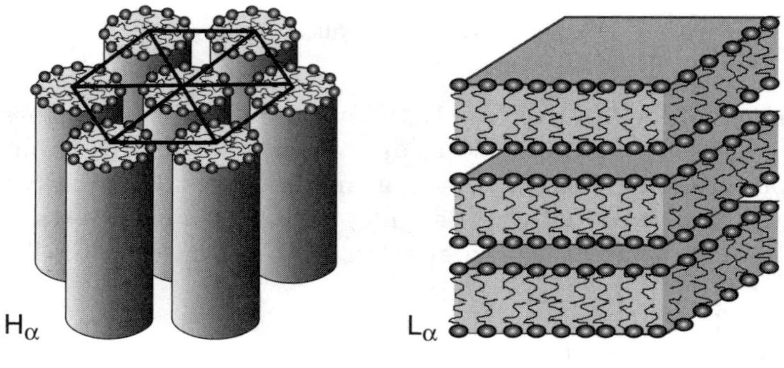

Fig. A.III.22 Hexagonal H_α and lamellar L_α phases. The index α indicates that the chains are "liquid." In this drawing, water is outside the cylinders and forms a continuous medium: such a phase is called "direct" (sometimes designed as H_1). In the opposite case (water inside the cylinders), it is called "inverted" (H_2 phase).

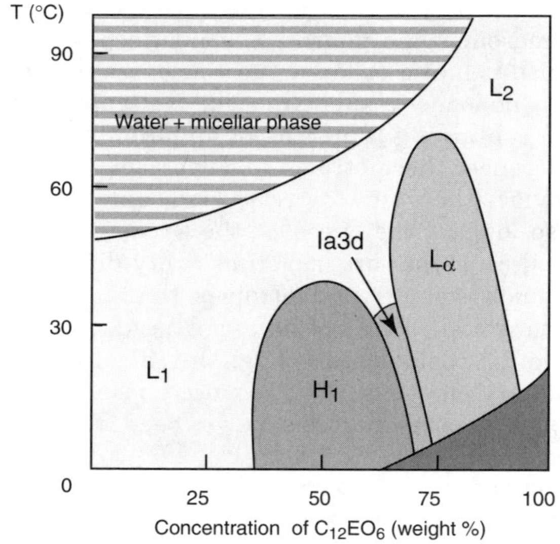

Fig. A.III.23 Phase diagram of the $C_{12}EO_6$/water binary mixture (from ref. [17a]).

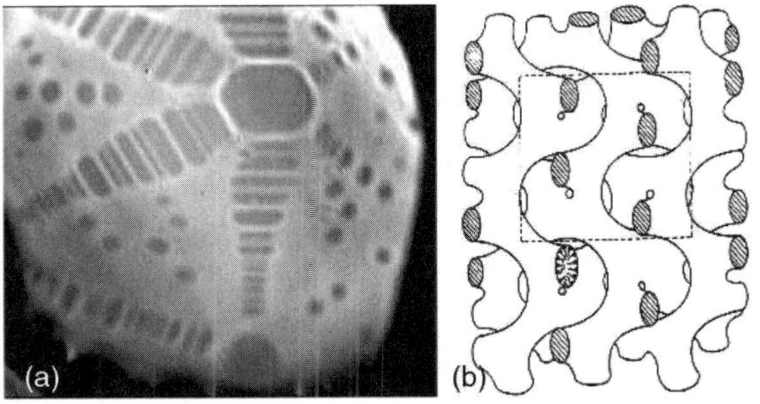

Fig. A.III.24 Cubic bicontinuous phase of symmetry Ia3d. a) Faceted surface of a droplet of the $C_{12}EO_6/H_20$ mixture; b) this structure was first proposed by Luzzati and coworkers [18].

It is worth noting that cubic phases can have various space groups and that one system can exhibit two such phases with different symmetries, like the $C_{12}EO_2$/water or DTACl/water mixtures, the phase diagrams of which are shown in figures A.III.25 and A.III.27.

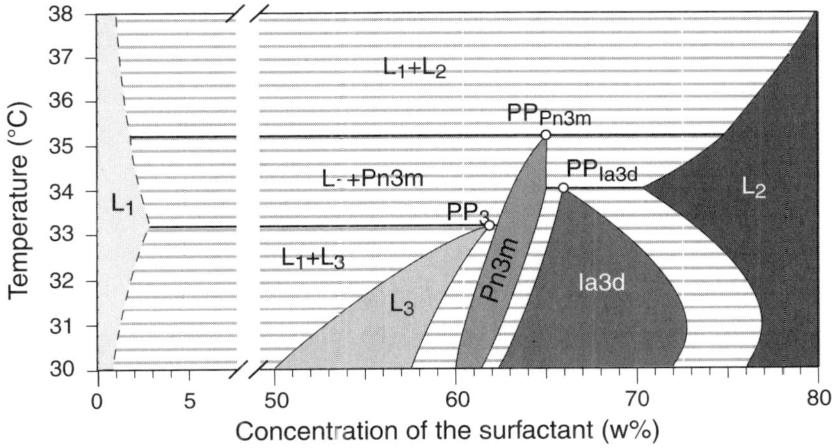

Fig. A.III.25 Phase diagram of the $C_{12}EO_2$/water binary mixture (from ref. [17b]). Besides the cubic Ia3d phase, let us note the presence of the second cubic phase of symmetry Pn3m (Fig. A.III.26). L_3 is the sponge phase.

The case of the DTACl/water mixture is different. Here, the first cubic phase (designated as Q'_α) is bicontinuous and situated between the hexagonal and lamellar phases, while the second (denoted Q''_α) consists of separate micelles forming an ordered lattice, a structure reminiscent of the adjacent L_1 phase.

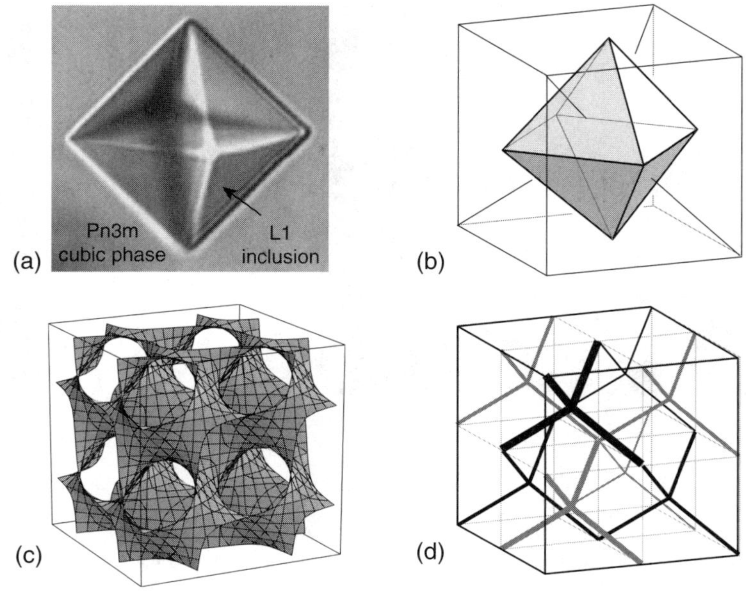

Fig. A.III.26 Cubic bicontinuous phase of symmetry Pn3m. a) Faceted inclusion of the micellar phase L1 in the Pn3m phase of the $C_{12}EO_2/H_20$ mixture. b) Orientation of the cubic unit cell with respect to the octahedral crystal habit. c) In the Pn3m structure, surfactant molecules are organized into a bilayer having the shape of the Infinite Periodic Minimal Surface D. d) The bilayer separates two interwoven labyrinths filled with water.

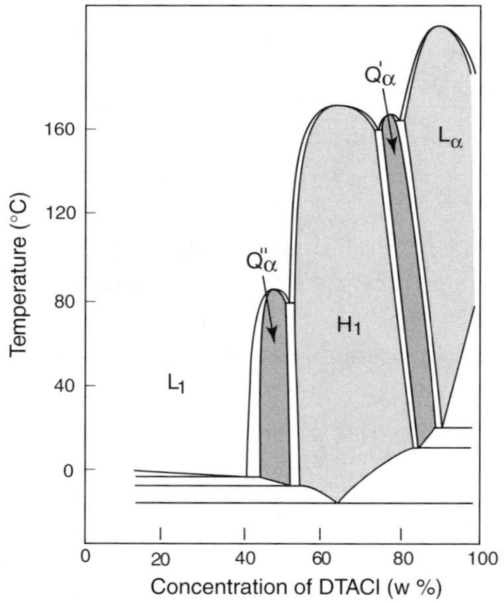

Fig. A.III.27 Phase diagram of the DTACl / water mixture. DTACl is dodecyltrimethyl-ammonium chloride (from ref. [19]).

Lastly, one can add a cosurfactant (e.g., an alcohol) to lyotropic mixtures. The example of the ternary system Aerosol OT/*para*-xylene/water is shown in figure A.III.28. We notice that lamellar and hexagonal phases are present over a large concentration range (the latter is an inverted phase, since the surfactant molecule has two bulky aliphatic chains).

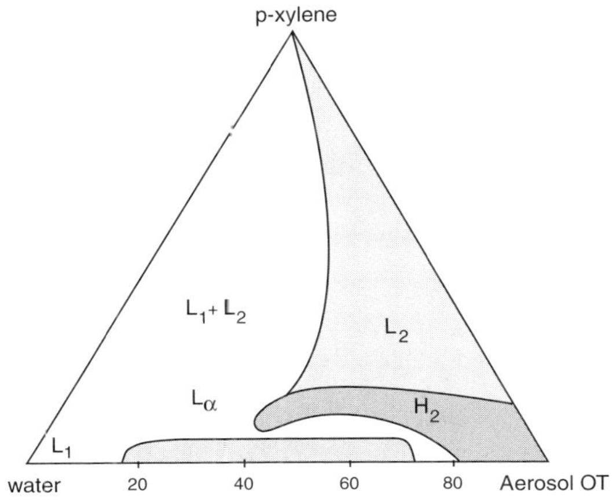

Fig. A.III.28 Simplified phase diagram of the ternary mixture Aerosol OT/p-xylene/water at room temperature. Aerosol OT is sodium bis(2-ethylhexyl) sulphosuccinate (from ref. [22]).

The point of adding a cosurfactant is not to render the phase diagram more complicated, but rather to enrich the "lyotropic mesomorphism" by varying the aggregates' shape. In this way, one can obtain new symmetries, as in the ribbon phases of Y. Hendriks and J. Charvolin, which are columnar biaxial phases on a rectangular lattice [20], or the biaxial nematic phase discovered in 1980 by Yu and Saupe [21].

Adding a salt to the solution can further diversify the phase diagram by modifying the electrostatic interactions between aggregates, as shown by the extensive literature on quaternary systems (brine/surfactant/cosurfactant) that are particularly rich in mesomorphic phases (see for instance the review article by P. Ekwall [22] containing many phase diagrams).

III.3 Liquid crystal diblock copolymers

These materials belong to the large family of polymer liquid crystals. They are a particularly good representative of the concept of molecular frustration which we have already encountered on several occasions. Consider two polymers A and B having little affinity for each other. When mixed, A and B separate just as oil and water do. The situation is radically changed if the A and B chains are covalently bonded to form the A-B copolymer. It is obvious that the system will try to reduce the contacts between A and B by forming domains rich in either A or B. The structure of the resulting phase strongly depends on the relative size of the two sequences. If A and B are close in size, one will have a lamellar phase of type -AA-BB-AA-BB- with A-B as the basic unit. An example of a copolymer (PS-PMMA) yielding a lamellar phase is given in figure A.III.29. If one of the two parts (say A) is smaller than the other, the A-B interface will curve and give rise to cylinders or even micelles. All three of these structures have been experimentally evidenced [23]. Thus, the situation strongly resembles that of lyotropic systems.

It is equally possible to change the interface curvature (and therefore the phase) by adding a homopolymer that prefers one of the domains [24]. The idea is similar to the addition of a cosurfactant to lyotropic systems.

(a)

(b)

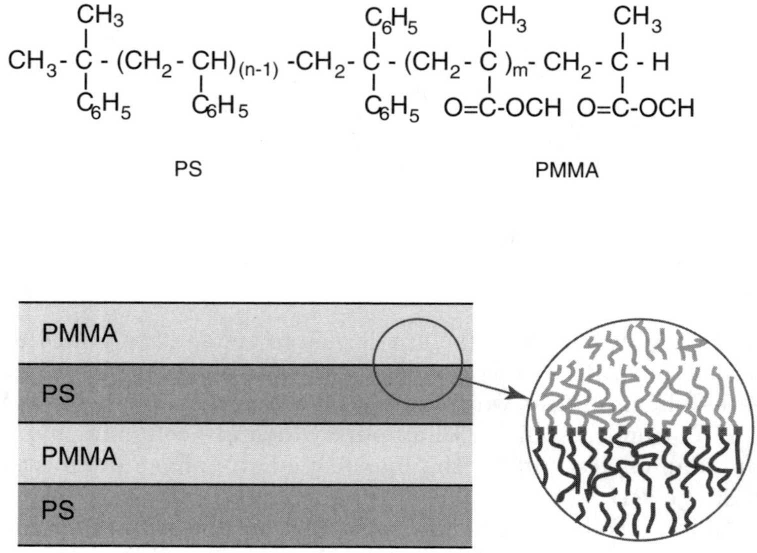

Fig. A.III.29 Expanded formula of PS (polystyrene)-PMMA (polymethyl methacrylate) (a) and its lamellar structure (b).

III.4 Colloidal liquid crystals

It has been known since 1936 [25] that the tobacco mosaic virus in aqueous suspension can have a nematic phase at sufficiently high concentration provided that the pH is kept between 7 and 8. Under these conditions, the virus has the appearance of stiff monodisperse cylinders (3000 Å long and 180 Å in diameter), negatively charged.

These experimental observations confirm the theoretical prediction of Onsager [26] on the behavior of a system of elongated rigid cylinders interacting uniquely by steric forces. In this case, the only relevant parameter is the entropy, which increases in the nematic phase with respect to the isotropic liquid. This result is at first glance puzzling, since the nematic phase seems more ordered than the isotropic one. The solution to this paradox lies in the distinction between translational and orientational entropy. Indeed, if the orientational entropy decreases in the nematic phase, it is nonetheless compensated for by the gain in translational entropy due to the reduction in excluded volume. Recent numerical simulations [27] show that other phases are also present in systems of hard, elongated spherocylinders at high concentration (they can acquire smectic, columnar, or crystalline order).

Finally, recent experiments on aqueous solutions of the semiflexible and chiral fd bacteriophage have revealed the existence of cholesteric and smectic phases [28]. Moreover, adding polystyrene spheres to the bacteriophage produces even more complex phase diagrams [29].

BIBLIOGRAPHY

[1] Gray G.W., *Molecular Structure and the Properties of Liquid Crystals*, Academic Press, New York, 1962.
In spite of its age, this work is still a valuable reference book on liquid crystals. Its author is a renowned English chemist, well known for synthesizing mesogenic molecules of the cyanobiphenyl family. These molecules are commonly employed in L.C. displays because of their remarkable chemical stability and their important dielectric anisotropy.

[2] Demus D., "100 years of liquid crystals chemistry," *Mol. Cryst. Liq. Cryst.*, **165** (1988) 45.
An excellent review article written by a German chemist from Halle, one of the hot spots in liquid crystal research.

[3] Gray G.W., Harrison K.J., Nash J.A., "Recent developments concerning biphenyl mesogens and structurally related compounds," in *Liquid Crystals*, Proc. Int. Conf. on Liq. Cryst., Bangalore Book Printers, Bangalore, 1973.

[4] Malthête J., Leclercq M., Dvolaitzky M., Gabard J., Billard J., Pontikis V., Jacques J., "Recherche sur les substances mésomorphes: III. Tolanes nématiques," *Mol. Cryst. Liq. Cryst.*, **23** (1973) 233.

[5] Roberts J.D., Caserio M.C., *Chimie Organique Moderne*, Interéditions, Paris, 1977.

[6] a) Deloche B., Charvolin J., Liébert L., Strzelecki L., *J. Physique (France)*, **36**, Coll. C1 (1975) 21.
b) Charvolin J., Deloche B., Chapter 15, in *Molecular Physics of Liquid Crystals*, Eds. Luckhurst G.R. and Gray G.W., Academic Press, London, 1979.

[7] a) Chandrasekhar S., Sadashiva B.K., Suresh K.A., *Pramana*, **9** (1977) 471.
b) Chandrasekhar S., Sadashiva B.K., Suresh K.A., Madhusudana N.V., Kumar S., Shashidhar R., Venkatesh G., *J. Physique (France)*, **40**, Coll. C3 (1979) 120.

[8] a) Nguyen H.T., Dubois J.C., Malthête J., Destrade C., *C. R. Acad. Sci., Paris*, **C286** (1978) 463.
b) Billard J., Dubois J.C., Nguyen H.T., Zann A., *Nouv. J. Chimie*, **2** (1978) 535.
c) Destrade C., Moncton M.C., Malthête J., *J. Physique (France)*, **40**, Coll. C3 (1979) 17.
d) Destrade C., Bernaud M.C., Nguyen H.T., *Mol. Cryst. Liq. Cryst. Lett.*, **49** (1979) 169.

e) Nguyen H.T., Destrade C., Gasparoux H., *Phys. Lett.*, **72A** (1979) 25.

[9] Destrade C., Malthête J., Nguyen H.T., Gasparoux H., *Phys. Lett.*, **78A** (1980) 82.

[10] Destrade C., Nguyen H.T., Gasparoux H., Malthête J., Levelut A.M., *Mol. Cryst. Liq. Cryst.*, **71** (1981) 111.

[11] Nguyen H.T., Malthête J., Destrade C., *J. Physique Lett.*, **42** (1981) L417.

[12] Cladis P.E., *Phys. Rev. Lett.*, **35** (1975) 48.

[13] a) Malthête J., Collet A., *Nouv. J. Chimie*, **9** (1985) 151.
b) Levelut A.M., Malthête J., Collet A., *J. Physique (France)*, **47** (1986) 351.
c) Malthête J., Collet A., *J. Am. Chem. Soc.*, **109** (1987) 7544.
d) Collet A., Dutasta J.P., Lozach B., Canceill J., *Topics in Current Chemistry*, **165** (1993) 103.

[14] Zimmermann H., Poupko R., Luz Z., Billard J., *Z. Naturforsch. A*, **40A** (1985) 149.

[15] Israelachvili J., *Intermolecular & Surface Forces*, Second Edition, Academic Press, London, 1996.

[16] Tanford C., *The Hydrophobic Effect*, Wiley, New York, 1973 and 1980.

[17] a) Mitchell D.J., Tiddy G.J., Waring L., Bostock T., McDonald M.P., *J. Chem. Soc. Faraday Trans. 1*, **79** (1983) 975.
b) Lynch M.L., Kochvar K.A., Burns J.L., Laughlin R.G., *Langmuir*, **16** (2000) 3537.

[18] Luzzati V., Tardieu A., Gulik-Krzywicki T., Rivas E., Reisshusson F., *Nature*, **220** (1968) 485.
For a review see Fontell K., *Colloid Polym. Sci.*, **268** (1990) 264.

[19] Balmbra R.R., Clunie J.S., Goodman J.F., *Nature*, **222** (1969) 1159.

[20] Hendriks Y., Charvolin J., *J. Physique (France)*, **42** (1981) 1427.

[21] Yu L., Saupe A., *Phys. Rev. Lett.*, **45** (1980) 1000.

[22] Ekwall P., in *Advances in Liquid Crystals*, Vol. 1, Ed. Brown G.H., Academic Press, London, 1975.

[23] a) Hasegawa H., Hashimoto T., *Macromolecules*, **18** (1985) 589.
b) Henkee H., Thomas E.L., Fetters L.J., *J. Mater. Sci.*, **23** (1988) 1685.

[24] a) Tanaka H., Hasegawa H., Hashimoto T., *Macromolecules*, **24** (1991) 240.
b) Winey K. I., Thomas E.L., Fetters L.J., *Macromolecules*, **25** (1992) 2645.

[25] Bawden F.C., Pirie N.W., Bernal J.D., Fankuchen I., *Nature*, **138** (1936) 1051.

[26] Onsager L., *Ann. N.Y. Acad. Sci.*, **51** (1949) 627.

[27] a) Stroobants A., Lekkerkerker H.N.W., Frenkel D., *Phys. Rev. A*, **36** (1987) 2929.
b) For a review, see Vroege G.J., Lekkerkerker H.N.W., *Rep. Prog. Phys.*, **55** (1992) 1241.

[28] Dogic Z., Fraden S., *Phys. Rev. Lett.*, **78** (1997) 2417.

[29] Adams M., Dogic Z., Keller S.L., Fraden S., *Nature*, **393** (1998) 349.

Part B

MESOPHASES WITH AN ORIENTATIONAL ORDER

Chapter B.I

Structure and dielectric properties of the nematic phase

This chapter is dedicated to the simplest of the orientational ordered mesophases: the uniaxial nematic. This phase appears usually directly below the isotropic liquid (except for a few cases when it is present between two more ordered phases, a feature known as *re-entrance*) and flows like a simple liquid, with macroscopic viscosities of the order of 1 poise for both thermotropic and lyotropic liquid crystals. Observation under the polarizing microscope of a free droplet placed on a slide reveals that the phase is birefringent and strongly diffuses light, as can easily be deduced from the characteristic "twinkling." Thermal fluctuations are therefore important. In larger drops, one can equally observe characteristic thread-like structures (see figure A.II.1) representing linear discontinuities of the optical axis. These defects, termed disclinations, are the analogue of dislocations in solids and shall be described in chapter B.III.

We start this introductory chapter by defining the concept of quadrupolar order (section I.1). We then describe the uniaxial nematic-isotropic liquid (section I.2) and biaxial nematic-uniaxial nematic (section I.3) phase transitions by the Landau theory. We subsequently show how the phenomenological Landau coefficients are experimentally determined, in particular by measuring the magnetic field-induced birefringence in the isotropic phase (Cotton-Mouton effect). The chapter ends with an overview of the dielectric properties of nematics, first at low frequency (dielectric anisotropy, flexo-electricity) (section I.4) and then at high frequency, in the optical range (birefringence, conoscopy, and Fermat-Grandjean principle) (section I.5).

I.1 Quadrupolar order parameter

I.1.a) Uniaxial nematics

In an ordinary nematic phase, all molecules tend to have the same direction, their centers of mass being distributed randomly, as in an ordinary liquid. Order is therefore purely **orientational.** It is equally **long range** because, in the absence of walls and external constraints, all molecules have at equilibrium the same average orientation, whatever the size of the sample (**collective behavior**). The so-called **director** is the unit vector, **n**, parallel to the average direction of molecular alignment, with the implicit assumption that molecules are free to turn about their axis. Such a medium is therefore optically **uniaxial**, the optical axes being parallel to the director **n**. Finally, **n** and −**n** are equivalent as nothing is changed by turning the molecule upside-down in an ordinary nematic (no ferroelectricity).

To describe the degree of molecular alignment parallel to the director **n**, let us model each molecule by a rigid rod and define a unit vector **a** parallel to the rod. Its components in the laboratory reference system (x, y, z) are, in terms of the polar angles θ and ϕ:

$$a_x = \sin\theta\cos\phi, \qquad a_y = \sin\theta\sin\phi, \qquad a_z = \cos\theta \tag{B.I.1}$$

The state of molecular alignment can be described by a distribution function $f(\theta,\phi)d\Omega$ giving the probability of finding the molecule in a small solid angle $d\Omega = \sin\theta d\theta d\phi$ around the direction (θ,ϕ).

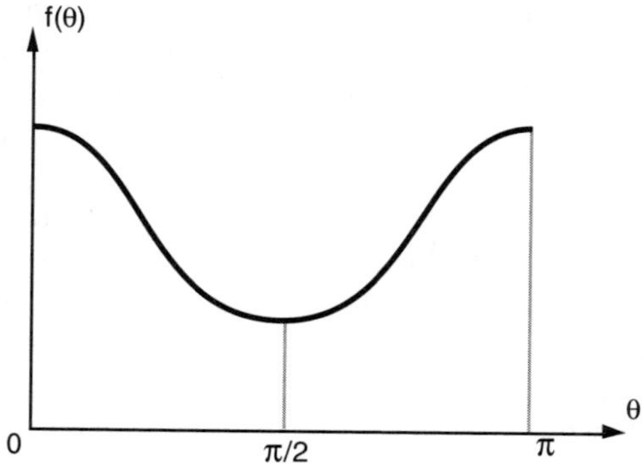

Fig. B.I.1 Generic shape of the distribution function f(θ).

Let us choose the z axis parallel to **n**. Because of the axisymmetry around **n**, the function $f(\theta,\phi)$ does not depend on ϕ. On the other hand, $f(\theta) = f(\pi-\theta)$ as the molecule has equal probabilities of pointing in one direction or the opposite way (**n** and −**n** are equivalent). The generic shape of f(θ) is given in figure B.I.1.

We shall now try to construct a *tensor* parameter that describes the average molecular orientation along **n** and that goes to zero in the isotropic phase. The vector $<\mathbf{a}> = \int f(\theta)\mathbf{a}d\Omega$ is not convenient, being zero by symmetry. On the other hand, the tensor S_{ij} defined by

$$S_{ij} = <a_i a_j> - \frac{1}{3}\delta_{ij} \tag{B.I.2}$$

is non-null in the nematic phase and goes to zero in the isotropic phase, as $<a_i^2> = 1/3$ and $<a_i a_j> = 0$ for $i \neq j$. This tensor is symmetric (hence diagonalizable) and traceless by construction. Furthermore, two of its eigenvalues must be equal for the nematic to be *uniaxial*. It must therefore have the following generic form

$$[S_{ij}] = S\begin{bmatrix} -\frac{1}{3} & 0 & 0 \\ 0 & -\frac{1}{3} & 0 \\ 0 & 0 & \frac{2}{3} \end{bmatrix} \tag{B.I.3}$$

in the eigenvector basis (**l**, **m**, **n**) (**n** being the director defined above) with an amplitude:

$$S = \frac{1}{2} < 3\cos^2\theta - 1 > = \int_0^{\pi} \frac{1}{2}(3\cos^2\theta - 1)\, 2\pi \sin\theta\, d\theta \tag{B.I.4}$$

In an arbitrary basis ($\mathbf{e}_1$, $\mathbf{e}_2$, $\mathbf{e}_3$), the components are

$$S_{ij} = S\left(n_i n_j - \frac{1}{3}\delta_{ij}\right) \tag{B.I.5}$$

with $n_i = \mathbf{n}.\mathbf{e}_i$. It is important to note that this tensor depends on three independent quantities, its amplitude S (which varies with temperature) and two of the three components of **n** Another important feature is that the tensor is invariant under the transformation $\mathbf{n} \rightarrow -\mathbf{n}$, an essential property of a nematic.

The scalar S is a measure of the degree of molecular alignment along the director. Its value is 1 when all the molecules are perfectly parallel to **n** for $\cos\theta = \pm 1$. Alternatively, it tends to −1/2 when $\cos\theta \rightarrow 0$, i.e., when the distribution function f(θ) is strongly peaked about π/2 (in this case, the molecules prefer a position perpendicular to **n**, Fig. B.I.2). Finally, S goes to zero in the isotropic liquid. This quantity can therefore be employed as an order parameter to describe the nematic-isotropic liquid phase transition. Note that in ordinary nematics, composed of elongated molecules, S is always positive.

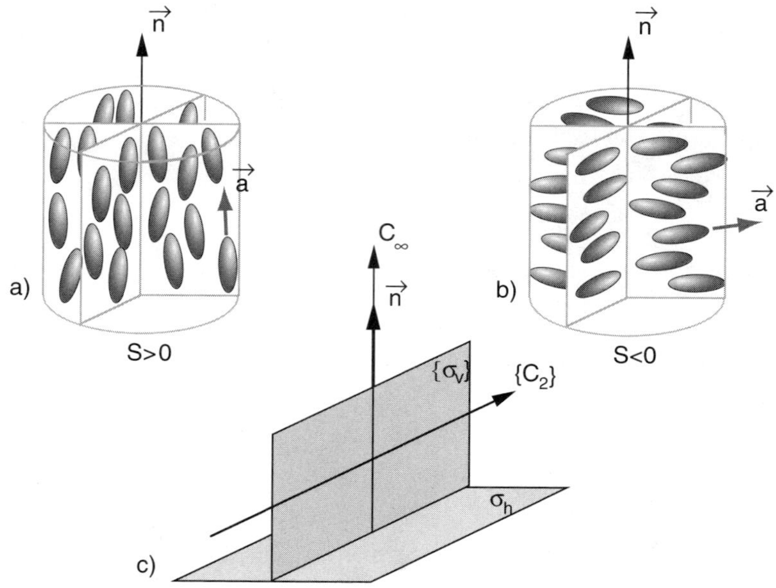

Fig. B.I.2 Two possible molecular configurations with positive (a) and negative (b) order parameter S. The director represents the symmetry axis of the system (c).

Rigorously speaking, this microscopic definition is only applicable to rigid, rod-like objects. However, molecules are usually more complex, being composed of rigid segments joined together by flexible parts. One must therefore find a more general approach, adapted to all molecular types. One such possibility consists of considering each molecule as a distribution $e_\alpha(\mathbf{r}_\alpha)$ of electric charges located by their position vectors $\mathbf{r}_\alpha$ with respect to the molecular center of mass. In electrostatics, one associates with this distribution the following multipolar moments (multipoles):

$$e = \sum_\alpha e_\alpha$$

$$p_i = \sum_\alpha e_\alpha r_{\alpha i} \tag{B.I.6}$$

$$q_{ij} = \frac{1}{2}\sum_\alpha e_\alpha \left(3r_{\alpha i}r_{\alpha j} - \mathbf{r}_\alpha^2\, \delta_{ij}\right)\;, \ldots$$

These quantities describe the anisotropy and the space orientation of the charge distribution $\{e_\alpha\}$ inside a molecule and allow for a precise definition of the fundamental concept of orientational order.

A system will be said to possess *long distance orientational order* if (at least) one of the multipole-multipole correlation functions tends to a finite value for $r \to \infty$.

By *definition*, order is *quadrupolar in a nematic*, meaning that the lowest-order multipole for which the correlation function does not go to zero in the limit $r \rightarrow \infty$ is the quadrupole ($<q_{ij}(\mathbf{0})q_{ij}(\mathbf{r})> \rightarrow$ Cst for $r \rightarrow \infty$).

To define an ***order parameter***, let us consider the average values of successive multipoles. The first is identically null, as the molecule is neutral. The second one is the dipolar moment borne by each molecule: it is generally non-null, but its average value $<\mathbf{p}>$ is zero (no ferroelectricity). The first moment whose average is not zero is the quadrupole $Q_{ij} = <q_{ij}>$.

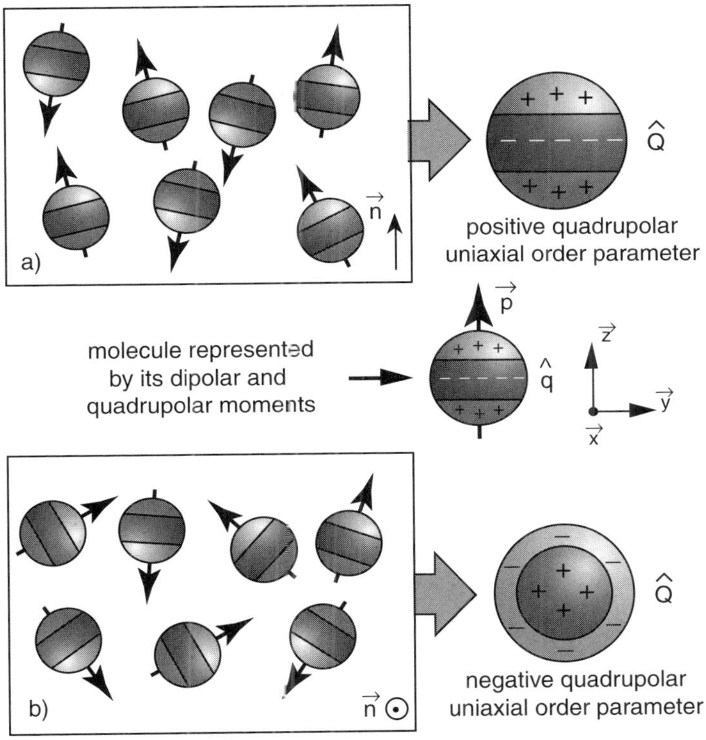

Fig. B.I.3 Definition of the quadrupolar order parameter Q_{ij}. a) $Q > 0$; b) $Q < 0$.

This second-order tensor is *symmetrical* (hence diagonalizable) and *traceless* by construction. For a **uniaxial** nematic, two of its eigenvalues are equal. Consequently, it can be written in an eigenvector basis $(\mathbf{l}, \mathbf{m}, \mathbf{n})$ in the general form

$$Q_{ij} = Q \begin{bmatrix} -\frac{1}{3} & 0 & 0 \\ 0 & -\frac{1}{3} & 0 \\ 0 & 0 & \frac{2}{3} \end{bmatrix} \qquad \text{(B.I.7)}$$

The unit vector **n**, parallel to the optical axis of the medium, is the **director**. It is also the isotropy axis of the $[Q_{ij}]$ tensor. The amplitude Q of the quadrupole depends on the temperature (and possibly on the density, but this latter quantity will be subsequently considered constant). In an arbitrary reference system ($\mathbf{e}_1$, $\mathbf{e}_2$, $\mathbf{e}_3$), the quadrupole components become:

$$Q_{ij} = Q(T)\,(n_i n_j - \frac{1}{3}\delta_{ij}) \qquad \text{(B.I.8)}$$

As previously indicated, this tensor depends on three independent quantities: its amplitude Q(T) and two of the three components of the director **n**.

As with S, it is important to note that the amplitude Q of the quadrupole can be **positive** as well as **negative**. In figure B.I.3, molecules are represented by their dipolar and quadrupolar moments. This representation makes no assumptions whatsoever about the molecular shape, thus being more general than the one in figure B.I.2, where each molecule is depicted as a rod. Nevertheless, as shown in figure B.I.2, the quadrupolar moments q_{ij} of each molecule can have two kinds of long-range correlation. In figure B.I.3a the average quadrupole $<Q_{ij}>$ is of positive amplitude ($Q > 0$), of axis **n** parallel to **z**, while in figure B.I.3b it is of negative amplitude ($Q < 0$), of axis **n** parallel to the **x** direction. One can therefore change the sign of Q (as for S) by differently orienting the same molecules. This fundamental property of Q (or S) shall be employed in section I.2, dedicated to the nematic-isotropic liquid phase transition. Observe that for rigid rods the two quantities Q and S only differ by a temperature-independent factor.

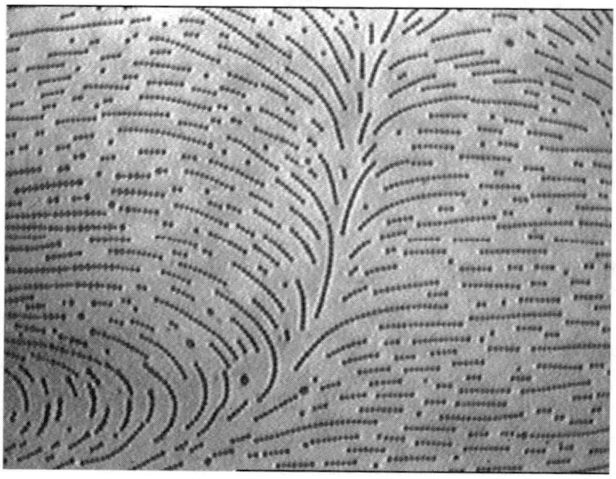

Fig. B.I.4 Decoration of the director field by water droplets forming long chains at the free surface of a nematic phase (see also figure B.I.30).

Finally, let us mention that the nematic director field can be revealed by dispersing an isotropic liquid (e.g., water) in the nematic phase. The droplets formed behave as macroscopic quadrupoles and form chains parallel or oblique with respect to the average orientation of the director **n** [1] (Fig. B.I.4).

I.1.b) Biaxial nematics

In polymers and lyotropic systems one can find optically **biaxial** nematics. In this case, the definition of the quadrupolar order parameter Q_{ij} remains unchanged, but its three eigenvalues are different. In the eigenvector basis (**l**, **m**, **n**) one can write:

$$Q_{ij} = \begin{bmatrix} \frac{-Q+P}{3} & 0 & 0 \\ 0 & \frac{-Q-P}{3} & 0 \\ 0 & 0 & \frac{2Q}{3} \end{bmatrix} \qquad \text{(B.I.8)}$$

which in an arbitrary reference system ($\mathbf{e}_1$, $\mathbf{e}_2$, $\mathbf{e}_3$) gives:

$$Q_{ij} = \frac{-1}{3}(Q-P)\, l_i l_j + \frac{-1}{3}(Q+P)\; m_i m_j + \frac{2}{3} Q\, n_i n_j \qquad \text{(B.I.9)}$$

The tensor now depends on five independent parameters: the two amplitudes Q(T) and P(T), and three out of the nine components of the basis vectors (**l**, **m**, **n**).

Remark that this tensor is invariant under the transformations $\mathbf{l} \rightarrow -\mathbf{l}$, $\mathbf{m} \rightarrow -\mathbf{m}$ and $\mathbf{n} \rightarrow -\mathbf{n}$. Finally, by setting $P = 0$ we are back in the uniaxial case.

In chapter B.VII we shall also see that in cholesterics the quadrupolar tensor is locally biaxial, the eigenvectors **l** and **m** turning with the molecules while the vector **n** (associated with the uniaxial part of the tensor) is parallel to the helicoidal axis.

I.2 The uniaxial nematic-isotropic liquid phase transition

To describe this transition we shall employ the Landau formalism. This phenomenological theory is applicable, as the quadrupolar symmetry is broken at the transition. We shall also see how the theory can be experimentally tested by means of the Cotton-Mouton effect and the action of a magnetic field on the nematic phase will be briefly reviewed.

I.2.a) Landau-de Gennes theory

For simplification, let us assume that the molecules are rigid rods and that the order parameter is the quadrupole S_{ij} (eq. B.I.5). Recall that it is oriented along **n** and its amplitude S is bounded between $-1/2$ and 1 (practically, between 0 and 1). Let us equally assume that the **free energy per unit volume $f(T,S_{ij})$ can be written as a power-law series of the order parameter** [2]. Strictly speaking, the free energy also depends on the density. However, we shall discard this variable as its variation at the transition is negligible. Retaining only the rotationally invariant terms of the development (the energy must not depend upon the referential choice), one finds the general form for a **uniaxial** nematic:

$$f_L(S_{ij},T) = f_0(T) + \alpha(T)\, S_{ii} + \frac{1}{2} a(T)\, S_{ij}S_{ji} - \frac{1}{3} b(T)\, S_{ij}S_{jk}S_{ki} + $$

$$+ \frac{1}{4} c(T)\, (S_{ij}S_{ji})^2 + \ldots \qquad \text{(B.I.10)}$$

Note that one can build one more fourth-order invariant of the form $S_{ij}S_{jk}S_{kl}S_{li}$; however, $S_{ij}S_{jk}S_{kl}S_{li} = (1/2)(S_{ij}S_{ji})^2$ if the S_{ij} tensor is symmetric, which justifies the adopted form B.I.10.

At each temperature T, the equilibrium state corresponds to the value of S_{ij} that minimizes f. If the nematic phase is stable at low temperature (which is usually the case), there is a temperature T* such that

$$(S_{ij}) = 0 \quad \text{if} \quad T > T^* \qquad \text{and} \qquad (S_{ij}) \neq 0 \quad \text{if} \quad T < T^* \qquad \text{(B.I.11)}$$

If $T > T^*$, the free energy is minimal for $S_{ij} = 0$: one therefore has $\alpha(T) \equiv 0$ (the development B.I.10 contains no first-degree term) and a(T) must be positive. If on the contrary $T < T^*$, the free energy must be minimal for a non-zero value of S_{ij} and a(T) must then be negative. Therefore the quantity a(T), measuring the concavity of the free energy in $S = 0$, must be zero at temperature T*. The simplest function a(T) to satisfy these conditions is:

$$a(T) = a_0\,(T - T^*) \qquad \text{(B.I.12)}$$

where a_0 is a positive constant. T* is termed the **spinodal temperature**. Below this temperature, the isotropic phase is absolutely unstable.

As to the other functions, $f_0(T)$, b(T), and c(T), they need not go to zero at the transition. Not too far from the transition temperature, they can therefore be considered constant and, as the free energy is only defined up to an additive constant, one can always take $f_0 = 0$.

It is important to note that the third-order term in the series cannot be discarded on symmetry grounds, as the states (S) and (–S) do not describe equivalent molecular configurations.

Also note that, if $S_{ij} = S(n_i n_j - \frac{1}{3}\delta_{ij})$, the free energy B.I.10 takes the following simple form:

$$f_L(S,T) = \frac{1}{2}A(T)\,S^2 - \frac{1}{3}B\,S^3 + \frac{1}{4}C\,S^4 + \ldots \qquad \text{(B.I.13)}$$

with

$$A(T) = A_o\,(T - T^*), \quad A_o = \frac{2}{3}a_o, \quad B = \frac{2}{9}b, \quad C = \frac{4}{9}c \qquad \text{(B.I.14)}$$

This result is easily proven by considering a reference frame with **z** // **n** (a legitimate choice, as the result must be independent of the system of reference).

In figure B.I.5 we have represented function f(S) for several characteristic temperatures taking B and C positive (in this case there is no minimum for $S < 0$). One can identify a **critical temperature** T_c for which the two phases coexist, having exactly the same free energy. The existence of an energy barrier between the two phases, due to the presence of the cubic term in the development, is the distinctive mark of a **first-order** phase transition. One can easily check that:

$$T_c - T^* = \frac{2}{9}\frac{B^2}{A_o C} \qquad \text{(B.I.15)}$$

and that

$$S(T{=}T_c) = S_c = \frac{2B}{3C} \qquad \text{(B.I.16)}$$

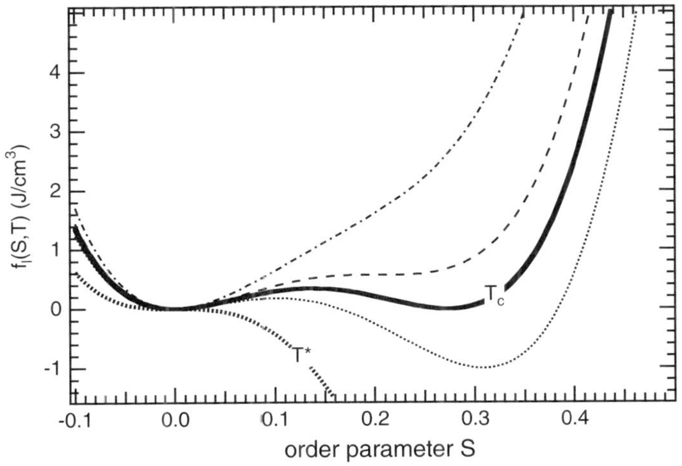

Fig. B.I.5 Landau free energy as a function of the quadrupolar order parameter S. The curves have been traced using experimental values for 5CB: $A_o = 0.13$ Jcm^{-3}K^{-1}, $B = 1.6$ Jcm^{-3}, $C = 3.9$ Jcm^{-3} [3]. From bottom to top: $T^* = 317.14$ K, $T_1 = 318.08$ K, $T_c = 318.26$ K, $T_2 = 318.4$ K, and $T_3 = 318.8$ K.

The transition is usually only weakly first order, so B and $T_c - T^*$ are small. In the following section, we shall see how the three coefficients A_o, B, and C are determined.

I.2.b) Determination of the Landau coefficients: the Cotton-Mouton and Kerr effects

When an isotropic fluid is exposed to a magnetic field **B**, it becomes slightly birefringent. On symmetry grounds, the optical axis of the medium is parallel to **B**. Experience shows that this effect is enhanced close to a nematic phase, because the molecules tend to align parallel to each other. We shall show how the measurement of magnetic birefringence in the isotropic liquid close to T_c allows verification of the Landau theory and determination of some of the coefficients.

First, let us briefly review the effects of a magnetic field **B**. Let **a** be the unit vector parallel to the major axis of the molecule and let $\chi^i_{//}$ and $\chi^i_\perp$ be the magnetic susceptibilities of the molecule in the directions parallel and perpendicular to **a**, respectively (the superscript i denotes an intrinsic quantity of the molecule, assumed to turn rapidly about its axis). Under the action of the field **B**, each molecule acquires a magnetization

$$\mathbf{m} = \frac{1}{\mu_o}[\chi^i_\perp \mathbf{B} + \Delta\chi^i(\mathbf{B.a})\,\mathbf{a}] \tag{B.I.17}$$

where $\Delta\chi^i = \chi^i_{//} - \chi^i_\perp$ is the magnetic susceptibility anisotropy of the molecule. This quantity is usually positive, but small (diamagnetic molecule), as we shall see later (section B.II.5.b). The torque exerted on the molecule by the magnetic field is

$$(\boldsymbol{\Gamma}^M)^i = \mathbf{m} \times \mathbf{B} \tag{B.I.18}$$

and the magnetic energy is

$$(f^M)^i = -\int_0^B \mathbf{m.dB} = -\frac{1}{2\mu_o}\chi^i_{//}B^2 - \frac{1}{2\mu_o}\Delta\chi^i(\mathbf{a.B})^2 \tag{B.I.19}$$

For $\Delta\chi^i > 0$, the energy is minimal when **B**//**a**. This means that each molecule tends to be aligned parallel to the magnetic field.

We can now determine the contribution f^M to the free energy of the isotropic liquid when it is subjected to a field **B**. Ignoring the constant term in equation B.I.19 and noting by N the number of molecules per unit volume:

$$f^M = -\frac{1}{2\mu_o} N\,\Delta\chi^i < (\mathbf{a.B})^2 > \tag{B.I.20}$$

As **B** is parallel to the optical axis of the medium, the formula can be expressed in terms of the order parameter S, since $(\mathbf{a}.\mathbf{B})^2 = B^2(\cos\theta)^2$. Up to a constant, one has:

$$f^M = -\frac{1}{3\mu_o} N\,\Delta\chi^i S B^2 \qquad \text{(B.I.21)}$$

Finally, the bulk free energy of the isotropic liquid to second-order terms (S being very small, of the order of 10^{-5}) is:

$$f = -\frac{1}{3\mu_o} N\,\Delta\chi^i S B^2 + \frac{1}{2} A_o(T - T^*)S^2 \qquad \text{(B.I.22)}$$

This energy is minimal for

$$S = \frac{N\,\Delta\chi^i B^2}{3\mu_o A_o(T - T^*)} \qquad \text{(B.I.23)}$$

One can show, by using the Vuks formulae [4], relating the principal polarizabilities of the medium and its principal indices (generalizing the Lorentz-Lorenz formula of isotropic media), that the field-induced birefringence Δn is related to S by the following relation:

$$\Delta n = \frac{1}{6\varepsilon_o} N \frac{2 + n^2}{n} \Delta\alpha^i S \qquad \text{(B.I.24)}$$

with n the optical index of the isotropic liquid and $\Delta\alpha^i = \alpha^i_{/\!/} - \alpha^i_{\perp}$ the anisotropy of the molecular polarizability at optical frequencies (defined by the usual relation $\mathbf{p}_{molecule} = (\underline{\alpha}^i)\mathbf{E}_l$ where $\mathbf{E}_l$ is the local field). One should keep in mind that this formula is only an approximation and that the linear dependence in the order parameter S is only justified for S << 1.

Combining the two equations B.I.23 and B.I.24 one has the Cotton-Mouton coefficient:

$$\frac{B^2}{\Delta n} = \frac{18\varepsilon_o\mu_o n}{(2+n^2)N^2\Delta\chi^i\Delta\alpha^i} A_o(T - T^*) \qquad \text{(B.I.25)}$$

According to the Landau theory, this coefficient must vary linearly with temperature and go to zero at the spinodal temperature T*.

The birefringence Δn is measured using the setup in figure B.I.6a. The sample of thickness d is placed in a temperature-controlled oven, itself situated in an electromagnet gap. A He-Ne laser polarized at 45° with respect to the field **B** defining the optical axis is directed through the sample. To measure the phase difference between the ordinary and the extraordinary rays propagating in the sample (see section I.5), an optical compensator of slow axis parallel to **B** and a second polarizer perpendicular to the first (analyzer) are placed in the beam. The transmitted intensity is reduced by adjusting the compensator. Total extinction is obtained when the phase difference $\Delta\varphi$ given by the compensator exactly cancels the one produced by the sample. Then

$$\Delta\varphi = -\frac{2\pi d\Delta n}{\lambda} \quad \text{(B.I.26)}$$

with λ the vacuum wavelength of the laser. An optical chopper and a lock-in amplifier can significantly improve the detection of the intensity minimum by the photodiode. With this setup and a good elliptic (Brace-Koehler) compensator, Δn variations down to the order of 10^{-6} can be detected.

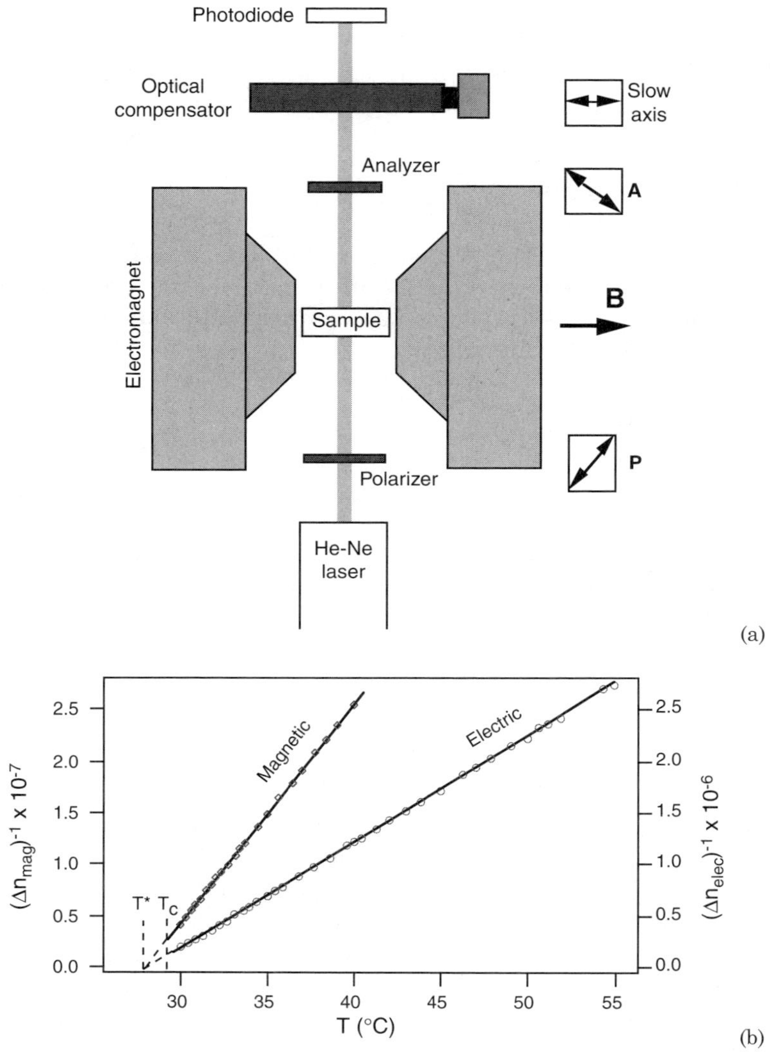

Fig. B.I.6 a) Possible setup for measuring the magnetic field-induced birefringence; b) reciprocal of the birefringence Δn induced by a magnetic or electric field as a function of temperature. The material is 6CB. The Kerr effect coefficient is practically temperature independent for this molecule (β = 0), but this is not generally the case (from ref. [5a, b]).

Figure B.I.6b shows the results obtained by Ratna et al. [5a,b] for 6CB. Similar measurements in different materials such as MBBA [5c] show that the reciprocal of the magnetic birefringence exhibits very clear linear behavior with temperature down to T_c. The same behavior is observed under an electric field when the Kerr constant (to be discussed later) does not depend on the temperature (this is the case for 6CB, as one can see from figure B.I.6b). The slopes of the curves give A_o while their zero intercepts give the spinodal temperature T*. The critical (coexistence) temperature being directly measured under the microscope, one obtains the difference $T_c - T^*$ (of the order of one degree in nematics) and then B^2/C by use of eq. B.I.15.

One further measurement is needed for determining the three Landau coefficients. The latent heat per unit volume at the transition ΔH is easily obtained. Employing the thermodynamic relations $\Delta H = T_c\Delta s$ and $s = -\partial f/\partial T$, one has

$$\Delta H = \frac{1}{2} A_o T_c S_c^2 = \frac{2}{9}\frac{A_o T_c B^2}{C^2} \qquad \text{(B.I.27)}$$

where the density jump at the transition has been neglected. In 5CB, $\Delta H \approx 1.55\ \mathrm{Jcm^{-3}}$, $T_c = 318.26$ K, $T^* = 317.14$ K, and $A_o \approx 0.13\ \mathrm{Jcm^{-3}K^{-1}}$, yielding $B \approx 1.6\ \mathrm{Jcm^{-3}}$ and $C \approx 3.9\ \mathrm{Jcm^{-3}}$. For MBBA, Vertogen and de Jeu [6] give $A_o \approx 0.028\ \mathrm{Jcm^{-3}K^{-1}}$, $B \approx 0.14\ \mathrm{Jcm^{-3}}$, and $C \approx 0.15\ \mathrm{Jcm^{-3}}$.

When the experiment is repeated under an electric field, the electrical birefringence is measured as a function of temperature (Kerr effect). One has a relation similar to B.I.21 between the electric free energy and the order parameter S:

$$f^E = -C_{Kerr} S E^2 \qquad \text{(B.I.28)}$$

If the molecule bears no permanent dipole (see section I.4 for details), the Kerr constant C_{Kerr} is determined as in the magnetic case and is proportional to the dielectric susceptibility anisotropy κ_a^i of the molecule at low frequency, which is temperature independent. If the molecule has a permanent dipole, it interacts strongly with the electric field, giving an additional contribution to the Kerr constant. This term can be shown to be proportional to $(3\cos^2\beta - 1)/T$ (with β the angle made by the dipole with the major axis and T the temperature) [7]. The Kerr constant can therefore be positive or negative (when the dipole is perpendicular to the molecular major axis). In certain materials such as PAA, C_{Kerr} can even change sign at a certain temperature ($T_c + 5$°C in PAA [8]).

Before addressing a detailed description of the dielectric properties of mesogenic molecules and uniaxial nematics (sections I.4 and I.5) we shall say a few words about the uniaxial nematic-biaxial nematic phase transition.

I.3 The uniaxial nematic-biaxial nematic phase transition

In section I.1 we mentioned the existence of biaxial nematics in polymers and lyotropic systems. The phase diagram of the ternary mixture potassium laurate-decanol-water is shown in figure B.I.7. This is the system where Yu and Saupe discovered in 1980 the first biaxial nematic phase [9]. The diagram exhibits three nematic phases, denoted by N_D, N_C, and N_{BX}. The first two are uniaxial; the third, **biaxial**, is sandwiched between the first two and only exists in a very narrow temperature range. The three nematic phases form a "tongue" surrounded by isotropic liquid.

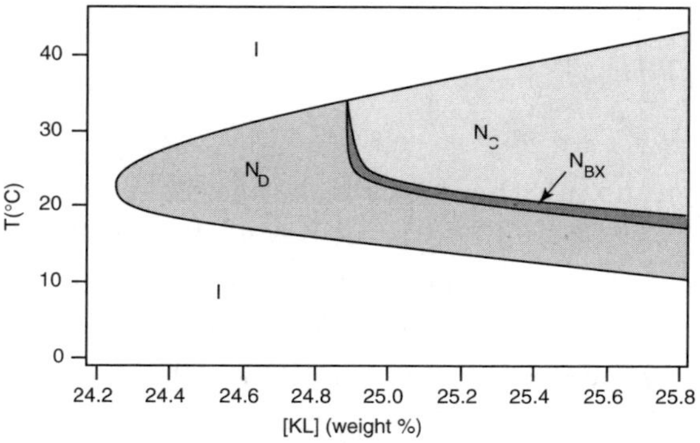

Fig. B.I.7 Phase diagram of the ternary mixture potassium laurate (LK)-decanol-D_2O. The N_c and N_D phases are uniaxial and surrounded by the isotropic phase. The N_{BX} phase is biaxial. Decanol represents 6.25% of the total weight (from ref. [9]).

In the N_D phase, the micelles (aggregates composed of several soap molecules and behaving as the thermotropic molecules, see figure B.I.8) have the shape of a **disk** perpendicular to the director **n**.

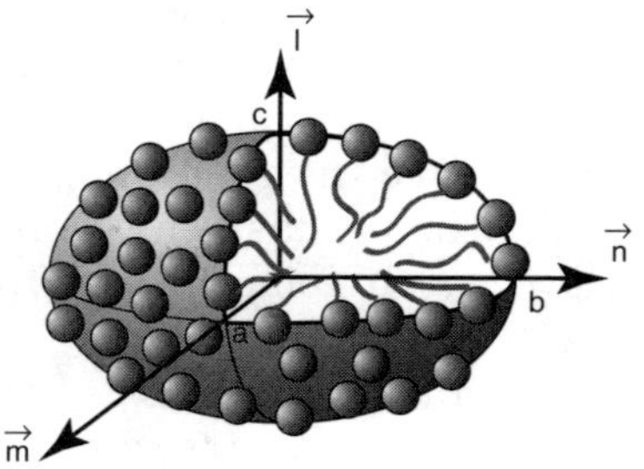

Fig. B.I.8 Schematic representation of a micelle in a nematic phase. The filled circles covering the micelle surface are the polar heads of the soap molecules; flexible segments pointing inwards are the aliphatic chains. In the N_D phase, a = b > c; in the N_C phase, a > b = c; finally, in the biaxial N_{BX} phase, a ≠ b ≠ c.

In this phase, molecules are on the average parallel to **n** and the birefringence is positive (see section I.5).

In the N_C phase, micelles have the shape of a revolution **cylinder** of axis parallel to the director **l**: this time, the molecules are on average perpendicular to **l** and the phase has negative birefringence.

In the intermediate phase N_{BX}, the micelles have the shape of a **generic ellipsoid** whose three principal axes are parallel to the three directors **l**, **m**, **n**. This phase is optically biaxial, as proven by NMR spectra as well as by conoscopic figures of magnetic field-oriented samples [9].

To study the order of the uniaxial nematic-biaxial nematic phase transition, Galerne and Marcerou [10] have precisely measured the differences between the optical indices n_1, n_2, n_3 of the three phases when the light is polarized along each of the three axes **l**, **m**, **n** (Fig. B.I.9).

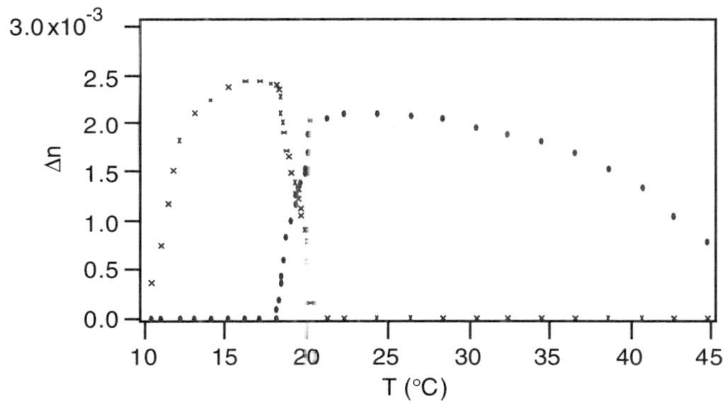

Fig. B.I.9 Refractive index difference in the three nematic phases as a function of temperature. Crosses represent the difference $n_3 - n_2$ and dots the difference $n_2 - n_1$ (from ref [10]).

In the N_D phase the birefringence $\Delta n = n_3 - n_2$ is positive and $n_1 = n_2$, while in the N_C phase $\Delta n = n_1 - n_2 < 0$ and $n_3 = n_2$. In the biaxial phase, all three indices are different. The birefringences vary continuously at the N_D-N_{BX} and N_{BX}-N_C transitions. These two transitions are therefore **second order**, contrary to the nematic-isotropic transition which is always first order.

The result is easily explained in terms of the Landau theory. One should first notice that this transition is analogous to a nematic-isotropic liquid transition **in two dimensions**. As in the uniaxial nematic phase the order is isotropic in the (**l**,**m**) plane normal to the director **n**, while it is nematic in the same plane of the biaxial phase. Orientational order in the (**l**,**m**) plane of the biaxial phase is quantified by the amplitude P of the biaxial component of the quadrupolar tensor Q_{ij} (eq. B.I.9). As it goes to zero in the uniaxial phase, this amplitude is the ideal order parameter to describe the uniaxial-biaxial phase transition. The following step is to develop the free energy in a Taylor series with respect to P. The first relevant term is the quadratic one, going to zero at the transition temperature T^* and that we will take as $(1/2)a_o(T-T^*)P^2$ in a mean-field theory. The following term is the quartic $(1/4)bP^4$, the cubic term being absent as **states P and –P are equivalent** (P changes to –P upon swapping **l** and **m**, while the system state remains unchanged). We can therefore write

$$f_L = \frac{1}{2} a_o(T - T^*)P^2 + \frac{1}{4} bP^4 + \ldots \qquad \text{(B.I.29)}$$

If the coefficient b is positive, the order parameter varies continuously at the transition, which is then **second order**, in agreement with the experiment described. In this case $T_c = T^*$ and

$$P = 0 \quad \text{for} \quad T > T_c \quad \text{and} \quad P = \sqrt{\frac{a_o}{b}(T_c - T)} \quad \text{for} \quad T < T_c \qquad \text{(B.I.30)}$$

In conclusion, the Landau theory shows that the uniaxial nematic-biaxial nematic phase transition **can** be second order, while the nematic-isotropic liquid phase transition is always first order. Obviously, this theory does not rule out a first-order transition if the coefficient b is negative.

In the following sections we summarize the dielectric properties of the uniaxial nematic phase, first at "low" frequency (from 0 to about 10^6 Hz) then at "very high frequency," in the optical domain (10^{15} Hz).

I.4 Low-frequency dielectric properties

Molecules usually react to (local) electric fields by changing their shape and orientation, in mesophases as well as in isotropic liquids or crystalline solids. However, the response can be stronger in a mesophase than in an isotropic liquid as a result of collective molecular behavior.

I.4.a) Potential energy of a molecule in an electric field

Let $e_\alpha(\mathbf{r}_\alpha)$ be the charge distribution of the molecule and Φ the potential describing the electric field:

$$\mathbf{E} = -\,\mathbf{grad}\Phi \qquad \text{(B.I.31)}$$

Electrostatic theory shows that the molecular potential energy is

$$U(\mathbf{E}) = e\Phi(\mathbf{0}) - \mathbf{p}.\mathbf{E}(\mathbf{0}) - \frac{1}{3} Q_{ij}E_{ij}(\mathbf{0}) - \ldots \qquad \text{(B.I.32)}$$

where quantities e, $\mathbf{p}$, Q_{ij}, ... are the successive multipoles defined in eq. B.I.6

and

$$E_{ij} = -\frac{\partial^2 \Phi}{\partial r_i \partial r_j} \qquad \text{(B.I.33)}$$

In a uniform electric field, the potential energy of a neutral molecule (e = 0) is only given by the dipolar moment **p**:

$$U(\mathbf{E}) = -\,\mathbf{p}.\mathbf{E} \qquad \text{(B.I.34)}$$

This energy varies with the angle between the dipole and the field, so there is a torque $\mathbf{\Gamma}^{Elec}$ acting on the molecule:

$$\mathbf{\Gamma}^{Elec} = \mathbf{p} \times \mathbf{E} \qquad \text{(B.I.35)}$$

Let us now discuss the value of the dipolar electric moment of a molecule.

I.4.b) Permanent and induced dipoles of an isolated molecule: role of the molecular symmetry

Certain mesogenic molecules possess a permanent dipolar moment, due to the permanent displacement of electrons from one atom to another at the level of intramolecular bonds, such as –C≡N or –CH=N–, etc. The cyano group –C≡N is known for its particularly strong dipolar moment.

In cyanobiphenyls, widely used in displays, the –C≡N group is attached to the end of the rigid biphenyl body, so the dipolar moment is usually directed along the molecule (Fig. B.I.10).

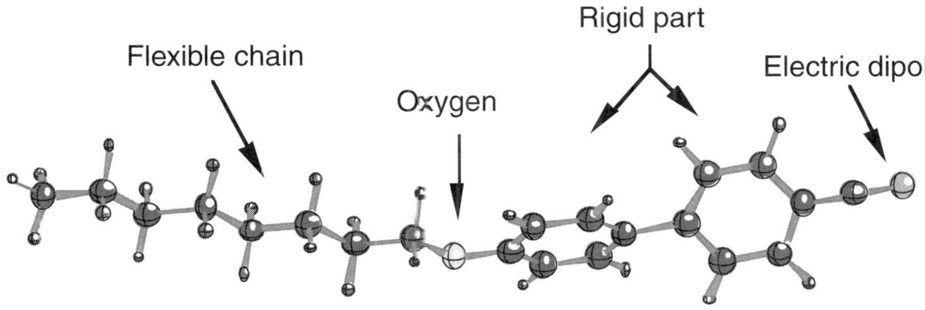

Fig. B.I.10 8OCB molecule.

In other molecules, the same –C≡N group is transversely attached, as in figure B.I.11.

Obviously, molecules are neither rigid nor motionless and they have access to an ensemble of states (configurations) that they explore as a result of thermal fluctuations. On the other hand, the energies of accessible states depend on the intramolecular structure and the intermolecular interactions. As a consequence, the average value of the dipolar moment and its orientation will depend on the symmetry of the molecule itself, as well as on that of the system it belongs to.

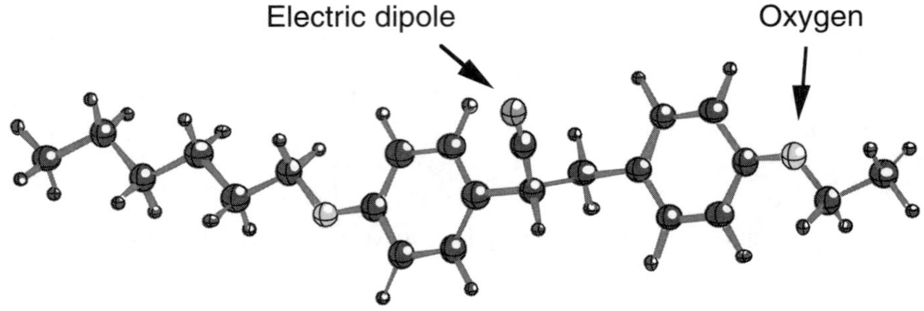

Fig. B.I.11 Molecule bearing a transverse dipole.

Let us start by considering the fluctuations of an isolated molecule (in vacuum). If the molecule has a mirror symmetry, the accessible states can be grouped in pairs, as two symmetric states have the same energy. Clearly, the average dipolar moment must be parallel to the mirror plane.

Fig. B.I.12 Molecule without a permanent dipole.

If the molecule has, besides a symmetry plane, a second-order axis normal to it, the average dipole must be zero (in the absence of external fields). An example of such a molecule is given in figure B.I.12.

When the molecule is subjected to an electric field, an additional dipole appears as the field deforms the charge distribution of the molecule. This **induced dipolar moment** is a function of the electric field and can be developed in a Taylor series with respect to the electric field:

$$p_i = a_{ij}E_j + b_{ijk}E_jE_k + c_{ijkl}E_jE_kE_l + \ldots \qquad \text{(B.I.36)}$$

where a_{ij} is the linear molecular polarizability tensor, while the two following terms correspond to non-linear molecular polarizabilities.

I.4.c) Absence of spontaneous polarization and flexo-electricity in the nematic phase

The existence of a permanent molecular dipole does not necessarily imply a spontaneous macroscopic polarization defined by:

$$\mathbf{P} = \mathrm{N} < \mathbf{p}(\mathbf{E} = \mathbf{0}) > \qquad \text{(B.I.37)}$$

where N is the number of molecules per unit volume and **p(E=0)** is the dipolar moment in the absence of the field. Indeed, the $\mathbf{D}_{\infty\mathbf{h}}$ symmetry of the undistorted uniaxial nematic phase is such that the spontaneous macroscopic polarization must be zero (Fig. B.I.13a). However, in the presence of a distortion breaking this symmetry, a polarization can appear. In the example of figure B.I.13b, the symmetry "left intact" by the deformation is the reflection in the (x, y) plane of the figure. Consequently, only the component P_z must be zero. Hence, **in nematics, distortions can induce a macroscopic polarization**. R. B. Meyer, who discovered this phenomenon, named it **flexo-electricity** (by analogy with piezo-electricity) [11]. Conversely, distortions can appear in a nematic subjected to an electric field.

The microscopic explanation of flexo-electricity is simple if the molecules are "wedge like" and bear a permanent dipole (Fig. B.I.13). Indeed, under a splay distortion, the molecules will rather place themselves "heads up" instead of "heads down" to improve the packing. As a consequence, the average value of the longitudinal dipolar moment will be non-null. One then speaks of **dipolar flexo-electricity**

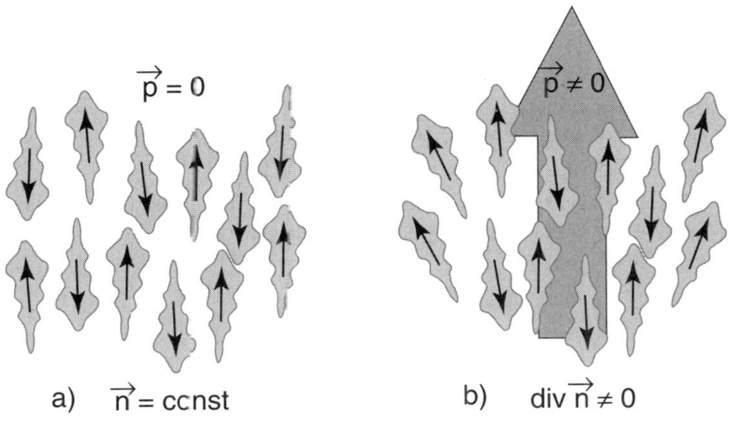

Fig. B.I.13 Flexo-electricity: a) without distortion the dipolar moment is zero; b) with a splay distortion, more molecules are "heads up" the result is a macroscopic polarization **P**.

Formally, to first order, the polarization must be **proportional** to the distortion $\partial n_i/\partial x_j$. The most general form of P that is linear in $\partial n_i/\partial x_j$ and **even** in **n**, is:

$$\mathbf{P} = e_1 \mathbf{n}\, \mathrm{div}\mathbf{n} - e_3 (\mathbf{n} \cdot \mathbf{grad})\mathbf{n} = e_1 \mathbf{n}\, \mathrm{div}\mathbf{n} + e_3\, \mathbf{n} \times \mathbf{curl}\mathbf{n} \qquad \text{(B.I.38)}$$

The e_1 term is associated with a splay deformation (div**n** $\neq 0$) and the e_3 one to a bend deformation (**curl n** $\perp$ **n**). As to twist deformations, they do not generate macroscopic polarization. Beware that two different sign conventions for e_3 are employed in the literature. In formula B.I.38 we have used that of Rudquist and Lagerwall [12], differing by a change of sign from the one first adopted by Meyer [11], and then by de Gennes in his book on liquid crystals.

The coupling between electric field and flexo-electric polarization leads to an additional term in the free energy:

$$f^F = -e_1 (\mathbf{n}\, \mathrm{div}\mathbf{n}) \cdot \mathbf{E} - e_3 (\mathbf{n} \times \mathbf{curl}\mathbf{n}) \cdot \mathbf{E} \qquad \text{(B.I.39)}$$

The first experimental value for the flexo-electric coefficient e_3 was given by Schmidt et al. [13]. In their experiment, a homeotropically aligned MBBA sample (molecules normal to glass plates) is placed in a horizontal electric field (Fig. B.I.14). If the molecular anchoring on the plates is weak enough (meaning that the molecules can easily change their direction at the glass surface), the experiment shows that the director field lines become curved in the presence of the field.

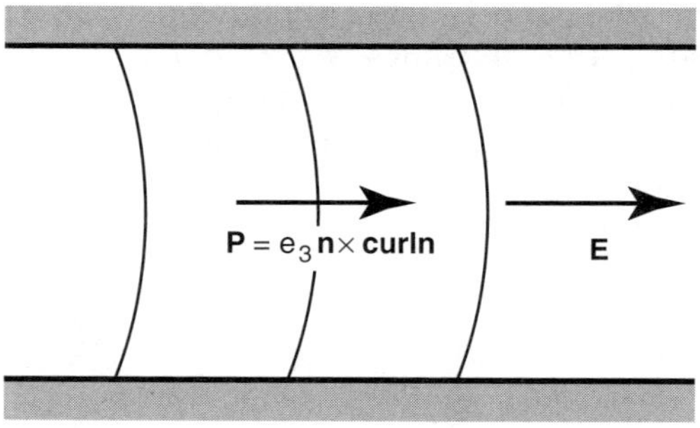

Fig. B.I.14 Measuring principle for the e_3 flexo-electric coefficient (from the experiment of Schmidt et al. [13]). Under the action of an applied field **E**, the director field lines get curved, generating a macroscopic polarization **P**. Molecules are weakly anchored at the surfaces.

Their curvature is determined by the competition between the gain in electric energy $-\mathbf{P} \cdot \mathbf{E}$ per unit volume, and the loss in elastic energy, of density $1/2 K_3 (\mathbf{n} \times \mathbf{curl}\, \mathbf{n})^2$ with K_3 the bend elastic constant (formula to be proven in the following chapter). By minimizing the total energy, the authors show that the curvature of director field lines (optically measurable) is proportional to the applied electric field and to the e_3 coefficient and inversely proportional to K_3 (neglecting the dielectric anisotropy term described in the previous section). This experimentally confirmed prediction allowed them to obtain the first estimate for the e_3 flexo-electric coefficient [13]:

$e_3 \approx 1.2\times10^{-12}$ C/m (MBBA, 22°C)

Strictly speaking, this measurement does not yield the sign of e_3. However, Schmidt et al. showed that a flow is induced in the sample parallel to the field and that its direction depends upon the sign of e_3. Thus they find $e_3 > 0$ in MBBA (note that Schmidt et al. use our sign convention for e_3 and not that of Meyer).

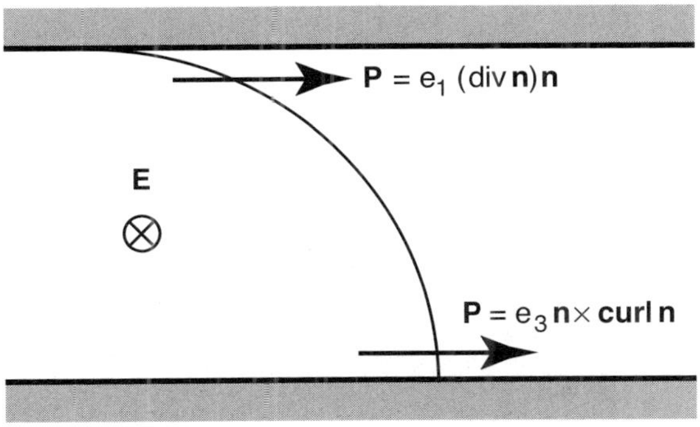

Fig. B.I.15 Hybrid cell used by Dozov et al.. [14] to measure the sum of the flexo-electric coefficients $e^* = e_1 + e_3$. The initial director configuration and the deformation-induced polarization are shown. The director turns about an axis normal to the plates (twist) upon applying the field E.

The difficulty in this experiment is obtaining a weak anchoring of the molecules on the glass plates (this concept will be further discussed in chapter B.V). The use of other geometries, better adapted to the measurement of flexo-electric coefficients and employing better controlled strong anchoring conditions, has subsequently been proposed, as in the experiment of Dozov et al. [14] where the nematic sample is sandwiched between two plates treated in planar and homeotropic anchoring, respectively (hybrid cell). These two strong and antagonistic boundary conditions deform the director field as shown in figure B.I.15, inducing splay distortion ($\mathrm{div}\,\mathbf{n} \neq 0$) at the planar anchoring plate, and bend distortion ($\mathbf{curl}\,\mathbf{n} \perp \mathbf{n}$) close to the homeotropically treated one. These deformations induce a macroscopic polarization represented in figure B.I.15 close to the plates. By applying a DC electric field **E** normal to the director, Dozov et al. showed first theoretically and then experimentally that the director (more precisely its projection to the plane of the plates) rotates around an axis normal to the plates by an angle $\Delta\Phi$ proportional to E and to the sum of the two flexo-electric coefficients $e^* = e_1 + e_3$ (with our definition of e_3). From the optical determination of $\Delta\Phi$ as a function of E, they obtain:

$e^* = e_1 + e_3 \approx 2.7\times10^{-12}$ C/m (MBBA, 20°C)

Combined with the experiment of Schmidt et al., this result allows for separate determination of e_1 and e_3.

A more direct and visual demonstration of flexo-electricity can be achieved by polarizing microscope observation of a thin nematic layer (5CB) deposed on the surface of an isotropic liquid (e.g., polyisobutylene). This experiment, performed by Lavrentovich and Pergamenshchik [15], shows that the nematic wetting of the liquid surface is not uniform, the liquid crystal forming a multitude of droplets resembling very flat convex lenses (Fig. B.I.16a). Observation between crossed polarizers shows that the director field inside the droplets is distorted, as shown in figure B.I.16b). This configuration, termed "boojum," will be described in chapter B.VI. The important fact is that all these droplets, circular in shape, spontaneously "orient" along the same direction. This collective effect shows that each droplet bears a dipolar moment **p** of flexo-electric origin. At equilibrium and if the thermal agitation is not too high, all these dipoles orient parallel to each other.

$$n_x = \cos 2\eta(x, y)$$
$$n_y = \sin 2\eta(x, y)$$

$$\eta(x, y) = \text{angle}(\mathbf{x}, \mathbf{r})$$

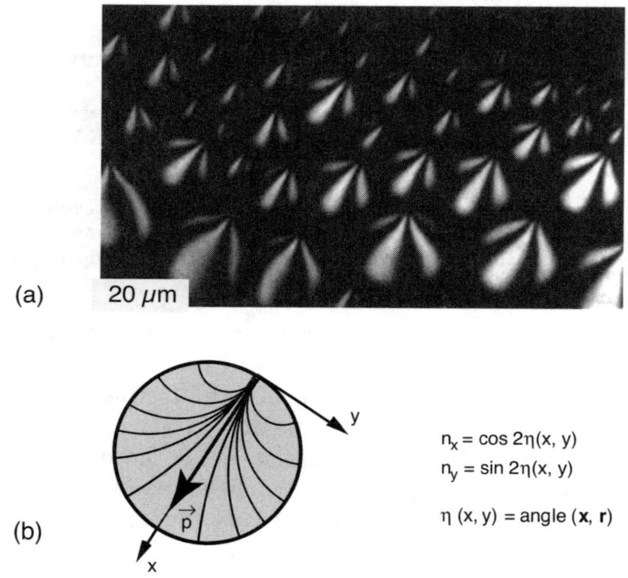

Fig. B.I.16 a) 5CB nematic droplets at the surface of an isotropic liquid (polyisobutylene). The droplets all orient in the same direction as they bear a flexo-electric dipole; b) schematic representation of the director field lines inside a droplet. The dipole is carried by the droplet symmetry axis (here taken along x). We also give the simplified analytical expression of the director field used to calculate the dipole **p** (formula B.I.40) (from ref. [15]).

This dipole moment can be determined for a droplet of radius R and of height h by using the formula B.I.38 and the expression of the director field given in figure B.I.16b [15]:

$$\mathbf{p} = \int \mathbf{P}\, dV = \pi e^* R h\, \mathbf{x} \tag{B.I.40}$$

Once again, the combination $e^* = e_1 + e_3$ is the relevant parameter. Theory shows that the dipoles are ordered if [16]:

$$\frac{p^2}{a^3 k_B T} \geq 60 \qquad \text{(B.I.41)}$$

with "a" the average distance between droplets and k_B the Boltzmann constant. This condition is largely satisfied in the described experiment.

Let us mention here the possibility of coupling between the director and the spatial derivatives of the electric field, yielding a free-energy term of the form $-e_2 n_i n_j \partial E_i/\partial x_j$. This kind of interaction, known as **quadrupolar flexo-electricity** (as it describes the interaction between a quadrupole and an inhomogeneous field [17]), is present in all systems, as opposed to the first one, which is dipolar in nature.

I.4.d) Dielectric anisotropy

We saw that distortions can induce a polarization of the nematic phase. This is however not the most common way of polarizing a medium. The "normal" method involves placing the nematic between the plates of a capacitor to which a potential difference is applied (Fig. B.I.17). Let **E** be the electric field inside the capacitor. This field acts upon the medium both by orienting the molecules parallel to itself and by inducing additional molecular polarization. The result is a polarization **P** proportional to the field **E**:

$$\mathbf{P} = \kappa \varepsilon_o \mathbf{E} \qquad \text{(B.I.42)}$$

with κ the **dielectric susceptibility** of the medium, supposed isotropic for the time being.

The field **E** is created by the charge density σ on the electrodes

$$E = \frac{\sigma_{ext} - \sigma_{pol}}{\varepsilon_o} \qquad \text{(B.I.43)}$$

These charges originate partly in the source employed to produce the electric potential difference (σ_{ext}), and partly in the polarization of the medium:

$$\sigma_{pol} = P = \kappa \varepsilon_o E \qquad \text{(B.I.44)}$$

One therefore has

$$E = \frac{1}{1+\kappa}\frac{\sigma_{ext}}{\varepsilon_o} = \frac{E_o}{1+\kappa} \qquad \text{(B.I.45)}$$

where

$$E_o = \frac{\sigma_{ext}}{\varepsilon_o} \tag{B.I.46}$$

is the electric field in the empty capacitor.

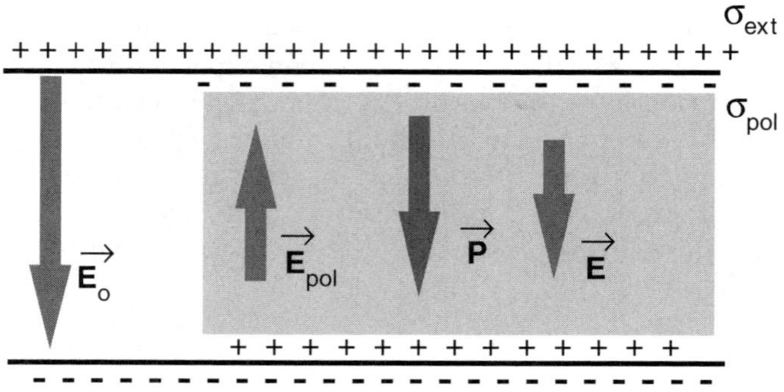

Fig. B.I.17 Dielectric medium inside a capacitor.

At a fixed external charge σ_{ext}, the dielectric reduces the electric field by a factor

$$\varepsilon = 1 + \kappa \tag{B.I.47}$$

known as the **dielectric constant** of the medium.

The field **E** is not usually uniform, so instead of eq. B.I.45 one has:

$$\mathrm{div}\,\mathbf{D} = \rho_{ext} \tag{B.I.48}$$

with

$$\mathbf{D} = (\varepsilon\,\varepsilon_o\mathbf{E}) = \varepsilon_o\mathbf{E} + \mathbf{P} \tag{B.I.49}$$

the electric displacement and ρ_{ext} the free charge density.

In nematics, the polarization **P** induced by a field **E** and the electric displacement **D** both depend on the orientation of **E** with respect to the director **n**. Equations B.I.42 and B.I.49 must then be modified such that $\mathbf{D}(\mathbf{E},\mathbf{n})$ and $\mathbf{P}(\mathbf{E},\mathbf{n})$ are linear in **E** and even in **n** (symmetry of the nematic phase). The most general expression of this kind for **D** is:

$$D_i = \varepsilon_o\varepsilon_{ij}E_j = \varepsilon_o(\varepsilon_1\delta_{ij} + \varepsilon_2 n_i n_j)E_j \tag{B.I.50}$$

or

$$D_i = \varepsilon_o(\varepsilon_{ij}^{iso} + \varepsilon_{ij}^{aniso})E_j \tag{B.I.51}$$

with

$$\varepsilon_{ij}^{iso} = \left(\varepsilon_1 + \frac{1}{3}\varepsilon_2\right)\delta_{ij} \tag{B.I.52a}$$

the isotropic part of the dielectric constant tensor and

$$\varepsilon_{ij}^{aniso} = \varepsilon_2\left(n_i n_j - \frac{\delta_{ij}}{3}\right) \tag{B.I.52b}$$

the anisotropic, traceless part of this tensor.

In a basis with the z axis parallel to **n**, the ε_{ij} tensor is diagonal:

$$\begin{bmatrix} \varepsilon_\perp & 0 & 0 \\ 0 & \varepsilon_\perp & 0 \\ 0 & 0 & \varepsilon_{//} \end{bmatrix} \tag{B.I.53}$$

By identification, $\varepsilon_1 = \varepsilon_\perp$ while ε_2 is the difference

$$\varepsilon_a = \varepsilon_{//} - \varepsilon_\perp \ (= \varepsilon_2) \tag{B.I.54}$$

between the dielectric constants $\varepsilon_{//}$ and $\varepsilon_\perp$ measured with **E** parallel and perpendicular to **n**, respectively. As for the coefficient ε^{iso}, it is the average of the ε_{ij} tensor eigenvalues:

$$\varepsilon^{iso} = \varepsilon_1 + \frac{1}{3}\varepsilon_2 = \frac{\varepsilon_{//} + 2\varepsilon_\perp}{3} \tag{B.I.55}$$

Among these quantities, the dielectric anisotropy ε_a is the most useful for applications as it gives the torque exerted by the electric field on the director in a TNC ("twisted nematic cell") displays (see next chapter)

$$\boldsymbol{\Gamma}^{Elec} = N\langle \mathbf{p} \times \mathbf{E} \rangle = \mathbf{P} \times \mathbf{E} \tag{B.I.56}$$

where the polarization **P** is generally given by a generalization of equation B.I.44 to anisotropic media:

$$P_i = \varepsilon_0 \kappa_{ij} E_j \tag{B.I.57}$$

By analogy with the dielectric constant, the dielectric susceptibility tensor κ_{ij} can be separated into two parts, isotropic and anisotropic, respectively:

$$\kappa_{ij} = \kappa_{ij}^{iso} + \kappa_{ij}^{aniso} \tag{B.I.58}$$

One therefore has:

$$\mathbf{P} = \mathbf{P}^{iso} + \mathbf{P}^{aniso} \tag{B.I.59}$$

As $\mathbf{P}^{iso}$ is parallel to **E**, only the anisotropic part of the susceptibility κ_{ij} can give a torque. Because

$$\kappa_{ij}^{aniso} = \varepsilon_{ij}^{aniso} = \varepsilon_a\left(n_i n_j - \frac{\delta_{ij}}{3}\right) \qquad \text{(B.I.60)}$$

hence

$$\mathbf{P}^{aniso} = \varepsilon_0 \varepsilon_a \mathbf{n}\,(\mathbf{n}.\mathbf{E}) \qquad \text{(B.I.61)}$$

and thus

$$\boldsymbol{\Gamma}^{Elec} = \varepsilon_0 \varepsilon_a (\mathbf{n} \times \mathbf{E})\,(\mathbf{n}.\mathbf{E}) \qquad \text{(B.I.62)}$$

For $\varepsilon_a > 0$, this torque tends to align the director with the field.
For $\varepsilon_a < 0$, the torque tends to align the director normal to the field (Fig. B.I.18).

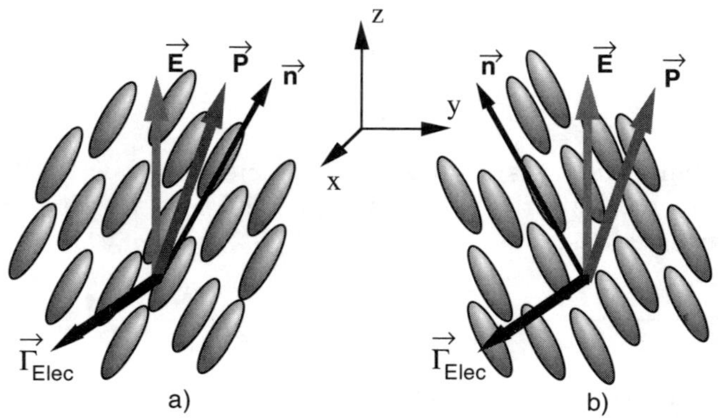

Fig. B.I.18 Action of an electric field **E** on the director **n**. a) Positive dielectric anisotropy: the polarization **P** is tilted with respect to **E**; the torque $\boldsymbol{\Gamma}^{Elec}$ tends to align the director along the field. b) Negative dielectric anisotropy: the polarization **P** is tilted with respect to **E**; the torque $\boldsymbol{\Gamma}^{Elec}$ tends to turn the director away from the field.

For completeness, the electric free energy f^E of the system (the quantity to minimize under fixed electric field) is given by:

$$f^E = \int_0^E -\mathbf{D}.d\mathbf{E} \qquad \text{(B.I.63)}$$

Keeping only the anisotropic part of the macroscopic polarization (eq. B.I.61), which amounts to ignoring the constant terms, leads to:

$$f^E = -\frac{1}{2}\varepsilon_0 \varepsilon_a (\mathbf{E}.\mathbf{n})^2, \qquad \text{(B.I.64)}$$

confirming that the energy is minimal when **E** // **n** or $\mathbf{E} \perp \mathbf{n}$ for $\varepsilon_a > 0$ or $\varepsilon_a < 0$, respectively.

The dielectric anisotropy ε_a depends on the molecular structure and on the frequency of the electric field.

In a DC field, the anisotropy in the dielectric susceptibility is mostly due to the permanent dipole(s) of the molecules. If the molecule carries a longitudinal dipole, the induced polarization in the **n** direction is strong, due to the alignment of the dipole along the field; on the other hand, polarization perpendicular to the director is weak, being only due to the internal polarizability of the molecule (eq. B.I.36). For transverse dipoles, the anisotropy changes sign. Anisotropy due to permanent dipoles is very strong. In cyanobiphenyls, it can be as high as +10 or even +15. Fairly strong negative anisotropies can be achieved with molecules bearing transverse permanent dipoles (see figure B.I.11).

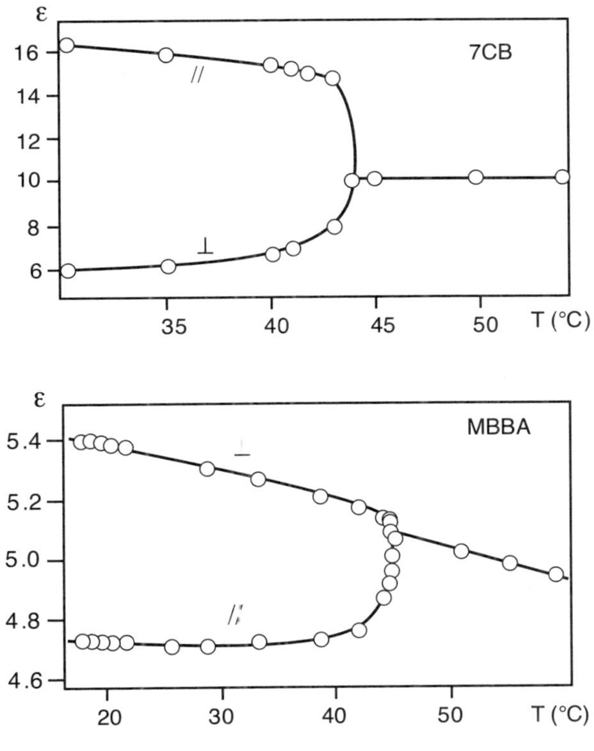

Fig. B.I.19 Dielectric constants as a function of temperature at 1kHz in two mesogenic products of opposite anisotropy (from ref. [18] for 7CB and ref. [19] for MBBA).

In an AC electric field, the response of the medium depends on the field frequency. At low enough frequency, permanent dipoles can adapt to the field variations (i.e., reach statistical equilibrium). This adaptation involves molecular reorienting, which takes a certain time. Consequently, when the frequency is too high, as in light waves, only intramolecular motions are fast enough to follow the field evolution. The dielectric anisotropy is therefore much weaker. In the nematic phase of 7CB, for instance, $\varepsilon_a \approx 10$ at 1 kHz but $\varepsilon_a \approx 0.5$ at optical frequencies (see next section). We shall note here the existence of nematic mixtures whose dielectric anisotropy changes sign at relatively low frequency (a few kHz). Finally, dielectric anisotropy ε_a varies with temperature and goes to zero discontinuously above T_c (Fig. B.I.19).

I.5 Optical properties

I.5.a) Preliminary experiment: double refraction of a nematic prism

The non-distorted (uniaxial) nematic phase is optically **birefringent and uniaxial**, exhibiting characteristic optical properties as illustrated by the following experiment. A good experimental example of birefringence is obtained by shining a depolarized laser beam through a nematic prism (Fig. B.I.20). The prism is in fact a wedge-shaped container filled with nematic (E9 from BDH). The container walls are treated by rubbing them to align the nematic molecules parallel to the wedge.

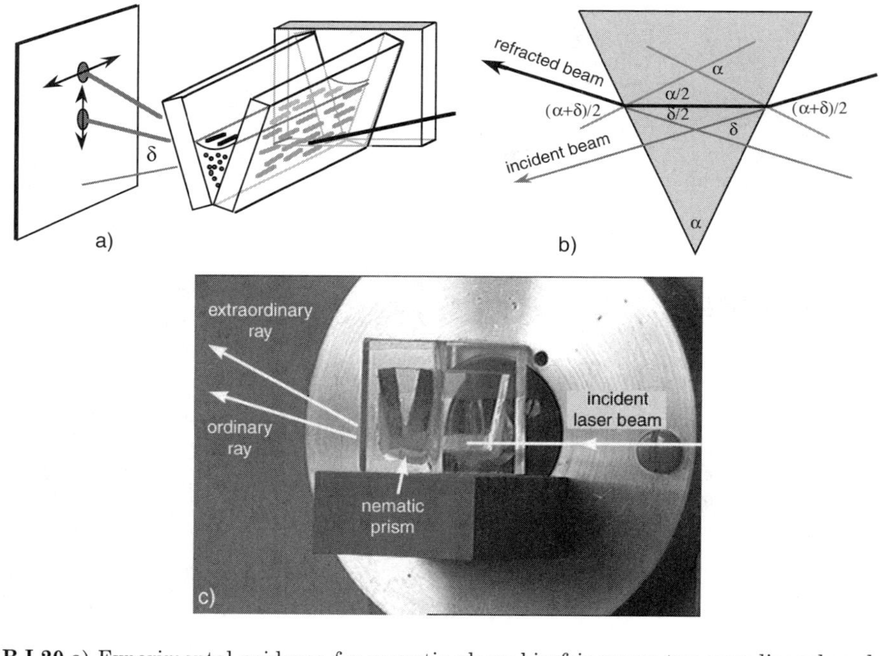

Fig. B.I.20 a) Experimental evidence for nematic phase birefringence; two rays linearly polarized with a 90° difference exit the prism; b) measuring the angle of least deviation; c) photograph of the experimental setup.

Two refracted rays exit the prism, both linearly polarized, but one normal to the refraction plane, and the other one in the plane. If the prism angle is $\alpha = 20°$, the angles of least deviation are $\delta_e = 13.5°$ for the first beam and $\delta_o = 9.1°$ for the second. With these data, one can compute the two refraction indices using the formula (see figure B.I.20b):

$$n_{e,o} = \frac{\sin\left(\frac{\alpha+\delta_{e,o}}{2}\right)}{\sin\left(\frac{\alpha}{2}\right)} \qquad \text{(B.I.65)}$$

yielding, for the E9 nematic, $n_e = 1.66$ and $n_o = 1.45$.

I.5.b) Ordinary and extraordinary rays: eigenmodes

The two rays exiting the prism correspond to the ordinary and extraordinary eigenmodes of electromagnetic waves propagating in a nematic. To find these modes, consider a wave with wave vector **k** and frequency ω:

$$\mathbf{B} = \mathbf{B}_o \exp i(\mathbf{k}.\mathbf{r} - \omega t), \qquad \mathbf{D} = \mathbf{D}_o \exp i(\mathbf{k}.\mathbf{r} - \omega t) \qquad \text{(B.I.66)}$$

The Maxwell equations

$$\operatorname{div}\mathbf{D} = 0 \qquad \text{and} \qquad \operatorname{div}\mathbf{B} = 0 \qquad \text{(B.I.67)}$$

imply that

$$\mathbf{k} \perp \mathbf{D} \qquad \text{and} \qquad \mathbf{k} \perp \mathbf{B} \qquad \text{(B.I.68)}$$

From the two remaining Maxwell equations,

$$\mathbf{curl}\ \mathbf{E} = -\,\partial\mathbf{B}/\partial t \qquad \text{(B.I.69)}$$

and

$$\mathbf{curl}\ \mathbf{H} = \partial\mathbf{D}/\partial t, \qquad (\mathbf{B} = \mu_o\mathbf{H}), \qquad \text{(B.I.70)}$$

one gets:

$$\mathbf{k}\times\mathbf{E} = \omega\mathbf{B} \Rightarrow \mathbf{B}\perp(\mathbf{k},\mathbf{E}) \qquad \text{and} \qquad \mathbf{k}\times\mathbf{H} = -\omega\mathbf{D} \Rightarrow \mathbf{D}\perp(\mathbf{k},\mathbf{H}) \qquad \text{(B.I.71)}$$

In short (Fig. B.I.21), vectors (**k**,**D**,**B**) form a direct trihedron and the vector **E** is contained in the plane (**k**,**D**).

By definition, the eigenmode amplitude does not depend on **r**. From eqs. B.I.70 and 71 one gets:

$$\mathbf{D} = -\frac{\mathbf{k}}{\omega} \times \left(\frac{\mathbf{k}}{\mu_0 \omega} \times \mathbf{E}\right) \tag{B.I.72}$$

which can be rewritten as

$$\frac{\mathbf{D}}{\varepsilon_0} = n^2\mathbf{E} - \mathbf{n}(\mathbf{n}.\mathbf{E}) \tag{B.I.73}$$

with

$$\mathbf{n} = \frac{\mathbf{k}c}{\omega} \qquad \text{(note that here } \mathbf{n} \text{ is not the director!)} \tag{B.I.74}$$

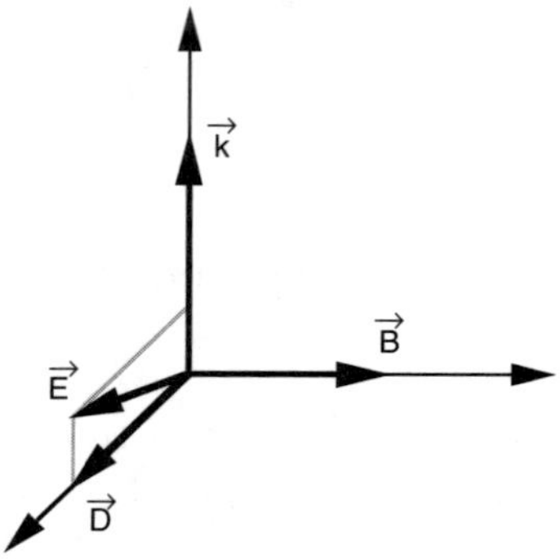

Fig. B.I.21 Relations between **B**, **D**, **E**, and **k** in an electromagnetic wave.

The eigenmodes are most easily found in the basis related to the propagation direction with axis **(3)** parallel to **n**. As **D** is normal to **n**, D_3 is zero and the third component of the vector equation B.I.72 is identically satisfied, reducing it to only **two equations** by use of the constitutive equation B.I.50:

$$(n^2\varepsilon_0)^{-1} D_1 = \varepsilon_0^{-1}(\bar{\varepsilon}_{11}D_1 + \bar{\varepsilon}_{12}D_2) \tag{B.I.75}$$

$$(n^2\varepsilon_0)^{-1} D_2 = \varepsilon_0^{-1}(\bar{\varepsilon}_{21}D_1 + \bar{\varepsilon}_{22}D_2) \tag{B.I.76}$$

where $[\bar{\varepsilon}_{ij}] = [\varepsilon_{ij}]^{-1}$ is the reciprocal tensor of $[\varepsilon_{ij}]$ and $n = |\mathbf{n}|$.

The desired eigenmodes (D_1, D_2) correspond to the eigenvectors of this equation system. The propagation velocities of the eigenmodes are given by the eigenvalues n.

Up till now, only the direction of axis **3** has been imposed. We can further simplify the calculations by choosing the orientation of axis **1**. Indeed, ε_{ij} being a uniaxial tensor, one of its eigenaxes is parallel to the director, while the other two are arbitrary. Equations B.I.75 and 76 can then be simplified by taking axis **(1)** perpendicular to the director **n**, as the restriction of tensor $[\bar{\varepsilon}_{ij}]$ to the plane (1, 2) is diagonal in this basis, which is shown denoting the angle between **k** and the director **n** by θ and explicitly writing the ε_{ij} tensor from eqs. B.I.50–54:

$$[\varepsilon_{ij}] = \begin{bmatrix} \varepsilon_\perp & 0 & 0 \\ 0 & \varepsilon_\perp + \varepsilon_a \sin^2\theta & \varepsilon_a \sin\theta\cos\theta \\ 0 & \varepsilon_a \sin\theta\cos\theta & \varepsilon_\perp + \varepsilon_a \cos^2\theta \end{bmatrix} \tag{B.I.77}$$

The reciprocal of this tensor is:

$$[\bar{\varepsilon}_{ij}] = \begin{bmatrix} 1/\varepsilon_\perp & 0 & 0 \\ 0 & (\varepsilon_\perp + \varepsilon_a \cos^2\theta)/(\varepsilon_\perp\varepsilon_{//}) & -(\varepsilon_a \sin\theta\cos\theta)/(\varepsilon_\perp\varepsilon_{//}) \\ 0 & -(\varepsilon_a \sin\theta\cos\theta)/(\varepsilon_\perp\varepsilon_{//}) & (\varepsilon_\perp + \varepsilon_a \sin^2\theta)/(\varepsilon_\perp\varepsilon_{//}) \end{bmatrix} \tag{B.I.78}$$

Its restriction to the (1,2) plane is diagonal, as advocated.

The eigenvectors and eigenvalues of the equation system B.I.75 and B.I.76 are therefore (Fig. B.I.22):

1. $(D_1,0)$ with $\dfrac{1}{n^2} = \dfrac{1}{n_o^2} = \dfrac{1}{\varepsilon_\perp}$ (B.I.79)

2. $(0,D_2)$ with $\dfrac{1}{n^2} = \dfrac{1}{n_{eff}^2} = \dfrac{1}{\varepsilon_{//}} + \dfrac{\varepsilon_a\cos^2\theta}{\varepsilon_{//}\varepsilon_\perp} = \dfrac{\sin^2\theta}{n_e^2} + \dfrac{\cos^2\theta}{n_o^2}$ (B.I.80)

where

$\dfrac{1}{n_o^2} = \dfrac{1}{\varepsilon_\perp}$ and $\dfrac{1}{n_e^2} = \dfrac{1}{\varepsilon_{//}}$ (B.I.81)

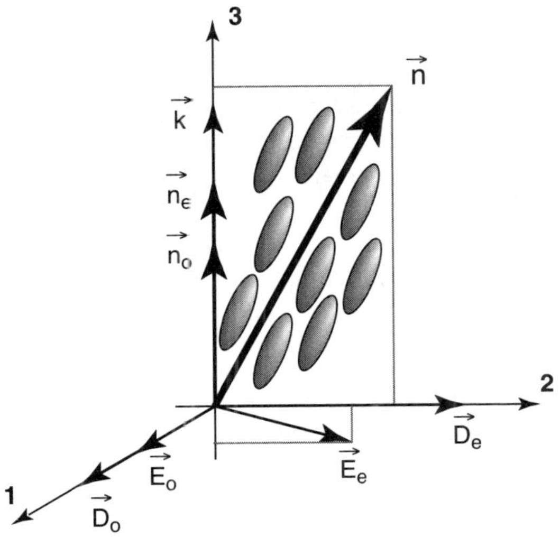

Fig. B.I.22 In a conveniently chosen basis, the eigenmodes for the Maxwell equations in a uniaxial medium are very simple. In the drawing, **n** is the director. The modulus of **n** (//**k**) gives the refraction index.

Consequently, there are two eigenmodes: the first is the **ordinary ray, polarized perpendicularly to the director**. The corresponding refraction index is n_o; the second one corresponds to the **extraordinary ray**. It is **polarized in the plane** defined by **k** (wave vector) and **n** (director); its refraction index n_{eff} depends on the angle θ between **k** and **n** and takes values between n_e and n_o. When **k** is parallel to **n**, $n_{eff}=n_o$ and the refractive index does not depend on the polarization direction. This defines the **optical axis** of a uniaxial nematic.

The ordinary and extraordinary rays have "physical existence," as under certain conditions they follow separate trajectories in the liquid crystal. This is illustrated in figure B.I.23. In the experiment, a lecithin-treated capillary tube is filled with a nematic (6CB, for instance). The lecithin treatment induces a molecular anchoring normal to the walls (homeotropic anchoring). It turns out that the lowest-energy director configuration compatible with the boundary conditions has axial symmetry. It will be described in detail in chapter B.IV, dedicated to defects. In cylindrical coordinates, the director remains in the plane (**r**, **z**) and varies continuously, from a radial orientation (parallel to **r**) at the walls to an axial orientation (parallel to **z**) on the **z** axis (Fig. B.I.24).

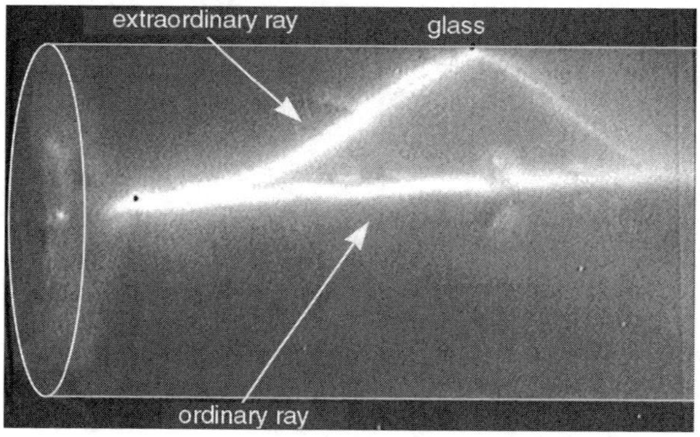

Fig. B.I.23 Trajectory of the ordinary and extraordinary rays in a nematic-filled tube. The laser beam is invisible in air, which diffuses light much less than the nematic phase.

A (depolarized) laser beam, parallel to **z**, but shifted with respect to the symmetry axis, enters the tube and splits into two rays. The one keeping the straight trajectory is the ordinary one: it is polarized normally to the i(**r**,**z**) plane and its index n_o is constant (so it is not deviated). The other ray curves as if it has tried to follow the director field lines. It is the extraordinary ray, polarized in the (**r**, **z**) plane: its refraction index n_{eff} depends on r, causing the deviation. It is then reflected by the tube walls and approaches the symmetry axis without crossing it. The rays are very apparent due to the director orientation fluctuations that strongly diffuse light. In the following section we shall show how to determine the trajectory of the extraordinary ray.

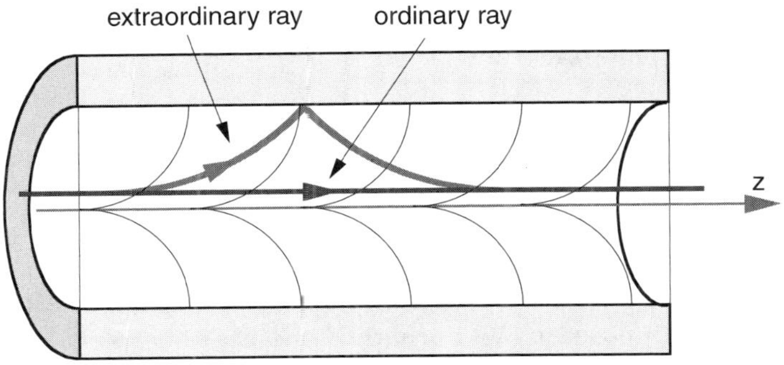

Fig. B.I.24 Director configuration and identification of the rays appearing in figure B.I.23.

I.5.c) Calculation of the extraordinary ray trajectory: generalization of the Fermat principle (Grandjean 1919)

It is common knowledge in optics that the direction of the wave vector **k** does not always coincide with that of the experimentally observed ray, whose real trajectory is given by the Poynting, or "energy" vector **S**, defined as

$$\mathbf{S} = \mathbf{E} \times \mathbf{H} \tag{B.I.82}$$

From figure B.I.22 it is obvious that a distinction should be made between the ordinary ray, for which $\mathbf{S} = \mathbf{S}_o = \mathbf{E}_o \times \mathbf{H}$ is parallel to **k**, and the extraordinary ray for which $\mathbf{S} = \mathbf{S}_e = \mathbf{E}_e \times \mathbf{H}$ is **not** parallel to **k**.

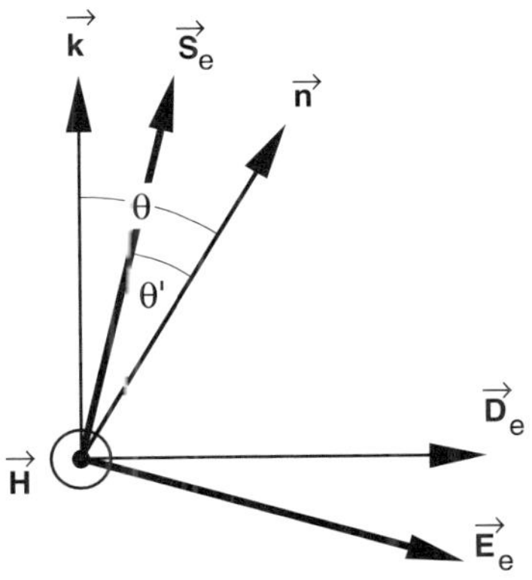

Fig. B.I.25 Position of the different vectors in the radial plane (extraordinary ray).

In this last case, the ray makes with the optical axis **n** an angle θ' differing from the angle θ between **k** and **n** (Fig. B.I.25).

It is readily obtained that:

$$\tan\theta' = \left(\frac{n_o}{n_e}\right)^2 \tan\theta \qquad \text{(B.I.83)}$$

This effect causes the splitting at the capillary entrance (double refraction). As the incident ray has its **k** vector parallel to the capillary axis, the same holds for both the ordinary and extraordinary rays in the nematic (the tangential component of **k**, supposed to be normal to the capillary axis, is continuous at the air-nematic interface). The ordinary ray propagates along the capillary axis, as its Poynting vector is parallel to **k** and the medium has the same index n_o everywhere. On the other hand, the extraordinary ray is deviated upon entering the capillary because its Poynting vector is no longer parallel to **k**, and its trajectory is not straight because of the variation of the n_{eff} index.

The deviation once explained, how can one calculate the ray trajectories? Directly solving the Maxwell equations is usually not feasible, except for some particular cases, an example of which shall be given in the following section. However, one can write down a variational principle allowing for approximate determination of the trajectory. This method, which amounts to **generalizing the Fermat principle to anisotropic media**, was introduced in 1919 by F. Grandjean [20]. Let us first recall the Fermat principle (as it can be proven, it is strictly speaking a theorem). Suppose the medium index n(**r**) only varies significantly over distances much larger than the light wave λ employed. The trajectory of the light ray between two points A and B is the extremum of the optical path $\int_A^B \mathbf{k}.\mathbf{dl}$, a condition mathematically expressed by a zero variation criterion:

$$\delta \int_A^B \mathbf{k}.\mathbf{dl} = 0 \qquad \text{with} \qquad \mathbf{k} = \frac{\omega}{c}\mathbf{n} \qquad \text{(B.I.84)}$$

In an isotropic medium, **k** and **dl** (// **S**) are parallel. The trajectory is therefore given by:

$$\delta \int_A^B n(\mathbf{r})\mathrm{d}l = 0 \qquad \text{(B.I.85)}$$

Consider now the case of a birefringent medium. If the director orientation changes are smooth enough ($\lambda|\nabla\cdot\mathbf{n}|<<1$), both the ordinary and extraordinary rays can be considered locally as propagating in a homogeneous medium having as the optical axis the one existing in that point. Grandjean assumes that, in spite of the continuous direction change, the wave propagates essentially in a straight line (proof for a particular case will be given in the next section) and at the velocity it would have in a homogeneous medium. These approximations are experimentally satisfied: the rays exiting the sample are usually well extinguished by a polarizer, showing that the wave polarizations at the exit point are the same as if the medium were homogeneous and with the same orientation of the optical axis as at the exit point.

Within these approximations, one can write for the **ordinary ray**:

$$\delta \int_A^B \mathbf{k}_o.\mathbf{dl} = 0 \qquad \text{with} \qquad \mathbf{k}_o = \frac{\omega}{c}\mathbf{n}_o \tag{B.I.86}$$

For this ray, $\mathbf{k}_o // \mathbf{dl}$ and n_o is a constant. Its trajectory is therefore a **straight line** joining the two points A and B.

For the **extraordinary ray**, the principle becomes:

$$\delta \int_A^B \mathbf{k}_e.\mathbf{dl} = 0 \qquad \text{with} \qquad \mathbf{k}_e = \frac{\omega}{c}\mathbf{n}_{eff} \tag{B.I.87}$$

Since vectors $\mathbf{k}_e$ and $\mathbf{dl}$ make a non-null angle $\alpha = \theta' - \theta$, the previous equation can be written as:

$$\delta \int_A^B n_{eff} \cos\alpha \, dl = 0 \tag{B.I.88}$$

or, from eqs. B.I.80 and B.I.83:

$$\delta \int_A^B \sqrt{\cos^2\theta' \, n_o^2 + \sin^2\theta' \, n_e^2} \, dl = 0 \tag{B.I.89}$$

This simple equation allows for determining the trajectory of the extraordinary ray. Solutions can be searched in parametric form x(u), y(u), z(u) where u is the parameter. Thus

$$dl = \sqrt{x'^2 + y'^2 + z'^2} \, du \tag{B.I.90}$$

and

$$\cos^2\theta' = \frac{(n_x x' + n_y y' + n_z z')^2}{x'^2 + y'^2 + z'^2} \tag{B.I.91}$$

where n_x, n_y, n_z are known functions for a given director configuration. Substituting in eq. B.I.89 yields:

$$\delta \int_A^B \sqrt{N^2(x'^2 + y'^2 + z'^2) - (N^2 - 1)(n_x x' + n_y y' + n_z z')^2} \, du = 0 \tag{B.I.92}$$

where $N = n_e/n_o$. Writing the Euler equations for the integrand yields a system of three second-order differential equations involving the three unknown functions x, y, and z and the parameter u.

One should note that equation B.I.89 can be obtained without employing variational calculus, by dividing the medium in many thin slices of uniform optical orientation. The real structure is recovered when their number goes to infinity. The extraordinary ray is then a broken line propagating from one slice to the other according to the Huygens construction. This technique shall be applied in the next section to describe light propagation in a twisted nematic. Grandjean solved the equation for the extraordinary ray trajectory in simple geometries, thus accounting for many optical properties of liquid crystals and their textures. As an application, we shall use this principle to calculate the extraordinary ray trajectory in the previously described capillary tube. As the director field lines are needed, they will be taken as circular arcs of radius R, centered on the capillary inner wall (Fig. B.I.26).

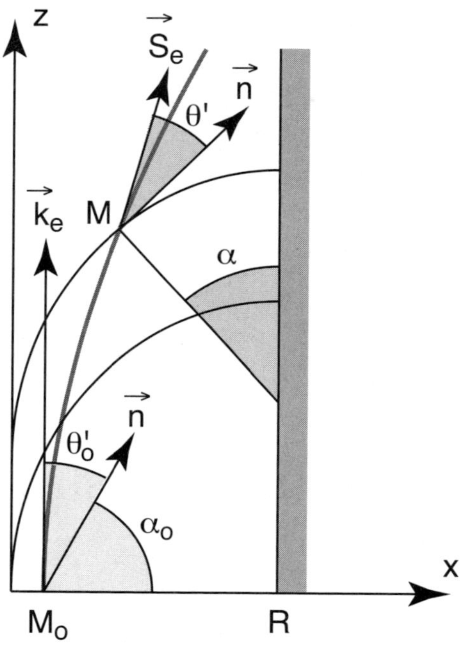

Fig. B.I.26 Notations used for calculating the extraordinary ray trajectory.

If the incident ray is parallel to the z axis, the extraordinary ray, described by the equation $z = f(x)$, remains in the radial plane (x, z). Let x_0 be the abscissa of the entrance point M_0 where one has:

$$f(x_0) = 0 \tag{B.I.93}$$

On the other hand, the tangential component of the wave vector is conserved on entering the capillary. The wave vector of the extraordinary ray $\mathbf{k}_e$ is therefore parallel to the z axis in M_0 and makes an angle $\pi/2 - \alpha_0$ with the director. Let θ'_0 be the angle made by the extraordinary ray with the director in this point.

From eq. B.I.83

$$\tan\theta_o' = \frac{1}{N^2 \tan\alpha_o} \qquad \text{(B.I.94)}$$

giving the slope of the trajectory at point M_o:

$$f'(x_o) = \tan(\theta_o' + \alpha_o) \qquad \text{(B.I.95)}$$

which can be written as a function of $X_o = x_o/R$ using the fact that for an arbitrary point M of coordinates (x, z) one has

$$n_z = \sin\alpha = \frac{R-x}{R}, \qquad n_x = \cos\alpha = \sqrt{\frac{2x}{R} - \frac{x^2}{R^2}} \qquad \text{(B.I.96)}$$

Finally,

$$f'(x_o) = \frac{X_o(2-X_o) + N^2(1-X_o)^2}{(N^2-1)(1-X_o)\sqrt{X_o(2-X_o)}} \qquad \text{(B.I.97)}$$

Let us now write the Fermat-Grandjean principle. From eq. B.I.89

$$\delta\int \sqrt{N^2 + (1-N^2)\cos^2\theta'}\,\sqrt{1+f'^2}\,dx = 0 \qquad \text{(B.I.98)}$$

yielding, after some algebra,

$$\delta\int \sqrt{N^2(1+f'^2) + (1-N^2)(\cos\alpha + f'\sin\alpha)^2}\,dx = 0 \qquad \text{(B.I.99)}$$

As the integrand L only depends on f' and x, the Euler equation is simply $\partial L/\partial f' = \text{Cst}$, yielding:

$$\frac{N^2 f' + (1-N^2)\sin\alpha\,(\cos\alpha + f'\sin\alpha)}{\sqrt{N^2(1+f'^2) + (1-N^2)(\cos\alpha + f'\sin\alpha)^2}} = \text{Cst} \qquad \text{(B.I.100)}$$

The constant is given by the initial condition B.I.97. Replacing $\cos\alpha$ and $\sin\alpha$ by their expressions B.I.96, one gets

$$f' = \frac{-C(A - \text{Cst}^2) + \sqrt{\Delta}}{A(A - \text{Cst}^2)} \qquad \text{(B.I.101)}$$

with

$$X = \frac{x}{R}$$

$$A = N^2 + (1-N^2)(1-X)^2$$

$$B = N^2 + 1 - A \qquad \text{(B.I.102)}$$

$$\Delta = C^2(A - Cst^2) - A(A - Cst^2)(C^2 - B\,Cst^2)$$

$$Cst = \frac{A_o f'_o + C_o}{\sqrt{A_o f'^2_o + 2C_o f'_o + B_o}}$$

We have solved equation B.I.101 numerically with the initial conditions B.I.93 and B.I.97 taking $X_o = 0.1$, $n_o = 1.45$, and $n_e = 1.66$. The trajectory thus determined is shown in figure B.I.27 (the calculation is analogous after reflection on the capillary wall). We obtain good agreement with the experiment, considering the approximation made on the director field lines. Asymptotically, the extraordinary ray propagates along the z axis.

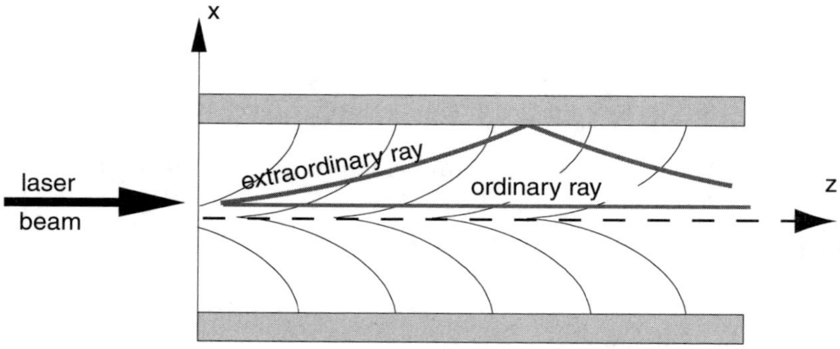

Fig. B.I.27 Calculated trajectory of the extraordinary ray when the director field lines are circular arcs centered on the capillary wall. Note the qualitative agreement with the real trajectory.

I.5.d) Eigenmodes of a twisted nematic

A particular director configuration, known as twisted, is optically very interesting, being widely employed in displays (see next chapter for the operation principle). In this configuration, the director turns uniformly, remaining perpendicular to a fixed direction that we shall take as the z axis (Fig. B.I.28). Let p be the pitch of this helical structure and $q = d\phi/dz = 2\pi/p$ its twist, with ϕ the angle between the director and the x axis.

In this section we shall only treat the case where the wavelength λ is very small in comparison with p. We shall equally assume that light is propagating along the z axis.

To find the eigenmodes propagating in this helical structure, let us model the twisted nematic as a stack of uniaxial birefringent slices each one having the optical axis turned by an angle $d\phi$ with respect to that of the adjacent slices (Fig. B.I.29). The rotation angle $d\phi$ between two slices of thickness dz is

$$d\phi = q\,dz \tag{B.I.103}$$

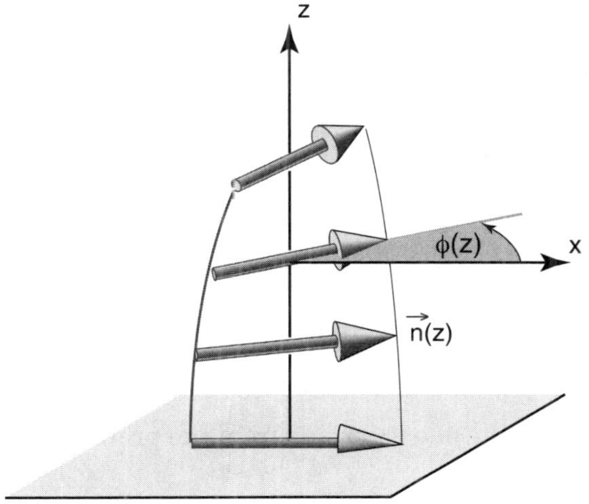

Fig. B.I.28 Twisted director configuration employed in nematic liquid crystal displays.

We are looking for eigenmodes of wave vector **k** parallel to **z**. The eigenmodes of each slice are linearly polarized; let $\mathbf{e}_o(z)$ and $\mathbf{e}_e(z)$ be the unit vectors parallel to the electric field of these modes (Fig. B.I.29). The first mode, polarized perpendicularly to the director ($\mathbf{e}_o(z) \perp \mathbf{n}(z)$), corresponds to the **ordinary ray**, while the second one, polarized parallel to the director ($\mathbf{e}_e(z) \,//\, \mathbf{n}(z)$), corresponds to the **extraordinary ray**. The eigenmodes of the entire stack can be written as

$$\mathbf{E}(z) = a(z)\mathbf{e}_o(z) + b(z)\mathbf{e}_e(z) \tag{B.I.104}$$

Let

$$\mathbf{E}_{in1} = \mathbf{E}(z) = a(z)\mathbf{e}_o(z) + b(z)\mathbf{e}_e(z) \tag{B.I.105}$$

be the electric field of the wave incident on the slice (z, z+dz) which will induce a dephasing between the ordinary and extraordinary rays such that the electric field emerges as:

$$\mathbf{E}_{s1} = a(z)\mathbf{e}_o(z)\exp(-\,ik_o dz) + b(z)\mathbf{e}_e(z)\exp(-\,ik_e dz) \tag{B.I.106}$$

where

$$k_o = 2\pi n_o/\lambda \qquad \text{and} \qquad k_e = 2\pi n_e/\lambda \tag{B.I.107}$$

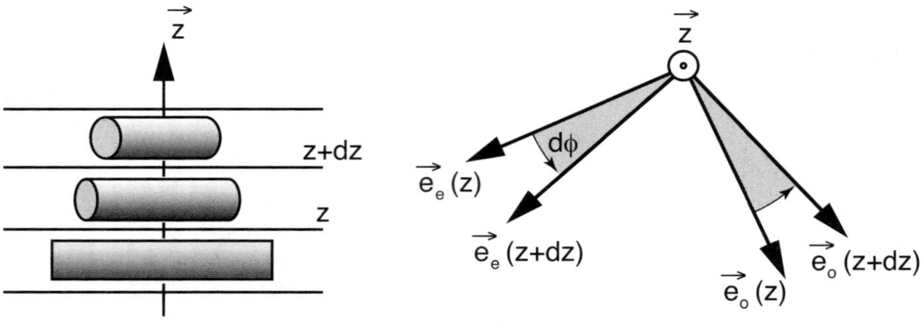

Fig. B.I.29 Twisted nematic modeled as a stack of uniaxial slices of thickness dz.

Following equation B.I.105, the field incident on the next slice reads

$$\mathbf{E}_{in2} = \mathbf{E}(z+dz) = a(z+dz)\mathbf{e}_o(z+dz) + b(z+dz)\mathbf{e}_e(z+dz) \qquad \text{(B.I.108)}$$

The field is identical to the one emerging from the previous slice, hence:

$$\mathbf{E}_{s1} = \mathbf{E}_{in2} \qquad \text{(B.I.109)}$$

Express $\mathbf{e}_o(z+dz)$ and $\mathbf{e}_e(z+dz)$ in terms of $\mathbf{e}_o(z)$ and $\mathbf{e}_e(z)$ of the previous slice:

$$\mathbf{e}_o(z+dz) = \mathbf{e}_o(z) - d\phi\, \mathbf{e}_e(z) \qquad \text{(B.I.110)}$$

and

$$\mathbf{e}_e(z+dz) = \mathbf{e}_e(z) + d\phi\, \mathbf{e}_o(z) \qquad \text{(B.I.111)}$$

and develop to first order a(z+dz), b(z+dz) and the exponentials in equation B.I.106. Using equation B.I.109, one has:

$$\frac{da}{dz} = -ik_o a - qb \qquad \text{(B.I.112)}$$

$$\frac{db}{dz} = qa - ik_e b \qquad \text{(B.I.113)}$$

The solutions of this system of equations can be written as:

$$a(z) = a_o \exp(\alpha z), \qquad b(z) = b_o \exp(\alpha z) \qquad \text{(B.I.114)}$$

yielding the eigenvalue equation:

$$\alpha^2 + i(k_o + k_e)\alpha - k_o k_e + q^2 = 0 \qquad \text{(B.I.115)}$$

with discriminant

$$\Delta = -\frac{4\pi^2(\Delta n)^2}{\lambda^2} - \frac{16\pi^2}{p^2} \qquad \text{(B.I.116)}$$

where $\Delta n = n_e - n_o$ is the nematic birefringence. In the limit of

$$\lambda << p\Delta n \qquad \text{(B.I.117)}$$

the roots of equation B.I.115 are simply:

$$\alpha_{+,-} = -ik_{o,e} \qquad \text{(B.I.118)}$$

Consequently, the two modes are given in this limit by:

$$\mathbf{1.}\ a(z) = a_o \exp(-ik_o z), \qquad b(z) = 0 \qquad \text{(B.I.119)}$$

$$\mathbf{2.}\ a(z) = 0, \qquad b(z) = b_o \exp(-ik_e z) \qquad \text{(B.I.120)}$$

The electric field of these two modes follows the rotation of the director for λ << pΔn.

This important result explains the optical properties of the twisted nematic display described in the next chapter.

Let us mention the existence of alternative approaches to the optical properties of liquid crystals, such as the Poincaré sphere [21] or the Jones matrices [22].

I.5.e) Polarized light contrast

The birefringence of the nematic phase is also apparent in its characteristic textures, visible in polarized microscopy and which are exemplified throughout this book. We shall be more particularly concerned with figures B.I.30, B.IV.3, and B.IV.45, displaying the so-called “plages à noyaux” textures, composed of linear (disclinations) or point (umbilics) defects.

When these textures are photographed in monochromatic light (with wavelength λ) between crossed polarizers, two extinction areas can be distinguished.

The most visible are called **isogyres;** they remain dark when the wavelength of the lighting changes but turn with the polarizers and cross at distinctive 90° angles. As the name indicates, **isogyres are curves of equal director orientation**. This is illustrated in figure B.I.30, where chains of water droplets reveal the director orientation. Also notice that this texture exhibits singularities of the director field, here a pair of $+2\pi$ and -2π disclinations where the isogyres meet at 90° angles.

More specifically, isogyres are places where the director is either in the plane parallel to the polarization P of the incident light, or in the plane normal to the polarization P. In the first case, only the extraordinary mode propagates in the liquid crystal, such that the exit polarization remains unchanged and normal to the analyzer A, hence the extinction. In the second case, the ordinary mode is the only one to propagate, also leading to extinction.

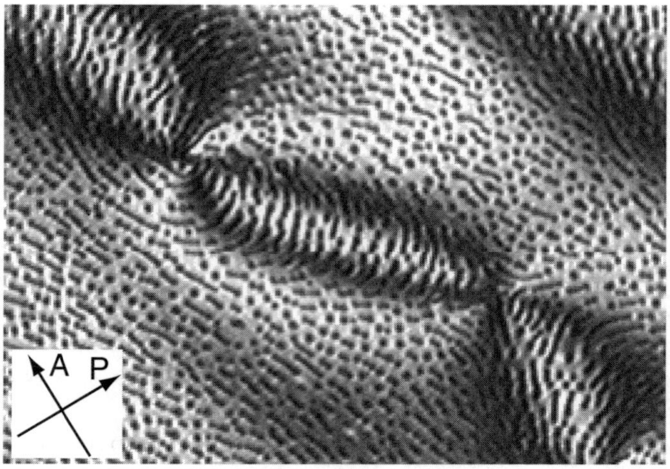

Fig. B.I.30 Pair of $+2\pi$ and -2π disclinations seen under the polarizing microscope between crossed polarizers. The chains of water droplets being parallel to the director, one can conclude that the extinction areas are **isogyres** – curves of equal director orientation.

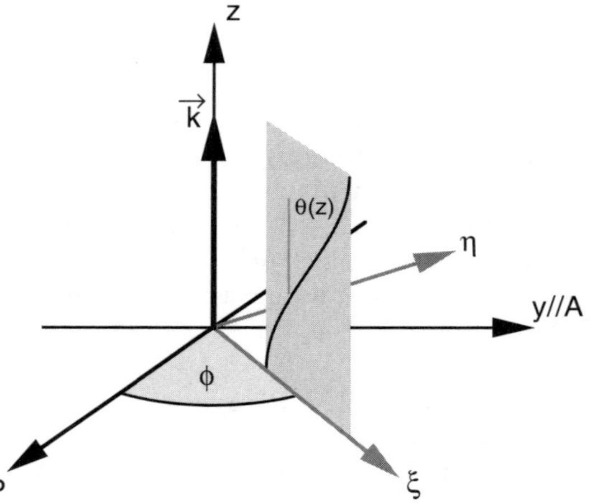

Fig. B.I.31 Propagation geometry for the ordinary and extraordinary modes.

With the director in the $(\boldsymbol{\xi},\mathbf{z})$ plane, tilted with respect to $\mathbf{P}//\mathbf{x}$ (see figure B.I.31), the two modes (ordinary and extraordinary) simultaneously propagate along the $\mathbf{z}$ direction. In the $(\boldsymbol{\xi},\boldsymbol{\eta},\mathbf{z})$ frame, comoving with the director, the electric field components of the linearly polarized light entering the liquid crystal are:

$$\mathbf{E}_{in}(z=0) = E[\cos\phi\exp(-i\omega t), -\sin\phi\exp(-i\omega t), 0] \qquad \text{(B.I.121)}$$

Upon leaving the liquid crystalline sample of thickness d, the electric field of the transmitted wave becomes:

$$\mathbf{E}_{tr}(z=d) = E[\cos\phi \exp(i\varphi_{eff} - i\omega t), -\sin\phi \exp(i\varphi_o - i\omega t), 0] \qquad \text{(B.I.122)}$$

where

$$\varphi_o = k_o d \qquad \text{(B.I.123)}$$

and

$$\varphi_{eff} = \int_A^B k_{eff}(z)dz \qquad \text{(B.I.124)}$$

are the respective phases of the ordinary and extraordinary modes after crossing the liquid crystal.

The polarization of the wave transmitted across the sample (eq. B.I.122) depends on the phase shift $\Delta\varphi = \varphi_{eff} - \varphi_o$. To be specific,

– for $\Delta\varphi = 2p\pi$ (integer p), it remains linear and parallel to x;

– for $\Delta\varphi = (2p+1)\pi$, it remains linear, but it is oriented along direction x', the symmetric of x with respect to the (ξ, z) plane;

– for any other value of the phase shift, it is elliptical.

Using the E_{ytr} component of the transmitted wave along the direction of the analyzer $\mathbf{A} // \mathbf{y}$, written as

$$E_{ytr} = E \sin\phi \cos\phi [\exp(i\varphi_{eff}) - \exp(i\varphi_o)] \exp(-i\omega t) \qquad \text{(B.I.125)}$$

one can determine the light intensity transmitted through the analyzer:

$$I_{tr} = I_{in} 2\sin^2\phi \cos^2\phi [1 - \cos(\Delta\varphi)] \qquad \text{with} \qquad \Delta\varphi = (\varphi_{eff} - \varphi_o) \qquad \text{(B.I.126)}$$

For $\phi = 0$ and $\phi = \pi/2$, this formula predicts the characteristic isogyre extinction.

However, the extinction can also occur for different orientations ϕ, when the phase shift between the ordinary and extraordinary modes is an integral multiple of 2π:

$$\Delta\varphi = \varphi_{eff} - \varphi_o = 2\pi\, p, \qquad p = 0, \pm 1, \pm 2, \ldots \qquad \text{(B.I.127)}$$

or, explicitly

$$\Delta\varphi = 2\pi \frac{d}{\lambda} \left(\left(\int_0^1 n_{eff}(\tilde{z}) d\tilde{z} \right) - n_o \right) = 2\pi p \qquad \text{(B.I.128)}$$

with $\tilde{z} = z/d$. As this new extinction condition depends on light color *via* the wavelength λ, it defines extinction areas termed **isochromes, over which the phase shift between the ordinary and the extraordinary waves is an**

integral multiple of 2π.

Isochromes are illustrated by the photos in figure B.I.32, showing a detail of the Néel points decorating the Bloch wall in figure B.IV.45. Besides the cross-shaped isogyres one can also see two isochromes encircling the -2π and $+2\pi$ singularities. Experiment shows that isochrome behavior is the exact opposite of that of the isogyres, since they do not move when turning the polarizers but shift when the wavelength is changed. To be more precise, isochromes move away from the center of the defect with increasing wavelength.

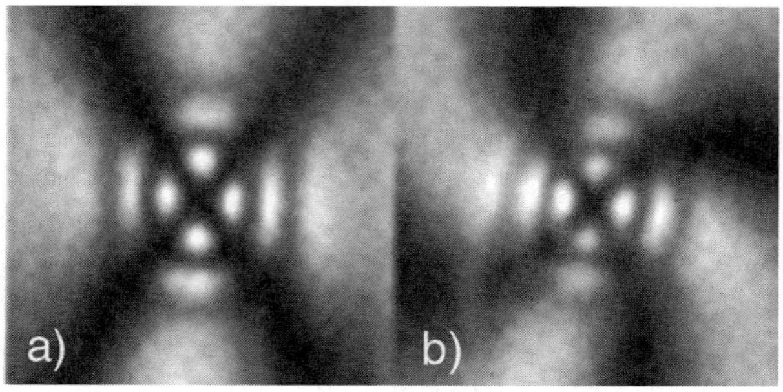

Fig. B.I.32 Detail of the -2π (a) and $+2\pi$ (b) singular Néel points on the Bloch wall in figure B.IV.45. Besides the cross-shaped isogyres one can also see two isochromes encircling the singularity centers.

Referring to §B.IV.8c, this behavior is easily explained knowing that: 1. In the center of the Néel singularities the director is parallel to the optical axis z, so $n_{eff} = n_0$ and the phase shift $\Delta\delta$ is zero; and 2. The $\theta(z,r)$ distortion due to the action of the electric field increases away from the center of the singularities. If $\theta(z,r)$ is known, $n_{eff}(z,r)$ can be determined from equation B.I.80. For $n_e > n_0$, the effective refractive index of the extraordinary wave increases with θ. Consequently, the average value $<n_{eff}(\widetilde{z})>$ appearing in the definition of the isogyres B.I.128 increases with r. Therefore, the radius r of isochrome p increases with the wavelength λ (which allows the interference order p to remain constant).

I.5.f) Conoscopy

The birefringence of the nematic phase is also expressed in the so-called conoscopy experiments. This relatively old optical method is used for investigating the optical anisotropy of crystalline media and yields the orientation of the optical axes. Throughout this book we shall see that conoscopy is extensively employed in liquid crystal experiments, starting with the Frederiks transition to be discussed in chapter B.II.

The principle of conoscopy is depicted in figure B.I.33a. A divergent beam of monochromatic light linearly polarized by the polarizer P is shone onto the plane-parallel sample of thickness d. It can simply be a laser beam diffused by tracing paper. The light transmitted by the sample goes through the analyzer A (perpendicular to P) before it is collected on a translucent screen (tracing paper).

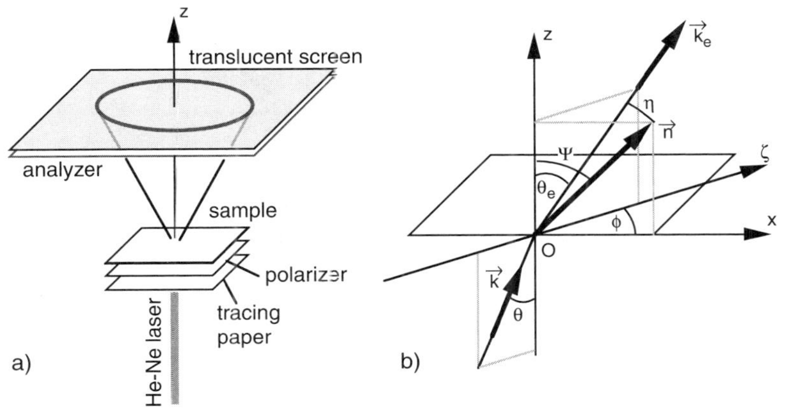

Fig. B.I.33 Conoscopy. a) Experimental setup. The analyzer and the polarizer are crossed. b) Refraction geometry on entering the nematic sample.

In this setup, every point on the screen is associated with a plane wave component with wavevector **k** of the incident beam. In an anisotropic sample the two modes, ordinary and extraordinary, propagate at different velocities along most directions **k** of the incident light and acquire a phase shift $\Delta\delta$ as they leave the sample. This phase shift $\Delta\varphi(\mathbf{k},\mathbf{n})$ depends on the direction of the incident wave and on the orientation of the liquid crystal along the light path. When the cone of incident light has a very small aperture, equation B.I.126 is a good approximation for the intensity of the light collected on the screen. Thus, the conoscopic figure that appears on the screen is composed of extinction areas similar to the isogyres and the isochromes discussed in the previous section.

To find the layout of the isochromes, we need to find the phase shift $\Delta\varphi(\mathbf{k},\mathbf{n})$. Let us assume that the director **n** (uniformly oriented) is contained in the (x, z) plane and makes an angle Ψ with the optical axis z of the setup. We then have

$$\mathbf{n} = (\sin\Psi, 0, \cos\Psi) \tag{B.I.129}$$

Let us equally assume that the polarizer and the analyzer are at a 45° angle with axes x and y. Consider an incident plane wave Σ with wavevector

$$\mathbf{k} = k(\sin\theta\cos\phi, \sin\theta\sin\phi, \cos\theta) \tag{B.I.130}$$

As it enters the liquid crystal, this wave gives rise to two plane waves, ordinary and extraordinary, with wave vectors $\mathbf{k}_o$ and $\mathbf{k}_e$ contained in the (ζ, z) plane just like $\mathbf{k}$. To simplify the diagram in figure B.I.33b, we only show the refraction conditions for the extraordinary wave.

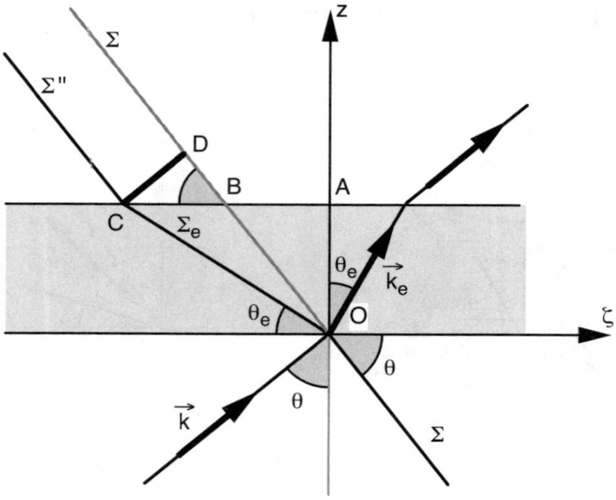

Fig. B.I.34 Determining the retardation of a plane wave after going through a plane-parallel plate.

The diagram in figure B.I.34 gives the positions of the incident Σ, refracted Σ_e, and emergent Σ'' wave fronts at the same moment. After going through the sample, the emergent wave Σ'' is obviously delayed with respect to the position of the incident wave in the absence of the nematic plate. The difference in optical path between Σ and Σ'' is:

$$CD_e = d(n_{eff}\cos\theta_e - \cos\theta) \tag{B.I.131a}$$

The same reasoning applied to the ordinary wave gives a retardation:

$$CD_o = d(n_o\cos\theta_o - \cos\theta) \tag{B.I.131b}$$

different from CD_e. The difference $CD_e - CD_o$ yields the phase shift:

$$\Delta\varphi = 2\pi\frac{d}{\lambda}(n_{eff}\cos\theta_e - n_o\cos\theta_o) \tag{B.I.132}$$

The phase shift must now be expressed as a function of the incident wave direction (θ, ϕ). For the second term of the equation, this is straightforward, the ordinary refractive index being a constant

$$n_o = \frac{\sin\theta}{\sin\theta_o} \tag{B.I.133}$$

whence

$$n_o \cos\theta_o = n_o \sqrt{1 - \frac{\sin^2\theta}{n_o^2}} \qquad \text{(B.I.134)}$$

The first term, due to the extraordinary wave, is more complicated because the effective refractive index depends on the angle η between the optical axis **n** and the wave vector $\mathbf{k}_e$ (see Fig. B.I.33b). We have:

$$\cos\eta = \mathbf{n}.\mathbf{k}/k = \cos\Psi \cos\theta_e + \sin\Psi \sin\theta_e \cos\phi \qquad \text{(B.I.135)}$$

Using equation B.I.80 (with η instead of θ), tedious algebra yields:

$$n_{eff} \cos\theta_e = \frac{n_o^2 - n_e^2}{n^2} \sin\theta \cos\phi \sin\Psi \cos\Psi + \frac{n_o n_e}{n} \sqrt{1 - \sin^2\theta \left(\frac{\sin^2\phi}{n_e^2} + \frac{\cos^2\phi}{n^2}\right)}$$

(B.I.136a)

with

$$n^2 = n_e^2 \cos^2\Psi + n_o^2 \sin^2\Psi \qquad \text{(B.I.136b)}$$

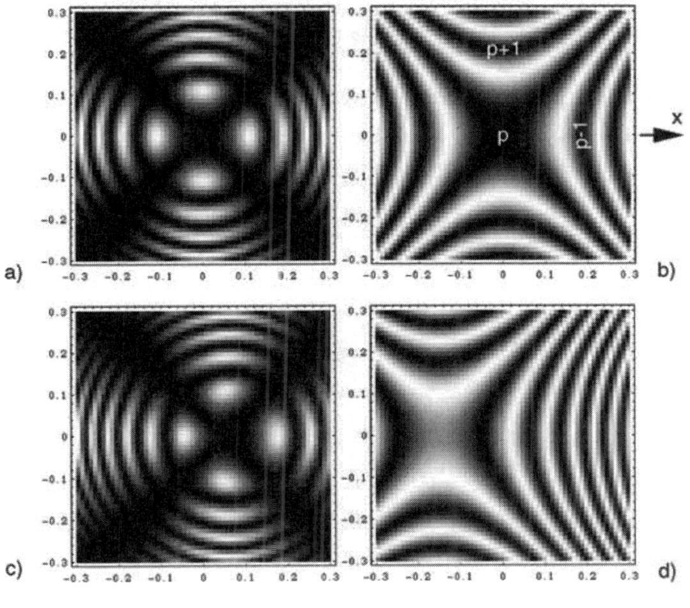

Fig. B.I.35 Conoscopic figures computed for a) $\Psi = 0$ (homeotropic geometry), b) $\Psi = \pi/2$ (planar geometry), c) $\Psi = 0.05$, and d) $\Psi = \pi/2 - 0.05$.

In homeotropic geometry (Fig. B.I.33a), the optical axis **n** is perpendicular to the nematic plate, Ψ=0, and the expression of the phase shift is much simpler:

$$\Delta\varphi = 2\pi \frac{d}{\lambda} n_o \left(\sqrt{1 - \frac{\sin^2\theta}{n_e^2}} - \sqrt{1 - \frac{\sin^2\theta}{n_o^2}} \right) \qquad \text{(B.I.137)}$$

Owing to revolution symmetry about the axis **z**//**n**, the phase shift $\Delta\varphi$ does not depend on the azimuthal angle ϕ and the isochromes, defined by $\Delta\varphi = 2\pi p$ ($p = 1,2,...$), are concentric circles as shown in figure B.I.35a (see also figure B.II.15a). In the limit of small incidence angle θ we have

$$\Delta\varphi = 2\pi \frac{d}{\lambda} \frac{n_o}{2} \left(\frac{1}{n_o^2} - \frac{1}{n_e^2} \right) \theta^2 \qquad \text{(B.I.138)}$$

so the diameters of successive isochromes vary as $p^{1/2}$. Notice that the conoscopic figure for the homeotropic sample also exhibits cross-shaped isogyres, as for **k** directions contained in the (P, z) and (A, z) planes, parallel and perpendicular to the incident polarization, respectively; only one of these modes can propagate and the beam remains linearly polarized.

In planar geometry (Fig. B.I.35b), the optical axis **n** is parallel to the sample surface, $\Psi = \pi/2$, and the formula for the phase shift is again simplified:

$$\Delta\varphi = 2\pi \frac{d}{\lambda} \left[n_e \sqrt{1 - \sin^2\theta \left(\frac{\sin^2\phi}{n_e^2} + \frac{\cos^2\phi}{n_o^2} \right)} - n_o \sqrt{1 - \frac{\sin^2\theta}{n_o^2}} \right] \qquad \text{(B.I.139)}$$

In the limit of small incidence angle,

$$\Delta\varphi = 2\pi \frac{d}{\lambda} (n_e - n_o) \left[1 + \frac{1}{2n_o} \left(\frac{\sin^2\phi}{n_e} - \frac{\cos^2\phi}{n_o} \right) \theta^2 \right] \qquad \text{(B.I.140)}$$

Writing the phase shift $\Delta\varphi$ as a function of the coordinates of the rays incident on the translucent screen placed at a distance L

$$x = L\theta \cos\phi \quad \text{and} \quad y = L\theta \sin\phi \qquad \text{(B.I.141)}$$

one has

$$\Delta\varphi(x,y) = 2\pi \frac{d}{\lambda} (n_e - n_o) \left[1 + \frac{1}{2n_o L^2} \left(\frac{y^2}{n_e} - \frac{x^2}{n_o} \right) \right] \qquad \text{(B.I.142)}$$

The isochromes, defined by $\Delta\varphi = 2\pi p$, of a planar nematic sample are therefore hyperbolae (see figure B.I.33c) of asymptotes

$$y = \pm \sqrt{\frac{n_e}{n_o}}\, x \qquad \text{(B.I.143)}$$

Due to the fact that $n_e > n_o$, the asymptotes of the conoscopic figure cross at an angle slightly larger than 90°. This dissymmetry allows determination of the director orientation (given here by the x axis). An alternative way of finding out the director orientation is to study the evolution of the isochromes upon heating the sample, which leads to a decrease in birefringence $\Delta n = n_e - n_o$. It is easily shown that the isochromes on the x axis converge towards the center of the figure while those on the y axis move away from it.

In conclusion, let us emphasize that conoscopy is a very convenient way of detecting and measuring a variation in the average director orientation **n** in nematics. To be more specific, we shall see in this book that the technique was widely employed in experiments on the Frederiks transition, on hydrodynamic instabilities, and on anchoring transitions. For instance, the conoscopic figure shifts, as in figure B.I.35c, when the homeotropic geometry is disturbed by a field or flow. Similarly, a deviation from the planar geometry leads to a tilt of the isochrome system (Fig. B.I.35d). Conoscopy is also very useful for determining the orientation of the average optical axis in smectic phases. In particular, it is by this technique that R.B. Meyer demonstrated the spontaneous polarization of the SmC* phase (see chapter C.IV in volume 2 on smectic and columnar liquid crystals).

BIBLIOGRAPHY

[1] a) Cladis P.E., Kléman M., Pieranski P., *C. R. Acad. Sci. (Paris)*, **273B** (1971) 275.
b) Poulin P., Cabuil. V., Weitz D.A., *Phys. Rev. Lett.*, **79** (1997) 4862.
c) Poulin P., Weitz D.A, *Phys. Rev. E*, **57** (1998) 626.
d) Lubensky T.C., Pettey D., Currier N., Stark H., *Phys. Rev. E*, **57** (1998) 610.

[2] De Gennes P.-G., *Mol. Cryst. Liq. Cryst.*, **12** (1971) 193.

[3] Coles H.J., *Mol. Cryst. Liq. Cryst. Lett.*, **49** (1978) 67.

[4] Vuks M.F., *Opt. Spektrosk.*, **20** (1966) 361.

[5] a) Ratna B.R., Vijaya M.S., Shashidar R., Sadashiva B.K., *Pramana*, Suppl. I (1973) 69.
b) Ratna B.R., Shashidar R., *Pramana*, **6** (1976) 278.
c) Stinson T.W., Litster J.D., *Phys. Rev. Lett.*, **25** (1970) 503.

[6] Vertogen G., de Jeu W.H., in *Thermotropic Liquid Crystals, Fundamentals*, Vol. 45, Springer-Verlag, Heidelberg, 1988, p. 227.

[7] Tsvetkov V.N., Ryumtsev E.I., *Sov. Phys. Crystallogr.*, **13** (1968) 225.

[8] Madhusudana N.V., Chandrasekhar S., in *Liquid Crystals and Ordered Fluids*, Vol. 2, Eds. Johnson J.F. and Porter R.S., Plenum, New York, 1967, p. 657.

[9] Yu L.J., Saupe A., *Phys. Rev. Lett.*, **45** (1980) 1000.

[10] Galerne Y., Marcerou J.P., *Phys. Rev. Lett.*, **51** (1983) 2109.

[11] Meyer R.B., *Phys. Rev. Lett.*, **22** (1969) 918.

[12] Rudquist P., Lagerwall S.T., *Liq. Cryst.*, **23** (1997) 503.

[13] Schmidt D., Schadt M., Helfrich W., *Z. Naturforsch.*, **27a** (1972) 277.

[14] Dozov I., Martinot-Lagarde Ph., Durand G., *J. Physique Lett.*, **43** (1982) L365.

[15] Lavrentovich O.D., Pergamenshchik V.M., in *Liquid Crystals in the Nineties and Beyond*, Ed. Kumar S., World Scientific, Singapore, 1995, p. 275.

[16] Bedanov V.M., Gadiyak G.V., Lozovik Yu E., *Phys. Lett. A*, **92** (1982) 400.

[17] Prost J., Marcerou J.P., *J. Physique (France)*, **38** (1977) 315.

[18] Lippens D., Parneix J.P., Chapoton A., *J. Physique (France)*, **38** (1977) 1465.

[19] Diquet D., Rondelez F., Durand G., *C. R. Acad. Sci. (Paris)*, Série B, **271** (1970) 954.

[20] Grandjean F., *Bull. Soc. Fr. Minér.*, **42** (1919) 42.

[21] Poincaré H., *Théorie Mathématique de la Lumière*, Vol. 2, Gauthier Villars, Paris, 1892.

[22] Jones R.C., *J. Opt. Soc. Am.*, **31** (1941) 488; ibid., **31** (1941) 493. See also chapter 4, in S. Chandrasekhar, *Liquid Crystals*, Second Edition, Cambridge University Press, Cambridge, 1992.

Chapter B.II

Nematoelasticity: Frederiks transition and light scattering

In the previous chapter we assumed that all the molecules were parallel to a fixed direction. This is the minimal energy configuration and it will be obtained if the nematic is free of external constraints. This is not always the case, as boundary conditions imposed by the container are not necessarily compatible with perfect alignment of the sample. An example is the case of a nematic between two glass plates with different molecular anchoring (e.g., planar and homeotropic), inducing bulk distortion of the director field. The resulting increase in free energy is the integral

$$F = \int f \, dv \qquad \text{(B.II.1)}$$

of the elastic deformation free energy density $f(\mathbf{n}, \nabla\mathbf{n}, \nabla^2\mathbf{n}, ...)$ over the entire nematic sample. In this chapter we describe this functional and its properties and show that, though fluid, a nematic can transmit torques.

To start with an illustration (section II.1), we describe an experiment that directly proves the existence of elastic torques (J. Grupp, 1983). The general form of the elastic distortion free energy f for a **uniaxial** nematic **(Frank-Oseen formula)** is then discussed (section II.2). We show the existence of several independent terms, each of them associated with a particular type of deformation. The equilibrium equations are formally established and the fundamental concepts of **molecular field** and surface and volume **elastic torques** are introduced (section II.3), allowing for interpretation of the Grupp experiment (section II.4). The action of a magnetic field is analyzed, and the **Frederiks instability** is subsequently described (section II.5) as well as its applications to **displays** (section II.6). We continue with a brief overview of elastic light scattering and its application to the measurement of the elastic constants (section II.7). The chapter concludes with a short introduction to non-linear optics (autofocusing of a laser beam and optical Frederiks transition, section II.8).

II.1 Grupp experiment

To prove that a nematic liquid crystal can transmit a torque, J. Grupp [1] used the setup described in figure B.II.1.

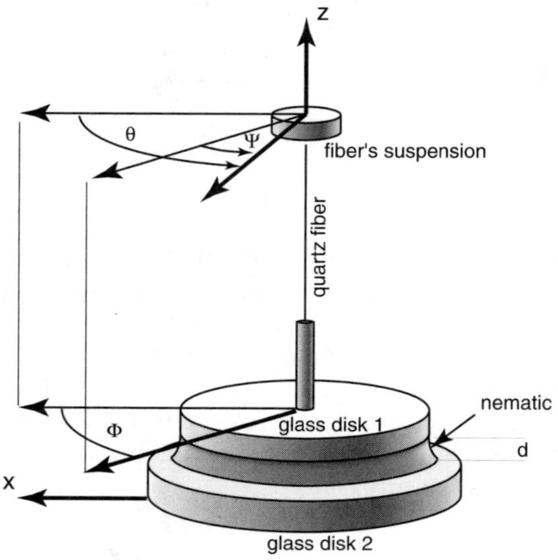

Fig. B.II.1 Setup used in the Grupp experiment to measure the twist nematic constant.

It consists of two parallel glass disks. The lower one (2) is fixed, while the upper one, of radius R = 15 mm, is suspended from a quartz fiber (diameter between 20 and 30 μm). The thickness of the nematic layer between the two disks can vary between 0 and about 200 μm with a precision of 1 μm. Both disks are surface treated to induce planar anchoring (**n** parallel to the x,y plane). On the fixed lower disk, the nematic liquid crystal, MBBA, is parallel to the x axis. Initially, the nematic orientation on the upper disk is the same and the quartz fiber is not under stress. One then rotates by an angle θ the fiber suspension mount and the mobile disk turns in the same direction, but by a smaller angle Φ. The fiber is therefore under a $\Psi = \theta - \Phi$ twist, associated with the elastic torque

$$\Gamma_{fiber} = D\,\Psi \tag{B.II.2}$$

with $D = 9.6\times10^{-10}$ Nm the torque constant of the fiber. The equilibrium configuration can only be explained if the nematic exerts on the mobile disk a compensating torque Γ_{nem}. One must have:

$$\Gamma_{fiber} + \Gamma_{nem} = 0 \tag{B.II.3}$$

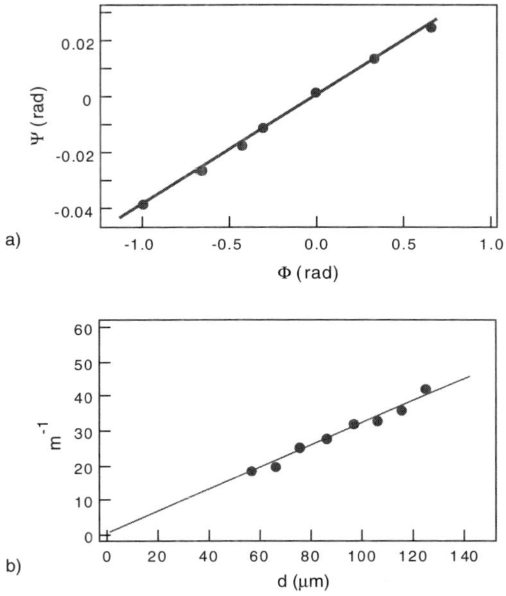

Fig. B.II.2 a) Twist angle Ψ of the quartz fiber as a function of the nematic layer twist Φ; b) m^{-1} coefficient as a function of the nematic layer thickness (from ref. [1]).

J. Grupp measured the torque as a function of the angle θ and showed that the twist angle Ψ of the quartz fiber varies linearly with the twist angle Φ of the nematic layer (Fig. B.II.2a). This dependence ($\Psi = m\Phi$), combined with the equilibrium equation B.II.3 indicates that Γ_{nem} is proportional to Φ. J. Grupp equally checked that the proportionality constant m is inversely proportional to the thickness d of the nematic layer (Fig. B.II.2b). Consequently,

$$\Gamma_{nem} \propto -\frac{\Phi}{d} \tag{B.II.4}$$

The surface torque exerted by the nematic on the disks is therefore proportional to the rotation angle of the director per unit length along the z axis (as we shall see later, this quantity is the opposite of the director field twist **n.curl n**). The associated twist elastic constant K_2 is given by

$$\Gamma_{nem} = -K_2\frac{\Phi}{d} \tag{B.II.5}$$

This elastic constant has force units, measured in Newtons in the international system MKSA and in dynes in CGS (1 N = 10^5 dyn). Grupp obtained $K_2 \approx 10^{-6}$ dyn for MBBA at room temperature.

In the next section we show the existence of two other fundamental bulk deformations, which can be characterized by two new elastic constants, K_1 and K_3.

II.2 Frank-Oseen free energy

Throughout this chapter, the discussion will be limited to the case of **uniaxial** nematics. Let **n** be the director (unit vector tangent to the direction of average molecular alignment). In the absence of external constraints, **n** is everywhere parallel to a fixed direction: in this reference state, $\nabla\mathbf{n} = 0$ and f = 0.

Consider now distortions with respect to this configuration, assuming the local order parameter S remains constant (fixed temperature). If $a|\nabla\mathbf{n}| << 1$, where a is a typical molecular distance, the free energy f associated with the elastic distortion can be expanded in a power series of the successive spatial derivatives of **n**.

First to be considered are the linear terms in $\nabla\mathbf{n}$. Among them, only div**n** and **n.curln** are invariant with respect to rotations of the reference system axes. The first term is forbidden, as the energy must remain invariant under $\mathbf{n} \rightarrow -\mathbf{n}$ (no ferroelectricity). The second one is also to be discarded in ordinary nematics, because it changes sign under a coordinate transformation of the form $x \rightarrow -x$, $y \rightarrow -y$, and $z \rightarrow -z$ (existence of an inversion center among the symmetry elements of a nematic). It is however allowed in a cholesteric liquid crystal.

Quadratic terms in $\nabla\mathbf{n}$ are more difficult to find and classify. The most obvious second-order invariants are the squares of the first-order scalars:

$$(\mathrm{div}\,\mathbf{n})^2, \qquad (\mathbf{n}.\mathbf{curl}\,\mathbf{n})^2$$

One must equally consider scalar products of vectors generated from **n** and its first derivatives $\nabla\mathbf{n}$. There are three independent vectors of this type:

$$\mathbf{curl}\,\mathbf{n}, \qquad \mathbf{n}\,\mathrm{div}\,\mathbf{n}, \qquad \mathbf{n}\times\mathbf{curl}\,\mathbf{n} = -(\mathbf{n}\nabla)\mathbf{n} \qquad \text{(B.II.6a)}$$

yielding three additional scalar invariants:

$$(\mathbf{curl}\,\mathbf{n})^2, \qquad (\mathbf{n}\times\mathbf{curl}\,\mathbf{n})^2, \qquad (\mathbf{n}.\mathbf{curl}\,\mathbf{n})\mathrm{div}\,\mathbf{n} \qquad \text{(B.II.6b)}$$

The combination $(\mathbf{n}.\mathbf{curl}\,\mathbf{n})\mathrm{div}\,\mathbf{n}$ can be eliminated from the start, as it changes sign under $\mathbf{n} \rightarrow -\mathbf{n}$. Furthermore, the term $(\mathbf{curl}\,\mathbf{n})^2$ is not independent of the others, since $(\mathbf{curl}\,\mathbf{n})^2 = (\mathbf{n}.\mathbf{curl}\,\mathbf{n})^2 + (\mathbf{n}\times\mathbf{curl}\,\mathbf{n})^2$. One is left with the three fundamental invariants

$$(\mathrm{div}\mathbf{n})^2, \qquad (\mathbf{n}.\mathbf{curl}\,\mathbf{n})^2, \qquad (\mathbf{n}\times\mathbf{curl}\,\mathbf{n})^2 \qquad \text{(B.II.7)}$$

Two more invariants can be built starting from vectors B.II.6:

$$\mathrm{div}(\mathbf{n}\,\mathrm{div}\,\mathbf{n}), \qquad \mathrm{div}(\mathbf{n}.\mathbf{curl}\ \mathbf{n}) \qquad \text{(B.II.8)}$$

These quantities, depending on second derivatives of the form $\partial^2 n_i/\partial x_j \partial x_k$, cannot be expressed as functions of the invariants B.II.7. Varying like the square k^2 of the perturbation wave vector, they are in principle of the same order of magnitude as the previous terms and cannot be eliminated.

The most general expression for the distortion free energy f is, to order k^2, the following [2]:

$$f = \frac{1}{2}K_1\,(\mathrm{div}\mathbf{n})^2 + \frac{1}{2}K_2\,(\mathbf{n}\,.\,\mathbf{curl}\ \mathbf{n})^2 + \frac{1}{2}K_3\,(\mathbf{n}\times\mathbf{curl}\ \mathbf{n})^2$$

$$+\frac{1}{2}K_4\,\mathrm{div}(\mathbf{n}\,\mathrm{div}\mathbf{n} + \mathbf{n}\times\mathbf{curl}\ \mathbf{n}) + K_{13}\,\mathrm{div}(\mathbf{n}\,\mathrm{div}\mathbf{n}) \qquad \text{(B.II.9)}$$

The term in K_4 groups together the two invariants B.II.8, because the second derivatives $\partial^2 n_i/\partial x_j \partial x_k$ disappear from the sum. As to the term in K_{13}, neglected by Frank [3], it is of second order in the derivatives.

One can almost always neglect the K_4 and K_{13} terms, of the form div(...), as their integral over the nematic volume V can be written as a surface integral over the entire boundary S:

$$\int_V \mathrm{div}(\ldots)\,dV = \int_S (\ldots)\,d\boldsymbol{\upsilon} \qquad \text{(B.II.10)}$$

with unit vector $\boldsymbol{\upsilon}$ the outward normal to the surface. Surface terms only appear when considering the effect of the sample boundaries, and even then they can be neglected in the case of **geometric surface anchoring conditions**. The molecular anchoring energy is then infinite or, physically speaking, very large compared to K/d, with d the smallest dimension of the sample. Within this approximation, K_4 and K_{13} terms can be omitted and the elastic distortion free energy can be written in the simplified Frank form:

$$f = \frac{1}{2}K_1\,(\mathrm{div}\mathbf{n})^2 + \frac{1}{2}K_2\,(\mathbf{n}.\mathbf{curl}\ \mathbf{n})^2 + \frac{1}{2}K_3\,(\mathbf{n}\times\mathbf{curl}\ \mathbf{n})^2 \qquad \text{(B.II.11)}$$

which we will generally use from now on (and which will be understood to imply strong surface anchoring conditions).

Let us go back to describing the different kinds of distortion. The easiest way to handle the problem is to write the free energy B.II.9 in a local reference system (x_1, x_2, x_3) such that axis 3 coincides with the director **n** at the considered point P. Following Sir C. Frank, the following local deformations can be defined [3]:

$$s_1 = n_{1,1}\,, \qquad s_2 = n_{2,2} \qquad \textbf{(splay)} \qquad \text{(B.II.12a)}$$

$$t_1 = -\,n_{2,1}, \qquad t_2 = n_{1,2} \qquad \textbf{(twist)} \tag{B.II.12b}$$

$$b_1 = n_{1,3}, \qquad b_2 = n_{2,3} \qquad \textbf{(bend)} \tag{B.II.12c}$$

$$d_1 = n_{1,13}, \qquad d_2 = n_{2,23} \qquad \textbf{(second-order splay-bend)} \tag{B.II.12d}$$

with $n_{i,j} = (\partial n_i/\partial x_j)_P$ and $n_{i,jk} = (\partial^2 n_i/\partial x_j \partial x_k)_P$. In this local reference system, $n_{3,i} = 0$ as **n** is a unit vector of components (close to P)

$$\mathbf{n} = \left(n_1, n_2, 1 - \frac{1}{2}n_1^2 - \frac{1}{2}n_2^2\right) \tag{B.II.13}$$

Then $\delta n_3 = -n_1\delta n_1 - n_2\delta n_2 = 0$ ($\delta\mathbf{n} \perp \mathbf{n}$) close to P, where $n_1 = n_2 = 0$. From B.II.13 one has:

$$f = \frac{1}{2}K_1(s_1 + s_2)^2 + \frac{1}{2}K_2(t_1 + t_2)^2 + \frac{1}{2}K_3(b_1^2 + b_2^2) + K_4\,(s_1 s_2 + t_1 t_2)$$

$$+ K_{13}\,[(s_1 + s_2)^2 - (b_1^2 + b_2^2) + d_1 + d_2] \tag{B.II.14}$$

Let us begin the discussion with the volume terms.

The first one, proportional to K_1, corresponds to a **twist** deformation characterized by div**n** $\neq 0$:

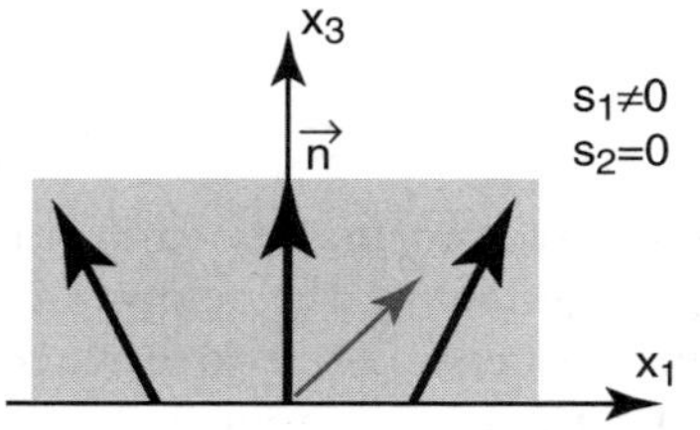

The K_2 term is associated with the **twist** mode already familiar from the Grupp experiment:

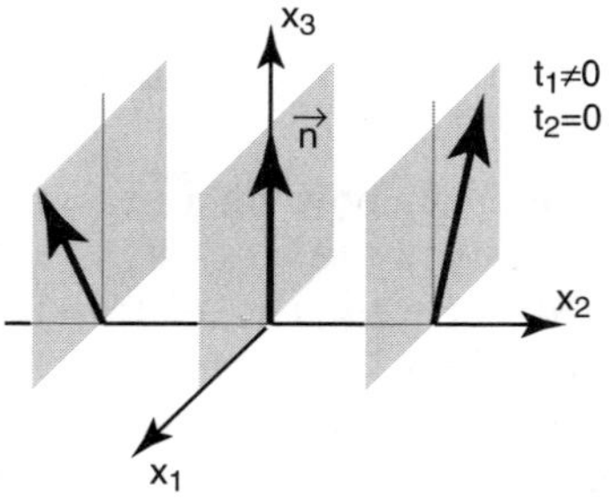

In the figure, twist only appears in the 2 direction and **n**// **curl n**.

The last fundamental term, the one in K_3, describes the **bend** in the director field lines.

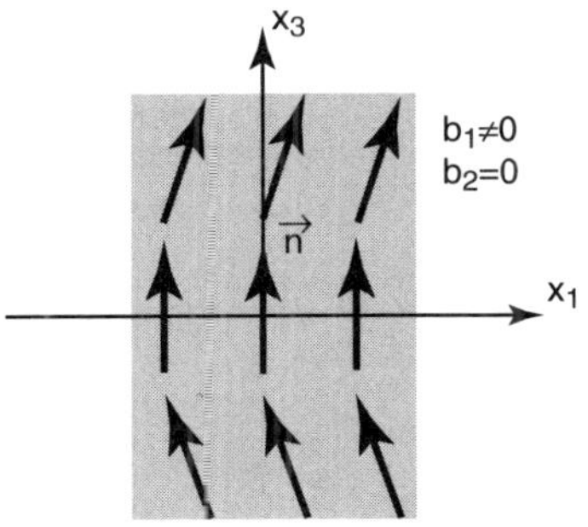

In the figure, the field lines are bent in the (x_1, x_3) plane and $\mathbf{n} \perp \mathbf{curl\,n}$.

Figure B.I.4 shows an example of experimental configuration involving both splay and bend deformations.

The K_4 surface term will not appear in the following experiments, but it plays a fundamental role in the understanding of the blue phases to be described in chapter B.VIII. This terms combines twist and splay deformations and is zero in the three examples already presented. However, there are configurations where this term is finite, while the K_1, K_2, and K_3 terms are zero (Fig. B.II.3).

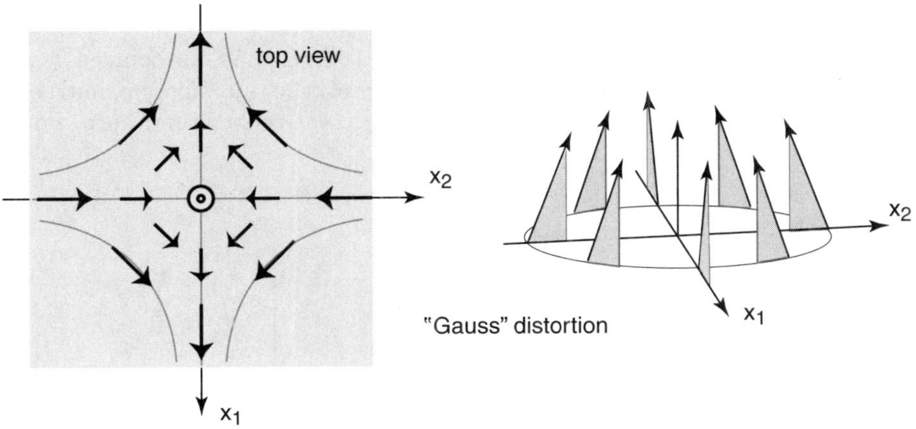

Fig. B.II.3 Example of a distortion involving the K_4 term.

In the center of the figure, $s_1 + s_2 = 0$ and $t_1 + t_2 = 0$ while $s_1 s_2 + t_1 t_2 = 0$. For a geometrical interpretation of this new type of distortion, consider the surface in each point perpendicular to the director. It can be defined here, because the total twist $t_1 + t_2$ is equal to zero. Let R_1 and R_2 be its **principal curvature radii** at point P (Fig. B.II.4). In this point, **total curvature** is also zero, since

$$\frac{1}{R_1} + \frac{1}{R_2} = \mathrm{div}\,\mathbf{n} = s_1 + s_2 = 0 \tag{B.II.15}$$

while the **Gaussian curvature**, given by

$$\frac{1}{R_1 R_2} = s_1 s_2 + t_1 t_2 \neq 0 \qquad \text{(B.II.16)}$$

is non-null.

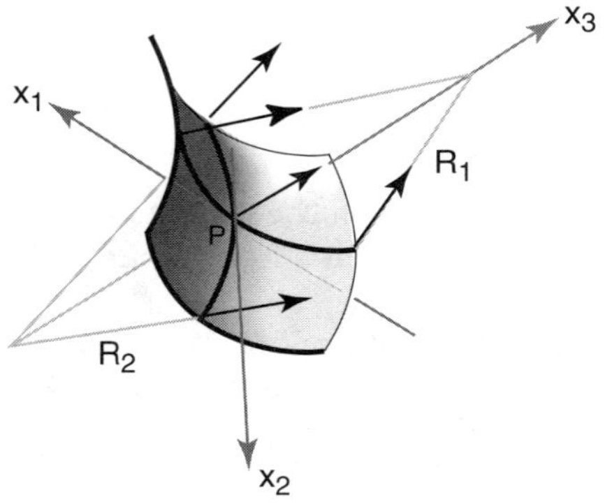

Fig. B.II.4 Geometrical interpretation of the K_4 term.

The K_4 term is therefore called the **Gauss term**; it is associated (in this example) with a "saddle-splay" deformation of the surface normal to the director. This term appears not only when the total twist is not zero, but also when it is maximum (Fig. B.II.5).

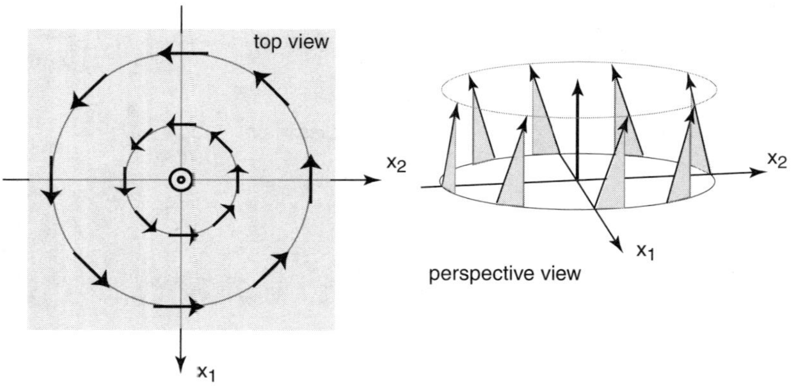

Fig. B.II.5 Example of deformation maximizing the total twist.

In this example, twist appears not just around one space direction, but rather around all axes originating in P. Then $t_1 = t_2 = t \neq 0$ and the total twist is the **double** of the one-directional twist already described. The free energy density at point P is then

$$f = (2K_2 + K_4)t^2 \tag{B.II.17}$$

As pointed out by M. Kléman, such a deformation is energetically more favorable than simple twist (i.e., $(1/2)K_2t^2$) if $K_4 < -(3/2)K_2 < 0$, an essential fact for the understanding of Blue Phases (see Chapter B.VIII).

Nothing will be said about the K_{13} term, as its relevance is under heavy controversy. This second-order term causes mathematical difficulties (the "Oldano-Barbero paradox") not yet solved. We will therefore neglect this term, especially since very recent results from a microscopic density functional theory seem to show that $K_{13} = 0$ [4].

Finally, let us briefly discuss director fields satisfying $\mathbf{curl\,n} = 0$ over the entire volume. One can then find a function $\phi(x,y,z)$ such that $\mathbf{n} = \mathbf{grad}\,\phi$ (i.e., a family of **parallel surfaces** everywhere normal to **n**). One such surface is defined by $\phi(x,y,z)$ = constant, and is characterized by

$$\mathrm{div}\,\mathbf{n} = \frac{1}{R_1} + \frac{1}{R_2} \tag{B.II.18}$$

and

$$\frac{1}{2}\mathrm{div}(\mathbf{n}\,\mathrm{div}\,\mathbf{n}) = \frac{1}{R_1R_2} \tag{B.II.19}$$

with R_1 and R_2 the two principal curvature radii of the surface in the point under discussion (Fig. B.II.4). The bulk free energy can then be rewritten in the simple form:

$$f = \frac{1}{2}K_1\left(\frac{1}{R_1} + \frac{1}{R_2}\right)^2 + (K_4 + 2K_{13})\frac{1}{R_1R_2} \tag{B.II.20}$$

This expression allows one to obtain the elastic deformation energy of a **thin plate** of thickness d (delimited by the two parallel surfaces ϕ = Cst and ϕ = cst + d) [5], by simply integrating over the thickness, the variation of the curvature radii with thickness being neglected. One has

$$f_{plate} = \frac{1}{2}K\left(\frac{1}{R_1} + \frac{1}{R_2}\right)^2 + \bar{K}\frac{1}{R_1R_2} \quad \text{(per unit surface area)} \tag{B.II.21}$$

with $K = K_1d$ and $\bar{K} = (K_4 + 2K_{13})d$. This formula was employed by Helfrich [6] and Canham [7] for describing elastic properties of biological membranes. It is equally used to model the curvature elasticity of lamellar phases.

Bear in mind that in cholesterics the $\mathbf{n.curl\,n}$ term cannot be neglected. The Frank-Oseen free energy is therefore written (omitting the surface terms):

$$F = k\,\mathbf{n.curl\,n} + \frac{1}{2}K_1(\mathrm{div}\,\mathbf{n})^2 + \frac{1}{2}K_2(\mathbf{n.curl\,n})^2 + \frac{1}{2}K_3(\mathbf{n}\times\mathbf{curl\,n})^2 \tag{B.II.22}$$

or it can be cast in an equivalent form (up to an additive constant) by using the spontaneous twist $q = k/K_2 = 2\pi/p$, with p the preferred cholesteric pitch:

$$f = \frac{1}{2} K_1 \, (\mathrm{div}\, \mathbf{n})^2 + \frac{1}{2} K_2 \, (\mathbf{n}.\mathbf{curl}\, \mathbf{n} + q)^2 + \frac{1}{2} K_3 \, (\mathbf{n} \times \mathbf{curl}\, \mathbf{n})^2 \tag{B.II.23}$$

Throughout the rest of this chapter, we will assume $q = 0$.

Let us conclude this section by giving the order of magnitude for the elastic constants K_i. In ordinary nematics, composed of short molecules, they have very similar values (up to anisotropy factors of the order of 2 or 3). One can estimate:

$$K_i \approx U/a$$

with U the typical energy of molecular interaction and "a" a molecular distance. With $U \approx 0.1$ eV and $a \approx 30$ Å one has $K_i \approx 5\times10^{-7}$ dyn, which is the same order of magnitude as in the Grupp experiment. In polymer nematics, the anisotropy of the elastic constants can be much larger (an order of magnitude or more) and no longer negligible.

II.3 Free energy minimization: molecular field and elastic torques

Equilibrium equations are obtained by writing the variation in the total free energy of the system

$$\delta F = \int \delta f \, dV \tag{B.II.24}$$

Using the following transformations

$$\delta f = \frac{\partial f}{\partial n_i} \delta n_i + \frac{\partial f}{\partial n_{i,j}} \delta n_{i,j} = \frac{\partial f}{\partial n_i} \delta n_i + \left(\frac{\partial f}{\partial n_{i,j}} \delta n_i \right)_{,j} - \frac{\partial}{\partial x_j} \left(\frac{\partial f}{\partial n_{i,j}} \right) \delta n_i \tag{B.II.25}$$

and integrating by parts eq. B.II.24, one gets:

$$\delta F = \int_V -h_i \, \delta n_i \, dV + \int_S \frac{\partial f}{\partial n_{i,j}} \delta n_i \, dS_j \tag{B.II.26}$$

with the notation

$$h_i = -\frac{\partial f}{\partial n_i} + \frac{\partial}{\partial x_j} \left(\frac{\partial f}{\partial n_{i,j}} \right) \tag{B.II.27}$$

For reasons soon to become apparent, vector **h** is called the **molecular field**.

Suppose now that the molecules turn in position, their centers of mass being fixed (translations will be discussed in the next chapter). As **n** is a unit vector, a **real** local rotation of the director is given by:

$$\delta\mathbf{n} = \delta\boldsymbol{\omega} \times \mathbf{n} \tag{B.II.28}$$

with $\delta\mathbf{n} \perp \mathbf{n}$. $\delta\boldsymbol{\omega}$ (normal to **n**) is the local director rotation vector. In index notation

$$\delta n_i = e_{ijk}\,\delta\omega_j\, n_k \tag{B.II.29}$$

yielding, after substitution in eq. B.II.26:

$$\delta F = \int_V - h_i e_{ijk}\delta\omega_j n_k dV + \int_S \frac{\partial f}{\partial n_{i,l}} e_{ijk}\delta\omega_j n_k dS_l \tag{B.II.30}$$

Recall that the e_{ijk} tensor is 1 or −1 when ijk is a circular permutation of 123 or 213, respectively, and 0 when (at least) two of the three indices are equal. Since $e_{ijk} = -e_{jik}$, one can also write

$$\delta F = \int_V \delta\boldsymbol{\omega}.(\mathbf{h} \times \mathbf{n})\, dV) + \int_S (\underline{C}\boldsymbol{\upsilon}).\delta\boldsymbol{\omega}\, dS, \tag{B.II.31}$$

with $\boldsymbol{\upsilon}$ the outward normal to the boundary S of the considered volume V and $\underline{C}$ the second-order tensor with components

$$C_{jl} = \frac{\partial f}{\partial n_{i,l}} e_{ijk} n_k \tag{B.II.32}$$

Physically speaking, δF is the work done on the system by the external torques when the director turns by $\delta\boldsymbol{\omega}$. This is therefore a "**torque**" elasticity, $\mathbf{h} \times \mathbf{n}$ and $\underline{C}\boldsymbol{\upsilon}$ having the dimensions of torque per unit volume and torque per unit surface, respectively. More precisely, $(\mathbf{h} \times \mathbf{n}).\delta\boldsymbol{\omega}$ is the work done on the system by the applied bulk torque. At equilibrium, one has:

$$\mathbf{h} \times \mathbf{n} = \boldsymbol{\Gamma}^V \text{(ext.)} \tag{B.II.33}$$

while $\underline{C}\boldsymbol{\upsilon}$ is the external surface torque applied to the system:

$$\underline{C}\boldsymbol{\upsilon} = \boldsymbol{\Gamma}^S \text{(ext.)} \tag{B.II.34}$$

Conversely, the quantity $\boldsymbol{\Gamma}^E = \mathbf{n} \times \mathbf{h}$ can be seen as the bulk elastic torque exerted by the medium on the director **n**. The bulk equilibrium equation is then:

$$\boldsymbol{\Gamma}^E + \boldsymbol{\Gamma}^V \text{(ext.)} = 0 \tag{B.II.35}$$

In practice, $\boldsymbol{\Gamma}^V$ (ext.) is a magnetic or electric torque. In the absence of an applied external torque, **h** is parallel to **n**. The director tends to align to the local field **h**, which explains the denomination of **molecular field**.

It is useful to know that **h** can generally be decomposed into three terms, each one corresponding to an elementary deformation, $\mathbf{h} = \mathbf{h}_s + \mathbf{h}_t + \mathbf{h}_b$ with:

$$\mathbf{h}_s = K_1\, \mathbf{grad}(\mathrm{div}\, \mathbf{n}) \qquad \text{(splay)}$$

$$\mathbf{h}_t = -K_2\{C\, \mathbf{curl}\, \mathbf{n} + \mathbf{curl}\, (C\mathbf{n})\} \qquad \text{(twist)} \qquad \text{(B.II.36)}$$

$$\mathbf{h}_b = K_3\{\mathbf{B} \times \mathbf{curl}\, \mathbf{n} + \mathbf{curl}\, (\mathbf{n} \times \mathbf{B})\} \qquad \text{(bend)}$$

with $C = \mathbf{n}.\mathbf{curl}\, \mathbf{n}$ and $\mathbf{B} = \mathbf{n} \times \mathbf{curl}\, \mathbf{n}$.

Also note that, taking $K_1 = K_2 = K_3 = -K_4 = K$ ("isotropic" elasticity) and $K_{13} = 0$ in the general formula B.II.19, f and **h** have very simple expressions:

$$f = \frac{1}{2} K\, \nabla_i n_j \nabla_i n_j \qquad \text{(B.II.37)}$$

and

$$\mathbf{h} = K\, \Delta \mathbf{n} \qquad \text{(B.II.38)}$$

These formulae will be very useful in the study of defects (Chapter B.IV).

II.4 Interpretation of the Grupp experiment

The formalism presented here is fundamental from the theoretical point of view, but not very useful in applications. In practice, the solution is often more easily found starting directly from the free energy. The Grupp experiment provides a good illustration.

The director is parallel to the x axis in z = 0 (Fig. B.II.6) and it turns in the sample but one can reasonably assume that it remains in the horizontal plane. Let d be the sample thickness and $\phi(z)$ the angle between the director and the x axis at height z. Boundary conditions impose

$$\phi(0) = 0 \quad \text{and} \quad \phi(d) = \Phi \quad (\mathrm{mod}\ \pi) \qquad \text{(B.II.39)}$$

The director distortion can be found by minimizing the free energy. The only non-null term is the one in K_2, corresponding to pure twist deformation. One immediately obtains:

$$\mathbf{n}.\mathbf{curl}\, \mathbf{n} = -\frac{\partial \phi}{\partial z} \qquad \text{(B.II.40)}$$

whence

$$F = \frac{1}{2} K_2 \int \left(\frac{\partial \phi}{\partial z} \right)^2 dz \qquad \text{(B.II.41)}$$

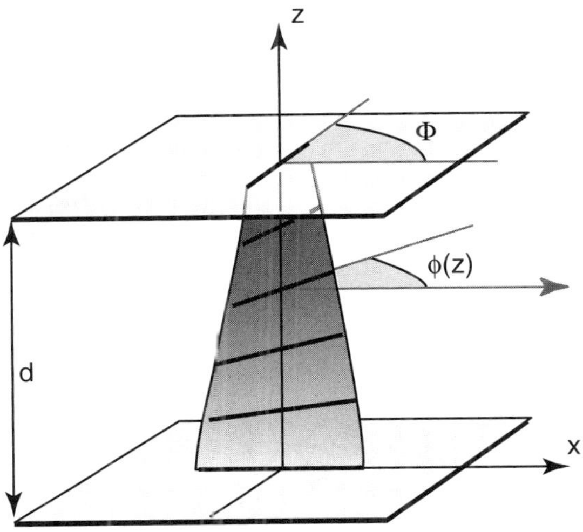

Fig. B.II.6 Director configuration in the Grupp experiment. The molecules are assumed parallel to the plates throughout the entire sample.

The bulk minimization equation is easily written:

$$K_2 \frac{d^2\phi}{dz^2} = 0 \qquad \text{(B.II.42)}$$

This is nothing more than the equilibrium equation B.II.35 with $\mathbf{\Gamma}^V(\text{ext.}) = 0$, as no external torque is applied. The bulk torque $\mathbf{\Gamma}^E$ exerted by the medium on its director is therefore in the z direction and its value is:

$$\Gamma_z^E = K_2 \frac{d^2\phi}{dz^2} \qquad \text{(B.II.43)}$$

Solving equation B.II.42 leads to the following solutions

$$\phi = \frac{\Phi + N\pi}{d} z \qquad N = 0, \pm 1, \pm 2, \ldots \qquad \text{(B.II.44)}$$

These possible director configurations (Fig. B.II.7) have an energy per unit surface in the (x, y) plane:

$$F = \frac{1}{2} K_2 \frac{(\Phi + N\pi)^2}{d} \qquad \text{(B.II.45)}$$

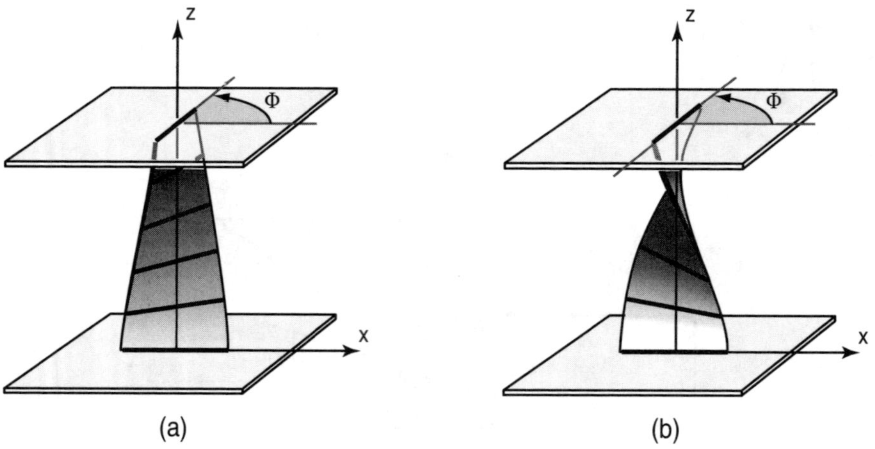

Fig. B.II.7 Two different director configurations respecting the boundary conditions on the plates. For $\Phi < \pi/2$, the elastic energy is lower in configuration (a) than in (b).

For $0 < \Phi < \pi/2$, the possible configurations can be classified in the order of increasing energy:

$$N = 0, -1, +1, -2, +2, \dots \tag{B.II.46}$$

The lowest energy configuration is therefore given by N = 0. The variational method allows for finding all possible director configurations, including those for which F has a local minimum. Obtaining the global minimum therefore demands explicit calculation of F. **The configuration corresponding to the global minimum of F is thermodynamically stable** while the other configurations are topologically unstable, a 2π twist being removable by a continuous transformation. However, they can be energetically **metastable**, depending on the elastic anisotropy of the material. The concepts of topological and energetic stability will be thoroughly discussed in the case of topological defects (section B.IV.7).

The surface torque exerted by the nematic on the upper plate is given by the energy variation δF when the plate turns by an angle $\delta\Phi$ at constant thickness d:

$$\delta F = K_2 \frac{\Phi}{d} \delta\Phi \tag{B.II.47}$$

It is obvious that for N = 0 the z component of the torque exerted by the nematic on the upper plate is

$$\Gamma_{nem} = -K_2 \frac{\Phi}{d} \tag{B.II.48}$$

in agreement with the experimental result B.II.5.

II.5 Magnetic field action

II.5.a) Physical interpretation of the bulk torque

As shown in the previous section, uniform twist distortion of the form $d\phi/dz = \text{Cst}$ transmits torques without modification, from one nematic surface to the other. As a matter of fact, the torque one needs to apply to the lower disk to keep it from turning is **completely** transmitted by the twisted nematic to the upper disk, where it produces the torsion of the quartz fiber. Consider now in the distorted nematic a layer of thickness dz, delimited by two planes parallel to (x, y) at heights z and z + dz (Fig. B.II.8).

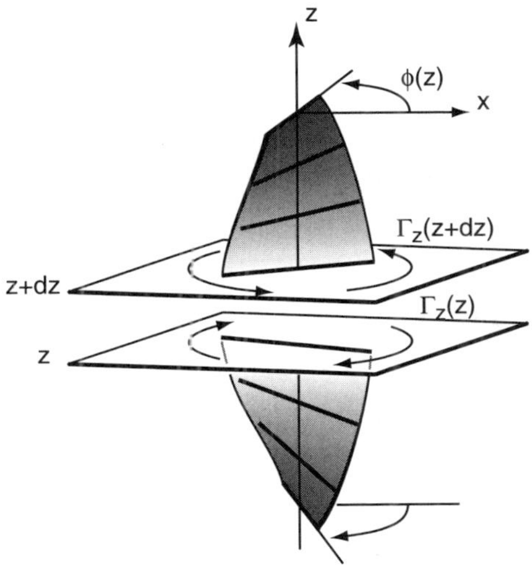

Fig. B.II.8 A layer of thickness dz in the distorted nematic is submitted to the torques Γ_z exerted upon its two boundaries at heights z and z + dz.

The twisted nematic on both sides of this layer exerts upon it the following torques (per unit surface):

– on the lower surface: $$\Gamma_z(z) = -K_2 \frac{d\phi}{dz}(z) \qquad \text{(B.II.49)}$$

– on the upper surface: $$\Gamma_z(z+dz) = K_2 \frac{d\phi}{dz}(z+dz) \qquad \text{(B.II.50)}$$

For uniform distortion, these two torques cancel each other, as in this case $d\phi/dz$ does not depend on z.

Let us now consider an equilibrium configuration where the twist deformation $\phi(z)$ is no longer given by a linear law. Still assuming that **n** is parallel to the (x, y) plane, the two torques sum up to (per unit surface):

$$\Gamma_z(z) + \Gamma_z(z+dz) = K_2 \frac{d^2\phi}{dz^2} dz \tag{B.II.51}$$

and, per unit volume:

$$\Gamma_z^E = K_2 \frac{d^2\phi}{dz^2} \tag{B.II.52}$$

We have thus recovered the previous expression B.II.43 for the bulk torque exerted by the medium on its director. To achieve mechanical equilibrium, this torque due to inhomogeneous distortion of the director field must in each point be balanced by another torque Γ_z^V (ext.) induced, for instance, by applying an electric or magnetic field:

$$\Gamma_z^E + \Gamma_z^V(\text{ext.}) = 0 \tag{B.II.53}$$

Before illustrating our discussion with some examples of field-induced distortions, we shall briefly review the action of a magnetic field upon a nematic.

II.5.b) Molecular magnetic field

In nematics, the magnetization per unit volume induced by a field **B** can be written as:

$$\mathbf{M} = \frac{1}{\mu_o}[\chi_\perp \mathbf{B} + (\chi_{//} - \chi_\perp)(\mathbf{B}.\mathbf{n})\,\mathbf{n}] \tag{B.II.54}$$

with $\chi_{//}$ and $\chi_\perp$ the magnetic susceptibilities of the medium, parallel and normal to the director, respectively. The magnetic torque (acting on the magnetization **M**) reads:

$$\mathbf{\Gamma}^M = \mathbf{M} \times \mathbf{B} = \frac{\chi_a}{\mu_o}(\mathbf{n}.\mathbf{B})\,\mathbf{n} \times \mathbf{B} \tag{B.II.55}$$

The medium reaches equilibrium when magnetic and elastic torques compensate:

$$\mathbf{\Gamma}^M + \mathbf{\Gamma}^E = \mathbf{\Gamma}^M + \mathbf{n} \times \mathbf{h} = 0 \tag{B.II.56}$$

or, alternatively,

$$\mathbf{n} \times (\mathbf{h} + \mathbf{h}^M) = 0 \tag{B.II.57}$$

with $\mathbf{h}^M = \chi_a/\mu_o)(\mathbf{n}.\mathbf{B})\mathbf{B}$. The $\mathbf{h}^M$ vector can thus be considered as the magnetic contribution to the molecular field.

Balancing the torques is often more easily obtained by direct minimization of the total free energy of the system $F = \int (f + f^M)dV$, with f^M the magnetic contribution to the free energy:

$$f^M = -\int_0^B \mathbf{M}\, d\mathbf{B} = -\frac{1}{2\mu_o}\chi_a(\mathbf{B}.\mathbf{n})^2 + \text{Const} \tag{B.II.58}$$

Note that if the anisotropy χ_a is positive, f^M is a minimum when $\mathbf{B}//\mathbf{n}$: the molecule tends to align parallel to the magnetic field. In the opposite case, the molecule will be perpendicular to the magnetic field: $\mathbf{B}\perp\mathbf{n}$.

Magnetic susceptibility of molecules crucially depends on their chemical nature. In liquid crystals, it is often negative, being dominated by the **diamagnetic** contribution of the benzene rings. On the other hand, the sign of the magnetic anisotropy depends on the way these benzene rings are oriented with respect to **n**. In molecules such as nCB, nOCB, nO.m, etc., the plane of the benzene rings is almost parallel to the molecular major axis, and hence to **n**. In these materials, χ_a is positive $(-\chi_\perp > -\chi_{//})$ and of the order 10^{-6}.

II.5.c) Magnetic coherence length

We now have all the elements to analyze the action of a magnetic field on a semi-infinite nematic slab limited by a plane wall on which molecular anchoring is parallel to the x direction (Fig. B.II.9).

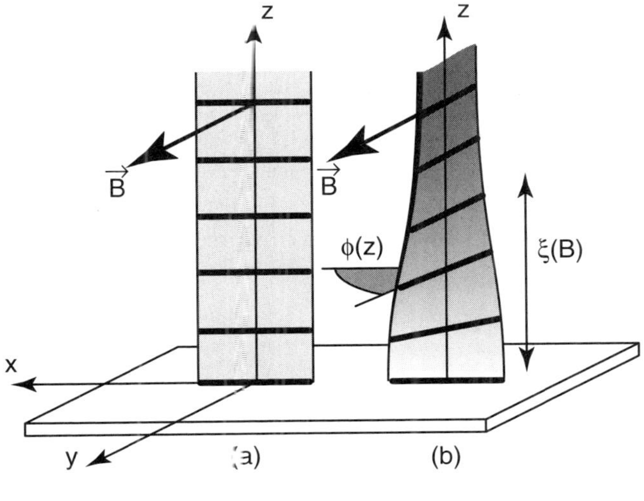

Fig. B.II.9 Definition of the magnetic coherence length ξ(B).

If the magnetic field is applied in the y direction, it will cause the molecules to turn ($\chi_a > 0$). Far from the wall, they are free to align along **B**. Close to the wall, however, the anchoring will impose the existence of a twist boundary layer of thickness ξ.

Assume that the molecules remain orthogonal to z and let $\phi(z)$ be the angle between **n** and the x axis. The torque exerted by the magnetic field is given by eq. B.II.55:

$$\Gamma_z^M = \frac{\chi_a}{\mu_o} B^2 \sin\phi \cos\phi \qquad \text{(B.II.59)}$$

and the equilibrium condition B.II.53 becomes:

$$K_2 \frac{d^2\phi}{dz^2} + \frac{\chi_a}{\mu_o} B^2 \sin\phi \cos\phi = 0 \qquad \text{(B.II.60)}$$

with boundary conditions:

$$\phi = 0 \quad \text{in} \quad z = 0 \qquad \text{(B.II.61)}$$

$$\phi = \pi / 2 \quad \text{in} \quad z = +\infty \qquad \text{(B.II.62)}$$

We can define the length $\xi(B)$ as

$$\xi(B) = \sqrt{\frac{\mu_o K_2}{\chi_a}} \frac{1}{B} \qquad \text{(B.II.63)}$$

which is called the **magnetic coherence length**. One easily finds the first integral:

$$\xi^2 \left(\frac{d\phi}{dz}\right)^2 = \cos^2\phi + \text{Const} \qquad \text{(B.II.64)}$$

The integration constant is zero from the $z = +\infty$ boundary condition, so:

$$\xi \frac{d\phi}{dz} = \pm \cos\phi \qquad \text{(B.II.65)}$$

The sign is not imposed, as the molecules can turn in one direction or the other. Thus, choosing the plus sign, integration gives:

$$\phi = \frac{\pi}{2} - 2 \arctan\left[\exp - \left(\frac{z}{\xi}\right)\right] \qquad \text{(B.II.66)}$$

The coherence length ξ can be seen as the thickness of the twist boundary layer. As an example, with a 1 tesla field, ξ is a few tens of micrometers in cyanobiphenyls.

II.5.d) Frederiks instability in a magnetic field

Consider now a nematic sample of thickness d between two parallel plates. As in the previous section, the anchoring is planar and the magnetic field along the y axis. The ϕ equation is the same, but, this time, the boundary conditions are given by $\phi = 0$ in $z = 0$ and d (Fig. B.II.10). For $d >> \xi$, we recover the case already discussed: the director is parallel to the field in the middle of the sample and two twisted boundary layers are formed at the plates. The situation is more complicated when the thickness d is of the order of, or less than, the penetration length ξ [8]. The "orienting" effects of the two surfaces add up and can even dominate the destabilizing action of the field, as we shall see in the following. Start by writing the solution as a Fourier series:

$$\phi = \sum_m \phi_m \sin\left[(2m+1)\frac{\pi z}{d}\right] \qquad \text{(B.II.67)}$$

Suppose ϕ remains small and develop the $\sin\phi \cos\phi$ term to third order in ϕ in the equilibrium equation B.II.60, which gives:

$$K_2 \frac{d^2\phi}{dz^2} + \frac{\chi_a}{\mu_o} B^2 \left(\phi - \frac{2}{3}\phi^3\right) = 0 \qquad \text{(B.II.68)}$$

An equation for ϕ_o can be obtained by replacing this expression for ϕ in the previous equation:

$$- K_2 \phi_o \frac{\pi^2}{d^2} + \frac{\chi_a}{\mu_o} B^2 \left(\phi_o - \frac{\phi_o^3}{2}\right) = 0 \qquad \text{(B.II.69)}$$

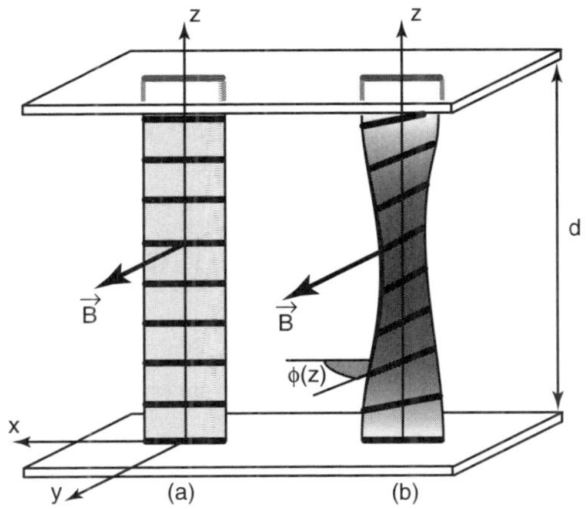

Fig. B.II.10 The geometry of the Frederiks transition. a) Director configuration for $B < B_c$; b) director configuration for $B > B_c$.

The angle ϕ_1 being proportional to ϕ_0^3, it has a negligible contribution near the threshold. The equation has two solutions, $\phi_0 = 0$ and a non-trivial one given by:

$$1 - \frac{\phi_0^2}{2} = \frac{B_c^2}{B^2} \tag{B.II.70}$$

where

$$B_c = \frac{\pi}{d}\sqrt{\frac{\mu_0 K_2}{\chi_a}} \tag{B.II.71}$$

This equation has no (real) solution unless $B > B_c$. Otherwise, the only solution is $\phi_0 = 0$. B_c is therefore the threshold (or critical) field above which the nematic starts twisting (Fig. B.II.10b). In particular, close to the instability threshold

$$\phi_0 \approx \pm 2\sqrt{\frac{B - B_c}{B_c}} \tag{B.II.72}$$

This is therefore a **normal bifurcation** (also called **supercritical**), as the twist angle in the middle of the sample increases continuously from 0 above the critical field as $(B-B_c)^{1/2}$ (Fig. B.II.11), behavior typical for the order parameter of a second-order phase transition in Landau theory. Note that the original mirror symmetry of the configuration is broken at the transition, when the molecules start turning.

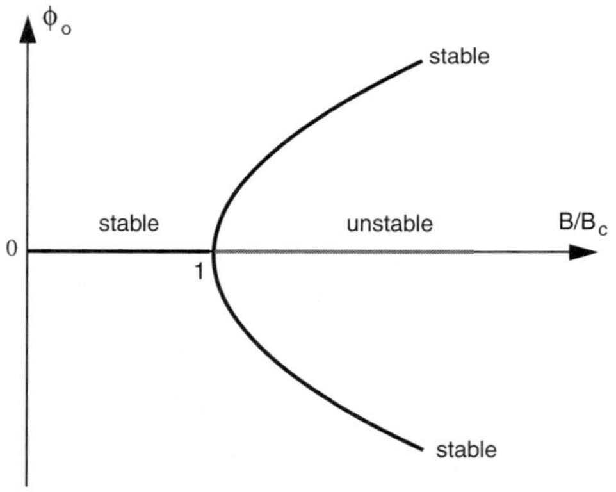

Fig. B.II.11 Tilt angle ϕ_0 as a function of B/B_c. When $B < B_c$, the undistorted configuration is stable. Above the threshold, the system breaks its symmetry by choosing one of the two (stable) distorted configurations.

The experimental determination of the threshold field can be achieved using the setup in figure B.II.12. The instability is detected by monitoring the rotation of the conoscopic figure produced by a He-Ne laser, with the sample between crossed polarizers making a 45° angle with the initial direction of

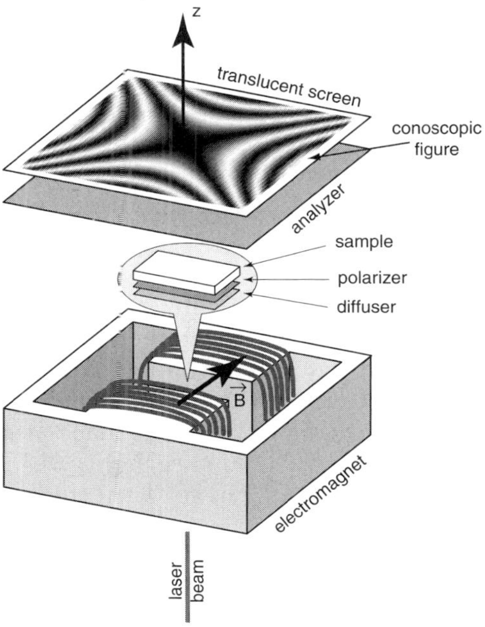

Fig. B.II.12 Setup used to investigate the Frederiks transition. The nematic is between two glass plates treated to ensure planar anchoring (with a polyimide layer rubbed in one direction). The d = 200 μm thickness is obtained using mylar spacers. A small piece of tracing paper is used as light diffuser. The "sample + diffuser + polarizer" sandwich is placed in the electromagnet gap. The conoscopic figure (interference fringes at infinity) is projected on a semi-opaque screen (tracing paper). Here, the fringes are hyperbolae of asymptotes parallel to the polarizer and analyzer. The Frederiks transition is signaled by the rotation of the conoscopic figure.

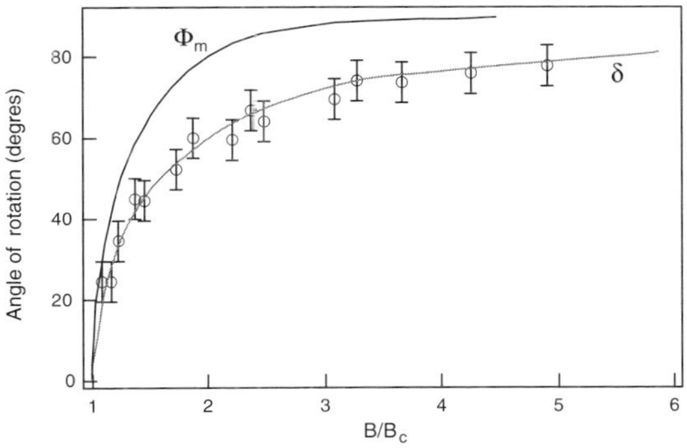

Fig. B.II.13 Rotation angle δ of the conoscopic figure asymptotes as a function of the reduced field B/B_c. The angle is zero below the critical field and increases continuously from zero above the threshold (supercritical bifurcation) (from ref. [10]). The upper curve is the maximum rotation angle Φ_m in the middle of the sample. It is related to δ by:

$$\tan 2\delta = \frac{2\sin\Phi_m}{2E(\pi/2,\sin\Phi_m) - F(\pi/2,\sin\Phi_m)}$$

where E and F are the complete elliptic integrals of the second and first kind, respectively.

molecular alignment [9]. Figure B.II.13 gives the rotation angle of the asymptotes to the interference fringes as a function of the field for MBBA [10], confirming supercriticality. For a 100 μm thick sample, the critical field is about $B_c \approx 530$ gauss, already demanding a fairly strong electromagnet.

II.5.e) Application to the determination of the elastic constants

In the previous experiment, the director field only suffered twist deformation. The critical field allows then for direct determination of the twist constant K_2. By modifying the geometry of the setup and the anchoring on the glass plates, the other elastic constants can be equally determined. Thus, with planar anchoring and a field perpendicular to the glass plates (Fig. B.II.14a), one has "splay" deformation above the threshold. As before, the equilibrium configuration is given by the balance between the magnetic and elastic torques: $\Gamma^E_y + \Gamma^M_y = 0$, yielding:

$$K_1 \frac{d^2\theta}{dz^2} + \frac{\chi_a}{\mu_o} B^2 \sin\theta\cos\theta = 0 \qquad \text{(B.II.73)}$$

Deformation appears above a threshold,

$$B_c = \frac{\pi}{d}\sqrt{\frac{\mu_o K_1}{\chi_a}} \qquad \text{(B.II.74)}$$

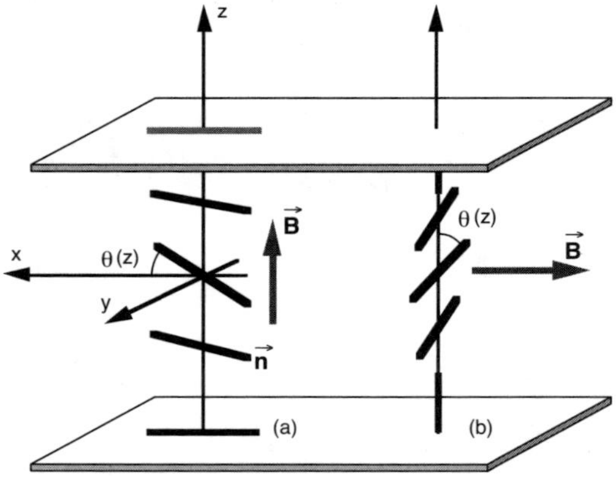

Fig. B.II.14 Another two geometries used for the study of the Frederiks transition.

With this setup, the transition is detected optically, with the sample between crossed polarizers that make a 45° angle with the average molecular alignment. As the molecules tilt, the phase shift $\Delta\varphi = 2\pi d\Delta n/\lambda$ between the ordinary and extraordinary rays crossing the sample varies, changing the intensity of the transmitted laser beam proportionally with the interference factor ($1 - \cos\Delta\varphi$). Experiments once again prove that the transition is supercritical, in agreement with the theoretical prediction.

In the geometry of figure B.II.14b, the field **B** is perpendicular to the molecules, themselves perpendicular to the plates. The critical field is given by a similar formula:

$$B_c = \frac{\pi}{d}\sqrt{\frac{\mu_o K_3}{\chi_a}} \qquad \text{(B.II.75)}$$

the deformation being now of the "bend" type.

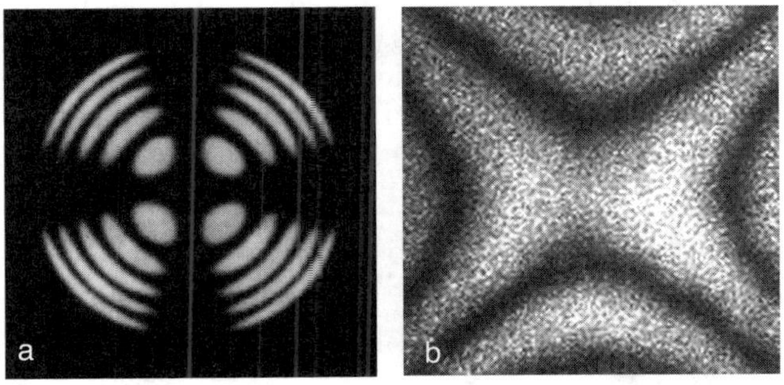

Fig. B.II.15 Conoscopic figures: a) homeotropic sample, b) planar sample.

The instability can be detected using polarized light conoscopy. The conoscopic figure is now composed of concentric circular interference fringes and an extinction cross (Fig. B.II.15). When the distortion appears, the fringes start drifting on the screen.

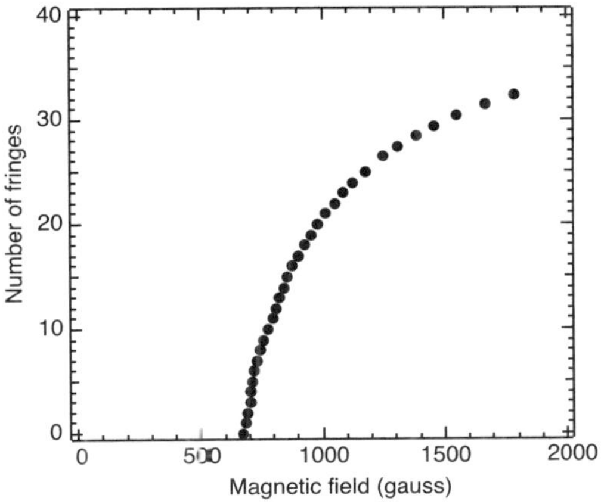

Fig. B.II.16 Number of interference fringes drifting through the center of the screen as a function of the magnetic field.

Figure B.II.16 shows the number of interference fringes having passed at the center of the screen as a function of the field B, allowing for an accurate determination of the critical field (about 670 gauss in this case: a 200 μm thick sample of the E8 mixture marketed by BDH).

One can equally use other techniques to determine the instability threshold. One of them, well adapted to the two previous geometries (Fig. B.II.14), consists of measuring the sample's electrical capacity for a given B field.

Throughout the preceding discussion we have assumed (with no justification other than facility in calculation) that the distortion of the director field above the instability threshold was translationally invariant in the plane of the sample. This hypothesis is experimentally confirmed in ordinary nematics consisting of short molecules, where all elastic constants are of the same order of magnitude. The situation can be quite different in polymers, where a stable stripe texture sometimes appears close to the instability threshold.

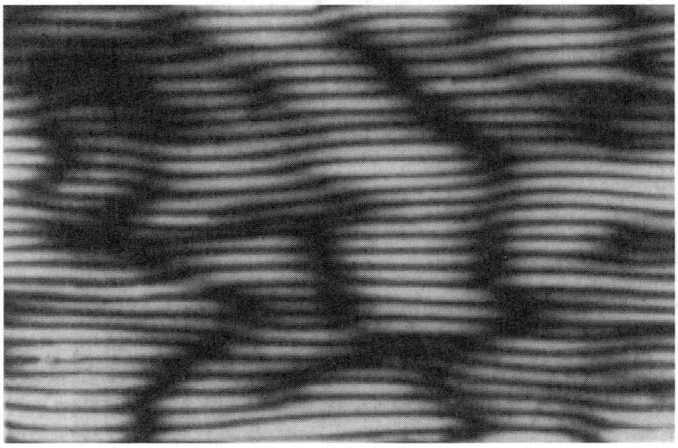

Fig. B.II.17 Stable periodic structure observed in PBG close to the Frederiks instability threshold. The director is initially parallel to the black and white stripes and the magnetic field is normal to the sample plane (photo by Lonberg and Meyer [11]).

This phenomenon was recently observed by Lonberg and Meyer [11] in PBG, or poly(γ-benzylglutamate) (Fig. B.II.17) in planar geometry (Fig. B.II.14a). This periodic structure is a result of the important elastic anisotropy, the K_1 constant being much larger than the two others. The anisotropy originates in the very high aspect ratio length/width, which is of the order of 70 for PBG, as opposed to 5 in ordinary nematics. One can show that, for K_1/K_2 larger than 3.3, a periodic deformation of the director field perpendicular to the initial molecular direction, and combining twist and splay, becomes energetically favorable. In this case, the instability threshold B_c is lower than the one given by eq. B.II.74 and must be computed numerically. For more details, see the article by Lonberg and Meyer [11].

II.6 Action of an electric field: displays

II.6.a) Employing an electric field

In the previous chapter we saw that an applied electric field **E** exerts on the director a torque:

$$\Gamma^{Elec} = \mathbf{P} \times \mathbf{E} = \varepsilon_a \varepsilon_0 \, (\mathbf{n}.\mathbf{E})\mathbf{n} \times \mathbf{E} \qquad \text{(B.II.76)}$$

where ε_a is the dielectric anisotropy of the medium. The electric contribution f^E to the free energy reads:

$$f^E = -\frac{1}{2}\varepsilon_0 \varepsilon_a \, (\mathbf{E}.\mathbf{n})^2 \qquad \text{(B.II.77)}$$

The critical electric field in a Frederiks experiment under an electric field can then be easily obtained by formally replacing χ_a/μ_0 by $\varepsilon_a\varepsilon_0$ in the previous formulae (for $\varepsilon_a > 0$). To compare the effectiveness of the electric and magnetic fields, let us look for the value of E corresponding to a magnetic field B of 1 gauss = 10^{-4} tesla. As in cyanobiphenyls $\chi_a \approx 10^{-6}$ and $\varepsilon_a \approx 10$, the equality of electric and magnetic free energy density yields the equivalence

$$B \approx 1 \text{ gauss} \leftrightarrow E \approx 10 \text{ V/m}$$

The typical electric field needed to induce the Frederiks instability in a sample of thickness 100 μm is then readily determined: from $B_c \approx 1000$ gauss, one can estimate $E_c \approx 20{,}000$ V/m.

Are such fields practically attainable? To answer the question, let us first discuss the problem of creating a uniform field over the entire sample volume. In the geometry of figure B.II.10, one only needs to place the sample between two plane parallel electrodes a distance L_y apart (Fig. B.II.18a). If the dimensions L_x and L_z of the electrodes are much larger than L_y, the result will be a uniform field, of intensity (in vacuum) U/L_y, where U is the applied tension. The cell containing the liquid crystal is a parallelepiped of size $l_x \times l_y \times l_z$, where the sample, of thickness d, is sandwiched between two glass plates. If the sample thickness is too small, the critical field becomes too high. If, on the contrary, it is too large, the alignment can be a problem and the response time of the system becomes too long. In laboratory experiments, $d \approx 30$ μm is a good compromise (the response time is then of the order of a few seconds). The sample thickness has a negligible contribution to the total cell thickness, since the thickness of the glass plates is millimetric. One therefore has, say, $l_z \approx 2$ mm. The dielectric constant of glass is different from that of air, so the electric field inside the cell is perturbed with respect to the uniform field inside the empty capacitor, but the perturbation is negligible if

$$l_x \approx l_y >> l_z$$

This requires a large enough cell,

$$l_x \approx l_y > 20 \text{ mm}$$

so the distance between the electrodes should be

$$L_y > 20 \text{ mm}$$

For this gap, the critical tension for the Frederiks instability is $U_c \approx 400$ V.

This is a high voltage, to be manipulated with caution.

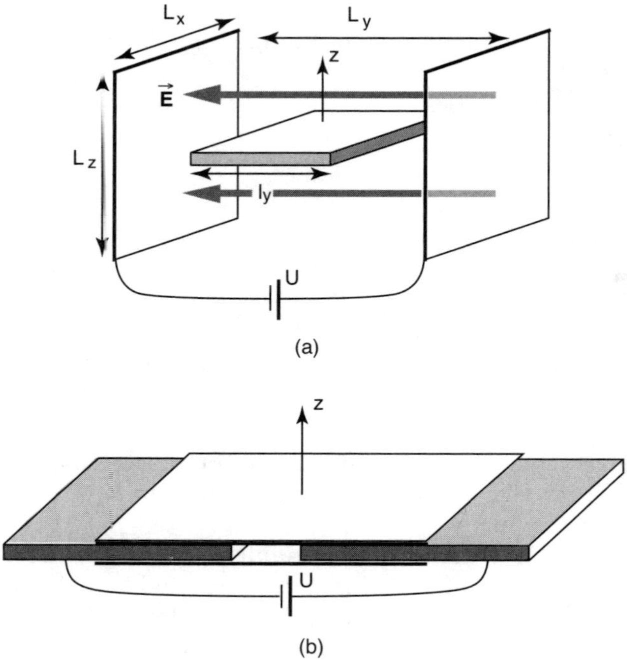

Fig. B.II.18 Two possible configurations for the study of the Frederiks transition under electric field in the geometry of figure B.II.10.

Two solutions can be adopted to reduce the necessary voltage:

– bring the electrodes closer, as in figure B.II.18b; however, this is not satisfactory, as the field becomes inhomogeneous;

– change the geometry by applying the field across the nematic layer (Figs. B.II.14a when $\varepsilon_a > 0$ and B.II.14b when $\varepsilon_a < 0$). The glass plates confining the liquid crystal can also serve as electrodes if covered by a semi-transparent film of ITO (mixture of indium and tin oxides). The gap between the electrodes is then equal to the nematic thickness d and the electric tension needed to induce the Frederiks instability is only a few volts.

Thus, the electric field appears to be more effective than the magnetic field in the two geometries. However, two comments (or even warnings) concerning its use are in order.

The first one bears upon the frequency of the electric field (DC or AC). Experiment shows that for a DC or low frequency ($f < 200$ Hz) field, the Frederiks instability is hidden by another instability, of electrohydrodynamic origin, consisting of the appearance of convective rolls provoked by the field-induced motion of the ions always present in the samples. This instability will be discussed in the next chapter. One can partially avoid this problem by using an AC field of high enough frequency (typically above 1 kHz), as the free energy is proportional to E^2 (because of the **n, −n** equivalence) and the sign of the field has no relevance. If the field frequency is much higher than the viscous relaxation frequency of the system (to be determined in chapter B.III, dedicated to nematodynamics) the above relations still hold, with E replaced by the effective (rms) value E_{eff}.

Another complication with respect to the magnetic case comes from the strong dielectric anisotropy: above the instability threshold, the distortion of the director field can significantly modify the electric field inside the sample. The non-linear analysis of the Frederiks instability is therefore much simpler in the case of an applied magnetic field; the latter is hardly modified by the presence of the material, contrary to the electric field. In particular, it can be shown that, under an electric field, a distinction must be made between conducting and isolating media [12]. Thus, the Frederiks bifurcation is first order (or **subcritical**) in a conducting sample if the dimensionless parameter $1 - (K_3-K_1) / K_3 - (\sigma_{//}-\sigma_{\perp}) / \sigma_{\perp}$ is negative, with $\sigma_{//}$ and $\sigma_{\perp}$ the electric conductivity along the director and perpendicular to it, respectively. The threshold above which the instability can develop is then smaller than the one given by the preceding formulae, obtained from a linear stability calculation. The bifurcation is supercritical in the isolating case and the preceding calculations still apply (see section II.8 to find out how the calculations are done in the subcritical case).

To illustrate our discussion, we shall show how a Twisted Nematic Cell (TNC) display works. This is one of the most important applications of nematic liquid crystals, as the technology is employed in most of the commercial displays.

II.6b) Setting up a 16-pixel display

The diagram of a 16-pixel display is shown in figure B.II.19. The setup, made in the laboratory by a graduate student of the École Centrale in Paris, consists of a thin layer of the E9 liquid crystal (trade name for a mixture of nCBs, nOCBs, and cyanoterphenyls created by BDH) sandwiched between two glass plates. Specially conceived for use in displays, this nematic has a very strong dielectric anisotropy: $\varepsilon_a = 15$. Transparent ITO strip-like electrodes are vacuum evaporated on the inner sides of the glass cell, parallel to the x direction on the lower plate and to the y direction on the upper one. The plates are also covered by a thin polymer film (polyvinyl alcohol, thickness 1μm) rubbed in the direction of the electrodes. This mechanical treatment (called "à la Chatelain" after the name of the inventor) ensures planar molecular alignment in the x direction on the lower plate and in the y direction on the upper one. The lower electrodes can be switched (Fig. B.II.20) between ground (null potential) and an AC source yielding

$$V_{column} = V_{oc} \cos \omega t \tag{B.II.78}$$

at a 1 kHz frequency (to avoid electrohydrodynamic instabilities) and an amplitude V_{oc} adjustable between 0 and 20 V. Similarly, the upper electrodes can be grounded or connected to a source yielding

$$V_{row} = - V_{oc} \cos \omega t \tag{B.II.79}$$

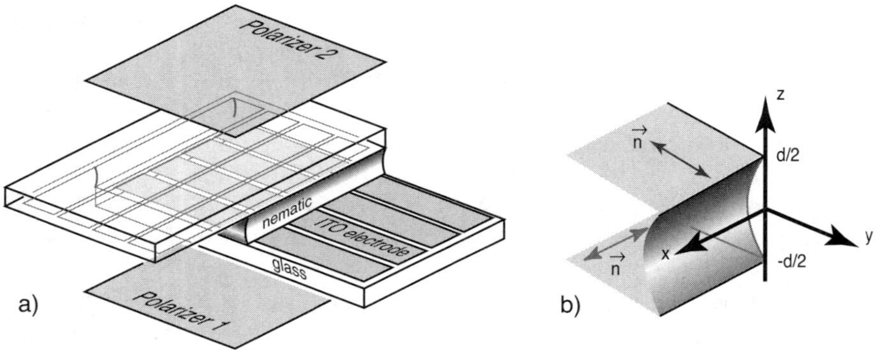

Fig.B.II.19 a) Diagram of a twisted nematic display; b) orientation of the director on the two electrodes.

By modifying the position of the switches, three different configurations can be obtained for a pixel, as the applied voltage between the upper and lower electrodes can be:

- $V_{row} - V_{column}$ for the pixels denoted by "2,"

- V_{row} or $- V_{column}$ for those indicated by "1,"

- 0 for the pixels indicated by "0."

Let us now consider each of these configurations.

Start with the **"0"** pixels. As the applied field is zero, the director field is obviously twisted, as in the Grupp experiment. More precisely, the director components are:

$$n_x = \cos\phi(z) \sin\theta(z), \quad n_y = \sin\phi(z) \sin\theta(z), \quad n_z = \cos\theta(z) \tag{B.II.80}$$

with boundary conditions:

$$\theta\left(\pm\frac{d}{2}\right)=\frac{\pi}{2},\qquad \phi\left(-\frac{d}{2}\right)=0,\qquad \text{and}\qquad \phi\left(+\frac{d}{2}\right)=\frac{\pi}{2} \tag{B.II.81}$$

In the absence of the field, the lowest-energy equilibrium configuration satisfying the condition of zero bulk torque is given by (Fig. B.II.21a):

$$\theta(z)=\frac{\pi}{2},\qquad \phi(z)=\frac{\pi z}{2d}+\frac{\pi}{4} \tag{B.II.82}$$

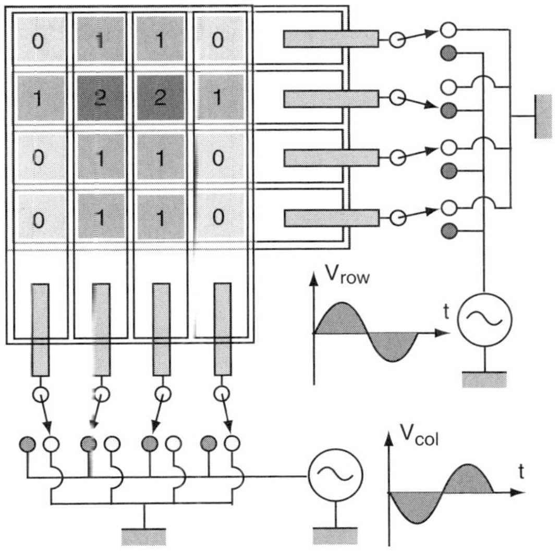

Fig. B.II.20 Electrical diagram of the 16-pixel display.

How does light propagate in such a pixel? We have seen in the previous chapter that, when the wave vector **k is parallel to the z axis**, and for **λ >> d**, two linearly polarized eigenmodes exist:

– the **extraordinary mode**, of polarization locally parallel to the director:

$$\mathbf{E}_e = E\,\mathbf{n}(z)\exp i(\mathbf{k}_e.\mathbf{r}-\Omega t) \tag{B.II.83}$$

– the **ordinary mode**, polarized perpendicularly to **n**:

$$\mathbf{E}_o = E\,\mathbf{m}(z)\exp i(\mathbf{k}_o.\mathbf{r}-\Omega t) \tag{B.II.84}$$

where **m** is a unit vector normal to both **n** and **z** while:

$$\mathbf{k}_o=\frac{2\pi n_o}{\lambda}\mathbf{z}\qquad \text{and}\qquad \mathbf{k}_e=\frac{2\pi n_e}{\lambda}\mathbf{z} \tag{B.II.85}$$

are the wave vectors defined by the ordinary (n_o) and extraordinary (n_e) refraction indices, respectively.

Eqs. B.II.83–85 show that **the polarization of light follows the director**. To put it differently, **the nematic layer turns the polarization through 90°**. Consequently, the **"0"** pixels **transmit light** when placed between crossed polarizers as shown in figure B.II.20.

As for the pixels of type **"2,"** the electric field **E** created by the applied AC voltage $V_{row}-V_{column}$ induces a torque tending to align the molecules along the z direction. If the rms value of the voltage is weak enough, this torque has no influence and the helical director configuration remains unchanged (Fig. II.21a). Above a critical voltage, the molecules will start to align in the direction of the field (Frederiks instability). This critical value is [13]

$$U_c = \pi \sqrt{\frac{K_1 + \frac{K_3 - 2K_2}{4}}{\varepsilon_0 \varepsilon_a}} \qquad \text{(B.II.86)}$$

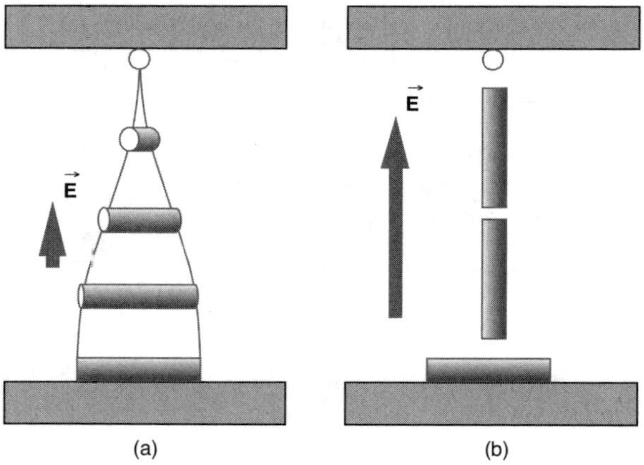

Fig. B.II.21 Director configuration: (a) weak field, (b) strong field.

Above the threshold, the helix gets distorted and light propagation becomes more difficult to describe, as the dielectric tensor $\varepsilon(\mathbf{r})$ appearing in the Maxwell equations varies with z in a complicated manner. However, if the applied tension is considerably higher than U_c, the molecules are aligned parallel to the electric field, itself almost parallel to the z axis (Fig. B.II.21b). In this **strong-field limit**, the medium behaves as if it were isotropic, the optical axis being normal to the glass plates. As a result, **the polarization of the incident light is not modified** in crossing the medium, because only the ordinary ray propagates:

$$\mathbf{E}_0 = \mathbf{E}_{inc} \exp i(\mathbf{k}_o.\mathbf{r} - \Omega t) \qquad \text{(B.II.87)}$$

with $\mathbf{E}_{inc}$ the polarization of the incident light. Such a pixel will not transmit light and it will appear as dark.

For the "1" pixels one can achieve a voltage lower than the Frederiks threshold. The helical director configuration remaining unchanged, they will transmit light, as the "0" pixels.

Also notice that contrast inversion can be obtained by placing the two polarizers parallel to the x or y axes (Fig. B.II.22).

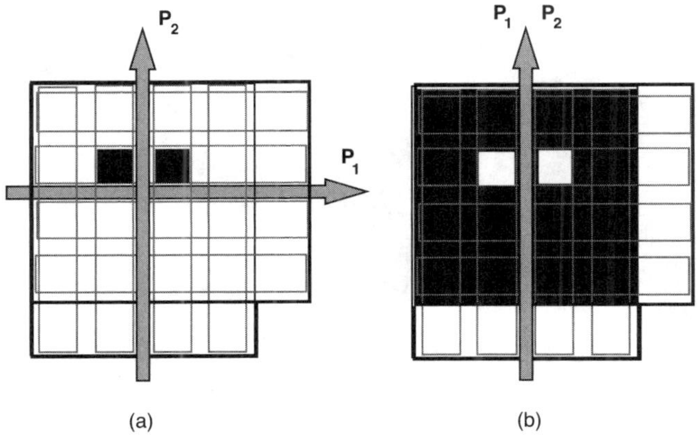

Fig. B.II.22 Contrast inversion of the pixels between crossed (a) or parallel (b) polarizers.

By flipping the switches, the individual pixels can be turned on and off. Nevertheless, multiplexing is needed to turn on a chosen pixel configuration. The display is limited by the pixel switching time, of about 30 s for the laboratory device. As discussed in the next chapter (on nematodynamics) this time is proportional to the thickness of the nematic layer. Satisfactory switching times (about 0.1 s) therefore demand nematic layers a few micrometers thick, imposing considerable constraints on the manufacturing process (constant thickness, high threshold field, etc.). The device cannot achieve video rate and is mostly used in wrist watches and pocket calculators ("sea-green" displays). Multiplexing can be avoided by using so-called active matrix displays, where each pixel is addressed separately by means of transistors placed close to the transparent electrodes [22].

II.7 Elastic light scattering and the determination of the Frank constants

In the previous chapter we saw that the trajectory of a laser beam in a nematic-filled capillary is perfectly visible as the light is strongly scattered by the director thermal fluctuations. The trajectory is still visible above the clearing point but less so, since the diffusion in the isotropic phase is much weaker than in the nematic. However, it can be followed over a much longer distance (a few centimeters compared to a few millimeters in the nematic phase). This effect is apparent in the two photos of figure B.II.23, presenting a very simple experiment where a He-Ne laser is shone from the side on a cell filled with the liquid crystal 6CB. While in the isotropic phase the beam is visible throughout the centimeter-thick sample, it only enters the nematic for a few millimeters before being completely smeared out.

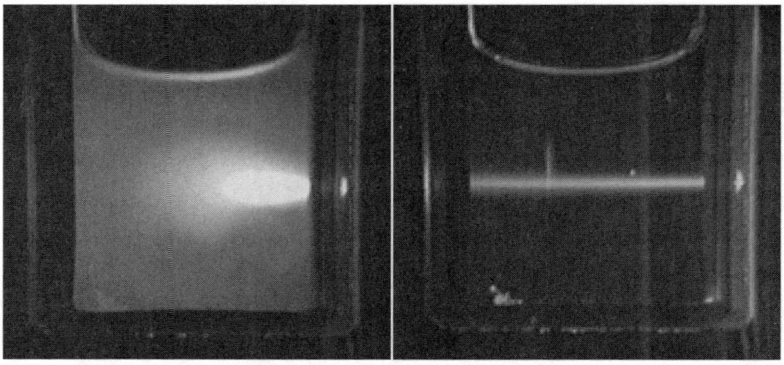

T = 27.1°C; nematic phase T = 28.6°C; isotropic phase

Fig. B.II.23 Laser beam penetrating a parallelepipedic cell filled with the liquid crystal 6CB. In the nematic phase, the beam is strongly diffused and it becomes rapidly smeared out (a few millimeters). In the isotropic phase, the beam is very clearly visible throughout the cell (especially close to the transition temperature).

To explain these features, let us first recall some classic results on elastic (Rayleigh) light scattering. Consider a homogeneous medium of dielectric tensor $\underline{\varepsilon}(\mathbf{r})$, an incident field of wave vector $\mathbf{k}_o$ and polarization $\mathbf{i}$ and a scattered field of wave vector $\mathbf{k}$ and polarization $\mathbf{f}$ (Fig. B.II.24), where $\mathbf{i}$ and $\mathbf{f}$ are unit vectors representing the chosen polarizations.

One can prove (see Appendix 1) that, in the limit of low dielectric anisotropy, the scattering cross-section σ_d per unit volume, given by

$$\sigma_d = \frac{1}{I}\frac{dP_d}{d\Omega} \tag{B.II.88}$$

with I the intensity of the incident wave and $dP_d/d\Omega$ the power per solid angle of the wave scattered in the $\mathbf{k}$ direction is:

$$\sigma_d(\mathbf{q}) = \left(\frac{\omega^2}{4\pi c^2}\right)^2 <|\mathbf{f}.\underline{\varepsilon}(\mathbf{q})\mathbf{i}|^2> \tag{B.II.89}$$

where $\mathbf{q}$ is the scattering vector $\mathbf{k} - \mathbf{k}_o$ (corresponding to a scattering angle θ such that $q = 2k\sin(\theta/2)$) and $\varepsilon(\mathbf{q})$ the Fourier transform of $\varepsilon(\mathbf{r})$:

$$\underline{\varepsilon}(\mathbf{q}) = \int_{\Omega} \underline{\varepsilon}(\mathbf{r}) \exp(-i\mathbf{q}.\mathbf{r})\, d^3\mathbf{r} \qquad \text{(B.II.90a)}$$

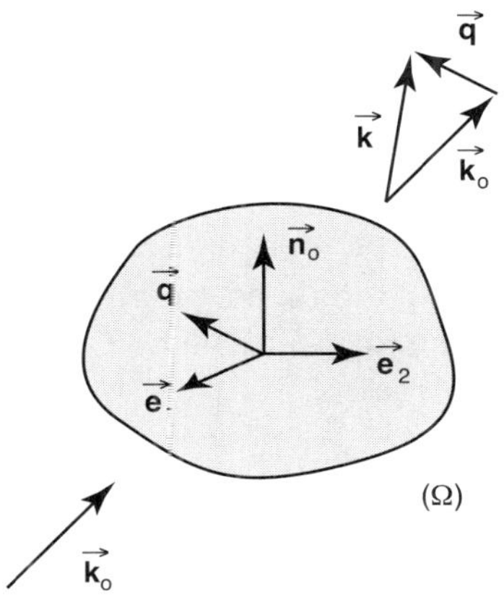

Fig. B.II.24 Scattering volume and the scattering vector **q**.

Note that $\underline{\varepsilon}(\mathbf{r})$ is the sum of all statistically independent modes $\underline{\varepsilon}(\mathbf{q})$:

$$\underline{\varepsilon}(\mathbf{r}) = \frac{1}{\Omega}\sum_{\mathbf{q}} \underline{\varepsilon}(\mathbf{q}) \exp(i\mathbf{q}.\mathbf{r}) \qquad \text{(B.II.90b)}$$

and that <...> represents a thermal average. In a nematic, the dielectric tensor can be written as:

$$\underline{\varepsilon} = \varepsilon_\perp \underline{I} + \varepsilon_a (\mathbf{n} : \mathbf{n}) \quad \text{with} \quad \varepsilon_a = \varepsilon_{//} - \varepsilon_\perp \quad \text{and} \quad (\mathbf{n} : \mathbf{n})_{ij} = n_i n_j \qquad \text{(B.II.91)}$$

and $\underline{I}$ the unit matrix. If the nematic is oriented, the director has everywhere the same average orientation $\mathbf{n}_o$. Because of the thermal fluctuations,

$$\mathbf{n}(\mathbf{r}) = \mathbf{n}_o + \delta\mathbf{n} \qquad \text{(B.II.92)}$$

where $\delta\mathbf{n}$ has the components n_1 and n_2 in the reference system given by $\mathbf{e}_1$, $\mathbf{e}_2$, $\mathbf{n}_o$ (Fig. B.II.24). We shall neglect the fluctuations in the amplitude S, and hence in $\varepsilon_{//}$ and $\varepsilon_\perp$. As we shall see later, this hypothesis is very well verified.

Simple algebra yields:

$$\mathbf{f}.\underline{\varepsilon}(\mathbf{q})\mathbf{i} = \varepsilon_a\{[(\mathbf{f}.\delta\mathbf{n}(\mathbf{q})](\mathbf{n}_o.\mathbf{i}) + (\mathbf{f}.\mathbf{n}_o)[(\delta\mathbf{n}(\mathbf{q}).\mathbf{i})]\} \qquad \text{(B.II.93)}$$

and hence

$$\sigma_d = \left(\frac{\varepsilon_a\omega^2}{4\pi c^2}\right)^2 \sum_{\alpha=1}^{2} <|n_\alpha(q)|^2> (i_\alpha f_z + i_z f_\alpha)^2) \qquad \text{(B.II.94)}$$

The thermal averages $<|n_\alpha(\mathbf{q})|^2>$ are determined from the Frank free energy:

$$F = \frac{1}{2}\int_\Omega \left[K_1\left(\frac{\partial n_1}{\partial x_1} + \frac{\partial n_2}{\partial x_2}\right)^2 + K_2\left(\frac{\partial n_1}{\partial x_2} - \frac{\partial n_2}{\partial x_1}\right)^2 + K_3\left(\frac{\partial n_1}{\partial x_3} - \frac{\partial n_2}{\partial x_3}\right)^2\right] d^3\mathbf{r} \qquad \text{(B.II.95)}$$

which, after Fourier transforming (and with the shorthand notation $|n_i(\mathbf{q})| = n_i$), can be written as (see Appendix 2):

$$F = \frac{1}{2}\frac{1}{\Omega}\sum_q K_1(q_1n_1+q_2n_2)^2 + K_2(q_2n_1 - q_1n_2)^2 + K_3q_z{}^2(n_1^2 + n_2^2)) \qquad \text{(B.II.96)}$$

This quadratic form can be diagonalized by an appropriate choice of the unit vectors $\mathbf{e}_1$ and $\mathbf{e}_2$, with $\mathbf{e}_1$ in the $(\mathbf{q}, \mathbf{n}_o)$ plane and $\mathbf{e}_2$ normal to it. One has then

$$\mathbf{q} = q_z\mathbf{n}_o + q_\perp\mathbf{e}_1 \qquad \text{(B.II.97)}$$

and

$$F = \frac{1}{2}\frac{1}{\Omega}\sum_q\sum_{\alpha=1}^{2}\{|n_\alpha(\mathbf{q})|^2(K_3q_z^2 + K_\alpha q_\perp^2)\} \qquad \text{(B.II.98)}$$

The two degrees of freedom $n_\alpha(\mathbf{q})$ are now totally decoupled so, by applying the energy equipartition theorem, one gets:

$$<|n_\alpha(\mathbf{q})|^2> = \Omega\frac{k_BT}{K_3q_z^2 + K_\alpha q_\perp^2} \qquad \text{(B.II.99)}$$

Replacing in eq. B.II.94, the final result is [13]:

$$\sigma_d = \Omega\left(\frac{\varepsilon_a\omega^2}{4\pi c^2}\right)^2\sum_{\alpha=1}^{2}\frac{k_BT\,(i_\alpha f_z + i_z f_\alpha)^2}{K_3q_z^2 + K_\alpha q_\perp^2} \qquad \text{(B.II.100)}$$

Thus, σ_d is directly proportional to the scattering volume and the scattered intensity is particularly high for small values of **q**. In principle, one only needs to measure σ_d as a function of **q** to determine the elastic constants. Such measurements have been performed by Langevin and Bouchiat [14], yielding values for the elastic constants close to those obtained by more direct methods, such as determining the Frederiks threshold.

In conclusion, light scattering in nematics is practically temperature independent, especially close to the transition to the isotropic liquid phase, due to the fact that ε_a is directly proportional to S (the quadrupolar order parameter) while K essentially varies like S^2 (as we shall show in chapter B.IV). Hence, σ_d does not depend on S (eq. B.II.100) and varies very little with temperature.

It is interesting to compare the nematic phase to an **ordinary** isotropic liquid, where the fluctuations in dielectric permittivity are induced by density fluctuations. With $\theta(\mathbf{r})$ the local dilation, the dielectric constant at **r** is given by

$$\varepsilon(\mathbf{r}) = \bar{\varepsilon} + \varepsilon' \theta(\mathbf{r}) \tag{B.II.101}$$

where $\bar{\varepsilon}$ is the average value of the liquid dielectric constant. The differential cross-scattering section σ_d^{iso} is computed as above, yielding:

$$\sigma_d^{iso} = \Omega \left(\frac{\varepsilon' \omega^2}{4\pi c^2}\right)^2 \frac{k_B T}{W} (\mathbf{f}.\mathbf{i})^2 \tag{B.II.102}$$

with W^{-1} the isothermal compressibility. We can now compare the intensities scattered by a nematic and a classical isotropic liquid. Equation B.II.100 gives the order of magnitude for a nematic:

$$\sigma_d \approx \Omega \left(\frac{\varepsilon_a \omega^2}{4\pi c^2}\right)^2 \frac{k_B T}{K q^2} \tag{B.II.103}$$

such that

$$\frac{\sigma_d}{\sigma_d^{iso}} \approx \frac{\varepsilon_a}{\varepsilon'} \frac{W}{K q^2} \tag{B.II.104}$$

One typically has $\varepsilon_a/\varepsilon' \approx 1$, $W \approx U/a^3$ and $K \approx U/a$ with U a binding energy and "a" a molecular distance, yielding:

$$\frac{\sigma_d}{\sigma_d^{iso}} \approx \frac{1}{q^2 a^2} \tag{B.II.105}$$

$2\pi/q$ is of the order of the optical wavelength (or even larger for small angle scattering) so $qa \approx 10^{-3}$. Consequently,

$$\sigma_d >> \sigma_d^{iso} \tag{B.II.106}$$

so the nematic scatters light much (several million times!) more than an ordinary isotropic liquid.

The result no longer holds when the liquid itself is the isotropic phase of a liquid crystal at a temperature close to the nematic-isotropic transition, in which case the pretransitional nematic fluctuations in the isotropic phase become important. These fluctuations of the quadrupolar order parameter S_{ij} engender fluctuations of the dielectric tensor ε_{ij} leading to anomalous light scattering (with respect to an ordinary liquid). To understand this phenomenon better, let us first recall the relation between ε_{ij} and S_{ij}. As liquids are rotationally invariant,

$$\varepsilon_{ij} = \bar{\varepsilon}\delta_{ij} + \frac{2}{3}\Delta\varepsilon\, S_{ij} \tag{B.II.107}$$

with $\bar{\varepsilon}$ the dielectric constant of the isotropic liquid (whose optical index $n = \bar{\varepsilon}^{1/2}$) and $\Delta\varepsilon$ the anisotropy of ε_{ij} if the liquid were completely oriented ($S = 1$). From eq. B.II.89, the scattering cross-section is:

$$\sigma_d = \left(\frac{4}{9}\frac{\Delta\varepsilon\omega^2}{4\pi c^2}\right)^2 <|\mathbf{f}.\underline{S}(\mathbf{q})\mathbf{i}|^2> \tag{B.II.108}$$

To obtain $<|\mathbf{f}.\ \underline{S}(\mathbf{q})\mathbf{i}|^2>$, one must return to the Landau free energy. Since S is small, one can employ the quadratic approximation and write:

$$F = \int_\Omega \frac{1}{2} a_o(T-T^*)\ S_{ij}S_{ji}\ d^3\mathbf{r}$$

$$= \int_\Omega \frac{1}{2} a_o\ (T-T^*)\ (S_{xx}^2 + S_{yy}^2 + S_{zz}^2 + 2S_{xy}^2 + 2S_{xz}^2 + 2S_{yz}^2)\ d^3\mathbf{r} \tag{B.II.109}$$

From this expression one can obtain σ_d by employing the energy equipartition theorem. Start by considering the polarization directions of the incident and scattered waves form a 90° angle (with, say, **i** along x and **f** along z). Then,

$$(\sigma_d)_{xz} = \frac{4}{9}\left(\frac{\Delta\varepsilon\omega^2}{4\pi c^2}\right)^2 <|S_{xz}(\mathbf{q})|^2> \tag{B.II.110}$$

where $<|S_{xz}(\mathbf{q})|^2>$ is given by the equipartition theorem:

$$\frac{1}{2\Omega} a_o(T - T^*) <|S_{xz}(\mathbf{q})|^2> = \frac{1}{2} k_B T \tag{B.II.111}$$

yielding immediately:

$$(\sigma_d)_{xz} = \frac{2\Omega}{9}\left(\frac{\omega^2}{4\pi c^2}\right)^2 \frac{(\Delta\varepsilon)^2 k_B T}{a_o(T-T^*)} \tag{B.II.112}$$

When the incident and scattered waves have parallel polarizations (along z for instance), the scattering cross-section can be obtained in a similar manner and reads

$$(\sigma_d)_{zz} = \frac{4}{9}\left(\frac{\Delta\varepsilon\omega^2}{4\pi c^2}\right)^2 <|S_{zz}(\mathbf{q})|^2> \tag{B.II.113}$$

To determine $<|S_{zz}(\mathbf{q})|^2>$ one also uses the equipartition theorem, keeping in mind that the S_{ij} tensor is traceless, so:

$$\frac{1}{2\Omega} a_o(T - T^*) <|S_{zz}(\mathbf{q})|^2> = \frac{1}{3} k_B T \tag{B.II.114}$$

and, by replacing in eq. B.II.113:

$$(\sigma_d)_{zz} = \frac{8\Omega}{27}\left(\frac{\omega^2}{4\pi c^2}\right)^2 \frac{(\Delta\varepsilon)2 k_B T}{a_o(T-T^*)} \tag{B.II.115}$$

In conclusion, the scattering sections $(\sigma_d)_{zz}$ and $(\sigma_d)_{xz}$ vary in $a + b(T - T^*)^{-1}$ and fulfill the notable relation:

$$\frac{(\sigma_d)_{zz}}{(\sigma_d)_{xz}} = \frac{4}{3} \tag{B.II.116}$$

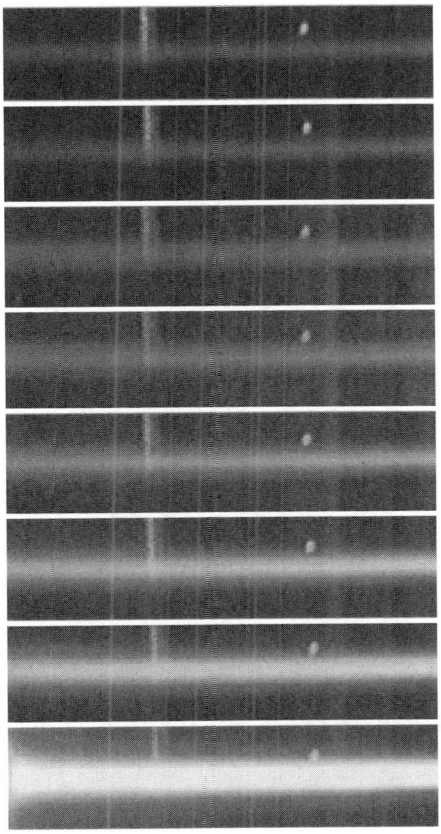

Fig. B.II.25 The trajectory of the laser beam becomes brighter when approaching the nematic phase (6CB sample).

These theoretical results are easily verified in experiments. Qualitatively, one only needs to watch a laser beam crossing a capillary filled with 6CB in the isotropic phase to realize that the scattered intensity strongly increases when approaching the nematic phase (Fig. B.II.25). Quantitatively, measuring the intensity scattered at an angle of 90° with respect to the incident wave using a photodiode shows that it varies as $a + b(T - T^*)^{-1}$, in agreement with theory (Fig. B.II.26). Finally, the relation B.II.116 was experimentally verified by Stinson et al. [15].

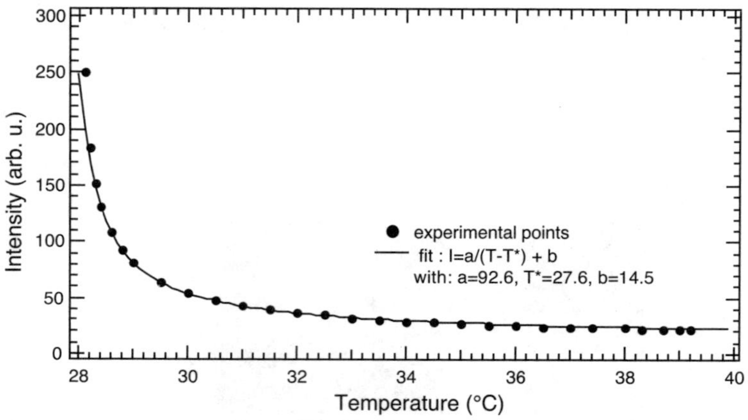

Fig. B.II.26 Scattered intensity in the isotropic phase as a function of temperature for a sample of 6CB. Note the strong increase on approaching the nematic phase.

We also emphasize that the scattered intensity does not change with the scattering vector **q**, fixed by the observation direction, because the correlation length (setting the scale of nematic fluctuations) is always much smaller than the wavelength of visible light.

Finally, let us compare the scattering cross-sections at the critical temperature T_c for the nematic phase and its isotropic liquid. From equations B.II.103 and B.II.115, one gets for an order of magnitude:

$$\frac{\sigma_d^{nem}}{\sigma_d^{iso}} \approx \frac{a_o(T_c - T^*)}{Kq^2} \qquad \text{(B.II.117)}$$

With $a_o \approx 0.2$ J/cm^3/K, $T_c - T^* \approx 1$°C, $K \approx 10^{-7}$ dyn and $q \approx 10^5$ cm^{-2}, the ratio is about 1000. The isotropic liquid at T_c scatters light much less than the nematic phase, but much (1,000 to 10,000 times) more than an ordinary liquid.

II.8 Nonlinear optics

II.8.a) Optical Frederiks transition

Under the action of a laser, the molecules of a fluid change their orientation, giving rise to optical nonlinearities. This effect is weak in common fluids and its experimental investigation requires very strong lasers. In nematic liquid crystals, the nonlinearity is much stronger because of the high anisotropy of the molecules and their collective behavior. To give an idea of this difference, we recall that a laser beam of intensity 100 W/cm^2 induces in nematics a birefringence variation of a few percent, while in an isotropic liquid like CS_2, the induced birefringence is of the order of 2×10^{-11}, a billion times weaker. It is therefore easy to understand that a strong enough polarized laser beam can cause the Frederiks transition in nematics. The instability is brought about by the electric field of the light wave, to which the molecules tend to align for positive birefringence ($\Delta n = (\varepsilon_{//})^{1/2} - (\varepsilon_{\perp})^{1/2} > 0$, with $\varepsilon_{//}$ and $\varepsilon_{\perp}$ the dielectric constants at the optical frequency of the laser). The instability threshold can be measured as shown in figure B.II.27.

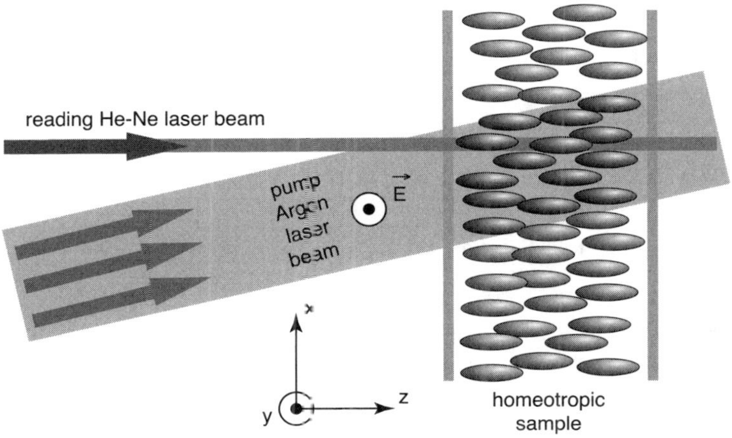

Fig. B.II.27 Light wave-induced Frederiks instability.

An Argon laser of linear polarization normal to the director is shone on a homeotropic sample under a very small incidence angle α. This beam, representing the external source, is adjustable in intensity and its cross-section is larger than the sample thickness to reduce finite-size effects. When the instability develops, the molecules tilt in the (y,z) plane (perpendicular to the incidence plane) by an angle θ(z) with respect to the sample normal. This angle is measured using a second, low intensity beam (He-Ne laser) arriving on the sample at normal incidence and polarized at 45° with respect to the (x, z) plane. The phase lag angle Δφ between the ordinary and extraordinary components of the "probe" beam can be easily shown to depend on the tilt angle θ(z) through the integral relation:

$$\Delta\varphi = \frac{2\pi}{\lambda}\int_{-d/2}^{+d/2}\left[\frac{n_o n_e}{\sqrt{n_e^2\cos^2\theta + n_o^2\sin^2\theta}} - n_o\right]dz \qquad \text{(B.II.118)}$$

with λ the vacuum wavelength of the He-Ne laser. This phase lag is zero below the instability threshold and was measured as a function of the power I of the source laser by Durbin et al. [16]. The experiment was performed on a homeotropic 5CB sample 250 μm thick and the results are displayed in figure

B.II.28; for clarity, only the experimental curve corresponding to $\alpha \approx 0$ is shown. Above a critical intensity $I_c \approx 155$ W/cm^2, molecules inside the sample start tilting. As the phase lag above the threshold increases continuously from zero, the bifurcation is supercritical (as almost all Frederiks transitions).

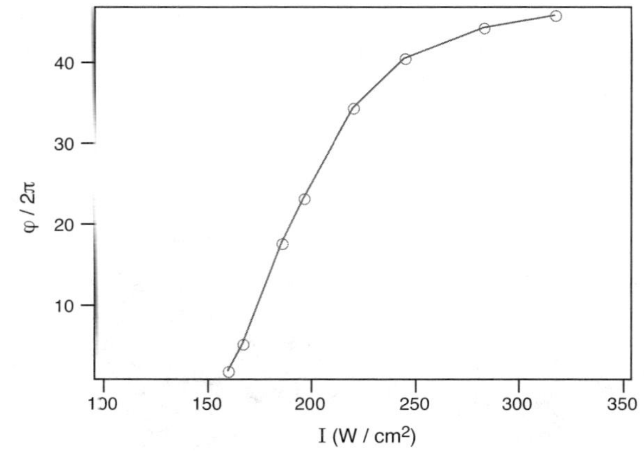

Fig. B.II.28 Phase lag angle $\varphi/2\pi$ as a function of the power I of the source beam (5CB homeotropic sample 250 μm thick) (from ref. [16]).

The threshold value I_c can be estimated from the free energy of the electromagnetic wave in the nematic sample [17]. Let $\theta(z)$ be the director tilt angle with respect to the **z** axis and **k** the wavevector of the source beam, taken as parallel to the **z** axis (normal incidence, $\alpha = 0$) (Fig. B.II.29). The nematic is a non-magnetic dielectric. The electromagnetic free energy consists of two terms: the electric energy

$$f^E = \frac{\mathbf{E}.\mathbf{D}}{2} \tag{B.II.119}$$

and the magnetic energy (not to be neglected, even if the molecular torque is mainly due to the electric field):

$$f^M = \frac{\mathbf{B}.\mathbf{H}}{2} \tag{B.II.120}$$

These contributions can be shown to be equal. The Maxwell equation $\mathbf{curl}(\mathbf{B}/\mu_o) = \partial\mathbf{D}/\partial t$ yields $\mathbf{D} = -(1/\mu_o\omega)(\mathbf{k}\times\mathbf{B})$, hence

$$f^E = -\frac{1}{2\mu_o\omega}\mathbf{E}.(\mathbf{k}\times\mathbf{B}) = \frac{1}{2\mu_o\omega}(\mathbf{k},\mathbf{E},\mathbf{B}) \tag{B.II.121}$$

Similarly, $\mathbf{B} = (1/\omega)(\mathbf{k}\times\mathbf{E})$ and, with the Maxwell equation $\mathbf{curl}\,\mathbf{E} = -\partial\mathbf{B}/\partial t$, one has:

$$f^M = \frac{1}{2\mu_o\omega}\mathbf{B}.(\mathbf{k}\times\mathbf{E}) = \frac{1}{2\mu_o\omega}(\mathbf{k},\mathbf{E},\mathbf{B}) \qquad \text{(B.II.122)}$$

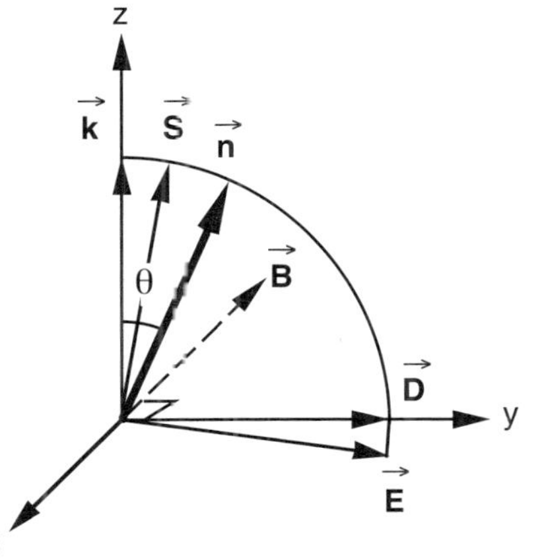

Fig. B.II.29 Relative position of the various vectors describing the electromagnetic wave.

The electric and magnetic energies are thus equal, so the overall electromagnetic energy is:

$$f^{EM} = \frac{1}{\mu_o\omega}(\mathbf{k},\mathbf{E},\mathbf{B}) \qquad \text{(B.II.123)}$$

where the wave vector **k** of the light wave, of expression:

$$\mathbf{k} = \frac{\omega}{c}n\,\mathbf{z} \qquad \text{(B.II.124)}$$

makes an angle θ with the optical axis, with n the effective index of the medium, taking values between n_o and n_e. Its value was determined in chapter B.I (eq. B.I.80):

$$n = \frac{n_o n_e}{\sqrt{n_o^2\sin^2 q + n_e^2\cos^2 q}} \qquad \text{(B.II.125)}$$

Defining the Poynting vector $\mathbf{S} = \mathbf{E} \times (\mathbf{B}/\mu_o)$, the electromagnetic free energy can be written in the simple form:

$$f^{EM} = \frac{1}{c}n\,S_z \qquad \text{(B.II.126)}$$

with S_z the z-component of the Poynting vector. The relevant value for the Frederiks instability is the time average of S_z, as the director cannot follow the electromagnetic field oscillations. Let us prove that this average is a constant (some authors, such as Durbin et al. [16], made the incorrect hypothesis of constant $|\mathbf{S}|$). The result is obtained from the law of energy conservation in the medium:

$$\mathrm{div}\,\mathbf{S} + \frac{\partial f^{EM}}{\partial t} = 0 \qquad \text{(B.II.127)}$$

by time averaging (denoted by <..>), knowing that translational invariance in the sample plane ensures $\partial/\partial x = \partial/\partial y = 0$. This implies

$$\frac{\partial \langle S_z \rangle}{\partial z} = 0 \qquad \text{hence} \qquad \langle S_z \rangle = \mathrm{Cst} \qquad \text{(B.II.128)}$$

Experimentally, $\langle S_z \rangle$ corresponds to the intensity I of the incident beam, so

$$\langle f^{EM} \rangle = \frac{1}{c} n I \qquad \text{(B.II.129)}$$

where n is the effective index, depending on the angle $\theta(z)$, yet to be determined. The same formula was obtained by Zel'dovich using the approximations of geometrical optics [18]. This angle and the Frederiks threshold can be determined by minimizing the total free energy of the system:

$$F = \int \left[\frac{1}{2} K_1 (\mathrm{div}\mathbf{n})^2 + \frac{1}{2} K_3 (\mathbf{n} \times \mathbf{curl}\ \mathbf{n})^2 + \frac{1}{c} n I\right] dz \qquad \text{(B.II.130)}$$

leading to the Euler equation:

$$(1 - K \sin^2\theta) \frac{d^2\theta}{dz^2} - K \sin\theta\cos\theta \left(\frac{d\theta}{dz}\right)^2 + \frac{I n_o}{c K_3} \frac{N \cos\theta \sin\theta}{(1 - N \sin^2\theta)^{3/2}} = 0 \qquad \text{(B.II.131)}$$

where

$$K = \frac{K_3 - K_1}{K_3} \qquad \text{and} \qquad N = 1 - \frac{n_o^2}{n_e^2} \qquad \text{(B.II.132)}$$

The absolute instability threshold and the corresponding critical intensity I_{Fr} can be obtained by searching for a solution of the simple form:

$$\theta = \theta_m \sin\left(\frac{\pi z}{d}\right) \qquad \text{(B.II.133)}$$

ensuring homeotropic anchoring on the glass plates, in $z = 0$ and $z = d$. Substituting this expression for θ in equation B.II.131 one gets, to first order in θ_m:

$$\left(-\frac{\pi^2}{d^2} + \frac{N I_{Fr} n_o}{c K_3}\right)\theta_m = 0 \qquad \text{(B.II.134)}$$

and therefore:

$$I_{Fr} = \frac{\pi^2}{d^2}\frac{c K_3}{n_o N} \tag{B.II.135}$$

In 5CB at 26°C, $n_o \approx 1.53$ and $n_e \approx 1.70$ (at $\lambda = 5890$ Å), $K_3 \approx 8.5\times10^{-12}$ N yielding, for a sample thickness d = 250 μm, a critical intensity $I_{Fr} \approx 133$ W/cm^2, close to the one measured by Durbin.

II.8.b) Nonlinear analysis

The bifurcation order can be found by looking for the general solution to equation B.II.131, subject to the condition

$$\theta(z) = \theta(d - z) \tag{B.II.136}$$

and letting θ_m be the maximum tilt angle of the director in the center of the cell, at z = d/2. Equation B.II.131 has the first integral:

$$\left(\frac{d\theta}{dz}\right)^2 = \frac{2I}{cK_3}\frac{n(\theta_m) - n(\theta)}{1 - K\sin^2\theta} \tag{B.II.137}$$

yielding, for 0 < z < d/2 and taking into account the boundary conditions:

$$z = \sqrt{\frac{cK_3}{2I}}\int_0^\theta \sqrt{\frac{1 - K\sin^2\theta}{n(\theta_m) - n(\theta)}}\,d\theta \tag{B.II.138}$$

Introducing (formally) an angle ψ such that

$$\sin\psi = \frac{\sin\theta}{\sin\theta_m} \tag{B.II.139}$$

Equation B.II.138 can be written as:

$$z\sqrt{\frac{2n_oI}{cK_3}} = \sin\theta_m(1 - N\sin^2\theta_m)^{1/4} \times$$

$$\int_0^\psi \sqrt{\frac{1 - K\sin 2\theta_m \sin 2\psi}{1 - \dfrac{\sqrt{1 - N\sin 2\theta_m}}{\sqrt{1 - N\sin 2\theta_m \sin 2\psi}}}}\cos\psi\,d\psi \tag{B.II.140}$$

The maximum distortion angle θ_m (playing the role of the order parameter for this Frederiks transition) is obtained by solving this equation at z = d/2 and $\psi = \pi/2$. Developing the equation obtained to second order in θ_m close to the absolute instability threshold of the homeotropic state, yields:

$$\theta_m^2 \approx \frac{1}{C}\left(\sqrt{\frac{I}{I_{Fr}}} - 1\right) \approx \frac{1}{2C}\left(\frac{I}{I_{Fr}} - 1\right) \qquad \text{(B.II.141)}$$

where I_{Fr} is the previously determined Frederiks threshold (eq. B.II.135) and C a material constant given by

$$C = \frac{1}{4}\left(1 - K - \frac{9N}{4}\right) \qquad \text{(B.II.142)}$$

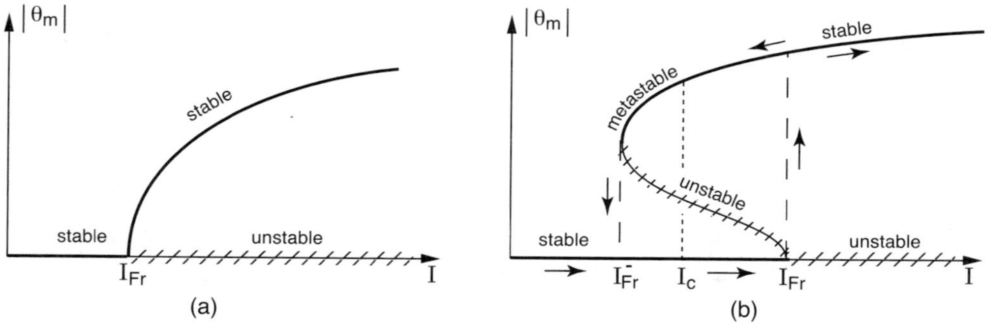

Fig. B.II.30 Maximum tilt angle as a function of the intensity of the source beam. The bifurcation is supercritical in (a) and subcritical in (b). The stability is indicated for each solution.

There are two possibilities, according to the sign of C:

– either $C > 0$ and the bifurcation is **supercritical**, the maximum angle varying as the square root of the distance to the threshold (Fig. B.II.30a);

– or $C < 0$ and the bifurcation is **subcritical**, as the solutions given by equation B.II.141 only exist for $I < I_{Fr}$ and they are unstable because $dI/d(\theta_m^2) < 0$. This solution branch appears hatched in figure B.II.30b. We shall subsequently prove the existence of an upper branch of stable solutions (thick line), having necessarily a finite value at the bifurcation threshold. The angle θ_m therefore undergoes a jump at the transition, contrary to the supercritical case.

In conclusion, two regions must be distinguished in the parameter plane (K_1/K_3, $\varepsilon_\perp/\varepsilon_{//}$): the one below the line

$$\frac{K_1}{K_3} + \frac{9}{4}\frac{\varepsilon_\perp}{\varepsilon_{//}} = \frac{9}{4} \qquad (C = 0) \qquad \text{(B.II.143)}$$

where the bifurcation is subcritical, and the one above it, where the bifurcation is supercritical. This limit, as well as experimental points corresponding to some known materials, are shown in figure B.II.31. Theory predicts a supercritical bifurcation for 5CB, 8CB and MBBA, in agreement with experimental observations of Csillag et al. [19], Durbin et al. [16], and Khoo [20]. On the other hand, it predicts a subcritical bifurcation for PAA, yet to be confirmed.

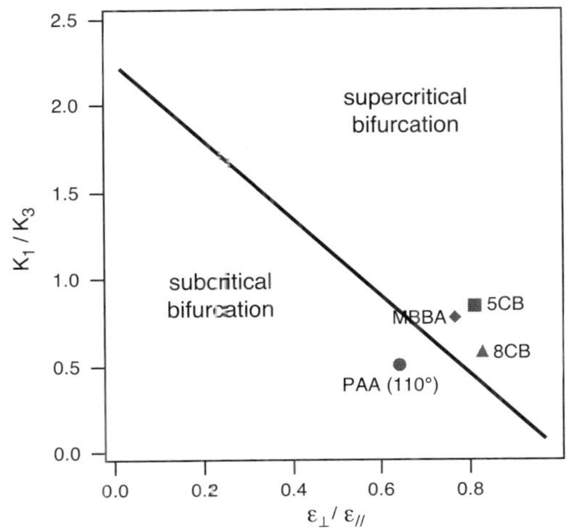

Fig. B.II.31 Theoretical prediction for the order of the optical Frederiks transition in the parameter plane K_1/K_3 and $\varepsilon_\perp/\varepsilon_{//}$ (from ref. [17]).

Let us go back to determining the two solution branches for the subcritical bifurcation. Developing equation B.II.140 to fourth order in θ_m leads to:

$$\theta_m^2 = \frac{-C \pm \sqrt{C^2 + 4D(\sqrt{I/I_{Fr}} - 1)}}{2D} \qquad \text{(B.II.144)}$$

where C is considered negative (as in PAA). The other constant reads:

$$D = \frac{1}{96}\left(\frac{11}{2} - K + \frac{9}{4}N + \frac{63}{4}KN - \frac{9}{2}K^2 - \frac{261}{32}N^2\right) \qquad \text{(B.II.145)}$$

and it will be taken as positive throughout this discussion, a condition fulfilled by all presented materials (if it were negative, the development should be taken to higher order). The formula yields the two already discussed solution branches:

– the unstable one, corresponding to the negative sign, reducing to the approximate formula B.II.141 for $I \rightarrow I_{Fr}$;

– the stable branch corresponding to the positive sign.

The two branches merge in the point (I_{Fr}^-, θ_m^-) where the square root in eq. B.II.144 goes to zero:

$$I_{Fr}^{-} = I_{Fr}\left(1 - \frac{C^2}{4D}\right)^2 \qquad \text{(B.II.146)}$$

$$\theta_m^{-} = \sqrt{-\frac{C}{2D}} \qquad \text{(B.II.147)}$$

Further insight into the thermodynamic significance of these two solution branches can be gained by calculating the total free energy of the system starting from equation B.II.130. Developing to sixth order in θ_m one has, up to an additive constant and a scaling factor:

$$F = -\left(\sqrt{\frac{I}{I_{Fr}}} - 1\right)\theta_m^2 + \frac{C}{2}\theta_m^4 + \frac{D}{3}\theta_m^6 + O(\theta_m^8) \qquad \text{(B.II.148)}$$

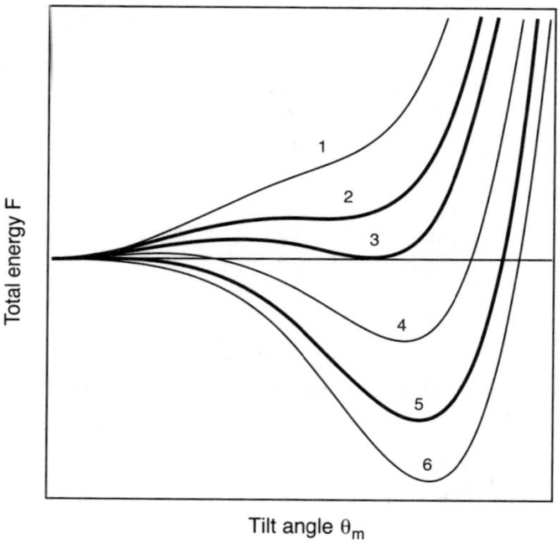

Fig. B.II.32 Total energy as a function of the maximum deformation angle for increasing intensity: 1) $I < I_{Fr}^{-}$; 2) $I = I_{Fr}^{-}$; 3) $I = I_c$; 4) $I_c < I < I_{Fr}$; 5) $I = I_{Fr}$; 6) $I > I_{Fr}$.

Figure B.II.32 displays some curves $F(\theta_m)$. Intensities I_{Fr} and I_{Fr}^{-} correspond to the spinodal limits of the homeotropic and distorted state, respectively. The two states have equal energy for a critical intensity I_c between these two values, easily obtained as:

$$I_c = I_{Fr}\left(1 - \frac{3C^2}{16D}\right)^2 \qquad \text{(B.II.149)}$$

with a critical angle:

$$\theta_m^c = \sqrt{-\frac{3C}{4D}} \qquad \text{(B.II.150)}$$

which completes the analogy with phase transitions. The existence of two distinct spinodal limits leads to a hysteresis cycle when progressively increasing and decreasing the beam intensity. However, its area is always smaller than that of the theoretical cycle depicted in figure B.II.30b, as thermal fluctuations (or sample defects) facilitate the jump from one branch to the other.

II.8.c) Autofocusing of a laser beam

Above the Frederiks threshold, molecular tilt modifies the local refraction index and thus the light ray propagation. In homeotropic geometry, increasing the light intensity leads to a corresponding increase in the refraction index. In nonlinear optics, this coupling is known to engender autofocusing; nevertheless, the phenomenon will only be noticeable if it dominates diffraction. One then expects the autofocusing threshold I_{AF} to be larger than the Frederiks threshold I_{Fr} for a supercritical bifurcation.

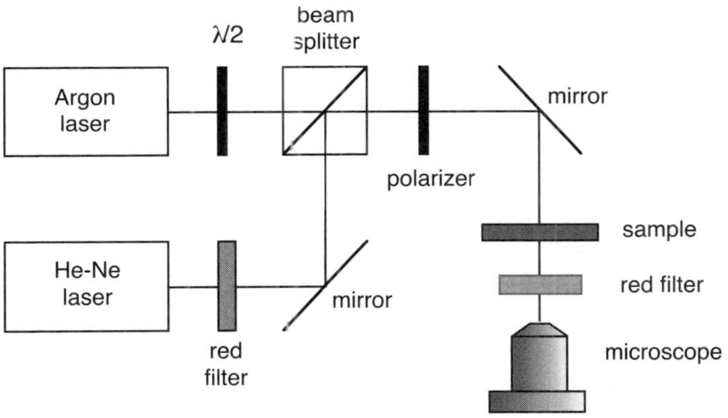

Fig. B.II.33 Experimental setup for measuring the autofocusing threshold of a laser beam (from ref. [21]).

To illustrate the experimental measurement of I_{AF} we present in figure B.II.33 the setup employed by Braun et al. [21]. It comprises an Argon laser linearly polarized in the gaussian mode TEM_{00} directed at normal incidence upon a homeotropic sample placed on the stage of an inverted microscope. By means of a separator cube, this beam is combined with a much weaker one (produced by a 3 mW He-Ne laser). The two beams are close in diameter (about 50 μm) and are aligned with micrometric precision in order to have the same trajectory in the sample. After crossing the sample, the intensity of the source beam is cut by a red filter. The sample acts as a convex lens on the probe beam, whose plane of smallest cross-section is determined using the microscope. The distance between this plane and the lower sample surface gives the focal length. Figure B.II.34 shows the variation of the focal length with the source beam intensity. Obviously, a clear distinction must be made between thin samples (100 μm, typically) whose f^{-1} is linear in I, and thick samples (700 μm) for which the dependence is rather exponential.

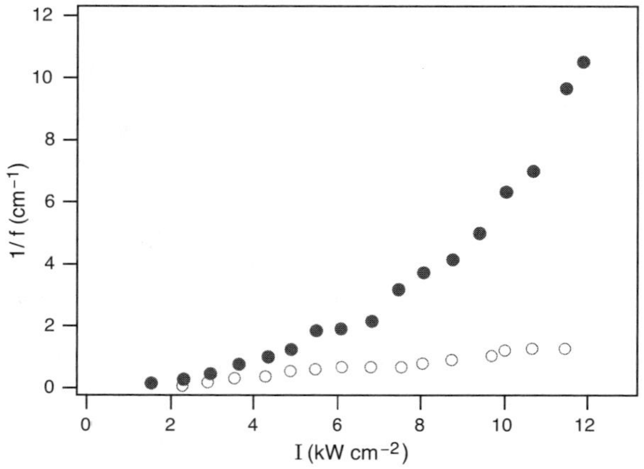

Fig. B.II.34 Focal length reciprocal as a function of the source beam intensity for two 6CB homeotropic samples, of thickness 100 μm (open circles) and 700 μm (solid circles) (from ref. [21]).

Experiment also shows that the minimal autofocusing intensity I_{AF} is much higher than the Frederiks threshold I_{Fr} (for a 6CB sample 100 μm thick, Braun et al. give $I_{Fr} \approx 870\ W/cm^2$ and $I_{AF} \approx 1250\ W/cm^2$, respectively).

Qualitative understanding of these results can be achieved by determining f for a thin sample. In a gaussian beam with diameter a, light intensity varies across the beam as $I\exp(-2r^2/a^2)$. Assume $I > I_{Fr}$ and neglect diffraction for the time being. Close to the beam center, the average refraction index n can be developed as:

$$n = n_o + n_2\,(I - I_{Fr})\exp\left(\frac{-2r^2}{a^2}\right) \qquad \text{(B.II.151)}$$

with n_2 the Kerr effect constant. This radial modulation deforms the wavefront, which acquires a curvature δC for a distance δz along the trajectory:

$$\delta C = \frac{\partial^2[n(r)\delta z]}{\partial r^2} = 4\,n_2\,(I - I_{Fr})\,a^{-2}\,\delta z \qquad \text{(B.II.152)}$$

This relation defines a typical autofocus length:

$$z_f = \frac{2}{a}\sqrt{n_2(I - I_{Fr})} \qquad \text{(B.II.153)}$$

A sample can be considered as thin if its thickness $d >> z_f$. The wavefront curvature when it exits the sample is then simply:

$$C = 4\, n_2\, (I - I_{Fr})\, a^{-2} d \qquad \text{(B.II.154)}$$

The actual wavefront curvature $1/f$ must take into account diffraction effects. It can be shown that

$$\frac{1}{f} = C - \frac{1}{ka^2} \qquad \text{(B.II.155)}$$

where k is the wave vector. By definition, the autofocusing threshold I_{AF} corresponds to $1/f = 0$ so that

$$I_{AF} = I_{Fr} + \frac{1}{4n_2dk} \qquad \text{(B.II.156)}$$

f^{-1} can then be written as:

$$\frac{1}{f} = 4\,\frac{n_2}{a^2}(I - I_{AF})\,d \qquad \text{(B.II.157)}$$

The linear proportionality with sample thickness is in agreement with experiment. Fitting the data allows for precise determination of I_{AF} as well as the Kerr effect constant n_2. For 6CB, one obtains $n_2 \approx 8.7\times10^{-8}\ \text{cm}^2/\text{W}$ and $I_{AF} \approx 1250\ \text{W/cm}^2$. However, the formula does not apply to thick samples, which still require satisfactory theoretical treatment.

BIBLIOGRAPHY

[1] Grupp J., *Phys. Lett. A*, **99** (1983) 373.

[2] Nehring J., Saupe A., *J. Chem. Phys.*, **54** (1971) 337 and **56** (1972) 5527.

[3] Frank F.C., *Disc. Faraday Soc.*, **25** (1958) 19.

[4] Yokoyama H., *Phys. Rev. E*, **55** (1997) 2938.

[5] Landau L., Lifchitz E., *Theory of Elasticity*, Mir, Moscow, 1967, p. 60.

[6] Helfrich W., *Z. Naturforsch.*, **28c** (1973) 693 and **33a** (1978) 305.

[7] Canham P.B., *J. Theor. Biol.*, **26** (1970) 61.

[8] Frederiks V., Zolina V., *Trans. Faraday Soc.*, **29** (1933) 919.

[9] Waton W., Ferre A., Candau S., Perbet J.N., Hareng M., *Mol. Cryst. Liq. Cryst.*, **78** (1981), 237.

[10] Cladis P., *Phys. Rev. Lett.*, **28** (1972) 1629.

[11] Lonberg F., Meyer R.B., *Phys. Rev. Lett.*, **55** (1985) 718.

[12] Deuling H., Helfrich W., *Appl. Phys. Lett.*, **25** (1974) 129.

[13] De Gennes P.-G., *Mol. Cryst. Liq. Cryst.*, **7** (1969) 325.

[14] Langevin D., Bouchiat M.J., *J. Physique (France)*, **36** Coll. C1 (1975) 197.

[15] Stinson T.W., Litster J.D., Clark N.A., *J. Physique (France)*, **33** Coll. C1 (1972) 69.

[16] Durbin S.D., Arakelian S.M., Shen Y.R., *Phys. Rev. Lett.*, **47** (1981) 1411.

[17] Ong H.L., *Phys. Rev. A*, **28** (1983) 2393.

[18] Zel'dovich B. Ya, Tabiryan N.V., Chilingaryan Yu., *Sov. Phys. J.E.T.P.*, **54** (1981) 32.

[19] a) Csillag L., Janossy I., Kitaeva V.F., Kroo N., Sobolev N.N., Zolotka A.S., *Mol. Cryst. Liq. Cryst.*, **78** (1981) 173.
b) Zolotka A.S., Kitaeva V.F, Kroo N., Sobolev N.N., Csillag L., *Sov. Phys. J.E.T.P. Lett.*, **34** (1981) 250.

[20] Khoo I.C., *Phys. Rev. A*, **23** (1981) 2078; *Phys. Rev. A*, **25** (1982) 1040 and 1637; *Phys. Rev. A*, **26** (1982) 1131; *Appl. Phys. Lett.*, **39** (1981) 937; *Appl. Phys. Lett.*, **40** (1982) 645.

[21] Braun E., Faucheux L.P., Libchaber A., *Phys. Rev. A*, **48** (1993) 611.

[22] Tsukada T., *Liquid Crystal Displays Addressed by Thin-Film Transistors*, Gordon and Breach, Amsterdam, 1996.

Appendix 1: Calculating the scattering cross-section

The exact calculation is fairly complicated for the general case of an anisotropic medium, as the refractive index depends on the direction of light propagation. A major simplification is to suppose that the rays propagate in an isotropic index of average index $\bar{n} = (\varepsilon_{//}\varepsilon_{\perp})^{1/2}$. Even though this assumption is only valid for small dielectric anisotropy $\varepsilon_a = \varepsilon_{//} - \varepsilon_{\perp}$ of the nematic (which is not always the case), we shall consider it true in the following.

The experiment is represented in Figure A.1.

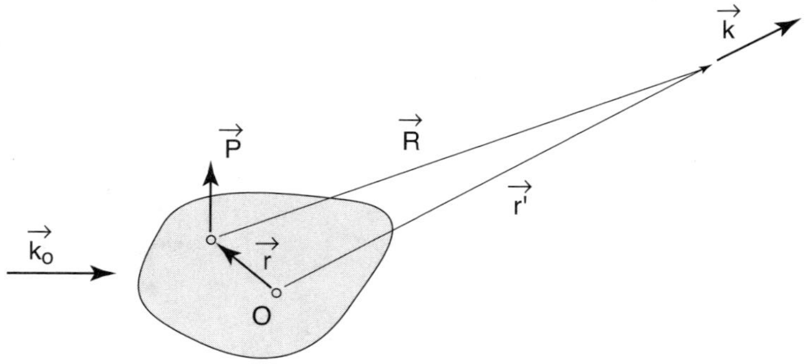

Fig. A.1 Diagram of a scattering experiment. The observation point **r'** is very far from the scattering volume. The incident wave, of wave vector $\mathbf{k}_0$, induces at each point **r** of the medium a polarization **P**(**r**) which sends out a field in all directions. The scattered field in **r'** is the sum of all these individual fields.

The incident wave, of wave vector $k_o = \omega\bar{n}/c$, linearly polarized along the direction **i**:

$$\mathbf{E}_{inc} = E_o\mathbf{i}\; e^{i\mathbf{k}_o\mathbf{r}} \quad (1.1)$$

induces around a point **r** the macroscopic polarization **P**(**r**):

$$\mathbf{P}(\mathbf{r}) = \varepsilon_o\,(\underline{\varepsilon} - \underline{I})\;\mathbf{E}_{inc}(\mathbf{r}) \quad (1.2)$$

This volume dipole itself re-emits a wave characterized at a point **r'** (for $r' \approx R >> r$) by the electric field:

$$\mathbf{E}_{diff}(\mathbf{r'}, \mathbf{r}) = \frac{\mu_o}{4\pi}\,\omega^2\,\frac{e^{ikR}}{R}\,\mathbf{P}_{\perp}(\mathbf{r}) \quad (1.3)$$

with $\mathbf{P}_{\perp}(\mathbf{r})$ the **P**-component normal to the $\mathbf{R} = \mathbf{r'} - \mathbf{r}$ direction:

$$\mathbf{P}_{\perp}(\mathbf{r}) = (\mathbf{P} \times \mathbf{u}) \times \mathbf{u} \quad (1.4)$$

where $\mathbf{u} = \mathbf{R}/R$, the unit vector parallel to the **R** direction and $k = \omega\bar{n}/c$. As the observation point is very far from the scattering volume, one can write:

$$kR \approx \mathbf{k}.\mathbf{R} = \mathbf{k}.(\mathbf{r}' - \mathbf{r}) \tag{1.5}$$

leading to the following expression for the total scattered field:

$$\mathbf{E}_{diff}(\mathbf{r}') = \int_{\Omega} \mathbf{E}_{diff}(\mathbf{r}', \mathbf{r})\, d^3\mathbf{r}$$

$$= \int_{\Omega} \frac{\mu_o}{4\pi}\omega^2 \frac{e^{i\mathbf{k}.(\mathbf{r}'-\mathbf{r})}}{R}[\varepsilon_o(\underline{\varepsilon} - \underline{I})\,\mathbf{E}_{inc}(\mathbf{r})]_{\perp}\, d^3\mathbf{r} \tag{1.6}$$

Its **f** component (perpendicular to **R** and **r'**) is then simply:

$$\mathbf{E}_{diff}(\mathbf{r}').\mathbf{f} = \frac{\varepsilon_o\mu_o}{4\pi}\frac{\omega^2}{R} e^{i\mathbf{k}.\mathbf{r}'}E_o \int_{\Omega} \mathbf{f}.(\underline{\varepsilon} - \underline{I}).\mathbf{i}\, e^{i(\mathbf{k}_o-\mathbf{k}).\mathbf{r}}\, d^3\mathbf{r} \tag{1.7}$$

or, for non-null scattering vector $\mathbf{q} = \mathbf{k} - \mathbf{k}_o$ (i.e., except for forward scattering):

$$\mathbf{E}_{diff}(\mathbf{r}').\mathbf{f} = \frac{\omega^2}{c^2}\frac{E_o}{4\pi R} e^{i\mathbf{k}.\mathbf{r}'} \int_{\Omega} \mathbf{f}.\underline{\varepsilon}(\mathbf{r}).\mathbf{i}\, e^{-i\mathbf{q}.\mathbf{r}}\, d^3\mathbf{r} \tag{1.8}$$

Introducing the Fourier transform of $\underline{\varepsilon}$:

$$\underline{\varepsilon}(\mathbf{q}) = \int_{\Omega} \underline{\varepsilon}(\mathbf{r})\, e^{-i\mathbf{q}.\mathbf{r}}\, d^3\mathbf{r} \tag{1.9}$$

the formula can be simplified to:

$$\mathbf{E}_{diff}(\mathbf{r}').\mathbf{f} = \frac{\omega^2}{c^2}\frac{E_o}{4\pi R} e^{i\mathbf{k}.\mathbf{r}'}\, \mathbf{f}.\underline{\varepsilon}(\mathbf{q})\mathbf{i} \tag{1.10}$$

We can now calculate the differential scattering cross-section, defined as:

$$\sigma_d = \frac{1}{I}\frac{dP_d}{d\Omega} \tag{1.11}$$

with I the intensity of the incident wave (energy flux per unit area and time) and P_d, energy flux per unit solid angle and time of the **f**-component of the wave scattered in the **k** direction. With these definitions:

$$I = \frac{1}{2}\frac{E_o^2}{\mu_o c} \tag{1.12}$$

$$\frac{dP_d}{d\Omega} = \frac{R^2}{2\mu_o c} \ <|\mathbf{E}_{diff}(\mathbf{r'}).\mathbf{f}|^2> \tag{1.13}$$

so the desired formula reads:

$$\sigma_d(\mathbf{q}) = \left(\frac{\omega^2}{4\pi c^2}\right)^2 <|\mathbf{f}.\underline{\varepsilon}(\mathbf{q})\,\mathbf{i}|^2> \tag{1.14}$$

Appendix 2: Free energy expression in the Fourier space

We shall assume the following typical form for the energy:

$$F = \int_{\Omega} \left[S^2 + \left(\frac{\partial S}{\partial x}\right)^2 \right] d^3\mathbf{r} \tag{2.1}$$

To determine its expression in the Fourier space, we introduce the Fourier transform of S(**r**):

$$S(\mathbf{q}) = \int_{\Omega} S(\mathbf{r})\, e^{-i\mathbf{q}.\mathbf{r}} d^3\mathbf{r} \tag{2.2}$$

It can be shown that the reverse transformation is:

$$S(\mathbf{r}) = \frac{1}{(2\pi)^3} \int_{\mathbf{q}} S(\mathbf{q})\, e^{i\mathbf{q}.\mathbf{r}} d^3\mathbf{q} \tag{2.3}$$

Replacing for S(**r**) in equation 2.1 yields:

$$F = \frac{1}{(2\pi)^6} \int_{\Omega} \int_{\mathbf{q}} \int_{\mathbf{q}'} [S(\mathbf{q})S(\mathbf{q}')(1 - q_x q'_x)\, e^{i(\mathbf{q}+\mathbf{q}').\mathbf{r}}]\, d^3\mathbf{q}\, d^3\mathbf{q}'\, d^3\mathbf{r} \tag{2.4}$$

and, applying the known properties

$$\int_{\Omega} e^{i(\mathbf{q}+\mathbf{q}').\mathbf{r}} d^3\mathbf{r} = (2\pi)^3\, \delta(\mathbf{q} + \mathbf{q}') \quad \text{and} \quad \int_{\mathbf{q}'} f(\mathbf{q}')\, \delta(\mathbf{q} + \mathbf{q}')\, d^3\mathbf{q}' = f(-\mathbf{q}) \tag{2.5}$$

to equation 2.4 one has:

$$F = \frac{1}{(2\pi)^3} \int_{\mathbf{q}} \int_{\mathbf{q}'} [S(\mathbf{q})S(\mathbf{q}')(1 - q_x q'_x)\, \delta(\mathbf{q}+\mathbf{q}')]\, d^3\mathbf{q} d^3\mathbf{q}'$$

$$= \frac{1}{(2\pi)^3} \int_{\mathbf{q}} [S(\mathbf{q})S(-\mathbf{q})\, (1 + q_x^2)]\, d^3\mathbf{q} \tag{2.6}$$

Since $S(-\mathbf{q}) = S^*(\mathbf{q})$, the final result is obtained by going back to a discrete sum (the modes are discretized by the boundary conditions):

$$F = \frac{1}{\Omega} \sum_{\mathbf{q}} [|S(\mathbf{q})|^2 + q_x^2 |S(\mathbf{q})|^2] \tag{2.7}$$

Chapter B.III

Nematodynamics and flow instabilities

In the previous chapter we determined the equilibrium equations for the director $\mathbf{n}(\mathbf{r})$ such that the total elastic energy is minimized. We shall now study the evolution of a (**uniaxial**) nematic taken out of thermodynamic equilibrium by an external perturbation. This is the field of nematodynamics, more complicated but also richer than conventional hydrodynamics, due to the coupling between the fields $\mathbf{n}(\mathbf{r})$ and $\mathbf{v}(\mathbf{r})$ (hydrodynamic velocity).

Therefore, we shall start this chapter not by expounding the general theory, but rather by describing simple dynamical phenomena illustrating some of the fundamental distinctions between a nematic and an isotropic fluid (section III.1). The first experiment, the Frederiks transition, should already be familiar. We shall see that, in order to determine the tilt time for the director, one must define the **rotational viscosity**, a physical quantity with no equivalent in ordinary fluids. We shall further see how this viscosity is associated with a new hydrodynamic mode, revealed at the macroscopic scale by the distinctive nematic scintillation. Finally, we shall present the experiments of Miesowicz and Gähwiller, proving that a nematic also behaves as an anisotropic fluid, its viscosity depending on director orientation with respect to the velocity and its gradient.

Next (section III.2), we shall take up the general Leslie-Ericksen theory. We shall see how to obtain the dynamical equations for the coupled fields $\mathbf{n}(\mathbf{r})$ and $\mathbf{v}(\mathbf{r})$ using the laws of mechanics and thermodynamics (Newton's laws, angular momentum theorem, the first and second principles of thermodynamics). These theoretical concepts will then be used to determine the hydrodynamic modes.

We shall subsequently study the Couette and Poiseuille laminary flows (section III.3). We shall show how, in a Poiseuille flow, transverse pressure gradients can develop: a surprising phenomenon, with no equivalent in ordinary fluids.

We shall also see that these flows exert hydrodynamic torques on the director (paragraph III.4). In turn, the resulting change in director orientation acts upon the flow. This feedback loop can sometimes destabilize the flow. We shall see that, in nematics, flow destabilization is related to a dimensionless parameter, the **Ericksen number**, more relevant than the Reynolds number, of major importance in hydrodynamics. We shall then describe the hydrodynamic instabilities that develop under simple or oscillatory (linear or elliptical) shear.

We shall conclude this chapter with a description of the convective instabilities under an electric field (section III.5) or in the presence of a vertical thermal gradient (Rayleigh-Bénard convection, section III.6).

III.1 Preliminary observations illustrating some fundamental differences between a nematic and an ordinary liquid

III.1.a) Frederiks instability and rotational viscosity

Simply by looking at a commercial grade twisted nematic display one notices that the switching time of a pixel is not instantaneous; it has a finite value, typically between 0.1 and 1 s. Laboratory investigation shows that this time is roughly proportional to the square of the thickness d of the nematic layer.

To determine this switching time, let us return to the example of the Frederiks transition in the simpler configuration of figure B.II.10. The – initially planar – sample is subjected to a magnetic field perpendicular to the director. We have shown that, above a certain value B_c of the magnetic field, the molecules in the middle of the sample tilt. This instability is driven by the torque exerted by the magnetic field on the director. At equilibrium, this torque is exactly compensated for by the elastic torque exerted by the medium upon the director. Out of equilibrium, the two torques are counteracted by a dynamical torque of viscous nature, originating in the "**friction**" exerted by the medium upon the director when this latter turns around an axis perpendicular to itself. We can assume this viscous torque to be proportional to the angular velocity of the director about an axis perpendicular to itself and write:

$$\mathbf{\Gamma}^v = -\gamma_1 \left(\mathbf{n} \times \frac{\partial \mathbf{n}}{\partial t}\right) \tag{B.III.1}$$

The coefficient γ_1, having the dimension of a dynamical viscosity, is the **rotational viscosity** of the medium. This positive quantity is measured in poise in the CGS system and in poiseuille (or Pa.s) in the international system (1 poiseuille = 10 poise). It is of the order of 1 poise in ordinary nematics.

The dynamics of the system are thus controlled by a new equation, expressing "torque balance":

$$\mathbf{\Gamma}^{v} + \mathbf{\Gamma}^{M} + \mathbf{\Gamma}^{E} = 0 \qquad \text{(B.III.2)}$$

where $\mathbf{\Gamma}^{M,E}$ are, respectively, the magnetic and elastic torque acting upon the director.

In the Frederiks experiment we considered, these torques are directed along the z axis, normal to the glass plates. Let ϕ be the angle between the director and the magnetic field; writing that the sum of their z-components is zero we find, using equations B.II.60 and B.III.1:

$$K_2 \frac{\partial^2 \phi}{\partial z^2} + \frac{\chi_a}{\mu_o} B^2 \sin\phi \cos\phi - \gamma_1 \frac{\partial \phi}{\partial t} = 0 \qquad \text{(B.III.3)}$$

This equation describes the dynamics of the Frederiks transition above the instability threshold. Let us search for the solution, supposing the angle ϕ remains small. In this limit, the preceding equation becomes:

$$\gamma_1 \frac{\partial \phi}{\partial t} = K_2 \frac{\partial^2 \phi}{\partial z^2} + \frac{\chi_a}{\mu_o} B^2 \left(\phi - \frac{2}{3}\phi^3\right) \qquad \text{(B.III.4)}$$

The solution satisfying the boundary conditions $\phi = 0$ in $z = 0$ and $z = d$ can be taken in the form:

$$\phi = \phi_o(t) \sin\left(\frac{\pi z}{d}\right) \qquad \text{(B.III.5)}$$

Replacing ϕ by its expression B.III.5 in eq. B.III.4 yields:

$$\left(1 - \frac{B_c^2}{B^2}\right)\phi_o - \frac{\phi_o^3}{2} = \frac{\gamma_1 \mu_o}{\chi_a B^2} \frac{d\phi_o}{dt} \qquad \text{(B.III.6)}$$

with B_c the critical field determined in the previous chapter $\left(B_c = \frac{\pi}{d}\sqrt{\frac{\mu_o K_2}{\chi_a}}\right)$. The general solution to this equation reads:

$$\phi_o^2(t) = \frac{\phi_o^2(\infty)}{1 + \left[\frac{\phi_o^2(\infty)}{\phi_o^2(0)} - 1\right] \exp\left[-\frac{2\chi_a}{\mu_o \gamma_1}(B^2 - B_c^2)\, t\right]} \qquad \text{(B.III.7)}$$

where $\phi_o(0)$ is the initial tilt angle, never rigorously zero because of cell imperfections and thermal fluctuations, and $\phi_o(\infty) \approx 2[(B - B_c)/B_c]^{1/2}$, the value of the angle ϕ_o in the infinite time limit for $B > B_c$. One can define a typical characteristic time for instability development:

$$\tau_b = \frac{\gamma_1 \mu_o}{\chi_a(B^2-B_c^2)} \approx \frac{d^2}{2\pi^2}\frac{\gamma_1}{K_2}\frac{B_c}{B-B_c} \qquad \text{(B.III.8)}$$

This tilt time, of the order of $d^2\gamma_1/2\pi^2K_2$ for $B = 2B_c$ (but the formula is not very accurate this far from the threshold), is typically 0.05 s for $d = 10\ \mu m$, $K_2 = 10^{-6}$ dyn and $\gamma_1 = 1$ poise. **Critical slowing down** occurs close to the instability threshold, as $\tau_b \to \infty$ for $B \to B_c$, typical behavior for a normal (supercritical) bifurcation.

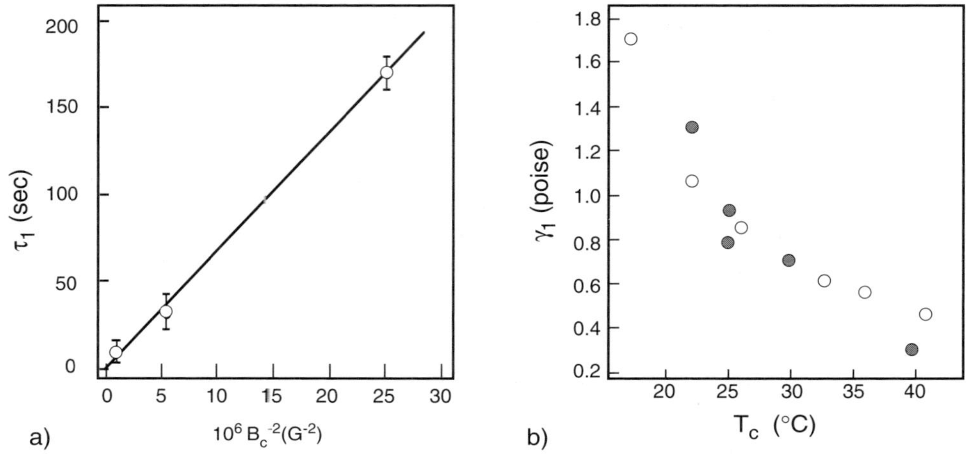

Fig. B.III.1 a) Relaxation times τ_r as a function of $1/B_c^2$ (from ref. [1]); b) viscosity γ_1 of MBBA as a function of temperature (from ref. [2]).

In principle, one only needs to measure this tilt time as a function of the applied field to determine the rotational viscosity γ_1. An alternative method consists of letting the system stabilize slightly above the instability threshold (about 10%) and then turning off the field. The system returns to equilibrium exponentially, with a characteristic time τ_r given by:

$$\tau_r = \frac{\gamma_1 \mu_o}{\chi_a B_c^2} = \frac{d^2}{\pi^2}\frac{\gamma_1}{K_2} \qquad \text{(B.III.9)}$$

Experiments confirm that, close to the threshold, the return to equilibrium is indeed monoexponential, with a characteristic time τ_r inversely proportional to B_c^2. For MBBA, this method yields $\gamma_1 \approx 0.8$ poise at room temperature [1]. This experiment also shows that γ_1 depends strongly on the temperature [2] (Fig. B.III.1b).

III.1.b) Scintillation and orientational fluctuations of the director

A thick sample of a nematic liquid crystal has an opalescent appearance, even when very well oriented. This scintillation is clearly visible when observing a planar sample under the polarizing microscope. These director fluctuations are thermally excited and decay with time due to the viscosity of the medium.

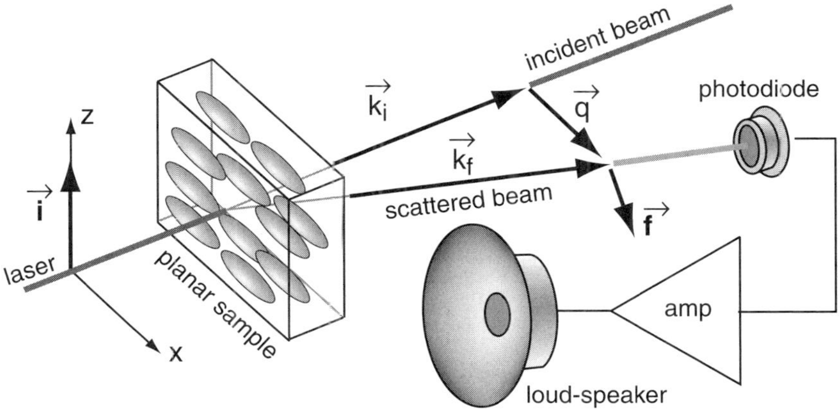

Fig. B.III.2 Experimental setup for "listening" to the orientation fluctuations of the director in a nematic phase.

What is the decay time for an orientational fluctuation? We can estimate it by writing that the sum of the viscous and elastic torques acting on the director is zero (eq. B.III.2). The underlying assumption – that the center of mass of the molecules remains immobile – is not always warranted, as we shall see later. In the previous chapter we saw that $\mathbf{\Gamma}^E = \mathbf{n} \times \mathbf{h}$. In this equation, $\mathbf{h}$ is the molecular field given by the simple expression $\mathbf{h} = K\Delta\mathbf{n}$ in the case of isotropic elasticity. Using the constitutive equation B.III.1, one has a dynamical equation for $\mathbf{n}$:

$$\gamma_1 \frac{\partial \mathbf{n}}{\partial t} = K\,\Delta\mathbf{n} \qquad \text{(B.III.10)}$$

This **diffusion equation for the orientation n** shows that a perturbation with wave vector $\mathbf{q}$ of the form $\mathbf{n} = \mathbf{n}_o\exp(i\mathbf{qr} - i\omega t)$ relaxes towards zero exponentially, with a characteristic time τ ($1/\tau = i\omega$) given by:

$$\frac{1}{\tau} = D_{orient}\,q^2 \qquad \text{(B.III.11)}$$

where:

$$D_{orient} = \frac{K}{\gamma_1} \qquad \text{(B.III.12)}$$

This quantity, with units of a diffusion constant, is **the orientational diffusivity** of the nematic. This coefficient is typically about 10^{-6} cm^2/s, much smaller than the thermal diffusivity (of the order of 10^{-3} cm^2/s) and the kinematic diffusivity (close to 1 cm^2/s).

These order of magnitude differences are important, showing that **an orientational fluctuation relaxes much more slowly than a fluctuation in temperature or vorticity**.

This diffusive mode of the director orientation does not exist in simple fluids. It can be investigated experimentally in inelastic light scattering, by performing the spectral analysis of the scattered light.

One can even "hear" the fluctuations in director orientation by using the simple setup depicted in figure B.III.2. In this experiment, devised by Piotr Pieranski, a photodiode receives the light scattered by a planar nematic sample exposed to the beam of a He-Ne laser. The output signal of the photodiode is then fed to a Hi-Fi amplifier which converts it to perfectly audible sound. In agreement with the theoretical predictions of section B.II.7 (see formula B.II.100), this experiment shows that the polarization of the scattered light makes a 90° angle with that of the incident beam and that its intensity decreases with q, i.e., with the angular distance between the photodiode and the incident beam.

We shall look at the hydrodynamic modes in greater detail in section B.III.2. Let us now briefly consider the rheological properties of the nematic phase under shear.

III.1.c) The Wahl and Fischer experiment: influence of shearing on the director orientation

Until now, we have only considered director dynamics in the absence of flow. Experiments show that a nematic flows like an ordinary fluid: tilting the recipient is enough to demonstrate it. While flowing, a nematic continually changes its appearance because of the changes in director orientation, proving the strong coupling between the director and the velocity field. The microscopic mechanisms that lead to this coupling are represented by the diagram in figure B.III.3a. In this drawing, the molecules are perpendicular to the velocity on the average. Under the action of a local shear, a given molecule undergoes collisions with molecules coming from the left and from the right. During these collisions the molecule is acted upon by forces $\mathbf{f}_{coll}$ and $-\mathbf{f}_{coll}$; their overall sum is zero, but the associate torques add up. Such a flow will then exert on the director a hydrodynamic torque proportional to the local shear rate and lead to a change in nematic orientation.

This effect is spectacularly illustrated by the Wahl and Fischer's experiment [3]. In principle (Fig. B.III.3.b), it resembles Grupp's experiment (see section B.II.1): a nematic liquid crystal layer (MBBA) is contained by capillarity between two glass disks (6 cm in diameter) parallel to the (x, y) plane and very close together (thickness $d \approx 0.2$ mm). The disks are treated with soybean lecithin to ensure homeotropic anchoring. One of the disks is fixed, while the other can turn about the vertical z axis. This system can produce a flow in which the local shear rate s = ∂v/∂y increases linearly with the distance r to the rotation axis. The orientation of the liquid crystal is detected between crossed polarizers by projecting the image of the nematic on a screen using a lens and mirror setup. Parallel lighting is used.

When the disks are at rest (s = 0), surface anchoring imposes a uniform orientation throughout the liquid crystalline sample: **n** = (0,0,1). In this case, the sample (if correctly oriented) appears uniformly black on the observation screen.

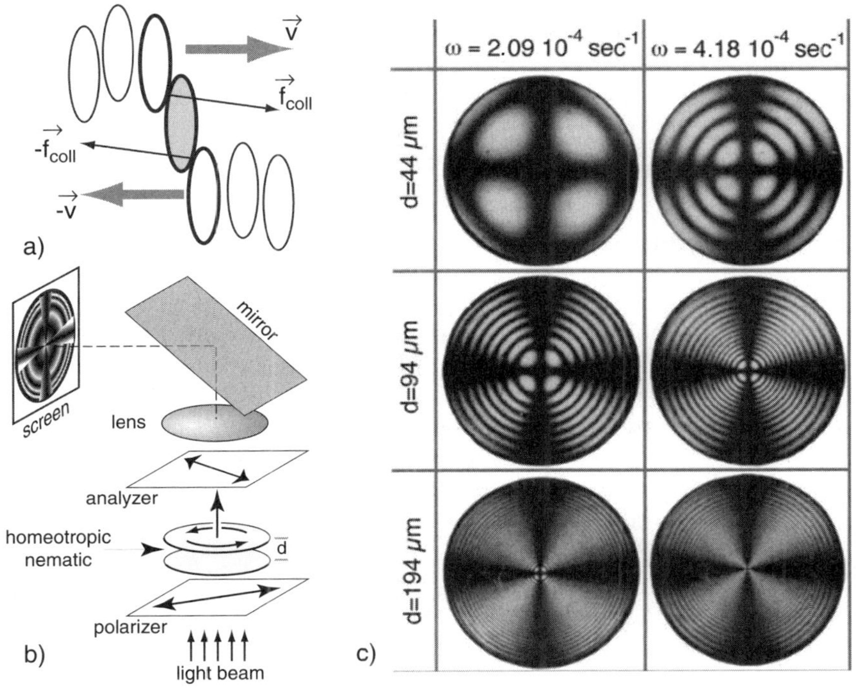

Fig. B.III.3 Wahl and Fischer experiment: a) Microscopic mechanism of the viscous torque; b) experimental setup; c) interference figure in polarized light (from ref. [3]).

Setting the mobile disk in motion (the typical angular velocity is of the order of 10^{-4} s^{-1}) has a spectacular effect: a black cross appears on the screen, whose axes are parallel to the polarizers, as well as a system of concentric rings, multicolored in white light and black in monochromatic light (see Fig. B.III.3c). Immediately after the rotation starts, the rings successively appear at the edge of the sample and move in to the center. The number of rings and their positions stabilize after a few minutes. These rings originate in the interference between the ordinary and extraordinary rays propagating across the sample. Thus, they are similar to those observed in the conoscopic figure B.II.15a, except that they appear in the image plane of the sample under parallel lighting. The obvious interpretation (by analogy with the magnetic field effects discussed in section B.II.5) is that the initially vertical molecules tilt in the orthoradial plane because of the shear. This experiment directly shows that the **flow exerts on the director a torque** that increases with the shear rate.

The analysis can be developed by studying the pattern of interference rings shown in figure B.III.3c. First of all, it shows that the flow-induced molecular tilt increases from the center of the sample towards the edge as a function of the local shear rate $s = \omega r/d$. Indeed, if we follow the reasoning of section B.I.5.e, the interference order p, i.e., the number of fringes counted from the center, depends on the difference δ in optical path between the ordinary and extraordinary rays according to the formula

$$p = \frac{\delta}{\lambda} \tag{B.III.13}$$

In the small distortion limit, $\Delta n \sin^2\theta << 1$, we have

$$\delta \approx d\Delta n \int_{-1/2}^{+1/2} \sin^2\theta(\tilde{y})\, d\tilde{y} \qquad \text{where} \qquad \tilde{y} = \frac{y}{d} \tag{B.III.14}$$

On the other hand, the images in figure B.III.3c show that, for d = 44 μm, the interference order at the edge of the sample increases (roughly) as the angular velocity squared; by doubling ω, one goes from p = 1 to p = 4. Thus, eq. B.III.14 suggests that, at low shear rates, the angle $\theta(\tilde{y})$ is proportional to the angular velocity. Assuming that the viscous torque acting on the director is proportional to the shear rate and in the small distortion limit, the equilibrium equation for the elastic and viscous torques reads:

$$K_3 \frac{d^2\theta}{dy^2} + \alpha \frac{\omega r}{d} = 0 \tag{B.III.15}$$

with α a phenomenological coefficient with viscosity units. In terms of the dimensionless variable $\tilde{y}$, this equation becomes

$$\frac{d^2\theta}{d\tilde{y}^2} + \frac{\alpha d \omega r}{K_3} = 0 \tag{B.III.16}$$

The obvious solution, satisfying the boundary conditions imposed by the homeotropic anchoring, is (Fig. B.III.4a):

$$\theta(\tilde{y}) = \theta_o \left(\frac{1}{4} - \tilde{y}^2\right) \qquad \text{with} \qquad \theta_o = \frac{\alpha d \omega r}{2K_3} \tag{B.III.17a}$$

which yields, using equations B.III.13 and B.III.14:

$$p \approx \frac{d\Delta n}{\lambda}\frac{\theta_o^2}{6} = \frac{\Delta n \alpha^2}{24\lambda K_3^2} d^3 \omega^2 r^2 \tag{B.III.17b}$$

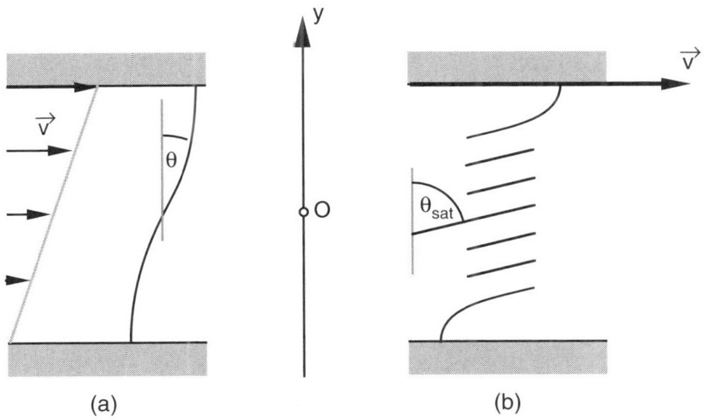

Fig. B.III.4 Distortion induced by shear flow. a) For low shear rates, the distortion angle θ remains small; b) the Wahl and Fischer experiment shows that, at high shear rates, the distortion "saturates," that is, the director tends to orient in such a way as to cancel the viscous torque.

These results confirm the proportionality between the tilt angle θ and the angular velocity and the ω^2 behavior of interference order p. The formula also predicts that, for the same ω and **r**, the interference order varies like d^3, in good agreement with the results of figure B.III.3c.

The solution B.III.17a has a deeper meaning, more easily perceived if we write it in the form:

$$\theta(\tilde{z}) = \theta_o\left(\frac{1}{4} - \tilde{y}^2\right) \quad \text{with} \quad \theta_o = \frac{1}{2}\frac{dv}{D} \qquad \text{(B.III.18)}$$

where v = ωr is the velocity and $D = K_3/\alpha$ a coefficient with diffusivity units. We now recognize it as being a similarity law, expressing the distortion as a function of the dimensionless quantity dv/D, similar to the Reynolds number, but for the fact that the vorticity diffusion constant is replaced by the diffusion constant of the orientation. We shall return to this similarity law in the section dedicated to the analysis of flow stability.

Finally, from the image taken at d = 194 μm and $\omega = 4.18\times10^{-4}\ s^{-1}$, we can infer that, at high shear rates, the interference order p saturates. Further on, in section B.III.4, we shall prove that this is an effect of the viscosity α decreasing with the tilt angle. More specifically, in the case of MBBA, there is an angle $\theta_{sat} < 90°$ such that $\alpha(\theta) \to$ Cst when $\theta \to \theta_{sat}$. It then ensues that in MBBA, **at high shear rates, the director assumes a specific orientation for which the viscous torque is zero**. At this limit, the director field has the configuration depicted in figure B.III.4b.

In section B.III.4 we shall also see that, in other circumstances, the viscous torque can, on the contrary, entail hydrodynamic instabilities radically different from those observed in isotropic fluids at high Reynolds number.

III.1.d) "Backflow" effect

The experiment we have just described illustrates a change in director orientation under the action of a mechanically induced flow. The reverse effect, called "backflow," also exists, corresponding to the appearance of a flow due to a change in director orientation [4]. For instance, this effect can show up in the Frederiks experiment with planar anchoring conditions when a magnetic field is applied in the direction perpendicular to the director [5].

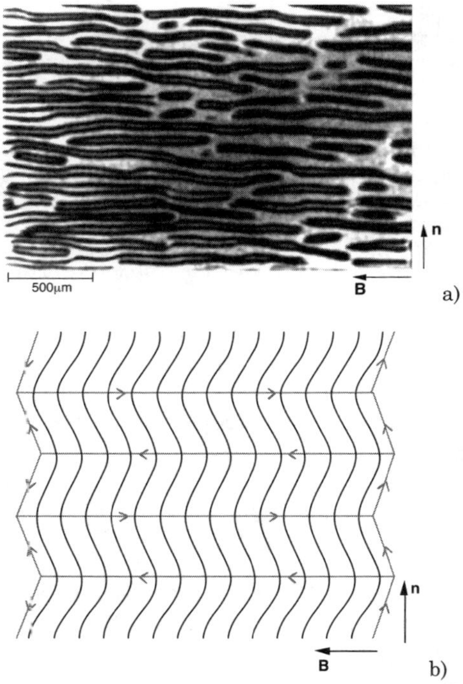

Fig. B.III.5 a) Periodic pattern observed in a planar MBBA sample immediately after applying a magnetic field of 10 $B_c^{(1)}$. Sample thickness is 90 μm (from ref. [5a]); b) director field deformation in the median plane of the sample and "backflow" in the transient regime.

Experiment shows that, for fields not much larger than the Frederiks critical field $B_c^{(1)}$, the director tilt is homogeneous over the entire sample; on the other hand, as shown in figure B.III.5, a periodic pattern spontaneously develops above a certain second threshold, $B_c^{(2)} > B_c^{(1)}$, of the order of 2–3 $B_c^{(1)}$. This pattern has a finite wave vector q and is the mark of a periodical director tilt. It is transitory and always ends up by completely disappearing to the advantage of the homogeneous solution (of wave vector q = 0) when the field stays on. This solution therefore has a higher energy than the homogeneous one, but also a higher growth rate for $B > B_c^{(2)}$, allowing it to develop transiently. It can be shown that this increase in growth rate originates in the "backflow" effect. This secondary flow, characterized (in the median plane of the sample) by the field lines represented as light gray lines in figure B.IIII.5b), favors director rotation, thus leading to a decrease in dissipation and an effective rotational viscosity γ_1^* lower than γ_1. The expression of this rotational viscosity will be given later, once the fundamentals of nematodynamics are presented (see eq. B.III.78b).

"Backflow" is important because it is employed in the design of new nematic liquid crystal displays where the pixels can be turned on or off by applying electric pulses of different shape [6]. In this new type of display, the electric field is only applied during the switch pulses, drastically reducing the use of electric energy.

Finally, the "backflow" effect is spectacularly illustrated in smectic C films, to be discussed in chapter C.VIII of the second volume.

III.1.e) Viscosities under shear: the experiments of Miesowicz and Gähwiller

To reduce the action of the viscous torque and the instabilities it leads to, and to keep nematic orientation uniform under shear, Miesowicz [7a, b] proposed as early as 1935 placing the sample in a strong magnetic field which orients the molecules. The use of suitable surfaces, treated to ensure planar or homeotropic anchoring, can equally favor molecular orientation along a given direction. Miesowicz also had the idea of determining the apparent viscosity of a nematic for different director orientations with respect to the velocity, by measuring the decay time of the oscillations of a pendulum composed of a rod and a plate immersed in the nematic (Fig. B.III.6). In this setup, dissipation is due to the viscous stress component in the direction of movement of the plate. In the Cartesian coordinate system (x, y, z) of axis y perpendicular to the plate surface and of axis x parallel to the flow direction, this is the σ^v_{xy} component of the viscous stress tensor $\underline{\sigma}^v$. Let

$$\mathbf{v} = (sy, 0, 0) \tag{B.III.19}$$

be the velocity close to the plate, with s the local shear rate. In classical hydrodynamics, the dynamical viscosity η relates the tangent stress σ^v_{xy} to the local shear rate:

$$\sigma^v_{xy} = \eta s \tag{B.III.20}$$

It is easily shown that the decay time for the pendulum oscillations is inversely proportional to the viscosity η.

Miesowicz measured this time for the three orientations of the director with respect to the velocity and the velocity gradient represented in figure B.III.6. He found different values for the viscosity and concluded that one must take into account (at least) three distinct viscosities for a nematic. These viscosity coefficients, denoted by η_a, η_b, and η_c, respectively, referring to the three diagrams in figure B.III.6, are such that:

$$\begin{aligned} &- \text{for } \eta_a: \quad && \mathbf{v} \perp \mathbf{n} \quad && \text{and} \quad && \mathbf{grad}\, v \perp \mathbf{n} \\ &- \text{for } \eta_b: \quad && \mathbf{v} \,//\, \mathbf{n} \quad && \text{and} \quad && \mathbf{grad}\, v \perp \mathbf{n} \\ &- \text{for } \eta_c: \quad && \mathbf{v} \perp \mathbf{n} \quad && \text{and} \quad && \mathbf{grad}\, v \,//\, \mathbf{n} \end{aligned} \tag{B.III.21}$$

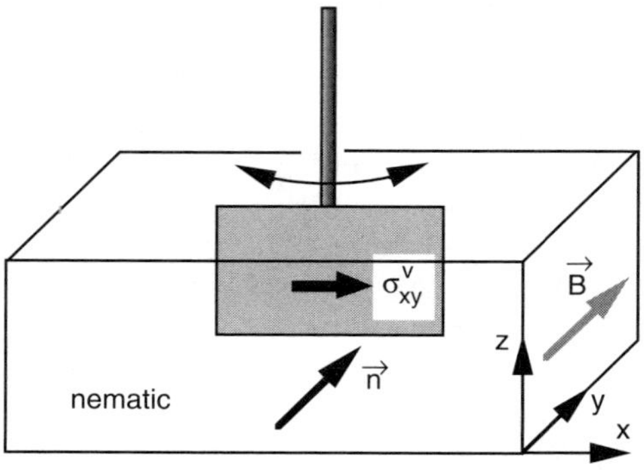

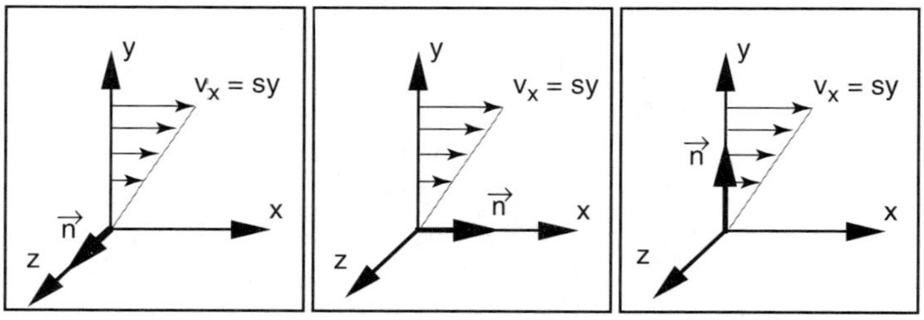

Fig. B.III.6 Principle of the Miesowicz experiment [7a, b]: a plate oscillates in the nematic oriented by a strong magnetic field. Dissipation is proportional to the σ^v_{xy} component of the stress tensor. The effective viscosity depends on the orientation a, b, or c of the director with respect to the flow.

The Miesowicz viscosities were measured several times by different authors. A remarkable study is that of Gähwiller [7c], who measured, among other things, the η_a, η_b and η_c viscosities in Poiseuille flows in a flat capillary tube (Fig. B.III.7a). Here, the nematic is also oriented using a strong magnetic field (of the order of 1 tesla).

As shown in figure B.III.7b, the viscosity anisotropy is significant. Most often, the viscosity is the lowest when the director is parallel to the velocity (η_b) and the highest when the director is parallel to the velocity gradient (η_c). Also note that the viscosities increase at lower temperature and are not equal at T_c, once again proving that the nematic-isotropic liquid transition is first order.

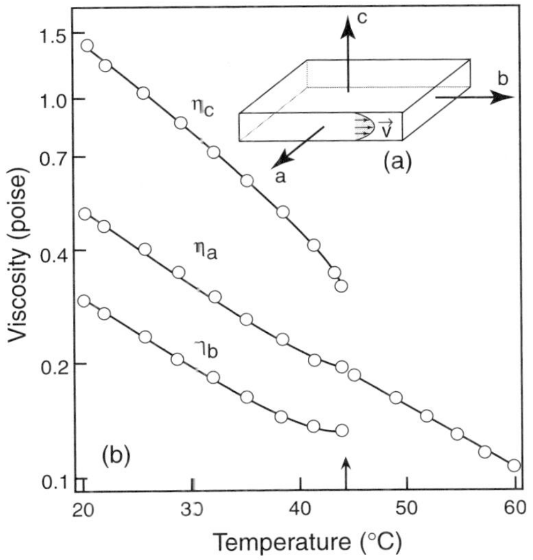

Fig. B.III.7 The Gähwiller experiment consists of determining the rate of flow of the nematic, oriented by a strong magnetic field, in a flat capillary. a) At fixed ΔP, the flow rate depends on the orientation of the nematic and is inversely proportional to the viscosity; b) viscosities of MBBA as a function of the temperature (from ref. [7c]).

We can thus conclude that a uniaxial nematic phase, first oriented by a strong magnetic field, behaves like a strongly anisotropic viscous fluid. Furthermore, this phase exhibits a rotational viscosity which governs the dynamics of the director in the absence of hydrodynamic flow. Finally, we note that the strong coupling between the director field and the velocity can lead to flow instability. To describe these phenomena, we need a theory of nematodynamics. Two different theoretical approaches were proposed in the literature: the macroscopic one, introduced by Ericksen [8] and Leslie [9] that we shall use in the following, and the one introduced by the Harvard group [10], at a more microscopic level, based on a study of the correlation functions, which we shall not discuss. Let us emphasize the role of de Gennes [11], who showed the equivalence of these two points of view.

III.2 Equations of the linear nematodynamics

III.2.a) Choice of the hydrodynamic variables

In an ordinary liquid, the hydrodynamic variables are the density ρ, the velocity $\mathbf{v}(\mathbf{r},t)$ or the momentum $\rho\mathbf{v}$, and the internal energy per unit volume e. A conservation law and a particular hydrodynamic mode are associated with each of these quantities. These modes can be propagating (e.g., the sound, associated to the density fluctuations), or damped, like the transverse shear waves governed by the kinematic viscosity $\nu = \eta/\rho$. These modes exhibit **macroscopic** relaxation times τ proportional to a certain power of the wavelength λ ($\lambda = 2\pi/q$); thus, in the $q \rightarrow 0$ limit, τ tends to infinity. These **long-lived modes** are termed **hydrodynamic modes**.

In a nematic, one must consider an additional variable, related to the orientational order of the molecules. In the previous chapter we showed that this order is characterized by three independent parameters: the amplitude of the order parameter Q and any two of the three components of the director **n**. The amplitude of the order parameter is not a hydrodynamic variable, as any perturbation of Q, be it of infinite wavelength, relaxes towards the equilibrium value over microscopic times. The director **n**, on the other hand, *is* a hydrodynamic variable, because the relaxation time of a distortion in orientation depends on the wavelength (τ varies like λ^2, from eq. B.III.11). This dependence stems from the fact that elastic torques are proportional to Kq^2; on the other hand, τ must diverge for $\lambda \rightarrow \infty$, since a uniform rotation of the director leaves unchanged the energy of the system, which has no reason to return to its initial configuration ($\tau \rightarrow \infty$).

In conclusion, a nematic has an additional hydrodynamic variable, the director **n**. This variable is not associated with any conservation law, but rather with the symmetry breaking at the isotropic liquid-nematic phase transition.

III.2.b) Surface torque field

Now that the hydrodynamic variables are defined, one must investigate the nature of the forces exerted by one part (I) of the medium on the adjoining part (II) (Fig. B.III.8).

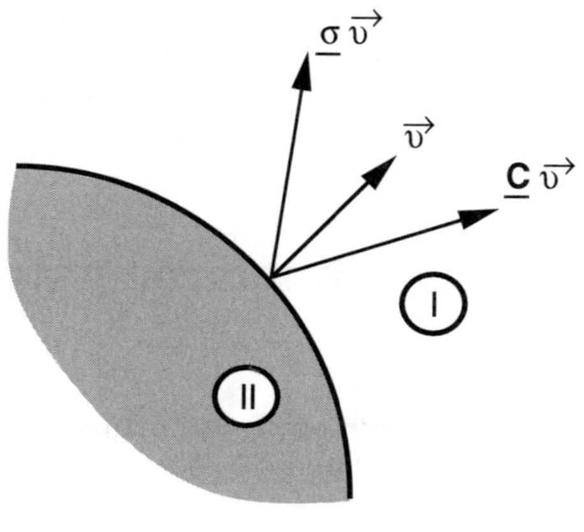

Fig. B.III.8 Definition of the surface torque field exerted by part (I) on part (II); $\underline{\sigma}\mathbf{v}$ is the overall resultant force and $\underline{C}\mathbf{v}$ is the surface torque.

We shall make the usual assumption in the mechanics of continuous media, namely that the action of (I) on (II) can be described by a **surface field**, which can be obtained from a second-order tensor $\underline{\sigma}$ termed the **total stress tensor**. Let $\boldsymbol{\upsilon}$ be the unit vector normal to the separation surface and directed from (II) towards (I). The force per unit surface exerted by (I) on (II) is then $\underline{\sigma}\boldsymbol{\upsilon}$. On the other hand, we know that a nematic also transmits torques (see chapter B.II). Thus, (I) exerts on (II) not only a force, but also a surface torque $\underline{C}\boldsymbol{\upsilon}$, where $\underline{C}$ is a second-order tensor we have already determined (eq. B.II.32).

This description is valid from a macroscopic point of view, because **molecular forces are short range** (a few molecular distances).

III.2.c) Conservation equations for the mass and the momentum

Suppose the nematic is in motion and let V_m denote a moving volume containing a group of molecules whose movement we follow (Fig. B.III.9). Let S_m be the moving surface delimiting the volume.

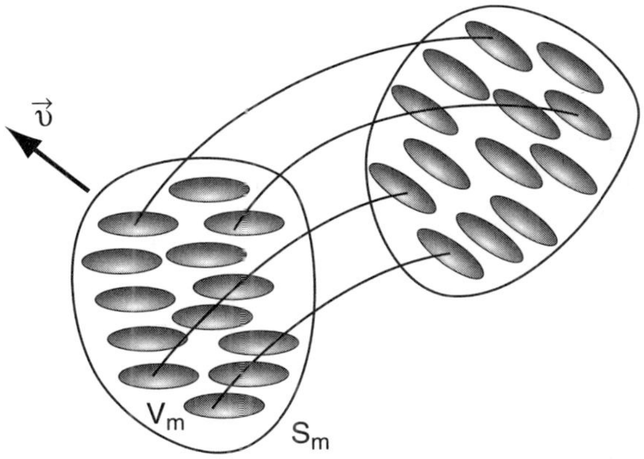

Fig. B.III.9 Moving volume V_m delimited by a moving surface S_m.

Applying Newton's law to these molecules yields:

$$\frac{d}{dt}\int_{V_m} \rho \mathbf{v}\, dV = \int_{S_m} \underline{\sigma}\, \boldsymbol{\upsilon}\, dS + \int_{V_m} \mathbf{F}\, dV \qquad \text{(B.III.22)}$$

In this equation, **F** is an external bulk force (gravity, for instance).

After transforming the surface integral in a volume integral and after "derivation under the integral" (see Appendix 1), one has:

$$\int_{V_m} \rho \frac{D\mathbf{v}}{Dt} dV = \int_{V_m} (\mathrm{div}\, \underline{\sigma} + \mathbf{F})\, dV \qquad \text{(B.III.23)}$$

In this expression, D/Dt represents the substantive derivative, i.e., a derivative taken over a group of molecules whose movement we follow. Let us recall that

$$\frac{D\mathbf{v}}{Dt} = \frac{\partial \mathbf{v}}{\partial t} + (\mathbf{v}.\nabla)\mathbf{v} \qquad \text{(B.III.24)}$$

Since the equality B.III.23 holds true irrespective of the integration volume V_m, this entails

$$\rho \frac{D\mathbf{v}}{Dt} = \mathrm{div}\, \underline{\sigma} + \mathbf{F} \qquad \text{(B.III.25)}$$

In this expression, $\mathrm{div}\,\underline{\sigma}$ is a vector with components $\sigma_{ij,j}$ (be careful: for Leslie, the components of this vector are $\sigma_{ij,i}$; de Gennes and Prost also employ this unusual convention in their book). Let us also recall that the mass conservation equation, cast in its conservative form, is written as for an ordinary fluid:

$$\frac{\partial \rho}{\partial t} + \mathrm{div} \rho \mathbf{v} = 0 \qquad \text{(B.III.26)}$$

If the nematic can be considered as **incompressible**, a valid assumption as long as the hydrodynamic velocity **v** remains small with respect to the speed of sound, this equation can be simplified to:

$$\mathrm{div}\mathbf{v} = 0 \qquad \text{(B.III.27)}$$

Throughout this chapter we shall assume that the nematic is incompressible. Let us now explicitly write down the stress tensor.

III.2.d) Constructing the stress tensor

In its most general form, this tensor can be written as

$$\underline{\sigma} = - P\, \underline{I} + \underline{\sigma}^E + \underline{\sigma}^v \qquad \text{(B.III.28)}$$

P is the **hydrostatic pressure**. It is the thermodynamic variable conjugated to the density ρ. In the case ρ = Cst that we consider, P corresponds to the Lagrange multiplier associated with the incompressibility condition $\mathrm{div}\mathbf{v} = 0$ (it is then no longer a state variable, but a parameter that must be calculated).

$\underline{\sigma}^E$ is the **elastic stress tensor** associated to the distortions of the director field.

Finally, $\underline{\sigma}^v$ is the **viscous stress tensor** leading to energy dissipation and irreversible entropy production.

One sometimes denotes $\underline{\sigma}^e = - P\,\underline{I} + \underline{\sigma}^E$. This is the **Ericksen tensor**.

The $\underline{\sigma}^E$ tensor can be expressed as a function of the Frank free energy f given in chapter B.II. To calculate its components, let us look for the variation in free elastic energy df when a molecule placed at **r** is displaced by an infinitesimal distance **u(r)**, the director **n remaining locally unchanged**, such that

$$\mathbf{n}'(\mathbf{r}') = \mathbf{n}'(\mathbf{r} + \mathbf{u}) = \mathbf{n}(\mathbf{r}) \qquad \text{(B.III.29)}$$

During the displacement **u**, the free energy (in an isothermal process) varies by:

$$df = \sigma^e_{ij} \frac{\partial u_i}{\partial x_j} = \sigma^E_{ij} \frac{\partial u_i}{\partial x_j} \qquad \text{(incompressible nematic)} \qquad \text{(B.III.30)}$$

This expression translates the first law of thermodynamics. The free energy f being a function of the n_i and their derivatives $\partial n_i/\partial x_j$, we can write, taking into account eq. B.III.29:

$$df = \frac{\partial f}{\partial n_{i,j}} dn_{i,j} \qquad \text{(B.III.31)}$$

as $dn_i = 0$.
By definition, we have

$$dn_{i,j} = \frac{\partial n_i'}{\partial x_j'} - \frac{\partial n_i}{\partial x_j} = \frac{\partial n_i}{\partial x_k}\frac{\partial x_k}{\partial x_j'} - \frac{\partial n_i}{\partial x_j} = \frac{\partial n_i}{\partial x_k}\left(\delta_{kj} - \frac{\partial u_k}{\partial x_j'}\right) - \frac{\partial n_i}{\partial x_j} = -\frac{\partial n_i}{\partial x_k}\frac{\partial u_k}{\partial x_j} \qquad \text{(B.III.32)}$$

where the last equality only holds in the $\mathbf{u} \to 0$ limit. Comparing expressions B.III.30 and B.III.31, taking into account eq. B.III.32, leads to:

$$\sigma^E_{ij}\frac{\partial u_i}{\partial x_j} = -\frac{\partial f}{\partial n_{i,j}}\frac{\partial n_i}{\partial x_k}\frac{\partial u_k}{\partial x_j} = -\frac{\partial f}{\partial n_{k,j}}\frac{\partial n_k}{\partial x_i}\frac{\partial u_i}{\partial x_j} \qquad \text{(B.III.33)}$$

whence, **u** being unspecified:

$$\sigma^E_{ij} = -\frac{\partial f}{\partial n_{k,j}}\frac{\partial n_k}{\partial x_i} \qquad \text{(B.III.34)}$$

Calculating this stress tensor for the Grupp experiment (Fig. B.II.1) is a particularly instructive exercise.

Indeed, the nematic exerts on the mobile plate not only a torque, but also a **normal force**. Physically, this is obvious since the total elastic energy F_{total} stored during a rotation of the upper plate by an angle Φ depends both on the thickness d of the nematic layer and on the sample surface area S. Let us determine this energy exactly. From eq. B.II.41, it is:

$$F_{total} = S\frac{1}{2}K_2 \int \left(\frac{\partial \phi}{\partial z}\right)^2 dz = \frac{1}{2}K_2 S \frac{\Phi^2}{d^2} d \qquad (B.III.35)$$

or, denoting by V = Sd the constant nematic volume (note that S is not constant and changes with d):

$$F_{total} = \frac{1}{2}K_2 V \frac{\Phi^2}{d^2} \qquad (B.III.36)$$

It is then clear that the mobile disk exerts on the nematic layer a normal force whose value is:

$$f_{normal} = \frac{\partial F_{total}}{\partial d} = -K_2 V \frac{\Phi^2}{d^3} = -K_2 S \frac{\Phi^2}{d^2} \qquad (B.III.37)$$

or, per unit surface:

$$\sigma = -K_2 \frac{\Phi^2}{d^2} \qquad (B.III.38)$$

This stress is negative, meaning that pressure must be exerted on the nematic layer if it is to maintain a constant thickness during rotation. Undoubtedly, the energy of the nematic decreases when the sample thickness increases.

The stress can also be determined by using the general formula B.III.34. For the Grupp experiment, one has:

$$f = \frac{1}{2}K_2 (n_{x,z}^2 + n_{y,z}^2) \qquad (B.III.39)$$

yielding:

$$\sigma_{zz}^E = \sigma = -K_2 (n_{x,z}^2 + n_{y,z}^2) = -K_2 \left(\frac{\partial \phi}{\partial z}\right)^2 = -K_2 \frac{\Phi^2}{d^2} \qquad (B.III.40)$$

Thus, we obtain the expected result.

An important point is that, in this experiment, the Ericksen stress σ_{zz}^E is a constant throughout the sample, the deformation being homogeneous. The resulting bulk force $div(\underline{\sigma}^E)$ associated with this stress is thus strictly zero.

The situation can change if the deformation of the director field is no longer homogeneous. This occurs, for instance, around a disclination line (see chapter B.IV) where the distortions $\partial n_i/\partial r$ vary with the distance from the center as 1/r. In this case, the bulk force $div(\underline{\sigma}^E)$ is no longer zero and results in an outward matter displacement in the least curved regions. At equilibrium, a radial pressure gradient sets in, coupled with a minute density gradient which exactly balances $div(\underline{\sigma}^E)$.

Let us return to the viscous stress tensor $\underline{\sigma}^v$.

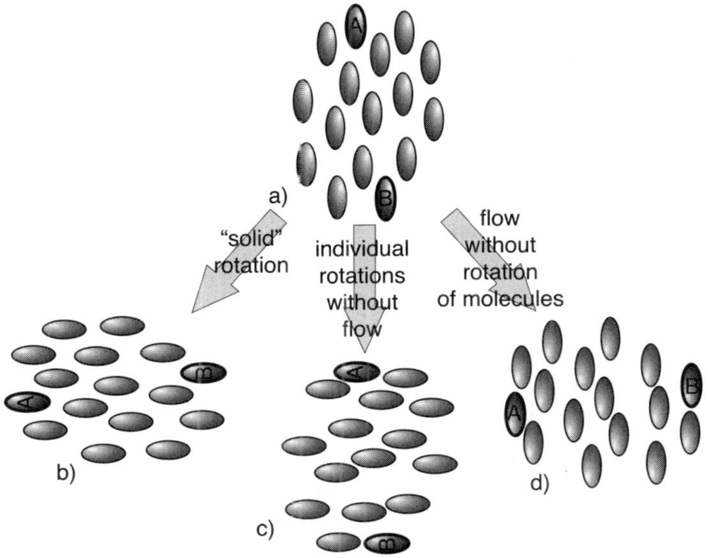

Fig. B.III.10 When a rotational flow of "solid rotation" type is imposed on a nematic, configuration (a) changes to configuration (b). The relative position of the molecules remaining unchanged after the rotation, there is no viscous dissipation However, if the molecules are turned about their (fixed) centers of mass, or if the fluid is turned at fixed molecular orientation, one goes from configuration (a) to configurations (c) or (d), respectively. In both cases, the relative position of the molecules changes and viscous dissipation occurs during relaxation towards thermodynamic equilibrium.

In an isotropic fluid, $\underline{\sigma}^v$ is a function of the local velocity gradients $\partial v_i/\partial x_j$. In a nematic, the situation is more complex, as energy dissipation also occurs when the molecules turn about their centers of mass without changing their positions ($\mathbf{v} = \mathbf{0}$), as proved by the Frederiks experiment. The $\underline{\sigma}^v$ tensor thus depends on the velocity gradients and on the local rotation rate of the molecules. This dependence can be justified by an explicit calculation of the dissipated energy, which we discuss in section III.2.f, for clarity's sake. At this point, one must keep in mind that dissipation, and hence $\underline{\sigma}^v$, must be zero for a "solid rotation" of the system (Fig. B.III.10). Consequently, the $\underline{\sigma}^v$ tensor cannot be a generic combination of the $\partial v_i/\partial x_j$ and $\partial n_i/\partial x_j$: it has to be a function of the A_{ij} components of the symmetric part of the shear rate tensor $\partial v_i/\partial x_j$ (as for an ordinary fluid) and of the N_i components of the **relative angular velocity N characterizing director rotation with respect to the fluid**:

$$A_{ij} = \frac{1}{2}\left(\frac{\partial v_i}{\partial x_j} + \frac{\partial v_j}{\partial x_i}\right) \tag{B.III.41}$$

$$\mathbf{N} = \frac{D\mathbf{n}}{Dt} - \boldsymbol{\Omega} \times \mathbf{n} \tag{B.III.42}$$

In this expression, the $\boldsymbol{\Omega}$ vector is the **local rotation rate** defined by the relation

$$2\Omega_{ij} = \frac{\partial v_i}{\partial x_j} - \frac{\partial v_j}{\partial x_i} = -2e_{ijk}\,\Omega_k \qquad \text{(B.III.43)}$$

We recall that $\mathbf{\Omega} = \boldsymbol{\omega}/2$, where $\boldsymbol{\omega} = \mathbf{curl\,v}$ is the vorticity. It is easily shown that A_{ij} and N_i are zero for solid rotations. To lowest order (linear approximation), the σ^v_{ij} components of the viscous stress tensor are then linear functions of the A_{ij} and N_i. On the other hand, σ^v_{ij} depends on the director orientation $\mathbf{n}$, so it must be invariant under a $\mathbf{n}$ to $-\mathbf{n}$ transformation. It can be shown that the most general form of this tensor still compatible with the symmetries of a uniaxial (and incompressible) nematic, is the following (see Appendix 4):

$$\sigma^v_{ij} = \alpha_1 (n_k A_{kl} n_l) n_i n_j + \alpha_2 n_j N_i + \alpha_3 n_i N_j + \alpha_4 A_{ij} + \alpha_5 n_j n_k A_{ki} + \alpha_6 n_i n_k A_{kj} \qquad \text{(B.III.44)}$$

The α_i constants, known as Leslie coefficients, are dynamic viscosities measured in poiseuille (Pa.s) in the international system and in poise in the CGS system. In this expression (only valid for an incompressible nematic), the α_4 term corresponds to the usual viscosity of an isotropic fluid. The physical signification of the other coefficients will be discussed in section III.3. Let us first write the equation of torque balance.

III.2.e) Equation of torque balance: application of the angular momentum theorem

By applying the angular momentum theorem to a mobile volume V_m of surface S_m inside the nematic one has:

$$\frac{d\mathbf{M}_o}{dt} = \int_{S_m} \mathbf{r} \times \underline{\sigma}\boldsymbol{\upsilon}\, dS + \int_{S_m} \underline{C}\boldsymbol{\upsilon}\, dS + \int_{V_m} \mathbf{r} \times \mathbf{F}\, dV + \int_{V_m} \boldsymbol{\Gamma}^M\, dV \qquad \text{(B.III.45)}$$

$\mathbf{M}_o$ is the angular momentum in O (origin of the laboratory frame of reference), $\mathbf{F}$ the density of external bulk forces and $\boldsymbol{\Gamma}^M$ the magnetic (or electric) bulk torque if a field is applied to the system. $\boldsymbol{\upsilon}$ is the unit normal pointing out of the volume V_m and $\mathbf{r} \times \underline{\sigma}\boldsymbol{\upsilon} + \underline{C}\boldsymbol{\upsilon}$ is the torque corresponding to the surface forces (internal stress) acting on the surface S_m. The $\underline{C}$ tensor is explicitly given by the equation B.II.32. As the moment of inertia of a molecule about its center of mass is very small we can write, to an excellent approximation

$$\mathbf{M}_o = \int_{V_m} \mathbf{r} \times \rho\mathbf{v}\, dV \qquad \text{(B.III.46)}$$

whence, after "derivation under the integral" (Appendix 1):

$$\frac{d\mathbf{M}_o}{dt} = \int_{V_m} \left(\mathbf{r} \times \frac{D\mathbf{v}}{Dt}\right) \rho \, dV \qquad \text{(B.III.47)}$$

The only thing left to do is transform the surface integral into a volume integral in eq. B.III.45. The tensor $\underline{\sigma}$ being asymmetric, one obtains

$$\int_{S_m} (\mathbf{r} \times \underline{\sigma}\boldsymbol{\upsilon} + \underline{C}\boldsymbol{\upsilon}) \, dS = \int_{V_m} (\mathbf{r} \times \mathrm{div}\underline{\sigma} + \boldsymbol{\Gamma}) \, dV \qquad \text{(B.III.48)}$$

where $\boldsymbol{\Gamma}$ is a bulk torque with components

$$\Gamma_i = -\, e_{ijk}\, \sigma_{jk} + C_{ij,j} \qquad \text{(B.III.49)}$$

(In Leslie's notation, also adopted by de Gennes, the torque associated with the asymmetric part of the σ_{ij} tensor reads $e_{ijk}\sigma_{jk}$, without the – sign.) But we already know that $\mathrm{div}\underline{\sigma} + \mathbf{F} = \rho\, D\mathbf{v}/Dt$ which, after introducing in B.III.48 and then in B.III.45 taking into account B.III.47, yields:

$$\int_{V_m} (\boldsymbol{\Gamma} + \boldsymbol{\Gamma}^M) \, dV = 0 \qquad \text{(B.III.50)}$$

This relation holds true for any volume V_m, so finally:

$$\boldsymbol{\Gamma} + \boldsymbol{\Gamma}^M = 0 \qquad \text{(B.III.51)}$$

$\boldsymbol{\Gamma}$ represents the bulk torque exerted by the medium on the director $\mathbf{n}$. It contains two terms, with the pressure not playing any role as $-\,\delta_{ij}P$ is a symmetric tensor:

$$\boldsymbol{\Gamma} = \boldsymbol{\Gamma}^v + \boldsymbol{\Gamma}^E \qquad \text{(B.III.52)}$$

In this equation, $\boldsymbol{\Gamma}^v$ is the **viscous torque**. From eq. B.III.44, its expression is:

$$\boldsymbol{\Gamma}^v = -\,\mathbf{n} \times (\gamma_1 \mathbf{N} + \gamma_2 \underline{A}\mathbf{n}) \qquad \text{(B.III.53)}$$

where $\gamma_1 = \alpha_3 - \alpha_2$ and $\gamma_2 = \alpha_6 - \alpha_5$. In this expression, $\underline{A}\mathbf{n}$ is the vector with components $(\underline{A}\mathbf{n})_i = A_{ij}n_j$.

Let us point out that we recovered the simplified expression B.III.1 of the viscous torque employed in the first section for the case of no hydrodynamic flow ($\mathbf{v} = \mathbf{0}$).

To assign a precise physical signification to the two viscosity constants γ_1 and γ_2, consider in a generic point of the flow a mobile frame of reference comoving with the molecules at that point. The velocity field close to this point can be decomposed into a rotational part (its field lines being concentric circles, Fig. B.III.11a) and its irrotational part (the field lines of which are hyperbolae, Fig. B.III.11b).

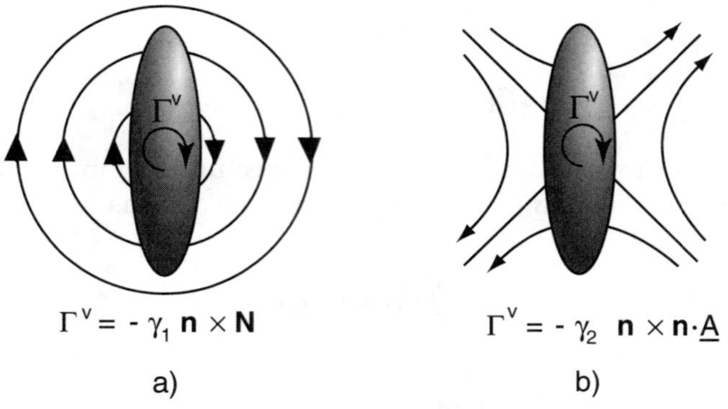

Fig. B.III.11 Viscous torque in a (a) rotational and (b) irrotational flow.

In the rotational flow represented in figure B.III.11a, the viscous torque is proportional to γ_1 and vanishes for $\mathbf{N} = \mathbf{0}$, i.e., when the director rotates at the same angular velocity as the fluid.

In the irrotational flow shown in figure B.III.11b, the torque is proportional to γ_2 and vanishes when the director is aligned along the eigenaxes of the A_{ij} tensor (in the drawing, the asymptotes to the four hyperbola branches).

In fact, the four viscosity coefficients α_2, α_3, α_5, and α_6 are not independent, but connected by the relation:

$$\alpha_2 + \alpha_3 = \alpha_6 - \alpha_5 \tag{B.III.54}$$

This equality was proven by Parodi [12] using the Onsager reciprocity relations. Thus, an incompressible nematic only has five independent viscosity coefficients.

$\mathbf{\Gamma}^E$ is the **elastic torque** given by equations B.III.49, B.III.34, and B.II.32 as:

$$\Gamma_i^E = e_{ijk}\frac{\partial f}{\partial n_{l,k}}\frac{\partial n_l}{\partial x_j} + \frac{\partial}{\partial x_l}\left(\frac{\partial f}{\partial n_{k,l}}e_{ijk}n_j\right) \tag{B.III.55}$$

At first glance, this expression bears little resemblance to the one found in chapter B.II:

$$\Gamma_i^E = e_{ijk}n_j h_k = e_{ijk}n_j\left(-\frac{\partial f}{\partial n_k} + \frac{\partial}{\partial x_l}\frac{\partial f}{\partial n_{k,l}}\right) \tag{B.III.56}$$

One can nevertheless show that these two formulae are in fact identical (see Appendix 2) on grounds of rotational invariance of the Frank free energy, namely $f(n_i, \partial n_i/\partial x_j)$ remains unchanged if both the director and the molecular centers of mass are rotated by the same angle ω ($\mathbf{u}(\mathbf{r}) = \boldsymbol{\omega}\times\mathbf{r}$ and $\boldsymbol{\delta}\mathbf{n}(\mathbf{r}) = \boldsymbol{\omega}\times\mathbf{n}$).

To sum up this section, let us explicitly write the torque balance equation B.III.51:

$$\mathbf{n} \times (\gamma_1 \mathbf{N} + \gamma_2 \underline{A}\,\mathbf{n}) = \mathbf{n} \times \mathbf{h} + \mathbf{n} \times \mathbf{h}^M \qquad \text{(B.III.57)}$$

$\mathbf{h}^M$ being the magnetic contribution to the molecular field: $\mathbf{h}^M = (\chi_a/\mu_0)(\mathbf{n}.\mathbf{B})\mathbf{B}$.

Before describing some specific flows, we shall first give the general expression of energy dissipation and obtain some thermodynamic inequalities concerning the viscosities and the thermal diffusivities.

III.2.f) Irreversible entropy production

Let us apply the first law of thermodynamics to a moving volume V_m (of surface S_m) inside the nematic. Let $e(\mathbf{r},t)$ be the internal energy per unit volume. According to the first law of thermodynamics:

$$\frac{d}{dt}\int_{V_m}\left(\frac{1}{2}\rho v^2 + e\right) dV = \int_{V_m} (\underline{\sigma}\boldsymbol{\upsilon}).\mathbf{v}\, dS + \int_{S_m} (\underline{C}\boldsymbol{\upsilon}).\left(\mathbf{n} \times \frac{D\mathbf{n}}{Dt}\right) dS$$

$$+ \int_{V_m} \mathbf{F}.\mathbf{v}\, dV + \int_{V_m} (\mathbf{n} \times \mathbf{h}^M).\left(\mathbf{n} \times \frac{D\mathbf{n}}{Dt}\right) dV + \int_{S_m} -\mathbf{q}.\mathbf{v}\, dS \qquad \text{(B.III.58)}$$

The left-hand side of this equation represents the variation per unit time of the total energy of the nematic contained inside the moving volume V_m.

The first term on the right-hand side is the work of the external stress $\underline{\sigma}\,\boldsymbol{\upsilon}$ acting on S_m.

The second term corresponds to the work of the surface torque $\underline{C}\boldsymbol{\upsilon}$ acting on S_m. This term can also be written in the equivalent form:

$$\underline{C}\boldsymbol{\upsilon}.\left(\mathbf{n} \times \frac{D\mathbf{n}}{Dt}\right) = (\underline{\pi}\boldsymbol{\upsilon}).\frac{D\mathbf{n}}{Dt} \qquad \text{(B.III.59)}$$

where $\underline{\pi}$ is the tensor with components $\pi_{ij} = \partial f/\partial n_{i,j}$.

The third term corresponds to the work of the bulk forces $\mathbf{F}$.

The fourth term gives the work of the magnetic torque when the director rotates at an angular velocity $\mathbf{n} \times D\mathbf{n}/Dt$. Note that this term can also be simplified to:

$$(\mathbf{n} \times \mathbf{h}^M).\left(n \times \frac{D\mathbf{n}}{Dt}\right) = \mathbf{h}^M.\frac{D\mathbf{n}}{Dt} \qquad \text{(B.III.60)}$$

as $\mathbf{n}$ is a unit vector.

Finally, the last term corresponds to the heat exchange by conduction across the surface, $\mathbf{q}$ being the heat flux vector. We neglected here all bulk heat intake (microwave heating, for instance). In a nematic, **q** depends on the relative orientation of the temperature gradient and the director **n**. If we let $k_{//}$ and $k_{\perp}$ be the thermal conductivities along the director and perpendicular to it, respectively, we get

$$\mathbf{q} = -k_{\perp}\,\mathbf{grad}\,T - (k_{//} - k_{\perp})\,\mathbf{n}\,(\mathbf{n}.\mathbf{grad}\,T) \qquad \text{(B.III.61)}$$

In the following step, we write eq. B.III.58 as a volume integral. This equation holding for all volumes V_m, the integrand is null, so:

$$\frac{D}{Dt}\left(\frac{1}{2}\rho v^2 + e\right) = (\sigma_{ij}v_i)_{,j} + \left(\pi_{ij}\frac{Dn_i}{Dt}\right)_{,j} + \mathbf{F}.\mathbf{v} + \mathbf{h}^M.\frac{D\mathbf{n}}{Dt} - \mathrm{div}\,\mathbf{q} \qquad \text{(B.III.62)}$$

This equation can be further simplified by using the equation of momentum conservation projected along **v**:

$$\rho\mathbf{v}.\frac{D\mathbf{v}}{Dt} = \mathbf{v}.\mathrm{div}\,\underline{\sigma} + \mathbf{F}.\mathbf{v} \qquad \text{(B.III.63)}$$

By subtracting from the previous equation one gets the following relation for the energy:

$$\frac{De}{Dt} = \sigma_{ij}\frac{\partial v_i}{\partial x_j} + \frac{\partial}{\partial x_j}\left(\pi_{ij}\frac{Dn_i}{Dt}\right) + \mathbf{h}^M.\frac{D\mathbf{n}}{Dt} - \mathrm{div}\,\mathbf{q} \qquad \text{(B.III.64)}$$

The same kind of equation can be written for the entropy per unit volume s. Indeed:

$$T\,ds = de - dW \qquad \text{(B.III.65)}$$

and, from equations B.II.26 and B.III.30, the work of the internal stress reads:

$$dW = \sigma^E_{ij}\,d\left(\frac{\partial u_i}{\partial x_j}\right) + \frac{\partial}{\partial x_j}(\pi_{ij}\,dn_i) - \mathbf{h}.d\mathbf{n} \qquad \text{(B.III.66)}$$

where σ^E_{ij} is the elastic stress tensor and **u** the displacement vector such that $\mathbf{v} = D\mathbf{u}/Dt$. In this expression, the first term stands for the work of the elastic stress $\underline{\sigma}^E$, while the next two terms represent, respectively, the work of surface torques and bulk torques acting on the director. We then have:

$$T\frac{Ds}{Dt} = \frac{De}{Dt} - \sigma^E_{ij}\frac{\partial v_i}{\partial x_j} - \frac{\partial}{\partial x_j}\left(\pi_{ij}\frac{Dn_i}{Dt}\right) + \mathbf{h}.\frac{D\mathbf{n}}{Dt} \qquad \text{(B.III.67)}$$

After replacing De/dt by its expression in the energy equation B.III.64, one gets an equation for the entropy s:

$$T\frac{Ds}{Dt} = \sigma^v_{ij}\frac{\partial v_i}{\partial x_j} + (\mathbf{h} + \mathbf{h}^M).\frac{D\mathbf{n}}{Dt} - \mathrm{div}\,\mathbf{q} \qquad \text{(B.III.68)}$$

We shall denote by $T\overset{c}{s}$ the **irreversible entropy production**, of general expression T(Ds/Dt) + T div($\mathbf{q}$/T) (see Appendix 3). We then can show that:

$$T\overset{\circ}{s} = \sigma^{v}_{ij}\frac{\partial v_i}{\partial x_j} + (\mathbf{h} + \mathbf{h}^{M}).\frac{D\mathbf{n}}{Dt} - \frac{\mathbf{q}}{T}.\mathbf{grad}\,T \qquad \text{(B.III.69a)}$$

or, by using eqs. B.III.53 and B.III.57:

$$T\overset{\circ}{s} = \sigma^{v}_{ij}\frac{\partial v_i}{\partial x_j} - \mathbf{\Gamma}^{v}.\left(\mathbf{n}\times\frac{D\mathbf{n}}{Dt}\right) - \frac{\mathbf{q}}{T}.\mathbf{grad}\,T \qquad \text{(B.III.69b)}$$

This quantity, always positive according to the second law of thermodynamics, is also the energy dissipated as heat. The three terms are easily recognized: the first one corresponds to the dissipation under shear, the second is associated with the dissipation during director rotation, while the third one is due to the heat flux. In Appendix 4, we provide a different expression for the dissipation and justify the constitutive laws for the nematic, that we have so far taken for granted.

From the condition that $T\overset{\circ}{s}$ be always positive, the following inequalities ensue:

$$\begin{aligned}
&k_{//}>0, \qquad k_{\perp}>0\\
&\alpha_4 > 0\\
&\alpha_1 + \alpha_4 + \alpha_5 + \alpha_6 > 0\\
&\gamma_1 > 0\\
&\gamma_1(\alpha_4 + \alpha_5) > \alpha_2\gamma_2
\end{aligned} \qquad \text{(B.III.70)}$$

III.2.g) Summary of the equations and description of the hydrodynamic modes

Let us rewrite the equations of nematodynamics. There are three fundamental equations:

$$\mathrm{div}\,\mathbf{v} = 0 \qquad \text{(incompressibility)} \qquad \text{(B.III.71a)}$$

$$\rho\frac{D\mathbf{v}}{Dt} = \mathrm{div}\,\underline{\sigma} + \mathbf{F} \qquad \text{(linear momentum equation)} \qquad \text{(B.III.71b)}$$

$$\mathbf{n}\times(\mathbf{h} + \mathbf{h}^{M} - \gamma_1\mathbf{N} - \gamma_2\underline{A}\,\mathbf{n}) = 0 \qquad \text{(angular momentum equation)} \qquad \text{(B.IIII.71c)}$$

where $\underline{\sigma} = -P\underline{I} + \underline{\sigma}^{E} + \underline{\sigma}^{v}$ is the total stress tensor, $\mathbf{h}$ is the molecular field, $\underline{A}$ is the symmetric part of the velocity gradient tensor, and $\mathbf{N}$ is the relative angular velocity of the director with respect to the fluid. We recall that

$$\sigma_{ij}^{E} = -\frac{\partial f}{\partial n_{k,j}}\frac{\partial n_k}{\partial x_i} \tag{B.III.72a}$$

$$h_i = -\frac{\partial f}{\partial n_i} + \frac{\partial}{\partial x_j}\left(\frac{\partial f}{\partial n_{i,j}}\right) \tag{B.III.72b}$$

$$\mathbf{h}^M = \frac{\chi_a}{\mu_o}(\mathbf{n}.\mathbf{B})\mathbf{B} \tag{B.III.72c}$$

$$\mathbf{N} = \frac{D\mathbf{n}}{Dt} - \boldsymbol{\Omega} \times \mathbf{n} \quad \text{with} \quad \boldsymbol{\Omega} = \frac{1}{2}\,\mathbf{curl\,v} \tag{B.III.72d}$$

$$(\mathbf{n}\underline{A})_i = n_j A_{ji} \quad \text{with} \quad A_{ij} = \frac{1}{2}\left(\frac{\partial v_i}{\partial x_j} + \frac{\partial v_j}{\partial x_i}\right) \tag{B.III.72e}$$

One must add to these equations the constitutive relations for the material:

$$\mathbf{q} = -k_\perp\,\mathbf{grad}T - (k_{//} - k_\perp)\,\mathbf{n}\,(\mathbf{n}.\mathbf{grad}T) \tag{B.III.73a}$$

$$\sigma_{ij}^{v} = \alpha_1(n_k A_{kl} n_l) n_i n_j + \alpha_2 n_j N_i + \alpha_3 n_i N_j + \alpha_4 A_{ij} + \alpha_5 n_j n_k A_{ki} + \alpha_6 n_i n_k A_{kj} \tag{B.III.73b}$$

These equations yield exact solutions for the director fluctuation eigenmodes. We determine the solution to eqs. B.III.71a–c using the form

$$\mathbf{v} = \delta\mathbf{v}\exp(i\omega t + i\mathbf{qr}) \tag{B.III.74a}$$

$$\mathbf{n} = \mathbf{n}_o + \delta\mathbf{n}\exp(i\omega t + i\mathbf{qr}) \tag{B.III.74b}$$

with $\delta\mathbf{v}$ and $\delta\mathbf{n}$ small enough to allow for equation linearization. The most convenient reference frame is given by ($\mathbf{e}_1$,$\mathbf{e}_2$, $\mathbf{e}_3 = \mathbf{n}_o$), so that the δn_1 and δn_2 components are decoupled. In the absence of the magnetic field, equations B.III.71 become:

$$\mathbf{q}.\mathbf{v} = 0 \tag{B.III.75a}$$

$$i\omega\rho\mathbf{v} = \mathrm{div}\underline{\sigma} \tag{B.III.75b}$$

$$\mathbf{n} \times (\mathbf{h} - \gamma_1\mathbf{N} - \gamma_2\underline{A}\,\mathbf{n}) = 0 \tag{B.III.75c}$$

The next step is to eliminate the v_2 component using the incompressibility equation. Two formally identical homogeneous equation systems are then obtained, involving v_1, δn_1 and v_3, δn_2, respectively. The compatibility equations for each system give the eigenmodes for director vibration.

Calculation shows that the structure of both compatibility equations depends on a parameter $\mu = K\rho/\eta^2$ where K and η represent an average of the elastic constants K_i, and an average of the viscosity coefficients α_i, respectively. This parameter is the ratio between the orientational diffusivity introduced in section III.1 and the kinematic viscosity, which is also the diffusivity of vorticity. Usually, $K \approx 10^{-6}$ dyn, $\eta \approx 0.1$ poise and $\rho \approx 1$ g/cm^3, so that $\mu \approx 10^{-4} << 1$. In such conditions, the two solutions of each compatibility equation are purely imaginary and can be written in the simple form (with $i\omega = -1/\tau$):

$$\tau_S = \frac{\eta}{Kq^2}, \qquad \tau_F = \frac{\rho}{\eta q^2} \tag{B.III.76}$$

These relaxation times are very different, since their ratio $\tau_F/\tau_S = \mu$ is much smaller than 1. The "S" modes are **slow**, while the "F" modes are **fast**. The first correspond to the director orientation modes already described at the beginning of the chapter. They only appear in nematics. The second modes, also present in ordinary liquids, describe the vorticity damping and correspond to overdamped shear waves (Fig. B.III.12). These results show that **the velocity fluctuations (or, more precisely, the vorticity fluctuations) relax much more quickly than the fluctuations in director orientation.**

For all practical purposes, we provide here the two pairs of slow and fast modes given by a complete calculation.

For the first pair, one must take in eq. B.III.76:

$$Kq^2 = K_1 q_\perp^2 + K_3 q_{//}^2 \tag{B.III.77a}$$

$$\eta = \gamma_1 - \frac{(q_\perp^2\alpha_3 - q_{//}^2\alpha_2)^2}{q_\perp^4\eta_b + q_\perp^2 q_{//}^2(\alpha_1+\alpha_3+\alpha_4+\alpha_5) + q_{//}^4\eta_c} \tag{B.III.77b}$$

where $q_{//}^2 = q_3^2$ and $q_\perp^2 = q_1^2 + q_2^2$. These modes involve splay and bend deformations.

For the second pair, the result is:

$$Kq^2 = K_2 q_\perp^2 + K_3 q_{//}^2 \tag{B.III.78a}$$

$$\eta = \gamma_1 - \frac{\alpha_2 q_{//}^2}{q_\perp^2\eta_a + q_{//}^2\eta_c} \tag{B.III.78b}$$

These modes involve twist and bend deformations. The slow mode of pure twist $q_{//} = 0$ is the only one to remain uncoupled to a hydrodynamic "backflow;" for this mode, $K = K_2$ and $\eta = \gamma_1$. For the other two modes, the corrections in γ_1 to eqs. B.III.77b and B.III.78b are due to the "backflow."

In these expressions, η_a, η_b and η_c are the Miesowicz viscosities, which can be expressed as a function of the Leslie viscosities α_i, as shown in the next section (eqs. B.III.84–86).

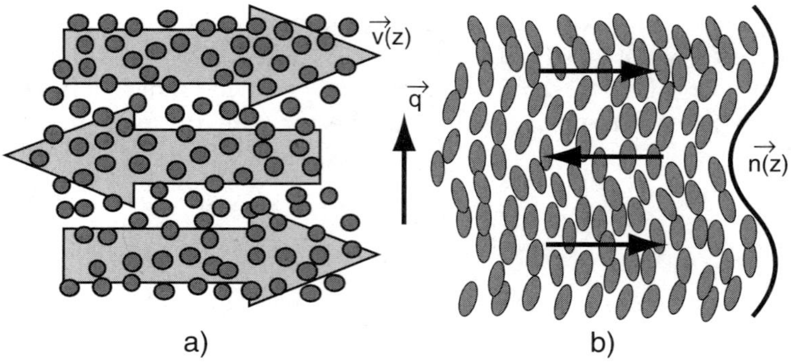

Fig. B.III.12 Pictorial description of a fast and a slow mode: (a) shows a shear wave and (b) an orientation wave. These two modes are overdamped (diffusive modes). Note that mode (b) induces in the fluid a secondary flow, termed "backflow," represented by arrows. This flow renormalizes the rotational viscosity γ_1 (see eq. B.III.77b).

These modes can be revealed using inelastic light scattering. As they are overdamped, the spectral density of the scattered light has a lorentzian shape, centered at the frequency of the incoming light and of width $1/\tau$. It is then obvious that the slow modes will have the main contribution to this spectral broadening. The experiment, based on optical mixing spectroscopy, was performed by L. Léger [13]. In principle, it should give access to various viscosities, but it is not very precise.

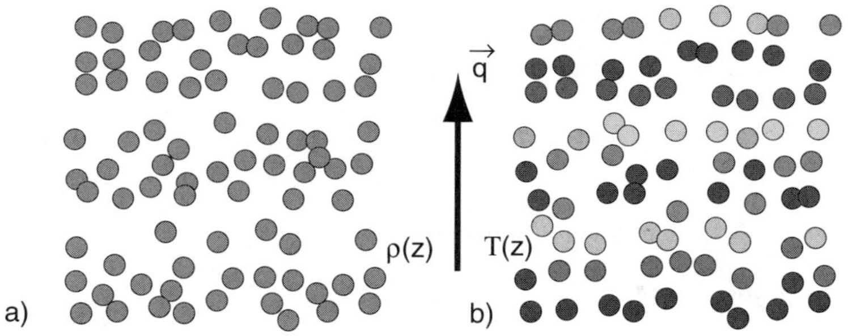

Fig. B.III.13 Pictorial description of a sound wave (a) and a thermal wave (b). Here, the shades of gray represent the temperature. These modes are also present in ordinary liquids.

To conclude this discussion on the hydrodynamic modes, let us point out that two propagating modes appear in nematics (corresponding to ordinary sound waves, Fig. B.III.13a), associated with the density fluctuations, and of dispersive relation

$$\omega = \pm\, cq \qquad \text{where} \qquad c = \sqrt{\left(\frac{\partial P}{\partial \rho}\right)_s} \tag{B.III.79}$$

as well as a diffusive thermal mode associated with temperature fluctuations (Fig. B.III.13b), of dispersive relation given by

$$i\omega = -\Lambda_{//}q_{//}^2 - \Lambda_{\perp}q_{\perp}^2 \quad \text{(B.III.80)}$$

where $\Lambda_{//,\perp} = k_{//,\perp}/\rho C_p$ are the thermal diffusivities parallel and perpendicular to the director, respectively.

We now turn to the analysis of the continuous Couette and Poiseuille flows described in the first section.

III.3 Laminary Couette and Poiseuille flows

III.3.a) Simple shear under strong magnetic field (Couette) and determining the Miesowicz viscosities

In this section we shall consider an even simpler experiment than the one performed by Miesowicz. Suppose that the liquid crystal is sheared at constant thickness between two parallel plates (Couette flow). Under shear, the director is subjected to a hydrodynamic torque of the order of γA (from eq. B.III.53) where γ is an average viscosity depending on γ_1 and γ_2. This torque is balanced by an elastic torque (if the molecules are firmly anchored at the surfaces) and a magnetic torque $(\chi_a/\mu_0)B^2\sin\theta$, with θ the angle between **n** and **B**. Under a strong magnetic field, the restoring magnetic torque is prevalent and the molecular tilt is given by:

$$\sin\theta \approx \frac{\gamma\mu_0 A}{\chi_a B^2} \quad \text{(B.III.81)}$$

The molecular alignment is only slightly disturbed if the angle $\theta \ll 1$. This is the case for a sufficiently low shear rate A:

$$A \ll \frac{\chi_a B^2}{\gamma\mu_0} \quad \text{(B.III.82)}$$

In the following, we shall assume this condition is fulfilled. For each of the three fundamental geometries represented in figure B.III.14 one can define an apparent viscosity η such that

$$\eta = \frac{\text{shear stress}}{\text{velocity gradient}} = \frac{\sigma_{ij}^v}{2A_{ij}} \quad \text{(B.III.83)}$$

In geometry (a), **n** is perpendicular to the flow and the velocity gradient, so:

$$A_{xy} = \frac{1}{2}\frac{\partial v}{\partial y}$$

$$\sigma_{xy}^{v} = \alpha_4 A_{xy} \tag{B.III.84}$$

$$\eta_a = \frac{\sigma_{xy}^{v}}{2A_{xy}} = \frac{1}{2}\alpha_4$$

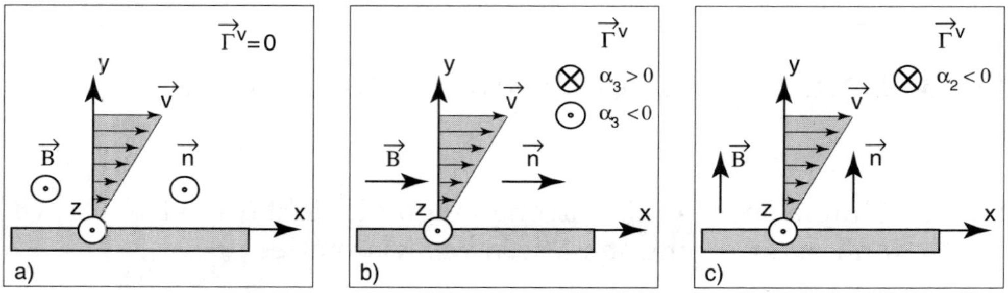

Fig. B.III.14 The three fundamental geometries allowing definition of the Miesowicz viscosities.

In geometry (b), **n** is parallel to the flow and perpendicular to the velocity gradient, thus:

$$A_{xy} = \frac{1}{2}\frac{\partial v}{\partial y} = N_y$$

$$\sigma_{xy}^{v} = \alpha_3 N_y + (\alpha_4 + \alpha_6)\, A_{xy} \tag{B.III.85}$$

$$\eta_b = \frac{1}{2}(\alpha_3 + \alpha_4 + \alpha_6)$$

Finally, in geometry (c), **n** is perpendicular to the flow and parallel to the velocity gradient, yielding:

$$A_{xy} = \frac{1}{2}\frac{\partial v}{\partial y} = -N_x$$

$$\sigma_{xy}^{v} = \alpha_2 N_x + (\alpha_4 + \alpha_5)\, A_{xy} \tag{B.III.86}$$

$$\eta_c = \frac{1}{2}(-\alpha_2 + \alpha_4 + \alpha_5)$$

Generally, if the director **n** makes an angle θ with the velocity while remaining in the plane defined by the velocity and its gradient, then the apparent viscosity of the sample depends on θ as:

$$\eta = \eta_b \cos^2\theta + \eta_c \sin^2\theta + \alpha_1 \sin\theta\cos\theta \qquad \text{(B.III.87)}$$

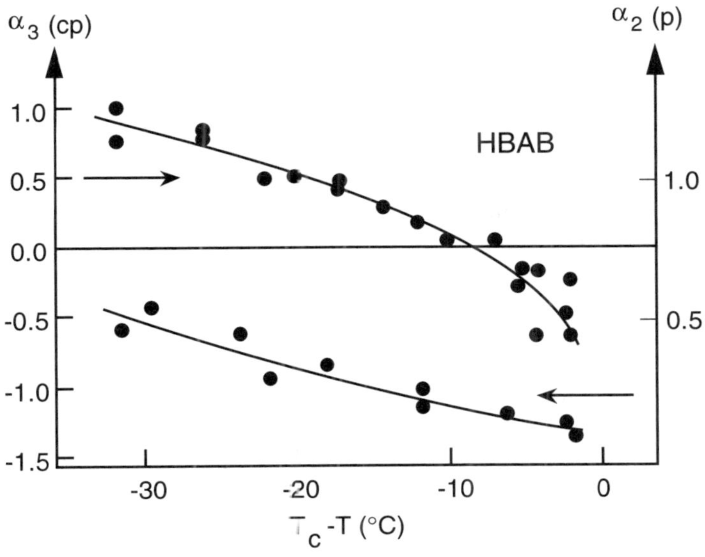

Fig. B.III.15 Temperature variation of the α_2 and α_3 viscosities of HBAB. T_c is the transition temperature to the isotropic liquid, close to 102° (from refs. [7c] and [15a]).

One can see that, by changing the flow geometry and providing γ_1 is known (it can easily be determined from the rise time for the Frederiks instability), five independent relations can be obtained between the six Leslie coefficients α_i. Comparing these data to those obtained in inelastic light scattering, the six Leslie coefficients can be independently determined without using the Parodi relation. For MBBA at room temperature, the experiment yields [14]:

$$\alpha_1 = 6.5\,\text{cp},\ \alpha_2 = -77.5\,\text{cp},\ \alpha_3 = -1.2\,\text{cp},$$

$$\alpha_4 = 83.2\,\text{cp},\ \alpha_5 = 46.3\,\text{cp},\ \alpha_6 = -34.4\,\text{cp} \qquad \text{cp} = \text{centipoise}$$

One can thus verify *ex post facto* that, for this material, the Parodi relation B.III.54 holds.

In MBBA, α_2 and α_3 are negative, but this is not a general property. For instance, in HBAB, α_3 changes sign with the temperature, whereas α_2 always remains negative (Fig. B.III.15) [7c, 15a].

We shall see later, in subsection (III.4.c), what is the interpretation of this sign change for α_3. Before that, let us consider the case of Poiseuille flow in a capillary.

III.3.b) Poiseuille flow and "transverse" viscous stress

We just saw that the viscosity of a nematic depends on its orientation with respect to the velocity and the velocity gradient. This behavior, fairly easy to predict, was indeed experimentally discovered by Miesowicz several decades before the Leslie-Eriksen theory was published.

The hydrodynamic behavior of nematics is however much richer than this simple anisotropy and can sometimes lead to surprising effects. Some of them were predicted from the analysis of the viscous tensor σ_{ij}^{v}, then experimentally confirmed. Other, more complex, effects were first experimentally revealed and only afterwards was the theoretical interpretation found.

As an example, consider the case when the director is tilted with respect to the shear plane (Fig. B.III.16). Intuitively, the effective viscosity should have a value between η_a and η_b. Calculating the σ_{xy} component of the Leslie-Ericksen tensor gives a result that supports this intuition:

$$\sigma_{xy} = \eta_{eff} s \tag{B.III.88}$$

where

$$\eta_{eff} = \eta_a \cos^2\phi + \eta_b \sin^2\phi \tag{B.III.89}$$

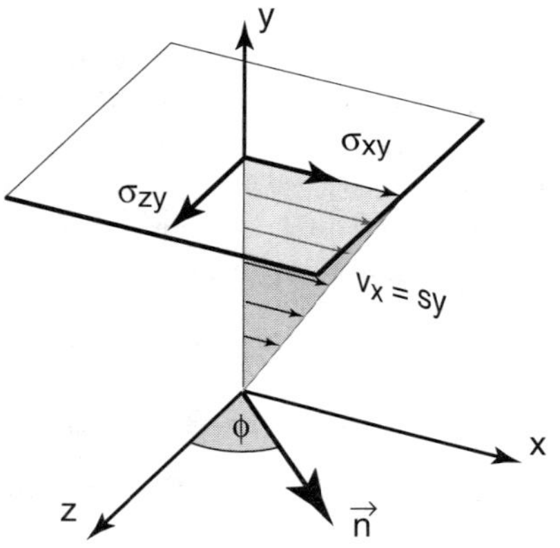

Fig. B.III.16 When the director is tilted with respect to the shear plane (x,y), the transverse stress σ_{zy} is not zero.

Let us now determine the stress σ_{zy} perpendicular to the shear plane:

$$\sigma_{zy} = [\,(\eta_a - \eta_b)\sin\phi\cos\phi\,]\,s \tag{B.III.90}$$

so this component also has a finite value when the director is tilted with respect to the shear plane. This much less intuitive result shows that if a nematic is sheared in the tilted geometry depicted in figure B.III.16, the displaced plate is not only hindered in its movement by the effective viscosity η_{eff}, but also pushed sideways by the force σ_{zy}.

In figure B.III.17 we show the setup for Poiseuille flow in a **narrow** rectangular pipe of dimensions $l_y << l_z << l_x$. In this geometry, the flow is along x, so:

$$v_x = v_x(y), \qquad v_y = v_z = 0 \tag{B.III.91}$$

Equations B.III.25, 28, and 44 yield:

$$0 = -\frac{\partial p}{\partial x} + \eta_{eff}(\phi)\frac{\partial^2 v_x}{\partial y^2} \tag{B.III.92}$$

and

$$0 = -\frac{\partial p}{\partial z} + (\eta_a - \eta_b)\sin\phi\cos\phi\frac{\partial^2 v_x}{\partial y^2} \tag{B.III.93}$$

whence, by eliminating $\partial^2 v_x/\partial y^2$, we get the ratio between the transverse and longitudinal pressure gradients:

$$\frac{\partial p/\partial z}{\partial p/\partial x} = \frac{(\eta_a - \eta_b)\sin\phi\cos\phi}{\eta_a\cos^2\phi + \eta_b\sin^2\phi} \tag{B.III.94}$$

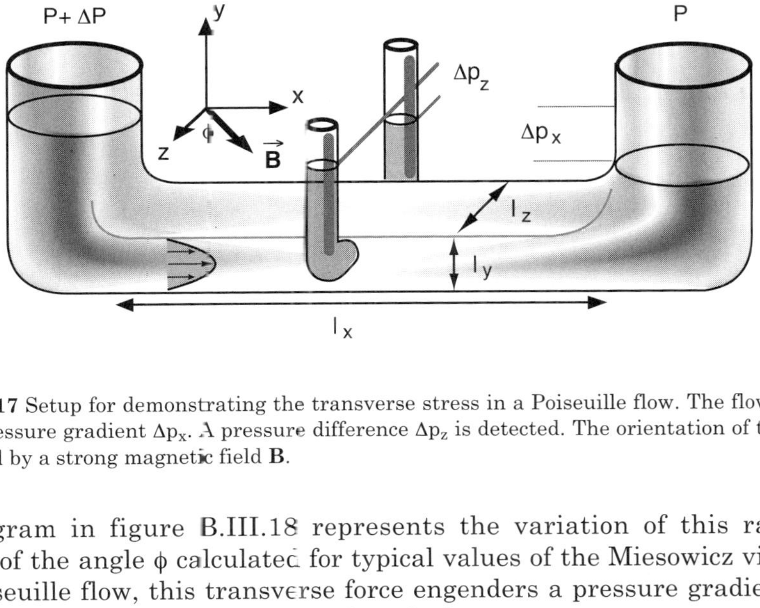

Fig. B.III.17 Setup for demonstrating the transverse stress in a Poiseuille flow. The flow is along x under a pressure gradient Δp_x. A pressure difference Δp_z is detected. The orientation of the nematic is **imposed** by a strong magnetic field **B**.

The diagram in figure B.III.18 represents the variation of this ratio as a function of the angle ϕ calculated for typical values of the Miesowicz viscosities. In a Poiseuille flow, this transverse force engenders a pressure gradient in the direction perpendicular to the shear plane [16].

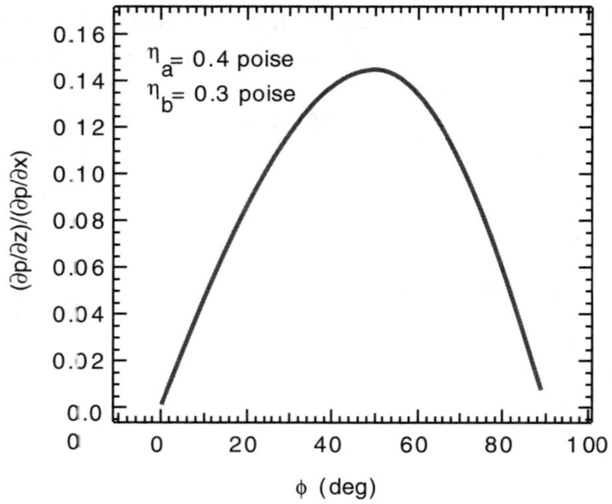

Fig. B.III.18 Variation in transverse pressure gradient as a function of the nematic orientation with respect to the flow direction (from ref. [16]).

Experiments performed on MBBA completely confirmed these theoretical predictions.

In the following section, we shall analyze the stability of these flows when the magnetic field is cut off.

III.4 Laminary flows and their stability

Let us first form some dimensionless numbers characteristic for the considered flow.

III.4.a) Reynolds and Ericksen numbers

In fluid mechanics one usually introduces a dimensionless number, called the **Reynolds number**, quantifying the relative importance of the inertial and viscous forces:

$$R_e = \frac{\text{inertial forces}}{\text{viscous forces}} \qquad \text{(B.III.95)}$$

In the case of stationary flow around a body of linear dimension L, this number is given by

$$R_e = \frac{VL}{\nu} \tag{B.III.96}$$

with V the velocity of the fluid far from the considered body and ν its kinematic viscosity ($\nu = \eta/\rho$, where η is the dynamic viscosity of the fluid). ν is also the diffusion constant of the vorticity.

The **Reynolds number** is important because it describes the family of flows around bodies of similar form for which

$$\frac{\mathbf{v}}{V} = f\left(\frac{\mathbf{r}}{L}, R_e\right) \qquad \text{(Reynolds similarity law)} \tag{B.III.97}$$

The Reynolds number also describes the loss of stability for a stationary flow. The critical value of the Reynolds number depends on the system investigated, but it is generally of a few tens in ordinary fluids.

The situation is completely different for a nematic, which can be destabilized even for very low Reynolds numbers. For instance, $R_e \approx 10^{-4}$ in the experiments on hydrodynamic instabilities or electrohydrodynamics described at the end of the chapter. As we shall see, this loss of stability manifests itself by a spontaneous and sudden disorientation of the director field. The Reynolds number is not relevant for this instability, at least when it is much smaller than 1. In the case of nematics, the instability is produced by the competition between elastic and viscous torques, described by the **Ericksen number**:

$$E_r = \frac{\text{viscous torque}}{\text{elastic torque}} \tag{B.III.98}$$

An order of magnitude estimation of the viscous and elastic torques gives $\alpha V/L$ and K/L^2, respectively, where L is a typical length (for instance, the sample thickness) and V a typical flow velocity (for instance, the velocity of the mobile plate in a shear experiment). The Ericksen number is then given by:

$$E_r = \frac{VL}{D} \tag{B.III.99}$$

where $D = K/\alpha \approx 10^{-5}$ cm^2/s is the diffusion constant of director orientation.

We have already employed the Ericksen number, without specifically defining it, when describing the Wahl and Fischer experiment at the beginning of this chapter. In this case, the tilt angle of the director under shear depends on the Ericksen number $\theta = \theta(z/d, Er)$ (eq. B.III.18). This formula defines **a family of stationary flows, all having similar distortions, parameterized by the Ericksen number**.

We shall show later that the Ericksen number, like the Reynolds number, can also describe the **stability loss of a flow**. It will be shown that in hydrodynamic instabilities (as well as in the electroconvective instability to be discussed later), the critical Ericksen number is typically 1, while the Reynolds number is still very small ($\approx 10^{-6} - 10^{-4}$). This order of magnitude difference stems from the fact that the vorticity diffuses much more quickly than the director orientation ($D \ll \nu$).

III.4.b) Definition of the hydrodynamic torque

To determine the behavior of the nematic under continuous shear in the three Miesowicz geometries (Fig. B.III.14), it is convenient to calculate first the viscous torque $\mathbf{\Gamma}^v$ (eq. B.III.53) exerted on the director.

In the (a) Miesowicz geometry (Fig. B.III.19a), one has

$$\mathbf{N} = (\mathbf{n}\underline{A}) = 0 \qquad \text{and} \qquad \mathbf{\Gamma}^v_{(a)} = 0 \tag{B.III.100}$$

The viscous torque being zero, one could expect this director orientation with respect to the flow to be **stable**. However, we shall see later that, for negative α_3, there is a critical shear rate above which the director spontaneously leaves its original orientation.

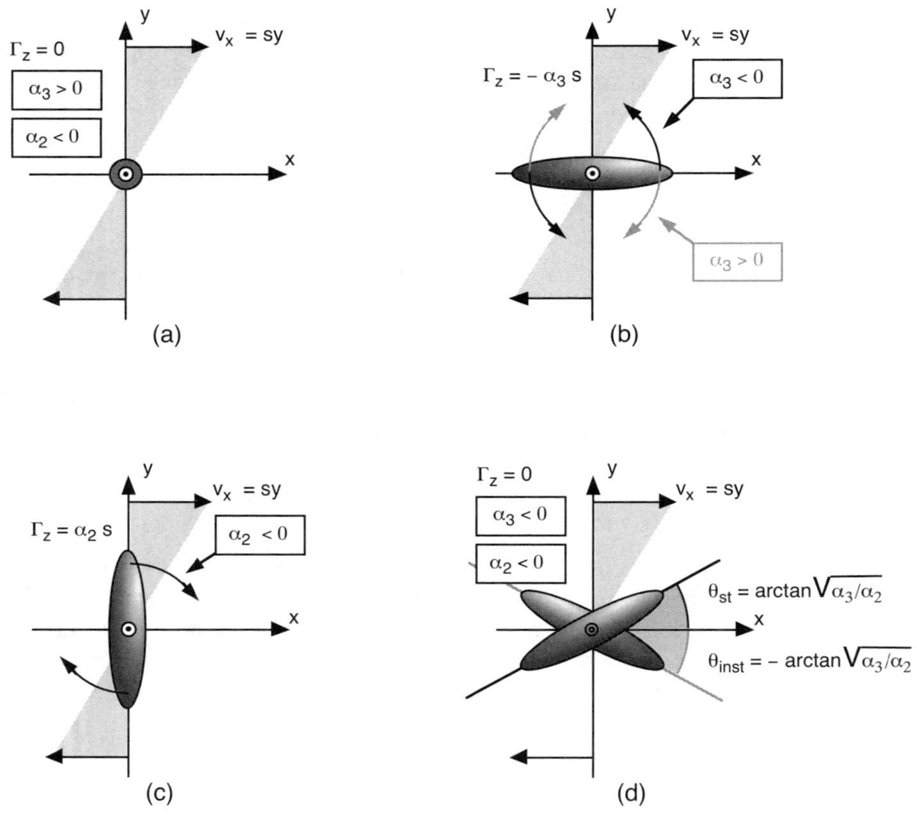

Fig. B.III.19 The viscosities α_2 and α_3 define the viscous torques per unit volume exerted upon the director when this is oriented either along (b), or perpendicular (c) to the simple shear flow. For $\alpha_3 < 0$, the nematic is **aligned by the flow**, the director being oriented in the direction θ_{st} for which the viscous torque is zero (d). This director orientation is stable. On the other hand, when α_3 is positive, the only stable orientation of the director is perpendicular to the shear plane (a).

In the (b) Miesowicz geometry (Fig. B.III.19b), using the Parodi relation yields:

$$\mathbf{N} = (\mathbf{n}\underline{A}) = (s/2)\,\mathbf{y} \quad \text{and} \quad \boldsymbol{\Gamma}^{v}_{(b)} = -(s/2)\,(\gamma_1 + \gamma_2)\,\mathbf{z} = -\,\alpha_3\, s\,\mathbf{z} \qquad \text{(B.III.101)}$$

where **y** and **z** (as well as **x**) are unit vectors parallel to the reference axes. Hence, the director tends to rotate perpendicular to its axis in the shear plane, as soon as the shear is applied. We should emphasize right away that the direction of rotation depends on the sign of the viscosity α_3 which, as we already know (see figure B.III.15), can change with the temperature.

In the (c) Miesowicz geometry (Fig. B.III.19c), which is also that of the Wahl and Fischer experiment, one has

$$(\mathbf{n}\underline{A}) = -\,\mathbf{N} = (s/2)\mathbf{x} \quad \text{and} \quad \boldsymbol{\Gamma}^{v}_{(c)} = (s/2)(\gamma_2 - \gamma_1)\,\mathbf{z} = \alpha_2\, s\mathbf{z} \qquad \text{(B.III.102)}$$

As previously, the director starts to turn as soon as the shear is applied.

In conclusion, formulae B.III.101 and B.III.102 confer a precise significance to the viscosity coefficients α_2 and α_3, that we summed up in the sketches of figures B.III.19 b and c.

The diagram in figure B.III.19d corresponds to the case, intermediate between b and c, when the director is in the shear plane and forms an angle θ with the velocity. In this geometry, $\mathbf{v} = (sy, 0, 0)$ and $\mathbf{n} = (\cos\theta, \sin\theta, 0)$. The expression B.III.53 of the viscous torque then gives

$$\boldsymbol{\Gamma}^{v} = (\alpha_2 \sin^2\theta - \alpha_3 \cos^2\theta)s\mathbf{z} \qquad \text{(B.III.103)}$$

Thus, the viscous torque depends on the director orientation, which begs the following question: **"Are there director orientations in the shear plane for which the viscous torque is null?"**

The answer to this question depends on the relative signs of the viscosity coefficients α_2 and α_3. Indeed, the equation

$$\alpha_2 \sin^2\theta - \alpha_3 \cos^2\theta = 0 \qquad \text{(B.III.104)}$$

has solutions

$$\theta_{st} = +\arctan[(\alpha_3/\alpha_2)^{1/2}] \quad \text{and} \quad \theta_{unst} = -\arctan[(\alpha_3/\alpha_2)^{1/2}] \qquad \text{(B.III.105)}$$

only if α_3/α_2 is positive. Is this condition fulfilled in real-life experiments?

As shown by the viscosity measurements, the α_2 coefficient is always negative, which is easily understood from the microscopic interpretation of the viscous torque given in figure B.III.3a, corresponding to the (a) Miesowicz geometry. On the other hand, the sign of the α_3 coefficient depends on the material and can change with temperature, a feature we have already mentioned (Fig. B.III.15) and to which we shall return later. Thus, in MBBA, the α_3 coefficient is negative over the entire existence range of the nematic phase: one can always find director orientations for which the viscous torque is zero. However, in the previously discussed case of HBAB, α_3 is negative at high temperature, close to the isotropic phase, but becomes positive when approaching the transition to the smectic phase: in this latter case, the viscous torque can no longer be annulled.

First we consider the case $\alpha_3 < 0$, for which eq. B.III.104 has the two solutions B.III.105, graphically represented in figure B.III.19d. For these two orientations, the viscous torque is zero, but are these orientations stable? To answer this question, let us first see what happens when the director moves away from the position $\theta_{st} > 0$ by an angle $|\delta\theta| << 1$. If $\delta\theta$ is positive, the emerging viscous torque has the same direction as in figure B.III.19c. This torque tends to reduce the fluctuation $\delta\theta$ and thus brings the director back towards the position θ_{st}. A similar result is obtained for a negative deviation $\delta\theta$, as the viscous torque engendered by the fluctuation changes sign along with $\delta\theta$. In conclusion, the $\theta_{st} > 0$ orientation is stable with respect to orientation fluctuations in the (x, y) plane. The same analysis applied to the θ_{inst} position leads to the opposite conclusion, namely that any fluctuation $\delta\theta$ grows. It is therefore towards the θ_{st} orientation that the director tends in the Wahl and Fischer experiment at high shear rates.

In the case $\alpha_3 > 0$, the amplitude of the torque given by eq. B.III.103 varies with the angle θ, but its sign remains constant irrespective of θ. Therefore, eq. B.III.104 has no solution: the director is not aligned by the flow, as Gähwiller noticed for the first time in HBAB [15a]. One could then imagine that, under the action of a very strong shear, the director starts to turn, accumulating the stress in the nematic slab. We shall see later that this reasoning is erroneous, as it assumes that the director remains in the shear plane (x, y). In fact, when the shear rate increases above a certain threshold, the director leaves the shear plane and orients along axis **z**, perpendicular to the shear plane, where the viscous torque annuls. It is also by leaving the (x, y) plane that the director field can relax excessive distortion.

III.4.c) Origin of the hydrodynamic torque

This preliminary analysis of shear flow stability emphasizes the highly important role played by the viscosity coefficient α_3. It is therefore useful to present Helfrich's explanation [17] of the sign of α_3 and the origin of the viscous torque in the (b) Miesowicz geometry.

The viscous torque exerted upon a molecule is due to its collisions with its neighbors. In figure B.III.20, identical to the one in figure B.III.3 except for the molecular orientation, the molecules arrive from the left, at the top, and from the right, at the bottom. Each collision imparts to the molecule a momentum $\mathbf{f}_{coll}dt$ whose direction depends on the details of the collision. If we represent the molecules by **"smooth" ellipsoids**, the forces are orthogonal to the surface of contact between the two molecules. These forces are oblique with respect to the flow velocity. The total force exerted on the molecule is null on average; as to the torque, its sign depends on the direction of the forces $\mathbf{f}_{coll}$.

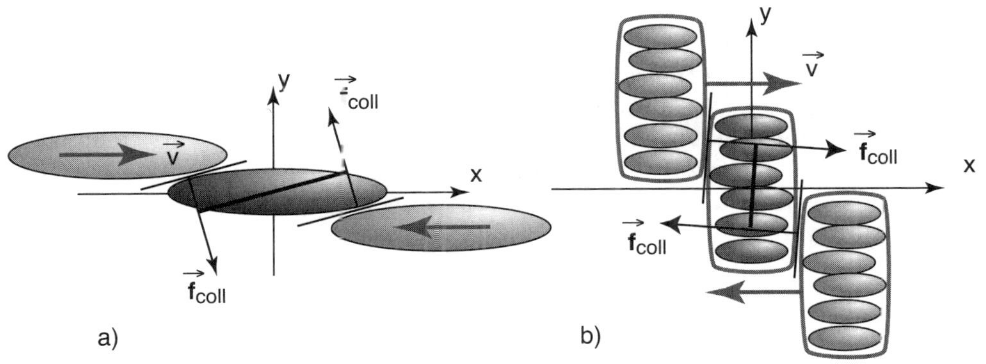

Fig. B.III.20 The sign of the viscous torque (and hence of α_3) depends on the shape of the objects undergoing a collision under the effect of the shear. In (a), these objects are smooth ellipsoids; in (b), they are cybotactic groups (from ref. [17]).

In figure B.III.20a, the torque is **positive**. How then can it change sign? We shall see in the second volume that, when cooling the nematic phase, a nematic → smectic phase transition occurs in certain materials. Close to the transition, the precursors of the smectic order appear in the nematic phase: the molecules form bunches, called "**cybotactic groups.**" The collisions between these **square-shaped bunches** generate forces whose moment of force with respect to the center is **negative** (Fig. B.III.20b). This phenomenon can qualitatively explain the sign change of α_3.

III.4.d) Measuring the α_3 viscosity

The α_3 viscosity coefficient can be determined in several ways. For instance, as Gähwiller did [7c], one can measure the effective shear viscosity $\eta(\mathbf{n})$ by the capillary method (Fig. B.III.7) for four well-chosen director orientations $\mathbf{n}_1,...,\mathbf{n}_4$ (imposed by the magnetic field), yielding four linear equations

$$\eta(\mathbf{n}_i) = \eta(\alpha_1,...,\alpha_6,\mathbf{n}_i), \qquad i = 1,...,4 \tag{B.III.106}$$

with the six unknowns $\alpha_1,...,\alpha_6$. By independently measuring the γ_1 viscosity, we have the additional equation:

$$\gamma_1 = \alpha_3 - \alpha_2 \tag{B.III.107}$$

Including the Parodi relation

$$\alpha_2 + \alpha_3 + \alpha_5 - \alpha_6 = 0 \qquad \text{(B.III.108)}$$

we finally have a system of six linear equations allowing us to find the six Leslie-Ericksen viscosities, including α_3.

This procedure is somewhat indirect, so we also present a much simpler method directly giving the ratio α_3/γ_1 [18]. The experimental setup is described in figure B.III.21. The liquid crystal sample is maintained by capillarity between two glass plates P_b and P_h treated in planar anchoring along the x axis. The two plates are fixed (one directly and the other by using the spacer C) on the two arms of a U-shaped metal piece. The two plates are parallel and separated by a gap d of about 150 μm. The B1 arm is fixed, while the other, B2, is attached to the spring R. The right end of the spring oscillates horizontally along the x axis with an amplitude X_o of a few millimeters, at a frequency of a few Hz. By carefully choosing the stiffness of the spring and the U-shaped piece, the amplitude of the oscillation can be reduced by a factor of about 1000, bringing it to only a few micrometers at the position of the sample. The displacement of the B2 arm is then given by $x(t) = 10^{-3}\, X(t)$. The distortion in the liquid crystal sample is detected by conoscopy.

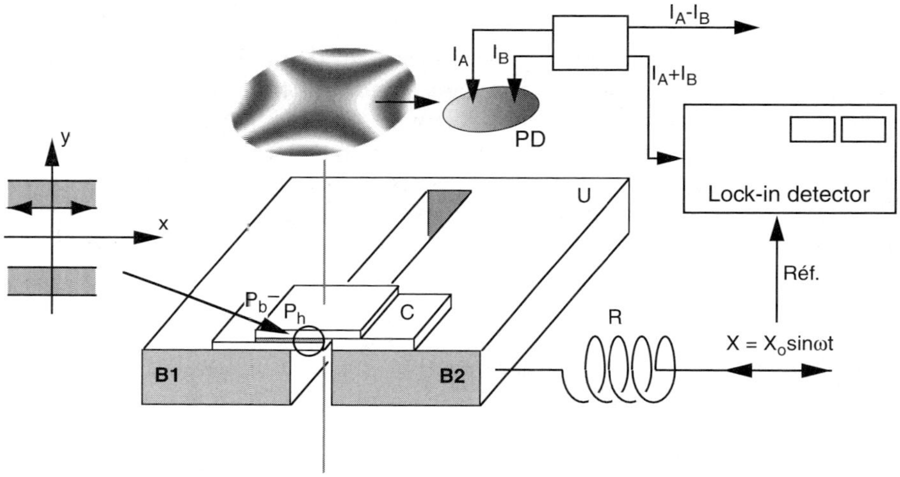

Fig. B.III.21 Experimental setup for measuring the α_3/γ_1 ratio. The liquid crystal, of orientation **n**//**z** is sheared between the glass plates P_h and P_b. Under the alternative shear, the director and, as a consequence, the conoscopic figure, start to oscillate (from ref. [18a]).

The shear geometry is the one in figure B.III.19b, only now it is the upper plate that oscillates. The thickness d, of the order of 150 μm, and the frequency ω, of a few s^{-1}, are chosen such that

$$\delta_{el} = (D/\omega)^{1/2} << d << \delta_v = (\nu_b/\omega)^{1/2} \qquad \text{(B.III.109)}$$

where δ_{el} and δ_v are the thicknesses of the elastic and viscous boundary layers, respectively. Here, $\nu_b = \eta_b/\rho \approx 1\ cm^2s^{-1}$ is the kinematic viscosity of the nematic and $D = K_1/\gamma_1 \approx 10^{-6}\ cm^2s^{-1}$ the orientational diffusivity. The velocity profile remains linear, as $d << \delta_v$, while the elastic torque can be neglected in the calculations because $\delta_{el} << d$.

Let

$$x = x_0 \sin\omega t \tag{B.III.110}$$

be the displacement of the upper plate. The velocity profile being linear, one has

$$s = \frac{dv_x}{dy} = \frac{\omega x_0 \cos\omega t}{d} \tag{B.III.111}$$

The equation of motion for the director B.III.51 then reads

$$\Gamma_z^v = -\gamma_1 \frac{d\theta}{dt} - \alpha_3 s = 0 \tag{B.III.112}$$

whence

$$\theta = \theta_0 \sin\omega t \qquad \text{with} \qquad \theta_0 = -\frac{x_0}{d}\frac{\alpha_3}{\gamma_1} \tag{B.III.113}$$

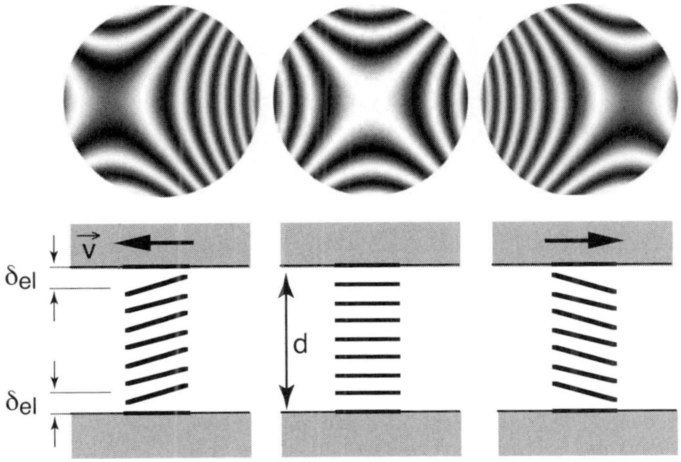

Fig. B.III.22 The distortion and oscillation of the conoscopic figure induced by the oscillating shear of a nematic sample with $\alpha_3 > 0$; the conoscopic figure moves in the direction of the shear. For negative α_3, the movement of the conoscopic figure is in the direction opposing the shear.

This director motion is reflected by the conoscopic figure (Fig. B.III.22) whose oscillations are detected using two photodiodes to measure the light intensities I_A and I_B in two very close points, separated by a distance Δx. An analogue circuit determines the sum $I = (I_A + I_B)/2$ and the difference $\Delta I = I_A - I_B$. The first signal I is measured under oscillating shear using a lock-in amplifier synchronized to the displacement $x = x_o \sin\omega t$. The second one, ΔI, is measured directly with the sample at rest. Under shear, the conoscopic figure turns by an angle θ' (proportional to θ), leading to an oscillation of the average intensity I as

$$\delta I = -\frac{\Delta I}{\Delta x} L\theta' = -\frac{\Delta I}{\Delta x} LC\theta = \frac{\Delta I}{\Delta x} LC \frac{x_o}{d}\frac{\alpha_3}{\gamma_1} \sin\omega t = \delta I_o \sin\omega t \qquad \text{(B.III.114)}$$

where L is the distance between the sample and the observation screen and C the proportionality constant between θ' and θ. The lock-in amplifier gives the amplitude δI_o of this signal, proportional to the ratio α_3/γ_1:

$$\frac{\alpha_3}{\gamma_1} = \frac{\delta I_o}{\Delta I}\frac{d}{x_o}\frac{\Delta x}{LC} = \frac{\delta I_o}{\Delta I} C' \qquad \text{(B.III.115)}$$

where C' is a constant to be determined.

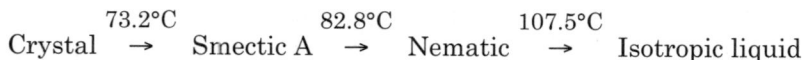

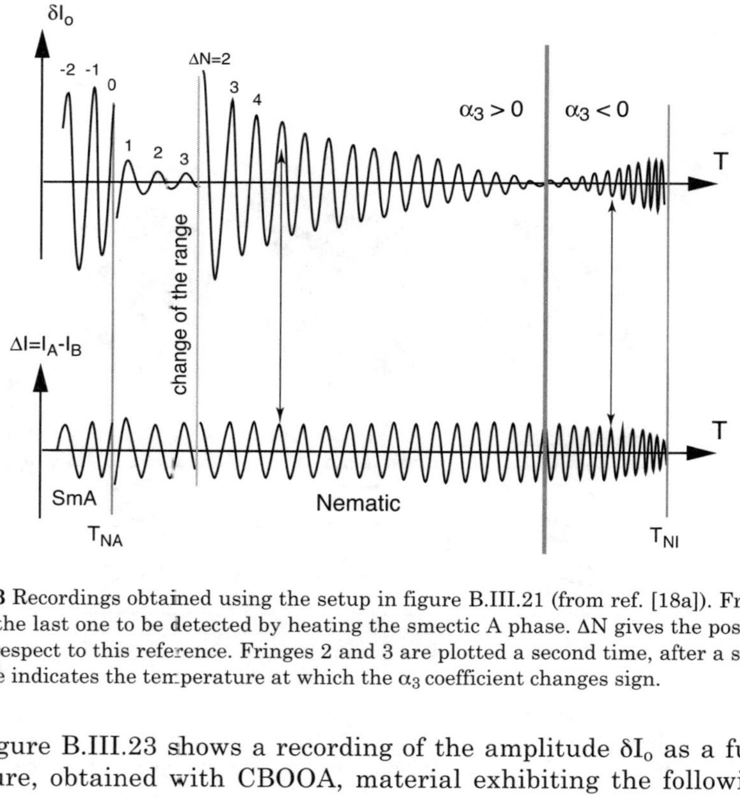

Fig. B.III.23 Recordings obtained using the setup in figure B.III.21 (from ref. [18a]). Fringe 0 is, by convention, the last one to be detected by heating the smectic A phase. ΔN gives the position of each fringe with respect to this reference. Fringes 2 and 3 are plotted a second time, after a scale change. The gray line indicates the temperature at which the α_3 coefficient changes sign.

Figure B.III.23 shows a recording of the amplitude δI_o as a function of temperature, obtained with CBOOA, material exhibiting the following phase sequence:

$$\text{Crystal} \xrightarrow{73.2^\circ C} \text{Smectic A} \xrightarrow{82.8^\circ C} \text{Nematic} \xrightarrow{107.5^\circ C} \text{Isotropic liquid}$$

In the same figure, we also show the recording of the intensity difference ΔI, a quantity needed to determine the ratio α_3/γ_1 from eq. B.III.115. Note that $\Delta I(T)$ oscillates and changes sign many times while the sample temperature increases from the SmA phase. This effect is due to the shift of interference fringes accompanying the change in interference order at the center of the conoscopic figure

$$N = \frac{(n_e - n_o)d}{\lambda} \tag{B.III.116}$$

Thus, besides the viscosity measurements, this experiment also shows that the birefringence $\Delta n = n_e - n_o$ of CBOOA decreases with increasing temperature.

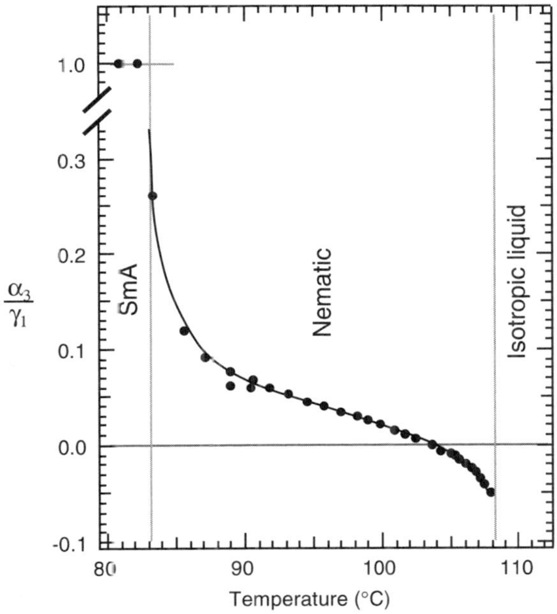

Fig. B.III.24 Value of the α_3/γ_1 ratio as a function of temperature in CBOOA (from ref. [18]).

The oscillations in $\Delta I(T)$ are of course reflected in the recording of the amplitude $\delta I_o(T)$, in agreement with eq. B.III.114. Note however that, close to the smectic phase, the maxima of ΔI coincide with the maxima of δI_o, while at high temperature, close to the isotropic phase, the maxima of ΔI coincide with the minima of δI_o. According to eq. B.III.114, this inversion can only be due to a sign change of the α_3/γ_1 ratio. To determine the value of the α_3/γ_1 ratio from eq. B.III. 115, one must first divide $\delta I_o(T)$ by $\Delta I(T)$, then multiply the result by C'. To determine the value of the C' coefficient, one need only notice that in the SmA phase the ratio α_3/γ_1 is 1; indeed, the molecules remain orthogonal to the layers which, in the absence of permeation, tilt by x/d. Finally:

$$\frac{\alpha_3}{\gamma_1} = \frac{\dfrac{\delta I_o(T)}{\Delta I(T)}}{\dfrac{\delta I_o(SmA)}{\Delta I(SmA)}} \tag{B.III.117}$$

The final result is plotted in figure B.III.24.

III.4.e) Destabilization under continuous shear when $\alpha_3 > 0$ in the (b) Miesowicz geometry (**n** // **v** and **n** ⊥ **grad**v)

In this paragraph, we continue the discussion started in section B.III.4b on the stability of a **continuous** simple shear flow. This time, however, the elastic torques will be considered.

We have seen that the viscous torque only goes to zero when the director **n** is orthogonal to the shear plane (x,y) ((a) Miesowicz geometry depicted in figure BIII.19a). On the other hand, what happens in a planar sample when the director is initially in the shear plane, oriented along the flow direction ((b) Miesowicz geometry)?

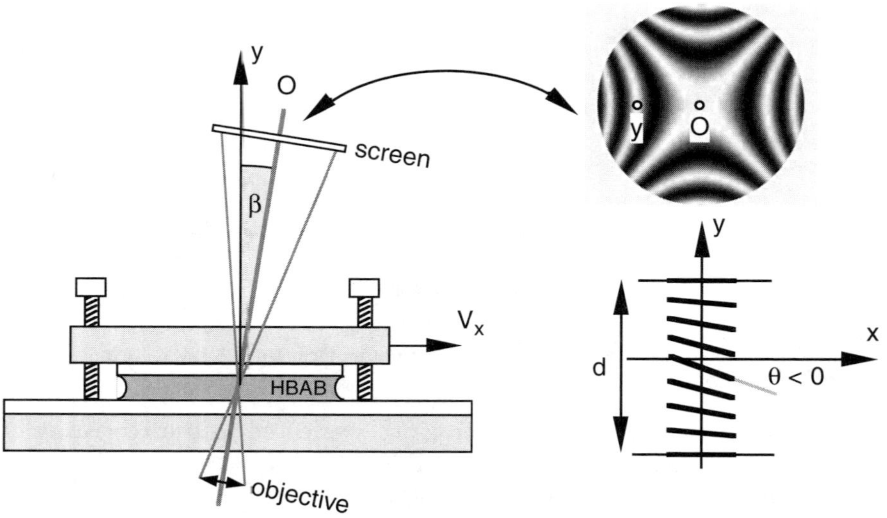

Fig. B.III.25 Setup for the study of hydrodynamic instabilities. The flow-induced distortion is detected by measuring the angle β by which the conoscopic figure tilts (from ref. [18a]).

This configuration can be obtained with the setup represented in figure B.III.25. The liquid crystal, HBAB in the nematic phase, is maintained by capillarity between two elongated glass plates, one of them shorter and narrower than the other. The two plates are treated to ensure planar anchoring (by SiO evaporation at oblique incidence) so that the molecules align on the plates along the shear direction x. The lower glass plate is fixed. The upper one is glued on a metallic piece bearing three teflon screws which allow adjustment of the spacing between the plates while their tips can slide on the lower plate. The upper plate is pulled at a constant velocity V in the x direction. The entire setup is temperature regulated to control the α_3 viscosity.

The average director orientation $<\theta>$ is detected by measuring the angle β by which the center of the conoscopic figure tilts. In the small θ limit, it can be shown that:

$$\beta \approx (n_o + n_e) <\theta> \qquad \text{(B.III.118)}$$

The values of the average angle $<\theta>$ as a function of the velocity V are plotted in figure B.III.26.

The upper curve was obtained at 94°C, i.e., in the temperature range where the α_3 coefficient is negative. As expected, the average distortion angle tends to its saturation value θ_{st}, of the order of 10°, for high shear rates.

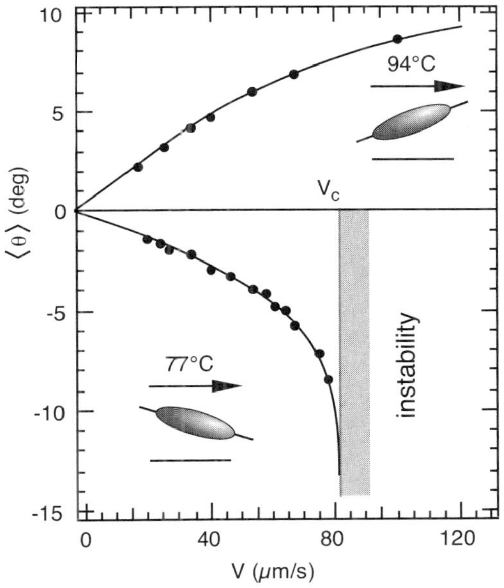

Fig. B.III.26 Values of the average distortion as a function of the shear velocity in HBAB. The sample thickness is 80 μm (from ref. [18a]).

The curve obtained at 77°C, a temperature at which α_3 is positive, is very different. It shows the existence of a critical shear velocity $V_c \approx 80$ μm/s, below which the "planar" conoscopic figure shifts in the shear direction maintaining its shape, while above this value it completely vanishes. More specifically, the experiment shows that for low velocities $V << V_c$, the tilt angle β of the conoscopic figure, and hence the angle $<\theta>$ (given by eq. B.III.118), are proportional to the shear velocity; however, the conoscopic figure shifts more and more quickly as the critical velocity V_c is approached, and completely disappears above V_c. Direct observation of the sample under the polarizing microscope shows that the director field is then very complex and non-stationary, up to the longest observation time t_{max}; indeed, in this setup the mobile plate can only move for at most 2 cm. Consequently, the observation time after applying the shear is limited to $t_{max} \approx 20$ s, for velocities of the order of 0.1 mm/s.

Thus, the flow becomes unstable above the critical velocity V_c. Additional measurements show that V_c depends on the sample thickness d (Fig. B.III.27) and on the temperature. Assuming the instability threshold is defined by a certain critical value Er_c of the Ericksen number $Er = Vd/D$, we predict that the product $V_c d$ must be constant. This prediction is well verified experimentally, as shown in figure B.III.27 where we find $V_c d \approx 7\times10^{-5}$ cm^2/s, irrespective of the sample thickness.

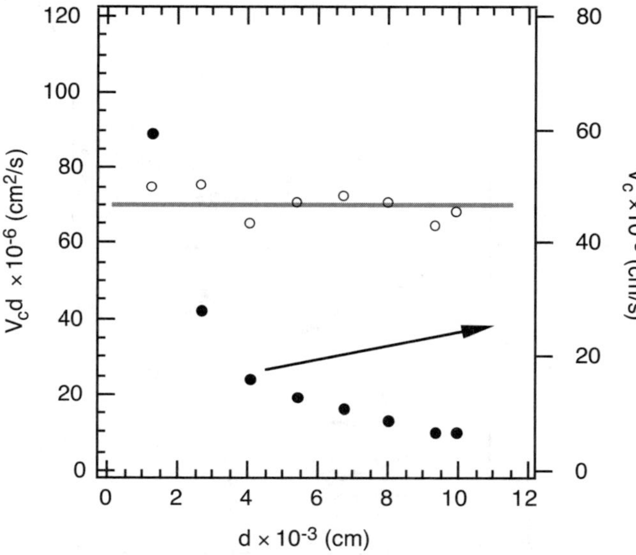

Fig. B.III.27 Critical velocity as a function of the sample thickness in HBAB. The product $V_c d$ is constant (from ref. [18a]).

A detailed analysis of flow stability in the (b) Miesowicz geometry was made by S.A. Pikin [19]. To find the instability threshold, consider the balance of the viscous and elastic torques. To simplify the calculations, let us assume that the shear viscosity η is independent of the angle θ, such that the velocity profile remains linear in spite of the director field distortions. Let us further assume that $K_1 = K_3$. The equation of torque balance then reduces to the simple form:

$$\Gamma_z = K_1 \frac{d^2\theta}{dy^2} + (\alpha_2 \sin^2\theta - \alpha_3 \cos^2\theta)\frac{V}{d} = 0 \qquad \text{(B.III.119)}$$

We define the dimensionless variables:

$$\tilde{y} = \frac{y}{d}, \qquad \varepsilon = \frac{\alpha_3}{\alpha_2} \qquad \text{and} \qquad \tilde{\theta} = \frac{\theta}{|\varepsilon|^{1/2}} \qquad \text{(B.III.120)}$$

as well as the Ericksen number:

$$Er' = \frac{Vd}{K_1/|\alpha_2|}|\varepsilon|^{1/2} \qquad \text{(B.III.121)}$$

In the small distortion approximation, eq. B.III.119 assumes the following form:

$$\frac{d^2\tilde{\theta}}{d\tilde{y}^2} = Er'(-I + \tilde{\theta}^2) \qquad \text{(B.III.122)}$$

$$\text{where } I = \text{sign}(\varepsilon) = \begin{cases} +1 \text{ for } \alpha_3 < 0 \\ -1 \text{ for } \alpha_3 > 0 \end{cases}$$

with the boundary conditions

$$\tilde{\theta}(\pm 1/2) = 0 \qquad \text{(B.III.123)}$$

This non-linear equation was numerically solved. The results for $\tilde{\theta}(0) = \tilde{\theta}_{max}$ and $<\tilde{\theta}>$ are plotted as a function of the number Er' in figure B.III.28. One can see that, for $\alpha_3 > 0$, the solution disappears above a certain critical value of the Ericksen number:

$$Er'_c = \frac{V_c d}{K_1/|\alpha_2|}|\varepsilon|^{1/2} = 4.75 \qquad \text{(B.III.124)}$$

With $V_c d \approx 7\times10^{-5}$ cm^2/s (as previously determined, see Fig. B.III.26), $\alpha_2 \approx 0.3$ poise, $\alpha_3 \approx 0.3$ cp (Fig. B.III.15), and $K_1 \approx 5\times10^{-7}$ dyn, we get $Er'_c = 4.2$, in fairly good agreement with the theoretical value.

In the case of HBAB, the parameter $\varepsilon = \alpha_3/\alpha_2$ tends to zero when the temperature approaches temperature T_i at which α_3 changes sign (Fig. B.III.15). As a consequence, according to eq. B.III.124 the critical velocity should diverge as $|\varepsilon|^{-1/2}$ close to T_i. This theoretical prediction, resulting from the definition of the Ericksen number Er', was verified experimentally, as one can see in figure B.III.29.

Another interesting feature is that, below the instability threshold, the numerical calculation yields two solutions, the first one corresponding to the distortion effectively observed. Is it possible to obtain the other solution experimentally, corresponding to stronger distortion? The answer is no. Indeed, the stability analysis of the two solutions with respect to fluctuations of the angle θ (which we shall not delve into here) shows that the experimentally obtained solution is stable, while the other is unstable.

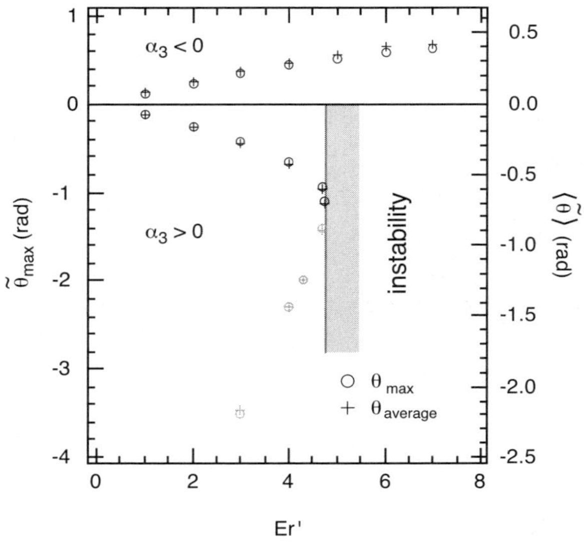

Fig. B.III.28 Numerical calculation of the distortion as a function of the Ericksen number under continuous shear (from ref. [18a]).

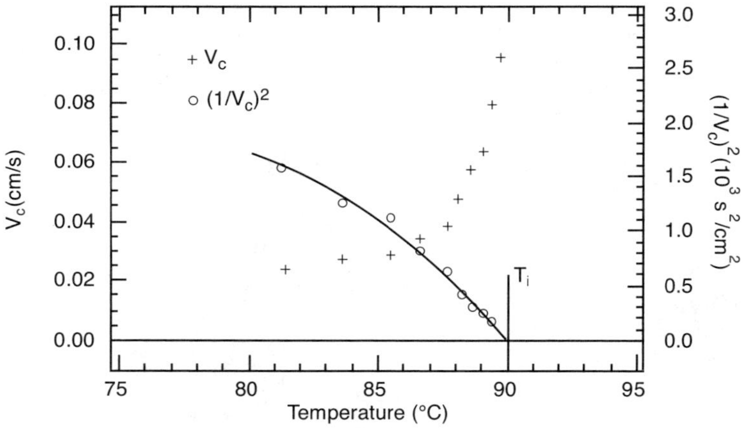

Fig. B.III.29 Values of the critical velocity V_c as a function of temperature in HBAB (from ref. [18a]).

Rigorously, one should also analyze the stability of the solutions obtained with respect to director fluctuations out of the shear plane (x, y). We sketch here the beginning of this analysis in the small θ (distortion) limit for which the solution of eq. B.III.119 can be found analytically. Indeed, for $\theta^2 << 1$

and $\alpha_3 > 0$, this equation has the simplified form

$$\frac{d^2\theta}{d\tilde{y}^2} \approx Er_{13} \qquad \text{where} \qquad Er_{13} = \frac{Vd}{K_1/\alpha_3} \tag{B.III.125}$$

with the obvious solution

$$\theta(\tilde{y}) = Er_{13}\frac{1}{2}\left(\tilde{y}^2 - \frac{1}{4}\right) \tag{B.III.126}$$

Assume now that to this distortion in the shear plane is superposed a small azimuthal fluctuation ϕ, which takes the director out of the shear plane, as shown in figure B.III.30.

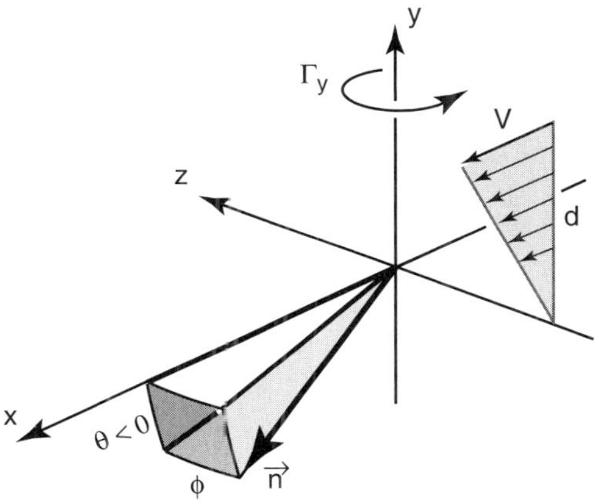

Fig. B.III.30 Asymmetric instability mechanism: in the presence of the shear-induced distortion $\theta < 0$, an azimuthal fluctuation ϕ creates a torque Γ_y, tending to increase the initial fluctuation.

In the presence of this fluctuation, a new component Γ_y of the viscous torque appears (eq. B.III.53), leading to a new torque balance equation:

$$\Gamma_y \approx K_2\frac{d^2\phi}{dy^2} + \alpha_2\,\phi\,\theta\frac{V}{d} = 0 \tag{B.III.127}$$

or

$$\frac{d^2\phi}{d\tilde{y}^2} = Er_{22}\,\phi\,\theta \qquad \text{with} \qquad Er_{22} = \frac{Vd}{K_2/|\alpha_2|} \tag{B.III.128}$$

Replacing for θ its expression B.III.126 yields the eigenvalue equation:

$$\frac{d^2\phi}{d\tilde{y}^2} = A\left(\tilde{y}^2 - \frac{1}{4}\right)\phi \qquad \text{where} \qquad A = \frac{1}{2}Er_{22}\,Er_{13} \tag{B.III.129}$$

The eigenvalue of the fundamental state is 45, giving an asymmetric instability threshold:

$$Er_{22}\,Er_{13} = 90 \qquad \text{(B.III.130)}$$

or

$$Er'^2 = 90\,\frac{K_1}{K_2} \qquad \text{(B.III.131)}$$

With the typical value $K_1/K_2 \approx 3/2$ for the elastic anisotropy:

$$Er'_{asym} \approx 11 > 4.75 = Er'_{sym} \qquad \text{(B.III.132)}$$

In this case, the asymmetric instability (out of the shear plane) occurs after the symmetric instability.

III.4.f) Destabilization under continuous shear for $\alpha_3 < 0$ in the (a) Miesowicz geometry (**n** ⊥ **v** and **n** ⊥ **grad** v)

Let us now return to the (a) Miesowicz geometry. The viscous torque being zero in symmetry terms, this flow should be stable, a least for low shear rates. Nevertheless, this is the geometry in which the first hydrodynamic nematic instability was observed.

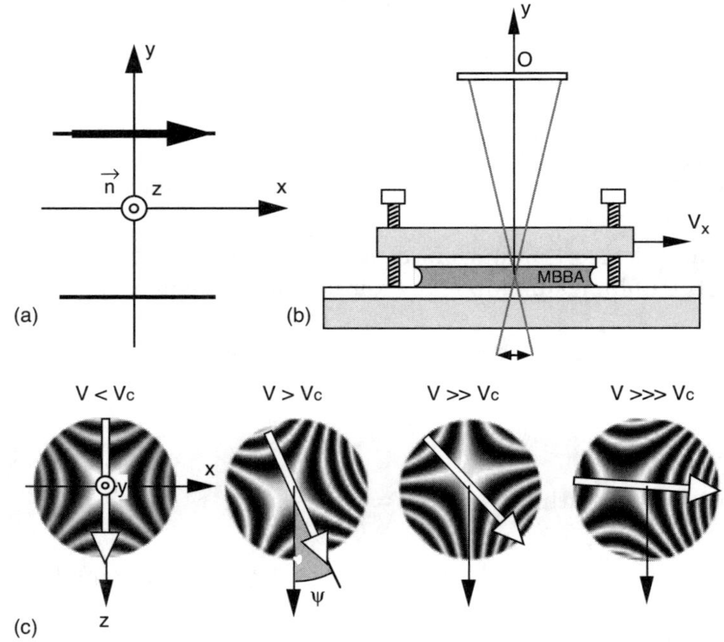

Fig. B.III.31 Hydrodynamic instability in the (a) Miesowicz geometry. a) Geometry. b) Experimental setup. c) Conoscopic figures: the initial director orientation is **n**//**z**. Above the threshold, the conoscopic figure is slightly tilted and turns around its axis. The rotation angle ψ of the conoscopic figure can be measured.

This instability was revealed using an experimental setup identical to the one in figure B.III.25, except for the fact that the anchoring direction is **n**//**z** instead of **n**//**x**. Observation of the conoscopic figure shows that at low shear rates the initial orientation **n**//**z** remains stable. Above a certain shear rate V_c/d (where d is the sample thickness), the conoscopic figure is slightly tilted "backwards" and at the same time turns around its axis, as shown in figure B.III.31. The rotation angle ψ increases (in absolute value) with the shear velocity and tends to 90° for high shear rates. In this limit, the conoscopic figure is tilted "backwards" by a few degrees, showing that, above a critical shear value, the director orientation changes as it progressively aligns in the shear plane along the direction defined by the angle θ_{st} (eq. B.III.105) for which the hydrodynamic torque is zero. Experimentally, one can observe in the same sample coexisting domains separated by walls and in which the ψ angles have opposite signs. This observation reflects the fact that the initial configuration possesses a mirror symmetry with respect to the (x,y) plane. The values of the rotation angle ψ of the conoscopic figure are plotted as a function of the shear velocity in figure B.III.32. The very distinct instability threshold is obtained for a critical value $V_c \approx 11.5$ μm/s.

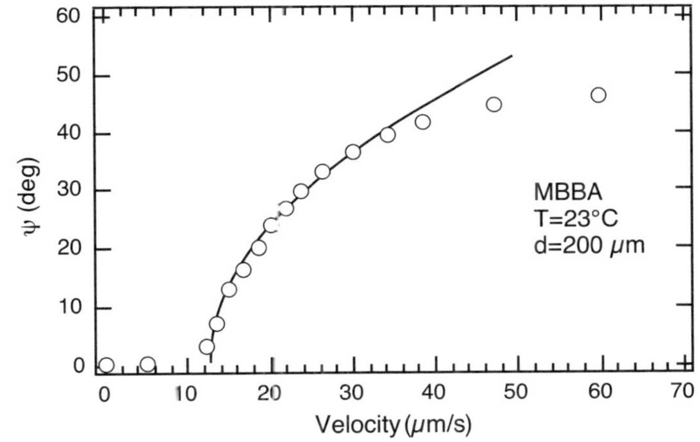

Fig. B.III.32 Values of the rotation angle of the conoscopic figure as a function of the shear velocity (from ref. [20a]).

These experimental results suggest that the hydrodynamic instability in the (a) Miesowicz geometry is similar to a Frederiks transition in a magnetic field (Fig. B.II.12), but for the fact that the twist distortion ($\partial\phi/\partial y$) is accompanied here by a splay distortion ($\partial\theta/\partial y$), where ϕ is the director rotation angle in the (x,z) plane and θ its tilt angle in the (y, z) plane.

Let us now investigate in more detail the origin of the instability. More specifically, we shall prove that above a critical shear rate s_c the director field becomes unstable with respect to a splay distortion in the vertical plane (of component $n_x(y)$) [20a]. We assume that the two Leslie coefficients α_2 and α_3 are negative (as for MBBA) and that no magnetic field is applied.

The origin of the instability can be qualitatively understood in the following way (Fig. B.III.33a, ref. [20]): suppose a $n_y > 0$ fluctuation develops. The director is then subjected to a torque $\Gamma_y^v > 0$ (if $\alpha_2 < 0$) proportional to n_y. This torque generates a new component $n_x > 0$, creating a torque Γ_x^v of the same sign as α_3 and proportional to n_x. For $\alpha_3 < 0$, this torque tends to amplify the initial fluctuation n_y, so it is destabilizing. This mechanism leads to a **homogeneous** deformation (not depending on x or z) of the director field if the destabilizing hydrodynamic torques (themselves depending on the shear rate s) overtake the elastic restoring torques (which do not depend on s) originating in the strong anchoring of the molecules on the surfaces limiting the sample. Hence, there exists a finite critical value of the shear rate, s_c, above which the instability develops. From this point of view, the situation is similar to a Frederiks transition.

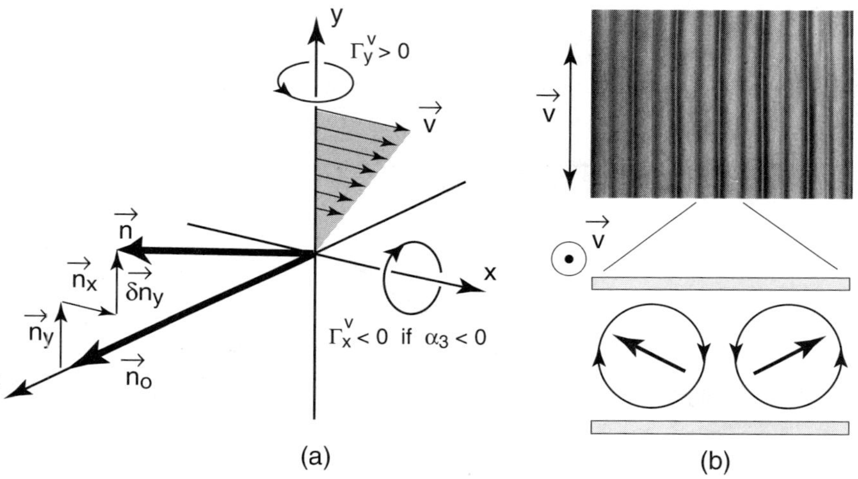

Fig. B.III.33 Hydrodynamic instabilities under shear: a) mechanism of the homogeneous instability. Suppose a $n_y > 0$ fluctuation develops. The flow induces a viscous torque $\Gamma_y^v > 0$ which generates a $n_x > 0$ component, then a torque $\Gamma_x^v < 0$ (for $\alpha_3 < 0$) tending to amplify n_y. b) Convective instability under shear flow. The secondary flow, $v_y(z)$ for instance, is generated by the shear in the presence of distortions $\partial n_y / \partial z$ (see eq. B.III.135).

This analogy can be taken even further, as the instability mechanism and the threshold value s_c remain the same when the sign of the distortions n_y and n_x is reversed. It is then obvious that, above the critical shear s_c, the domains corresponding to the two distortion signs can simultaneously appear. We shall see later that under oscillating shear (or in a stabilizing magnetic field parallel to z), a domain system systematically develops which is periodical in the z direction, the **homogeneous** instability is replaced by a "**roll**" instability [20b] (Fig. B.III.33b).

In order to find the instability threshold, we shall write the linearized movement equations for the stationary regime in the general case [21]. In the following, n_x, n_y, v_x, v_y, and v_z represent the **fluctuations** of the director and of the hydrodynamic velocity, respectively.

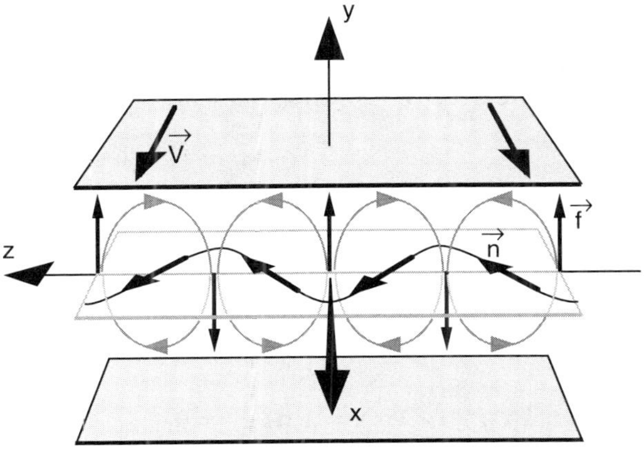

Fig. B.III.34 Viscous forces **f**//**y** induced by the flow **v** = (Vy/d, 0, 0) in the presence of the distortion **n** ≈ (n_{xo}sinqz,0,1).

The balance of the viscous and elastic **torques** writes:

$$\Gamma_x = 0 = -\left(K_1\frac{\partial^2}{\partial y^2} + K_3\frac{\partial^2}{\partial z^2}\right) n_y + \alpha_3 s\, n_x + \alpha_3 \frac{\partial v_z}{\partial y} + \alpha_2 \frac{\partial v_y}{\partial z} \tag{B.III.133}$$

$$\Gamma_y = 0 = \left(K_2\frac{\partial^2}{\partial y^2} + K_3\frac{\partial^2}{\partial z^2}\right) n_x - \alpha_2 s\, n_y - \alpha_2 \frac{\partial v_x}{\partial z} \tag{B.III.134}$$

while the equations for the bulk **forces** expressing momentum conservation are:

$$\rho s v_y = \left(\eta_a\frac{\partial^2}{\partial y^2} + \eta_c\frac{\partial^2}{\partial z^2}\right) v_x + \frac{\alpha_5 - \alpha_2}{2}\frac{\partial}{\partial z}(s\, n_y) \tag{B.III.135}$$

$$\frac{\partial P}{\partial y} = \left(g\frac{\partial^2}{\partial y^2} + \eta_c\frac{\partial^2}{\partial z^2}\right) v_y + \frac{\alpha_5 + \alpha_2}{2}\frac{\partial}{\partial z}(s\, n_x) \tag{B.III.136}$$

$$\frac{\partial P}{\partial z} = \left(b\frac{\partial^2}{\partial z^2} + \eta_b\frac{\partial^2}{\partial y^2}\right) v_z + \frac{\alpha_6 + \alpha_3}{2}\frac{\partial}{\partial y}(s\, n_x) \tag{B.III.137}$$

with

$$b = \frac{1}{2}(2\alpha_1+\alpha_3+\alpha_4+2\alpha_5+\alpha_6) \quad \text{and} \quad g = \frac{1}{2}(\alpha_4-\alpha_5-\alpha_2) \tag{B.III.138}$$

In eqs. B.III.135–7, the last terms in $s\partial n_i/\partial x_j$ correspond to the bulk forces created by the shear $s = \partial v_x/\partial y$ in the presence of director field distortions.

These forces, with no equivalent in isotropic liquids, create a secondary roll-shaped flow (Figs. B.III.33b and 34).

The **incompressibility** condition gives

$$\frac{\partial v_y}{\partial y} + \frac{\partial v_z}{\partial z} = 0 \tag{B.III.139}$$

We still need to write the boundary conditions on the plates in $y = \pm d/2$:

$$n_x(y = \pm d/2) = n_y(y = \pm d/2) = 0 \tag{B.III.140a}$$

$$v_x(y = \pm d/2) = v_y(y = \pm d/2) = v_z(y = \pm d/2) = 0 \tag{B.III.140b}$$

For the **homogeneous instability**, the solution is **translation invariant** in the horizontal plane (x, z). In addition to $\partial/\partial x = 0$, we also have has $\partial/\partial z = 0$. From the incompressibility condition B.III.139 and the boundary conditions it is obvious that $v_y = 0$ whence, from eq. B.III.136, P = constant. Finally, eq. B.III.135 gives $v_x = 0$. We shall seek for a solution under the form:

$$\frac{n_x}{n_{xo}} = \frac{n_y}{n_{yo}} = \cos q_y y \qquad \frac{v_z}{v_{zo}} = \cos q_y y \tag{B.III.141}$$

with $q_y = \pi/d$. We emphasize that a distortion (n_x, n_y) generates a secondary flow in the z direction (an effect of the "transverse" stress, see section III.3.b).

After eliminating v_z in the torque equations B.III.133 and 134, two equations are obtained for n_{xo} and n_{yo}:

$$K_1 q_y^2 n_{yo} + \alpha_3 \frac{\eta_a}{\eta_b} s n_{xo} = 0 \tag{B.III.142a}$$

$$\alpha_2 s n_{yo} + K_2 q_y^2 n_{xo} = 0 \tag{B.III.142b}$$

Note that in the first equation, the η_a/η_b correction comes from the additional viscous torque generated by the secondary flow in the z direction. The "transverse" viscous stresses, discussed in section III.3.b, are at the origin of this secondary flow. The homogeneous equation system B.III.142 only has non-trivial solutions if its determinant is zero which, replacing q_y by its value, gives:

$$\pi^2 = s\, d^2 \sqrt{\frac{\eta_a \alpha_2 \alpha_3}{\eta_b K_1 K_2}} \tag{B.III.143}$$

This relation defines the critical threshold s_c for the appearance of a **homogeneous** deformation of the director field. Clearly, this solution only exists if the product $\alpha_2\alpha_3 > 0$, viz. $\alpha_3 < 0$ for $\alpha_2 < 0$. For $\alpha_2\alpha_3 < 0$, the homogeneous equation system B.III.142 only has the trivial solution $n_{xo} = n_{yo} = 0$. The nematic is thus stable with respect to the homogeneous instability. The relation B.III.143 also determines the value of the critical Ericksen number above which the flow is destabilized

$$Er_c'' = V_c d \sqrt{\frac{\eta_a \alpha_2 \alpha_3}{\eta_b K_1 K_2}} = \pi^2 \qquad \text{(B.III.144)}$$

where $s_c = V_c/d$.

This relation predicts that for a 200 μm thick MBBA sample, the critical velocity of one plate with respect to the other is of the order of 10 μm/s, in good agreement with the experimental results (see Fig. B.III.32).

Experience also shows that the rotation angle of the conoscopic figure varies as the square root of the distance from the threshold ($\psi \propto (V - V_c)^{1/2}$) (Fig. B.III.32) [20a,d]. The bifurcation is thus normal (also termed supercritical). To prove it theoretically, one needs to take into account the nonlinear terms in the equations of motion and write the corresponding amplitude equation.

This analysis can be easily extended to the situation when a stabilizing magnetic field is applied along the z axis. Under these conditions, the critical shear rate is given by a formula similar to eq. B.III.143, where the K_i are replaced by $K_i(1 + \chi_a B^2 d^2/\mu_o \pi^2 K_i)$ with i = 1,2. At stronger fields, the critical shear rate becomes proportional to B^2, a result confirmed experimentally in MBBA for B < 1200 gauss. If, on the other hand, the applied field is above this limit, experiment shows that **another instability, of convective origin** appears, involving bend distortions in the direction z, perpendicular to the shear plane (Figs. B.III.33b and 34). By generalizing eqs. B.III.133–138 to take into account the magnetic field it can be shown that the threshold for the appearance of these convective rolls is below the threshold of the homogeneous solution previously calculated.

Instabilities in the geometry (a) of Miezowicz have also been observed in a channel (Poiseuille) flow [20c].

III.4.g) Destabilization under linear oscillating shear in the (a) Miesowicz geometry (**n** ⊥ **v** and **n** ⊥ **grad** v)

Hydrodynamic **roll** instabilities can also be produced under a sinusoidal (or square) oscillating shear of frequency f = 1/T. They can be studied using an experimental setup similar to the one in figure B.III.31, but where the moving piece is set in motion by a crankshaft system of adjustable eccentricity (Fig. B.III.35). The total oscillation amplitude D of the upper plate can thus be adjusted between 0 and a few millimeters. The system is placed in a magnetic field parallel to the initial orientation **z** of the liquid crystal. This field is stabilizing both for the n_x and the n_y fluctuations. Using glass plates covered by a thin semi-transparent layer of gold or ITO evaporated under oblique incidence (this method yields planar anchoring, see chapter B.V), an electric field can equally be applied along the vertical direction y. For MBBA, where the dielectric anisotropy is negative, this field is stabilizing for the n_y fluctuations, but does not act on the n_x fluctuations. The instabilities are detected by monitoring the diffraction figure of a He-Ne laser beam due to the periodic distortions of the director field, which can be approximately written as:

$$n_x(y,z,t) = n_{xo}(t)\cos(\pi y/d)\cos(qz) \quad \text{(B.III.145a)}$$

$$n_y(y,z,t) = n_{yo}(t)\cos(\pi y/d)\cos(qz) \quad \text{(B.III.145b)}$$

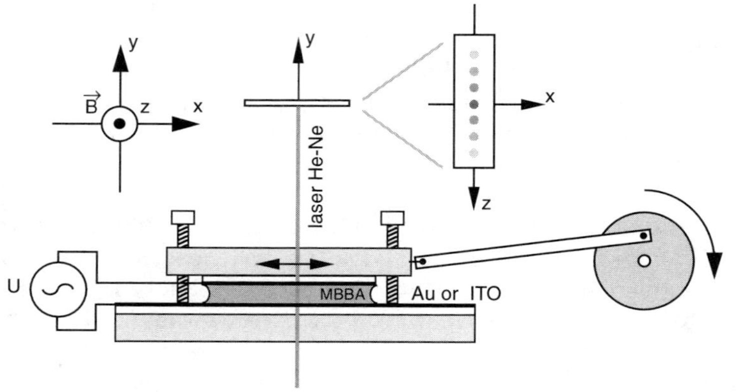

Fig. B.III.35 Setup used for the study of hydrodynamic instabilities under oscillating linear shear. The periodic distortion typical for the instability is detected by the diffraction of a laser beam (from ref. [22b]).

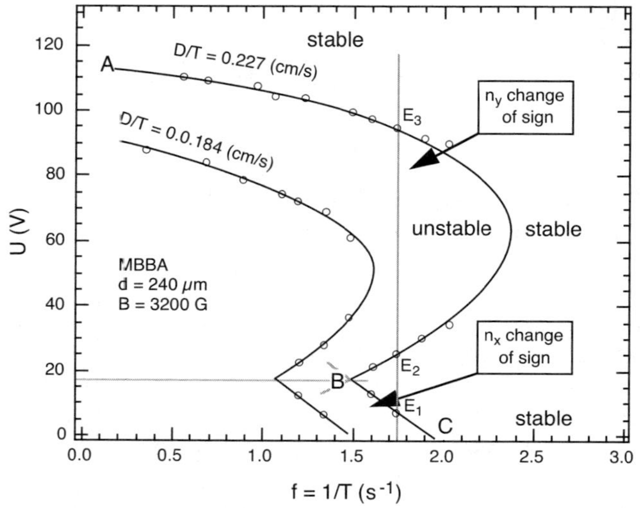

Fig. B.III.36 "Roll" instability diagram for MBBA under oscillating shear (from ref. [22b]).

In figure B.III.36, we present the experimental instability diagram obtained by varying the shear frequency f and the voltage U, at constant magnetic field B and shear velocity D/T.

Let us first consider the curve ABC, obtained at a velocity D/T = 0.227 cm/s. To the right of the curve, the initial orientation **n//z** remains stable, while to the left one can observe the formation of convection rolls, shown under the polarizing microscope in figure B.III.33b. Note that the curve exhibits two branches AB and BC, separated by the cusp B. Thus, by continuously increasing the electric field, at a fixed frequency (say, f = 1.75 Hz) from zero, the borderline ABC is crossed three times. More specifically, four electric field intervals can be defined, and the system has a different behavior in each of them. We shall describe them one by one:

– In the first interval $[0, E_1]$, the roll instability is observed. Its analysis under the polarizing microscope yields the following symmetry relations:

$$n_x(y,z,t+T/2) = -n_x(y,z,t) \qquad \text{(B.III.146a)}$$

$$n_y(y,z,t+T/2) = n_y(y,z,t) \qquad \text{(B.III.146b)}$$

This instability mode under oscillating shear shall be termed "**X-mode,**" as it is the x component of the distortion that changes sign when the shear direction is reversed (see Fig. B.III.37a).

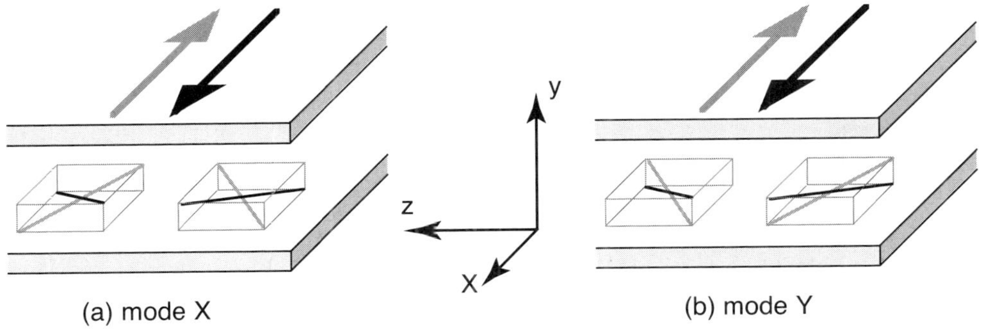

Fig. B.III.37 The two instability modes under oscillating shear. a) X-mode: the x component of the distortion changes sign with the shear; b) Y-mode: the y component of the distortion changes sign with the shear.

– In the second interval $[E_1, E_2]$, the rolls disappear. The system is **stable**.

– In the third interval $[E_2, E_3]$, rolls are again observed. Their analysis under the polarizing microscope leads to the following symmetry relations:

$$n_x(y,z,t+T/2) = n_x(y,z,t) \qquad \text{(B.III.147a)}$$

$$n_y(y,z,t+T/2) = -n_y(y,z,t) \qquad \text{(B.III.147b)}$$

In contrast to the first regime, we shall now speak of a **"Y-mode"** (see Fig. B.III.37b).

– In the last interval $[E_3, \infty]$, i.e., for strong fields, the rolls disappear: one can speak of **restabilization**.

The exact *ab initio* analysis, starting from the equations of nematodynamics, of the instabilities under oscillating shear is too complex to tackle here. Nevertheless, a very simple model which preserves the essential physics of the problem is edifying. This model is based upon the following evolution equations for the amplitudes $n_{xo}(t)$ and $n_{yo}(t)$ of the periodical distortions:

$$\frac{dn_{xo}}{dt} + \frac{n_{xo}}{T_x} + A_1\, s(t) n_{yo} = 0 \qquad \text{(B.III.148a)}$$

$$\frac{dn_{yo}}{dt} + \frac{n_{yo}}{T_y} + A_2\, s(t) n_{xo} = 0 \qquad \text{(B.III.148b)}$$

These equations are very similar to formulae B.III.142, except for the presence of the dissipative terms in $d\mathbf{n}/dt$.

The characteristic relaxation times T_x and T_y associated with the elastic, magnetic, and dielectric torques can be written as:

$$\frac{1}{T_x} = \frac{1}{T_x^e} + \frac{\chi_a B^2/\mu_o}{\gamma_x} \qquad \text{(B.III.149a)}$$

$$\frac{1}{T_y} = \frac{1}{T_y^e} + \frac{\chi_a B^2/\mu_o}{\gamma_y} - \frac{\varepsilon_o \varepsilon_a E^2}{\gamma_y} \qquad \text{(B.III.149b)}$$

with γ_x and γ_y some effective rotational viscosities.

The last terms in eqs. B.III.148 represent the action of shear on the director and introduce a coupling between its two components. These coupling terms form a feedback loop which induces the instability. One should note here that these terms stem from the roll flow engendered by the already discussed bulk forces. Thus, the phenomenological coefficients A_1 and A_2 depend on $(\alpha_5 - \alpha_2)/2$ and $(\alpha_5 + \alpha_2)/2$. Finally, $s(t)$ is the shear rate, which reads

$$s(t) = \frac{D\pi}{Td}\cos(2\pi f t) = s_o \cos(2\pi f t) \qquad \text{(B.III.150)}$$

in the case of sinusoidal excitation.

As the results of figure B.III.36 were obtained at constant magnetic field, the relaxation time T_x itself is a constant and can be used as a time unit to form the following dimensionless variables:

$$\tilde{t} = t/T_x, \qquad \tilde{f} = f T_x \quad \text{and} \quad r = (T_x/T_y)^{1/2} \qquad \text{(B.III.151)}$$

It is also convenient to rescale the distortions

$$\tilde{n}_x = n_{xo}\,(B/A)^{1/2} \qquad \text{and} \qquad \tilde{n}_y = n_{yo} \tag{B.III.152}$$

in order to simplify eqs B.III.148:

$$\frac{d\tilde{n}_x}{d\tilde{t}} + \tilde{n}_x + C\,s(\tilde{t})\,\tilde{n}_y = 0 \tag{B.III.153a}$$

$$\frac{d\tilde{n}_y}{d\tilde{t}} + r\tilde{n}_y + C\,s(\tilde{t})\,\tilde{n}_x = 0 \tag{B.III.153b}$$

where

$$C = (A_1A_2)^{1/2} s_0 T_x \qquad \text{and} \qquad s(\tilde{t}) = \cos(2\pi\tilde{f}\tilde{t}) \tag{B.III.154}$$

Numerically integrating eqs. B.III.153 yields the instability diagram in figure B.III.38. This latter shares some common features with the experimental diagram of figure B.III.36:

First and foremost, the presence of the two previously described instability modes X and Y. The X-mode, for which the n_x distortion changes sign when the shear is reversed, is obtained for $r < 1$, i.e., for $T_x < T_y$. Conversely, the numerical calculation gives the Y-mode for $T_x > T_y$. As for a zero field the X-mode is observed, and as $r < 1$ for this mode, one must have $T_x < T_y(E = 0)$. However, MBBA has negative dielectric anisotropy ($\varepsilon_a < 0$), so the relaxation time T_y decreases with increasing electric field (eq. B.III.149b) such that, for a certain critical field E_B, $T_x = T_y(E_B)$. We conclude that for $E > E_B$ the X-mode is replaced by the Y-mode, in agreement with the experimental observations.

The second common point between the experimental and theoretical instability diagrams is the presence of an upper boundary for the electric field, $E_{max} = U_{max}/d$, above which the **n**//**z** position of the director is again stable. The existence of this upper boundary is not surprising, as the electric field has a strong stabilizing effect.

The field value E_{max} can be analytically determined starting from the observation that, for $r > 1$, the $\tilde{n}_x$ distortion does not change sign, such that its average value $<\tilde{n}_x>$ is not zero. Replacing $\tilde{n}_x$ by its average value in eq. B.III.153b, one can integrate this equation analytically to obtain:

$$\tilde{n}_y = -C <\tilde{n}_x> \frac{2\pi\tilde{f}\,\sin(2\pi\tilde{f}\tilde{t}) + r\cos(2\pi\tilde{f}\tilde{t})}{r^2 + (2\pi\tilde{f})^2} \tag{B.III.155}$$

On the other hand, the average value $<\tilde{n}_x>$ can be determined from eq. B.III.153a:

$$<\tilde{n}_x> = -C < s(\tilde{t})\,\tilde{n}_y > \tag{B.III.156}$$

giving, after replacing $\tilde{n}_y$ by its expression B.III.155, the following self-consistency relation:

$$<\tilde{n}_x> = <\tilde{n}_x> \frac{1}{2} \frac{C^2 r}{r^2 + (2\pi\tilde{f})^2} \qquad \text{(B.III.157)}$$

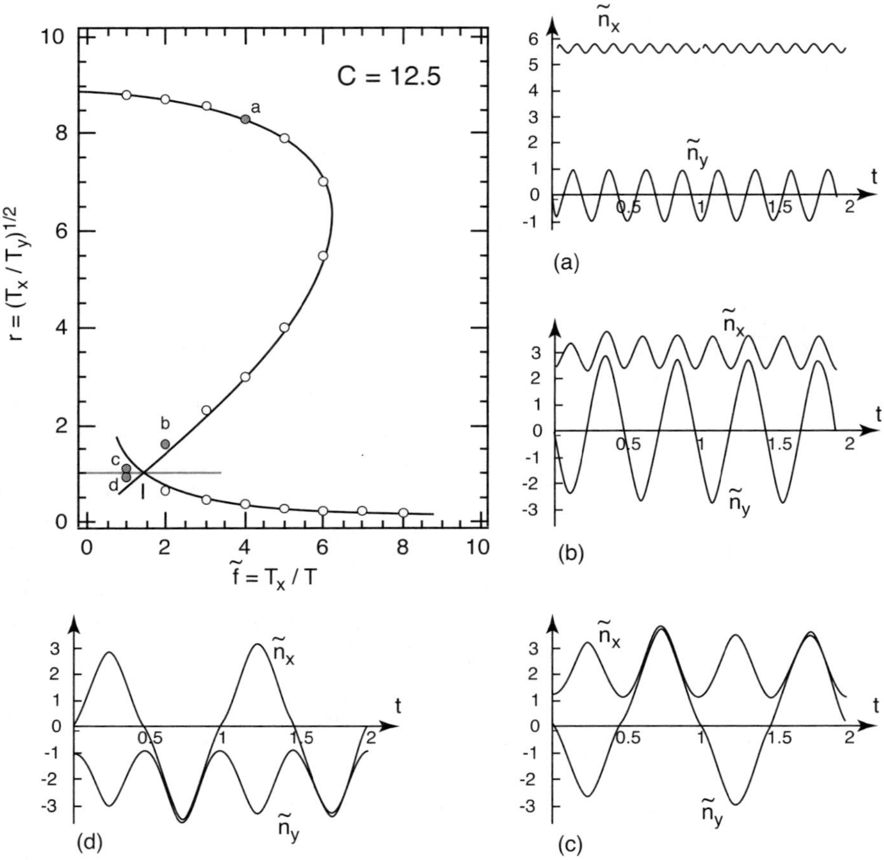

Fig. B.III.38 Theoretical instability diagram under sinusoidal oscillating shear. The open dots were obtained by numerical integration of equations B.III.153. The shape of the solutions is illustrated by graphs a,b,c (for the Y-mode) and d (for the X-mode). Solid lines in the instability diagram have been plotted from the approximate analytical solutions.

Finally:

$$\tilde{f}^2 = \frac{r(C^2 - 2r)}{2(2\pi)^2} \qquad \text{(B.III.158)}$$

This approximate solution, represented in the theoretical diagram, has the same upper limit as the exact numerical solution:

$$\left(\frac{T_x}{T_y}\right)_{max} = \frac{C}{2^{1/2}} = 8.84 \qquad \text{(B.III.159)}$$

with C = 12.5.

Following exactly the same procedure, one can get an approximate solution for the X-mode and the corresponding self-consistency relation:

$$<\tilde{n}_y> = <\tilde{n}_y> \frac{1}{2r} \frac{C^2}{1+(2\pi\tilde{f})^2} \qquad \text{(B.III.160)}$$

Thus, the stability limit of the X-mode is:

$$r = \frac{1}{2} \frac{C^2}{1+(2\pi\tilde{f})^2} \qquad \text{(B.III.161)}$$

This function is plotted in the diagram of figure B.III.38.

The analytical approximations for the X- and Y-modes cross at the point I with coordinates

$$\tilde{f} = \left(\frac{C^2 - 2}{8\pi^2}\right)^{1/2}, \qquad r = 1 \qquad \text{(B.III.162)}$$

Its position in the stability diagram is close to that of the cusp B in the experimental diagram. Unfortunately, this resemblance is misleading, as the numerical integration of the equation system B.III.153 for r = 1 yields the solution $\tilde{f} = 0$, which means that the equation system B.III.153 is not sufficient to describe the instabilities close to r = 1.

III.4.h) Elliptical shear: director precession

An elliptical shear being only the superposition of two oscillating linear shear deformations with a certain phase shift, one could imagine that it will bring nothing new with respect to the situation already discussed. Nevertheless, we shall see that this is not the case and that nematics still have something in store for us. The phenomena discussed in this section were first observed experimentally (by chance, *nota bene*) before being explained by the theory of nematodynamics.

The experimental setup described in figure B.III.39 [22a] can create elliptical shear. The liquid crystal (MBBA or another nematic with negative dielectric anisotropy) is held by capillarity between two transparent electrodes treated for homeotropic anchoring. The electrodes are mounted on organic glass rods. Each rod is fixed to the membrane of a speaker and supported by a flexible strip. One rod oscillates along x and the other along y.

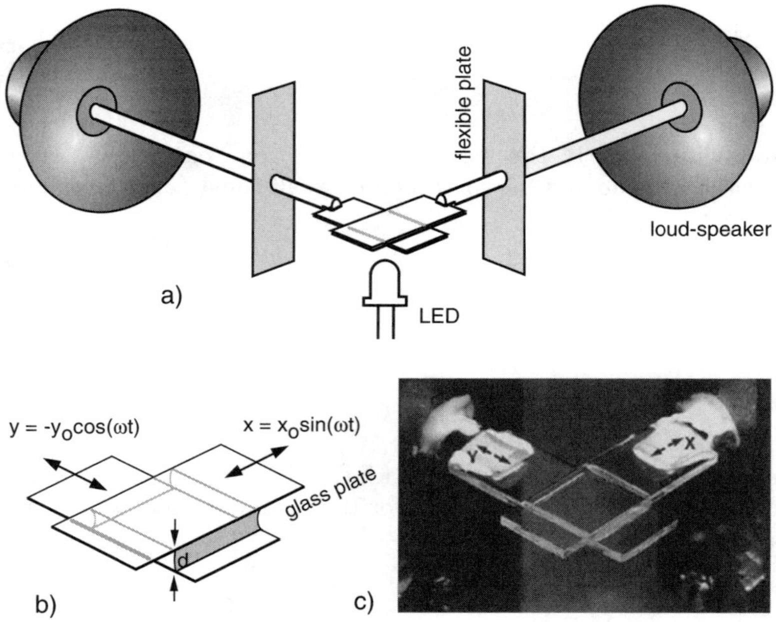

Fig. B.III.39 Setup used for studying the action of an elliptic shear on a nematic. The distortion of the director field induced by the shear can be detected by stroboscopic conoscopy. The stroboscopic effect is obtained using a light-emitting diode (LED) which provides monochromatic lighting and is powered by current pulses at a frequency $\omega + \Delta\omega$ (from ref. [22a]).

The speakers are powered at the same frequency, but with a $\pi/2$ phase shift and variable voltage amplitudes, and the motion of the glass plates is given by:

$$x = x_o \cos(\omega t) \qquad \text{(upper plate)} \tag{B.III.163a}$$

$$y = -\, y_o \sin(\omega t) \qquad \text{(lower plate)} \tag{B.III.163b}$$

At frequencies low enough so that the viscous penetration length

$$\delta = \sqrt{\frac{2\nu}{\omega}} \qquad \text{where} \qquad \nu = \eta / \rho \tag{B.III.164}$$

is much larger than the distance d between the glass plates, the velocity profiles are linear and their superposition results in elliptic shear. To put it differently, with respect to the lower plate, the motion of a point P (described by the position vector **r**) on the upper plate is elliptical, of equation:

$$\mathbf{r}(t) = (x_o\cos(\omega t),\ y_o\sin(\omega t),\ 0) \tag{B.III.165}$$

In the absence of an applied electric field, the initial orientation of the liquid crystal is $\mathbf{n}_o = (0,0,1)$. Applying a low amplitude circular shear ($r = x_o = y_o$) with velocity

$$\mathbf{v}(z,t) = \frac{\dot{r}z}{d} \qquad \text{(B.III.166)}$$

perturbs this orientation. By conoscopy and using a stroboscope the director can be shown to describe a cone of axis **z**, with an opening angle $\theta_0 \approx r/d$.

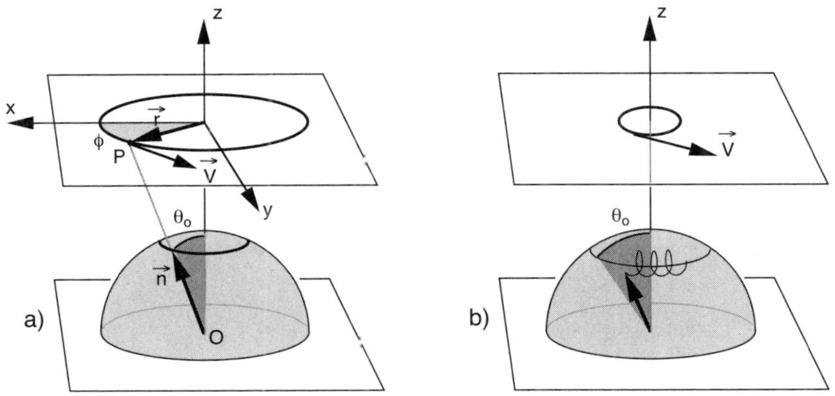

Fig. B.III.40 Director trajectories under circular shear. a) Without an electric field, the director describes a cone with axis **z** at an angular velocity $d\phi/dt = \omega$; b) when an electric field above the Frederiks transition threshold is applied, the average director orientation (defined by $\theta_0 = \langle\theta(z = d/2, E, t)\rangle$ where <...> denotes a time average) slowly precesses with an angular velocity $\Omega \ll \omega$.

Representing the successive director orientations by points on the unit sphere (see chapter B.IV), the trajectory follows a parallel. This director motion **n**(t) is in phase with the **r**(t) trajectory of the upper plate, as shown in the diagram of figure B.III.40a. Consequently,

$$\mathbf{n} = (\sin\theta_0\cos\phi, \sin\theta_0\sin\phi, \cos\theta_0) \quad \text{with} \quad \phi = \omega t \qquad \text{(B.III.167)}$$

where θ_0 and ϕ are, respectively, the polar and azimuthal angles of the director. The director is then always oriented along the OP line and its average orientation <**n**> remains vertical.

When an electric field above the Frederiks transition threshold is applied, it entails a distortion of the director field which can be described by the polar and azimuthal angles $\theta(z,E,t)$ and $\phi(x,y,t)$ (Fig. B.III.41a). This distortion can be determined under the polarizing microscope by observing the texture of the sample between crossed polarizers (Fig. B.III.41b). Under circular shear with a low amplitude r, the texture in figure B.III.41b is deformed as shown in figures B.III.41c and d. The motion of the isogyres shows that the director **c** (the projection of the director **n** onto the (x,y) plane) rotates slowly and regularly with an angular velocity Ω. One can thus conclude that, under circular shear, the director precesses about the **z** axis with the angular velocity Ω. The vibration at frequency ω adds to this slow motion, so the director trajectory on the unit sphere is complex, as illustrated in figure B.III.40b.

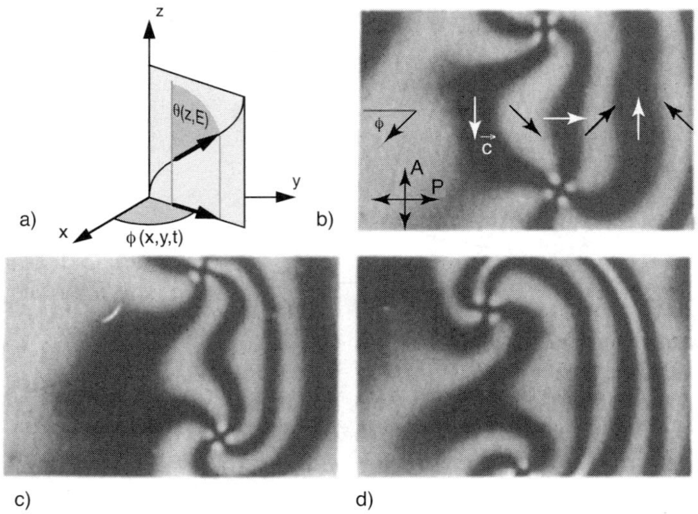

Fig. B.III.41 Effect of the director precession under circular shear. a) The distortion induced by the electric field. b) The texture, as seen under the polarizing microscope. The motion of the isogyres, from b) to d) shows that the azimuthal angle $\phi(x,y,t)$ of the average distortion at an angular velocity $\Omega = d\langle\phi\rangle/dt$ (from ref. [22c]).

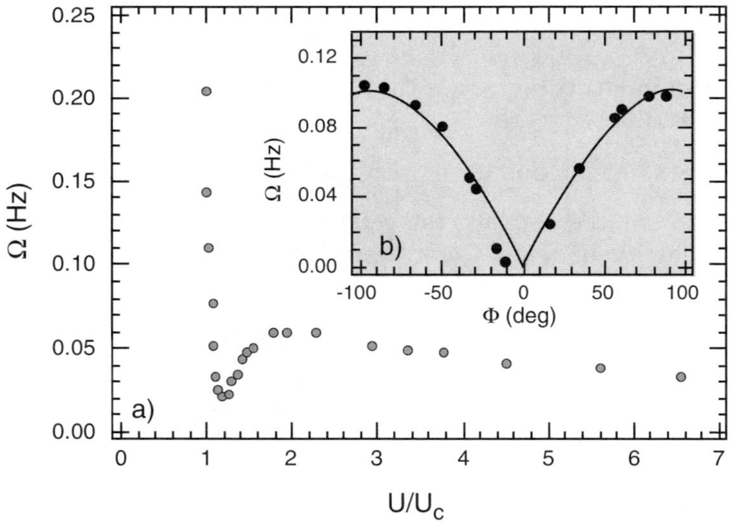

Fig. B.III.42 Measuring the precession velocity: a) as a function of the applied voltage under circular shear, b) as a function of the phase shift Φ between the motions along x and y: at $\Phi = 0°$ the shear is linear and becomes circular for $\Phi = \pm 90°$. Merck phase 5A, $f = \omega/2\pi = 155$ Hz (from ref. [22d]).

T. Börzsönyi et al. [22d] measured the precessional angular velocity Ω as a function of the parameters of the elliptic shear (ω, x_o, y_o) and as a function of the applied voltage. The angular velocity Ω strongly varies with the voltage close to the threshold of the Frederiks transition (see Fig. B.III.42), but tends towards a constant in the strong field limit $U >> U_c$. These authors also measured Ω, at constant shear amplitudes along x and y as a function of the phase shift Φ, defined by: $\mathbf{r} = A(\cos(\omega t), \cos(\omega t - \Phi), 0)$. The experimental curve in figure B.III.42b shows that $\Omega = 0$ for linear shear ($\Phi = 0$).

Finally, the authors showed that Ω is proportional to the shear frequency f (Fig. B.III.43a) and to the square of the shear amplitude A (Fig. B.III.43b).

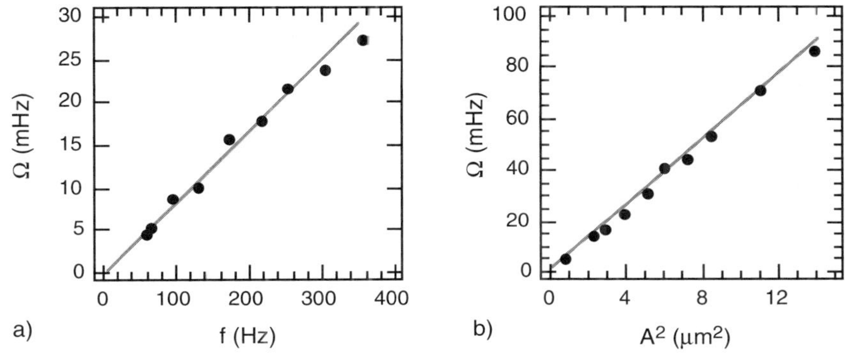

Fig. B.III.43 Measuring the precession velocity: a) as a function of the shear frequency f; $x_o = 0.8\ \mu m$, $y_o = 1\ \mu m$, $d = 20\ \mu m$; b) as a function of the amplitude $A = x_o = y_o$ of the circular shear; $f = 122$ Hz, $U/U_c = 2.3$ (Merck phase 5A) (from ref. [22d]).

All these experimental results are well described by the following theoretical formula:

$$\Omega \approx -\frac{\omega x_o y_o}{2d^2}\frac{\alpha_2\alpha_3}{\gamma_1} \tag{B.III.168}$$

found by T. Börzsönyi et al. [22d]. For a simplified proof of eq. B.III.168, we shall only consider the strong field limit, when the director is confined to the (x, y) plane. Its initial orientation is $\mathbf{n} = (1, 0, 0)$. Let us also assume that the frequency ω is much larger than the reciprocal of the characteristic time of elastic relaxation $\tau = \gamma_1 d^2/K$. In this limit, the elastic torques can be neglected and the equation of motion for the director can be simply written as:

$$\Gamma_y = 0 = \gamma_1\frac{\partial n_z}{\partial t} + \alpha_3\frac{\partial v_x}{\partial z} \tag{B.III.169a}$$

$$\Gamma_z = 0 = -\gamma_1\frac{\partial n_y}{\partial t} - \alpha_2\frac{\partial v_y}{\partial z}n_z \tag{B.III.169b}$$

Obviously, these equations are only valid to first order in the perturbations $\delta\mathbf{n} = (0,n_y,n_z)$, with $|n_y|$, $|n_z| << 1$. We shall see later that in order to fulfill this condition the amplitude of motion for the two plates must be small with respect to the sample thickness d.

Suppose first that only the shear along x is applied to the nematic, initially oriented along the same direction x. The motion of the director is given by eq. B.III.169a, describing the balance between the hydrodynamic torque exerted on the molecules by the shear (see section B.III.4d) and the viscous torque opposing director rotation. In these conditions

$$n_z = -\frac{\alpha_3}{\gamma_1}\frac{x_o}{d}\sin(\omega t) \qquad \text{(B.III.170)}$$

The director oscillates around the average orientation $\mathbf{n} = (1,0,0)$.

Let us now add the shear along y. By replacing B.III.170 in eq. B.III.169b, we have:

$$\frac{\partial n_y}{\partial t} = -\frac{\omega y_o}{d}\frac{x_o}{d}\frac{\alpha_2\alpha_3}{\gamma_1^2}\sin^2(\omega t) \qquad \text{(B.III.171)}$$

whence

$$n_y = -\frac{\omega y_o x_o}{2d^2}\frac{\alpha_2\alpha_3}{\gamma_1^2}\left[t - \frac{\sin(2\omega t)}{2\omega}\right] \qquad \text{(B.III.172)}$$

This formula shows that, due to the $\pi/2$ shift between the two shear motions, the n_y component acquires an average angular velocity proportional to $\alpha_2\alpha_3/\gamma_1^2$. Furthermore, during the precession, the director trajectory on the unit sphere oscillates about the equator (Fig. B.III.44).

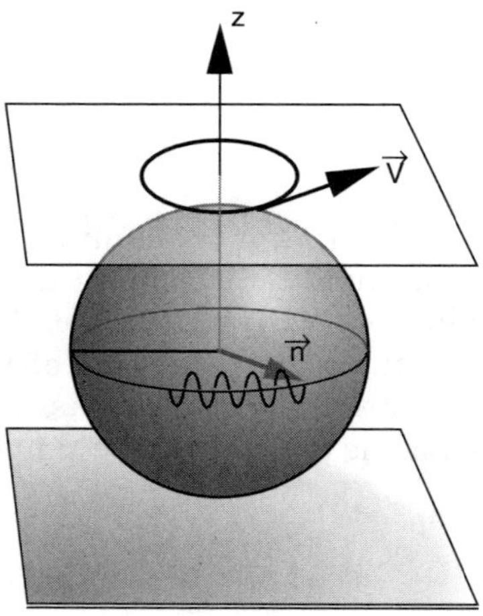

Fig. B.III.44 Director precession under elliptic shear in strong field. When α_3 is negative, the director precession follows the shear.

Let us point out that, in the materials employed, $|\alpha_3| << |\alpha_2|$, such that $\gamma_1 \approx -\alpha_2$ and that Ω is proportional to the α_3/α_2 ratio (up to a sign change). Measuring Ω then determines α_3/α_2, and thus the sign of α_3. Also, as the precession angular velocity Ω depends on the thickness, the phase will become twisted (see also chapter C.VIII of volume 2) if the two glass plates are not perfectly parallel. Indeed, suppose there is a thickness gradient along x:

$$d = d_0 + \alpha x \tag{B.III.173}$$

If at the moment t = 0, the azimuthal angle ϕ is uniformly null ($\phi = 0$), then, after a time t, one has:

$$\phi(x,t) = \Omega(x,t)\,t \approx -\frac{\omega x_0 y_0}{2}\frac{\alpha_2\alpha_3}{\gamma_1}\frac{t}{(d_0+\alpha x)^2} \tag{B.III.174}$$

so the distortion

$$\frac{d\phi}{dx} \approx 2\alpha t\,\frac{\omega x_0 y_0}{2d_0^3}\frac{\alpha_2\alpha_3}{\gamma_1} \tag{B.III.175}$$

increases with time. Optically, this is shown by an accumulation of isogyres in the sample (Fig. B.III.45). However, the phase gradient does not increase indefinitely, for when the distortion exceeds a certain threshold, pairs of dislocations of opposite sign nucleate in the isogyre system and their movement ("climb") allows the distortion to relax.

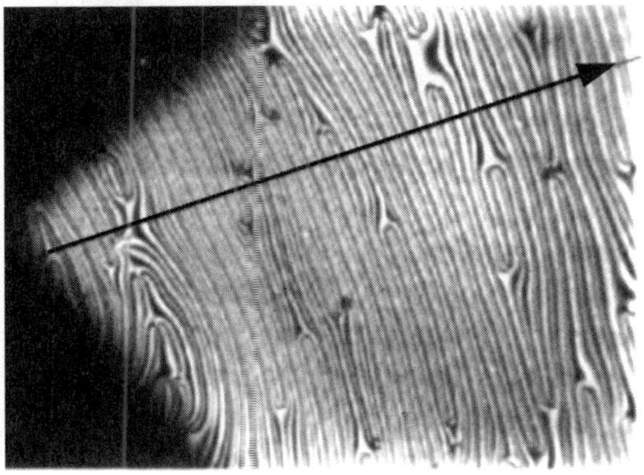

Fig. B.III.45 Winding of the phase ϕ in a wedge-shaped sample. The arrow indicates the direction of thickness gradient (from ref. [22c]).

III.4.i) Instabilities under elliptic shear

As we just saw, in the absence of an electric field and for low elliptic shear rates, the average director orientation remains the same as at equilibrium: < **n** > // **z**. This changes with increasing shear rate and, above a certain threshold, periodic rolls appear in the sample, optically similar to the ones observed under linear shear (Fig. B.III.33b).

These novel instabilities can be detected by their influence on the diffraction pattern of a laser beam. The photos in figure B.III.46 show the diffraction pattern for several shear geometries. The main feature of the rolls is that they make with the major axis of the shear ellipse an angle Ψ of about 70°. This angle changes sign when the ellipse is described in the opposite direction. The rolls then transform into a square two-dimensional pattern at higher shear rate. When the shear is circular, the rolls disappear and the system goes directly towards the square texture.

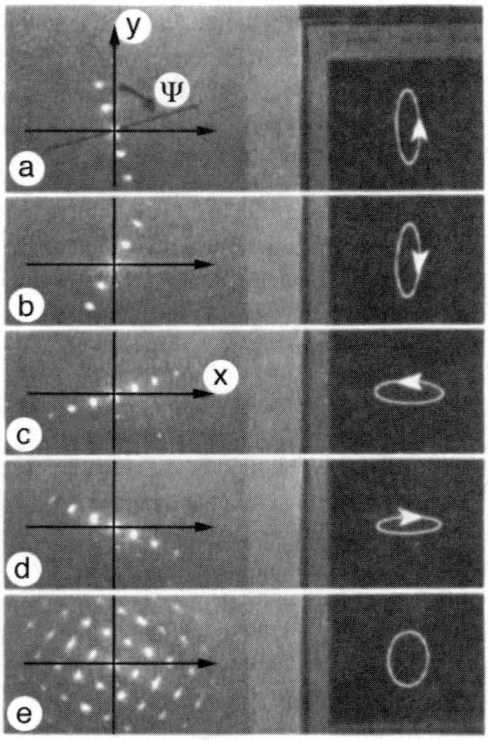

Fig. B.III.46 Hydrodynamic instabilities under elliptic shear (MBBA). Left: diffraction pattern of a laser beam produced by the periodic distortion. Right: shear geometry. In the case of elliptic shear (a – d) the rolls that appear make an angle $\pm\Psi$ with the major axis of the ellipse. The angle changes sign when the shear direction is reversed. When close to circular shear (e), two-dimensional instabilities develop (from ref. [22a]). The experimental phase diagram also shows that a second dimensionless parameter – the shear ellipticity defined by the ratio x_0/y_0 – plays an essential role, evidenced by the appearance of a cusp on the stability curve at the point $x_0/y_0 = 1$.

A typical phase diagram, obtained by E. Guazzelli and E. Guyon [22e] at constant thickness and shear frequency, is shown in figure B.III.47. Note that the stability limit of the homeotropic sample is close to a hyperbola of equation $x_0 y_0 = 137\ \mu m^2$. Other measurements, at different values of thickness and excitation frequency, show that this approximate instability threshold is given by a critical value of the Ericksen number

$$Er = \frac{\omega x_0 y_0}{D} \qquad \text{(B.III.176)}$$

with D the orientational diffusivity.

J. Sadik et al. [22f] theoretically showed that the position of this cusp on the diagonal of the stability diagram is related to this latter being invariant under an exchange of the x_0 and y_0 amplitudes (mirror symmetry), an operation that also preserves the Ericksen number (eq. B.III.176).

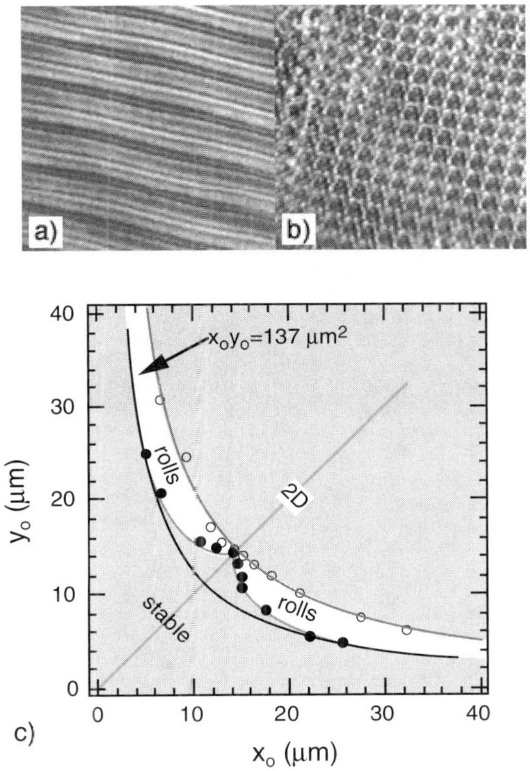

Fig. B.III.47 Measuring the instability threshold in an MBBA sample of thickness d = 50 μm at a frequency f = 150 Hz. The solid dots show the stable → rolls transition. The open dots represent the transition between rolls (a) and 2D structures (b) (from ref. [22e]).

Finally, let us note that E. Guazzelli and A.-J. Koch [22g] performed experiments under periodic (but no longer elliptic) shear, by superposing along the x and y axes oscillations with frequencies ω and nω, for the integral value of

n. The parametric equations x = x(t) and y = y(t) then define curves resembling flowers with n ± 1 petals.

III.5 Convective instabilities of electrohydrodynamic origin

In chapter B.II, we noted that under certain conditions an electric field (AC or DC), could induce convection rolls and hinder the observation of the Frederiks instability. Experiments also show that a similar flow can appear when the applied electric field stabilizes the initially selected molecular orientation. This happens for planar anchoring conditions (Fig. B.III.48) if the sample exhibits negative dielectric anisotropy ($\varepsilon_a < 0$ in MBBA, for instance). This is a novel instability, different in origin from the mechanical Frederiks instability.

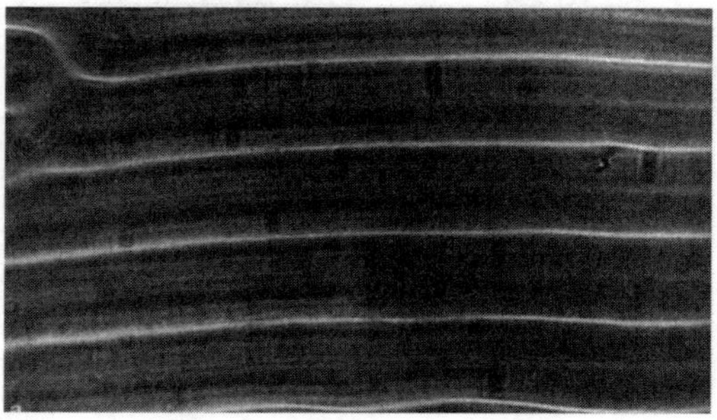

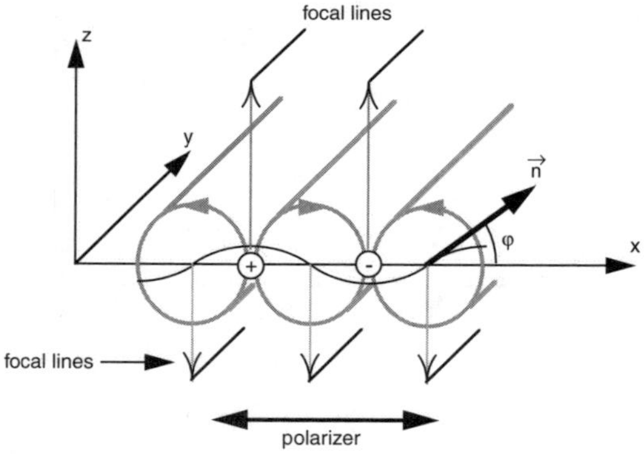

Fig. B.III.48 Convection rolls (Williams domains [26]) in a planar MBBA sample. They are perpendicular to the average director alignment. V = 6 V and f = 60 Hz. The lighting is polarized parallel to the director. Only the extraordinary ray propagates. Being sensitive to the variations of the local index, this ray is deflected by the director field distortions. If the focus is slightly above the sample, the bright lines where the rays converge (caustics) can be seen. It can be seen that the sample behaves like a phase grating. In natural light, the rolls disappear completely (photo by Y. Galerne).

In order to describe this instability we need to take into account a new ingredient, namely the existence of **bulk electric charges**. These charges, positive and negative ions, are always present in the samples. If the frequency of the electric field is lower than the frequency of charge relaxation, these can move (**conduction regime**) and accumulate in the space regions where the director field lines exhibit maximal curvature. This **focalization** effect is due to the anisotropy in electrical conductivity of the nematic. The ions then have a twofold destabilizing action, as their movement and the local field they create engenders dielectric and viscous destabilizing torques (Fig. B.III.49). These torques are countered by a restoring elastic torque tending to "smooth out" the distortions in the director field. The existence of a threshold field for the appearance of the convective structures is a consequence of the interplay between these torques. To understand these mechanisms better, let us now write down the equations of the problem. The following model is due to Carr [23], Helfrich [24], Dubois-Violette, de Gennes, and Parodi [25].

III.5.a) Theoretical model

For definiteness, we assume that the anchoring is planar and that the material has negative dielectric anisotropy ($\varepsilon_a < 0$). The electrodes are in $z = \pm d/2$ and the director is at rest, parallel to the x axis (Fig. B.III.48b). To keep the calculations as simple as possible, we further assume that all relevant quantities vary as $\exp(iq_x x)$ (one-dimensional model). Experimentally, the periodicity of the rolls is close to 2d ($q_x^c \approx \pi/d$) and we shall use this information when necessary. It can be proven using a two-dimensional model that takes into account thickness effects, but our goal here is to describe the origin of the convective phenomenon using the least algebra possible.

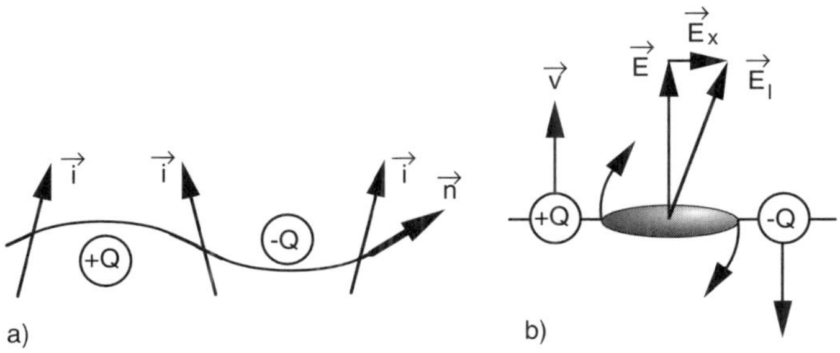

Fig. B.III.49 Instability mechanism in the conducting regime. a) A periodical variation in the curvature of the directory field lines leads to the accumulation of space charges ± Q; b) the periodical charge distribution engenders a destabilizing electric torque. This torque is amplified by the hydrodynamic shear created by the motion of the charges.

The important quantities are:

– the applied electric field $E(t) = V(t)/d$ of angular frequency ω;
– the local bend distortion of the field **n**, described by the local angle $\varphi = n_z(x,t)$, or by the curvature of the field lines $\psi(x,t) = \partial\varphi/\partial x$;
– the density $Q(x,t)$ of the space charge, taken as zero in the absence of the electric field. We shall also assume the electrodes are perfect and do not inject any charges in the medium;
– the velocity hydrodynamic $v_z(x,t)$.

First of all, these variables must satisfy the equation of torque balance (along y)

$$-K_3\frac{\partial^2\varphi}{\partial x^2} - \varepsilon_a\varepsilon_o E^2\left(\varphi + \frac{E_x}{E}\right) + \gamma\left(\dot{\varphi} - \frac{\partial v_z}{\partial x}\right) = 0 \qquad \text{(B.III.177)}$$

In this equation, the K_3 term represents the elastic restoring torque. The second term is the dielectric torque, the tilt angle $\varphi + E_x/E$ being measured with respect to the local field $\mathbf{E}_l$ resulting from the superposition of the applied field E and the transverse field E_x induced by the charges Q (Fig. B.III.50). The last term is the viscous torque acting on the director: its first φ component is stabilizing while the other is destabilizing and stems from the coupling between the director and the flow. For the sake of simplicity, we assumed that $\gamma_1 = -\gamma_2 = \gamma$.

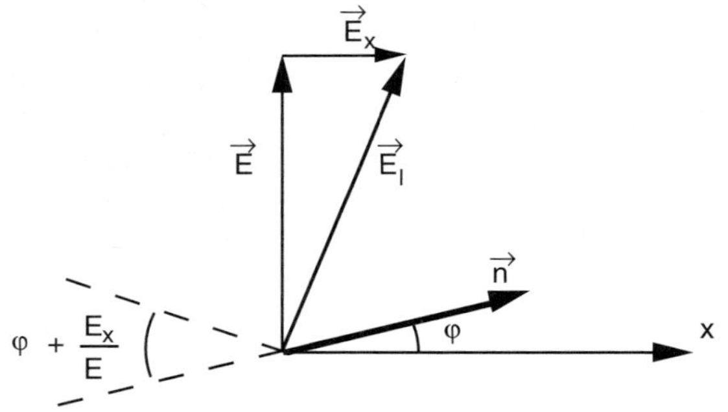

Fig. B.III.50 Position of the director with respect to the local field.

The field E_x is given by the Maxwell equation

$$\operatorname{div}\mathbf{D} = \varepsilon_{//}\varepsilon_o\frac{dE_x}{dx} + \varepsilon_a\varepsilon_o E\frac{d\varphi}{dx} = Q \qquad \text{(B.III.178)}$$

with **D** the electric displacement vector.

By derivatizing eq. B.III.177 with respect to x and using the previous equation, one can get an equation for the local curvature ψ of the director field lines:

$$K_3 q_x^2 \psi - \varepsilon_a \varepsilon_o E^2 \left(\frac{\varepsilon_\perp}{\varepsilon_{//}} \psi + \frac{1}{\varepsilon_o \varepsilon_{//}} \frac{Q}{E} \right) + \gamma(\dot{\psi} + q_x^2 v_z) = 0 \qquad \text{(B.III.179)}$$

which can also be written as

$$\dot{\psi} + \frac{\psi}{T_o} = \frac{\varepsilon_a}{\varepsilon_{//}} + \frac{QE}{\gamma} - q_x^2 v_z \qquad \text{(B.III.180)}$$

where T_o is the relaxation time of a bend deformation in the presence of an electric field:

$$T_o^{-1} = \frac{K_3 q_x^2 - \varepsilon_a \varepsilon_o \dfrac{\varepsilon_\perp}{\varepsilon_{//}} E^2}{\gamma} \qquad \text{(B.III.181)}$$

As there is an additional variable Q, one needs one more equation, which is the one of charge conservation:

$$\frac{\partial Q}{\partial t} + \operatorname{div} \mathbf{j} = 0 \qquad \text{(B.III.182)}$$

where $\mathbf{j} = \underline{\sigma}\mathbf{E}$ is the current density. We denote by $\sigma_{//}$ (resp. $\sigma_\perp$) the conductivity in the direction parallel (resp. perpendicular) to the director and $\sigma_a = \sigma_{//} - \sigma_\perp$. This anisotropy can be significant. For instance, in MBBA $\sigma_a > 0$ and $\sigma_a/\sigma_{//} \approx 50\%$.

We obtain:

$$\operatorname{div} \mathbf{j} = \sigma_{//} \frac{dE_x}{dx} + \sigma_a E \frac{d\varphi}{dx} \qquad \text{(B.III.183)}$$

which, after substitution in eq. B.III.182 and using eq. B.III.178 yields:

$$\dot{Q} + \frac{Q}{\tau} + \sigma_H E \psi = 0 \qquad \text{(B.III.184)}$$

with

$$\tau^{-1} = \frac{\sigma_{//}}{\varepsilon_o \varepsilon_{//}} \quad \text{and} \quad \sigma_H = \left(\frac{\varepsilon_\perp}{\varepsilon_{//}} - \frac{\sigma_\perp}{\sigma_{//}} \right) \qquad \text{(B.III.185)}$$

τ^{-1} represents a **charge relaxation frequency** (experimentally, of the order of a few hundred Hz), while in the expression of σ_H, the σ_a term is often dominant.

Now we only need to write the momentum conservation equation for the v_z component of the velocity:

$$\rho \dot{v}_z = -\frac{d}{dx} \Gamma_y^v + \eta_a \frac{\partial^2 v_z}{\partial x^2} + QE \qquad \text{(B.III.186)}$$

It is easy to recognize here the viscous coupling between the flow and the director rotation, the usual viscous force and the force acting on the bulk charges. Taking $\gamma_1 = |\gamma_2| = \gamma$, we can rewrite this equation as

$$\dot{v}_z + \frac{v_z}{T_v} = \frac{QE}{\rho} - \gamma\frac{\dot{\psi}}{\rho} \tag{B.III.187}$$

with

$$T_v^{-1} = \frac{(\gamma + \eta_a)\, q_x^2}{\rho} \tag{B.III.188}$$

T_v is the diffusion time of the vorticity across a convection roll, of the order of 10^{-5} to 10^{-3} s.

The problem is much simpler if the frequency $f = \omega/2\pi$ of the applied field is small before T_v^{-1}. Practically, this condition requires $f << 10$ kHz for a 100 μm sample. Thus, it is not very restrictive and usually fulfilled in experiments. One then has $v_z \propto (QE - \gamma\psi)$ and the equations B.III.184 and B.III.180 become, after eliminating v_z:

$$\dot{Q} + \frac{Q}{\tau} + \sigma_H E\,\psi = 0 \tag{B.III.189}$$

$$\dot{\psi} + \frac{\psi}{T_E} + \frac{1}{\eta_H} QE = 0 \tag{B.III.190}$$

with

$$T_E = \frac{T_o\eta_a}{\gamma + \eta_a} \qquad \text{and} \qquad \eta_H = \frac{\eta_a}{1 - \frac{\varepsilon_a}{\varepsilon_{//}}\left(1 + \frac{\eta_a}{\gamma}\right)} \tag{B.III.191}$$

Typical values for MBBA are $\tau^{-1} \approx 200$ Hz and $T_E^{-1} \approx 1$ Hz in the convective regime, the last value being calculated for a sample thickness of 100 μm and an applied electric field of 600 V/cm. Hence, the curvature elastic deformations relax much more slowly than the charges.

Eqs. B.III.189 and B.III.190 show that the space charge Q and the curvature ψ are parametrically coupled. Generally, we shall look for a solution of the following form (omitting the $\cos(q_x x)$ factor):

$$Q(t) = Q_o(t)\, e^{st}, \qquad \psi(t) = \psi_o(t)\, e^{st} \tag{B.III.192}$$

where Q_o and ψ_o are periodical functions with the same angular frequency ω as the field $E = E_o \cos\omega t$. The instability develops when the real part of the growth rate s goes to zero.

For $\omega = 0$ and close to the threshold, one obtains $Q = -\tau\sigma_H E\psi$ from eq. B.III.189, and eq. B.III.190 becomes:

$$\dot{\psi} + \psi\left(\frac{1}{T_E} - \tau\frac{\sigma_H}{\eta_H}E^2\right) = 0 \qquad \text{(B.III.193)}$$

The instability condition at zero frequency is straightforward

$$\frac{1}{T_E} - \tau\frac{\sigma_H}{\eta_H}E^2 < 0 \qquad \text{(B.III.194)}$$

or, knowing that T_E depends on the electric field (eqs. B.III.191 and B.III.181)

$$E^2 > E_c^2 = \frac{K_3 q_x^2}{\eta_B \lambda}\frac{1}{\xi^2 - 1} \qquad \text{(B.III.195)}$$

with $q_x \approx \pi/d$ and

$$\eta_B = \eta_a \frac{\gamma}{\gamma + \eta_a} \qquad \lambda = -\frac{1}{\eta_B}\varepsilon_a\varepsilon_0\frac{\varepsilon_\perp}{\varepsilon_{//}} \qquad \xi^2 = \frac{\sigma_H \tau}{\eta_H \lambda} \qquad \text{(B.III.196)}$$

ξ^2 is the Helfrich parameter. It must be above 1 for the instability to occur. For MBBA, $\xi^2 \approx 3$.

If the field frequency is finite, but lower than τ^{-1}, the threshold can be estimated by noting that ψ varies very little with time. Thus, we only need to keep the zero-frequency component in its Fourier expansion:

$$\psi(t) \rightarrow \bar{\psi} \qquad \text{(B.III.197)}$$

Q(t) can be taken in the form

$$Q = Q'\cos\omega t + Q''\sin\omega t \qquad \text{(B.III.198)}$$

which gives, after plugging in eq. B.III.189

$$Q'' = \omega\tau Q' \qquad \text{and} \qquad Q' = -\sigma_H\bar{\psi}E_o\tau\frac{1}{1+\omega^2\tau^2} \qquad \text{(B.III.199)}$$

In the first approximation, the charge Q is proportional to the applied electric field (as $\omega\tau << 1$). This is why one speaks of the **conduction regime**. Returning to eq. B.III.190, one can equally see that its action, proportional to QE, is always destabilizing. Indeed, after averaging this equation over a period, we have

$$\frac{\bar{\psi}}{T_E} + \frac{1}{2\eta_H}Q'E_o = 0 \qquad \text{(B.III.200)}$$

This relation fixes the critical instability threshold

$$\bar{E}_c^2 = \frac{1}{2} E_o^2 \approx \frac{K_3 q_x^2}{\eta_B \lambda} \frac{1 + \omega^2\tau^2}{\xi^2 - (1 + \omega^2\tau^2)} \qquad \text{(B.III.201)}$$

where the quantities η_B, λ and ξ were defined in eq. B.III.196. We emphasize that this approximate theory does not allow us to determine q_x, which will be taken as π/d. This threshold field diverges for a cut-off frequency given by $\xi^2 = 1 + \omega^2\tau^2$, i.e.,

$$f_{cutoff} \approx \frac{\sqrt{\xi^2 - 1}}{2\pi} \frac{1}{\tau} \qquad \text{(B.III.202)}$$

Above this value, the conduction regime is replaced by the dielectric regime. The existence of this cut-off is well documented experimentally.

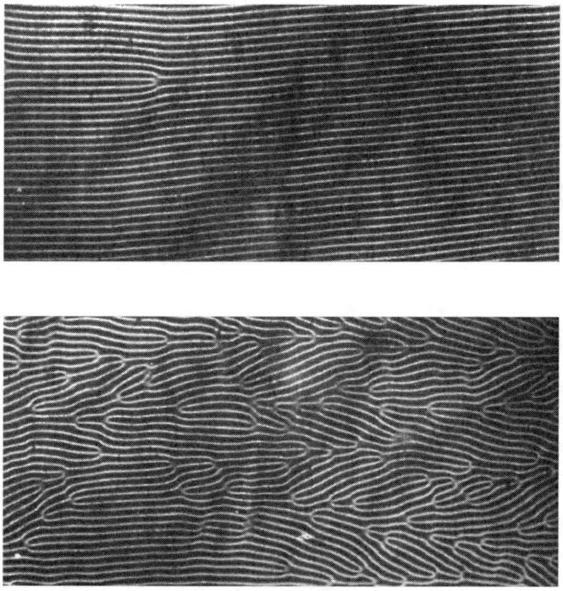

Fig. B.III.51 Stripes in the dielectric regime. Close to the threshold, the stripes are straight (a). By slightly increasing the voltage, an instability leading to a herringbone structure (b) is obtained (photos by Y. Galerne [27b]).

The **dielectric regime** corresponds to the case $\omega\tau >> 1$. Experimentally, it is characterized by much higher threshold fields (at least one order of magnitude) and by the appearance of periodical stripes, with a wavelength much smaller than the sample thickness (Fig. B.III.51) [27].

This regime can be theoretically analyzed using the same equations, B.III.189 and 190, as before, but the calculations are more involved. Most importantly, the wave vector q_x is no longer fixed by the sample thickness, becoming a free parameter one must determine.

We take this occasion to point out that this regime is much better described by the one-dimensional model than the conduction regime. It can be shown that, as the frequency is high, the Q charge is static (hence the name of dielectric regime), and that at the threshold the three terms in eq. B.III.179 are of the same order of magnitude:

$$\varepsilon\varepsilon_o E^2 \approx K_3 q_x^2 \approx \eta\omega \qquad \text{(B.III.203)}$$

where ε and η are known combinations of the dielectric constants and the viscosities. These two relations show that the threshold field is proportional to $\omega^{1/2}$, while the spatial periodicity of the stripes decreases as $\omega^{-1/2}$. These two quantities are also independent of the thickness, in contrast with the conduction regime. These theoretical predictions are well verified experimentally.

III.5.b) Nonlinear regime

To conclude this chapter by showing the wealth of behavior exhibited by nematic liquid crystals, let us qualitatively describe the nonlinear evolution of the system above the threshold [28]. Figure B.III.52 presents the different bifurcations leading to space-time chaos. Between the bifurcations, convective structures with well-defined topologies can be observed.

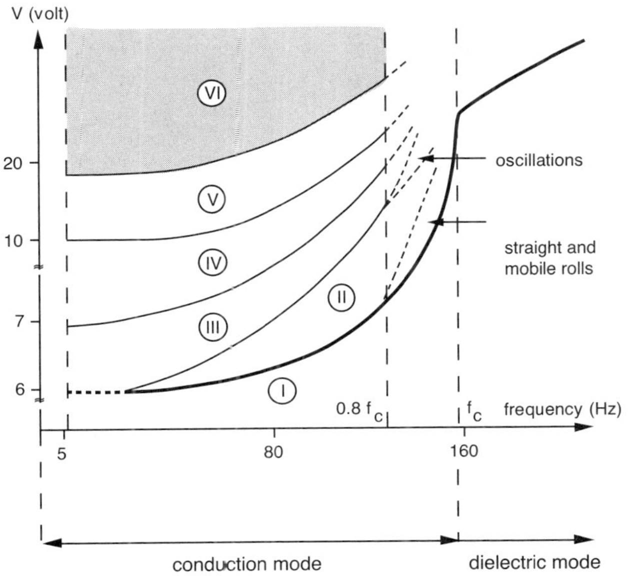

Fig. B.III.52 Bifurcation diagram in a nematic (from ref. [28]). (I) Stable nematic; (II) straight rolls; (III) zigzag; (IV) skewed varicose; (V) rectangular lattice; (VI) chaotic regime. The situation becomes more complicated at the transition between the conduction and dielectric regimes, where new instability modes appear.

Thus, region I of the diagram corresponds to the stable nematic, while region II sees the appearance of the straight rolls for which we determined the threshold. It is shown experimentally that the transition towards straight rolls is second order. This transition is not observed at very low frequency, where the system forms a zigzag structure as soon as the threshold is reached (Fig. B.III.53a). The transition is then first order, the zigzag angle being finite at the threshold.

Region III represents the domain where the zigzag structure exists. The rolls are locally tilted by an angle $\pm\theta$ with respect to their average direction. Each domain of tilt $+\theta$ is then separated from its neighbor, with tilt $-\theta$, by a bend wall. The transition between straight rolls and the zigzag structure is second order. The point on the phase diagram where the three phases (stable nematic, straight rolls and zigzag) coexist, is a multicritical triple point (Lifshitz point).

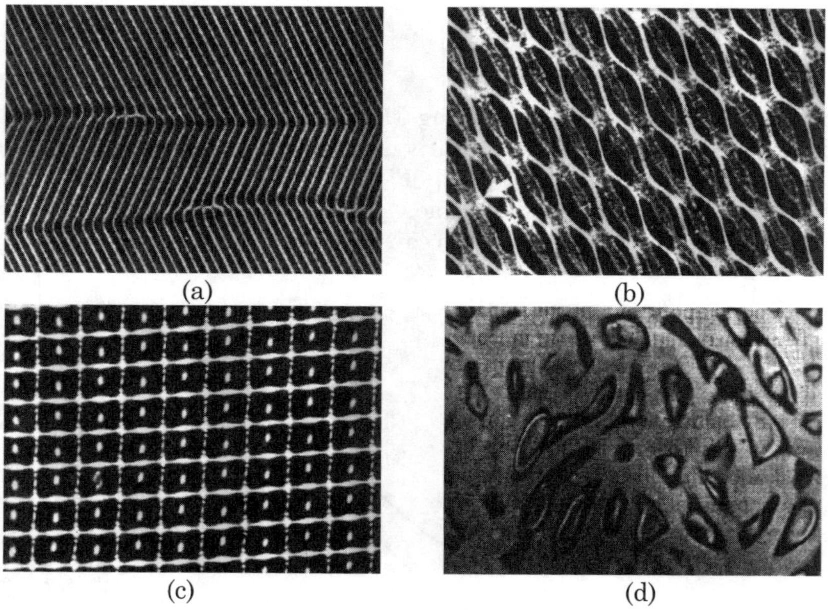

Fig. B.III.53 (a) Zigzag structure; (b) periodical pinching or "varicose"; (c) rectangular lattice; (d) chaotic regime (photos by A. Joets and R. Ribotta).

In region IV, the rolls are periodically pinched along their axis. This mode is termed "skewed varicose" (Fig. B.III.53b). The pinch amplitude increases with the applied voltage and the structure eventually becomes a rectangular lattice (Fig. B.III.53c). The structure is then bimodal. It can be found in region V of the diagram. The varicose-rectangle transition is clearly first order, with coexistence of the two phases at the threshold. This feature is confirmed by the sudden appearance of a higher system symmetry. The rectangles are stable over a voltage range of a few volts.

At higher voltage, the rectangular structure loses all apparent symmetry and rapidly fluctuates with time (Fig. B.III.53d). It is followed by the dynamic scattering mode (DSM), in which the sample strongly scatters light. This is region VI of the bifurcation diagram. The turbidity of DSM was employed in the past (the 1960s) to realize the first displays. This kind of display has been completely abandoned because of its low energy performance, compounded by degradation problems due to the presence of ions and electrochemical reactions in the samples.

Finally, in addition to the pinching, the numerous structural defects (dislocations and grain boundaries, Fig. B.III.54) can also transform a plane flow into a 3D one. These defects play a fundamental role in the evolution towards chaos. Thus, the 2D-3D transition can take place abruptly, by proliferation of defects, as in a martensitic transition.

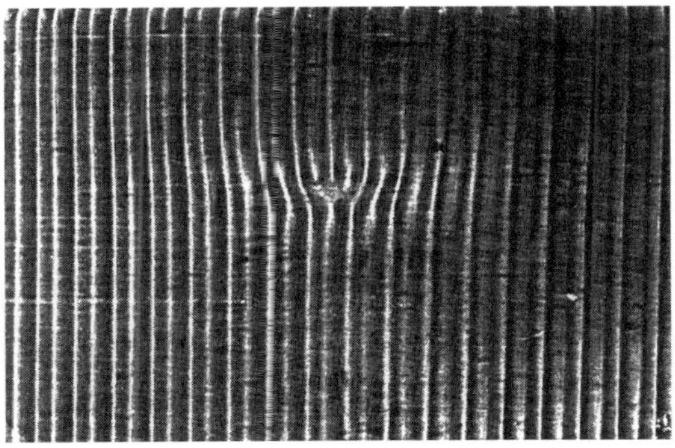

Fig. B.III.54 Wedge dislocation in a system of straight parallel rolls (photo by A. Joets).

For a more complete description of the spatial structures and of the defects, we refer the reader to the papers in ref. [29].

III.6 Thermal instabilities

III.6.a) Qualitative analysis, role of the thermal conductivity anisotropy

In the previous sections we saw that the coupling between the flow **v**(**r**,t) and the orientation **n**(**r**,t) leads to hydrodynamic instabilities appearing at much lower Reynolds numbers than in isotropic liquids.

The thermal Rayleigh-Bénard instabilities are another example of phenomena dramatically affected by this coupling. For an intuitive understanding, consider a horizontal nematic layer of thickness d, confined between two solid surfaces maintained at different temperatures which impose a vertical temperature gradient β. It is known that, in an ordinary fluid, convection rolls develop above a critical temperature gradient that can be determined as shown in Appendix 5.

Below the threshold, the liquid is at rest ($\mathbf{v} = 0$) and the isotherms $T(\mathbf{r}) = \text{const}$ are planar. As shown in the diagram of figure B.III.55, in a nematic liquid crystal one must specify not only the initial velocity and temperature distributions, but also the initial orientation of the liquid crystal. This latter depends on the anchoring conditions at the boundaries. For planar anchoring along the x axis, this initial orientation will be:

$$\mathbf{n} = [1,0,0] \tag{B.III.204}$$

In an isotropic liquid, convective currents develop in the boundary layer and the isotherms become deformed when the temperature gradient

$$\beta = \frac{\partial T}{\partial z} \tag{B.III.205}$$

exceeds the instability threshold, β_{crit}.

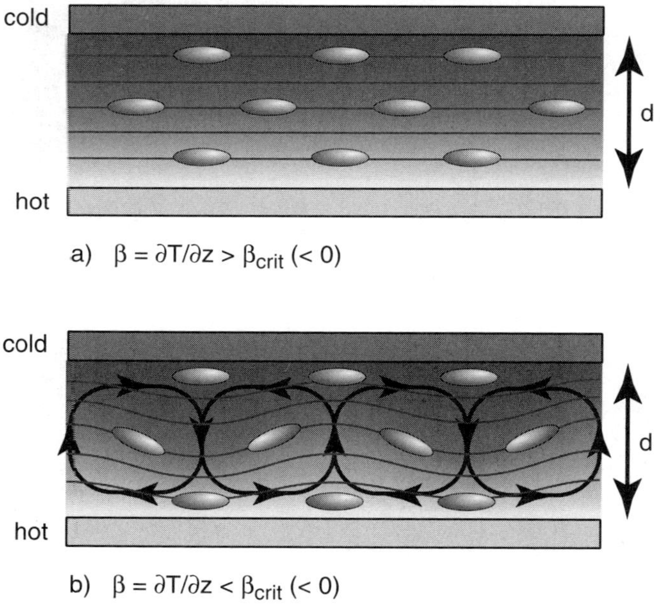

Fig. B.III.55 Rayleigh-Bénard instability in a nematic layer. Below the threshold (a), the molecules are everywhere parallel to the solid surfaces. Above the threshold (b), convective currents, a deformation of the isotherms and a distortion of the initial orientation of the liquid crystal, all appear simultaneously.

The two processes act simultaneously, as isotherm deformation is necessary to maintain the convection (via the buoyancy force), while convection is needed to deform the isotherms.

In a nematic, convective currents deform not just the isotherms, but also the director orientation (due to the hydrodynamic torques). The diagram in figure B.III.55 shows a realistic director configuration resulting from the action of convective rolls.

Because of the anisotropy of the liquid crystal, this orientation deformation reflects back not only on the flow (viscosity anisotropy) but also on the heat transfer, the thermal conductivity being also anisotropic (it is a tensor quantity). We shall see later that the thermal conductivity of nematics (MBBA, 5CB) is higher along the director than in the directions perpendicular to it. For the configuration shown in figure B.III.55, this conductivity anisotropy leads to an additional heating of the nematic in the places where the convective currents are already ascending. In conclusion, as the effect of thermal anisotropy adds to that of convection, the threshold of the Rayleigh-Bénard instability should be lower than in an isotropic liquid.

This qualitative analysis underlines the crucial role played by the thermal conductivity anisotropy so we shall give some details about it before turning to a detailed calculation of the threshold.

III.6.b) Thermal conductivity anisotropy

We saw that the thermal conductivity of a nematic is anisotropic, taking different values along the director and in the directions perpendicular to it. In a generic frame of reference, it is described by a second-order symmetric tensor relating the heat flux q_i to the temperature gradients $\partial T/\partial x_j$:

$$q_i = -k_{ij}\frac{\partial T}{\partial x_j} \tag{B.III.206}$$

This tensor is the sum of two terms

$$k_{ij} = k_{iso}\,\delta_{ij} + k_a\left(n_i n_j - \frac{\delta_{ij}}{3}\right) \tag{B.III.207}$$

where

$$k_{iso} = \frac{k_{//} + 2k_{\perp}}{3} \tag{B.III.208}$$

and

$$k_a = k_{//} - k_{\perp} \tag{B.III.209}$$

The coefficients $k_{//}$ and $k_{\perp}$ represent the thermal conductivities in directions parallel and perpendicular to the director, respectively.

The thermal conductivities $k_{//}$ and $k_{\perp}$ were first determined for MBBA [30] then, more recently, for 5CB [31]. In these two materials, the k_a anisotropy of the nematic phase is positive and the two thermal conductivities are of the same order of magnitude as the conductivity of the isotropic phase. As shown by the diagram in figure B.III.56, the ratio $k_{//}/k_{\perp}$ varies between 1.4 at the nematic-isotropic transition and 1.8 at the crystal-nematic transition.

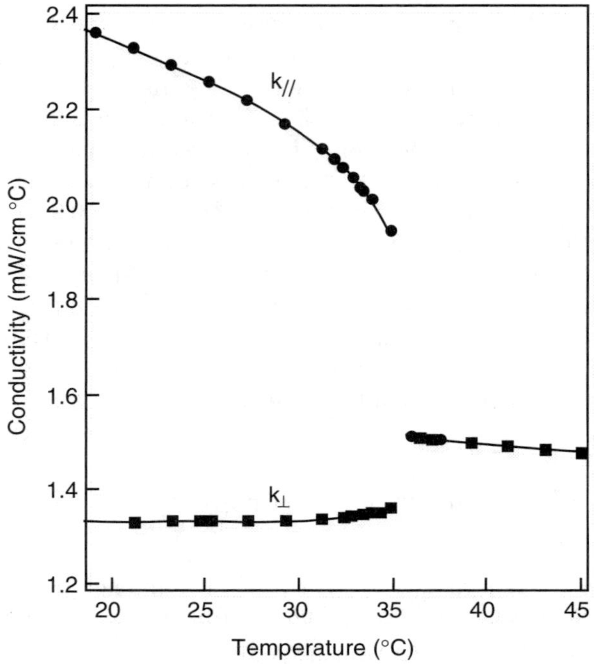

Fig. B.III.56 Thermal conductivities in 5CB (from ref. [31]).

III.6.c) Heat focusing

Since the experimentally determined anisotropy k_a of the thermal conductivity of thermotropic nematics is of the same order of magnitude as their average conductivity k_{iso}, it is very interesting to consider the effects of this anisotropy on the Rayleigh-Bénard instabilities [32, 33].

First of all, let us try to understand how the heat diffuses in a nematic distorted by convection (Fig. B.III.57).

To simplify the calculations, we shall adopt the one-dimensional model for the instabilities (see Appendix 5) and assume that the distortion, as well as the perturbations of the velocity v_z and the temperature δT, only depend on the horizontal coordinate x:

$$\mathbf{n} \approx [1, 0, \phi(t,x)] \qquad \text{(B.III.210)}$$

By only keeping the terms linear in ϕ, the heat flux $\mathbf{q}$ can be obtained from eq. B.III.206:

$$\mathbf{q} \approx -[k_a\beta\phi(x), 0, k_\perp\beta] \tag{B.III.211}$$

Besides the component $q_z = -k_\perp\beta$ directed along the thermal gradient, this flux has a component $q_x = -k_a\beta\phi$ orthogonal to the temperature gradient and linear in ϕ.

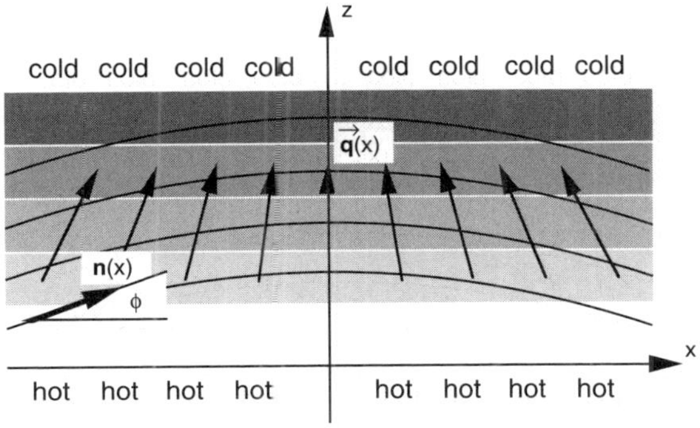

Fig. B.III.57 Effect of the thermal conductivity anisotropy: the heat flux $\mathbf{q}(x)$ induced by the vertical temperature gradient $\beta = \partial T/\partial z$ is deflected by the distortion $\phi(x)$.

In other words, the heat flux is no longer parallel to the temperature gradient as soon as the molecules are tilted with respect to the gradient. If $k_a > 0$, the flux $\mathbf{q}$ tends to be parallel to the director $\mathbf{n}$.

Thus, in the molecular configuration presented in figure B.III.57, the transverse heat flux varies with x like the angle of director tilt and changes sign in x = 0. The flux divergence is then no longer zero

$$\mathrm{div}\,\mathbf{q} = -k_a\beta\frac{\partial\phi}{\partial x} \neq 0 \tag{B.III.212}$$

This quantity goes through a maximum in $\phi = 0$ (i.e., in x = 0).

To put it differently, the heat flux is "**focused**" by the director field distortion ϕ towards the places where ϕ changes sign.

III.6.d) Calculating the instability threshold in planar geometry

We shall now determine the threshold of the convective instability. For ease of calculation, we shall make the one-dimensional approximation (Appendix 5) assuming that all perturbations only depend on x (Fig. B.III.58):

$$\mathbf{v} \approx [0, 0, v_z(t,x)] \tag{B.III.213}$$

$$\mathbf{T} \approx \beta z + \theta(t,x) \tag{B.III.214}$$

$$\mathbf{n} \approx [1, 0, \phi(x,t)] \tag{B.III.215}$$

In this approximation, the temperature evolution is described by the diffusion equation:

$$\frac{\partial \theta}{\partial t} = -\frac{1}{\rho C}\operatorname{div} q - \beta v_z \tag{B.III.216}$$

where, in contrast with eq. B.III.211, the heat flux

$$\mathbf{q} \approx -\left[k_{//}\frac{\partial \theta}{\partial x} + k_a\beta\phi(x), 0, k_\perp\beta\right] \tag{B.III.217}$$

comprises the additional term $k_{//}\partial\theta/\partial x$, due to the horizontal temperature gradient.

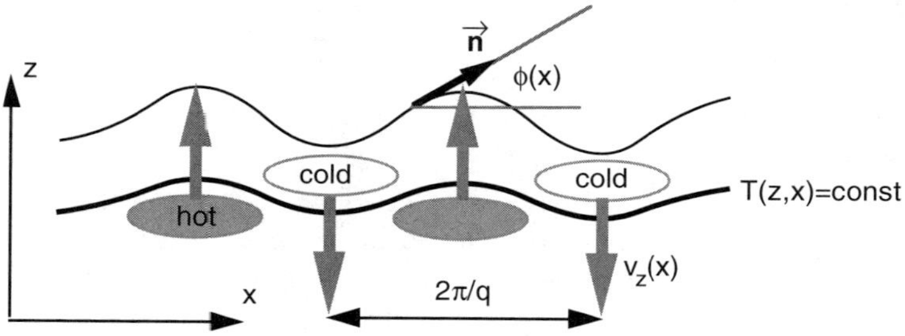

Fig. B.III.58 One-dimensional model for the convective instabilities.

By plugging this expression in the diffusion equation B.III.216, we have:

$$\frac{\partial \theta}{\partial t} = \Lambda_{//}\frac{\partial^2\theta}{\partial x^2} + \Lambda_a\beta\frac{\partial \phi}{\partial x} - \beta v_z \tag{B.III.218}$$

where $\Lambda = k/\rho C$ represents the heat diffusion coefficient.

To model the evolution of convection currents, the nematodynamics equation (see eq. B.III.71) must be used for the v_z velocity component taking into account the buoyancy force:

$$\rho\frac{Dv_z}{Dt} = \partial_x\sigma^v_{zx} + \rho g\alpha\theta \tag{B.III.219}$$

where α is the thermal dilation coefficient. In this equation, the nonlinear acceleration term $(\mathbf{v}.\boldsymbol{\nabla})\mathbf{v}$ is completely negligible, the convection currents being very slow at the instability threshold. As to the σ^v_{zx} component of the viscous stress tensor, it contains one term stemming from the velocity gradient and a second term resulting from the director rotation:

$$\sigma^{v}_{zx} = \eta_c \frac{\partial v_z}{\partial x} + \alpha_2 \frac{\partial \phi}{\partial t} \qquad \text{(B.III.220)}$$

Replacing this expression in eq. B.III.219 yields:

$$\rho \frac{\partial v_z}{\partial t} = \eta_c \frac{\partial^2 v_z}{\partial x^2} + \alpha_2 \frac{\partial^2 \phi}{\partial x \partial t} + \rho g \alpha \theta \qquad \text{(B.III.221)}$$

Finally, the distortion evolves according to the equilibrium equation for the viscous, elastic, and hydrodynamic torques:

$$\gamma_1 \frac{\partial \phi}{\partial t} = K_3 \frac{\partial^2 \phi}{\partial x^2} - \alpha_2 \frac{\partial v_z}{\partial x} \qquad \text{(B.III.222)}$$

Equations B.III.218, 221 and 222 form a system of three coupled differential equations that must be solved to find the evolution of the three perturbations ϕ, θ, and v_z on which the convective instabilities depend.

Solving these equations is facilitated by introducing a new variable:

$$\psi = \frac{\partial \phi}{\partial x} \qquad \text{(B.III.223)}$$

Derivatizing eq. B.III.222 with respect to x and dividing it by γ_1, we obtain:

$$\frac{\partial \psi}{\partial t} = D_3 \frac{\partial^2 \psi}{\partial x^2} + \frac{\partial^2 v_z}{\partial x^2} \qquad \text{(B.III.224)}$$

where

$$D_3 = \frac{K_3}{\gamma_1} \qquad \text{(B.III.225)}$$

and

$$\frac{-\alpha_2}{\gamma_1} \approx 1 \qquad \text{(B.III.226)}$$

Dividing eq. B.III.221 by ρ yields

$$\frac{\partial v_z}{\partial t} = \nu_c \frac{\partial^2 v_z}{\partial x^2} + \nu_2 \frac{\partial \psi}{\partial t} + g \alpha \theta \qquad \text{(B.III.227)}$$

with

$$\nu_c = \frac{\eta_c}{\rho} \qquad \text{(B.III.228)}$$

and

$$\nu_2 = \frac{\alpha_2}{\rho} \tag{B.III.229}$$

Using the following to solve equations B.III.218, 224, and 227:

$$\theta = \theta_o e^{st} \cos(qx) \tag{B.III.230}$$

$$\psi = \psi_o e^{st} \cos(qx) \tag{B.III.231}$$

$$v_z = v_{zo} e^{st} \cos(qx) \tag{B.III.232}$$

we obtain:

$$(s + D_3 q^2)\psi_o + q^2 v_{zo} = 0 \tag{B.III.233}$$

$$(s + \Lambda_{//} q^2)\, \theta_o - \Lambda_a \beta\, \psi_o + \beta v_{zo} = 0 \tag{B.III.234}$$

$$g\alpha\, \theta_o + s\nu_2 \psi_o - (s + \nu_c q^2)\, v_{zo} = 0 \tag{B.III.235}$$

Equating the determinant of this system to zero for s = 0 fixes the instability threshold. Explicitly, it yields:

$$g\alpha\beta(\Lambda_a q^2 + D_3 q^2) + D_3 q^2 \Lambda_{//} q^2 \nu_c q^2 = 0 \tag{B.III.236}$$

Solving this equation with respect to β gives the critical value of the temperature gradient at the instability threshold:

$$\beta_{crit} = \frac{\beta_{iso}}{\Lambda_a / D_3 + 1} \tag{B.III.237}$$

where

$$\beta_{iso} = -\frac{\nu_c \Lambda_{//} q^4}{g\alpha} \tag{B.III.238}$$

is the critical temperature gradient in the isotropic case ($\Lambda_a = 0$).

The expression B.III.237 of the threshold depends on the ratio between the two diffusion constants D_3 and Λ_a. We saw that the typical value for the orientation diffusivity is

$$D_3 = \frac{K_3}{\gamma_1} \approx 10^{-6}\ \mathrm{cm^2 s^{-1}} \tag{B.III.239}$$

As for the thermal diffusivity anisotropy, we have just seen that it is positive and its order of magnitude is

$$\Lambda_a = \frac{k_a}{\rho c} \approx 5\times10^{-4}\ \mathrm{cm^2 s^{-1}} \qquad \text{(B.III.240)}$$

Consequently, the ratio Λ_a/D_3 is of the order of 500.

Thus, the effect of the thermal diffusivity anisotropy on the Rayleigh-Bénard instability is very significant, leading to a decrease in the instability threshold by a factor of 500.

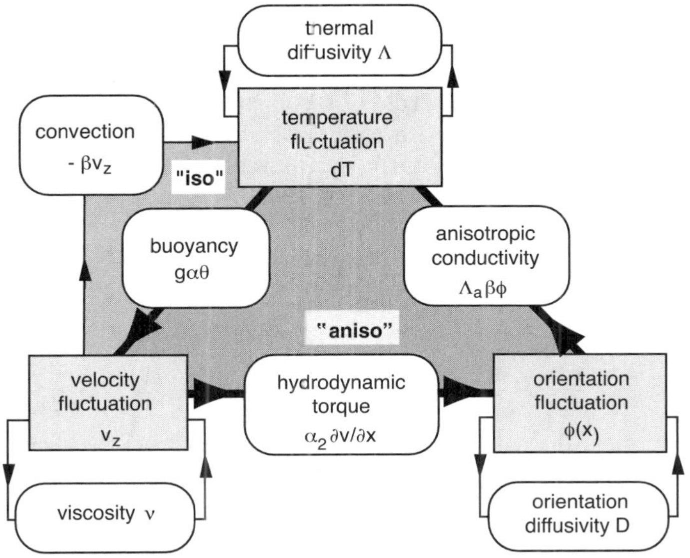

Fig. B.III.59 Rayleigh-Bénard instability in nematics. The chain of couplings temperature-velocity-orientation-temperature forms a feedback loop (represented in thick line and labeled "aniso"). Due to the slow distortion relaxation, this feedback loop is 500 times more effective than the classical loop (labeled "iso"). The ordinary mechanism of thermal convection only plays a minor role here.

In order to understand the ins and outs of the theoretical model better, we give in figure B.III.59 a graphical representation of the couplings between the hydrodynamic variables entering the equation system B.III.233–235. This diagram shows that the couplings between the variables form two feedback loops: one of them, labeled "iso," involves heat convection and is of the same nature as in isotropic liquids (see Fig. A.5.3). The other one, labeled "aniso" and drawn in a thick line, only exists in nematics. This second loop is 500 times more effective than the first, because the distortions it involves relax very slowly.

The noteworthy fact is that the $v_z\beta$ term, usually the source of thermal convection, only plays a minor role in nematics. To see it better, consider a nematic with negative thermal diffusivity anisotropy. Neglecting the "1" in the denominator of the threshold expression B.III.237, it is obvious that a change in the sign of the thermal diffusivity anisotropy leads to a change in the sign of the critical thermal gradient. In other words, the nematic would have to be heated from the top in order for the instability to occur if the thermal conductivity anisotropy were negative.

III.6.e) Instabilities in homeotropic geometry

In the absence of a nematic with negative anisotropy of the thermal conductivity, one can exchange the role of the $k_{//}$ and $k_\perp$ conductivities to trigger the instabilities by heating from the top. Such an exchange takes place when the initial orientation of the nematic is vertical instead of horizontal (Fig. B.III.60).

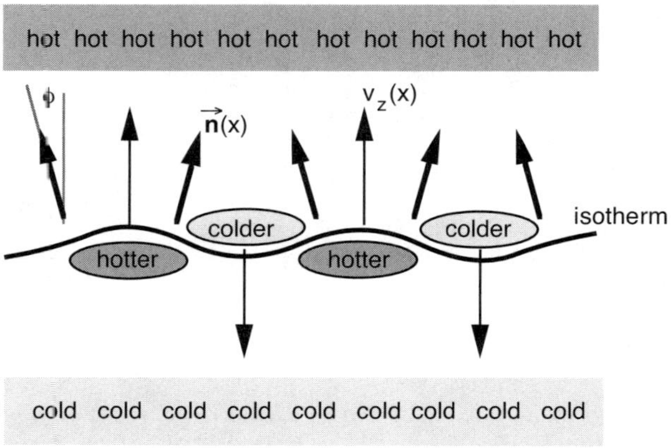

Fig. B.III.60 Thermal instability in homeotropic geometry.

In this geometry, in the presence of an orientation fluctuation φ(x), the temperature gradient

$$\mathbf{grad}\,T = \left[\frac{\partial \theta}{\partial x}, 0, b\right] \tag{B.III.241}$$

induces the heat flux

$$\mathbf{q} \approx -\left[k_\perp \frac{\partial \theta}{\partial x} - k_a \beta \phi(x), 0, k_{//}\beta\right] \tag{B.III.242}$$

where the role of the conductivities is indeed exchanged (cf. eq. B.III.217).

The change in geometry also has an effect on the viscous stresses and the torques. The viscosity entering the Navier-Stokes equation is no longer η_c but η_b (the velocity being parallel to the director). Finally, the hydrodynamic torque is proportional to α_3 and not α_2. To keep the same sign of the hydrodynamic torque and to be able to make the same approximation as for the planar geometry, we shall assume that

$$\alpha_3 > 0 \qquad \text{and} \qquad \frac{\alpha_3}{\gamma_1} \approx 1 \tag{B.III.243}$$

Taking all these modifications into account, the following expression is obtained for the threshold

$$\beta_{crit} = \frac{\beta_{iso}}{1 - \Lambda_a / D_1} \tag{B.III.244}$$

where

$$\beta_{iso} = -\frac{\nu_a \Lambda_\perp q^4}{g\alpha}, \qquad D_1 = \frac{K_1}{\gamma_1} \qquad \text{and} \qquad \nu_b = \frac{\eta_b}{\rho} \tag{B.III.245}$$

Neglecting the "1" in the denominator we reach the conclusion that the sign of the critical gradient in homeotropic geometry is reversed with respect to the planar case.

III.6.f) Experimental evidence

The Rayleigh-Bénard instabilities were observed in the two geometries: planar and homeotropic [33, 34].

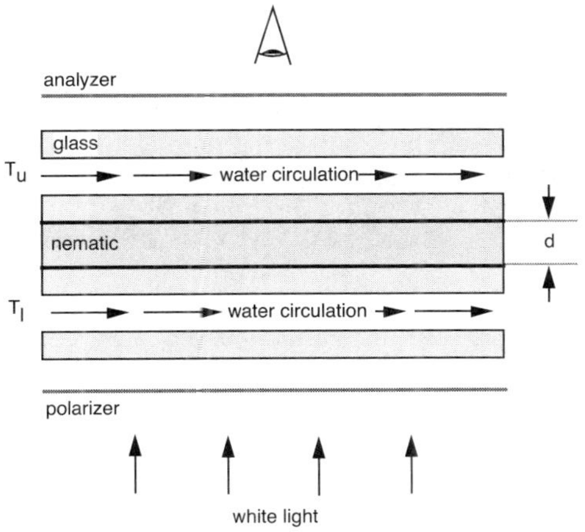

Fig. B.III.61 Experimental setup for the observation of the Rayleigh-Bénard instabilities in nematics.

The experimental setup is described in figure B.III.61. A layer of nematic liquid crystal is contained between two glass plates surface treated to ensure planar or homeotropic anchoring. The d distance between the plates is set at about 1 mm by glass spacers. The plates are temperature regulated using two independent water flows. The instabilities are optically detected by observation in white polarized light.

The photo in figure B.III.62 shows a typical image of the instabilities that develop in planar geometry. In agreement with the theoretical predictions, these instabilities appear when the nematic layer is heated from below ($T_d > T_u$). This image, taken using light polarized along the nematic orientation, shows the liquid crystal is distorted by the convection currents. Close to the threshold, these currents have the shape of rolls perpendicular to the orientation of the liquid crystal.

Fig. B.III.62 Thermal Rayleigh-Bénard instabilities in a planar nematic layer heated from below (photo by E. Plaut [33]).

Fig. B.III.63 Thermal Rayleigh-Bénard instabilities in a homeotropic nematic layer heated from above (from ref. [34]).

The structure of the convection currents is very different in the homeotropic geometry [34]. As shown in the photo of figure B.III.63, the convective currents now form a lattice of square cells. Furthermore, the orientation of the lattice varies over the nematic layer. This is remarkable in itself, but the most noteworthy feature is that, in agreement with the theoretical predictions, the instabilities develop when the nematic layer is heated from above.

In the two geometries, the temperature difference at the threshold is a few degrees for a 1 mm thick sample, confirming the threshold reduction by a factor of the order of 500 with respect to the isotropic case.

As discussed in section III 6.c, this reduction of the critical gradient with respect to the isotropic case is related to the exceptionally long-lived orientation fluctuations; the orientational diffusivity D being of the order of 10^{-6} cm^2/s, the relaxation time of a distortion with wavelength $\lambda = 10^{-1}$ cm is:

$$\tau_{dist} = D^{-1}\left(\frac{\lambda}{2\pi}\right)^2 = 10^6\left(\frac{10^{-1}}{6}\right)^2 \approx 300\ \text{s} \qquad \text{(B.III.246)}$$

This relaxation time determines the instability dynamics. Indeed, as for the Frederiks transition, the time needed for the instability to develop after suddenly applying a thermal gradient $\Delta T > \Delta T_{crit}$ or, conversely, for the distortion to disappear after the gradient is decreased, is

$$\tau = s^{-1} \approx \tau_{dist}\frac{\Delta T}{\Delta T - \Delta T_{crit}} \qquad \text{(B.III.247)}$$

(This result is obtained by analyzing the solutions of eqs. B.III.233–235 close to the threshold, where $s \neq 0$). This so-called critical slowing down of instability dynamics close to the threshold considerably increases the experimental times needed to determine the critical value of the gradient. For instance, after applying a temperature difference $\Delta T = 1.1\ \Delta T_{crit}$, one must wait for 1 h to see the instabilities develop.

III.6.g) Thermal instabilities in the presence of a magnetic field

A nematic layer of thickness d = 1 mm is very sensitive to a magnetic field, as even a modest field, of a few hundred gauss, is well above the critical Frederiks field. Indeed, for this thickness, the critical field in planar geometry is typically:

$$B_c = \frac{\pi}{d}\sqrt{\frac{\mu_0 K_1}{\chi_a}} \overset{\text{S.I. units}}{\approx} 3000\sqrt{\frac{4\pi\times10^{-7}\times5\times10^{-12}}{1.5\times10^{-6}}} \approx 6\times10^{-3}\,\text{tesla} \approx 60\ \text{gauss}$$

(B.III.248)

This estimate shows the relevance of experiments where the temperature gradient and the magnetic field are applied at the same time.

First consider the planar geometry. The action of the field introduces in the torque equation (see eq. B.III.222) a new term

$$\gamma_1 \frac{\partial \phi}{\partial t} = K_3 \frac{\partial^2 \phi}{\partial x^2} - \alpha_2 \frac{\partial v_z}{\partial x} \pm \frac{\chi_a}{\mu_o} B^2 \phi \qquad \text{(B.III.249)}$$

the sign of which depends on the orientation of the field with respect to the director. If the magnetic field is horizontal, i.e., parallel to **n**, it tends to stabilize the initial orientation of the liquid crystal: the magnetic term is to be taken with a " – " sign. Conversely, a vertical magnetic field will be destabilizing, leading to a " + " sign in the equation above.

If the field B is close to B_c, the instability threshold (given by the condition s = 0) satisfies the following equation:

$$\Delta T_{crit}(B) \approx \Delta T_{crit}(B{=}0) \left[1 \pm \frac{B^2}{B_c^2}\right] \qquad \text{(B.III.250)}$$

where the " + " and " – " signs correspond to a horizontal and vertical magnetic field, respectively. This variation in the critical gradient as a function of the magnetic field is summarized in the diagram of figure B.III.64. It should be noted that in the **B** ⊥ **n** case the instability threshold decreases and goes to zero for B = B_c. In the other case, the threshold increases with the field.

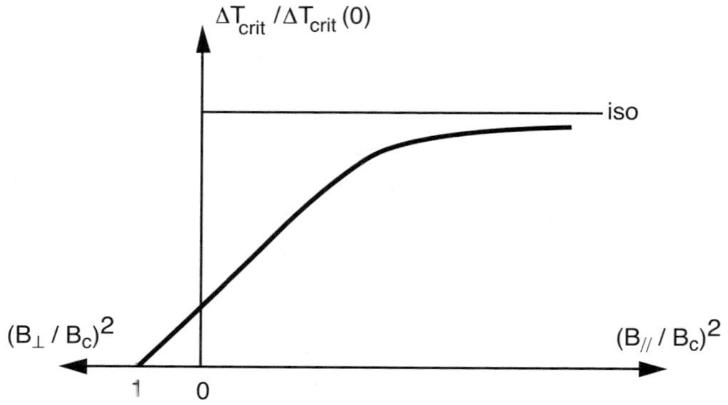

Fig. B.III.64 Theoretical variation of the R-B instability threshold as a function of the magnetic field.

If the magnetic field is very strong (B >> B_c), the magnetic torque dominates in eq. B.III.249 and the distortions in the director field become negligible. An essential link in the feedback loop is then destroyed (Fig. B.III.59) and the R-B instability threshold tends towards the value it would have in the isotropic case. This effect was recently demonstrated (Fig. B.III.65) [35].

Consider now the homeotropic geometry. In this case, the effects of the magnetic field on the thermal instabilities were studied by E. Guyon et al. [36]. We should remember that, in the absence of the field, convection can be triggered by heating the nematic from above (see the discussion in section

III.6.e) and figure B.III.60). By analogy with the planar geometry, it can be shown that applying a magnetic field leads to a reduction or an increase of the linear instability threshold when the field is perpendicular or parallel to the director, respectively. This can be seen by replacing in the expression B.III.244 of the linear instability threshold the orientational diffusivity D_1 only due to the elasticity, by an effective diffusivity in the presence of the magnetic field

$$D_1(B) = \frac{K_1 \pm \dfrac{\chi_a d^2 B^2}{\pi^2 \mu_o}}{\gamma_1} \tag{B.III.251}$$

In this formula d is the sample thickness.

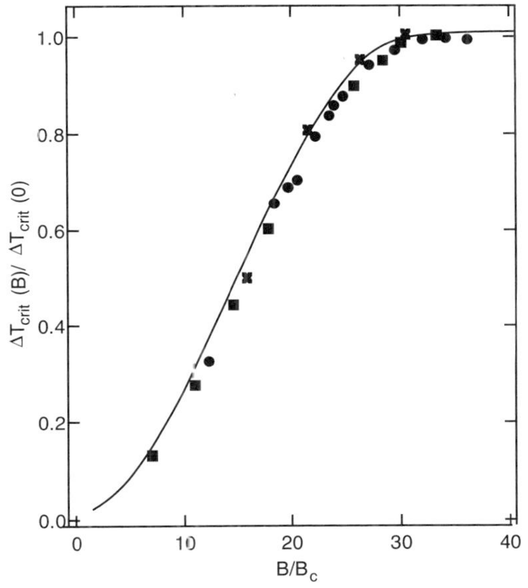

Fig. B.III.65 Variation of the R-B instability threshold as a function of the magnetic field in the experiment of Berge, Ahlers, and Cannel [35].

The " – " sign corresponds to a horizontal magnetic field: indeed, in this case, the field is perpendicular to the director and hence destabilizing, which slows down the distortion relaxation. The diffusivity $D_1(B)$ thus decreases and tends towards zero in the limit $B \rightarrow B_c = (\pi/d)(\mu_o/\chi_a)^{1/2}$ (B_c is the threshold of the Frederiks instability). As a result, the threshold temperature gradient β_{crit}, given by eq. B.III.244, goes to zero in this limit.

On the other hand, a magnetic field which is vertical, and so parallel to the director, increases the orientation diffusivity D_1 (" + " sign in eq. B.III.251) and leads to an increase of the instability threshold. According to eq. B.III.244, the instability threshold $\beta_{crit} \rightarrow +\infty$ when $\Lambda_a/D_1(B) \rightarrow 1$ from above. For even higher magnetic fields, the ratio $\Lambda_a/D_1(B)$ goes below 1 and the critical gradient

β_{crit} changes sign; to trigger the linear instability, the nematic layer must then be heated from below. In this situation, the instability threshold $\beta_{crit} \rightarrow -\infty$ when $\Lambda_a/D_1(B) \rightarrow 1$ from below.

Is this divergence of the linear instability threshold predicted by eq. B.III.244 observable? In particular, what happens when a homeotropic nematic layer is heated from below? The experiments of Guyon et al. [36] showed that the divergence of the linear instability threshold cannot be observed, being hidden for thermal gradients β of the order of β_{iso} by oscillating instabilities with the following characteristics:

1. The period of the oscillations, measured using a thermocouple, is of the order of one minute, and corresponds to the characteristic time $(\Lambda_\perp q^2)^{-1}$ of thermal diffusion over a distance comparable to the thickness ($q = \pi/d$).

2. The bifurcation is subcritical (as in the example of figure B.II.30b), such that above a critical threshold β_c the oscillations (which can be triggered by locally heating the sample using an electric resistor) grow ($|\beta| > |\beta_c|$), while below ($|\beta| < |\beta_c|$), they are exponentially damped. At $|\beta_c|$ the amplitude of the oscillations is finite.

3. β_c is comparable to β_{iso}.

4. The bifurcation being subcritical, there is a nonlinear branch of finite amplitude corresponding to $|\beta| > |\beta_0|$ with $|\beta_0| < |\beta_c|$.

Without going into a detailed analysis, let us try to understand the origin of the oscillations starting with the information that β_c is of the order of β_{iso}. In an isotropic liquid, the Rayleigh-Bénard instabilities (Appendix 5) correspond to a supercritical bifurcation (see for instance Fig. B.II.30a) such that the evolution of a velocity fluctuation is exponential (proportional to e^{st}) with s real and positive for $|\beta| > |\beta_{iso}|$, while s becomes negative for $|\beta| < |\beta_{iso}|$. Suppose that $|\beta| < |\beta_{iso}|$, so s is negative and the velocity and temperature fluctuations relax to zero. In a homeotropic nematic, the flow is accompanied by a distortion in the director field. When the sample is heated from below ($\beta < 0$) this distortion, because of the heat focusing, leads to a temperature decrease for the rising liquid (Fig. B.III.60 with inverted thermal gradient β). It so happens that the velocity fluctuation relaxes to zero faster than the $\phi(x)$ distortion of the director field. Consequently, the liquid still "feels" the effect of heat focusing so it gets cooler and starts to descend: this mechanism leads to an **oscillation** phenomenon.

It is then obvious that the growth rate s of the instability is no longer real (as we assumed in paragraph B.III.6d), but complex. The imaginary part of s fixes the oscillation frequency. The real part is positive for $|\beta| > |\beta_c|$ and negative for $|\beta| > |\beta_c|$.

BIBLIOGRAPHY

[1] Cladis P.E., *Phys. Rev. Lett.*, **28** (1972) 1629.

[2] a) Gähwiller Ch., *Phys. Lett.*, **36A** (1971) 311.
b) Pieranski P., D.Sc. Thesis, University of Paris XI, Orsay, 1972.
c) Prost J., D.Sc. Thesis, Bordeaux I University, Bordeaux, 1973 (N° 392).
d) Prost J., Gasparoux H., *Phys. Lett.*, **36A** (1971) 245.
e) Gasparoux H., Prost J., *J. Physique (France)*, **32** (1971) 953.

[3] Wahl J., Fischer F., *Mol. Cryst. Liq. Cryst.*, **22** (1973) 359.

[4] a) Brochard F., *Mol. Cryst. Liq. Cryst.*, **23** (1973) 51.
b) Svensek D., Zumer S., *Liq. Cryst.*, **28** (2001) 1389.

[5] a) Lonberg F., Fraden S., Hurd A.J., Meyer R.B., *Phys. Rev. Lett.*, **52** (1984) 1903.
b) Ong H.L., *Jpn. J. Appl. Phys. 1*, **31** (1992) 858.
c) Grigutsch M., Klöpper N., Schmiedel H., Stannarius R., *Phys. Rev. E*, **49** (1994) 5452.
d) Simões M., Palangana A.J., Arroteia A.A., Vilarim P.R., *Phys. Rev. E.*, **63** (2001) 041707.

[6] a) Martinot-Lagarde Ph., Dozov I., Polossat E., Giocondo M., Lelidis I., Durand G., Angele J., Pecout B., Boissier A., *SID '97 Digest*, (1997) 41.
b) Dozov I., Nobili M., Durand G., *Appl. Phys. Lett.*, **70** (1997) 1179.
c) Dozov I., Durand G., *Pramana*, **53** (1999) 25 (reprinted from *Liquid Crystals Today*, **8** (1998) 1).
d) Giocondo M., Lelidis I., Dozov I., Durand G., *Eur. Phys. J. App. Phys.*, **5** (1999) 227.
e) McIntosh J.G., Kedney P.J., Leslie F.M., *Mol. Cryst. Liq. Cryst. A*, **330** (Part III) (1999) 1769.
f) McIntosh J.G., Leslie F.M., *J. Eng. Math.*, **37** (2000) 129.

[7] a) Miesowicz M., *Nature*, **17** (1935) 261.
b) Miesowicz M., *Nature*, **158** (1946) 27.
c) Gähwiller Ch., *Mol. Cryst. Liq. Cryst.*, **20** (1973) 301.

[8] a) Ericksen J.L., *Arch. Ration. Mech. Analysis*, **4** (1960) 231.
b) Ericksen J.L., *Physics Fluids*, **9** (1966) 1205.

[9] a) Leslie F.M., *Quart. J. Mech. Appl. Math.*, **19** (1966) 357.
b) Leslie F.M., *Arch. Ration. Mech. Analysis*, **28** (1968) 265.

[10] Martin P.C., Parodi O., Pershan P.S., *Phys. Rev. A*, **6** (1972) 2401.

[11] De Gennes P.-G., *The Physics of Liquid Crystals*, Clarendon Press, Oxford, 1974.

[12] Parodi O., *J. Physique (France)*, **31** (1970) 581.

[13] a) Orsay Group on Liquid Crystals, *J. Chem. Phys.*, **51** (1969) 816.
b) Orsay Group on Liquid Crystals, *Mol. Cryst. Liq. Cryst.*, **13** (1971) 187.
c) Léger L., D.Sc. Thesis, University of Paris XI, Orsay, 1971.

[14] See table 5.1 on page 189 of ref. [11].

[15] a) Gähwiller Ch., *Phys. Rev. Lett.*, **28** (1972) 1554.
b) Pieranski P., Guyon E., *Phys. Rev. Lett.*, **32** (1974) 924.

[16] Pieranski P., Guyon E., *Phys. Lett. A*, **29** (1974) 237.

[17] a) Helfrich W., *J. Chem. Phys.*, **50** (1969) 100.
b) Helfrich W., *J. Chem. Phys.*, **53** (1970) 2267.
c) Helfrich W., *J. Chem. Phys.*, **56** (1972) 3167.

[18] a) Pieranski P., "Hydrodynamic instabilities in nematics. An experimental study," D.Sc. Thesis, University of Paris XI, Orsay, 1976.
b) Pieranski P., Guyon E., *Comm. on Phys.*, **1** (1976) 45.

[19] Pikin. S.A., *Sov. Phys. JETP*, **38** (1974) 1246.

[20] a) Pieranski P., Guyon E., *Solid St. Comm.*, **13** (1973) 435.
b) Pieranski P., Guyon E., *Phys. Rev. A*, **9** (1974) 1974.
c) Janossy I., Pieranski P., Guyon E., *J. Physique (France)*, **37** (1976) 1105.
d) Horn D., Guyon E., Pieranski P., *Rev. Phys. Appl.*, **11** (1976) 139.

[21] Dubois-Violette E., Durand G., Guyon E., Manneville P., Pieranski P., in *Solid State Physics*, Supp. 14, Ed. Liebert L., Academic Press, New York, 1978.

[22] a) Pieranski P., Guyon E., *Phys. Rev. Lett.*, **39** (1977) 1280.
b) Guazzelli E., "Two experimental studies of disorder in physical hydrodynamics," D.Sc. Thesis, University of Provence, Marseille, 1986.
c) Dreyfus J.-M., Pieranski P., *J. Physique (France)*, **42** (1981) 459.
d) Börzsönyi T., Buka A., Krekhov A.P., Scaldin O.A., Kramer L., *Phys. Rev. Lett.*, **84** (2000) 1934.
e) Guazzelli E., Guyon E., *J. Physique (France)*, **43** (1982) 985.
f) Sadik J., Rothen F., Bestgen W., Dubois-Violette E., *J. Physique (France)*, **42** (1981) 915.
g) Guazzelli E., Koch A.-J., *J. Physique (France)*, **46** (1985) 673.

[23] Carr E.F., in *Ordered Fluids and Liquid Crystals*, Adv. Chem. Series Am. Chem. Soc. Pub., 1967, p. 76.

[24] Helfrich W., *J. Chem. Phys.*, **51** (1969) 4092.

[25] Dubois-Violette E., de Gennes P.-G., Parodi O., *J. Physique (France)*, **32** (1971) 305.

[26] Williams R., *J. Chem. Phys.*, **39** (1963) 384.

[27] a) Galerne Y., Durand G., Veyssié M., *Phys. Rev. A*, **6** (1972) 484.
b) Galerne Y., D.Sc. Thesis, University of Paris XI, Orsay, 1973.

[28] Joets A., Ph.D. Thesis, University of Paris XI, Orsay, 1984.

[29] a) Joëts A., Ribotta R., *J. Physique (France)*, **47** (1986) 595.
b) New nonlinear regimes were recently observed in electroconvection. See for instance Plaut E., Decker W., Rossberg A. G., Kramer L., Pesch W., Belaidi A., Ribotta R., *Phys. Rev. Lett.*, **79** (1997) 2367.
c) For a review on the electrohydrodynamic instabilities in nematics see Kramer L., Pesch W., *Ann. Rev. Fluid Mech.*, **27** (1995) 515, and the chapter "Electrohydrodynamic instabilities in nematic liquid crystals," in *Pattern Formation in Liquid Crystals*, Eds. Kramer L. and Buka A., Springer-Verlag, New York, 1996.

[30] a) Pieranski P., Brochard F., Guyon E., *J. Physique (France)*, **33** (1972) 681.
b) Vilanove R., Guyon E., Mitescu C., Pieranski P., *J. Physique (France)*, **35** (1974) 153.

[31] Ahlers G., Cannel D.S., Berge L.I., Sakurai S., *Phys. Rev. E*, **49** (1994) 545.

[32] a) Dubois-Violette E., *C. R. Acad. Sci. (Paris)*, **21** (1971) 923.
b) Dubois-Violette E., Guyon E., Pieranski P., *Mol. Cryst. Liq. Cryst.*, **26** (1974) 193.

[33] a) Plaut E., Ph.D. Thesis, University of Paris XI, Orsay, 1996.
b) Plaut E., Ribotta R., *Europhys. Lett.*, **38** (1997) 441.

[34] Pieranski P., Dubois-Violette E., Guyon E., *Phys. Rev. Lett.*, **30** (1973) 736.

[35] Berge L.I., Ahlers G., Cannel D.S., *Phys. Rev. E*, **48** (1993) R3236.

[36] a) Guyon E., Pieranski P., Salan J., *C. R. Acad. Sci. (Paris)*, **287** Série B (1978) 41.
b) Guyon E., Pieranski P., Salan J., *J. Fluid. Mech.*, **93** (1979) 65.

Appendix 1: "Derivation under the integral" theorem

Let $f(\mathbf{r},t)$ be a generic function (which can be scalar, vectorial, or tensorial in nature) and V_m a mobile volume composed of particles to be followed in their motion. We shall prove the following identity:

$$\frac{d}{dt}\int_{V_m} f\,\rho\,dV = \int_{V_m} \frac{Df}{Dt}\rho\,dV \tag{1.1}$$

where D/Dt is the total (also called material) derivative with respect to time.

The simplest way to prove this theorem is to describe the motion of the fluid in the Lagrangian formalism. This amounts to giving the trajectory of every particle in the form:

$$\mathbf{r} = \mathbf{r}(\mathbf{a}, t) \tag{1.2}$$

with $\mathbf{a}$ the particle position at a given moment, for instance $t = 0$. Then $\mathbf{a} = \mathbf{r}(\mathbf{a}, 0)$.

This equation defines a coordinate change from $\mathbf{a}$-space to $\mathbf{r}$-space:

$$dx_i = \frac{\partial x_i}{\partial a_j} da_j\ , \tag{1.3}$$

by which a volume element $d^3\mathbf{a}$ around $\mathbf{a}$ is transformed into a volume element $d^3\mathbf{r} = |J|\,d^3\mathbf{a}$, with J the Jacobian (i.e., the determinant of the transformation matrix):

$$J = \det\left(\frac{\partial x_i}{\partial a_j}\right) \tag{1.4}$$

In the Lagrangian description, mass conservation reads:

$$\frac{D}{Dt}[\rho(\mathbf{r},t)d^3\mathbf{r}] = \frac{D}{Dt}[\rho(\mathbf{r},t)\,|J|]\,d^3\mathbf{a} = 0 \tag{1.5}$$

yielding:

$$\frac{D}{Dt}[\rho(\mathbf{r},t)\,|J|] = 0 \tag{1.6}$$

The theorem is proven by first transforming the integral over the mobile (time-dependent) volume in an integral over the fixed volume V_a taken in reference space:

$$\int_{V_m} f(\mathbf{r},t)\,\rho\,dV = \int_{V_a} f[\mathbf{r}(\mathbf{a},t),t]\,\rho\,|J|\,dV_a \tag{1.7}$$

then derivatizing this new integral with respect to time, using the mass conservation law (1.6) written in its Lagrangian form. The calculations are then straightforward, as the integration volume V_a is time independent:

$$\frac{d}{dt}\int_{V_m} f\,\rho\,dV = \frac{d}{dt}\int_{V_a} f\,\rho\,|J|\,dV_a = \int_{V_a} \frac{D}{Dt}(f\,\rho\,|J|)\,dV_a$$

$$= \int_{V_a} \frac{Df}{Dt}\rho\,|J|\,dV_a = \int_{V_m} \frac{Df}{Dt}\rho\,dV \tag{1.8}$$

q. e. d.

Appendix 2: Rotational identity

To find the rotational identity, we shall write that the energy is invariant ($\delta f = 0$) for all deformations of the type

$$\mathbf{u}(\mathbf{r}) = \boldsymbol{\omega} \times \mathbf{r} \qquad \text{and} \qquad \delta\mathbf{n}(\mathbf{r}) = \boldsymbol{\omega} \times \mathbf{n} \tag{2.1}$$

with **v** a generic rotation vector (but constant in space).

After the deformation

$$\delta f = \sigma_{ij}^{E} \frac{\partial u_i}{\partial x_j} + \frac{\partial f}{\partial n_i} \delta n_i + \frac{\partial f}{\partial n_{i,l}} \delta n_{i,l} \tag{2.2}$$

One immediately calculates

$$\frac{\partial u_i}{\partial x_j} = - e_{ijk}\omega_k \tag{2.3}$$

$$\delta n_i = e_{ijk}\, \omega_j\, n_k \tag{2.4}$$

which gives, using the expression B.III.28 of the elastic stress:

$$\delta f = e_{ijk} \left(- \frac{\partial f}{\partial n_{l,k}} \frac{\partial n_l}{\partial x_i} + n_k \frac{\partial f}{\partial n_i} + \frac{\partial f}{\partial n_{i,l}} \frac{\partial n_k}{\partial x_l} \right) \omega_j = 0 \tag{2.5}$$

This equation being satisfied for all ω_j, after rearranging the dummy indices we obtain the rotational identity:

$$e_{ijk} \left(\frac{\partial f}{\partial n_{l,k}} \frac{\partial n_l}{\partial x_j} + n_j \frac{\partial f}{\partial n_k} + \frac{\partial f}{\partial n_{k,l}} \frac{\partial n_j}{\partial x_l} \right) = 0 \qquad \forall\, i \tag{2.6}$$

Using this relation, it is obvious that the two expressions found for the elastic torque (eqs. B.III.55 and 56) are identical.

Appendix 3: Calculation of the irreversible entropy production

Consider a mobile volume V_m delimited by the surface S_m. Let S be the entropy of this system and $Q_{rév}$ the heat exchanged by conduction with the external environment. From the second law of thermodynamics, any transformation respects the following inequality:

$$dS \geq \frac{\delta Q_{rév}}{T} \tag{3.1}$$

Let s be the entropy per unit volume and **q** the heat flux across the mobile surface S_m. We assume here that the medium is incompressible (ρ = Cst). The previous inequality becomes (per unit time):

$$\frac{d}{dt}\int_{V_m} s\, dV \geq \int_{S_m} -\frac{\mathbf{q}.\mathbf{n}}{T}\, dS \tag{3.2}$$

which can also be written under the equivalent form (using the derivation theorem proven in Appendix 1):

$$\int_{V_m} \left(\frac{Ds}{Dt} + \operatorname{div}\frac{\mathbf{q}}{T}\right) dV \geq 0 \tag{3.3}$$

This inequality holding for any choice of the volume V_m, it ensues:

$$\frac{Ds}{Dt} + \operatorname{div}\frac{\mathbf{q}}{T} \geq 0 \tag{3.4}$$

We shall call **irreversible entropy production** the quantity

$$T\overset{\circ}{s} = T\frac{Ds}{Dt} + T\operatorname{div}\frac{\mathbf{q}}{T} \geq 0 \tag{3.5}$$

N.B.: if the fluid is not incompressible, the specific quantities (per mass unit) must be used. If s is the specific entropy, the general form of irreversible entropy production is written as:

$$\rho T\overset{\circ}{s} = \rho T\frac{Ds}{Dt} + T\operatorname{div}\left(\frac{\mathbf{q}}{T}\right) \geq 0 \tag{3.6}$$

Appendix 4: Energy dissipation and constitutive laws in the formalism of Leslie–Ericksen–Parodi

We shall first show that the dissipation can be written as:

$$T\overset{\circ}{s} = \sigma_{ij}^{s} A_{ij} + (\mathbf{h} + \mathbf{h}^{M}).\mathbf{N} - \frac{\mathbf{q}}{T}.\mathbf{grad}\,T \tag{4.1}$$

where σ_{ij}^{s} and A_{ij} are the symmetric parts of the viscous stress tensor σ_{ij}^{v} and of the distortion rate tensor $\partial v_i/\partial x_j$, respectively. Let us also define the antisymmetric parts of these tensors, σ_{ij}^{a} and Ω_{ij} (cf eq. B.III.43). Then, from eq. B.III.69a:

$$T\overset{\circ}{s} = \sigma_{ij}^{s} A_{ij} + \sigma_{ij}^{a} \Omega_{ij} + (\mathbf{h} + \mathbf{h}^{M}).\frac{D\mathbf{n}}{Dt} - \frac{\mathbf{q}}{T}.\mathbf{grad}\,T \tag{4.2}$$

But, from the torque balance theorem (eq. B.III.51), we know that:

$$\mathbf{\Gamma}^{v} = -\,\mathbf{n} \times (\mathbf{h} + \mathbf{h}^{M}) \tag{4.3}$$

where $\mathbf{\Gamma}^{v}$ is the viscous torque associated with the antisymmetric part of the viscous stress tensor $\underline{\sigma}^{v}$ (eq. B.III.49):

$$\Gamma_i^{v} = -e_{ijk}\,\sigma_{jk}^{a} \tag{4.4}$$

Using 4.3 and then 4.4 one successively gets:

$$(\mathbf{h} + \mathbf{h}^{M}).\frac{D\mathbf{n}}{Dt} = (\mathbf{h} + \mathbf{h}^{M}).\mathbf{N} + (\mathbf{h} + \mathbf{h}^{M}).(\mathbf{\Omega} \times \mathbf{n}) = (\mathbf{h} + \mathbf{h}^{M}).\mathbf{N} - \mathbf{\Omega}.\mathbf{\Gamma}^{v} \tag{4.5}$$

and

$$-\,\mathbf{\Omega}.\mathbf{\Gamma}^{v} + \sigma_{ij}^{a}\,\Omega_{ij} = e_{ijk}\sigma_{jk}^{a}\Omega_i - e_{ijk}\Omega_k\sigma_{ij}^{a} = 0 \tag{4.6}$$

From these two equalities and eq. 4.2, formula 4.1 is obtained. It should be noted that this expression makes it clear that dissipation remains unchanged under a "solid" rotation of the entire system, an essential property of the laws of thermodynamics.

Let us now see how the constitutive laws for the material can be obtained [1, 2]. We start by choosing $\underline{\sigma}^{a}$, $\mathbf{h} + \mathbf{h}^{M}$, and $\mathbf{q}$ as forces and $\underline{A}$, $\mathbf{N}$, and **grad**T as the associated fluxes. If the fluxes are weak on the molecular scale (as the hydrodynamic theory assumes), a linear relation can be written between the forces and the fluxes. Taking into account the presence of an inversion center among the symmetry elements of the nematic phase, it can be shown that these relations can be written in the following general form:

$$\sigma_{ij}^{s} = L_{ijkl}A_{kl} + M_{ijk}N_k \tag{4.7a}$$

$$h_i + h_i^M = M'_{ijk}A_{jk} + P_{ij}N_j \tag{4.7b}$$

$$q_i = k_{ij}T_{,j} \tag{4.7c}$$

where all the coefficients L, M, M', and P are in units of dynamic viscosity (1 poiseuille = 1 Pa.s = 10 poise) and where k_{ij} is the thermal conductivity matrix. To further simplify the equations, we use the Onsager theorem [3], according to which:

$$M_{ijk} = M'_{ijk} \tag{4.8}$$

This theorem expresses the "time-reversal invariance" of the equations of motion for the particles forming the system. On the other hand, matrices $\underline{L}$, $\underline{M}$, $\underline{P}$, and $\underline{k}$ must be compatible with the local $D_{\infty h}$ symmetry of the nematic. Consequently, the only vector that can appear in their expression is the director **n**. Finally, all quantities must be invariant when **n** is changed to −**n**. Note that under this transformation $\underline{A}$ and $\underline{\sigma}$ are invariant, while **N** and **h** change sign.

It can be shown that the most general structure of equations 4.7 compatible with these requirements is the following (for an incompressible nematic, $A_{ii} = 0$):

$$\sigma_{ij}^s = \alpha_1(n_k A_{kl} n_l)n_i n_j + \alpha_4 A_{ij} + \frac{1}{2}(\alpha_5 + \alpha_6)(n_i n_k A_{kj} + n_j n_k A_{ki}) + \frac{1}{2}\gamma_2(n_i N_j + n_j N_i) \tag{4.9a}$$

$$h_k + h_k^M = \gamma_2 n_i A_{ki} + \gamma_1 N_k \tag{4.9b}$$

$$\mathbf{q} = -k_\perp \,\mathbf{grad}\, T - (k_{//} - k_\perp)\,\mathbf{n}\,(\mathbf{n}.\mathbf{grad}\, T) \tag{4.9c}$$

To explicitly determine the complete viscous stress tensor $\underline{\sigma}^v$, one only needs use relations 4.3 and 4.4 which give its antisymmetrical part as a function of the molecular field. It then comes:

$$\sigma_{ij}^v = \alpha_1(n_k A_{kl} n_l)n_i n_j + \alpha_2 n_j N_i + \alpha_3 n_i N_j + \alpha_4 A_{ij} + \alpha_5 n_j n_k A_{ki} + \alpha_6 n_i n_k A_{kj} \tag{4.10}$$

with the additional relations

$$\gamma_1 = \alpha_3 - \alpha_2 \tag{4.11a}$$

and

$$\gamma_2 = \alpha_3 + \alpha_2 = \alpha_6 - \alpha_5 \tag{4.11b}$$

This last equality was established by Parodi [4] using the Onsager reciprocity relations [3].

BIBLIOGRAPHY

[1] Ericksen J.L., *Arch. Ration. Mech. Analysis*, **4** (1960) 231; *Phys. Fluids*, **9** (1966) 1205.

[2] Leslie F.M., *Quart. J. Mech. Appl. Math.*, **19** (1966) 357; *Arch. Ration. Mech. Analysis*, **28** (1968) 265.

[3] de Groot S.R., Mazur P., *Non-Equilibrium Thermodynamics*, Dover, New York, 1984.

[4] Parodi O., *J. Physique (France)*, **31** (1970) 581.

Appendix 5: The Rayleigh-Bénard instability in isotropic fluids

The typical experimental setup for studying the Rayleigh-Bénard instabilities [1,2] consists of a horizontal layer of an isotropic liquid (thickness d) confined between two solid surfaces imposing a vertical temperature gradient (Fig. A.5.1).

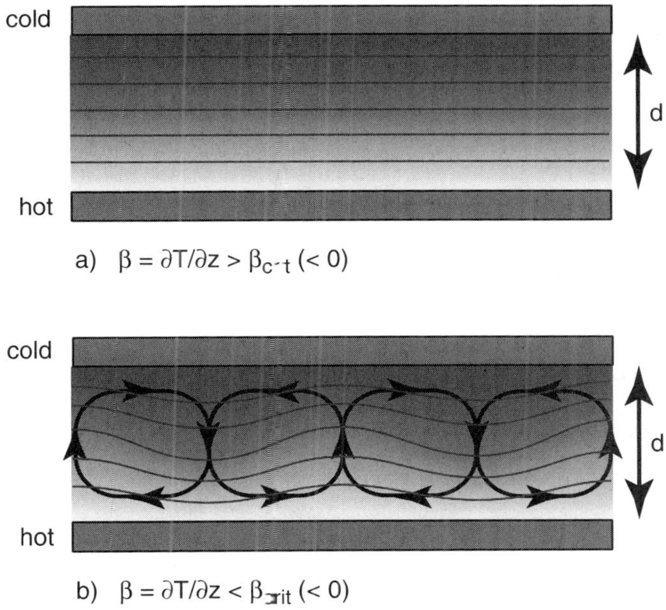

Fig. A.5.1 Rayleigh-Bénard instability in a layer of isotropic liquid. a) Below the threshold; b) above the threshold, the convective currents deform the isotherms.

As represented in figure A.5.1, the warm (and lighter) liquid is below, while the cooler (and usually heavier) liquid is above. Obviously, the situation is mechanically unstable. The experiment shows that convective currents appear when the temperature gradient

$$\beta = \frac{\partial T}{\partial z} \tag{5.1}$$

exceeds a certain threshold β_{crit}. The exact value of the threshold depends on the physical characteristics of the liquid and on the boundary conditions [1]. In most experimental situations, determining the threshold is very difficult since the flow $\mathbf{v}(\mathbf{r},t)$ and the temperature distribution $T(\mathbf{r},t)$ are complex due to the boundary conditions.

To estimate the value of the threshold for the convective instabilities, let us simplify the problem as much as possible by considering the figure A.5.2 diagram.

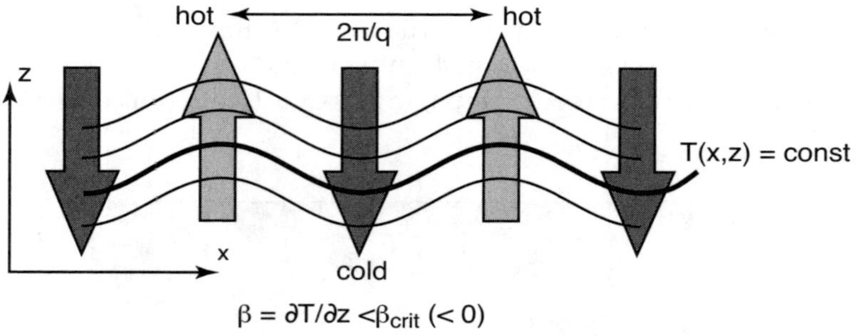

Fig. A.5.2 One-dimensional model for the convective instabilities.

Here, the convective currents are strictly vertical:

$$\mathbf{v} = [0,0,v_z(t)\cos(qz)] \tag{5.2}$$

and the temperature perturbation $\delta T(\mathbf{r},t)$ is one dimensional:

$$T = \beta z + \delta T = \beta z + \theta(t)\cos(qx) \tag{5.3}$$

The evolution of the temperature perturbation is given by the thermal diffusion equation:

$$\frac{\partial \theta}{\partial t} = \Lambda \frac{\partial^2 \theta}{\partial x^2} - \beta v_z \tag{5.4}$$

comprising, in addition to the relaxation term due to diffusion (Λ is the thermal diffusion coefficient), the convection term $-\beta v_z$. This term couples the temperature evolution to that of the current v_z.

As to the evolution of the convective current, it is described by the Navier-Stokes equation

$$\frac{\partial v_z}{\partial t} = \nu \frac{\partial^2 v_z}{\partial x^2} + g\alpha\theta \tag{5.5}$$

containing both the viscous relaxation term ($\nu = \eta/\rho$ is the kinematic viscosity) and the buoyancy force (g is the gravity acceleration and α the thermal expansion coefficient). This second term introduces a coupling between the flow and the temperature field.

As the figure A.5.3 diagram illustrates, the mutual coupling of the two perturbations δT and v_z determines a feedback loop whose action adds to the stabilizing effects of the thermal diffusivity and the viscosity.

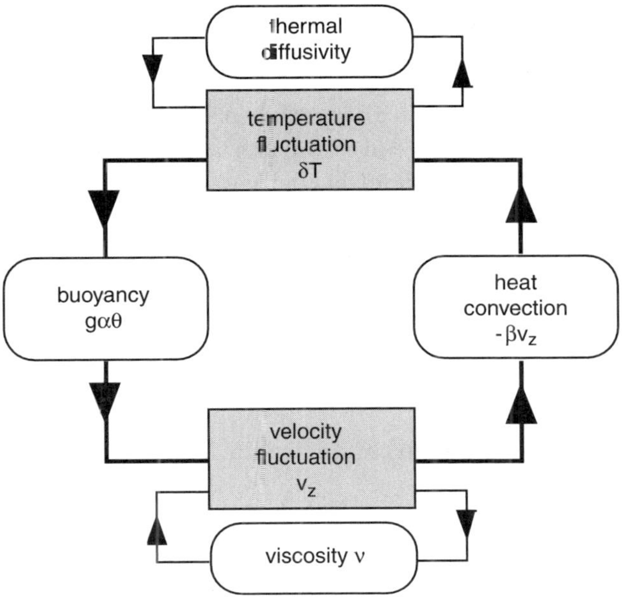

Fig. A.5.3 Rayleigh-Bénard instability: the coupling between the velocity and temperature fluctuations leads to a feedback loop.

Eqs. 5.4 and 5.5 admit solutions of the type:

$$\theta = \theta_0 e^{st} \cos(qx) \tag{5.6}$$

and

$$v_z = v_{zo} e^{st} \cos(qx) \tag{5.7}$$

where the amplitudes θ_o and v_{zo} can be finite, provided that the determinant is zero:

$$(s + \nu q^2)(s + \kappa q^2) + g\alpha\beta = 0 \tag{5.8}$$

The solutions

$$s_{\pm}(q) = \frac{-(\nu q^2 + \Lambda q^2) \pm \sqrt{(\nu q^2 - \Lambda q^2)^2 - 4g\alpha\beta}}{2} \tag{5.9}$$

of this last equation describe the temporal evolution of the perturbations (5.6) and (5.7).

Let us examine the nature of these solutions as a function of the argument

$$P = (\nu q^2 - \Lambda q^2)^2 - 4g\alpha\beta \tag{5.10}$$

of the square root in formula 5.9. The sign and amplitude of this parameter can be modified "at will" in the experiments by tuning the sign and amplitude of the thermal gradient β.

In the absence of a vertical thermal gradient ($\beta = 0$), this parameter is positive. It increases when the fluid is warmer at the bottom than on top ($\beta < 0$) as the thermal expansion coefficient is positive. Conversely, when the system is heated from above, P decreases and can become negative.

For $P < 0$, the solutions $s_\pm$ are complex, meaning that the perturbations oscillate. However, in this case the real part of the roots $s_\pm$

$$\mathrm{Re}[s_\pm] = -\frac{\nu q^2 + \Lambda q^2}{2} \tag{5.11}$$

is always negative, so the perturbation oscillations are damped. The system is thus stable.

In the opposite case, when $P > 0$, the solutions $s_\pm$ are purely real. For small P, they are both negative, but one of them, s_+, changes sign when P increases. The condition that the s_+ root is zero

$$-(\nu q^2 + \Lambda q^2) + \sqrt{(\nu q^2 - \Lambda q^2)^2 - 4g\alpha\beta} = 0 \tag{5.12}$$

defines the threshold of the Rayleigh-Bénard instability.

Introducing the perturbation wavelength $\lambda = 2\pi/q$, this expression for the threshold can be written as a function of a dimensionless quantity, called the Rayleigh number:

$$R = -\frac{g\alpha\beta\lambda^4}{\nu\Lambda} = (2\pi)^4 \tag{5.13}$$

In fact, the instability wavelength λ_{crit} at the threshold is related to the thickness d of the liquid layer:

$$\lambda_{crit} = \lambda'_{crit}\, d \tag{5.14}$$

therefore we shall rather use this thickness in the definition of the Rayleigh number, the threshold value of which becomes:

$$R = -\frac{g\alpha\beta d^4}{\nu\Lambda} = R_{crit} \tag{5.15}$$

The critical value of the Rayleigh number, as well as the scaled wavelength λ', depend on the boundary conditions.

For instance, when the limiting surfaces are completely rigid and perfectly conductive, in their vicinity the isotherms are planar and the fluid is at rest. With these boundary conditions, the critical Rayleigh number is

$$R_{crit} = 1708 \tag{5.16}$$

and the convection rolls are practically circular, as $\lambda'_{crit} = 2.016$.

In practice, in order to trigger the R-B instabilities, one only needs to apply a large enough temperature difference $\Delta T = \beta d$ between the two surfaces. We should point out right away that the critical value of this temperature difference varies as d^{-3}.

Let us consider the example of a water layer of thickness d = 1 cm, and with:

$$g = 981\ cm/s^2$$

$$\alpha = 2\times10^{-3} K^{-1}$$

$$\nu = 10^{-2} cm^2/s$$

$$\Lambda = 1.4\times10^{-3} cm^2/s$$

we have:

$$\Delta T_{crit} = 1708 \frac{\nu\Lambda}{\alpha g d^3} = 1.2\times10^{-2} K$$

BIBLIOGRAPHY

[1] Chandrasekhar S., *Hydrodynamic and Hydromagnetic Stability*, Dover, New York, 1961.

[2] Guyon E., Hulin J.-P., Petit L., *Hydrodynamique Physique*, Inter Editions/Editions du CNRS, Les Ulis, 1991.

Chapter B.IV

Defects and textures in nematics

Thus far we have only considered large-scale deformations of continuous director fields. We have shown that, in the absence of applied fields, the equilibrium director configuration must satisfy in the bulk the condition of zero elastic torque $\mathbf{\Gamma}^E$, while at the surfaces it must be compatible with the boundary conditions. In the case of strong anchoring on the walls limiting the sample, these conditions are geometric in nature and cannot be modified. As such, they can be **topologically incompatible** with the existence of a continuous director field.

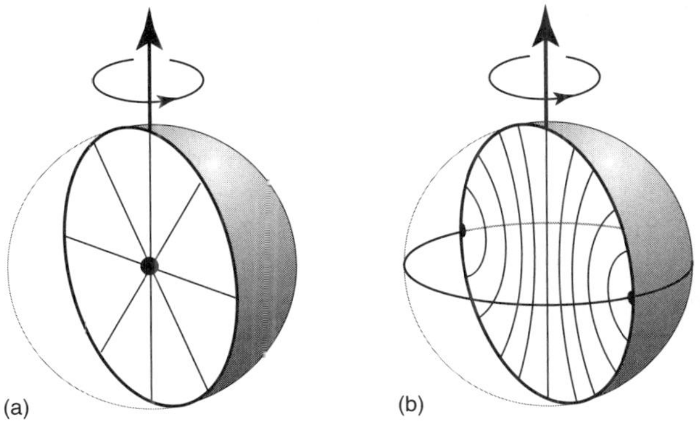

Fig. B.IV.1 Two possible ways of ensuring homeotropic anchoring for a nematic-filled sphere. In (a) there is a point singularity in the center of the sphere, while in (b) the line singularity follows the equator.

A simple example is that of a nematic-filled sphere imposing homeotropic anchoring on its inner surface. It is obvious that the cavity cannot be filled by a continuous director field. Two possible configurations are shown in figure B.IV.1:

The first case in (a) is that of a purely radial director field. The singularity is point-like and located in the center of the sphere. At this point div**n**, and therefore the elastic energy, diverge. Such singularities are termed **umbilics**.

In configuration (b), the central singularity is replaced by a line following the equator. It is easily seen that the director field is singular on the line. As we shall see later, this is a wedge disclination loop of rank 1/2.

This simple example proves the existence of **singular director fields**, sometimes imposed by the system geometry. The singularities can be point-like or linear (or, exceptionally, surface-like in nematics). This chapter treats such singular vector fields. The solutions of nematoelasticity equations are similar to those of the point defects and dislocations encountered in ordinary solids. However, while dislocations break the translational symmetry of a crystal, **these defects break the rotation symmetries of the nematic**. Termed **disclinations** by Frank [1], they are easily seen under a polarizing microscope and form textures (structures of defects) allowing for clear identification of nematic phases.

It is these characteristic structures that we shall discuss first (section IV.1). We shall then introduce the Volterra process allowing us to construct and classify the different topological defects (section IV.2). Their elastic and energetic properties will then be discussed (section IV.3), revealing the need for a separate treatment of dislocation cores, where elastic energy diverges. This divergence is only apparent and can be removed by taking into account the variation of the quadrupolar order parameter inside the core, which in turn requires that Ginzburg terms involving the gradient of the quadrupolar order parameter, S_{ij}, be included in the Landau-de Gennes free energy (section IV.4). Long-distance elastic interactions between two parallel disclinations will be studied (section IV.5). Their mobility will be estimated by a simplified calculation (section IV.6) and the result compared to experimental data. We shall subsequently discuss the "escape in the third dimension" (section IV.7), a topological process which can sometimes suppress the singular core of a disclination line. Finally, we shall tackle the issue of walls and their stability (section IV.8). As to surface disclinations, they will be treated in chapter B.V dedicated to anchoring transitions.

IV.1 Polarizing microscope observations

There are several ways of preparing a nematic liquid crystal sample. The easiest one consists of setting a nematic drop on a glass plate treated in homeotropic anchoring. This particular anchoring prevents the defects from "sticking" to the surface. The "**threadlike texture**" [2] appearing immediately after placing the liquid crystal drop (Fig. B.IV.2) is typical of this phase and allows for immediate identification. The threads resemble flexible filaments floating inside the drop. Two kinds can be clearly distinguished: **thin threads**, strongly scattering light and **thick threads**, whose diameter and contrast are not well defined; the latter can even be overlooked under certain polarization conditions. Each of them corresponds to a topologically distinct dislocation type. Appearing under the form of closed loops, these defects evolve and completely disappear in a few minutes (unless they get caught on dust particles or on the surfaces). The defects reappear when stirring the liquid crystal with a spatula, proving they are a consequence of high Ericksen number flows. Defect lines also appear when filling a homeotropic sample by capillarity, a feature to be discussed in section IV.6 on dynamic properties. They can also be obtained by rapid quenching from the isotropic to the liquid crystal phase. Defect generation during directional growth of the nematic phase into its isotropic liquid will also be discussed in section IV.6.

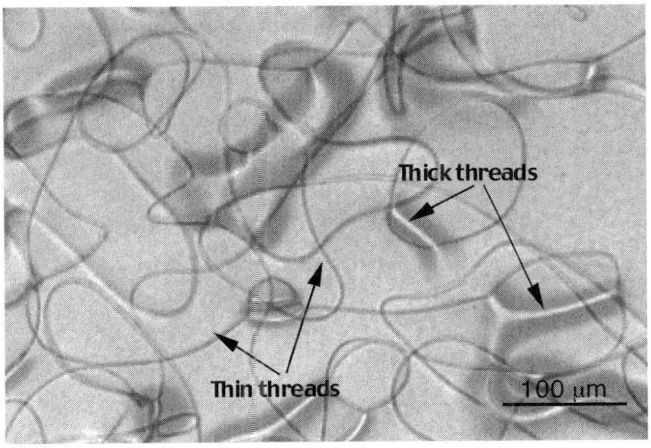

Fig. B.IV.2 Threadlike texture in a nematic droplet placed on a glass plate treated in homeotropic anchoring.

The liquid crystal can also be sandwiched between two untreated glass plates favoring weak planar anchoring. A new texture, called "**plage à noyaux**" by G. Friedel [2], can then be observed under the microscope. Its appearance (Fig. B.IV.3) largely depends on the observation conditions. In unpolarized light, one sees a bright background with isolated dark points corresponding to the dislocation cores, where the director exhibits an orientational discontinuity (or abrupt variation). This central part scatters light more or less strongly, hence the optical contrast. Between crossed polarizers, these points appear connected by black branches, indicating the local direction of the neutral lines of the sample (i.e., the optical axis orientation, tangent to the director). We recall that extinction is achieved when the optical axis is either parallel or perpendicular to the light polarization. Outside the core, molecular orientation varies slowly. Two defect types can be immediately distinguished: they have either four or two branches. We shall see later that the former correspond to thick threads and the latter to thin ones.

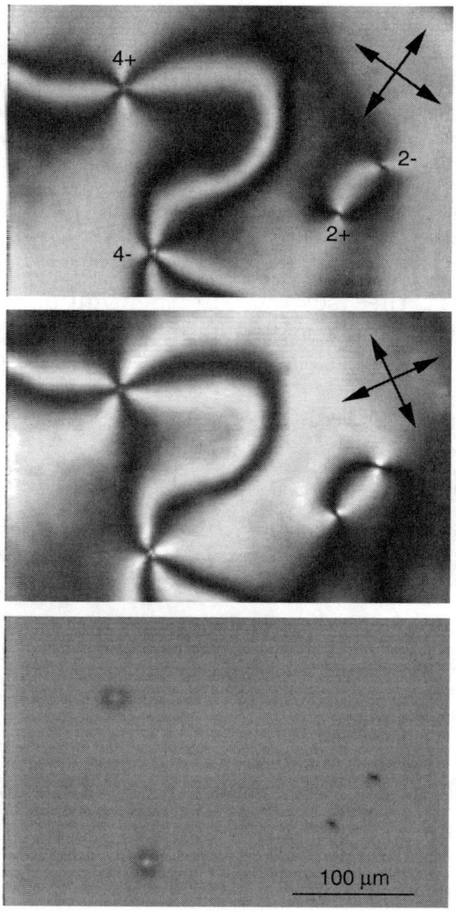

Fig. B.IV.3 "Plage à noyaux" observed between crossed polarizers and in natural light, showing all four defect types. The liquid crystal is 8CB in a 10 μm sample with no surface treatment at 40°C. In natural light, 2+ and 2– defects exhibit a thin and strongly scattering core (black dots) while the cores of 4+ and 4– defects are larger and much less contrasted.

The threads, normal to the glass plates, are observed along their axis. On turning the polarizers, the extinction branches can rotate one way or the other. A **sign** must therefore be associated with each defect, + if the branch rotation follows that of the polarizers and – if they turn the opposite way. One therefore has at least four different defect types in a "plage à noyaux." For the moment, we shall designate them as 4+, 4–, 2+, and 2– (Fig. B.IV.3).

Another way of creating defects is confining the nematic in a cavity such that the boundary conditions are incompatible with a continuous director field. The case of the sphere, already discussed in the introduction, can appear at the nematic-isotropic liquid transition when the two phases coexist [3], or when the nematic forms small droplets dispersed in a non-miscible fluid [4] or in a polymer matrix (PDLC or "polymer dispersed liquid crystals" [5]). For the latter situation anchoring at the polymer surface is usually planar so a binodal structure can appear, with two umbilics at the North and South poles (Fig. B.IV.4).

Fig. B.IV.4 Binodal structure induced by planar anchoring at the sphere surface. Two point defects appear at the North and South poles.

An electric field applied to a thin PDLC layer can induce a Frederiks transition inside the droplets, dramatically changing film transmittance [6].

The cylindrical configuration [7] is easily obtained in a tube filled by capillarity. Glass capillaries of internal diameter between 10 and 200 μm are usually employed. If the glass surface is treated in homeotropic anchoring, a thick thread along the capillary axis can be observed using polarized microscopy. For easier observation, the capillary should be immersed in a liquid of refractive index matching that of the glass. This thread is sometimes interrupted by one or several point defects [8] as shown in figure B.IV.5.

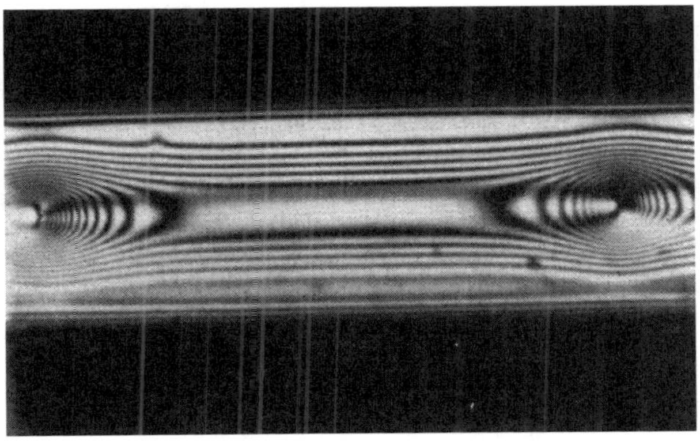

Fig. B.IV.5 Thick thread in a round capillary, interrupted by two point defects (umbilics). The dark lines are interference fringes between the ordinary and the extraordinary rays.

A semi-cylindrical configuration can also be obtained using flat capillaries like the one depicted in figure B.IV.6. With homeotropic anchoring, two thin threads appear on the sides [9]. Usually they are not straight, but rather develop a zigzag pattern parallel to the capillary axis and with a perfectly defined wavelength, showing that a disclination line can sometimes be unstable in a confined environment. We maintain that the zigzag instability is not the mark of a biaxial nematic, as it appears in products like MBBA or 8CB.

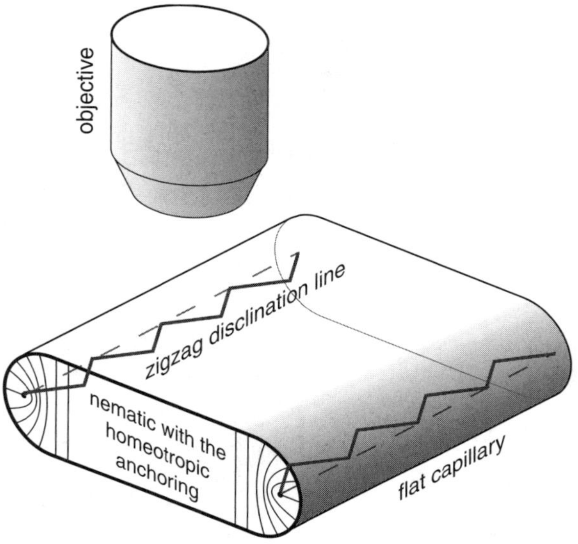

(a)

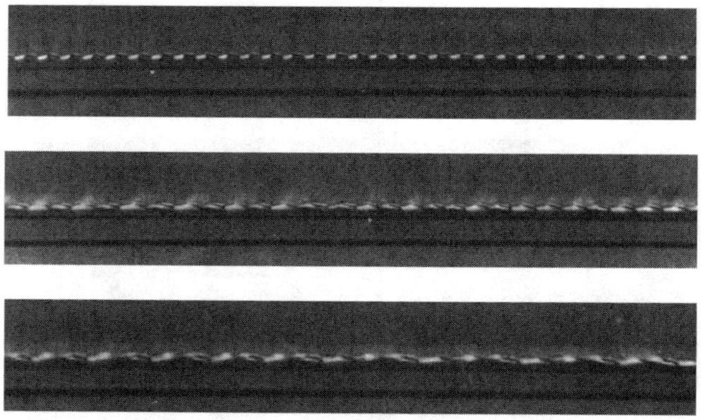

(b)

Fig. B.IV.6 Zigzag instability of a thin thread appearing on the side of a homeotropically treated flat capillary. a) Experimental setup; b) microscope observation between crossed polarizers of a 20 μm thick 8CB sample. The wavelength strongly decreases on approaching the transition temperature to the smectic A phase. From top to bottom, $\Delta T = T - T_{SmA} = 0.05$, 0.15, and 0.30°C (from ref. [9]).

Let us now describe the Volterra process allowing a rotational defect to be created.

IV.2 The Volterra-de Gennes-Friedel process

IV.2.a) Symmetry elements of a uniaxial nematic

Before describing the defects of a certain phase it is important to classify its symmetry elements. In uniaxial nematics, these are:

– all translations
– all rotations about **n**
– all rotations of an angle π or a multiple of π about all axes perpendicular to **n**
– all mirror planes parallel or perpendicular to **n**
– an inversion center

IV.2.b) Generating a disclination

We shall first describe the general principle of the Volterra construction. Let L be a contour and Σ a generic surface limited by L. Imagine cutting open the material along Σ. Let Σ_1 and Σ_2 be the two sides of the slit (Fig. B.IV.7). Without deforming them, pull Σ_1 and Σ_2 a distance $\mathbf{d}(\mathbf{r})$ apart. If the displacement is a solid rotation of angle Ω around the unit vector $\boldsymbol{\upsilon}$, then:

$$\mathbf{d}(\mathbf{r}) = 2\sin(\Omega/2)\,\boldsymbol{\upsilon}\times\mathbf{r} \qquad \text{(B.IV.I)}$$

If $(\Omega,\boldsymbol{\upsilon})$ is a symmetry operation of the undistorted material, it is possible to add in (or remove from) the space swept by the two edges of the slit during their displacement a slice of such material, so that the contact along Σ_1 and Σ_2 is made without a singularity. One then suppresses the stress imposed on the material by this procedure. When the system has relaxed, a singularity, called **disclination**, remains along L, accompanied by a persistent volume deformation. Before describing the nematic case, a few definitions are in order. Let **t** be the unit vector tangent to the disclination line. If $\boldsymbol{\upsilon}$ is parallel to **t**, the line is of the **wedge** type. If, on the other hand, $\boldsymbol{\upsilon}$ is perpendicular to **t** one speaks of a **twist** disclination. In all other cases, the line is **mixed**.

Figure B.IV.7b illustrates the application of the Volterra process to generating a straight wedge disclination in nematics with a vector $\boldsymbol{\upsilon}$ parallel to the line and $\Omega = \pi$. Notice that the operation $(\Omega,\boldsymbol{\upsilon})$ corresponds to a rotation of angle π about an axis normal to the director. The stages of the construction, (i) cut, (ii) turn, (iii) fill in with undistorted nematic, and (iv) relax, are easy to conceive and visualize.

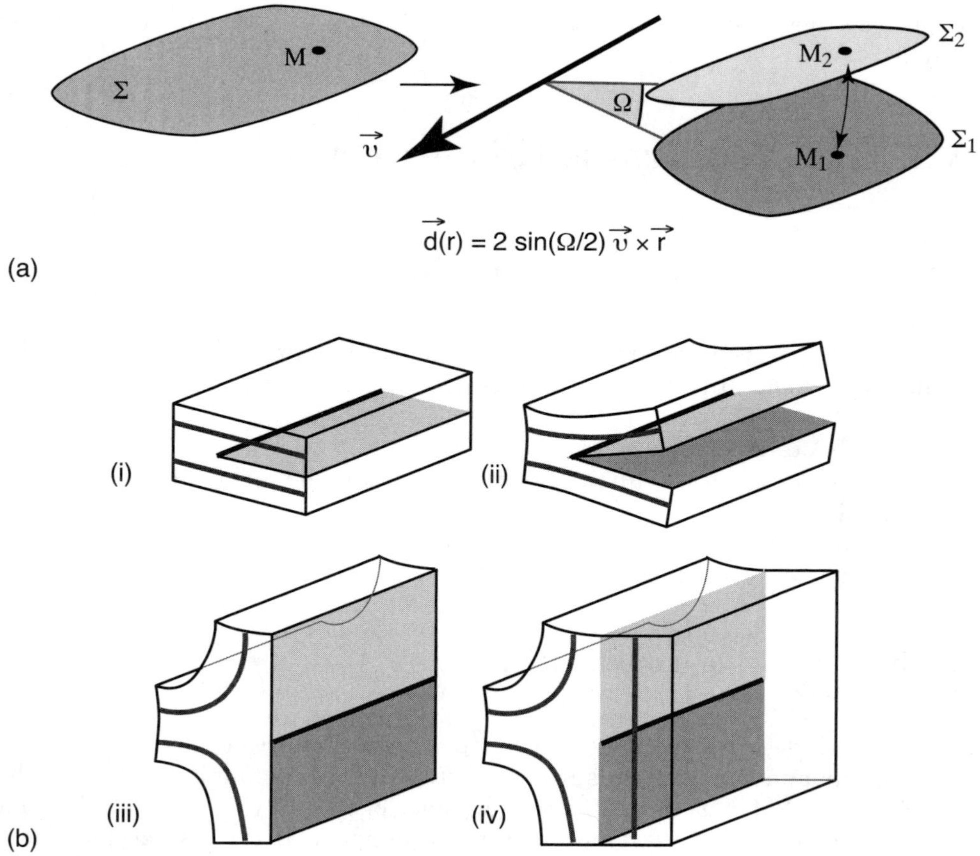

Fig. B.IV.7 a) Generic Volterra process [10]; b) Volterra process applied to generating a wedge line of angle π in a nematic. The dark stripes on the side represent the director field lines.

Things are completely different in the case of a twist disclination, when the rotation axis $\boldsymbol{\upsilon}$ is normal to the line (see the vector field in Figure B.IV.10b, where the rotation axis is horizontal). In the general case, the Volterra process is more difficult to observe, as it also engenders dislocations (a concrete example will be given in the case of columnar phases, treated in the second volume) [11]. In nematics, however, this difficulty disappears because each dislocation can be decomposed into a sum of infinitesimal dislocations. All translations being allowed, they disappear by viscous relaxation. This feature reduces the Volterra process, to simply rotating the molecules around their center of mass with a relative angle Ω between the two sides of Σ. First introduced by de Gennes, this mechanism is equivalent to the Volterra process in a nematic, as shown by J. Friedel. Figure B.IV.8 displays the most common director fields of wedge disclinations obtained by the Volterra-de Gennes-Friedel process.

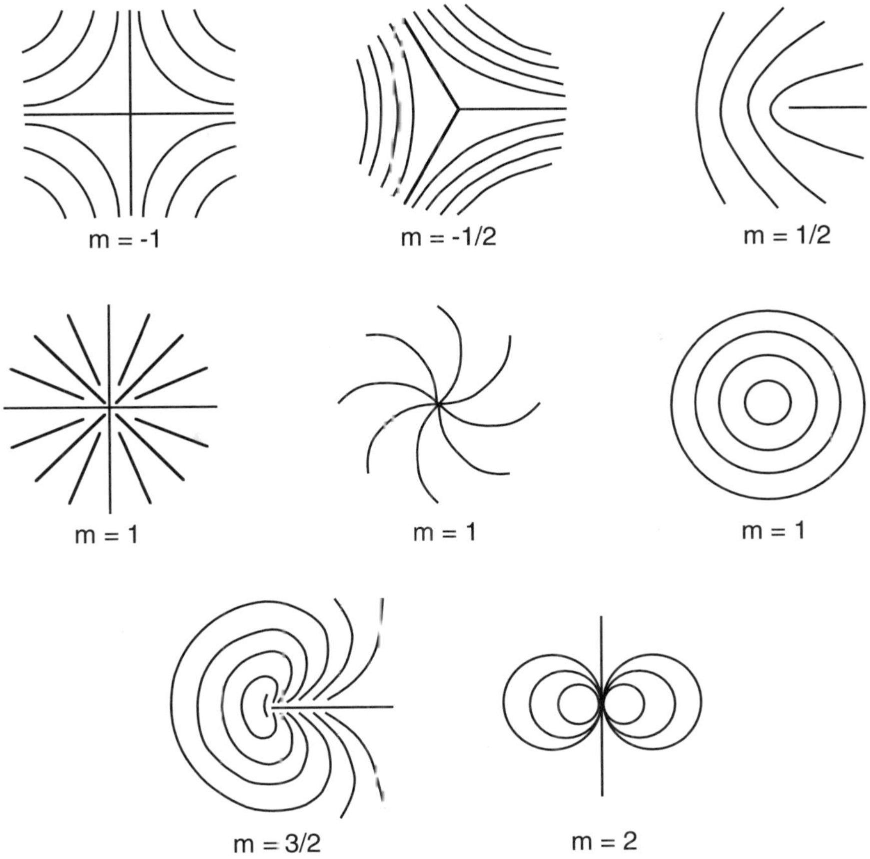

Fig. B.IV.8 Examples of planar wedge disclinations.

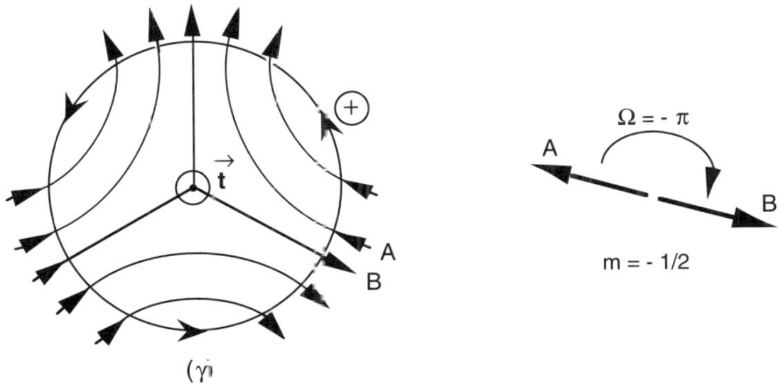

Fig. B.IV.9 Hodograph of a wedge line [11a].

To each wedge disclination is conventionally associated a number m termed **topological rank** and defined in the following manner: Let γ be a closed loop (sometimes called a **Burgers circuit** [11]) embracing the line L (Fig. B.IV.9). The circuit is oriented, defining at the same time a positive rotation sense (corkscrew rule). Starting from an initial point on this contour, one draws the encountered unit vectors tangent to the molecular direction (the director **n**). After a complete rotation, **n** will have turned by an angle Ω which can be positive or negative depending on the rotation sense. The rank m is defined by:

$$\Omega = 2\pi m \qquad \text{(B.IV.2)}$$

In nematics m can be an integer or a half-integer (because **n** and **−n** are equivalent). In a system of spins, on the other hand, defects with half-integer rank are forbidden. In Figure B.IV.8, the topological rank is indicated below each configuration.

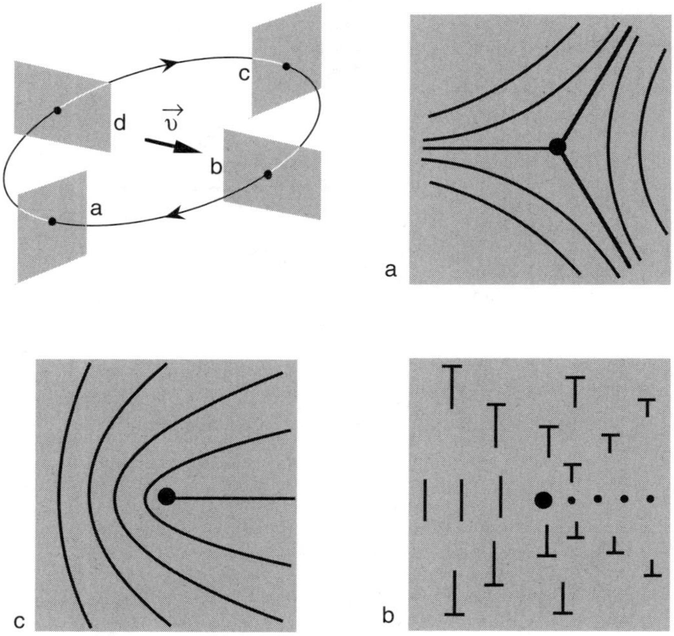

Fig. B.IV.10 Flat disclination loop. If the rotation vector **υ** is parallel to the loop plane, the disclination type changes along the line, from pure wedge in sections (a) and (c) to pure twist in (b) and (d) and mixed elsewhere. Far from the loop, the director is horizontal and parallel to the (ac) direction. Note that the director field distortions are localized near the loop. In section (b), the so-called "nail representation" has been used to describe the director field. Tilted molecules are represented by nails, proportional in length to the director projection in the plane of the drawing.

It is now obvious that 4+, 4−, 2+, and 2− defects in the previous section correspond to wedge disclinations of rank m = +1, − 1, +1/2, and −1/2, respectively. The "plages à noyaux" where all lines are normal to the sample plane are however exceptional. In most cases, the lines have a variety of shapes, sometimes closing back on themselves to form loops, clearly visible in the threadlike textures appearing in thick drops. The disclination type then varies along the line, since the angle between the line tangent t and the rotation vector $\boldsymbol{\upsilon}$ changes. For instance, when the vector $\boldsymbol{\upsilon}$ is in the plane of the loop (Fig. B.IV.10), the line is of wedge type in the (ac) section, twist in the (bd) section, and mixed elsewhere. It is worth noting that in the (ac) plane the line has topological rank 1/2 on one side and – 1/2 on the other, showing that the conserved quantity along the line is the vector $\Omega\boldsymbol{\upsilon}$, and not the rank m. This is a consequence of the fact that the nematic must remain undistorted far from the loop: the sum of the topological ranks in the (ac) section is then zero.

If, on the other hand, the rotation vector $\boldsymbol{\upsilon}$ is perpendicular to the loop plane, the line is everywhere of the twist type.

Finally, let us get back to Figure B.IV.1b, representing a disclination loop having the topology of a 1/2 wedge line in all sections, which might seem contradictory with the preceding discussion. In this case, however, the nematic is no longer undistorted far from the loop and acquires a radial orientation.

IV.3 Energy of a wedge planar line in isotropic elasticity

To simplify the calculations, let us suppose that the nematic elastic constants satisfy $K = K_1 = K_2 = K_3 = -2K_4$ and $K_{13} = 0$. The elastic energy then reduces to:

$$f = \frac{1}{2} K \partial_i n_j \partial_i n_j \qquad \text{(B.IV.3)}$$

We shall equally assume that the lines are of the wedge type (straight) and **planar**, meaning that the director is always contained in the plane normal to the line. This last hypothesis is not always verified close to the dislocation core, as we shall see at the end of the chapter.

In two dimensions (the (x, y) plane normal to the line) one has, denoting by $\varphi(x, y)$ the angle between the director **n** and the x axis:

$$f = \frac{1}{2} K (\nabla \varphi)^2 \qquad \text{(B.IV.4)}$$

The total energy of the system is a minimum when φ verifies the Euler equation:

$$\Delta\varphi = 0 \qquad \text{(B.IV.5)}$$

written in cylindrical coordinates as:

$$\frac{1}{r}\frac{\partial}{\partial r}\left(r\frac{\partial \varphi}{\partial r}\right)+\frac{1}{r^2}\frac{\partial^2 \varphi}{\partial \theta^2}=0 \qquad \text{(B.IV.6)}$$

The obvious solution is:

$$\varphi = m\theta + \text{Cst} \qquad \text{(B.IV.7)}$$

representing a disclination of topological rank m (Fig. B.IV.11). The director field has no singularity in the half-plane $\theta = 0$ for integer or half-integer m (as **n** and **−n** are equivalent). Note that, in a spin system, only integer rank defects are allowed.

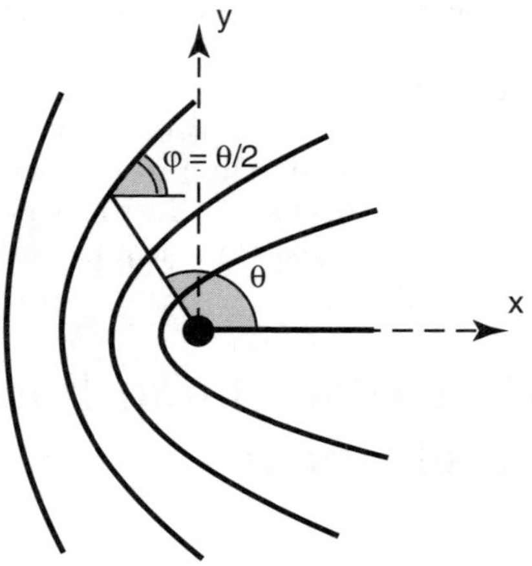

Fig. B.IV.11 Line of topological rank m = 1/2.

The energy per unit volume is:

$$f=\frac{1}{2}K\left(\frac{1}{r}\frac{\partial \varphi}{\partial \theta}\right)^2=\frac{1}{2}K\frac{m^2}{r^2} \qquad \text{(B.IV.8)}$$

or, per dislocation unit length along z:

$$E=\int_{r_c}^{R} f\,2\pi r\,dr=\pi K m^2 \ln\left(\frac{R}{r_c}\right) \qquad \text{(B.IV.9)}$$

The energy of a disclination line changes as m^2, the same way as in solids the dislocation energy is proportional to the Burgers vector squared. Planar lines of low topological rank are thus energetically favored, and only those with $m = \pm 1$ or $m = \pm 1/2$ are experimentally obtained.

In eq. B.IV.9, R is the sample size if the line is unique. Generally, the sample contains many lines of both signs and R must be taken as **the average distance between these lines**, a result to be proven in section IV.5.

As to r_c, it is a cut-off length called the **core radius**. Inside the core ($r < r_c$) energy density diverges as $1/r^2$. The core is usually of atomic size and we shall prove in the next section that the quadrupolar order parameter S goes to zero inside the core. Admitting this result without proof for the moment, let us try to make a rough estimate of the core energy.

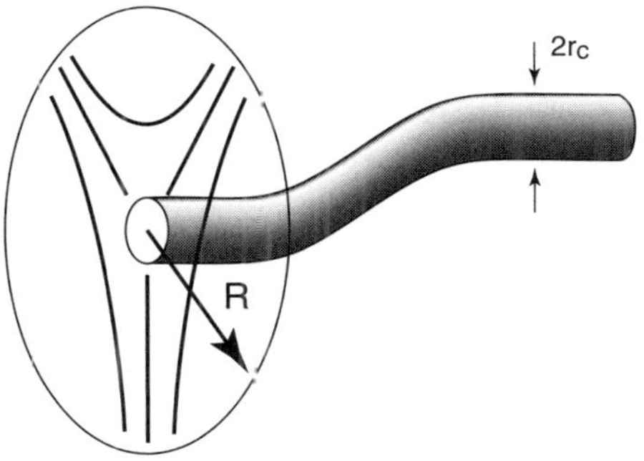

Fig. B.IV.12 In the two-phase model, the core of a disclination is represented by a tube of radius r_c filled by the isotropic liquid.

We shall equally assume that the core is filled with the isotropic liquid and that, in the nematic phase, the elastic constants and the order parameter take their equilibrium values (**"two-phase model"**) (Fig. B.IV.12). In the core, the isotropic liquid is undercooled and its energy per unit volume is $\Delta H \Delta T/T_c$, with ΔT the shift with respect to the transition temperature T_c and ΔH the transition latent heat per unit volume. Furthermore, an interface of energy γ per unit surface appears between the two phases, therefore:

$$E_{core} \approx \Delta H \frac{\Delta T}{T_c} \pi r_c^2 + 2\pi r_c \gamma \qquad \text{(B.IV.10)}$$

Minimizing with respect to r_c the total free energy of the line yields the value of the core radius:

$$r_c = d_o \frac{T_c}{2\Delta T}\left[\sqrt{1 + 2m^2 \frac{\Delta T}{T_c}\frac{d_s}{d_o}} - 1\right] \qquad \text{(B.IV.11)}$$

with the microscopic lengths:

$$d_o = \frac{\gamma}{\Delta H} \qquad \text{and} \qquad d_s = \frac{K}{\gamma} \qquad \text{(B.IV.12)}$$

The first one is the **capillary length**, while the second is given by the competition between surface tension and elasticity, being related to the penetration length of surface effects in the nematic bulk. The core radius remains very small at all temperatures, reaching a maximum at the critical temperature T_c:

$$r_c(T_c) = \frac{m^2}{2} d_s \qquad \text{(B.IV.13)}$$

and decreasing as $1/\sqrt{\Delta T}$ at sufficiently low temperature:

$$r_c \approx m\sqrt{\frac{KT_c}{2\,\Delta H\,\Delta T}} \qquad \text{if} \qquad \frac{\Delta T}{T_c} >> \frac{d_o}{d_s} \qquad \text{(B.IV.14)}$$

With typical values $T_c = 300$ K, $K = 10^{-12}$ J/m, $\Delta H = 10^6$ J/m^3, and $\gamma = 10^{-5}$ J/m^2 one has, for an $m = 1/2$ line at the critical temperature, $d_o = 0.1$ Å, $d_s = 1000$ Å, and $r_c = 500$ Å. Note that the simplified formula B.IV.14 is valid as soon as $\Delta T >> 0.03$ K and yields, for instance, $r_c = 120$ Å for $\Delta T = 1$ K and $r_c = 40$ Å for $\Delta T = 10$ K. **The core radius is therefore microscopic at all temperatures.** Using formula B.IV.14 the core energy is easy to estimate:

$$E_c \approx \frac{\pi K m^2}{2} \qquad \text{(B.IV.15)}$$

This contribution is rather small (about 10%) in comparison with the elastic deformation energy B.IV.9. In order to tackle the issue of the real order parameter profile inside the core, one must first generalize the Landau theory.

IV.4 Continuous core model: Landau-Ginzburg-de Gennes free energy

For an accurate description of the way S, the quadrupolar order parameter, varies in the core of a wedge line, the terms including gradients of the quadrupolar order parameter S_{ij} must be included in the Landau free energy $f_L(S_{ij},T)$ (eq. B.I.10). These terms are [12]:

$$f_G = \frac{1}{2} L_1\, \partial_i S_{jk} \partial_i S_{jk} + \frac{1}{2} L_2\, \partial_i S_{ij} \partial_k S_{kj} + \frac{1}{4} L_3 S_{ij} \partial_i S_{kl}\, \partial_j S_{kl} + \ldots \qquad \text{(B.IV.16)}$$

to third order in S. The previous expression seems completely different from that of Frank. However, the three constants L_1, L_2, and L_3 are intimately related to the Frank constants as formula B.IV.16 can be rewritten in a more familiar way using the definition of the tensor order parameter $S_{ij} = S\,[n_i n_j - (1/3)\delta_{ij}]$:

$$f_G = L_1 S^2\,[(\mathrm{div}\,\mathbf{n})^2 + (\mathbf{n}.\mathbf{curl\,n})^2 + (\mathbf{n}\times\mathbf{curl\,n})^2 - \mathrm{div}(\mathbf{n}\times\mathbf{curl\,n} + \mathbf{n}\,\mathrm{div}\,\mathbf{n})\,]$$

$$+ \frac{1}{2} L_2 S^2\,[(\mathrm{div}\,\mathbf{n})^2 + (\mathbf{n}\times\mathbf{curl\,n})^2]$$

$$+ \frac{1}{6} L_3 S^3\,[\,2\,(\mathbf{n}\times\mathbf{curl\,n})^2 - (\mathrm{div}\,\mathbf{n})^2 - (\mathbf{n}.\mathbf{curl\,n})^2 + \mathrm{div}(\mathbf{n}\times\mathbf{curl\,n} + \mathbf{n}\,\mathrm{div}\,\mathbf{n})\,]$$

$$+\frac{1}{3}L_1\,(\mathbf{grad}S)^2+\frac{1}{18}L_2[\,(\mathbf{grad}S)^2+3\,(\mathbf{n}.\mathbf{grad}S)^2]$$

$$+\frac{1}{3}L_2S\,\mathbf{grad}S\,(\,2\mathbf{n}\mathrm{div}\mathbf{n}+\mathbf{n}\times\mathbf{curl}\,\mathbf{n})+\text{terms}\,(L_3\,S\,\partial_iS\partial_jS) \qquad \text{(B.IV.17)}$$

In this expression, the terms of the first three lines describe the energy change with the variation of **n** (at fixed order parameter S), so they can be identified to the Frank terms, while the following terms contain the order parameter gradient. Identification leads to:

$$K_1=(2L_1+L_2)S^2-\frac{1}{3}L_3S^3+\ldots$$

$$K_2=2L_1S^2-\frac{1}{3}L_3S^3+\ldots$$

$$K_3=(2L_1+L_2)S^2+\frac{2}{3}L_3S^3+\ldots \qquad \text{(B.IV.18)}$$

$$K_4=-\,2L_1S^2+\frac{1}{3}L_3S^3+\ldots$$

explicitely showing how the Frank constants vary with the order parameter S. With the simplifying assumption $L_2 = L_3 = 0$ one has $K_1 = K_2 = K_3 = -K_4 = K = 2LS^2$, where $L_1 = L$. It is now clear that the divergence in elastic distortion energy can be eliminated by letting $S = 0$ inside the dislocation core. One can even go further and completely determine the radial profile of the order parameter S around a dislocation of rank m. Indeed, taking as reference the energy of the nematic at temperature T and assuming that the angle φ only depends on θ, the free energy density can be written as:

$$f=f_L[S(r)]-f_L[S_\infty]+L[S(r)]^2\left(\frac{1}{r}\frac{\partial\varphi}{\partial\theta}\right)^2+\frac{1}{3}L\left(\frac{dS}{dr}\right)^2 \qquad \text{(B.IV.19)}$$

where

$$f_L=\frac{1}{2}A(T)S^2-\frac{1}{3}BS^3+\frac{1}{4}CS^4 \qquad \text{(B.IV.20)}$$

is the usual Landau free energy and S_∞ the order parameter value at temperature T in the undistorted nematic (i.e., the one minimizing the Landau energy f_L):

$$S_\infty=\frac{B+\sqrt{B^2-4AC}}{2C} \quad \text{with} \quad A=A_0(T-T^*) \qquad \text{(B.IV.21)}$$

We recall that T* is the spinodal temperature of the isotropic phase. Minimizing the total free energy with respect to the two variables φ and S leads once again to $\varphi = m\theta$ and to a new equation describing the order parameter profile in the disclination core:

$$\frac{2}{3}L\frac{1}{r}\frac{d}{dr}\left(r\frac{dS}{dr}\right) - \left(A + 2L\frac{m^2}{r^2}\right)S + BS^2 - CS^3 = 0 \qquad \text{(B.IV.22)}$$

with boundary conditions:

$$S = 0 \quad \text{in} \quad r = 0 \qquad \text{and} \qquad S = S_\infty \quad \text{for} \quad r = +\infty \qquad \text{(B.IV.23)}$$

An approximate solution to this differential equation can be determined using an equation of the form [13]:

$$S = S_\infty\left[\frac{r}{\sqrt{r^2 + a^2}}\right]^\nu \qquad \text{(B.IV.24)}$$

which has the required behavior in r = 0 and r = ∞ on condition that we take $\nu = m\sqrt{3}$. Parameter "a," fixing the core radius, can subsequently be determined by line energy minimization. This method gives a result very close to the "exact" one obtained by direct integration of equation B.IV.22 (posing no particular problem).

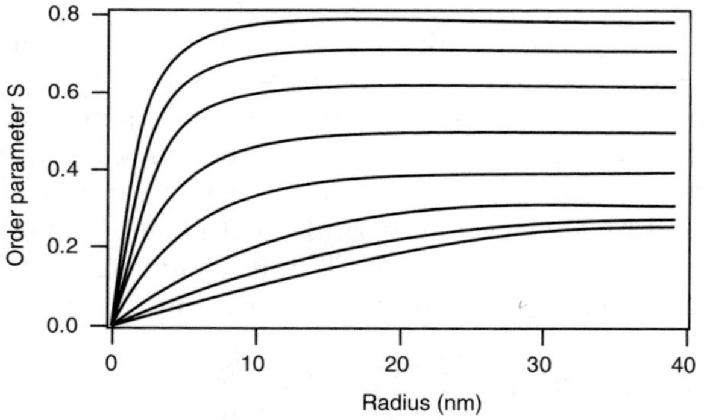

Fig. B.IV.13 Order parameter profile S(r) in the core of a wedge line of rank m = 1/2. The curves (from bottom to top) correspond to temperature shifts $T_c - T$ of 0, 0.05, 0.25, 1, 2.5, 7.5, and 10°C.

In figure B.IV.13, we plotted some order parameter profiles numerically computed by P. Ribière for a line of rank 1/2 taking for the Landau coefficients the values experimentally determined in 5CB [14]: $A_o = 0.13$ J/K/cm^3, $B = 1.6$ J/cm^3, $C = 3.9$ J/cm^3, $L = 10^{-13}$ J/cm, $T^* = 317.14$K and $T_c - T^* = 1.12$K. We have also computed, as a function of temperature, the total line energy (with an outer cut-off length R = 1 μm, Fig. B.IV.14a) and its core radius r_c (taking $S(r_c) = 0.9\,S_\infty$, Fig. B.IV.14b). These results are close to those given by the "two-phase model" of the previous section: the comparison has been made considering that outside the core $K = 2LS_\infty^2$ and using the following equations: $\Delta H = (1/2)A_oT_cS_c^2$ (see eq. B.I.27) and $\gamma = (2\sqrt{2}/81)B^3L^{1/2}C^{-5/2}$, to be proven in the chapter on interfaces. These calculations confirm that the "two-phase model" yields the correct order of magnitude for the core radius and the line energy.

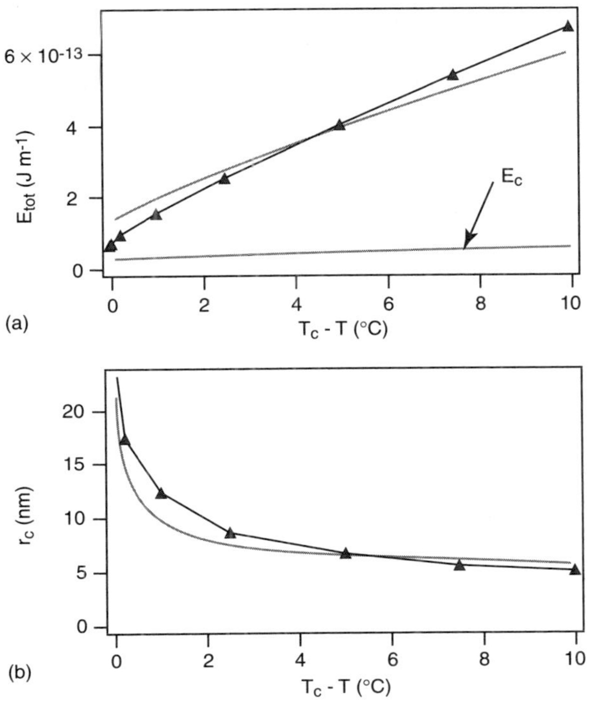

Fig. B.IV.14 Comparison between the "exact" model (solid triangles) and the "two-phase model" (dotted line) for an m = 1/2 wedge line. a) Total line energy as a function of temperature for an outer cut-off length R = 1 μm (the core energy in the two-phase model is also shown: its contribution is negligible with respect to the elastic line energy); b) core radius as a function of temperature.

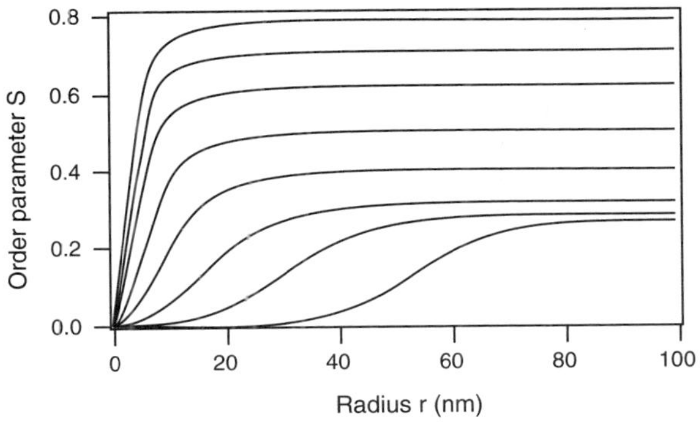

Fig. B.IV.15 Order parameter profile S(r) in the core of an m = 1 wedge line. From top to bottom, the curves correspond to temperature shifts $T_c - T$ of 0, 0.05, 0.25, 1, 2.5, 5, 7.5, and 10°C. As expected, the core is larger than for a 1/2 line.

The same calculations can be performed for a wedge line of rank m = 1. The results are qualitatively the same; even the order parameter profile, in spite of having a horizontal tangent in r = 0, globally preserves the same shape (Fig. B.IV.15).

We insist on the fact that all these calculations are only valid for planar lines. As we shall see in section IV.6, this hypothesis is only justified for lines of half-integer rank, while the others can relax the singularity in their structure by "escaping in the third dimension." Before discussing this feature, let us determine the elastic interaction between two parallel wedge lines. This calculation is important, as it imposes the outer cut-off length R when the sample contains more than one line.

IV.5 Interaction energy between two parallel wedge lines

We still assume that the lines are planar, the director being contained in the plane (x, y). The two lines, parallel to the z axis, are located by the position vectors a_1 and a_2 (Fig. B.IV.16). Angle φ between the director and the x axis must verify the equilibrium equation $\Delta\varphi = 0$. As the equation is linear, the solutions corresponding to each line can be superposed giving [15]:

$$\varphi = m_1\theta_1 + m_1\theta_2 + \text{Const} \qquad \text{(B.IV.25)}$$

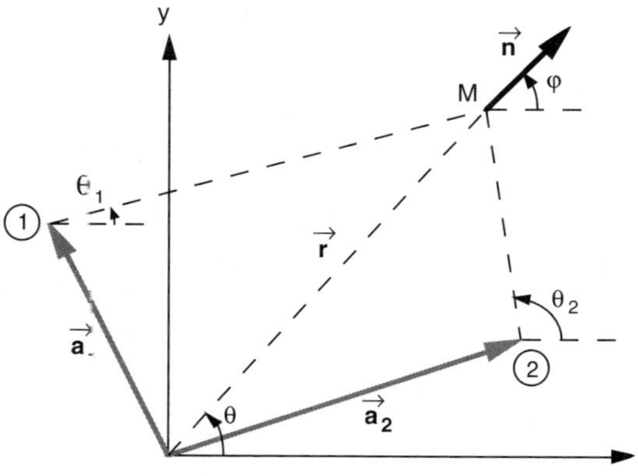

Fig. B.IV.16 Notations for two parallel disclination lines.

The bulk free energy can be easily obtained in isotropic elasticity employing equation B.IV.4:

$$f = \frac{1}{2}K\left[m_1\,\mathbf{grad}\,\theta_1 + m_2\,\mathbf{grad}\,\theta_2\right]^2 \qquad \text{(B.IV.26)}$$

but

$$\mathbf{grad}\,\theta_i = \frac{\mathbf{x}}{(\mathbf{r} - \mathbf{a}_i)^2} \tag{B.IV.27}$$

whence, after substitution

$$f = \frac{1}{2}K\left[\frac{m_1^2}{(\mathbf{r} - \mathbf{a}_1)^2} + \frac{m_2^2}{(\mathbf{r} - \mathbf{a}_2)^2} + 2m_1m_2\frac{(\mathbf{r} - \mathbf{a}_1)(\mathbf{r} - \mathbf{a}_2)}{(\mathbf{r} - \mathbf{a}_1)^2(\mathbf{r} - \mathbf{a}_2)^2}\right] \tag{B.IV.28}$$

Ericksen rewrites this equation in the form [16]:

$$f = \frac{1}{2}K\left[\left(\sum_{i=1}^{2} m_i\right)\left(\sum_{i=1}^{2}\frac{m_i}{(\mathbf{r}-\mathbf{a}_i)^2}\right) - m_1m_2\frac{r_{12}^2}{(\mathbf{r}-\mathbf{a}_1)^2(\mathbf{r}-\mathbf{a}_2)^2}\right] \tag{B.IV.29}$$

with $\mathbf{r}_{12} = \mathbf{a}_1 - \mathbf{a}_2$. This important formula can be generalized to a finite number of disclinations and shows that, if the sum $\sum m_i$ over the topological ranks of all disclinations is zero, the first term in equation B.IV.29 is equally null. As the remaining term behaves like $1/r^4$ for large r, total energy remains finite, even in an infinite system.

The total energy can be explicitly calculated for two disclinations of opposite sign, $m_1 = -m_2 = m$:

$$E_{12} = \int \frac{1}{2}Km^2\frac{r_{12}^2}{(\mathbf{r} - \mathbf{a}_1)^2(\mathbf{r} - \mathbf{a}_2)^2}2\pi r\,dr \tag{B.IV.30}$$

Introducing the core radius r_c as the lower cut-off length, integration yields:

$$E_{12} = 2\pi Km^2 \ln\left(\frac{r_{12}}{r_c}\right) \tag{B.IV.31}$$

Thus, the elastic energy of the system is twice the energy of an isolated disclination provided that the upper cut-off length R for the two lines is taken to be not the sample size, but rather their mutual distance r_{12}. E_{12} also represents the interaction energy between the two disclinations. Here, E_{12} diminishes with decreasing r_{12}; as expected, **two defects of opposite sign attract each other** with a force F_{12} inversely proportional to the distance:

$$F_{12} = 2\pi K\frac{m^2}{r_{12}} \tag{B.IV.32}$$

In the general case, the m_im_j term in B.IV.29 stands for the interaction energy between defects i and j. If the defects have the same sign, their interaction energy, proportional to $-m_im_j\ln(r_{ij})$, diminishes with increasing r_{ij}: **two defects of the same sign repel each other**.

We emphasize that the energy of a closed disclination loop in an infinite sample does not diverge, the distortions created by two diametrally opposed line elements exactly compensating each other at infinity. A loop of radius R therefore has typical energy:

$$E_{loop} = 2\pi KR \ln\left(\frac{R}{r_c}\right) + 2\pi RE_c \qquad \text{(B.IV.33)}$$

decreasing with the radius, which leads to finite-time collapse of the loop. Generally speaking, interactions between defects lead to their progressive disappearance during a sufficiently long annealing of the sample. This process, similar in nature to solid recrystallization, is pretty slow (minutes to hours) because of the finite line **mobility**. The following section shows how to calculate, and then experimentally measure, the mobility in the simplified case of a straight wedge line.

IV.6 Dynamics of a planar wedge line: calculating the friction force

Consider a planar wedge line of rank S moving with constant velocity V along the x axis. The line is straight and parallel to the z axis. Outside the core, line displacement engenders in place director rotation (possible "backflow" effects are neglected, in spite of the fact they can play a role, as shown recently by numerical simulations [17]). Inside the core, the situation is more complicated and will not be discussed here. Nevertheless, we shall suppose that dissipation in the core is negligible.

To find out the friction force exerted by the medium on the moving line, let us determine the energy dissipated in the system per unit time. Neglecting secondary flow (backflow) ($\mathbf{v} = 0$), one simply has:

$$\Phi = \gamma_1 \int \left(\mathbf{n} \times \frac{\partial \mathbf{n}}{\partial t}\right)^2 dxdy \qquad \text{(B.IV.34)}$$

with γ_1 the nematic rotational viscosity defined in the previous chapter. In the stationary regime:

$$\mathbf{n}(x, y, t) = \mathbf{n}(x - Vt, y) \qquad \text{(B.IV.35)}$$

whence, in the reference frame (X,Y) comoving with the line ($X = x-Vt$ and $Y = y$) using the fact that $\mathbf{n}$ is a unit vector:

$$\Phi = \gamma_1 V^2 \int \left(\frac{\partial \mathbf{n}}{\partial X}\right)^2 dXdY \qquad \text{(B.IV.36a)}$$

Note φ the angle between the director and the X axis. The integral becomes:

$$\Phi = \gamma_1 V^2 \int \left(\frac{\partial \varphi}{\partial X}\right)^2 dXdY \qquad \text{(B.IV.36b)}$$

where the integration is from the core radius to infinity. To perform it, one must first obtain the angle φ from the balance between viscous and elastic torques described by eq. B.III.51. In isotropic elasticity, the latter reads:

$$K\Delta\varphi = \gamma_1 V \frac{\partial \varphi}{\partial X} \qquad \text{(B.IV.37a)}$$

Expressing all lengths in terms of the core radius r_c, it can be cast in the dimensionless form:

$$\Delta\varphi = E_r \frac{\partial \varphi}{\partial X^*} \qquad \text{(B.IV.37b)}$$

with $X^* = X/r_c$, $Y^* = Y/r_c$ and $E_r = \gamma_1 V r_c/K$, the Ericksen number formed with the disclination core radius. Outside the core, the dissipation is necessarily of the form:

$$\Phi = \gamma_1 V^2 f(E_r) \qquad \text{(B.IV.38)}$$

where $f(E_r)$ depends on the Ericksen number. This function was calculated by Ryskin and Kremenetsky [18] for small E_r (practically, for line velocities less than 100μm/s). For a disclination of rank m, they find:

$$f(E_r) = \pi m^2 \ln\left(\frac{3.6}{E_r}\right) \qquad \text{(B.IV.39)}$$

To justify this law, let us return to the equation with dimension B.IV.37a, showing that two regions must be distinguished in the vicinity of the line (Fig. B.IV.17):

– the one inside a circle of radius given by the diffusion lengths of orientational changes ($r < K/\gamma_1 V$); here, the Laplacian dominates ($\Delta\varphi \approx 0$) and the solution is similar to the static one;

– the one outside this circle, where $\partial\varphi/\partial X \approx 0$ essentially (no dissipation) and φ can be taken as a constant, except in the "wake" ($\theta = \pi$) where φ changes by $2\pi m$.

In the first region (appearing shaded in the drawing), the dissipated energy is non-null and can be easily estimated using the static solution $\varphi = m\theta$ of the equation $\Delta\varphi = 0$. In the second region, dissipation is negligible since $\partial\varphi/\partial X \approx 0$. From B.IV.36b one gets:

$$\Phi \approx \gamma_1 V^2 m^2 \int \left(\frac{\partial \theta}{\partial X}\right)^2 dXdY \approx \gamma_1 V^2 m^2 \int_{r_c}^{K/\gamma_1 V} \int_0^{2\pi} \left(\frac{\sin\theta}{r}\right)^2 r dr d\theta \qquad \text{(B.IV.40a)}$$

yielding

$$\Phi \approx \pi\gamma_1 V^2 m^2 \ln\left(\frac{1}{E_r}\right) \qquad \text{(for an infinite sample)} \qquad \text{(B.IV.40b)}$$

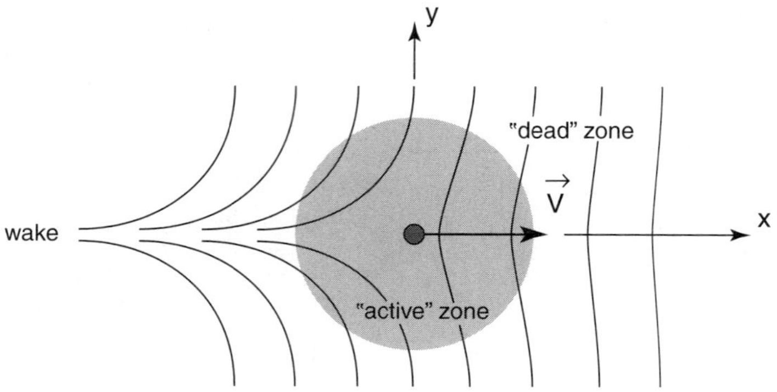

Fig. B.IV.17 Director field lines around a −1/2 disclination moving with velocity V in the X direction. The energy is essentially dissipated in the central area enclosing the line, of radius $K/\gamma_1 V$. Outside this region, dissipation is negligible as the director does not turn.

This approximate formula is similar to that of Ryskin and Kremenetsky (up to the numerical factor 3.6, not very relevant for a small Ericksen number). Note that this law is only accurate when the sample size R is larger than the diffusion length $K/\gamma_1 V$. Otherwise, the upper limit in the integral B.IV.40a should be set to R leading to:

$$\Phi \approx \pi\gamma_1 V^2 m^2 \ln\left(\frac{R}{r_c}\right) \qquad \text{if} \qquad R < \frac{K}{\gamma_1 V} \qquad \text{(B.IV.40c)}$$

We thus revive an old formula given by Imura and Okano as early as 1973 [19]. Let us also emphasize the strong analogy between this calculation and the friction force exerted by a stationary flow on a vorticity line source (practically, a cylindrical wire normal to the velocity). The Reynolds number then plays the role of the Ericksen number, vorticity the role of the angle φ, kinematic viscosity of the fluid is the equivalent of the diffusion coefficient K/γ_1 of director orientation changes, and the boundary layer appearing around the wire (where fluid vorticity is concentrated) corresponds to the central area enclosing the defect, where the dissipation takes place. The analogy is therefore complete.

In conclusion, let us determine the friction force acting on the line (again, this is a configurational force, and not a real one: to put it differently, everything happens as if the line were a real object acted upon by a friction force). In the stationary regime one has, from the dissipation theorem:

$$\Phi = -\mathbf{F}_v . \mathbf{V} \qquad \text{(B.IV.40d)}$$

where $\mathbf{F}_v$ is the desired friction force. Employing eqs. B.IV.40b, c one has:

$$\mathbf{F}_v = -\pi\gamma_1 m^2 \ln\left(\frac{3.6K}{\gamma_1 V r_c}\right)\mathbf{V} \qquad \text{in an infinite sample} \qquad \text{(B.IV.41a)}$$

or

$$\mathbf{F}_v = -\pi\gamma_1 m^2 \ln\left(\frac{R}{r_c}\right)\mathbf{V} \qquad \text{for} \qquad R < \frac{3.6K}{\gamma_1 V} \qquad \text{(B.IV.41b)}$$

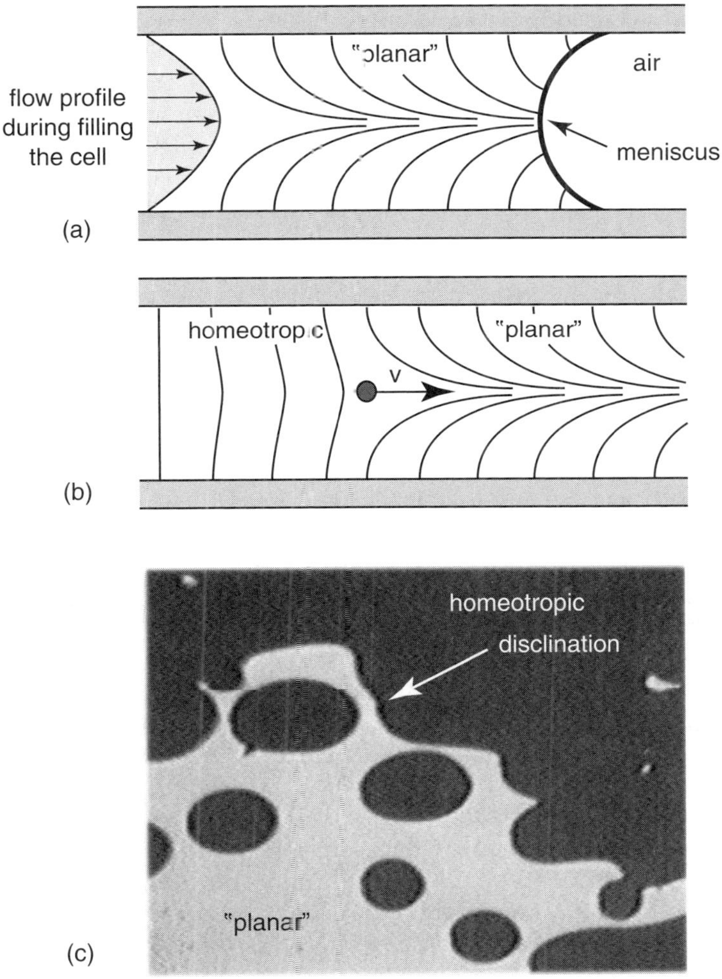

Fig. B.IV.18 a) Diagram of the director field lines while filling a homeotropic sample. Poiseuille flow, and molecular anchoring at the nematic-air meniscus, favor planar molecular orientation in the middle of the sample; b) fairly rapidly, disclination lines separating planar and homeotropic regions nucleate at sides of the sample or on dust particles; c) crossed polarizers photo of a sample after filling. Homeotropic domains nucleate and progressively invade the distorted planar area.

In order to measure the disclination mobility, two experimental problems have to be solved: first create a straight line, then control its movement.

A simple method for creating a –1/2 disclination line consists of filling by capillarity a homeotropic sample. Experiment shows that in the middle of the sample the molecules orient along the velocity, while at the surfaces they remain normal to the glass plates (Fig. B.IV.18a). The induced deformations relax towards the lower-energy homeotropic configuration by nucleation (on the sample boundaries or on dust particles) and propagation of one or more disclination lines (Fig. B.IV.18b and c).

When the filling is complete, this experiment allows for measuring the natural advancing velocity of a disclination line. Here, the driving force is given by the difference in elastic energy between the planar and homeotropic regions ($F_{elastic} \approx (1/2)(K/(d/2)^2)d = 2K/d$, with d the sample thickness). As to the friction force, it is given by formula B.IV.41. As the two forces balance each other one has, taking for the outer cut-off R half the sample thickness,

$$V_{line} \approx \frac{8K}{\pi\gamma_1 d}\left[\ln\left(\frac{d}{2r_c}\right)\right]^{-1} \approx \frac{K}{\gamma_1 d} \qquad \text{(B.IV.42)}$$

As an order of magnitude, $K = 5\ 10^{-7}$ dyn and $\gamma_1 = 0.1$ poise yielding for a 30 μm sample $V_{line} \approx 17$ μm/s, in good agreement with the experimental result. Furthermore, experiment confirms the inverse proportionality between line velocity and sample thickness.

We shall see in chapter B.VI that a –1/2 disclination line can get depinned from the nematic-isotropic liquid interface, forming metastable planar triangles in a homeotropic environment. It will be shown that the dynamics of these triangles, and in particular their survival, is conditioned by the mobility of the –1/2 line.

Nevertheless, velocities achieved in these configurations are very small. The lines can be accelerated by applying a magnetic or electric field. Such an experiment was performed by P. Cladis et al. [20]. Their setup is shown in figure B.IV.19a. The nematic (6CB) is placed between two glass plates treated in homeotropic anchoring. Two parallel metallic wires, equally treated in homeotropic anchoring, serve as both spacers and electrodes. They have a diameter d = 140 μm, are placed at a distance 1 mm apart, and allow application of a horizontal electric field. As the glass plates are covered with an I.T.O. layer, a vertical electric field can equally be applied. The boundary conditions on the plates and on the wire surface initially lead to the formation of two –1/2 wedge disclinations adhering to the wires. They can be displaced towards the center of the sample by applying a voltage between the wires (typically, 90 V at a 1 kHz frequency). Two "planar" regions are formed along each wire, separated from the homeotropic central region by the two –1/2 lines (Fig. IV.19b). In this context, "planar" implies that the molecules are parallel to the plates only at mid-distance between them. Clearly, in the presence of a horizontal field the two planar zones are energetically favored. Once the lines are displaced to the sample center, the Cladis experiment consists of replacing the initial horizontal field by a vertical field of intensity E = U/d (U being the voltage applied between the glass plates). Once switching is done, the two −1/2 lines return to the wires with a velocity V proportional to the applied voltage (Fig. B.IV.20).

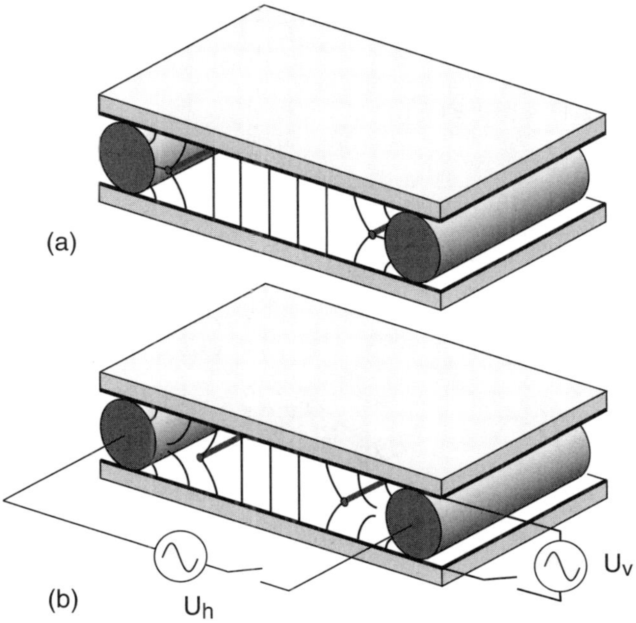

Fig. B.IV.19 a) Without an applied field, homeotropic anchoring on the glass plates and on the wires generate two −1/2 disclination lines along the wires; b) when a 90 V electric voltage is applied between the wires, the position of the disclination lines is shifted towards the center of the sample. Upon removing the applied voltage, the two lines return to their initial position, with a velocity controlled by the voltage U between the glass plates.

In order to explain this experimental result, let us first determine the line velocity as a function of E. The configurational force acting on each line depends on the elastic and electric excess energy E_{pl} stored in the (energetically metastable) planar domains. Let θ be the angle between the director and the x axis and L the width of a planar stripe. In isotropic elasticity and assuming a uniform electric field throughout the sample, one has:

$$E_{pl} = L\int_{-d/2}^{+d/2}\left[\frac{1}{2}K\left(\frac{\partial\theta}{\partial y}\right)^2 - \frac{1}{2}\varepsilon_o\varepsilon_a E^2(\sin^2\theta - 1)\right]dy \qquad \text{(B.IV.43)}$$

Minimization with respect to θ requires:

$$K\frac{\partial^2\theta}{\partial y^2} = -\frac{d}{d\theta}\left(\frac{\varepsilon_a\varepsilon_o}{2}E^2\sin^2\theta\right) \qquad \text{(B.IV.44)}$$

with boundary conditions:

$$\theta = \pm\frac{\pi}{2} \quad \text{in} \quad y = \pm\frac{d}{2} \qquad \text{(B.IV.45)}$$

The equation has a first integral:

$$\frac{1}{2}K\left(\frac{\partial\theta}{\partial y}\right)^2 + \frac{1}{2}\varepsilon_o\varepsilon_a E^2\sin^2\theta = \text{Cst} \qquad \text{(B.IV.46)}$$

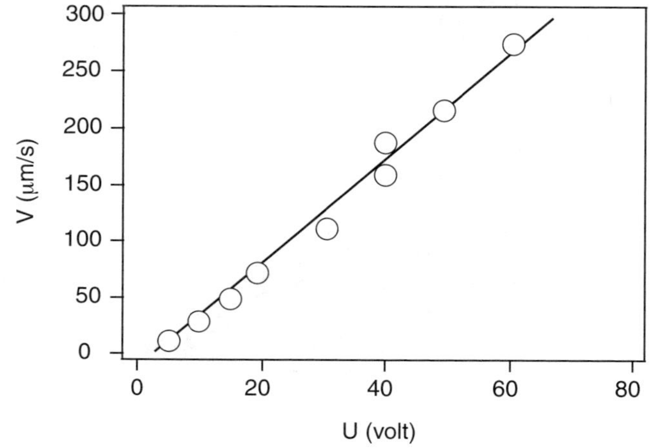

Fig. B.IV.20 Dislocation line velocity vs. the voltage applied between the electrodes. 140 μm thick 6CB sample at 36°C (from Cladis et al. [20]).

For a strong enough field, the magnetic coherence length $\xi = [K/(\varepsilon_a\varepsilon_o E^2)]^{1/2}$ is very small with respect to the sample thickness d and the integration constant is roughly: $\text{Cst} = (1/2)\varepsilon_o\varepsilon_a E^2$. E_{pl} is then obtained from equations B.IV.43 and B.IV.46:

$$E_{pl} = L\int_{-d/2}^{+d/2}\varepsilon_o\varepsilon_a E^2\cos^2\theta\, dy = 2L\sqrt{\varepsilon_o\varepsilon_a K}\, E \qquad \text{(B.IV.47)}$$

In the case $\xi << d$, the absolute value of the elastic force F_e acting on the disclination is therefore:

$$F_e = \frac{dE_{pl}}{dL} = 2\sqrt{\varepsilon_o\varepsilon_a K}\, E \qquad \text{(B.IV.48)}$$

This force is balanced by the friction force on the line, given by eq. B.IV.41a, so one finally has:

$$V = \frac{8\sqrt{\varepsilon_0 \varepsilon_a K}}{\pi \gamma_1 d \ln\left(\frac{3.6}{E_r}\right)} U \qquad \text{(B.IV.49)}$$

Note that the obtained V(U) dependence is not strictly linear, as the speed also appears in the definition of E_r. It is however easy to see that $\ln(3.6/E_r)$ does not vary much in the explored velocity range (between 6 and 9 if we assume a core radius $r_c \approx 30$ Å not depending on the velocity). On the other hand, this formula does not take into account dissipation in the core and the effects of the electric field on the line mobility, so it should not be trusted to provide more than an order of magnitude. Keeping this in mind, comparison with experiment gives satisfactory results: data in figure B.IV.20 yield a slope of 4.7 μm/V/s at 36°C for 6CB while theory gives about 2 μm/V/s with $K \approx 5\times10^{-7}$ dyn, $\varepsilon_a \approx 10$, $\gamma_1 \approx 0.5$ poise, $d = 140$ μm, and $\ln(3.6/E_r) \approx 8$. Taking into account the simplifications made and our approximate knowledge of material constants, the agreement is reasonable.

Another experiment deals with the movement of twist disclinations [21]. These lines are obtained in planar samples twisted at 90°, such as those employed in displays. Indeed, as the medium is not chiral, both left- and right-handed twist configurations of the director can be adopted to satisfy the boundary conditions on the glass plates. Two types of domains with opposite chiralities appear, separated by twist lines (Fig. B.IV.21).

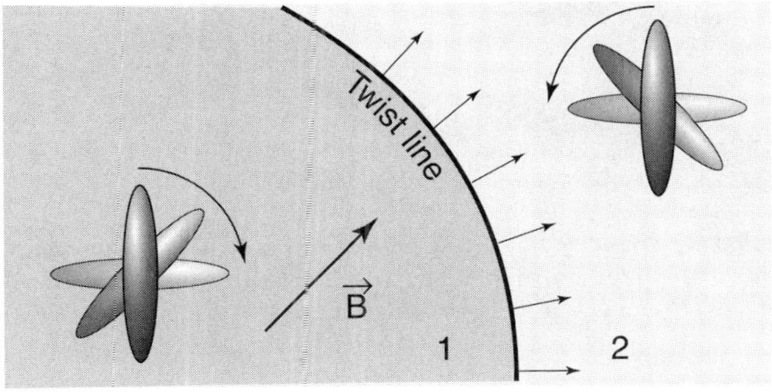

Fig. B.IV.21 Two domains with opposite chirality separated by a twist line. Applying the magnetic field **B** favors domain 1 over domain 2 and the line moves.

In the absence of applied fields, these domains have exactly the same twist energy, but one can favor one domain over the other by applying a magnetic field parallel to the plates at an angle of 45° with respect to the average molecular orientation on the plates. This energy imbalance between the domains gets the twist lines moving and the same experiment can be performed as in the case of wedge lines. The reader is referred to the original papers for a detailed presentation.

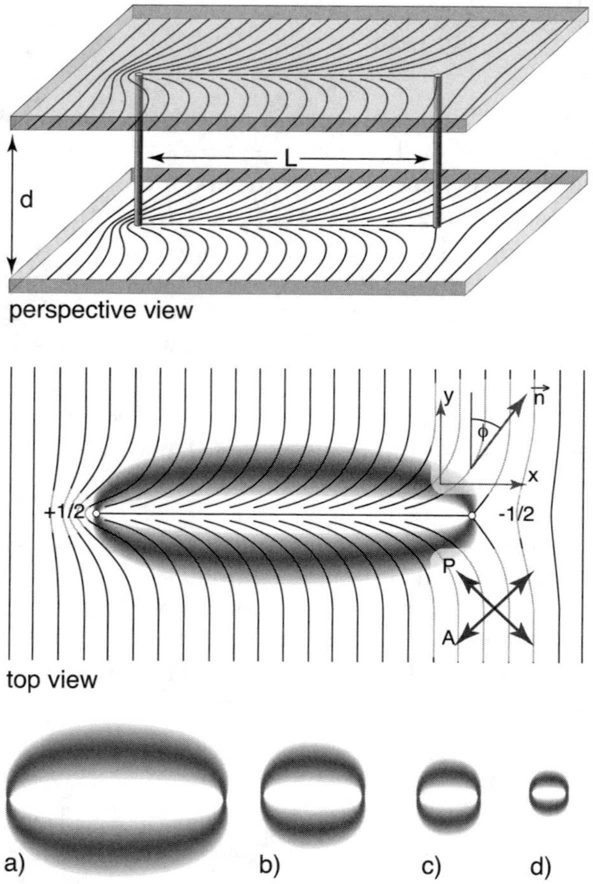

Fig. B.IV.22 Annihilation of two +1/2, –1/2 disclination lines in a planar sample. A π-wall forms between the two lines when they are far apart (L>d). Pictures a-d show the two lines observed between crossed polarizers under the microscope at times t = 1.5 s, 1.75 s, 1.78 s, and 1.82 s. (from ref. [22]).

Finally, we mention a recent experiment [22] on the dynamics of annihilation of two wedge disclinations lines of ranks +1/2 and –1/2 in a planar sample, which are oriented perpendicularly to the glass plates. Let L be the distance between the lines and d the sample thickness. The main result is that L decreases linearly in time as long as L > d, whereas at short distances (L < d), L varies as $(t_c - t)^{1/2}$, where t_c is the collapse time. The first regime results from the existence of a π wall connecting the lines (this wall forms because of the planar anchoring on the glass plates, Fig. B.IV.22), whereas at short distances (L < d), anchoring no longer plays any role, the lines having the same behavior as in an infinite medium. In this limit, a square-root law is expected, as the result of the balance between the elastic force (in 1/L) and the viscous force, proportional to dL/dt (within a ln(L) correction).

IV.7 Wedge line stability: escape in the third dimension

Two different kinds of lines are observed under the polarizing microscope: rather diffuse thick threads with a large core and sharper contrast thin threads. We shall show that thick threads correspond to integer rank wedge lines ($m = \pm 1$) where the core singularity has disappeared, while thin threads are half-integer rank singular lines ($m = \pm 1/2$).

Thus far, we have assumed the director field to be planar (**n** everywhere perpendicular to the line) and the medium to be infinite. Are these assumptions reasonable? To find out, we shall analyze the stability of a planar wedge line with respect to a three-dimensional perturbation, first in an infinite medium and then in a limited one, closer to the experimental situation. A distinction shall be made between topological and energetic stability, as these two concepts are not equivalent.

IV.7.a) Topological stability

Consider a disclination of rank m. **In the planar model**, it is always **singular**, i.e., it cannot be created or suppressed without using a scalpel. **Does the disclination remain singular when the director regains its three-dimensional orientational liberty?**

To answer this question, consider a Burgers circuit enclosing the line. Let **n**(s) be the director at point s on this circuit and assign to it a point P(s) on the unit S2 sphere such that **OP**(s) = **n**(s) (Fig. B.IV.23). The same reference frame is used in real space and on the S2 sphere. As s moves on the Burgers circuit, P(s) traces a contour on the S2 sphere.

For $m = \pm 1/2$ disclinations, the trajectory of P(s) is a semi-circle. In figure B.IV.23, this trajectory starts at the North pole and ends at the South pole. The starting and ending points need not be identical; as **n** and **–n** are equivalent, they can also be diametrically opposed. Consider now a **topologically continuous** deformation of the director field: the trajectory P(s) itself is deformed, but the two extremities remain diametrically opposed; thus, it can never be reduced to one point. On the other hand, it is obvious that a wedge disclination of rank $m = 1/2$ can be transformed in a wedge disclination of rank $m = -1/2$ or in a twist disclination by appropriately deforming the trajectory of P(s).

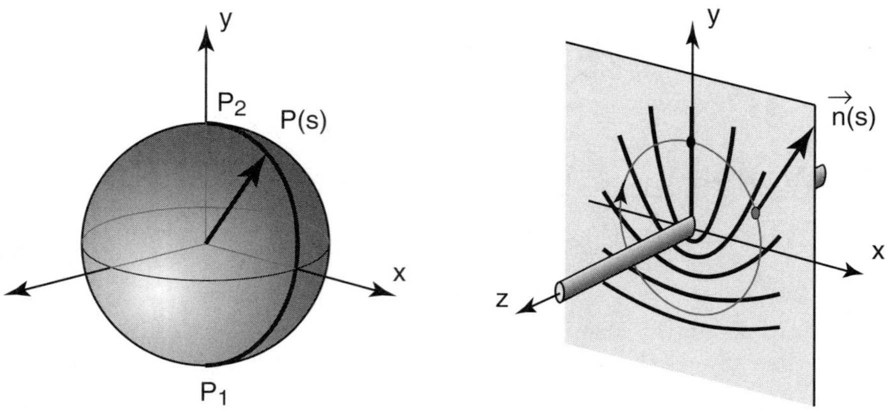

Fig. B.IV.23 Representing a wedge disclination by a trajectory on the S2 sphere.

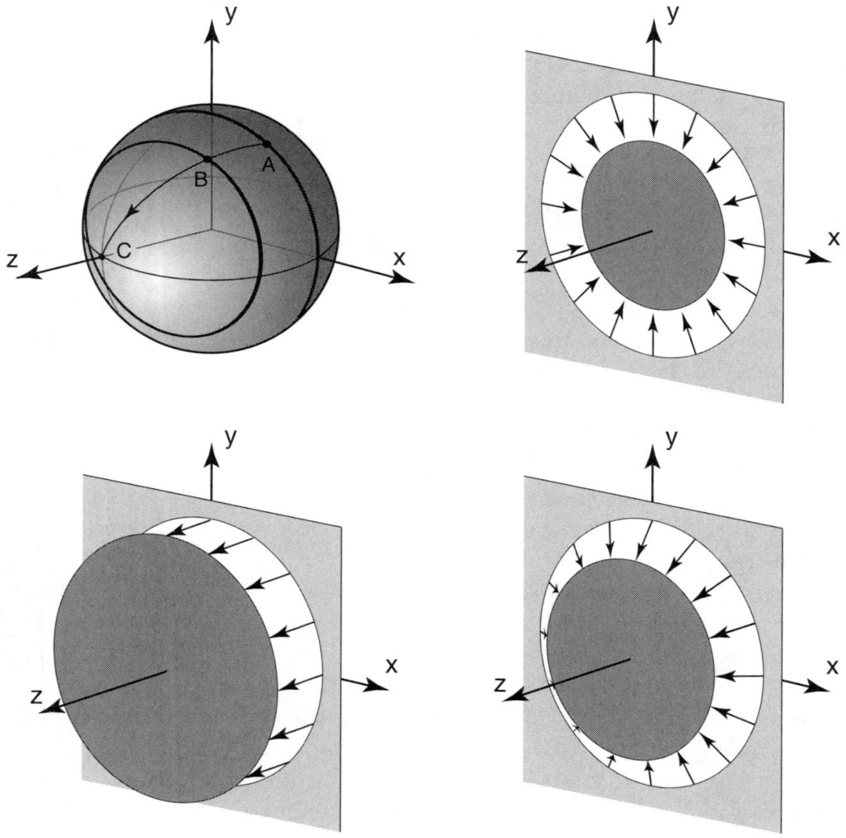

Fig. B.IV.24 Topological instability of an m = 1 disclination.

The case of $m = \pm 1$ planar disclinations is totally different, because the S2 image P(s) of a Burgers circuit enclosing these lines is a great circle (A), corresponding to $\pm 2\pi$ director rotation. As illustrated in figure B.IV.24, this great circle can be continuously reduced to the circle (B), then to the point (C). **Disclinations of rank $m = \pm 1$ are therefore topologically unstable in three dimensions**, being continuously removable. The real space transformation is described in the three diagrams of figure B.IV.24, showing the way the director leaves the plane along the Burgers circuit and "escapes" to the third dimension, parallel to the line axis. Clearly, all trace of the initial distortion is lost when **n** is everywhere parallel to the **z** axis. Reasoning along the same lines, one can generally show that:

1. All integer rank disclinations are topologically unstable.
2. Disclinations of arbitrary half-integer rank can be continuously transformed into a 1/2 disclination, which is topologically stable.

These two results can be obtained in the frame of the much more general theory of topological defects in systems with an order parameter arising as a result of broken symmetry. This theory, developed by Kléman and Toulouse, employs the mathematical concept of the homotopy group of the internal states manifold of the system [11a, b].

IV.7.b) Energetic stability (infinite medium)

Let us now discuss the energetic stability of a planar line with respect to a three-dimensional distortion [23]. Unless explicitly stated otherwise, the three main Frank constants will be taken as equal; let $\alpha(\theta)$ be the angle between the director and the (x,y) plane normal to the line ($-\pi/2 \leq \alpha \leq +\pi/2$) (Fig. B.IV.25). The director components become

$$n_x = \cos\varphi \cos\alpha , \qquad n_y = \sin\varphi \cos\alpha , \qquad n_z = \sin\alpha \tag{B.IV.50}$$

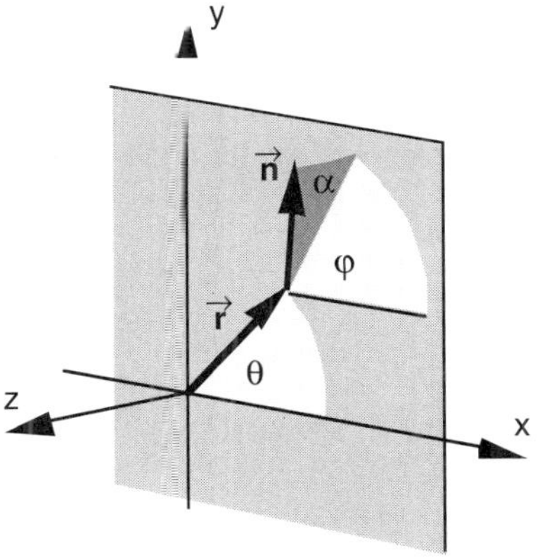

Fig. B.IV.25 Definition of the angles α, φ, and θ.

From eq. B.IV.4 the line energy for this director field is:

$$E = \frac{K}{2} \ln\left(\frac{R}{r_c}\right) \int_0^{2\pi} (\alpha'^2 + \varphi'^2 \cos^2\alpha)\, d\theta \qquad \text{(B.IV.51)}$$

where ' designates the θ derivative. Minimization with respect to φ again yields $\varphi'' = 0$ and auto-similar solutions of the type $\varphi = m\theta$ correspond to wedge disclinations of rank m. For these solutions, angle α must satisfy:

$$\alpha(\theta + 2\pi) = (-1)^{2m} \alpha(\theta) \qquad \text{(B.IV.52)}$$

otherwise a singularity would appear in the $\theta = 0$ semi-plane. In the limit of weak deviations (small α), the energy becomes:

$$E = K(\pi m^2 + \delta) \ln\left(\frac{R}{r_c}\right) \qquad \text{(B.IV.53)}$$

where

$$\delta = \frac{1}{2} \int_0^{2\pi} (\alpha'^2 - m^2\alpha^2)\, d\theta \qquad \text{(B.IV.54)}$$

The sign of δ determines line stability. If there are functions α(θ) rendering δ negative, the line is unstable with respect to three-dimensional perturbations; otherwise, the line is stable.

For wedge lines of integer rank m the answer is immediate. Indeed, $\delta < 0$ for $\alpha = \text{cst} \neq 0$ (solution satisfying B.IV.52). **Lines of rank $m = \pm 1, \pm 2, \ldots$ are therefore unstable with respect to director tilt in the direction of the line axis.** The half-integer case is less obvious, since $\alpha = \text{Cst}$ is no longer allowed by the condition B.IV.52, that can be written explicitly as

$$\alpha(\theta + 2\pi) = -\alpha(\theta) \qquad \text{(half-integer m)} \qquad \text{(B.IV.55)}$$

This relation implies that the function α is 4π-periodic, as $\alpha(\theta+4\pi) = -\alpha(\theta+2\pi) = \alpha(\theta)$ and can therefore be developed in a discrete Fourier series of the generic form:

$$\alpha(\theta) = \sum_{p=0}^{\infty} \left[a_p \cos\left(\frac{p\theta}{2}\right) + b_p \sin\left(\frac{p\theta}{2}\right)\right] \qquad \text{(integer p)} \qquad \text{(B.IV.56)}$$

Using eq. B.IV.55 it is easily shown that

$$\sum_{n=0}^{\infty} \left[a_{2n} \cos\left(\frac{2n\theta}{2}\right) + b_{2n} \sin\left(\frac{2n\theta}{2}\right)\right] = 0 \qquad \text{(integer n)} \qquad \text{(B.IV.57)}$$

whence

$$\alpha(\theta) = \sum_{n=0}^{\infty}\left[a_{2n+1}\cos\left(\frac{(2n+1)\theta}{2}\right) + b_{2n+1}\sin\left(\frac{(2n+1)\theta}{2}\right)\right] \qquad \text{(B.IV.58)}$$

From this formula and eq. B.IV.55 one has:

$$\delta = \frac{\pi}{2}\sum_{n=0}^{\infty}\left\{(a_{2n+1}^2 + b_{2n+1}^2)\left[\frac{(2n+1)^2}{4} - m^2\right]\right\} \qquad \text{(B.IV.59)}$$

Irrespective of the function α, the terms of this series are always positive for $m = \pm 1/2$, proving that **$m = \pm(1/2)$ lines are energetically stable** with respect to escape in the third dimension. For the other half-integer lines ($m = \pm 3/2,...$) one can always find a function α yielding negative δ. These lines (or at least their planar configurations) are therefore energetically unstable.

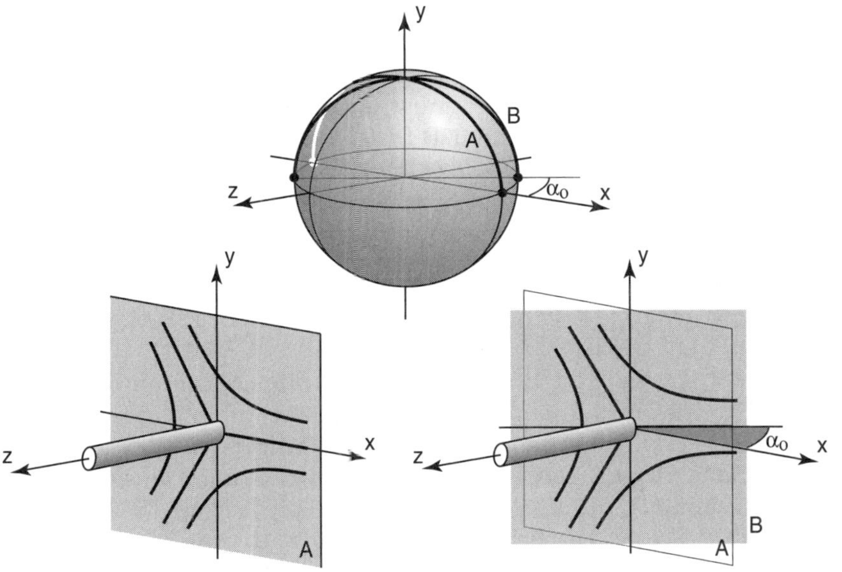

Fig. B.IV.26 Transformation of an $m = -1/2$ line in real space and on the S2 sphere when the director acquires an out-of-plane tilt given by $\alpha = \alpha_0 \sin(\theta/2)$.

However, lines of rank $\pm 1/2$ are marginally stable with respect to deformations of the form $\alpha = \alpha_0 \cos(\theta/2)$, giving $\delta = 0$. It is therefore plausible that the stability of these lines could be different in anisotropic elasticity. As a matter of fact, Anisimov and Dzyaloshinskii [24] proved that planar disclinations of rank $m = \pm(1/2)$ are stable only if

$$K_2 \geq \frac{1}{2}(K_1 + K_3) \qquad \text{(B.IV.60)}$$

Otherwise, these lines are unstable with respect to a three-dimensional deformation such as the one depicted in figure B.IV.26, representing a $-1/2$ wedge line parallel to the z axis after the director acquired a tilt $\alpha = \alpha_0 \cos(\theta/2)$. Before deformation, the S2 image of the Burgers circuit is the half-great circle (A); after tilt, the image of the circle is the half-great circle (B) obtained by turning (A) through a (small) angle α_0 about the y axis. In real space, this means that the planar director configuration is maintained, but in a plane (B) making an angle α_0 with plane (A) perpendicular to the disclination axis (still parallel to z). This "escaping" phenomenon explains the zigzag instability of 1/2 lines in flat capillaries.

To date, there is no general theory of the energetic stability of disclination lines, covering all ranks and taking into account anisotropic elasticity. Nevertheless, Anisimov and Dzyaloshinskii have once again shown that $m = \pm 1$ planar disclinations become energetically stable (metastable, to be exact) for

$$K_2 > 2K_3 \qquad \text{and} \qquad K_3 < K_1 \tag{B.IV.61}$$

but these last conditions are never satisfied experimentally, so that lines of rank ± 1 are generally unstable and must disappear in an **infinite** medium by third-dimension "escape." Condition B.IV.60 is less restrictive and both cases are experimentally accessible. One can therefore create either planar or three-dimensional $\pm 1/2$ lines.

IV.7.c) Disclinations in a confined medium (capillary tube)

As we have just shown, a ± 1 line necessarily disappears in an infinite medium by "escaping" to the third dimension. This process can be "blocked" by confining the nematic to a round capillary treated in homeotropic anchoring. The escape only takes place in the center, allowing suppression the core singularity (Fig. B.IV.27), but not the line itself, which is imposed by the topology of the medium. The "escape" is easily recognizable in homeotropically treated cylindrical capillaries (Fig. B.IV.5). For this particular configuration, the radius of curvature of the director field lines must be proportional to the capillary radius r, which is the only length scale of the problem. The bend energy is therefore of the order $\pi r^2 K_3 / r^2 = \pi K_3$, independent of the radius.

This result is confirmed by an exact calculation by Cladis and Kléman [7a] yielding the energy per capillary unit length:

$$E = \pi\left(K_1 + K_3 \frac{k}{\tan k}\right) \quad \text{with} \quad \tan^2 k = \frac{K_3 - K_1}{K_1} \quad \text{for} \quad K_3 > K_1 \tag{B.IV.62}$$

and

$$E = \pi\left(K_1 + K_3 \frac{k}{\text{th} k}\right) \quad \text{with} \quad \text{th}^2 k = \frac{K_1 - K_3}{K_1} \quad \text{for} \quad K_1 > K_3 \tag{B.IV.63}$$

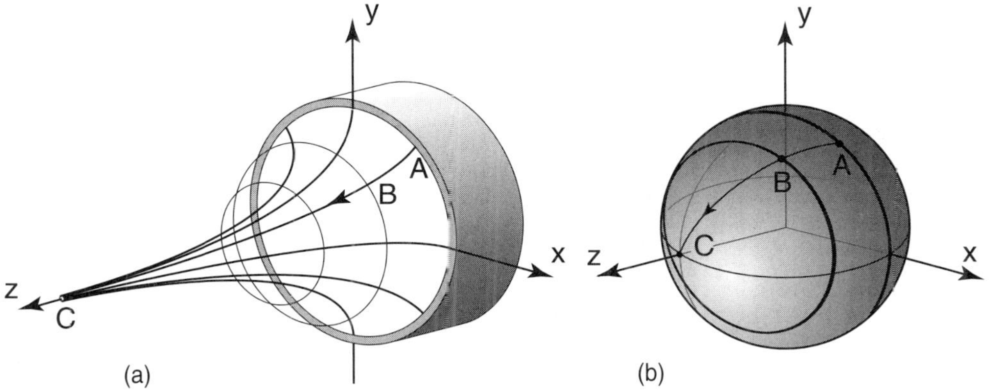

Fig. B.IV.27 m = 1 radial line with a regular core in a round capillary (a) and its representation on the S2 sphere (b). The core singularity can be suppressed by distorting the field lines.

The escape can take place either upwards or downwards, such that point defects can appear on the capillary axis, as shown in figure B.IV.5. These defects, termed **umbilics**, must necessarily have a singular core and their structure can be either radial or hyperbolic (Fig. B.IV.28).

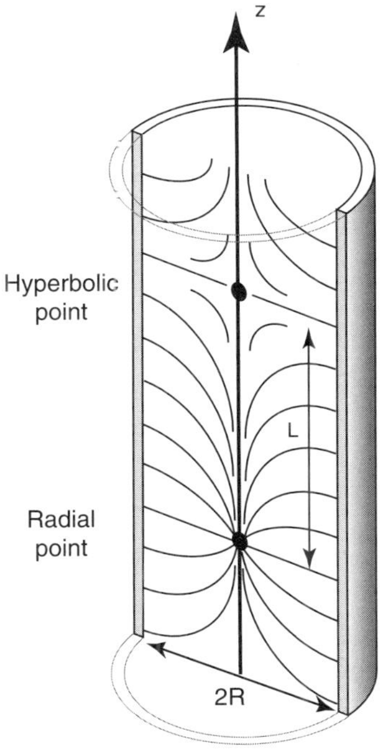

Fig. B.IV.28 Field line structure around the singular points appearing on the axis of a homeotropically treated capillary.

Their energy is of the order of Kr, with r the capillary radius. Experimentally, numerous umbilics are nucleated after a rapid quench from the isotropic phase. The configuration without defects being energetically favorable, an interesting question is how the structure relaxes. Intuitively, one can imagine that the defects will annihilate by pairs of opposite signs until complete disappearance. However, the real mechanism is more subtle, as shown by the NMR experiments in porous media of Crawford et al. [25a]. These authors use polycarbonate membranes with calibrated cylindrical pores of radius $0.05\ \mu m < R < 0.5\ \mu m$ filled with the nematic phase. The many umbilics nucleated at the isotropic-nematic transition only disappear when their mean distance is comparable to the pore radius. This anomaly was explained by Vilfan et al. [25b] who determined numerically (in the isotropic elasticity approximation) the energy difference ΔF between a perfect structure and a configuration with two defects, one radial and the other hyperbolic, a distance L apart. For a more realistic treatment, the authors have assumed the anchoring energy w_o on the inner surface of the capillary to be finite, satisfying the equation:

$$f_s = f_{so} - \frac{1}{2} w_o (\mathbf{n} \cdot \boldsymbol{\nu})^2 \qquad \text{(B.IV.64)}$$

with $\boldsymbol{\nu}$ the surface normal and f_s the surface energy. This assumption is not essential: the result remains qualitatively the same for infinite anchoring energy. The simulated interaction energy is shown in figure B.IV.29 for $w_oR/K = 10$. One can see that ΔF, which is also the interaction energy between the two umbilics, increases for $L < L_o \approx 0{,}25R$ and decreases when $L > L_o$, such that the two defects attract and finally annihilate if they are close enough. Otherwise, they counterintuitively repel each other. This calculation explains the metastability of "umbilical lattices" appearing in the capillaries for $L > L_o$.

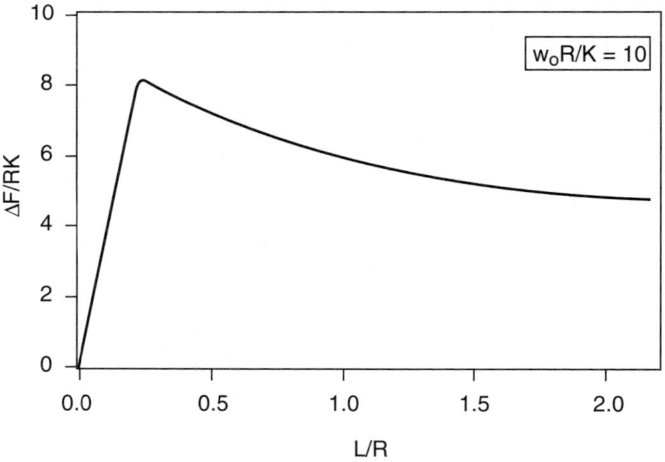

Fig. B.IV.29 Interaction energy between a radial and a hyperbolic defect vs. the separating distance (from ref. [25b]).

Circular m = +1 lines in a capillary with planar anchoring conditions can undergo a similar escape process (Fig. B.IV.30), by which the director acquires a radial twist. This deformation costs an extra twist energy but suppresses the divergence of the K_3 bend term. Calculating the exact energy of a +1 circular line leads to results resembling those obtained for the radial line (simply replace K_1 by K_2, tan by sin and th by sh). The choice between the two configurations (radial or circular) therefore depends on the elastic constants, their anisotropy, and the boundary conditions on the sample surface.

It is easily seen that escape in the third dimension allows for suppressing the singular core of an m = –1 line and, in general, of all integer-rank lines (positive or negative). Integer rank lines (for which Ω is a multiple of 2π) can thus be identified with the **thick threads** observed in the samples.

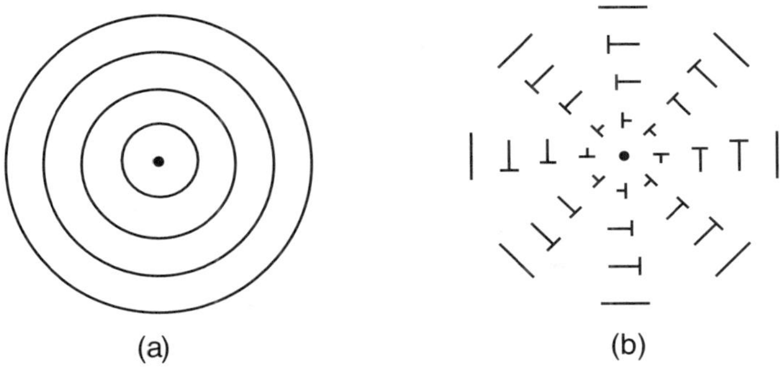

Fig. B.IV.30 a) m = 1 circular line with a singular core; b) the singularity can be removed by twisting the director field.

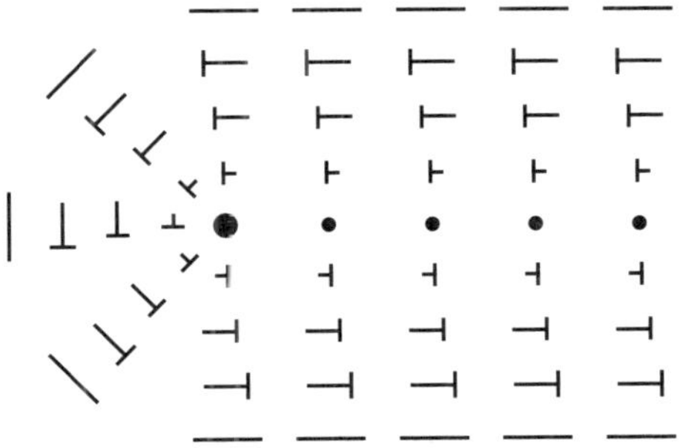

Fig. B.IV.31 Trying to relax the singularity for an m = 1/2 leads to the formation of a twist wall attached to the line.

One can obviously ask the same question in the case of half-integer lines. As illustrated by the S2 trajectory, they are topologically stable, meaning that their core singularity cannot be removed as in the case of integer rank lines (Fig. B.IV.31). Half-integer lines are therefore very contrasted in microscopy as their singular core strongly diffuses light: they are the **thin threads** often observed in thick samples. Note that thin threads are sometimes unstable, tending to escape partially in the third dimension in anisotropic elasticity. An already encountered example is that of the zigzag formed by a 1/2 lined on the side of a flat capillary (Fig. B.IV.6b). In this case, the molecular configuration is imposed by the homeotropic anchoring on the inner capillary wall, while the line itself must locally tilt with respect to the capillary axis in order to minimize its energy. The line cannot go too far from its average position, so it develops a zigzag pattern.

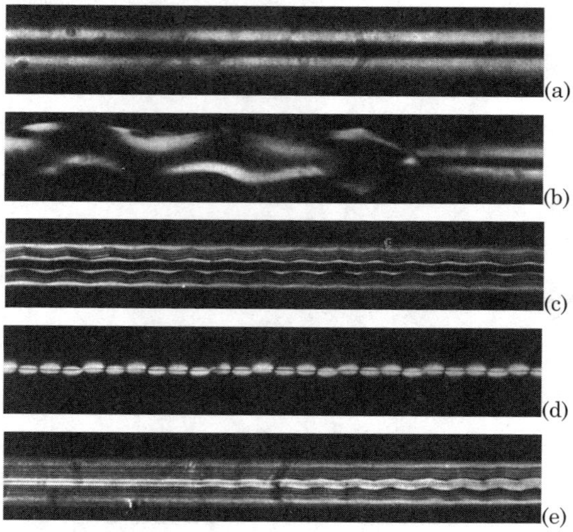

Fig. B.IV.32 Observation between crossed polarizers of the 8CB nematic phase in a homeotropically treated round capillary (radius R) a) +1 line with continuous core: R = 75 μm, $\Delta T \geq 0.5°C$; b) dissociation into two 1/2 lines: R = 75 μm and $\Delta T = 0.5°C$; c and d) same capillary (R = 25 μm) at the same temperature ($\Delta T = 0.1°C$). Turning the capillary by 90° around its axis, the two lines are observed to zigzag in the same radial plane; e) the zigzags progressively disappear on approaching the transition to the smectic phase (at the left extremity of the image). Here, the capillary (R = 25 μm) is placed in a lengthwise gradient, the temperature increasing by about 0.1°C from left to right (from ref. [9]).

Ultimately, the competition between the energy gained by tilting the line and the energy lost at the meeting point between a zig and a successive zag will determine the final wavelength of the structure.

A disclination line can also spontaneously dissociate into two lower-rank lines. This can happen in a capillary, as shown in figure B.IV.32. In this experiment, the nematic (8CB) is introduced in a round capillary internally treated in homeotropic anchoring. The adopted configuration strongly depends on the temperature, namely on the distance ΔT from the smectic A-nematic transition temperature: $\Delta T = T - T_{NA}$. Far from the transition ($\Delta T > 0.5°C$), a regular core +1 disclination appears, similar to the one described by Cladis and Kléman (Fig. B.IV.32a). When approaching the smectic phase, typically for

$\Delta T \approx 0.5°C$, the line becomes unstable and dissociates into two 1/2 lines (Fig. B.IV.32b, c, d). These lines zigzag, like the ones observed in flat capillaries (see Fig. B.IV.6) as a result of the partial director escape in the direction of the capillary axis (itself induced by the elastic constant anisotropy). This 3D distortion and the zigzag disappear very close to the smectic A transition as the K_2 and K_3 elastic constants diverge (see chapter C.I. of the second volume on smectics): the two 1/2 lines are then planar and straight (Fig. B.IV.32e).

This transition sequence is clearly induced by the temperature evolution of the elastic constants. It could only be explained by comparing the energies of different line types. These calculations are difficult in confined geometry and have not yet been rigorously performed (see [9] for more details). Therefore, we shall only emphasize that the diverging constants K_2 and K_3 (K_1 remaining "normal") favor planar radial configurations and the dissociation of a +1 line into two +1/2 lines, because of the m^2 energy variation; indeed, assuming low core energy (or, equivalently, large enough capillary radius):

$$2E_{planar}(m = 1/2) < E_{planar}(m = 1) \qquad \text{(B.IV.65)}$$

This inequality explains the appearance of two planar 1/2 lines when sufficiently close to the smectic phase.

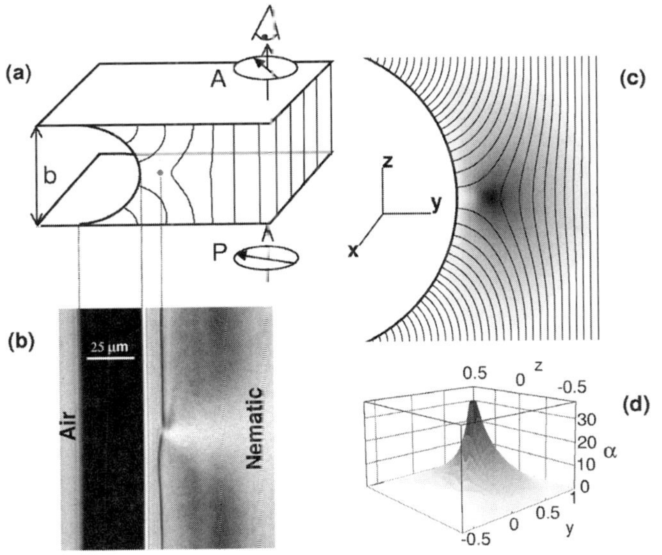

Fig. B.IV.33 a) Formation of a −1/2 disclination line near the edge of a homeotropic nematic sample; b) top view of the sample when observed under the microscope with the polarizer (P) and the analyzer (A) partially crossed. The black region corresponds to the meniscus with the air. A point defect is visible on the disclination line, separating two zones tilted in opposed directions; c) calculated projection of the director field on the (y, z) plane. The director is tangent to the solid lines. The intensity of the gray shading is an encoding for the magnitude of the escaped component of the director field along the line (x-component). The corresponding angle between the director and the (y, z) plane is show in (d) (from ref. [23b]).

Finally, we mention that similar observations have been made recently for a −1/2 line located near a free surface. The geometry is depicted in figure IV.33: the sample is homeotropic and forms a circular meniscus in contact with air at the edges. When the director is perpendicular to the free surface, a −1/2 line forms spontaneously in the vicinity of the tip of the meniscus [26a]. In this case, the distance L between the line and the meniscus depends on the penetration length of the anchoring at the free interface, which is defined as the ratio between the Frank elastic constant K and the anchoring energy W_a: $L_p=K/W_a$ (this notion will be defined more precisely in chapter B.V.6). If the sample thickness d is much larger than L_p ("strong anchoring" or "soft elasticity" limit), the line locates at a distance L that depends only on d: $L/d \approx 0.1$. By contrast, this distance decreases when L_p becomes comparable or larger than d, which can lead to the total disappearance of the line: it then becomes "virtual," which means that the core of the line lies in the air. In this case, there are no longer any singularities in the director field. This phenomenon occurs when approaching a smectic phase, because the elastic constant K diverges (more precisely, K_2 and K_3 diverge at the second order smectic phase transition, see chapter C.I of the second volume). The distance between the line and the tip of the meniscus also depends on the anchoring angle θ of the director at the free surface, defined as the angle between the director and the normal to the interface. When this angle increases, the distance L decreases until the line disappears completely. This effect has been observed in MBBA and in I52 (sold by Merck) and has revealed a temperature-induced-anchoring-angle transition in these materials.

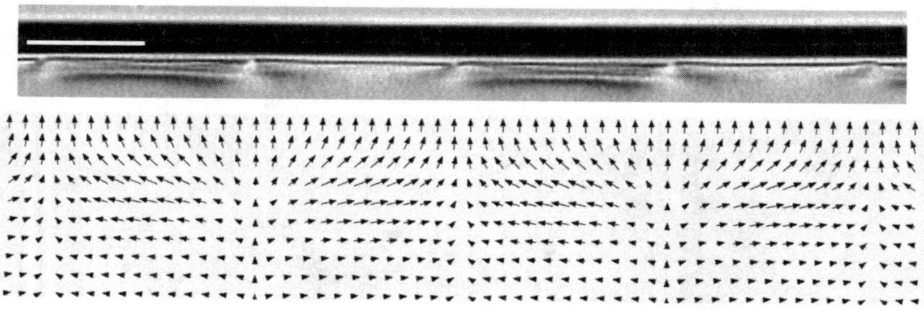

Fig. B.IV.34 Regular distribution of point defects along a −1/2 disclination line (the sample is 50 μm thick). Bottom: projection of the director field in the mid-plane of the sample (from ref. [26b]).

Another interesting observation is that the line is not planar. For that reason, zones tilted in opposed directions may form, which are separated by point-like defects (Fig. B.IV.34). The latter may be nucleated by various methods (thermal heating from the smectic phase, thermal quench from the isotropic liquid, or by using an electric field). The main result is that the point defects can be of two types (+−) or (−+) according to the fact that they separate a zone "−" from a zone "+," or vice versa. For that reason they can anneal. It is what they do when they are a short distance apart (less than 2 to 3b with b the sample thickness). Otherwise they repel each other: for that reason, they can form very regular (metastable) patterns (Fig. B.IV.34). The conditions of formation of these patterns have been studied in detail, numerically [26c] and experimentally, by performing thermal or electrical quenches [26b], and by directional solidification [26d].

IV.8 Bloch and Ising walls induced by the Frederiks instability

Thus far, we have only considered linear or point-like defects, the most frequent in nematics. In this section we show a way to obtain stable walls. These surface defects have been studied first theoretically by Brochard [27], then experimentally by Léger [28] in a Frederiks instability experiment. We begin this section describing a similar experiment performed by Gilli et al. [29] in Nice. We then survey briefly the theory on Ising and Bloch walls before discussing the experimental results.

IV.8.a) Experimental setup

As shown in figure B.IV.35, Gilli's setup consists of a homeotropic nematic sample of thickness d, prepared between two transparent ITO electrodes, and two magnets placed side by side on a turntable below the sample.

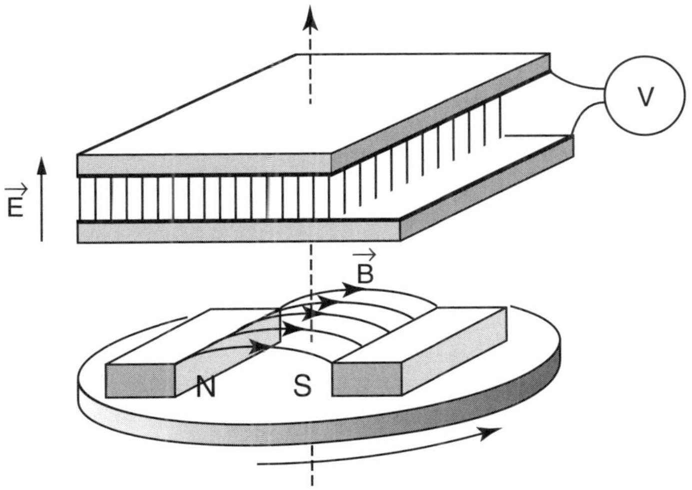

Fig. B.IV.35 Gilli's experimental setup, allowing for simultaneous application of a horizontal magnetic field and a vertical electric field (along z) to a homeotropic sample.

Thus, both a vertical electric field **E** and a horizontal magnetic field **B** can be applied to the sample. In Gilli et al.'s experiment [29], this latter field can turn with a angular velocity ω, but we shall take it as fixed, along the x axis. We shall equally assume negative dielectric anisotropy ($\varepsilon_a < 0$) and positive magnetic anisotropy ($\chi_a > 0$). Under these conditions, fields **E** and **B** are destabilizing and will favor the formation of a wall in the (y, z) symmetry plane, halfway between the magnets.

IV.8.b) Theoretical predictions

The response of the nematic slab to the applied fields is obtained by solving the dynamical equation:

$$\gamma_1^* \frac{\partial \mathbf{n}}{\partial t} = -\mathbf{h} - \mathbf{h}^M - \mathbf{h}^E \qquad \text{(B.IV.66)}$$

expressing the balance of viscous, elastic, magnetic, and electric torques. γ_1^* is the rotational viscosity corrected for possible "backflow" effects. As these corrections (calculated by Brochard [27]) are generally weak ($\approx$ 10%), we shall suppose in the following that $\gamma_1^* \approx \gamma_1$.

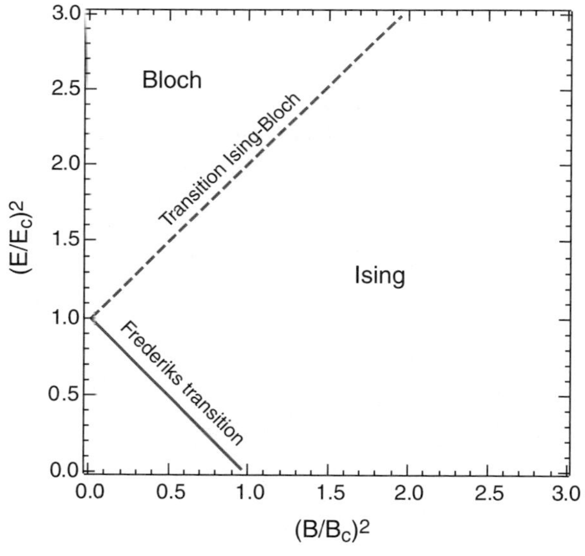

Fig. B.IV.36 Stability diagram in the presence of a magnetic and an electric field.

Let us first analyze the stability of the homeotropic nematic. This can be done by linearizing eq. B.IV.66 with respect to n_x and n_y. After some tedious algebra, one has:

$$\gamma_1 \frac{\partial n_x}{\partial t} = K_3 \frac{\partial^2 n_x}{\partial z^2} + \left(\frac{\chi_a}{\mu_o} - \varepsilon_o \varepsilon_a E^2\right) n_x \qquad \text{(B.IV.67a)}$$

$$\gamma_1 \frac{\partial n_y}{\partial t} = K_3 \frac{\partial^2 n_y}{\partial z^2} - \varepsilon_o \varepsilon_a E^2 n_y \qquad \text{(B.IV.67b)}$$

These equations have solutions of the type $n_i = n_i^o \sin(kz)\, e^{\sigma t}$ with $k = \pi/d$ to ensure homeotropic anchoring. The first one gives the growth rate σ_1 of the n_x component:

$$\sigma_1 = \frac{\frac{\chi_a}{\mu_o} B^2 - \varepsilon_o \varepsilon_a E^2 - K_3 k^2}{\gamma_1} \qquad \text{(B.IV.68a)}$$

while the second one yields, for the n_y component:

$$\sigma_2 = \frac{-\varepsilon_o \varepsilon_a E^2 - K_3 k^2}{\gamma_1} \qquad \text{(B.IV.68b)}$$

We introduce the usual critical fields (see chapter B.II):

$$B_c = k\sqrt{\frac{\mu_o K_3}{\chi_a}}, \qquad E_c = k\sqrt{\frac{K_3}{-\varepsilon_o \varepsilon_a}} \qquad \text{(B.IV.69)}$$

and find out that the nematic is unstable in the x direction when

$$\frac{B^2}{B_c^2} + \frac{E^2}{E_c^2} > 1 \qquad \text{(B.IV.70a)}$$

and along y when

$$\frac{E^2}{E_c^2} > 1 \qquad \text{(B.IV.70b)}$$

It is obvious that the absolute stability limit of the nematic is given by condition B.IV.70a, more restrictive than the second one B.IV.70b. This limit is represented in the stability diagram of figure B.IV.36 (solid line).

An important feature is the qualitative difference between the cases $\mathbf{B} \neq 0$ and $\mathbf{B} = 0$. Indeed, when the magnetic field is non-null, the distortion first appears along x, while for B = 0 it can appear in any direction, as the equations for n_x and n_y become identical. These conclusions are obvious on symmetry grounds, as shown in figure B.IV.37. Indeed, in the presence of an **applied magnetic field B**, the complete system "sample + electric field + magnetic field" has symmetry D_{2v} **below** the Frederiks threshold, possessing three second-order axes and three mirror planes. **Above** the threshold, **this symmetry is broken**, but the symmetry breaking depends on the direction adopted by the distortion.

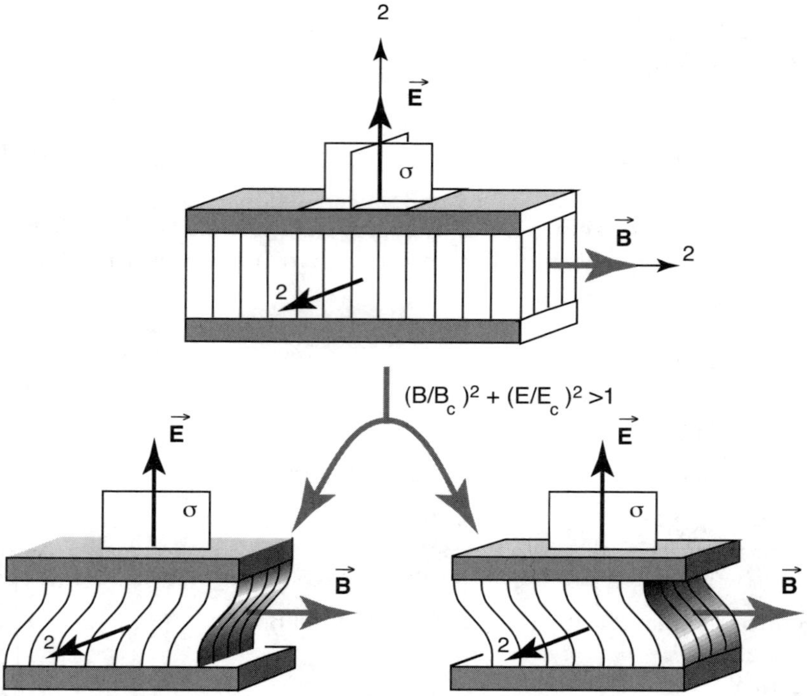

Fig. B.IV.37 Symmetry breaking at the Frederiks transition in a homeotropic sample under the combined action of a magnetic and an electric field. "2" indicates second-order axes.

In principle, two cases can appear:

– the system only develops one of the two components n_x and n_y, the other remaining zero. The final state therefore has C_{2v} symmetry, as it preserves the σ mirror plane parallel to the distortion plane, as well as the second-order axis normal to it. This symmetry breaking $D_{2v} \rightarrow C_{2v}$ leads to two pairs of equivalent states: $(n_x,0)$ and $(-n_x,0)$ or $(0,n_y)$ and $(0,-n_y)$. This is similar to the paramagnetic→ferromagnetic transition in magnetic systems yielding two energetically equivalent states, with magnetizations M and –M, respectively.

– both components n_x and n_y appear; symmetry breaking is then complete, all symmetry elements of the D_{2v} group being lost. This symmetry breaking leads to four equivalent states: (n_x,n_y), $(-n_x,n_y)$, $(n_x,-n_y)$, and $(-n_x,-n_y)$.

Calculation then shows that the pair $(n_x,0)$ and $(-n_x,0)$ has the lowest threshold and minimizes the energy (see Fig. B.IV.38).

We shall now determine the amplitude of the solutions above the Frederiks instability threshold [30]; this requires that the first nonlinear saturation terms be kept when writing eq. B.IV.66. One usually looks for solutions of the approximate form:

$$n_x = X(x,y,t)\sin(kz) \tag{B.IV.71a}$$

$$n_y = Y(x,y,t)\sin(kz) \tag{B.IV.71b}$$

where amplitudes X and Y in the center of the sample depend on x and y (only the first term of the Fourier series along z has been preserved). Returning to the starting equation B.IV.66 and integrating over the sample thickness leads to the following equations for amplitudes X and Y:

$$\gamma_1 \frac{\partial X}{\partial t} = (\mu + \gamma)X + K_2 \nabla^2 X + (K_1 - K_2)\left(\frac{\partial^2 Y}{\partial x \partial y} + \frac{\partial^2 X}{\partial y^2}\right) - aX^3 - aY^2 X \tag{B.IV.72a}$$

$$\gamma_1 \frac{\partial Y}{\partial t} = (\mu - \gamma)Y + K_2 \nabla^2 Y + (K_1 - K_2)\left(\frac{\partial^2 X}{\partial x \partial y} + \frac{\partial^2 Y}{\partial y^2}\right) - aX^2 Y - aY^3 \tag{B.IV.72b}$$

with the new parameters:

$$\mu = \frac{\chi_a}{2\mu_o} B^2 - \varepsilon_0 \varepsilon_a E^2 - K_3 k^2 \tag{B.IV.73a}$$

$$\gamma = \frac{\chi_a}{2\mu_o} B^2 \tag{B.IV.73b}$$

$$a = \frac{1}{2} K_1 k^2 \tag{B.IV.73c}$$

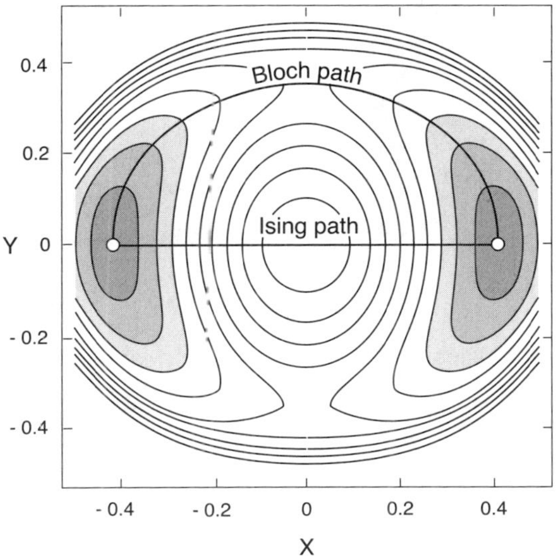

Fig. B.IV.38 Above the Frederiks threshold, the energy per unit sample surface has two minima with coordinates (–X,0) and (X,0). The passage from one minimum to the other can follow two paths: the one along the X axis corresponds to the Ising wall; the other one, skirting around the central maximum, corresponds to the Bloch wall.

In a similar way, one can obtain the energy per sample unit surface (i.e., integrated over the thickness) for homogeneous distortion (constant X and Y):

$$<F> = \frac{d}{4}\left[-(\mu+\gamma)X^2 - (\mu-\gamma)Y^2 + \frac{a}{2}(X^2+Y^2)^2\right] \qquad \text{(B.IV.74)}$$

This function is plotted in figure B.IV.38 above the Frederiks threshold.

Equations B.IV.72 generalize the linearized equations B.IV.67. They recall that the X deformation is the first to appear, as soon as $\mu + \gamma > 0$; the instability condition B.IV.70a is thus recovered. Furthermore, they give the amplitude X near the threshold for a homogeneous deformation in the (x,y) plane:

$$X^{\pm} = \pm\sqrt{\frac{\mu+\gamma}{a}}\,, \qquad Y = 0 \qquad \text{(B.IV.75)}$$

These two solutions correspond to the two minima of the energy landscape in figure B.IV.38. As they have the same energy, after the field is applied + and – domains often appear simultaneously in different spots throughout the sample. Each domain is then separated from its neighbor by a wall whose nature only depends on its orientation with respect to the field **B** and on the distance to the threshold. The walls themselves are solutions to the equations B.IV.72a and b.

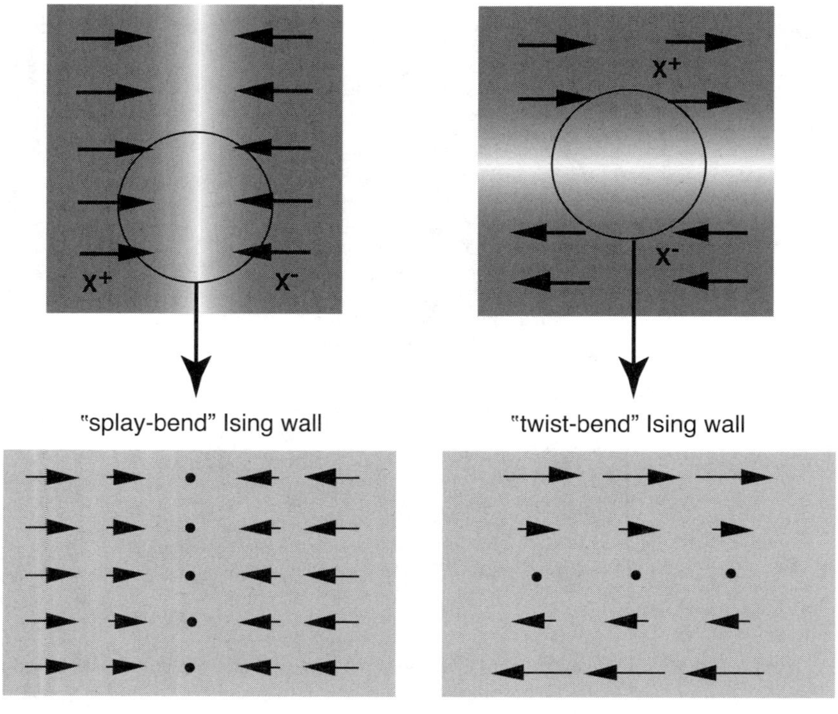

Fig. B.IV.39 Two possible structures for an Ising wall.

By analogy with magnetic systems, two types of wall can be considered: **Ising walls**, where the director remains in the (**E**, **B**) plane, and **Bloch walls**, where the director leaves this plane.

Let us begin with **Ising walls**, the simplest of the two because the n_y component is zero everywhere (Y = 0); furthermore, the amplitude X must be zero in the middle of the sample for symmetry reasons. In figure B.IV.38, this is described by the path parallel to the X axis, joining the two energy minima and passing through the central maximum. The structure (X profile) and energy of an Ising wall depend on its orientation **m** (the normal to the wall) with respect to the magnetic field **B** (Fig. B.IV.39). For **m** parallel to **B**, and therefore to x, the deformations involved are of the bend and splay types. The solution to eq. B.IV.72a is then:

$$X = \pm X^+ \mathrm{th}\left(\frac{x}{\xi_1}\right) \quad \text{with} \quad \xi_1 = \frac{2K_1}{\mu + \gamma}, \qquad Y = 0 \qquad \text{(B.IV.76a)}$$

where ξ_1 is the wall thickness.

When **m** is orthogonal to **B**, wall distortions are of the bend and twist types. In this case:

$$X = \pm X^+ \mathrm{th}\left(\frac{y}{\xi_2}\right) \quad \text{with} \quad \xi_2 = \frac{2K_2}{\mu + \gamma}, \qquad Y = 0 \qquad \text{(B.IV.76b)}$$

where ξ_2 is the new wall thickness.

In the general case (any **m**), the three deformation types are present and the wall energy takes a value between the two previous extremal values. This energy dependence on the **m** orientation with respect to **B** explains the elliptical shape of the closed walls separating two domains of opposite sign (Fig. B.IV.40). Clearly, ellipse lengthening in the direction of the field **B** implies $K_2 < K_1$, while shortening signifies $K_1 < K_2$.

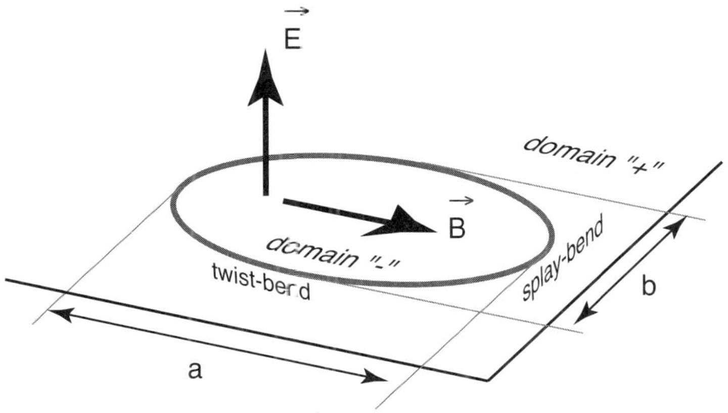

Fig. B.IV.40 Elliptical Ising wall separating two domains with different tilt.

In **Bloch walls**, very well known in magnetism, the sign change of X is accompanied by the appearance of the Y component. In the diagram of figure B.IV.35, the path joining the two minima skirts round the central maximum by one of the two valleys around it. As for the Ising walls, the structure of a Bloch wall depends on its orientation **m** with respect to the magnetic field **B**. To simplify the calculations we assume all elastic constants are equal. If the wall is normal to the field, deformation components X and Y only depend on x. The following field can be shown to satisfy the equations:

$$X = \pm\sqrt{\frac{\mu+\gamma}{a}}\ \mathrm{th}\left(x\sqrt{\frac{2\gamma}{K}}\right) \tag{B.IV.77a}$$

$$Y = \pm\sqrt{\frac{\mu-3\gamma}{a}}\left[\mathrm{ch}\left(x\sqrt{\frac{2\gamma}{K}}\right)\right]^{-1} \tag{B.IV.77b}$$

The graph of these functions is shown in figure B.IV.41. Formula B.IV.77b clearly shows that the Bloch wall can only exist if:

$$\mu \geq 3\gamma \tag{B.IV.78a}$$

or, explicitly, if

$$\left(\frac{E}{E_c}\right)^2 \geq 1 + \left(\frac{B}{B_c}\right)^2 \tag{B.IV.78b}$$

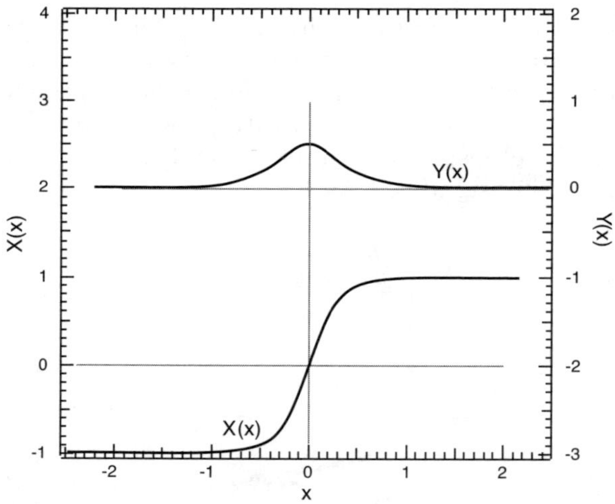

Fig. B.IV.41 X and Y distortions in a Bloch wall.

This inequality separates the parameter plane (B^2,E^2) in two distinct domains (Fig. B.IV.36); below the dotted line given by $(E/E_c)^2 = 1+(B/B_c)^2$ only Ising walls can exist, while above it Bloch walls can also appear. Inequality B.IV.78b also has a very simple physical interpretation in the energy landscape shown in figure B.IV.38. Indeed, a Bloch wall can only appear if there are two "valleys" around the central maximum or, equivalently, if two saddle-points are present on the Y axis. This condition is only achieved with an electric field sufficiently larger than the critical threshold E_c. On the other hand, the presence of two different passageways signifies in real space that, upon crossing the wall, the vector $\mathbf{p} = (X, Y)$ can turn either clockwise or counterclockwise. Hence, Bloch walls possess a **chirality**, which can change sign along the wall (Fig. B.IV.42): in this case, at the inversion (Néel) point appears a disclination of rank $m = \pm 1$ whose internal structure shall be discussed later.

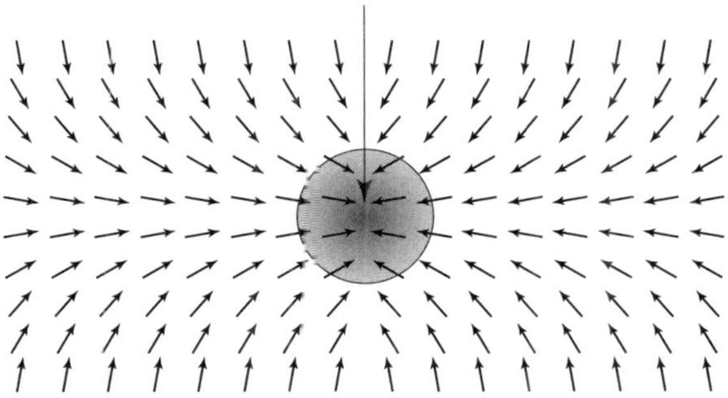

Fig. B.IV.42 The two possible chiral structures of a Bloch wall.

To conclude this theoretical discussion we point out that, in anisotropic elasticity, the internal structure and energy of the Bloch walls depend on their orientation **m** with respect to the direction **B** of the magnetic field. In particular, the slope of the Ising-Bloch transition line in figure B.IV.36 depends on the angle between **m** and **B**. The line equation can be shown to be:

$$\left(\frac{E}{E_c}\right)^2 = 1 + \frac{1-\alpha}{2\alpha}\left(\frac{B}{B_c}\right)^2 \qquad \text{(B.IV.79c)}$$

where constant α depends on the elastic anisotropy and on the angle $(\mathbf{m}, \mathbf{B})$. For instance [30]:

$$\alpha = \frac{K_2\left(\sqrt{1+8K_1/K_2}-1\right)^2}{16K_1 - K_2\left(\sqrt{1+8K_1/K_2}-1\right)^2} \qquad \text{when} \qquad \mathbf{m} \,//\, \mathbf{B} \qquad \text{(B.IV.80a)}$$

and

$$\alpha = \frac{K_1\left(\sqrt{1+8K_2/K_1}-1\right)^2}{16K_2 - K_1\left(\sqrt{1+8K_2/K_1}-1\right)^2} \qquad \text{when} \qquad \mathbf{m} \perp \mathbf{B} \qquad \text{(B.IV.80b)}$$

IV.8.c) Comparison with experiment

Let us now return to the Gilli experiment, schematized in figure B.IV.35. In this experiment, wall formation depends on the (deliberate) curvature of field lines **B** and on their change in tilt (with respect to the horizontal) from one side to the other of the symmetry plane Σ (Fig. B.IV.43). Indeed, no matter how small, a tilt of the **B** field favors one or the other of the configurations (X, 0) and (–X, 0). As a natural consequence, a wall appears in the Σ plane.

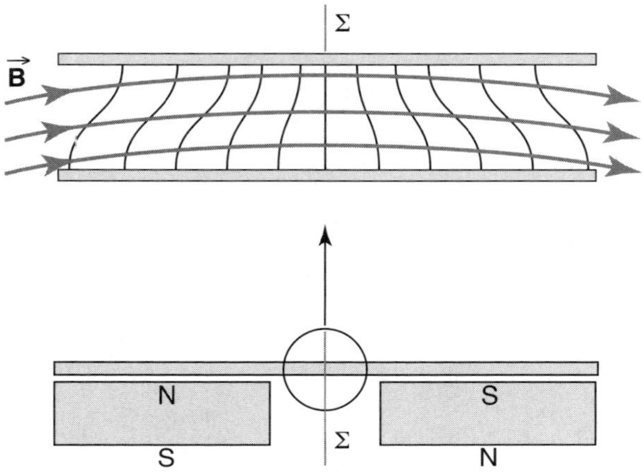

Fig. B.IV.43 Controlled generation of an Ising wall in the presence of a slightly curved magnetic field.

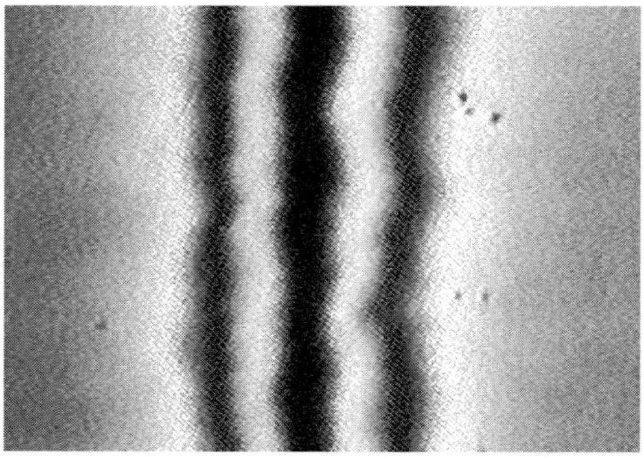

Fig. B.IV.44 Polarizing microscope image of a wall created by a slightly curved magnetic field (cf. Fig. B.IV.43). The wall is of the Ising type because at the center the director field is not distorted (extinction between crossed polarizers).

For weak enough (or zero) electric field, microscope observation between crossed polarizers reveals the existence of a black stripe at the center of the wall (Fig. B.IV.44). As this stripe is observed irrespective of the polarizer orientation, the (X, Y) deformation is zero in that area, and the wall is therefore of the "Ising" type. Note that its thickness ξ in figure B.IV.44 is about 500 μm, larger than the sample thickness d = 120 μm. When measured as a function of magnetic field intensity, ξ is found to diverge like $(B - B_c)^{-1/2}$, in agreement with equation B.IV.76.

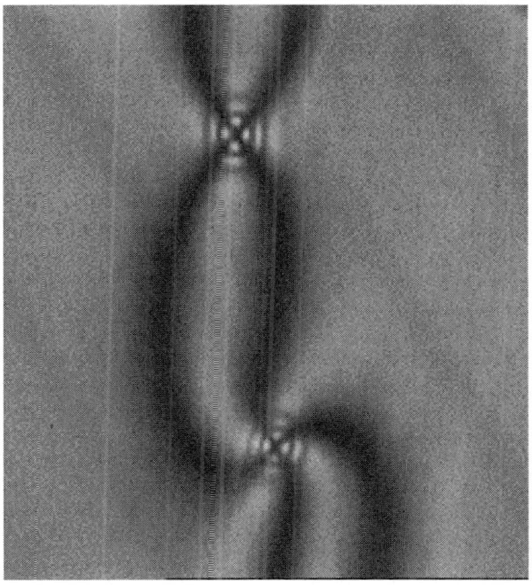

Fig. B.IV.45 Polarizing microscope image of a wall induced by a slightly curved magnetic field in the presence of a high enough electric field. As the distortion amplitude (X,Y) is everywhere non-null, the wall is of the Bloch type. The director projection onto the horizontal plane turns by 180° when crossing the wall, as indicated by the two extinctions in polarized light. Wall chirality changes sign twice, leading to the formation of two disclination lines, of rank +1 and –1.

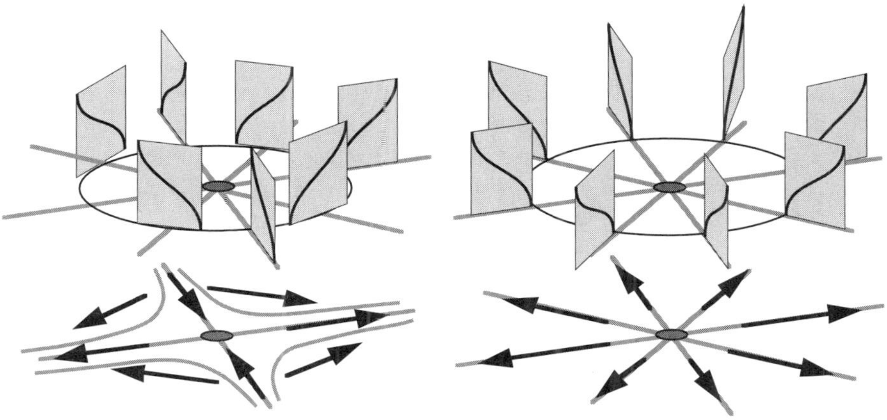

Fig. B.IV.46 Three-dimensional diagram of m = –1 and m = +1 disclination lines.

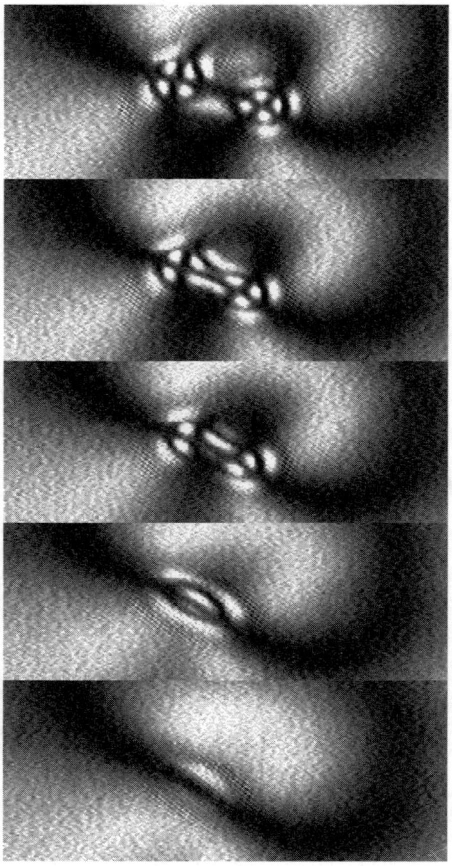

Fig. B.IV.47 Annihilation of two lines with ranks +1 and − 1. The time lapse between two images is 1 s.

The appearance of the magnetic field-induced wall changes when the electric field intensity reaches a certain threshold. As illustrated in figure B.IV.45, the transmitted light intensity is no longer zero at the center of the wall, but two new extinction areas appear, showing that the director projection onto the horizontal plane turns by 180° upon crossing the wall, which is therefore of the Bloch type. Note that this wall exhibits two disclination lines, of rank +1 and −1, respectively (Fig. B.IV.46), each one indicating a change in chirality along the wall.

The width of a Bloch wall depends on the magnetic field intensity and diverges when **B** goes to zero. In this limit, the four extinction branches of the isolated disclination lines are 90° apart. One can equally observe extinction rings around the line centers in monochromatic light. These interference rings between the ordinary and extraordinary rays indicate a decrease in distortion amplitude on approaching the center.

When the lines approach each other, their director fields are perturbed by the elastic interactions. For instance, one can see in figure B.IV.47 the way two lines of opposite sign attract and ultimately annihilate.

As for walls, the appearance of lines breaks the symmetry of the system "sample + field." This symmetry breaking can be induced in a controlled way, for instance, by provoking a radial symmetry flow or by employing fields with the appropriate geometry. In the Gilli setup, fast enough rotation ($\approx$ 1 rps) of the magnetic field induces lines parallel to the rotation axis. This is how the lines appearing in the figure B.IV.45 photo were produced.

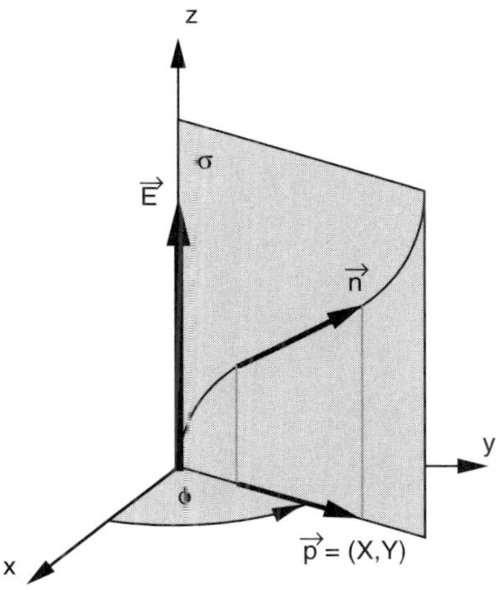

Fig. B.IV.48 Order parameter for the Frederiks transition induced by an electric field normal to the sample plane: the vector order parameter **p** = (X,Y) is characterized by its modulus and its direction ϕ.

We conclude by pointing out that these lines can be seen as topological defects of the vector field **p** = (X,Y) appearing when the symmetry of the system "sample+vertical electric field" is broken, going from $D_{\infty h}$ to C_{2v} above the Frederiks threshold (Fig. B.IV.48). This situation is obviously richer than for $\mathbf{B} \neq 0$, the initial group being here continuous ($D_{\infty h}$) rather than discrete (D_{2v}) as in the magnetic case. The vector order parameter **p** can thus have any orientation in the (x, y) plane, allowing for the appearance of disclination lines above the threshold.

The vector order parameter **p** can equally be represented by the complex number

$$\Psi = X + iY = |\Psi|\, e^{i\phi} \qquad \text{(B.IV.81)}$$

of phase ϕ equal to the angle between **p** and the x axis and amplitude $|\Psi|$ equal to the modulus of **p**. We shall see later that the same kind of order parameter is used in superconductors (or superfluids). Clearly, only integer rank defects m = ±1, ±2, ... are allowed, as the phase can only vary by a multiple of 2π when going round a closed circuit C:

$$\delta = \int_C d\phi = 2\pi m \qquad \text{(m a rational integer)} \qquad \text{(B.IV.82)}$$

This originates from the fact that **p** and **–p** are not physically equivalent.

BIBLIOGRAPHY

[1] Frank F.C., *Disc. Faraday Soc.*, **25** (1958) 19.

[2] Friedel G., *Annales Phys.*, **19** (1922) 273.

[3] a) Golemme A., Zumer S., Allender D.W., Doane J.W., *Phys. Rev. Lett.*, **61** (1988) 2937.
b) Vilfan I., Vilfan M., Zumer S., *Phys. Rev. A*, **40** (1989) 4724.
c) Kralj S., Zumer S., Allender D.W., *Phys. Rev. A*, **43** (1991) 2943.
d) Huang W., Tuthill G.F., *Phys. Rev. E*, **49** (1994) 570.

[4] a) Volovik G.E., Lavrentovich O.L., *Sov. Phys. J.E.T.P.*, **58** (1983) 1159.
b) Drzaic P., *Mol. Cryst. Liq. Cryst.*, **154** (1988) 289.

[5] a) Doane J.W., Golemme A., West J.L., Whitehead J.B., Jr., Wu B.-G., *Mol. Cryst. Liq. Cryst.*, **165** (1988) 51.
b) Ondris-Crawford R., Boyko E.P., Wagner B.G., Erdmann J.H., Zumer S., Doane J.W., *J. Appl. Phys.*, **69** (1991) 6380.
c) for a review see Drzaic P.S., *Liquid Crystal Dispersions*, Liquid Crystals Series, Vol. 1, World Scientific, Singapore, 1995.

[6] a) Dubois-Violette E., Parodi O., *J. Physique (France)*, **30** (1969) C4-57.
b) Williams R.D., *J. Phys. A: Math Gen.*, **19** (1986) 3211.
c) Kralj S., Zumer S., *Phys. Rev. A*, **45** (1992) 2461.
d) Xu F., Kitzerow H.-S., Crooker P.P., *Phys. Rev. E*, **49** (1994) 3061.

[7] a) Cladis P., Kléman M., *J. Physique (France)*, **33** (1972) 591.
b) Cladis P., Kléman M., *Mol. Cryst. Liq. Cryst.*, **16** (1972) 1.
c) Meyer R.B., *Phil. Mag.*, **27** (1973) 405.
d) Williams C., Pieranski P., Kléman M., *Phys. Rev. Lett.*, **29** (1972) 90.

[8] Saupe A., *Mol. Cryst. Liq. Cryst.*, **21** (1973) 211.

[9] Mihaïlovic M., Oswald P., *J. Physique (France)*, **49** (1988) 1467.

[10] Volterra V., *Ann. Ecol. Norm. Suppl.*, **24** (1907) 401.

[11] a) Kléman M., *Points, Lines, and Walls in Liquid Crystals, Magnetic Systems, and Various Media*, John Wiley & Sons, Chichester, 1983.
b) Chaikin P.M., Lubensky T.C., *Principles of Condensed Matter Physics*, Cambridge University Press, Cambridge, 1995.

[12] a) De Gennes P.-G., *Mol. Cryst. Liq. Cryst.*, **12** (1971) 193.
b) Vertogen G., de Jeu W.H., *Thermotropic Liquid Crystals, Fundamentals*, Springer-Verlag, Heidelberg, Vol. 45, 1988.

[13] Fan C., *Phys. Lett.*, **34A** (1971) 335.

[14] Madhusudana N.V., Pratibha R., *Mol. Cryst. Liq. Cryst.*, **89** (1982) 249.

[15] Dafermos C.M., *Quart. J. Mech. Appl. Math.*, **23** (1970) 549.

[16] Ericksen J.L., *Liquid Crystals and Ordered Fluids*, Eds. Johnson J.F. and Porter R.S., Plenum, New York, 1970, p. 189.

[17] a) Tòth G., Denniston C., Yeomans J., *Phys. Rev. Lett.*, **88** (2002) 105504.
b) Svensek D., Zumer S., *Phys. Rev. E, 66* (2002) 021712.

[18] Ryskin G., Kremenetsky M., *Phys. Rev. Lett.*, **67** (1991) 1574.

[19] Imura H., Okano K., *Phys. Lett.*, **42A** (1973) 403.

[20] Cladis P., van Saarloos W., Finn P.L., Kortan A.R., *Phys. Rev. Lett.*, **58** (1987) 222.

[21] Gerritsma C.J., Geurst J.A., Spruijt A.M.J., *Phys. Lett.*, **43A** (1973) 356.

[22] Bogi A., Martinot-Lagarde P., Dozov I., Nobili M., *Phys. Rev. Lett.*, **89** (2002) 225501.

[23] Dzyaloshinskii I.E., *Sov. Phys. J.E.T.P.*, **31** (1970) 773.

[24] Anisimov S.I., Dzyaloshinskii I.E., *Sov. Phys. J.E.T.P.*, **36** (1973) 774.

[25] a) Crawford G.P., Allender D.W., Doane J.W., *Phys. Rev. A*,**45** (1992) 8693.
b) Vilfan I., Vilfan M., Zumer S., *Phys. Rev. A*, **43** (1991) 6875.
c) For a more detailed calculation see Kralj S., Zumer S., *Phys. Rev. E*, **51** (1995) 366.

[26] a) Ignes-Mullol J., Baudry J., Lejcek L., Oswald P., *Phys. Rev. E*, **59** (1999) 568.
b) Ignes-Mullol J., Baudry J., Oswald P., *Phys. Rev. E*, **63** (2001) 031701.
c) Holyst R., Oswald P., *Phys. Rev. E*, **65** (2002) 041711.
d) Ignes-Mullol J., Scurtu L., Oswald P., *Eur. Phys. J. E.*, **10** (2003) 281.

[27] Brochard F., *J. Physique (France)*, **33** (1972) 607.

[28] Léger L., *Mol. Cryst. Liq. Cryst.*, **24** (1973) 33.

[29] Gilli J.M., Morabito M., Frisch T., *J. Phys. II (France)*, **4** (1994) 319.

[30] Frisch T., Ph.D. thesis, “Ondes Spirales dans les Cristaux Liquides et Théorie de Ginzburg-Landau,” University of Nice Sophia-Antipolis, 1994.

Chapter B.V

Anchoring and anchoring transitions of nematics on solid surfaces

In the absence of surfaces and applied forces, the director field **n**(**r**) of a nematic at equilibrium is uniform (**r** independent) and its orientation is arbitrary. This translational and rotational freedom recalls that of a ship on a flat sea. Surface-imposed orientation of the liquid crystal is similar to dropping anchor, which fixes the ship position with respect to the sea bottom. Thus, one speaks of nematic **anchoring** at a surface.

This chapter, as well as the following one (dedicated to the nematic-isotropic liquid interface), will show that the contact of a nematic with any surface influences its director field. Anchoring is considered **strong** when it permanently fixes molecular orientation ("geometric" anchoring condition), and **weak** otherwise. It can be **monostable** (only one possible anchoring) or **multistable** (several possibilities) and its character can change with experimental conditions such as temperature, chemical environment, etc., sometimes leading to **anchoring transitions**. These are the phenomena we shall discuss.

The chapter is organized as follows. We start with a historical overview of the anchoring concept (section V.1). The notion of interface is then discussed, first in general terms (section V.2) and then using concrete examples, allowing us to define the fundamental concept of "**interface symmetry**" (section V.3). The existence of monostable and multistable anchoring conditions lead us to the problem of anchoring **selection** by wetting (section V.4). Some classical anchoring transitions on typical substrates are subsequently discussed (section V.5). At this point we shall insist on the analogies between these transitions and classical phase transitions. Finally, the experimental measurement of the anchoring energy is discussed, using the simple example of homeotropic anchoring (section V.6).

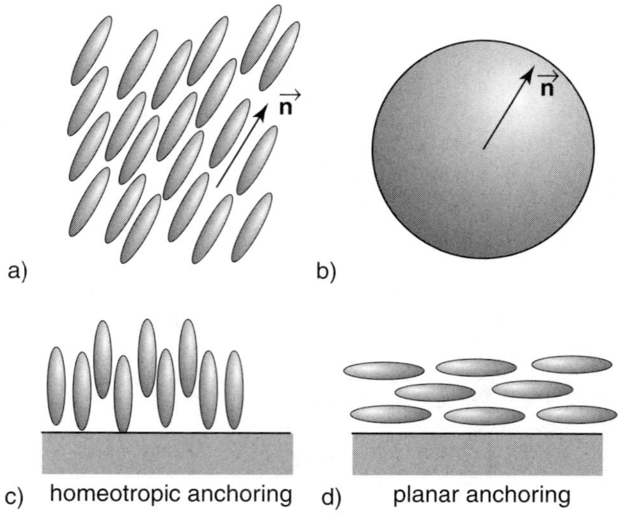

Fig. B.V.1 Anchoring concept: a) and b) in an isotropic space, the nematic can have any orientation **n**. Its image on the S2 sphere is an arbitrary point; c) and d) surfaces break spatial isotropy and remove this degeneracy, imposing the nematic orientation. The terms "homeotropic" and "planar" date back to the days of G. Friedel.

Be advised that the topic is too vast to be completely covered in an introductory textbook, so this chapter only presents part of the known results. More extensive information can be found in the review articles cited (references [1–4]).

V.1 Precursors

Nematic anchoring on crystal surfaces was already known in the times of G. Friedel (Fig. B.V.1). First Ch. Mauguin [5] and then M. Grandjean came to the conclusion that "*...well defined orientations are almost always obtained when placing a drop of anisotropic liquid on crystal cleavage surfaces or in cleavage cracks.*" They also mention that "*...the observed orientations usually have a simple relation with crystal symmetry*" [6]. In his famous work [7], G. Friedel specifies that "*orientation should be understood as referring not to the normal homeotropy, occurring in the absence of intimate contact, which is substrate-independent, but rather to orientations with respect to the crystalline structure, for which the optical axis, in most cases if not always, is placed parallel to the crystal surface and in this plane assumes well defined directions.*" When mentioning "normal homeotropy" in the cited passage, Friedel was alluding to "*the property of smectic bodies of orienting with their optical axis normal to the preparation surface.*"

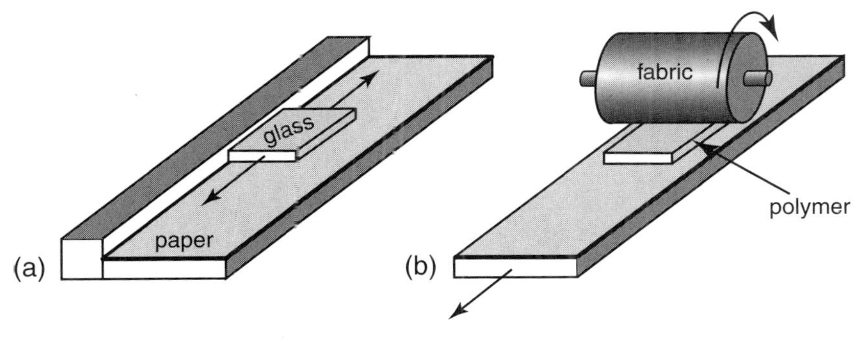

Fig. B.V.2 The Châtelain method: the surface obtained by rubbing a glass plate on paper induces planar anchoring (a). Anchoring quality is much better when the glass surface has first been covered with a thin polymer layer (polyvinyl alcohol, polyimides, etc.), and the paper replaced by cloth placed on a rotating cylinder (b).

Much later, Châtelain found out that paper-rubbed glass surfaces could align nematics in the rubbing direction [8] (Fig. B.V.2). This empirical method was then considerably improved following the practical demands related to display production. Other surface treatments, providing various types of high-quality anchoring, were equally devised [1–4].

The anchoring phenomenon is usually very complex and difficult to explain from a purely microscopic point of view. We shall therefore adopt a more phenomenological ("Landau-like") approach in order to classify different types of anchoring according to their common features.

V.2 On the notion of interface

A surface limiting a nematic volume is in fact an interface between the nematic phase and another phase. An interface is generally characterized by its shape, composition, and internal structure, all of which depend on thermodynamic parameters such as temperature, pressure, or the chemical potentials of the different components of the two phases in contact. Usually, medium structure and composition vary considerably across the interface (a few molecular lengths) on a distance ξ termed "**interface thickness**" (Fig. B.V.3).

In this chapter and the following one we shall see that this definition must be qualified according to the circumstances. Thus, the interface between a nematic and its isotropic phase is very different from the one between the nematic and a solid substrate (assumed insoluble in the mesogenic substance).

In the first case (thoroughly discussed in the following chapter), the quadrupolar order parameter varies continuously from zero (in the isotropic liquid) to a finite value (in the nematic phase) over a relatively large distance (several molecular lengths). Chemical components are easily exchanged across this interface which is also **rough** at the molecular scale (i.e., strongly fluctuating as a result of thermal agitation). Being easily deformed, the interface can take various shapes when subjected to magnetic or elastic torques.

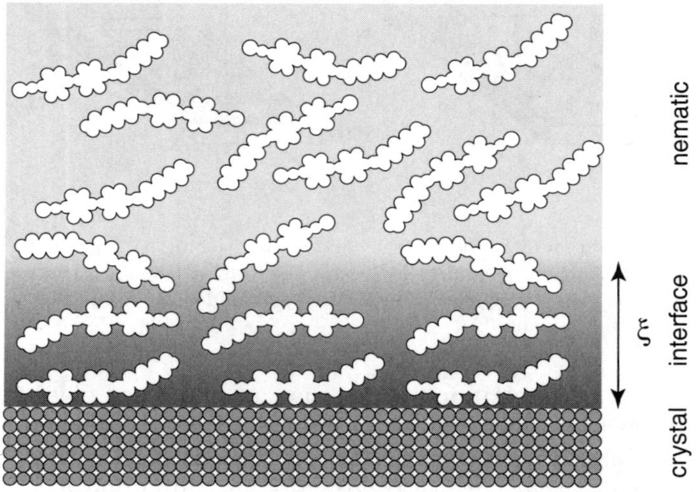

Fig. B.V.3 Schematic representation of an interface between the nematic phase and a crystal. Nematic molecules (e.g., 8CB) are adsorbed to the crystal surface where they form a regular pattern (2D crystal). Uniaxial nematic order is recovered beyond the ξ-thick perturbed layer.

For a nematic-solid substrate interface the situation is completely different. A good example is provided by the interface between a nematic and the cleavage surface of muscovite mica. Flat and smooth at the molecular scale, this interface retains its shape because of the crystal stiffness, simplifying the problem with respect to the case of the nematic-isotropic liquid interface. On the other hand, going from positional crystalline order to purely orientational order leads to deeper changes in structure and symmetry than for an isotropic liquid.

V.3 Interface symmetry and classification of the different types of anchoring

The relation between solid substrate symmetry and the anchoring it induces is particularly easy to discuss in the case of crystal surfaces obtained by cleavage (gypsum, mica, graphite) or, in certain cases, by mechanical cutting (silicon wafers). In all these cases, the surfaces possess two-dimensional periodic structures. Such an example, given in figure B.V.4, show the atomic force microscopy (AFM) image of a gypsum cleavage surface.

Although structural details have little relevance in a phenomenological approach, it is however interesting to know the **point symmetry of the surface**, as demonstrated by the following examples.

Let us start with a gypsum cleavage along its (010) crystalline plane. This material of monoclinic symmetry has two second-order symmetry axes along the [010] direction and orthogonal to the cleavage plane (010): This cleavage surface is thus invariant under 180° rotations (Fig. B.V.5). Furthermore, it has no symmetry plane, a feature that will later prove to be important. However, the two surfaces resulting from the same cleavage operation are mirror reflections of each other, as cleaving takes place along a symmetry plane.

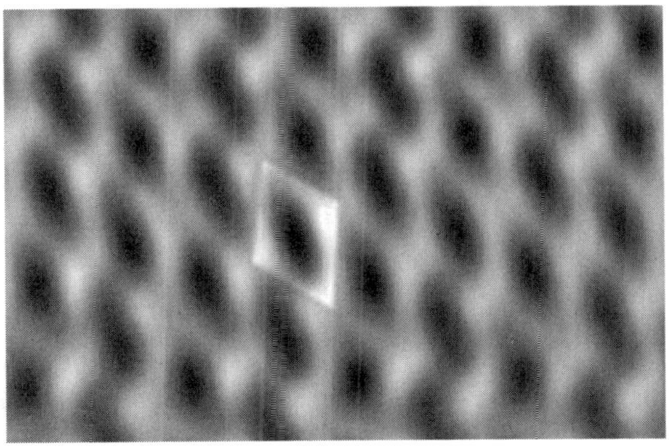

Fig. B.V.4 Atomic force microscopy (AFM) image of a gypsum cleavage surface.

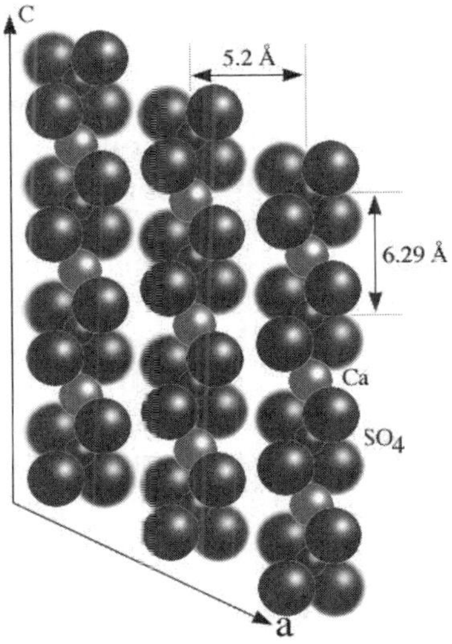

Fig. B.V.5 The cleavage surface (**a**, **c**) of gypsum, whose chemical formula is ($CaSO_4$ $2(H_2O)$) exhibits rows of $CaSO_4$ stoichiometry. The surface symmetry is C_2.

Cleavage surfaces of muscovite mica have totally different symmetry. As shown in figure B.V.6, the mica surface is essentially made up of oxygen atoms placed on a Kagomé lattice, each of them shared between the bases of two SiO_4 tetrahedra belonging to a layer of stoichiometry Si_2O_5. The adjacent layer has the same stoichiometry but is displaced by **l** with respect to the surface layer. As a consequence of this displacement, the C_{6v} symmetry of the Kagomé lattice is broken, the only preserved symmetry element being mirror plane σ parallel to **l**.

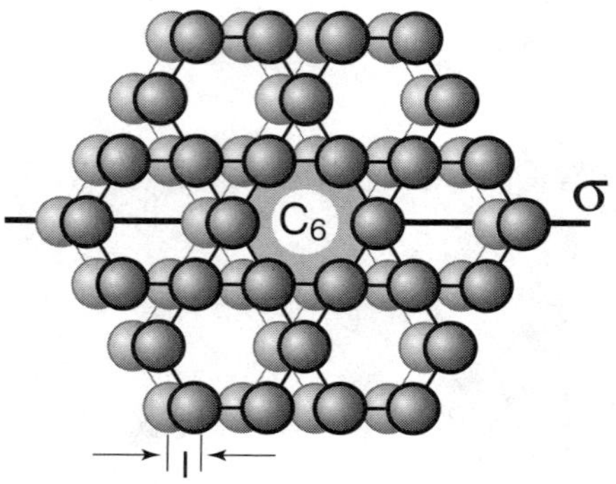

Fig. B.V.6 A muscovite mica cleavage surface is essentially made up of oxygen atoms placed on a Kagomé lattice. Surface symmetry is "almost" C_{6v}. Indeed, the next atomic layer (similar to the first) is shifted by **l** in the direction parallel to the mirror plane σ. This displacement breaks C_{6v} symmetry which is reduced to C_s.

It turns out that in muscovite mica the third equivalent oxygen layer, placed deeper under the surface, is translated with respect to the surface layer in a direction **l'** not contained in the σ plane. This new translation breaks the mirror symmetry with respect to the σ plane. Consequently, the surface of muscovite mica has no symmetry element but the identity. Nevertheless, this departure from mirror symmetry is so weak as to be generally neglected.

In conclusion, let us describe the surface of a silicon wafer. Obtained by mechanical cutting, its orientation with respect to the crystalline planes (hkl) is imperfect. It therefore consists of terraces separated by steps. The width l of the terraces depends on the step height h and on the angle α between the cutting surface and the crystal planes (hkl): $l \approx h/\alpha$. This is illustrated in figure B.V.7, showing a vicinal surface close to a (100) plane. It is made up of terraces of individual symmetry C_{2v}, but the direction of the mirror plane changes by 90° between two neighboring terraces. Terrace width is about 1000 Å for $\alpha \approx 10^{-3}$. At the macroscopic scale of a nematic sample (a few dozen microns for a display), the influence of individual terraces is "wiped out" by the nematic elasticity and the substrate behaves as if it had C_{4v} symmetry. There is however a boundary layer of thickness l (the terrace width) where the terraces still show their presence. This elastic boundary layer defines the real interface thickness between the liquid crystal and the substrate, sometimes much larger than in the case of smooth crystal surfaces.

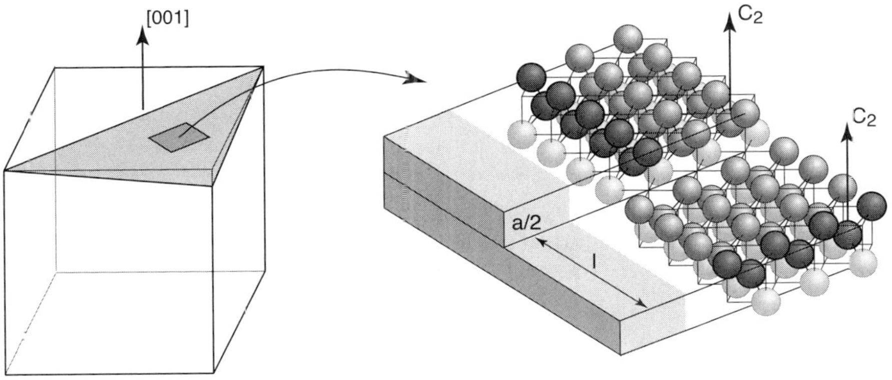

Fig. B.V.7 A vicinal surface close to the (001) plane of silicon is formed of terraces. The height of the step between adjacent terraces is a/2, i.e., half the cubic repeat distance. The silicon atoms connected by covalent bonds form rows. One such row is shaded on each of the adjacent terraces. The symmetry of an individual terrace is C_{2v}. Structures of the adjacent terraces are connected by the screw axis 4_2 symmetry operation.

The concept of interface becomes even more difficult in the case of another substratum, widely employed for its anchoring properties: SiO films evaporated on glass under oblique incidence [9–12]. These layers, a few hundred Ångstrom thick, have a very porous structure made up of "dendrites" tilted in the evaporation plane. As we shall see later, at a macroscopic scale this substrate behaves as if it had mirror symmetry with respect to the evaporation plane σ (Fig. B.V.8).

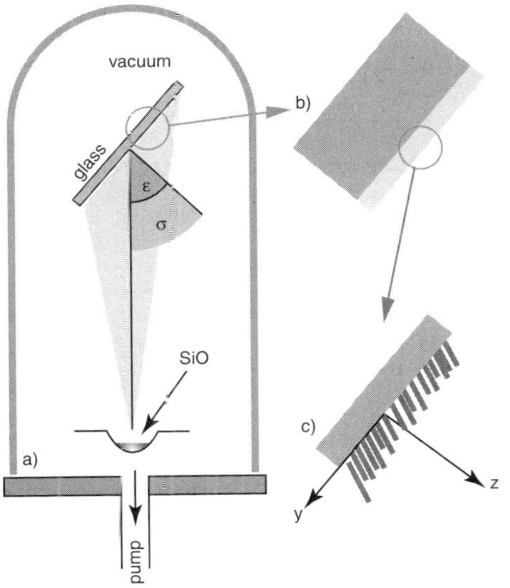

Fig. B.V.8 Obliquely evaporated SiO layers. a) Preparation method; b, c) porous dendritic structure of the evaporated layer.

Consider now the general case of a solid surface of symmetry G_s covered by the nematic phase. In the absence of external torques, the director **n** chooses a particular direction, $\mathbf{n}_a$, corresponding to the minimum of surface free energy. Obviously, the surface free energy is also a minimum for any other direction:

$$\mathbf{n}'_a = g\, \mathbf{n}_a\,, \qquad g \in G_s \tag{B.V.1}$$

obtained from $\mathbf{n}_a$ by a symmetry operation of the G_s group.

If $\mathbf{n}_a$ is invariant under all symmetry elements in the G_s group, the anchoring is called **monostable**, being unique. The interface structure preserves the symmetry of the solid surface.

Otherwise, at least one of the symmetry elements of the G_s group is broken, leading to the existence of at least two equivalent anchoring directions. This is the case with **multistable** anchoring. By definition, the order m of multistability is the number of equivalent distinct directions (up to a sign change) belonging to the set $\{g\mathbf{n}_a\}$. It depends both on the **surface symmetry G_s** and on the group of **interface symmetry G_a** defined as the set of all operations $g \in G$, leaving the orientation $\mathbf{n}_a$ unchanged:

$$g \in G_a \qquad \text{if} \qquad g\mathbf{n}_a = \pm\, \mathbf{n}_a \tag{B.V.2}$$

By definition, the G_a group is a sub-group of G_s so one can define the quotient group:

$$G_m = G_s / G_a \tag{B.V.3}$$

whose order m determines the multistability.

This anchoring classification according to the invariance of the anchoring direction under certain symmetry operations of the G_s group is illustrated in the diagrams of figures B.V.9 and 10 for the $G_s = \{I,\sigma\}$ symmetry typical of SiO layers evaporated under oblique incidence.

If $\mathbf{n}_a$ is a direction in the mirror plane σ, $\sigma\mathbf{n}_a = \mathbf{n}_a$ so $G_a = G_s$ and $G_m = \{I\}$. Similarly, for $\mathbf{n}_a$ orthogonal to the mirror plane σ, $\sigma\mathbf{n}_a = -\,\mathbf{n}_a$, $G_a = G_s$, and $G_m = \{I\}$. In both cases, the anchoring is monostable.

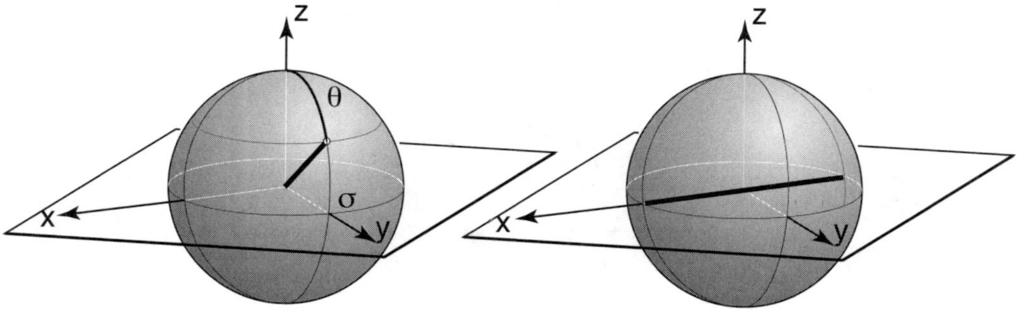

Fig. B.V.9 Monostable anchoring on a substrate of symmetry C_s. To the left, oblique anchoring; to the right, planar anchoring.

For all other orientations of $\mathbf{n}_a$ (see figure B.V.10), $\sigma\mathbf{n}_a \neq \pm\, \mathbf{n}_a$, such that $G_a = \{I\}$ and $G_m = \{I,\sigma\}$. Anchoring is therefore bistable.

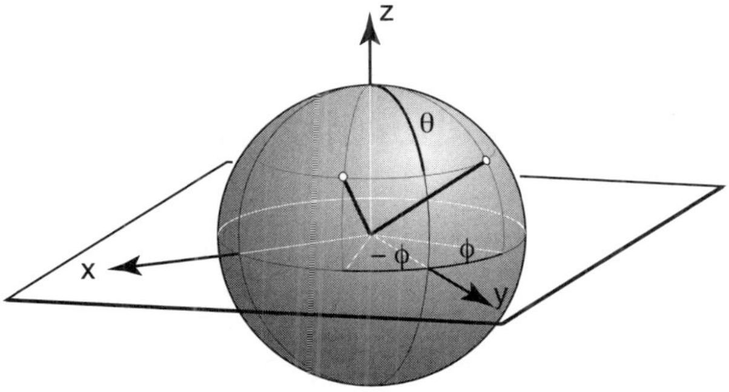

Fig. B.V.10 Bistable anchoring on a substrate of symmetry C_s. The two possible anchoring directions are symmetric with respect to the mirror plane σ (yz plane).

The same classification scheme, applied to the interfaces between the nematic phase and an isotropic phase, allows for defining the three types of anchoring shown in figure B.V.11. From left to right:

1. **Homeotropic** anchoring, monostable as it respects the symmetry G_s of the "substrate" ($G_s = G_a$ so $G_m = \{I\}$);

2. **Oblique degenerate** anchoring, respecting the mirror symmetry σ (and the identity I), for which $G_a = \{I,\sigma\}$. This anchoring is multistable, of order infinity since $G_m = C_\infty$ is a continuous group;

3. **Planar degenerate** anchoring, multistable of order infinity, respecting both the mirror symmetry σ and 180° rotation: the quotient group is also continuous.

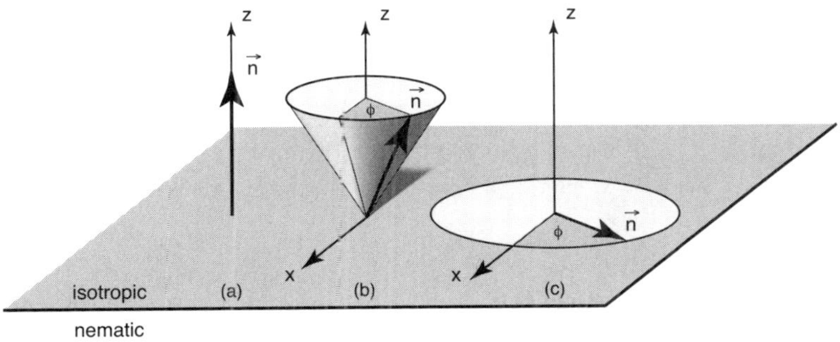

Fig. B.V.11 The three possible types of anchoring at the nematic-isotropic interface. a) monostable homeotropic, b) oblique degenerate, and c) planar degenerate.

We shall subsequently show that it is sometimes possible to achieve on the same surface the different types of symmetry-determined anchoring by modifying the thermodynamic parameters fixing the interface state. It will also be shown that the passage from one kind of anchoring to another has all the attributes of a phase transition. But before tackling this problem, let us see how the selection is done for multistable anchoring.

V.4 Wetting and anchoring selection

In the multistable case, the system must choose an anchoring direction in each point on the substrate. This need to make a local choice is a common feature of systems with broken symmetry. In magnetic systems, for instance, the direction of the magnetization **M** at the para- → ferro-magnetic transition is undetermined (with an infinite number of possible choices), but can be imposed by applying a weak magnetic field that breaks spatial isotropy. The same happens with anchoring, as we shall now show.

In the case of multistable anchoring on solid substrates, B. Jérôme [13] proved by a very elegant experiment that the choice is generally made at the moment the nematic wets the substrate. The experiment is very simple, consisting of the following steps: 1) place a nematic drop on the substrate (Fig. B.V.12); 2) let it spread, and 3) observe it under the polarizing microscope after spreading.

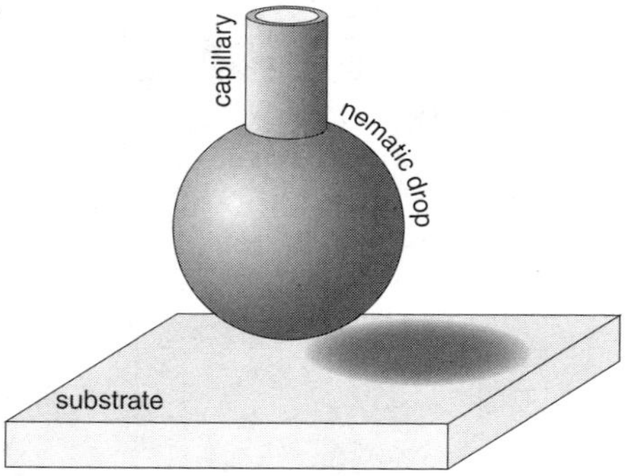

Fig. B.V.12 Placing a nematic droplet on an anisotropic substrate allows for rapid identification of the anchoring type.

This method was successfully used to identify the type of anchoring on SiO layers and on crystalline substrates such as graphite, mica, gypsum, or silicon. In the case of SiO, it demonstrated not only the three types of anchoring predicted by symmetry (see section V.3), but also yielded a relation between the evaporation angle ε and the corresponding anchoring. Figure B.V.13 shows different photos of the same drop of nematic E8 (eutectic mixture sold by BDH) on an evaporated SiO layer for multistable anchoring. The photos, taken between crossed polarizers for different drop orientations, are typical of bistable oblique anchoring: indeed, the drop is divided into four domains

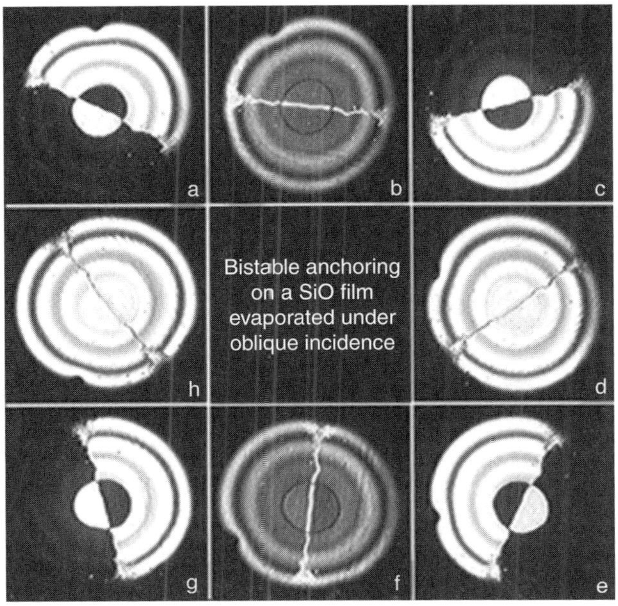

Fig. B.V.13 A drop of nematic E8 (volume 10^{-2} μl) spread on a SiO layer evaporated at a 72° oblique incidence. The photos were taken between crossed polarizers parallel to the sides of the image. Between two successive photos the drop has been turned by 22.5°.

separated by two walls. One of the walls is a straight line, parallel to the evaporation plane, which is also the symmetry plane σ. The other wall is closed and shifted with respect to the center of the drop in the evaporation direction. According to their optical contrast, the four domains can be grouped together into two distinct pairs, each representing one of the two possible anchoring transitions. This is summarized in figure B.V.14.

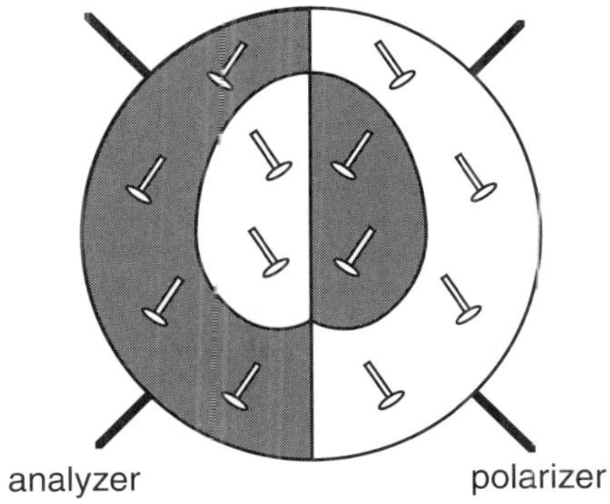

Fig. B.V.14 Choice of the two anchoring directions in the nematic drop. Oblique director orientation is represented by nails driven at an angle into the substrate plane (x, y).

To understand the final structure of the drop, B. Jérôme showed that one had to analyze the evolution of the air/SiO/nematic contact line during spreading (Fig. B.V.15). The argument goes like this: When the drop touches the substrate, the contact angle δ is worth π (point contact). As the drop spreads, the radius r of the contact line increases, while the contact angle decreases. Experiment shows that the anchoring direction is locally given by the contact angle δ and by the angle ψ between the normal to the contact line and the evaporation plane.

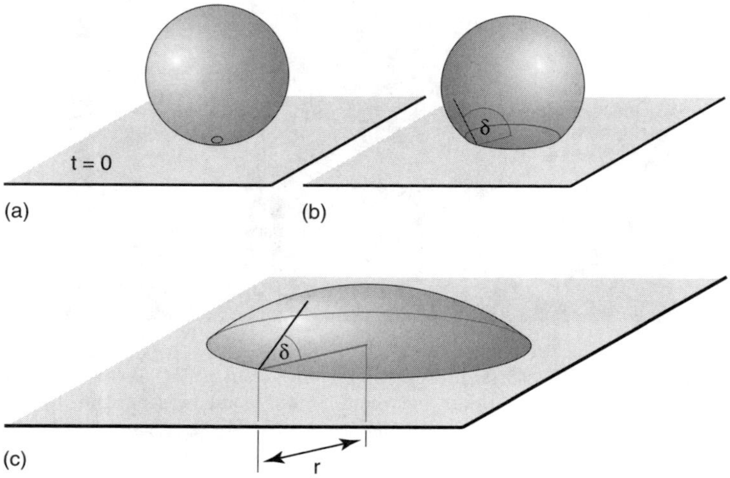

Fig. B.V.15 During spreading of the nematic drop, the instantaneous contact angle δ varies between π and 0 as a function of the radius r. a) $\delta = \pi$; b) $0 < \delta < \pi$; c) $\delta \approx 0$.

To be more precise, the choice of an anchoring direction at a given point on the substrate only depends on the orientation of molecules at the drop free surface when they touch the substrate.

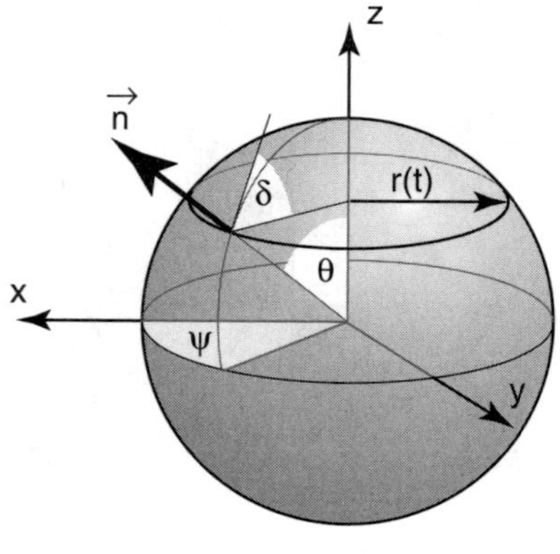

Fig. B.V.16 Relation between the molecular direction **n** at the free surface close to the contact line and the spreading parameters δ and ψ.

The molecules then choose between the two anchoring directions the one "closer" to their orientation at the free surface (this "closeness" will be later defined in terms of the thermodynamic surface potential, or "anchoring energy"). The free surface anchoring of the employed nematic E8 (cyanobiphenyl mixture marketed by BDH) is homeotropic, greatly simplifying the reasoning, as one only needs to know the angles δ and ψ to obtain the director orientation at the contact line on the drop free surface. To understand the phenomenon better, let us draw on the unit S2 sphere the molecular orientation at the free surface close to the contact line (Fig. B.V.16). The director **n** is homeotropically oriented at the free surface, having a polar angle:

$$\theta = \delta = \delta(r) \quad \text{with} \quad 0 \leq \theta \leq \pi \tag{B.V.4}$$

and the previously defined ψ ($0 \leq \psi \leq 2\pi$) as azimuthal angle. Representing these orientations on the S2 sphere, one gets each time a parallel of latitude $\theta = \delta(r)$ with r the instantaneous drop radius in the substrate plane. During spreading, this parallel "rises" from the South pole ($\theta = \delta = \pi$) of the S2 sphere to the North pole (for complete wetting), thus sweeping the entire sphere surface. Hence, the spreading mechanism creates a one-to-one relation:

$$(r,\psi) = R(\theta,\psi) \tag{B.V.5}$$

between the S2 sphere director coordinates (θ,ψ) (upon which depends the thermodynamic surface potential $W(\theta,\psi)$, also termed "anchoring energy") and the polar coordinates (r,ψ) of the spreading drop. This relation is mathematically defined, on the one hand by the inverse function of $r(\delta)$,

$$r = \delta^{-1}(\theta) \tag{B.V.6}$$

and, on the other hand, by the identity of the azimuthal angles on the S2 sphere and the spreading drop. To get the function $r = \delta^{-1}(\theta)$, one can assume as a first approximation that the drop maintains the shape of a spherical cap during spreading. The volume V being constant, the radius r of the base and the contact angle δ are related by the expression of the cap volume:

$$V = \frac{\pi}{6} h\,(3r^2 + h^2) \tag{B.V.7a}$$

with

$$h = \frac{r}{\sin\delta} - \frac{r}{\tan\delta} \tag{B.V.7b}$$

the spherical cap height. Solving these equations with respect to r yields the desired relation $r = \delta^{-1}(\theta)$.

Once relation R (B.V.5) is known, the general shape of the surface potential $W(\theta,\psi)$ can be obtained from the structure of the spreading drop (and vice versa). As a matter of fact, during the wetting process molecules pass from the free surface to the substrate surface with a tank-treading motion. Approaching the substrate, they interact with it and "enter" one of the basins of attraction of the surface potential W(θ,ψ), easily represented on the S2(θ,ψ) sphere by tracing its equipotential lines. Assuming that the subsequent director evolution is along the steepest descent towards the minimum of the chosen basin of attraction, the anchoring direction at point (r,ψ) is unambiguously determined.

A good example is provided by the bistable oblique anchoring corresponding to the photos of figure B.V.13, where the drop is divided into four separate domains. The surface potential W(θ,ψ), represented in figure B.V.17, corresponds to the drop structure of the spread drop. The S2 sphere is divided into four lunes by two great circles representing the geometrical place of points equidistant to the two favored anchoring directions. One of the circles, C_σ, is in the mirror plane σ (evaporation plane) while the other, C_c, is normal to σ and tilted with respect to the (x,y) plane of the substrate. In a relief map of the thermodynamic surface potential W(θ,ψ) on the S2 sphere, these two great circles become watersheds separating the four basins of attraction corresponding to the four minima defining the two favored anchoring directions.

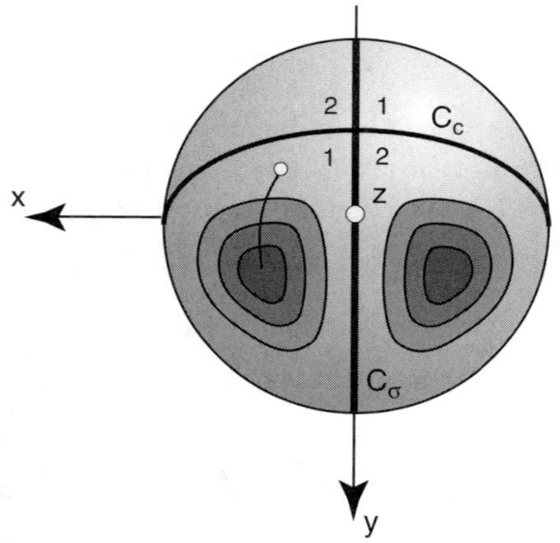

Fig. B.V.17 Division of the S2 sphere into four "basins of attraction" corresponding to the two anchoring directions (two diametrally opposed basins correspond to the same anchoring direction). The two great circles C_c and C_σ represent the "watersheds" between the basins of the four minima of the surface potential W(θ,ψ). Level lines are represented around the two visible minima.

Using relation R, the thermodynamic potential W(θ,ψ) of figure B.V.17 can be projected onto the drop plane (r,ψ). The resulting projection, W(r,ψ), is shown in relief in figure B.V.18. One realizes that the C_σ circle yields a watershed along the diameter of the drop parallel to the mirror plane σ. The other circle C_a gives a closed watershed, enclosed by the drop and shifted from the origin. These two contours divide the W(r,ψ) relief into four basins of attraction. Note that two diametrally opposed basins lead to the same anchoring direction on the substrate. The next two figures depict the R-projection of the surface potentials typical for monostable planar and oblique anchoring, respectively. For planar anchoring, the function $W(\theta,\psi)$ has two minima separated by watershed C_σ.

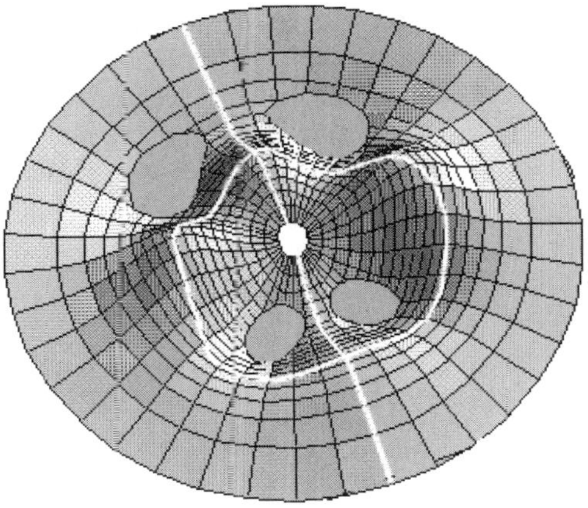

Fig. B.V.18 Projection $W(r,\psi)$ on the substrate plane of the surface potential $W(\theta,\psi)$ in figure B.V.17 using the wetting relation $\Re$.

In R-projection, the surface potential exhibits two basins of attraction separated by a watershed parallel to the great axis of the drop contained in the evaporation plane σ. In this case, the planar anchoring direction is the same on both sides of the "watershed," being perpendicular to the σ plane.

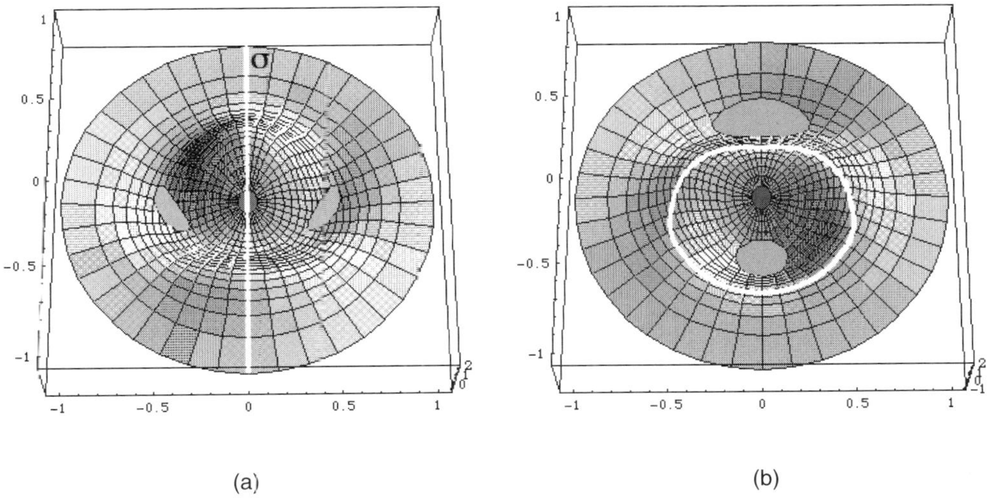

Fig. B.V.19 Surface potentials projected onto the plane of the spread drop for planar monostable (a) and oblique monostable (b) anchoring.

For oblique monostable anchoring, the two minima of the surface potential are separated on the S2 sphere by the great circle C_c perpendicular to the σ plane and to the anchoring direction. In R-projection, this circle appears as a watershed enclosing one of the two minima. Again, the anchoring direction is the same inside the two domains. In the case of monostable anchoring, a watershed necessarily separates two domains with identical substrate anchoring. However, inside each domain the director links up differently to the free surface of the drop, so that the two domains are separated by a visible wall along the watershed. These theoretical predictions of drop separation into domains are in agreement with experimentally observed structures.

Let us now describe some of the typical anchoring transitions observed on SiO layers or on crystal cleavage surfaces.

V.5 Anchoring transitions

V.5.a) Anchoring transitions on evaporated SiO layers

On evaporated SiO layers, the anchoring changes with the evaporation angle ε and the thickness d of the layer. Anchoring angles θ and ψ measured for a layer thickness of about 200 Å are shown in the diagram of figure B.V.20 versus the evaporation angle. This anchoring evolution can be represented on the S2 sphere with increasing evaporation angle (Fig. B.V.21).

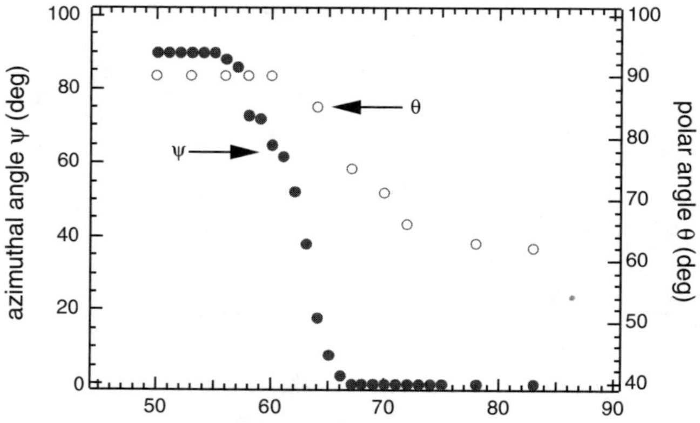

Fig. B.V.20 Variation in the anchoring direction on a 200 Å SiO layer as a function of the incidence evaporation angle (from ref. [12]).

At first ($\varepsilon < 55°$), anchoring is planar monostable, corresponding to two diametrally opposed points PM and PM' on the S2 sphere. These points are on the x axis perpendicular to the evaporation plane σ. For $55° < \varepsilon < 65°$, anchoring becomes oblique bistable: its S2 image is then a pair of trajectories OB and OB' symmetrical with respect to the σ plane and originating in points PM and PM', respectively. These two trajectories converge to the same point in the symmetry plane σ. For larger evaporating angles ($\varepsilon > 65°$), the anchoring becomes oblique monostable: its S2 image is the segment OM contained in the symmetry plane σ.

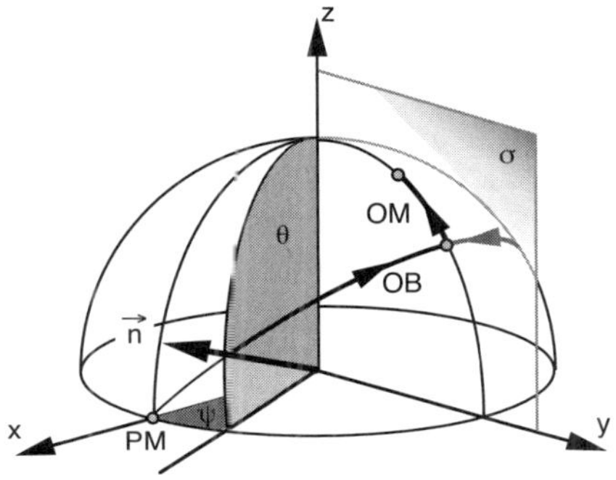

Fig. B.V.21 Evolution of the anchoring direction on a SiO layer with increasing evaporation angle. Planar monostable at first (PM point), the anchoring becomes oblique bistable (OB trajectory) and eventually oblique monostable (OM segment).

Based on the discussion in the previous section, this anchoring evolution can be interpreted in terms of the changes in surface potential $W(\theta,\psi)$ as a function of the evaporation angle ε. The simplest conceivable approach is to develop the potential $W(\theta,\psi)$ in a series of orthogonal functions with ε-depending coefficients. The real spherical harmonics $Y_l^m(\theta,\psi)$ are very convenient, as the potential W depends on the angular variables θ and ψ. As this function is invariant under **n** reversal, odd l harmonics must be discarded. For even l harmonics, invariance under reflection in the mirror plane $\sigma = (y,z)$ requires that certain values of m also be eliminated. The simplest development with the required features is:

$$W(\theta,\psi) = -X\sin^2(\theta)\cos(2\psi) - Y\sin^4(\theta)\cos(4\psi) - Z\sin(2\theta)\sin(\psi) \qquad \text{(B.V.8a)}$$

The shape of this potential depends on the three coefficients X, Y, and Z. When only interested in the existence of potential minima and their relative depth, one can introduce, without loss of generality, the normalization condition:

$$X^2 + Y^2 + Z^2 = 1 \qquad \text{(B.V.8b)}$$

reducing the number of independent coefficients to only two (X and Y in the following) and remain in the northern hemisphere $Z > 0$.

The study of $W(\theta,\psi)$ as a function of these two coefficients leads to the "anchoring diagram" in figure B.V.22.

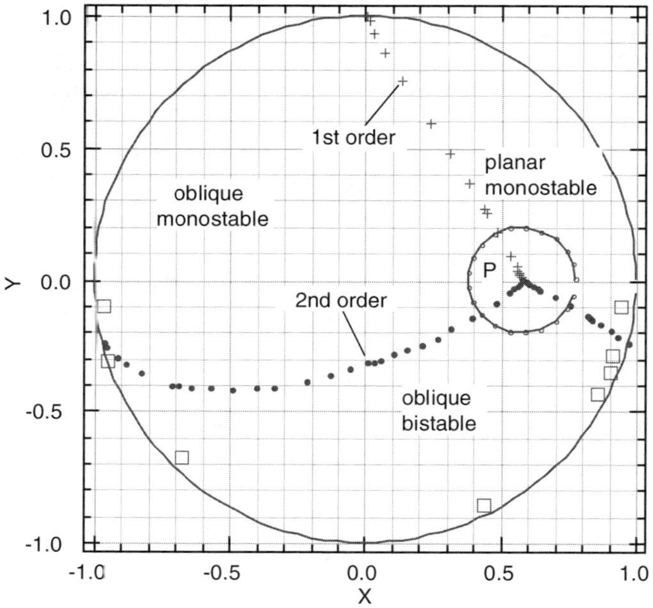

Fig. B.V.22 Theoretical anchoring diagram determined from the spherical harmonics development of the surface potential W (equation B.V.8a). The calculated potential shape in some points of the circular trajectory enclosing the bicritical point P is shown in figure B.V.23. The open squares mark the experimental trajectory corresponding to increasing evaporation angle.

The (X, Y) plane is divided into three domains corresponding to the three types of anchoring: planar monostable, oblique bistable, and oblique monostable. Two kinds of transitions can take place between these three anchoring conditions. Thus, the transition between planar and oblique monostable anchoring is necessarily discontinuous (first order), involving a discontinuous (90°) change in the anchoring direction. Conversely, transitions between oblique bistable and a monostable anchoring are continuous within this model. The evolution of the $W(\theta,\psi)$ potential at these transitions is illustrated in figure B.V.23 by a series of "potential maps," each of which corresponds to one point on the circular trajectory around point P with coordinates (0.577,0). This point, where a first-order transition line separates into two second-order lines, is called **bicritical** in phase transition terminology. At this point anchoring is degenerate, the $W(\theta,\psi)$ potential being minimal (in the sphere representation) along a great circle tilted with respect to the substrate plane (x, y) and orthogonal to the symmetry plane σ. On a map of potential $W(\theta,\psi)$ at P, these minima lie at the bottom of a constant depth valley. In a point close to P, this valley is still visible, but its depth is no longer constant (Fig. B.V.23).

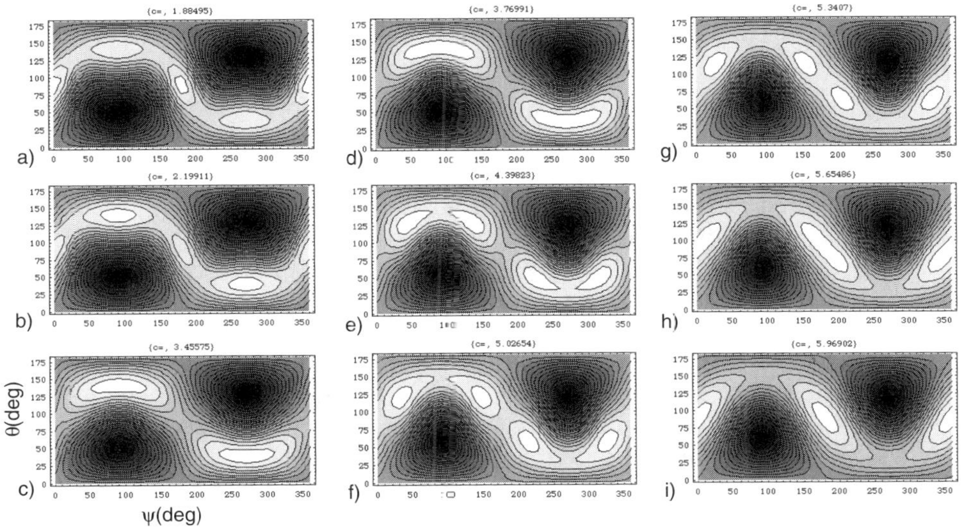

Fig. B.V.23 Evolution of the $W(\theta,\psi)$ potential during anchoring transitions. a, b, c) First-order transition between planar (a) and oblique (c) monostable anchoring; d, e, f) second-order transition between oblique monostable (d) and oblique bistable (e, f) anchoring; g, h, i) second-order transition between oblique bistable (g) and planar monostable (i) anchoring. Lighter areas correspond to potential minima.

In order to establish the correspondence between this simple theoretical model and experimental reality, one must find the relation between the X, Y, Z coefficients in development B.V.8 and the experimentally controlled parameters. For SiO layers, anchoring transitions are a function of the evaporation parameters: incidence angle ε and layer thickness d. The three coefficients X, Y, Z thus depend on ε and d:

$$X = X(\varepsilon, d)$$

$$Y = Y(\varepsilon, d) \qquad \text{(B.V.9)}$$

$$Z = Z(\varepsilon, d)$$

These equations represent an application A between the experimental parameter plane and the theoretical anchoring diagram of figure B.V.22. At constant d and with ε varying from 0 and $\pi/2$, this application gives the trajectory marked by open squares (□) in figure B.V.22, beginning at $\varepsilon = 0$ in the domain of planar monostable anchoring, ending at $\varepsilon = \pi/2$ in the domain of oblique monostable anchoring and passing through the domain of oblique bistable anchoring for $56° < \varepsilon < 63°$. When in the oblique bistable domain, the trajectory is in agreement with the relation between the experimentally measured angles θ and ψ.

V.5.b) Anchoring transitions on muscovite mica

From a theoretical viewpoint, the previously mentioned anchoring transitions would be completely similar to phase transitions if only one could control the surface potential $W(\theta,\psi)$ by changing the intensive thermodynamic parameters such as the temperature T, pressure P, or chemical potentials μ. Unfortunately, anchoring on SiO layers is hardly sensitive to these parameters.

On the other hand, anchoring on cleaved surfaces of crystals such as mica or gypsum turns out to be very sensitive to humidity or, more generally, to the chemical potentials of water, alcohols, or diols when they are dissolved in the nematic phase [14]. A practical way of controlling the chemical potential of these substances is to put the nematic phase in contact with a gaseous mixture of precisely controlled composition (Fig. B.V.24).

The anchoring of the E9 nematic mixture on muscovite mica is planar, but the azimuthal anchoring angle ψ depends strongly on the chemical potentials of water and ethylene glycol (EG). These chemical potentials are controlled experimentally by adjusting the composition of a $N_2/H_2O/EG$ gaseous mixture in contact with the nematic phase. A given composition is obtained by mixing three gas flows in appropriate proportions. The first one, Φ_A, contains pure nitrogen; the second, Φ_W, is water-saturated nitrogen; and the third, Φ_{EG}, is nitrogen saturated in ethylene glycol.

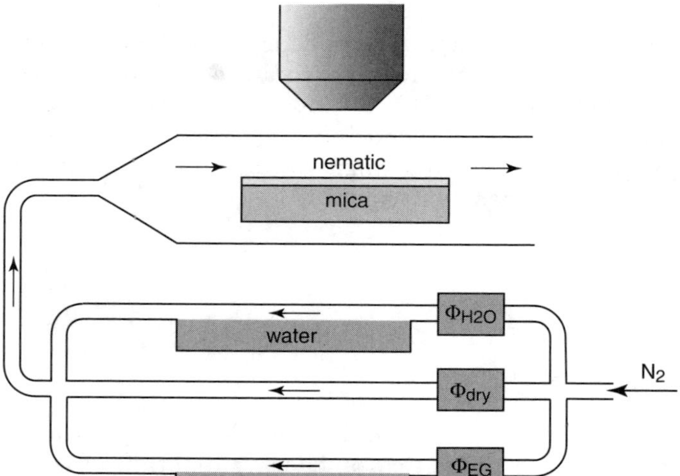

Fig. B.V.24 Experimental setup for studying water and/or ethylene glycol-induced anchoring transitions at the E9/muscovite mica interface. The nematic is spread in a thin layer onto the mica cleavage surface. Water and ethylene glycol vapor circulate in the nitrogen flow above the nematic. Partial pressures in the gas mixture are controlled by means of three flowmeters regulating the three flows: dry nitrogen, water-saturated nitrogen, and ethylene glycol-saturated nitrogen.

Measurements of the azimuthal anchoring angle as a function of the position in the ternary mixture diagram (Fig. B.V.25) reveal the presence of three types of anchoring. In domain I, anchoring is orthogonal to the symmetry plane σ, and therefore monostable. In domain II, anchoring is also monostable, but parallel to the σ plane. Finally, in domain III, anchoring is oblique with respect to the σ plane, and hence bistable. By changing the composition of the gas flow one can easily induce one of the three possible phase transitions we shall now describe.

The I→II transition, taking place between two types of anchoring with the same symmetry $G_a = C_{2v}$, is necessarily first order. As the director turns discontinuously by $\pm\pi/2$ at the substrate surface, this transition leads to the appearance of surface walls. Inside the walls, the azimuthal anchoring angle ψ varies continuously between the two values ψ_I and $\psi_{II} = \psi_I \pm \pi/2$. One speaks of a $+\pi/2$ wall when the director turns clockwise (upon crossing the wall from the inside to the outside of a closed domain), and of a $-\pi/2$ wall otherwise. The two photos in figure B.V.26, taken a few seconds apart, demonstrate the I→II transition. In agreement with the theoretical predictions, two sorts of elliptical walls appear, differing in shape and evolution. Thus, when two walls of the same type (say $+\pi/2$) touch, they coalesce. Conversely, when the walls have opposite sign, they stick together yielding a π wall.

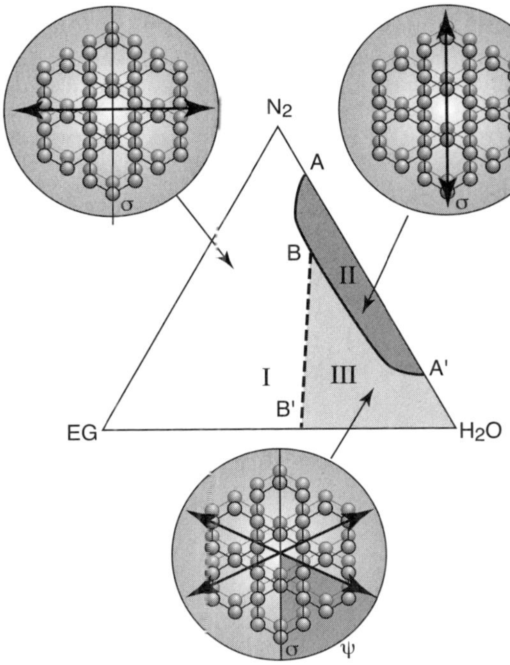

Fig. B.V.25 Anchoring diagram of the E9/muscovite mica interface in the presence of water and ethylene glycol vapor.

The II→III transition is also first order, although it results in symmetry breaking. It is visible in the figure B.V.28 photo, where an m = 1/2 disclination splits into three walls, each one separating two domains with opposite sign oblique anchoring. In fact, the bulk disclination first turns into a π surface wall (itself unstable) which separates into three walls. This phenomenon is interesting from the viewpoint of phase transitions, showing that a first-order transition can occur without nucleation. Indeed, when a π wall is present in the system, it already bears the seed of all possible anchoring transitions. At the transition, it only needs dissociate in the suitable way to engender the new anchoring directions.

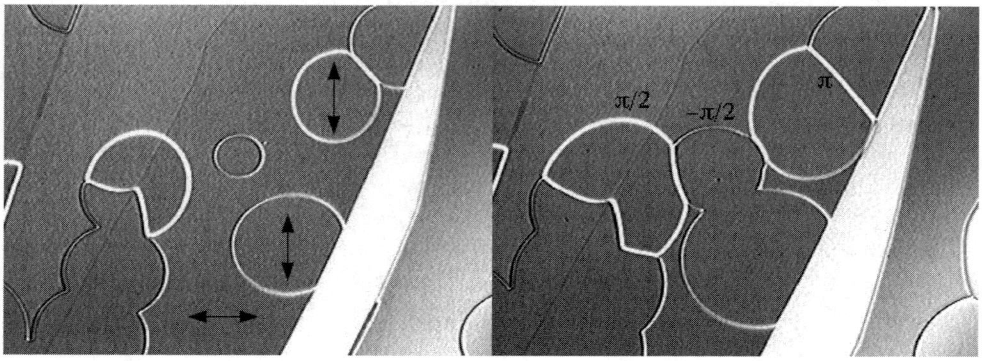

Fig. B.V.26 The I→II anchoring transition (Fig. B.V.25) between two anchoring conditions: parallel and orthogonal to the symmetry plane σ. This first-order transition takes place by domain nucleation and wall movement. The walls are of two types: +π/2 and −π/2. When in contact, two walls of the same type coalesce while two walls of different sign stick together yielding a π wall.

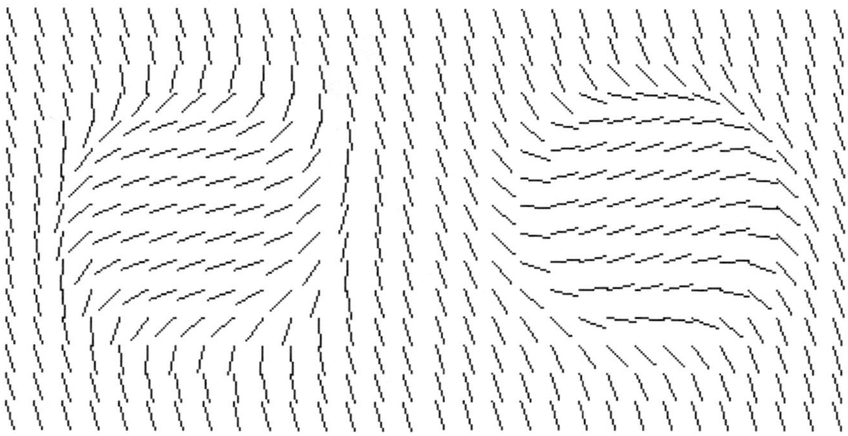

Fig. B.V.27 Structure of the surface walls +π/2 and −π/2 appearing at the I→II anchoring transition. These walls are visible in the figure B.V.26 photos.

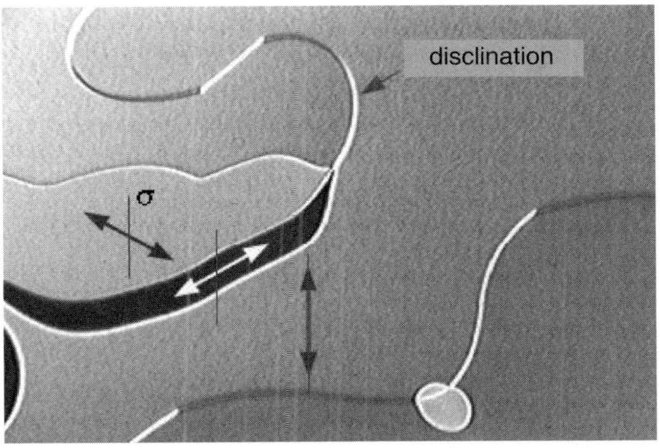

Fig. B.V.28 Transition between monostable anchoring parallel to the symmetry plane (II in figure B.V.25) and bistable anchoring oblique with respect to the symmetry plane (III in figure B.V.25).

The I→III transition is second order, having as order parameter the angle $\Psi = \pi/2 - \psi$ between the oblique anchoring directions and the symmetry plane. As illustrated in figure B.V.29, this angle increases as the square root of the difference $\Delta\Phi$ between the composition of the ternary mixture and the critical composition Φ_{crit}.

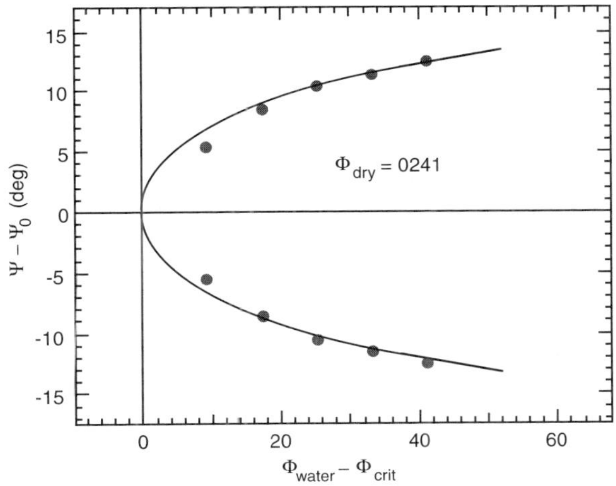

Fig. B.V.29 Variation in the anchoring direction with the gaseous mixture composition. Dry nitrogen flow is maintained constant while changing the proportions between the water- and ethylene glycol-saturated flows. In the ternary diagram of figure B.V.25, this corresponds to a horizontal trajectory across the transition line BB'.

In addition to the continuous variation in order parameter, this anchoring transition exhibits all the features of a second-order phase transition. In particular, strong orientational fluctuations – **critical fluctuations** – can be observed using polarizing microscopy very close to the transition line (Fig. B.V.30). These spontaneous thermal fluctuations are the sign of the diverging susceptibility related to the variations in order parameter. In other words, the anchoring becomes very "soft" and easily perturbed.

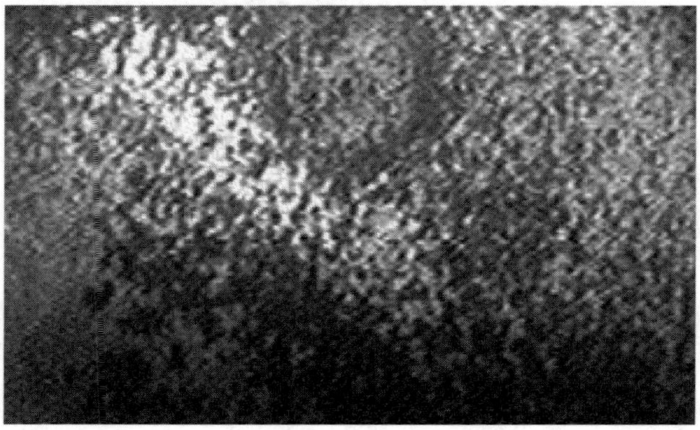

Fig. B.V.30 Transition between the monostable planar anchoring perpendicular to the plane of symmetry (I on Fig. B.V.25) and the oblique bistable anchoring (III). The second-order character of this transition manifests itself by the existence of very strong fluctuations of the anchoring direction.

Close to the critical line (BB' in the diagram of figure B.V.25), it is extremely easy to influence the choice the interface must make when entering domain III in the anchoring diagram. For instance, a slight deformation of the mica surface is enough to favor one anchoring direction over the other. This effect has been demonstrated in a very simple experiment, described in figure B.V.31 [15]. It consists of placing a tiny glass bead (≈ 20 μm in diameter) between a mica sheet and a solid transparent substrate. By evacuating the air between the mica sheet and the substrate, the sheet adheres to the substrate following as close as possible the shape of the bead. One can easily prove that the resulting strain tensor ε_{ij} at the mica surface has an elongational component (the symmetrical and traceless part of the tensor $\underline{\varepsilon}$) which can break the mirror symmetry of the mica surface. In polar coordinates (r,ϕ) the amplitude $\varepsilon(r)$ of this tensor decreases as $1/r^2$. Locally, its principal axes are, by symmetry, radial and orthoradial, respectively. In cartesian coordinates (x,y) with the x axis parallel to σ, the tensor can be expressed as:

$$\underline{\varepsilon} = \varepsilon(r)\cos(2\phi)\begin{pmatrix} 1 & 0 \\ 0 & -1 \end{pmatrix} + \varepsilon(r)\sin(2\phi)\begin{pmatrix} 0 & 1 \\ 1 & 0 \end{pmatrix} \quad \text{with} \quad \varepsilon(r) = \frac{\alpha}{r^2} \qquad \text{(B.V.10)}$$

It is obvious that for $\phi = 0, \pi/2, \pi$ and $3\pi/2$, one of the eigenaxes of tensor ε_{ij} is parallel to the mirror plane: the symmetry of the mica surface is then preserved along these axes. Everywhere else, the σ symmetry is broken and one of the two anchoring orientations is thus favored.

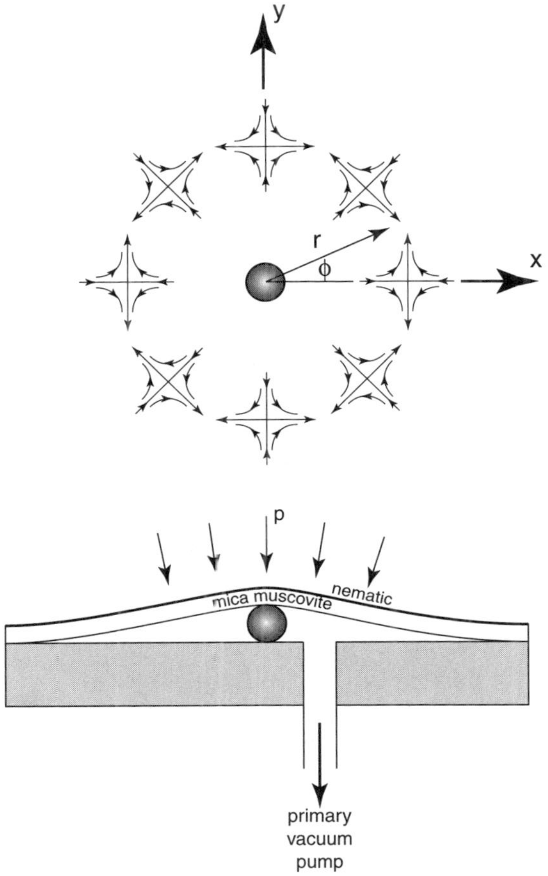

Fig. B.V.31 Deformation of a mica muscovite sheet caused by a tiny glass bead.

Fig. B.V.32 The "butterfly texture" around the glass bead shows the influence of surface deformation on a second-order anchoring transition.

The outcome is a "butterfly texture," centered on the position of the bead and consisting of two pairs of domains with different signs of the order parameter $\Delta\Psi$ (Fig. B.V.32).

Let us now return to the general anchoring diagram in figure B.V.25. The diagram can be easily interpreted by developing the anchoring energy $W(\Psi)$ in a Fourier series ("Landau model"):

$$W(\Psi) = X\cos(2\Psi) + Y\cos(4\Psi) + Z\cos(6\Psi) + \Sigma\sin(2\Psi) + \ldots \qquad \text{(B.V.11)}$$

Without deformation, $\Sigma = 0$ as a result of mirror symmetry. Studying the minima of this expression as a function of X, Y, and Z (with the normalization condition $X^2 + Y^2 + Z^2 = 1$) leads to the anchoring diagram in figure B.V.33.

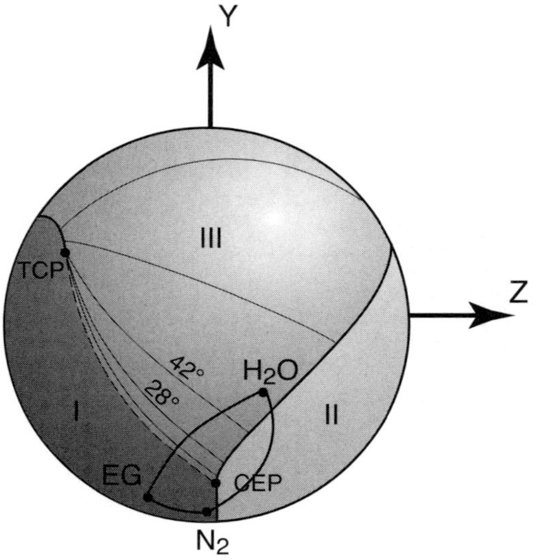

Fig. B.V.33 Theoretical anchoring diagram obtained from development B.V.11. The sphere is divided into three domains by first- or second-order transition lines. TCP is the tricritical point where the II→III transition changes order. The second-order line beginning at TCP ends on a first-order line at the critical point CEP. The variation in anchoring angle Ψ in the domain III corresponding to bistable oblique anchoring is indicated by several level lines $\Psi(X,Y,Z) = 0°, 14°, 28°, 42°, \ldots$

The sphere $X^2 + Y^2 + Z^2 = 1$ is divided into three domains corresponding to the three experimentally observed anchoring conditions. The I→II transition is first order in this model, in agreement with experiment. However, transitions I→III and II→III can be either first or second order as the boundaries between the III domain and the two others contain tricritical points where the transition order changes.

In order to establish the relation between this theoretical diagram and experiment, we have represented on the sphere $X^2 + Y^2 + Z^2 = 1$ the spherical triangle N_2-H_2O-EG reproducing the topology of the experimental anchoring diagram in figure B.V.25. The experiment can be seen to cover only a small part of the theoretical phase diagram.

In conclusion, let us explain the effect of a mechanical deformation of the mica surface. In the presence of a deformation, the surface potential W(Ψ) is perturbed, the development coefficients in B.V.11 depending on the spatial coordinates (r, ϕ):

$$X = X_c + a\, S_1(r,\phi)$$

$$Y = Y_o + b\, S_1(r,\phi) \qquad \text{(B.V.12)}$$

$$\Sigma = c\,S(r,\phi) = c\,S(r)\sin(2\phi)$$

The most important effect of the perturbation is to break the mirror symmetry by the Σ term which changes sign four times for a complete tour of the bead, explaining the experimentally observed quadrant texture.

V.5.c) Anchoring transitions on gypsum

Anchoring transitions on the cleavage surface of gypsum were studied using the same setup as for mica (Fig. B.V.24).

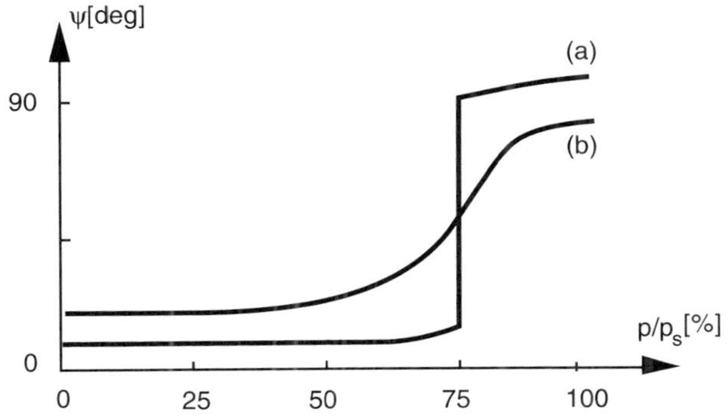

Fig. B.V.34 Anchoring at the E9/gypsum interface. a) First-order transition induced by the presence of water vapor; b) continuous variation of the anchoring angle above the critical diethylene glycol concentration.

The C_2 symmetry of the gypsum surface (and the absence of mirror symmetry) impose monostable planar anchoring. As the C_2 symmetry is preserved irrespective of the planar molecular orientation at the surface, the angle ψ between the anchoring direction and a particular crystallographic direction (e.g., the **c** axis of the $CaSO_4$ rows) can vary with the gaseous mixture composition. The presence of water vapor in the flow induces a variation in the anchoring angle ψ represented in figure B.V.34.

The variation presents a jump of almost 90° when the partial pressure p_w of water vapor reaches 75% of the saturated vapor pressure p_s. This first-order anchoring transition, visible with polarizing microscopy, is similar to the one on mica aside from the fact that the walls around the new phase have a less regular contour. Anchoring at the E9/gypsum interface is also sensitive to the presence of other substances, such as diols. In particular, adding diethylene glycol vapor in the gas flow reduces the Δψ discontinuity and ultimately eliminates it: this point where the transition disappears is the **critical point** C in the ternary phase diagram of figure B.V.35.

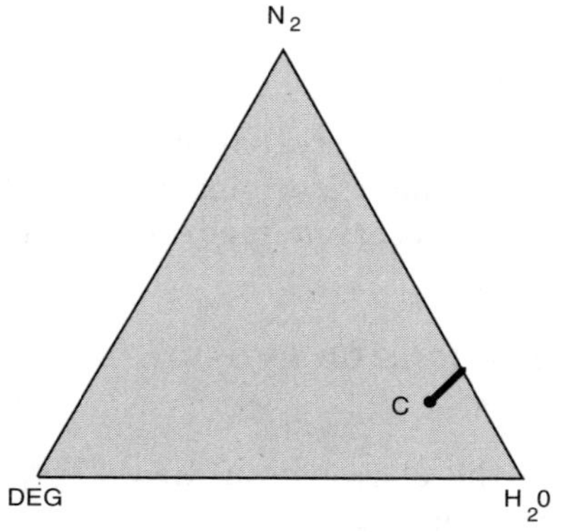

Fig. B.V.35 Anchoring diagram of the E9/gypsum interface in the presence of water vapor and diethylene glycol.

This last example emphasizes the analogy between anchoring transitions and classical phase transitions.

V.6 Measuring the anchoring energy in the homeotropic case

Thus far, we have only described anchoring and the related transitions in a phenomenological manner, based upon symmetry arguments only. Unfortunately, such considerations cannot provide information on the anchoring force, depending mostly on the chemistry of the interacting substrate and liquid crystal. Nevertheless, this problem is of considerable practical import, as the anchoring force plays an essential role in applications. The lack of theoretical predictions increases the necessity to measure this quantity experimentally. We shall illustrate this discussion with the simple example of the Frederiks transition between two plates treated in homeotropic anchoring.

It was shown in chapter B.II that the director field becomes unstable above a certain critical field B_c given by:

$$B_c^\infty = \frac{\pi}{d}\sqrt{\frac{\mu_o K_3}{\chi_a}} \tag{B.V.13}$$

where d is the sample thickness, K_3 the bend constant, and χ_a the diamagnetic susceptibility anisotropy. This expression is obtained assuming that the anchoring remains homeotropic whatever the value of the applied field. To put it differently, the anchoring energy is assumed infinite, or at least large enough so that the molecules do not tilt on the plates.

In this section we shall determine the limits of this "geometrical" approximation. For finite anchoring energy, the experimentally measured critical field B_c is of course less than its theoretical limit B_c^∞. In order to calculate it, one needs an explicit form of the anchoring energy. For small angle θ, the most current form is that of Rapini and Papoular [16]:

$$W(\theta) = \frac{1}{2} W_a \,\theta^2 \tag{B.V.14}$$

The critical field is obtained by minimizing the total system energy, including the surface energy. With the notations of figure B.II.14b, the total energy is:

$$F = \frac{1}{2}\int_{-d/2}^{d/2}\left[K_3\left(\frac{d\theta}{dz}\right)^2 - \frac{\chi_a}{\mu_o}B^2\sin^2\theta\right]dz + 2W(\theta_s) \tag{B.V.15}$$

In the previous formula, $\theta(z)$ is the angle between the director and the z axis and θ_s, the anchoring angle on the plates. For this particular geometry the surface K_4 term need not be considered, which simplifies the discussion. The distortion being symmetric at the threshold, the solution can be looked for in the form:

$$\theta(-z) = \theta(z) \qquad \text{with} \qquad \left(\frac{d\theta}{dz}\right)_{z=d/2} = -\left(\frac{d\theta}{dz}\right)_{z=-d/2} \tag{B.V.16}$$

Minimization yields:

$$K_3\frac{d^2\theta}{dz^2} + \frac{\chi_a}{\mu_o}B^2\sin\theta\cos\theta = 0 \qquad \text{in the bulk} \tag{B.V.17}$$

$$K_3\frac{d\theta}{dz} + \frac{dW}{d\theta} = 0 \qquad \text{in } z = +\frac{d}{2} \tag{B.V.18a}$$

$$-K_3\frac{d\theta}{dz} + \frac{dW}{d\theta} = 0 \qquad \text{in } z = -\frac{d}{2} \tag{B.V.18a}$$

The first equation expresses the equilibrium of elastic and magnetic torques in the bulk. The last two state the equilibrium of surface torques at the walls. Note that the substrate acts upon the director with a restoring torque proportional to the anchoring force W_a. Rapini and Popoular solved these equations, showing that the critical field is given by the following implicit equation [16]:

$$\cot\left(\frac{\pi}{2}\frac{B_c}{B_c^\infty}\right) = \frac{\pi L_{eff}}{d}\frac{B_c}{B_c^\infty} \tag{B.V.19}$$

where the new length L_{eff} is given by:

$$L_{eff} = \frac{K_3}{W} \tag{B.V.20}$$

and can be considered as an "effective anchoring penetration" length (Fig. B.V.36).

These formulae yield the difference between the critical field B_c and its asymptotic value B_c^∞. Two cases must be distinguished:

Either $d \gg L_{eff}$: in this "strong anchoring" limit, the critical field is close to its asymptotic value. To first order in L_{eff}/d:

$$B_c = B_c^\infty\left(1 - \frac{2L_{eff}}{d}\right) \tag{B.V.21}$$

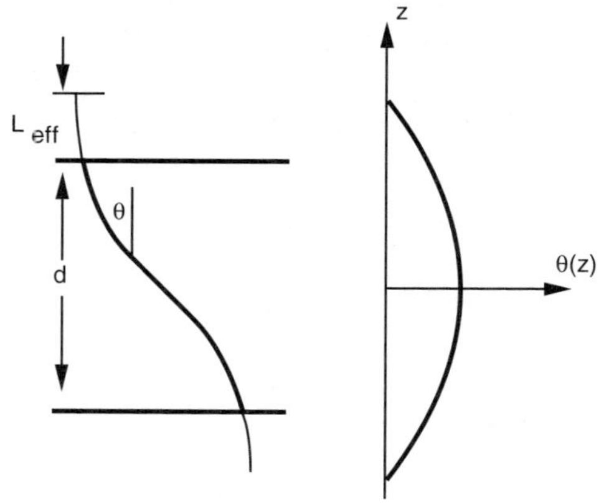

Fig. B.V.36 Graphical representation of the penetration length.

Or $d \approx L_{eff}$: the anchoring is "weak" and the correction is important.

Thus, the "anchoring force" depends on the experimental conditions, in our case the sample thickness. Figure B.V.37 represents the dimensionless critical field B_c/B_c^{∞} as a function of the ratio d/L_{eff}. This curve shows that one only needs to measure the critical field as a function of the sample thickness in order to obtain the anchoring energy and the K_3/χ_a ratio. Instead of using several samples of different thickness, one can also use a wedge-like sample. This method, recently introduced by Gu, Uran, and Rosenblatt [17] is very convenient: only the wedge angle counts, and not the absolute sample thickness.

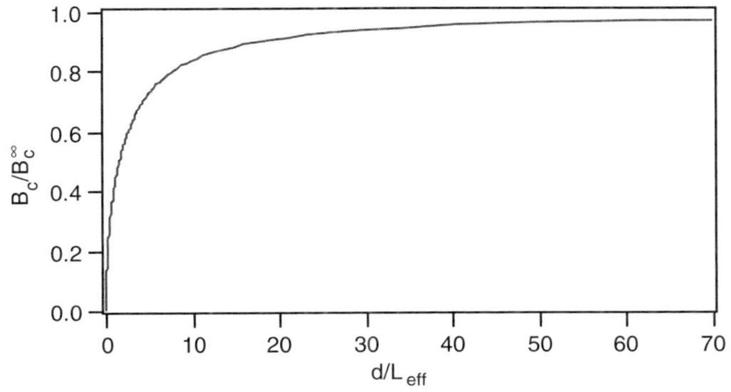

Fig. B.V.37 Critical field normalized by its "infinite anchoring" value vs. the ratio between sample thickness and anchoring penetration length.

This experiment shows that the anchoring energy typically varies between 10^{-4} and 10^{-1} erg/cm^2 depending on the applied surface treatment (silanes, lecithin, etc.). The corresponding anchoring penetration length ranges from about 100 Å (strong anchoring) to several micrometers (weak anchoring).

Measurements of the anchoring energy in planar orientation on SiO layers evaporated under oblique incidence are reported in refs. [18,19].

BIBLIOGRAPHY

[1] Cognard J., *Mol. Cryst. Liq. Cryst.*, **78** Suppl.1 (1982) 1.

[2] Jérôme B., *Rep. Prog. Phys.*, **54** (1991) 391.

[3] Sonin A.A., *The Surface Physics of Liquid Crystals*, Gordon and Breach, Luxembourg, 1995.

[4] a) Yokoyama H., *Mol. Cryst. Liq. Cryst.*, **165** (1988) 265.
b) Yokoyama H., in *Handbook of Liquid Crystal Research*, Eds. Collings P.J. and Patel J.S., Oxford University Press, Oxford, 1997, p. 179.

[5] Mauguin Ch., *Bull. Soc. Franç. Cryst.*, **34** (1911) 71.

[6] Grandjean F., *Bull. Soc. Franç. Cryst.*, **39** (1916) 164.

[7] Friedel G., "Etats mésomorphes de la matière," *Annales de Physique*, **18** (1922) 273.

[8] Châtelain P., *Bull. Soc. Franç. Minéral.*, **66** (1943) 105.

[9] Janning J.L., *Appl. Phys. Lett.*, **21** (1972) 173.

[10] Guyon E., Pieranski P., Boix M., *Lett. Appl. Eng. Sci.*, **1** (1973) 19.

[11] Monkade M., Boix M., Durand G., *Europhys. Lett.*, **5** (1988) 697.

[12] Jérôme B., Pieranski P., Boix M., *Europhys. Lett.*, **5** (1988) 693.

[13] Jérôme B., Pieranski P., *J. Physique (France)*, **49** (1988) 1601.

[14] a) Bechhoefer J., Jérôme B., Pieranski P., *Phys. Rev. A*, **41** (1990) 3187.
b) Bechhoefer J., Duvail J.-L., Masson L., Jérôme B., Hornreich R.M., Pieranski P., *Phys. Rev. Lett.*, **64** (1990) 1911.

[15] Kitzerow H.-S., Jérôme B., Pieranski P., *Physica A*, **174** (1991) 163.

[16] Rapini A., Papoular M., *J. Physique (France)*, **30** Coll. C4 (1969) 54.

[17] Dong-Feng Gu, Uran S., Rosenblatt Ch., *Liq. Cryst.*, **19** (1995) 427.

[18] Faetti S., Nobili M., Schirone A., *Liq. Cryst.*, **10** (1991) 95.

[19] Yokoyama H., Kobayashi S., Kamei H., *J. Appl. Phys.*, **61** (1987) 4501.

Chapter B.VI

The nematic-isotropic liquid interface: static properties and directional growth instabilities

After having studied nematic anchoring on solid substrates, we shall now study the interface between the nematic phase and its melted isotropic liquid. Because the material is the same on both sides of the interface, the two phases can be described by the same order parameter, the quadrupolar moment S_{ij}. This parameter describes symmetry breaking at the transition. Unlike density or impurity concentration, it is not conserved; its equilibrium profile across the interface will therefore be easily determined. Nevertheless, complications appear when submitting the interface to chemical or mechanical constraints, as it can deform or become unstable. In addition to the well-known Mullins-Sekerka instability, developing when the nematic phase grows into its isotropic liquid, instabilities of elastic origin can appear when the nematic is confined to a restricted volume.

These diverse phenomena are the scope of this chapter. We start with an overview of the techniques employed to characterize the interface and its static properties such as the director anchoring angle, interface surface tension, and width (section VI.1). The interface structure will then be analyzed in terms of the "Landau-Ginzburg-de Gennes" model. Although it yields the interface intrinsic width as well as its energy, this phenomenological model must be improved to describe oblique molecular anchoring. For example, it must take into account ordo-electricity, analogous to the flexo-electricity already presented in chapter B.I (section VI.2). We then show that the interface is unstable in confined geometry as a result of elastic torques, and we illustrate it by describing umbilic formation at the interface (section VI.3). The Gibbs-Thomson relation will be generalized by including elastic corrections, allowing us to consider the problem of the equilibrium configurations of a nematic domain in equilibrium with its isotropic liquid. In particular, we shall insist on the "boojum" configuration encountered in thin smectic C* films and in certain Langmuir layers behaving as two-dimensional nematics (section VI.4). In conclusion, we discuss interface destabilization in directional growth (Mullins-Sekerka instability) and the secondary instabilities of the cellular front in the strongly nonlinear regime. We shall also show that a disclination line can become depinned from the growth front under certain experimental conditions to be specified (section VI.5).

VI.1 Anchoring angle, surface tension, and width of the nematic-isotropic interface

The first experiment demonstrating oblique molecular anchoring at the nematic-isotropic interface was performed by Vilanove and coworkers [1]. Their experimental setup is shown in figure B.VI.1.

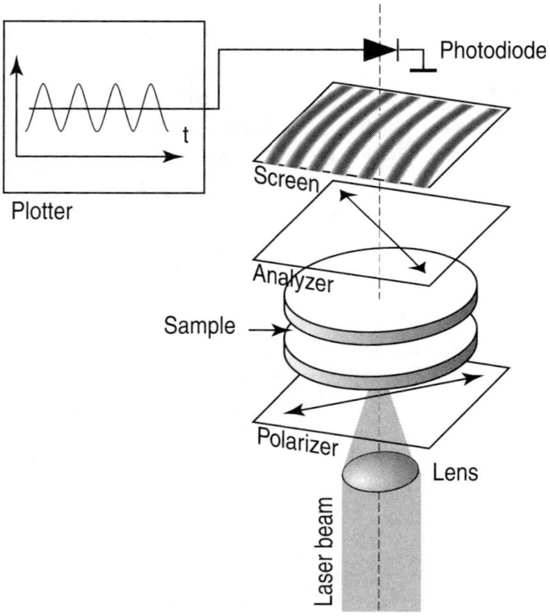

Fig. B.VI.1 Experimental setup for measuring the anchoring angle at the nematic-isotropic liquid interface.

It consists of a planar MBBA sample (of thickness d = 500 μm), oriented parallel to the x axis and sandwiched between two parallel glass plates, each one independently regulated in temperature. The sample is observed in conoscopy, with the polarizers at 45° from the direction of molecular alignment (x axis). A photodiode placed at the center of the conoscopic figure measures the local light intensity.

First, the sample is completely nematic, and its conoscopic figure is made up of two families of hyperbolae, centered on the z axis.

The upper side of the sample is then progressively heated until the first isotropic droplets are obtained. A thin isotropic layer then appears between the nematic phase and the upper plate. Experiment shows that the conoscopic figure abruptly shifts along the x axis (either to the left or to the right) when the nematic-isotropic liquid interface forms. This sudden shift shows that the director acquires a tilt angle $\pm \theta_a \neq 0$ at the new interface interface (Fig. B.VI.2). The tilt angle can be determined by measuring using the photodiode variation Δp of the interference order at the center of the conoscopic figure. A simple calculation gives:

$$\Delta p \approx \frac{1}{\lambda} \int_0^d \Delta n \sin^2\theta(z)\, dz \qquad \text{(B.VI.1)}$$

with Δn the average birefringence of the nematic phase in the temperature range under consideration and $\theta(z)$ the angle between the director and the x axis at height z. The formula is only valid under the (experimentally verified) condition $\Delta n \sin^2\theta << 1$. As $\theta = \theta_a z/d$ in the nematic layer, the integral in equation B.VI.1 can be shown to yield:

$$\frac{\sin(2\theta_a)}{2\theta_a} = 1 - \frac{2\Delta p\,\lambda}{\Delta n\, d} \qquad \text{(B.VI.2)}$$

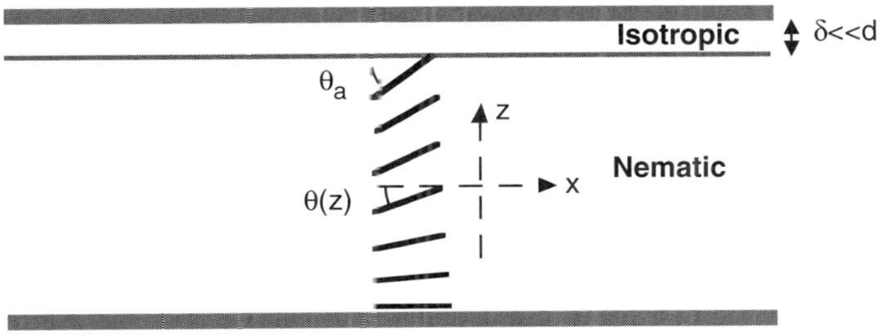

Fig. B.VI.2 Director configuration in the presence of a thin isotropic layer.

Experimentally, $\Delta p \approx 6$ in an MBBA sample of thickness $d \approx 500$ µm, so $\theta_a \approx 23°$ for $\lambda/\Delta n = 4.3$ µm. This angle is very close to the equilibrium value, as the elastic torque on the director at the interface, exerted by the nematic because of director field deformation, is extremely weak in a 500 µm sample.

Let us now see how the surface tension is determined.

One possible method is the sessile drop. First employed by R. Williams [2] in nematics, it consists of measuring the equilibrium height h_{eq} of a nematic drop placed on a horizontal substrate immersed in the isotropic liquid (Fig. B.VI.3). One can show [3] that the height of a large enough drop depends on the gravity g, the density difference $\Delta\rho$ between the two phases, the contact angle φ between drop surface and substrate, and the surface tension to be determined. Two situations can arise:

– either $\varphi < \pi/2$; the drop has the form of a flattened spherical cap, with total height:

$$h_{eq} = 2\cos\frac{\phi}{2}\sqrt{\frac{\gamma}{g\,\Delta\rho}} \qquad \text{(B.VI.3a)}$$

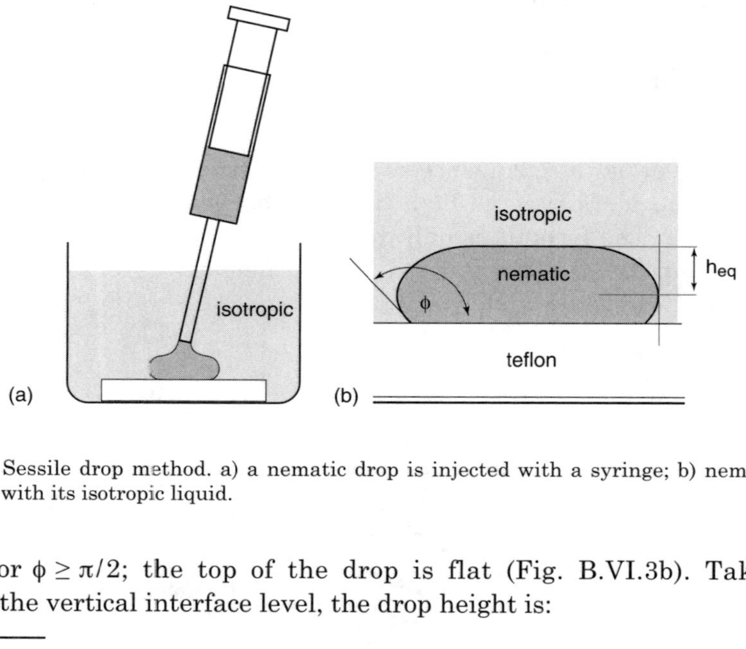

Fig. B.VI.3 Sessile drop method. a) a nematic drop is injected with a syringe; b) nematic drop in equilibrium with its isotropic liquid.

– or $\phi \geq \pi/2$; the top of the drop is flat (Fig. B.VI.3b). Taking as a reference the vertical interface level, the drop height is:

$$h_{eq} = \sqrt{\frac{2\gamma}{g\,\Delta\rho}} \qquad \text{(B.VI.3b)}$$

irrespective of the wetting angle ϕ.

This second case is experimentally convenient, since there is no need to measure the contact angle. With nematics, it can be achieved by placing the drop on a teflon plate, itself situated on the bottom of a transparent container filled with the isotropic liquid, with the entire setup maintained at a temperature slightly higher than the transition temperature T_c by about 0.1°C typically. Once injected, the nematic drop (typical volume 0.25 cm^3) falls to the bottom of the container and spreads on the teflon substrate, which it wets only partially. Although the drop slowly melts (thermochemical equilibrium takes a long time to reach), its height is stable after about 1 min, showing that mechanical equilibrium was achieved. For MBBA, this height is $h_{eq} \approx 1.45$ mm, so, from formula B.VI.3b and using $\Delta\rho \approx 1.9\times10^{-3}$ g/cm^3 [4] one has:

γ (MBBA) $\approx 1.9\times10^{-2}$ erg/cm^2

Similar values (between 10^{-2} and 2×10^{-2} erg/cm^2) were obtained for the cyanobiphenyls of the nCB series [5].

However, this method gives no information as to the interface width or the director anchoring energy at the interface. More elaborate techniques must therefore be employed, such as measuring interface reflectivity in polarized light under normal (or quasi-normal) incidence [6]. This parameter depends on the molecular orientation at the interface and on its "effective width," a concept to be discussed in detail, but which can be roughly defined as the typical distance over which the refractive index of the medium goes from n_{nem} to n_{iso}.

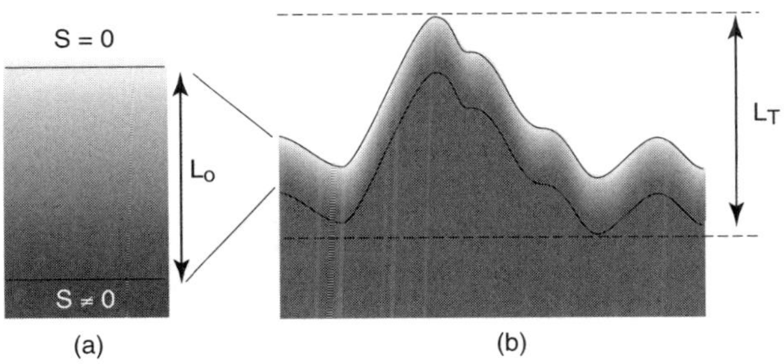

Fig. B.VI.4 Schematic representation of the two effects reducing the intensity of reflected light: a) Intrinsic interface profile; b) diffuse profile due to thermal fluctuations.

Without thermal fluctuations, this width would be just the distance L_o over which the quadrupolar order parameter goes to zero (Fig. B.VI.4a). In the next section we shall determine this intrinsic width as a function of the Landau-Ginzburg parameters. In fact, the interface fluctuates due to thermally excited capillary waves (Fig. B.VI.4b). These waves, of strong amplitude for low surface tension, diffuse light, thus reducing the measured reflection coefficient.

Taking into account both these effects, Meunier and Langevin-Cruchon [7] proved that the nematic-isotropic interface behaves like a thin layer with an error function variation of the refractive index:

$$n(z) = \frac{1}{2}\left[n_{nem} + n_{iso} + (n_{iso} - n_{nem})\,\mathrm{erf}\left(\frac{z}{L_{eff}}\right)\right] \qquad \text{(B.VI.4)}$$

and effective width

$$L_{eff} \approx \sqrt{L_T^2 + \frac{\pi^2}{24} L_o^2} \qquad \text{(B.VI.5)}$$

where L_T is the "thermal width," given by

$$L_T^2 \approx \frac{k_B T}{2\pi\gamma} \ln\left[\frac{\lambda}{(n_{nem} + n_{iso})\alpha L_o}\right] \qquad \text{(B.VI.6)}$$

In this formula, α is the angular aperture of the measuring system (in the experiment to be described, this angle is fixed by the aperture diaphragm at $\alpha = 0.005$ rd).

The reflectivity is then given by the classical formula [8]:

$$R \approx \left[\frac{1}{n_{nem} + n_{iso}} \int_{-\infty}^{+\infty} \frac{\partial n(z)}{\partial z} \exp\left(\frac{4\pi i\, n(z)\, z}{\lambda}\right) dz\right]^2 \qquad \text{(B.VI.7)}$$

and yields in this case:

$$R \approx R_o \exp\left(-\frac{\pi^2 \bar{k}^2 L_{eff}^2}{12}\right) \qquad \text{(B.VI.8)}$$

with

$$\bar{k} = \pi \frac{n_{iso} + n_{nem}}{\lambda} \qquad \text{(B.VI.9a)}$$

and

$$R_o = \left(\frac{n_{iso} - n_{nem}}{n_{iso} + n_{nem}}\right)^2 \qquad \text{(B.VI.9b)}$$

Here, λ is the light wavelength (assumed to be much larger than L_{eff}), and R_o the theoretical reflectivity for an infinitely thin interface.

In conclusion, measuring the interface reflectivity gives access to its effective width L_{eff}. If the surface tension is known, relation B.VI.6 yields the thermal width L_T, so that the intrinsic interface width L_o can be obtained.

Moreover, the reflectivity depends on light polarization through n_{nem} when the director is tilted with respect to the interface normal. In principle, measuring R for two different polarizations would then yield the anchoring angle θ_a.

Finally, the interface anchoring energy can be determined by applying a magnetic field and measuring the variation in anchoring angle as a function of the applied field. We shall return to this measurement after describing the experimental setup.

The setup, represented in figure B.VI.5, consists of a 2 cm thick nematic cell (C), situated in a vertical temperature gradient and a horizontal magnetic field allowing one to impose planar molecular orientation. A laser beam polarized at 45° from the neutral lines of the sample (parallel and perpendicular to the field **B**) arrives at quasi-normal incidence on the nematic-isotropic interface. The reflected beam crosses first a diaphragm (D) of angular aperture 0.005 rd, then a Wollaston prism (W) which separates the two components polarized in the directions parallel and perpendicular, respectively, to the magnetic field **B**. The intensities $I_{//}$ and $I_{\perp}$ of these two beams are then measured using two photodiodes (PH2) and (PH3). The intensity of the incident beam is also measured using photodiode (PH1). When rigorously calibrated [6b], this setup allows measurement of the reflectivities $R_{//}$ and $R_{\perp}$.

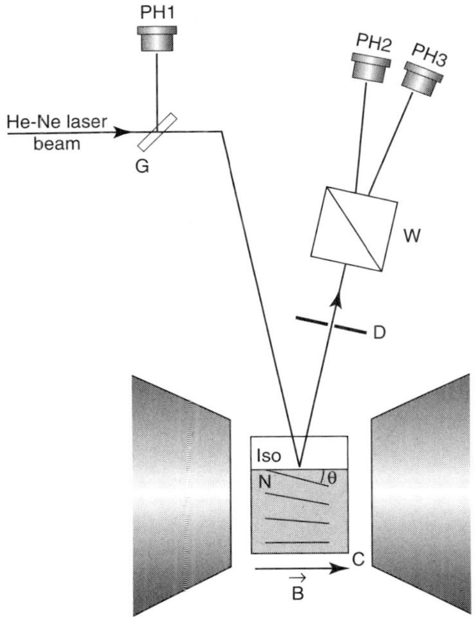

Fig. B.VI.5 Experimental setup for measuring the reflectivity of the nematic-isotropic liquid interface. (from ref. [6]). C: cell containing the sample; D: diaphragm; W: Wollaston prism; G: glass plate; PH1-3: photodiodes; N: nematic phase; Iso: isotropic liquid.

Theoretically, these coefficients are obtained from eq. B.VI.8 replacing n_{nem} by the index corresponding to the chosen polarization. More precisely,

$$n_{nem} = n_o \tag{B.VI.10}$$

for "perpendicular" polarization ($R_{\perp}$), whatever the value of the field B, while:

$$n_{nem} = n(\theta) \quad \text{with} \quad \frac{1}{n(\theta)^2} = \frac{\cos^2\theta}{n_o^2} + \frac{\sin^2\theta}{n_e^2} \tag{B.VI.11}$$

for parallel polarization ($R_{//}$). In this case, n_{nem} depends on the field B through the variation of the anchoring angle θ (with respect to its zero-field value θ_a).

The experiment, performed on nCBs, confirms that $R_\perp$ is B independent between 0 and 8000 gauss. From this result, Faetti and Palleschi obtain an effective width L_{eff} between 300 and 700 Å depending on the particular substance. Having measured the surface tension by the sessile drop method they determine the thermal width L_T and roughly estimate the intrinsic interface width L_o at a few hundred Ångstroms, about a dozen molecular lengths.

Finally, let us show how the anchoring angle and the anchoring energy W_a can be measured. We shall consider that the anchoring law is of the form:

$$W(\theta) = \frac{1}{2} W_a (\theta - \theta_a)^2 \qquad \text{(B.VI.12)}$$

In the presence of an applied field **B**, the director makes with the interface an angle θ smaller than θ_a. This angle can be obtained by writing the torque equilibrium equations in the bulk and at the surface. The derivation is similar to the one developed in section B.V.8 of the previous chapter, so only the final result will be given here. We shall assume that the magnetic field is weak enough so that $\theta_a - \theta << 1$ and we shall take $K = K_1 = K_3$

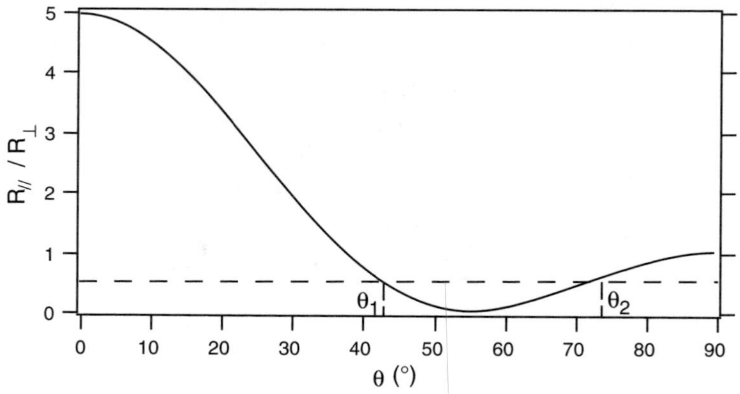

Fig. B.VI.6 Reflectivity ratio $R_{//} / R_\perp$ as a function of the molecular anchoring angle θ at the interface.

(this condition is well verified experimentally at the nematic-isotropic liquid transition [9]). The angle θ is then given by the following implicit formula [6a]:

$$\theta \approx \theta_a + \frac{\sqrt{K\chi_a/\mu_o}\cos\theta\, B}{W_a} \qquad \text{(B.VI.13)}$$

In principle, the anchoring energy W_a can then be obtained by measuring θ as a function of B. This can be achieved experimentally by determining the reflectivity $R_{//}$ or, better still, the $R_{//}/R_{\perp}$ ratio. This ratio has been traced in figure B.VI.6 as a function of the anchoring angle θ. The values employed are those for 7CB: L_o = 450 Å, λ = 6328 Å, n_o = 1.5335, n_e = 1.635, and n_{iso} = 1.564 [6]. In general, two possible values of θ (θ_1 and θ_2 on the graph) exist for a given reflectivity. An easy way of finding out which is the actual angle is to increase the field B, thus reducing the angle θ. θ_1 is the valid solution if the reflectivity $R_{//}$ increases, otherwise one must choose θ_2. The anchoring angle and its evolution with the field B can thus be unambiguously determined, and the anchoring energy W_a can then be deduced from the theoretical law B.VI.13. For 7CB, for instance, Faetti and Palleschi find $\theta_a \approx 53°$ and $W_a \approx 1.6\times10^{-3}$ erg/cm^{2}rad^2 for a surface tension $\gamma \approx 1.8\times10^{-2}$ erg/cm^2. The surface tension anisotropy turns out to be important, of the order of 10%.

Let us now seek a theoretical interpretation.

VI.2 Landau-Ginzburg-de Gennes theory

We shall make the simplifying assumption that the director has everywhere the same orientation, making an angle θ with the interface normal (the z axis). The bulk free energy then depends only on the amplitude S of the tensor order parameter S_{ij} and on its derivatives (here, dS/dz).

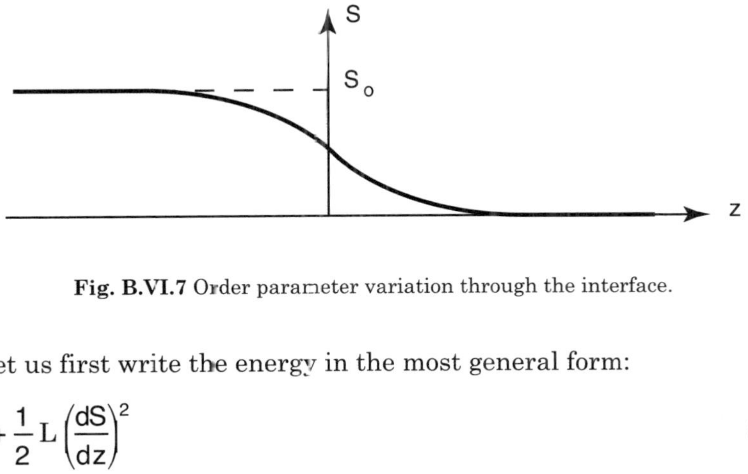

Fig. B.VI.7 Order parameter variation through the interface.

Let us first write the energy in the most general form:

$$f = f_L(S) + \frac{1}{2} L \left(\frac{dS}{dz}\right)^2 \qquad \text{(B.VI.14)}$$

where:

$$f_L(S) = \frac{1}{2} A_o (T - T^*) S^2 + \ldots \qquad \text{(B.VI.15)}$$

is the Landau free energy and T* the spinodal temperature of the isotropic phase.

Thermodynamic equilibrium imposes:

$$-\frac{\delta f}{\delta S}=\frac{d}{dz}\frac{\partial f}{\partial S_{,z}}-\frac{\partial f}{\partial S}=0 \qquad \text{(B.VI.16)}$$

whence, using equation B.VI.14:

$$L\frac{d^2S}{dz^2}=\frac{df_L}{dS} \qquad \text{(B.VI.17)}$$

The boundary conditions are (Fig. B.VI.7):

$$S=S_0 \quad \text{in} \quad z=-\infty \qquad \text{(B.VI.18a)}$$

$$S=0 \quad \text{in} \quad z=+\infty \qquad \text{(B.VI.18b)}$$

with S_0 the value of the order parameter in the nematic phase.

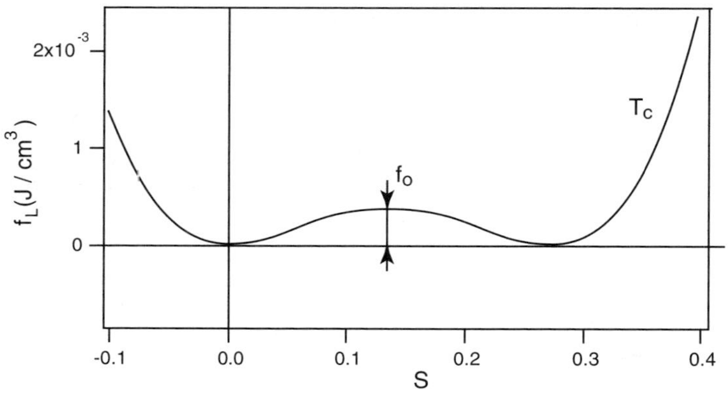

Fig. B.VI.8 Profile of the Landau free energy at the critical temperature. The curve has been traced employing experimental values for 5CB. The barrier height is about 0.5 J/cm^3.

First, let us show that the system is necessarily at the coexistence temperature T_c. Multiplying both sides of equation B.VI.17 by dS/dz and integrating over the z axis:

$$\int_{-\infty}^{+\infty}\frac{d^2S}{dz^2}\frac{dS}{dz}dz=\int_{-\infty}^{+\infty}\frac{df_L}{dS}\frac{dS}{dz}dz \qquad \text{(B.VI.19)}$$

which, after integration by parts and using the boundary conditions gives:

$$0=\Delta f=f_L(0)-f_L(S_0) \qquad \text{(B.VI.20)}$$

The two phases have the same energy, so the system must be at the transition temperature T_c.

The excess free energy of the system, or surface tension γ is, by definition:

$$\gamma = \int_{-\infty}^{+\infty} \left[f_L(S) + \frac{L}{2}\left(\frac{dS}{dz}\right)^2 \right] dz \qquad \text{(B.VI.21)}$$

Using the first integral given by eq. B.VI.17:

$$\frac{L}{2}\left(\frac{dS}{dz}\right)^2 = f_L \qquad \text{(B.VI.22)}$$

one gets the following simplified formula:

$$\gamma = \int_{-\infty}^{+\infty} L\left(\frac{dS}{dz}\right)^2 dz \qquad \text{(B.VI.23)}$$

To continue the calculations the exact form of $f_L(S)$ is needed, and then one must solve the nonlinear equation B.VI.17. We shall do this later for the nematic-isotropic liquid interface, but let us first establish qualitatively how the interface width L_0 and its energy γ depend on the order parameter jump S_0 and on the height f_0 of the energy barrier between the two phases at the transition temperature (Fig. B.VI.8). Eqs. B.VI.22 and B.VI.21 show that:

$$L_0 \approx S_0 \sqrt{\frac{L}{f_0}} \qquad \text{(B.VI.24)}$$

and, respectively,

$$\gamma \approx f_0 L_0 \approx S_0 \sqrt{L f_0} \qquad \text{(B.VI.25)}$$

Thus, for higher stiffness L, the interface is wider and of higher energy. On the other hand, the energy increases and the width decreases for a higher energy barrier f_0.

We conclude this section by presenting the complete calculations for the nematic case. With the previously introduced notations:

$$f_L(S) = \frac{1}{2} A_0 (T - T^*) S^2 - \frac{1}{3} B S^3 + \frac{1}{4} C S^4 \qquad \text{(B.VI.26)}$$

while L depends on the molecular anchoring angle at the interface (see eq. B.IV.17):

$$L = \frac{2}{3} L_1 + \frac{1}{9} L_2 + \frac{1}{3} L_2 \sin^2\theta \qquad \text{(B.VI.27)}$$

Note that $\theta = 0$ corresponds to planar molecular anchoring at the interface and that the stiffness L must always be positive, requiring $L_1 + (2/3)L_2 > 0$. Solving the differential equation B.VI.17 and using eq. B.VI.26, one gets a hyperbolic tangent profile:

$$S(z) = \frac{1}{2}\left[1 - \mathrm{th}\left(\frac{z}{L_0}\right)\right] \qquad \text{(B.VI.28)}$$

of width:

$$L_0 = \frac{3\sqrt{2LC}}{B} \qquad \text{(B.VI.29)}$$

with $S_0 = 2B/3C$. Up to a multiplicative constant, this width is just the correlation length ξ_c at the transition temperature T_c, a parameter accessible in X-ray scattering experiments.

The surface tension γ can be obtained from expression B.VI.23 and is given by:

$$\gamma = \frac{2\sqrt{2}}{81}\frac{B^3}{C^2}\sqrt{\frac{L}{C}} \qquad \text{(B.VI.30)}$$

One can see that measuring L_0 or γ allows determination of the value of the stiffness constant L, provided parameters B and C are known (and they are for some materials, see chapter B.I). In the case of 5CB, where $B \approx 1.6\ \mathrm{J/cm^3}$ and $C \approx 3.9\ \mathrm{J/cm^3}$, Faetti and Palleschi obtain $\gamma \approx 1.5\times10^{-2}\ \mathrm{erg/cm^2}$, yielding $L \approx 10^{-6}$ erg/cm from eq. B.VI.30. Employing eq. B.VI.29, one can also derive the interface width $L_0 \approx 160$ Å, which is totally compatible with the reflectivity determinations.

This model, proposed by de Gennes [10], does not explain the oblique molecular anchoring at the interface. Indeed, from the complete expression B.VI.27 for the stiffness, the surface tension is minimal either for $\theta = 0$, or for $\theta = \pi/2$, according to the sign of L_2. This prediction is very seldom confirmed in experiments.

To explain oblique anchoring, one must take into account the presence of a finite interface polarization, like the one induced by the gradient dS/dz of the quadrupolar order parameter at the interface. This phenomenon, known as ordo-electricity, was discovered by Prost and Marcerou [11] and applied to the nematic-isotropic interface by Barbero et al. [12]. It is of the same nature as the flexo-electric effect. Thus, in the same way that splay or bend distortions induce a polarization, it can be shown that a gradient of the quadrupolar order parameter creates a polarization of the general form:

$$\mathbf{P}_0 = r_1\,(\mathbf{n}.\mathbf{grad}S)\,\mathbf{n} + r_2\,\mathbf{grad}S \qquad \text{(B.VI.31)}$$

To prove that this polarization brings about a non-negligible correction to the surface stiffness L, let us determine the electric energy. We suppose that the director has a uniform orientation θ. In the purely dielectric regime (a reasonable assumption, since the Debye screening length $\approx 1\ \mu m$ is much larger than the correlation length $\xi_c \approx 100$ Å), the z component of the electric displacement vector is zero (as $\mathrm{div}\mathbf{D} = \partial D_z/\partial z = 0$ and $D_z = 0$ "far" in the isotropic and nematic phases), yielding:

$$D_z = \varepsilon_0\, \varepsilon_{zz} E_z + P_{oz} = 0 \qquad \text{(B.VI.32)}$$

with

$$\varepsilon_{zz} = \varepsilon_{//}\sin^2\theta + \varepsilon_{\perp}\cos^2\theta \qquad \text{(B.VI.33a)}$$

and

$$P_{oz} = (r_2 + r_1 \sin^2\theta)\,\frac{dS}{dz} \qquad \text{(B.VI.33b)}$$

Note that, in eq. B.VI.32, the field-induced polarization is contained in the term $\varepsilon_0\varepsilon_{zz}E_z$, while P_{oz} is the permanent polarization related to the order parameter gradient (ordo-electricity). This relation allows determination of the E_z component of the electric field. As the electric contribution to the free energy is:

$$f_E = -\int_0^{E_z} \mathbf{D}\,d\mathbf{E} = -\frac{1}{2}\varepsilon_0\,\varepsilon_{zz}\,E_z^2 - P_{oz}E_z \qquad \text{(B.VI.34)}$$

one has (with $\varepsilon_a = \varepsilon_{//} - \varepsilon_{\perp}$):

$$f_E = \frac{1}{2\varepsilon_0\varepsilon_{\perp}}\,\frac{(r_2 + r_1\sin^2\theta)^2}{1 + (\varepsilon_a/\varepsilon_{\perp})\sin^2\theta}\left(\frac{dS}{dz}\right)^2 \qquad \text{(B.VI.35)}$$

Finally, everything works as if the interface had an effective rigidity:

$$L^* = \frac{2}{3}L_1 + \frac{1}{9}L_2 + \frac{1}{3}L_2\sin^2\theta + \frac{1}{\varepsilon_0\varepsilon_{\perp}}\,\frac{(r_2 + r_1\sin^2\theta)^2}{1 + (\varepsilon_a/\varepsilon_{\perp})\sin^2\theta} \qquad \text{(B.VI.36)}$$

The surface tension being proportional to $(L^*)^{1/2}$ (eq. B.VI.30 remains applicable under the assumption that r_1 and r_2 are S independent, which is a fairly crude approximation [12]) one only needs to minimize L^* with respect to θ in order to find the anchoring angle θ_a. In conclusion, ordo-electricity allows us to explain why the angle θ_a is usually different from 0 or $\pi/2$.

An alternative explanation of the oblique molecular anchoring at the interface is to assume that ferroelectric order (and therefore a non-null polarization) appears at the interface and exponentially decreases away from it in the two phases [13]. This can only arise if the two molecular extremities are of different types, an assumption not needed in the ordo-electricity argument. The $\mathbf{n}_s \leftrightarrow -\mathbf{n}_s$ equivalence is then broken at the interface (with $\mathbf{n}_s$ the director at the interface) and the surface free energy can be formally developed as

$\gamma = \gamma_{\pi/2} - \gamma_1(\mathbf{n}_s.\boldsymbol{\nu}) + (1/2)\gamma_2(\mathbf{n}_s.\boldsymbol{\nu})^2$ including a new term, proportional to $\mathbf{n}_s.\boldsymbol{\nu} = \sin\theta$ ($\boldsymbol{\nu}$ is the outer unit normal at the interface of the nematic domain). If γ_1 and γ_2 are both positive, the preferred angle is $\theta_a = \pi/2$ for $\gamma_1 > \gamma_2$ and $\theta_a = \arcsin(\gamma_1/\gamma_2)$ for $\gamma_1 < \gamma_2$: in this case, γ can be developed in $(\theta - \theta_a)$ giving: $\gamma \approx \gamma_{eq} + (1/2)\,W(\theta - \theta_a)^2$ where $\gamma_{eq} = \gamma_{\pi/2} - (1/2)(\gamma_1^2/\gamma_2)$ and $W = \gamma_2 - \gamma_1^2/\gamma_2$.

VI.3 Instabilities in confined geometry

Thus far we have implicitly assumed (and experiment has confirmed) that the nematic-isotropic liquid front was stable. In fact, an instability can develop below a certain threshold thickness. This instability is very difficult to observe in planar samples, leading to a "hills and valleys" structure consisting of surface disclinations normal to the planar anchoring direction. However, a readily observed similar instability appears in homeotropic samples when the nematic layer is very thin (a few micrometers) and the vertical temperature gradient is not too strong [1]. Under these conditions, the interface is decorated with singular points, called **umbilics**, easily seen under the microscope between crossed polarizers (figure B.VI.9a). The defect core always sends out four black branches, turning either with the polarizers or the other way and showing that defect rank is $m = \pm 1$. By changing the focus one can also show that these defects deform the interface, +1 defects forming peaks, while −1 defects correspond to valleys. Figure B.VI.9b shows a diagram of the director field in a vertical plane connecting two umbilics of opposite sign.

In trying to explain the origin of this instability we shall make the simplifying assumption that the director remains in the (x, z) plane being invariant in the y direction. Furthermore, we shall assume that the destabilized interface takes the shape of a zigzag with wavelength Λ and angle α with respect to the horizontal (Fig. B.VI.10). Finally, the anchoring will be assumed to be strong at both interfaces, imposing:

$$\theta = \frac{\pi}{2} \qquad \text{in} \qquad z = 0 \qquad \text{(B.VI.37a)}$$

and

$$\theta = \alpha + \theta_a \qquad \text{in} \qquad z = \xi = h + \alpha x \qquad \left(-\frac{\Lambda}{4} \leq x \leq \frac{\Lambda}{4}\right) \qquad \text{(B.VI.37b)}$$

Here, θ is the angle between the director and the z axis and θ_a is the anchoring angle at the nematic-isotropic interface. This amounts to assuming the anchoring penetration lengths K/W at the two interfaces are small with respect to the average thickness h of the nematic layer. Let G be the vertical temperature gradient and neglect gravity effects.

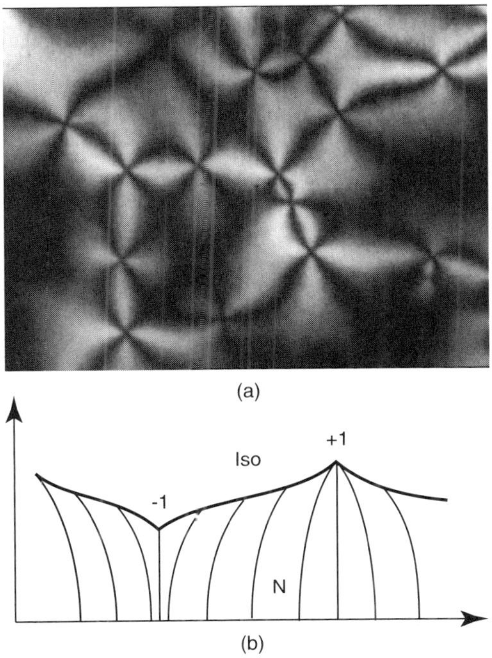

Fig. B.VI.9 a) Umbilics at the nematic-isotropic liquid interface observed between crossed polarizers in a homeotropic 8CB sample of total thickness 20 μm. The nematic layer is a few micrometers thick; b) diagram of the director field in a vertical plane connecting two umbilics of opposite sign.

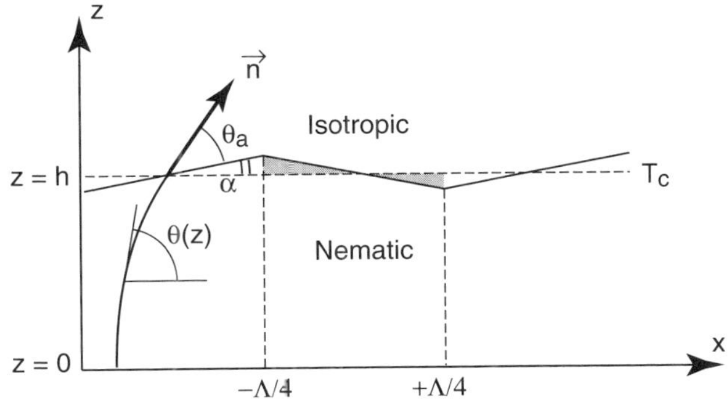

Fig. B.VI.10 Unstable nematic layer. The gray areas are oversaturated because of the temperature gradient.

To explain the instability, let us calculate the system energy F in a stripe of width Λ/2 ($-(\Lambda/4) \leq x \leq (\Lambda/4)$). On unit length along y, it reads:

$$F = \int_{-\Lambda/4}^{+\Lambda/4} \int_0^{\xi} \frac{1}{2} K \left(\frac{\partial \theta}{\partial z}\right)^2 dxdz + \int_{-\Lambda/4}^{+\Lambda/4} \gamma \sqrt{1 + \left(\frac{d\xi}{dx}\right)^2} dx + F_G + F_D \qquad \text{(B.VI.38)}$$

In this expression, the first term stands for the Frank elastic energy (we neglect the x variation of θ) and the second for the surface energy. The third one, F_G, corresponds to the undercooling energy of the isotropic liquid below the average front position (situated in the gradient at the critical temperature T_c) and the overheating energy of the nematic above this position. As this energy is $\Delta H \Delta T / T_c$ per unit volume, one can easily check that:

$$F_G = \frac{\Delta H \, G \, \Lambda^3}{192 \, T_c} (\tan \alpha)^2 \qquad \text{(B.VI.39)}$$

The last term corresponds to the average energy of a peak and a valley which, from a topological viewpoint, are surface disclinations. The order of magnitude is:

$$F_D \approx \varepsilon K \qquad \text{(B.VI.40)}$$

where ε is a constant of order unity.

Before explicitly calculating the energy B.VI.38, let us first note that minimization with respect to θ leads to a linear z dependence or, taking into account the boundary conditions:

$$\theta(z) = \frac{\pi}{2} + \frac{\theta_a + \alpha - \pi/2}{\xi} z \qquad \text{(B.VI.41)}$$

Replacing θ by this expression in eq. B.VI.38 one has, to lowest order in α:

$$F = F_0 - \frac{K\Lambda}{h} \alpha \left(\frac{\pi}{2} - \theta_a\right) + \frac{1}{4} \gamma \, \alpha^2 \Lambda + F_G + F_D \qquad \text{(B.VI.42)}$$

where F_0 is the system energy for a horizontal interface ($\alpha = 0$). Clearly, tilting the interface reduces the elastic energy by "straightening" the director field lines. The thinner the nematic layer, the stronger this effect (1/h dependence). On the other hand, the surface energy increases and so does the bulk energy, because of the oversaturation in the two phases. Finally, costly surface disclinations must be created.

The overall system energy excess per unit sample area is:

$$\Delta F = -2 \frac{K}{h} \alpha \left(\frac{\pi}{2} - \theta_a\right) + \frac{1}{2} \gamma \, \alpha^2 + \frac{\Delta H \, G \, \Lambda^2}{96 \, T_c} \alpha^2 + \frac{2\varepsilon K}{\Lambda} \qquad \text{(B.VI.43)}$$

The interface is unstable if there are values of α and Λ rendering $\Delta F < 0$. To find them out, let us minimize ΔF with respect to the two variables. Starting with Λ yields:

$$\Lambda^3 = \frac{96\,\varepsilon K\,T_c}{G\,\Delta H\,\alpha^2} \tag{B.VI.44}$$

and, replacing in eq. B.VI.43:

$$\Delta F = -2\frac{K}{h}\alpha\left(\frac{\pi}{2} - \theta_a\right) + \frac{1}{2}\gamma\,\alpha^2 + 3\left(\frac{G\,\Delta H\,(\varepsilon K)^2}{96\,T_c}\right)^{1/3}\alpha^{2/3} \tag{B.VI.45}$$

This expression must then be minimized with respect to α and, writing that $\Delta F = 0$ at the threshold, one easily obtains the critical thickness h_c **below** which the instability appears:

$$h_c = \left(\frac{\pi}{2} - \theta_a\right)\left(\frac{12\,T_c\,\varepsilon^2\,K^2}{\gamma\,G\,\Delta H}\right)^{1/4} \tag{B.VI.46}$$

For this thickness, the angular amplitude of the zigzag is:

$$\alpha_c = \left(\frac{\varepsilon^2 G\,\Delta H\,K^2}{12\,T_c\,\gamma^3}\right)^{1/4} \tag{B.VI.47}$$

at a wavelength:

$$\Lambda_c = 4\sqrt{3}\sqrt{\frac{T_c}{G}\frac{\gamma}{\Delta H}} \tag{B.VI.48}$$

The threshold amplitude is finite, indicating a subcritical bifurcation.

With $\varepsilon = 1$, $K = 2\times10^{-7}$ dyn, $\gamma = 2\times10^{-2}$ erg/cm^2, $\Delta H = 2\times10^7$ erg/cm^3, $T_c = 300$ K, and $\theta_a = 45°$, one typically gets:

$$h_c(\mu m) \approx 1.1\,G^{-1/4} \tag{B.VI.49a}$$

$$\alpha_c = 0.07\,G^{1/4} \tag{B.VI.49b}$$

$$\Lambda_c\,(\mu m) = 38\,G^{-1/2} \tag{B.VI.49c}$$

where G is given in K/cm. Thus, extremely weak gradients, of order $G < 10^{-2}$ K/cm, are needed for the instability to develop and be visible under the microscope. Indeed, one has $h_c = 6$ µm, $\alpha_c = 0.01$ rd, and $\Lambda_c = 0.1$ mm for $G = 10^{-3}$ K/cm, a very small gradient.

This crude estimation shows that an instability of elastic origin can appear at the N-I interface for different anchoring conditions on the glass plate and at the interface. It also shows that the instability can only develop in a very thin nematic layer (a few micrometers) with a very small vertical gradient. Experimentally, however, one never observes parallel disclination lines, but rather a lattice of umbilics. This is not surprising, as the energy of the point defects must be much lower than that of lines.

Moreover, some predictions can be made in this case. Imagine a square lattice of umbilics is formed and let Λ be the average distance between two defects of opposite sign.

The energy excess ΔF is given by a formula obtained from the previous one by replacing the line energy $-2\varepsilon K/\Lambda$ by the umbilic energy $\varepsilon Kh/\Lambda^2$. The typical umbilic energy is taken as εKh, with h the thickness of the nematic layer and ε a constant of order unity. The system energy then becomes (up to numerical factors of order unity):

$$\Delta F \approx -\frac{K}{h}\alpha + \gamma\,\alpha^2 + \frac{\Delta H\, G\, \Lambda^2}{100\, T_c}\alpha^2 + \frac{\varepsilon K h}{\Lambda^2} \qquad \text{(B.VI.50)}$$

A first minimization with respect to Λ gives:

$$\Lambda = \left(\frac{100\,\varepsilon\, K\, h\, T_c}{\alpha^2\, G\, \Delta H}\right)^{1/4} \qquad \text{(B.VI.51)}$$

and

$$\Delta F = \left(-\frac{K}{h} + \frac{1}{5}\sqrt{\frac{\varepsilon\, K\, h\, G\, \Delta H}{T_c}}\right)\alpha + \gamma\alpha^2 \qquad \text{(B.VI.52)}$$

It is immediately apparent that the instability develops ($\Delta F < 0$) when the coefficient of the linear α term changes sign, namely at a critical thickness:

$$h_c \approx \left(\frac{25\, K\, T_c}{\varepsilon\, G\, \Delta H}\right)^{1/3} \qquad \text{(B.VI.53)}$$

For this thickness, the angle α is zero and Λ is infinite. Unlike previously, the transition is then continuous (or supercritical). It must then be much easier to observe, as there is no longer an energy barrier to cross. Also, the critical thickness now decreases as $G^{-1/3}$ (instead of $G^{-1/4}$) and its value is much more acceptable than before. Indeed, one has $h_c \approx 10\ \mu m$ in a vertical gradient of 0.1 K/cm with the same numerical values as before. One can equally determine the average distance between two umbilics and the average interface tilt when the thickness of the nematic layer is, say, half the critical thickness. Minimizing ΔF, one gets:

$$\alpha \approx 0.65\frac{K}{\gamma h_c} \qquad (h = h_c/2) \qquad \text{(B.VI.54)}$$

and, using eqs. B.VI.51 and 53:

$$\Lambda \approx 7.4\sqrt{\frac{T_c}{G}\frac{\gamma}{\Delta H}} \qquad \text{(B.VI.55)}$$

Up to a numerical factor, this formula is similar to the one found in the two-dimensional case. For G = 0.1 K/cm, one has $\alpha \approx 0.004$ rd and $\Lambda \approx 140$ μm, in good agreement with the experimentally observed values.

VI.4 Elastic correction to the Gibbs-Thomson relation

In the previous section we assumed that the interface was locally flat, except for the singular points. This is in fact a crude approximation: experiment shows that the surface shape is more complicated. To determine its morphology, one must correctly minimize the **total** system energy, including both surface energy and the bulk elastic energy. The minimization is done in two steps: first at "fixed surface," by varying the director orientation, and then at "fixed texture," by displacing the interface. The first step leads to the equilibrium equation for bulk torques and a similar surface equation; the second gives the Gibbs-Thomson relation determining interface temperature as a function of its curvature and of the local director field deformations. This relation is also the differential equation giving the exact interface shape.

This section shows that the Gibbs-Thomson equation is more complicated in nematics than in solids, due to the presence of new terms directly related to the curvature elasticity of the medium. This equation will then be solved to give the equilibrium shape of a nematic germ in contact with its isotropic liquid. We shall see that the problem has more than one solution, some of which can be determined analytically and compared to the experimental results.

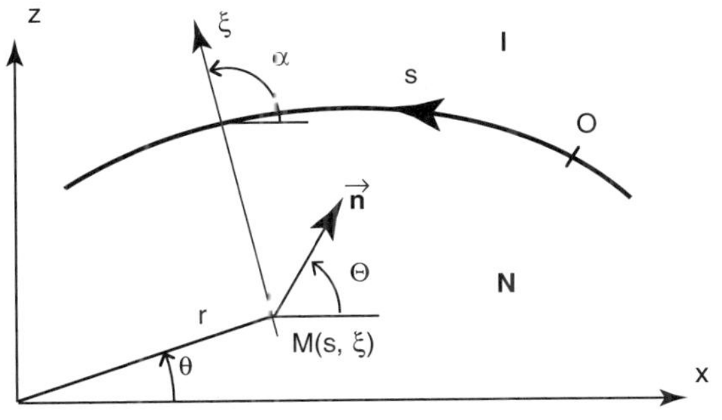

Fig. B.VI.11 Nematic-isotropic liquid interface. Definition of the (s, ξ) coordinates and of angles α and θ. In the drawing, the interface radius of curvature and angles α and θ are positive.

To simplify the calculations as much as possible, let us assume the nematic is two dimensional, the director remaining in the (x, z) plane. As we shall see later, this situation has experimental relevance for Langmuir layers or smectic C* films. The following demonstration is from J. B. Fournier [14]. Let Θ be the angle between the director and the x axis and α the angle between the interface normal and the x axis. The interface is parameterized by its arc length s while a point M in the bulk is given by coordinates (s, ξ), with ξ the distance to the surface measured along the interface normal.

The total system energy is:

$$F = F_s + F_v = \int \gamma\, ds + \int \left[\Delta f + \frac{1}{2} K (\nabla \theta)^2\right] dV \qquad \text{(B.VI.56)}$$

assuming $K_1 = K_3 = K$ and where Δf is the condensation free energy of the nematic phase ($\Delta f = f_N - f_I = \Delta H \delta T / T_c$ with $\delta T = T - T_c$). The first term is the surface energy, where γ is an even function $\gamma(\varphi)$ of the angle $\varphi = \alpha - \theta$ between the director and the interface normal. The second term is the bulk energy, including oversaturation and elastic deformation.

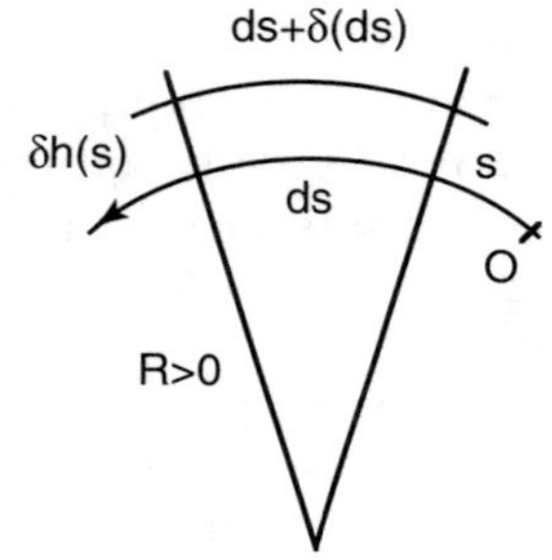

Fig. B.VI.12 Variation δ(ds) of a surface element as a result of a "normal" interface displacement δh(s).

Let us now consider an interface displacement $\delta h(s)$ along its normal, at fixed texture $\theta(s, \xi)$. The interface energy varies by:

$$\delta F_s = \int [\delta\gamma\, ds + \gamma\, \delta(ds)] = \int \left(\delta\gamma + \frac{\gamma}{R}\delta h\right) ds \qquad \text{(B.VI.57)}$$

The change in γ is related to the variation $\delta\alpha = -\, d(\delta h)/ds$ in interface tilt and to the modification $\delta\theta = (\partial\theta/\partial\xi)\delta h$ of the molecular orientation. By analogy with the interface curvature $1/R = d\alpha/ds$, Fournier defines the molecular orientation curvatures $1/R_{//} = \partial\theta/\partial s$ and $1/R_{\perp} = \partial\theta/\partial\xi$ in the directions parallel and perpendicular to the interface, respectively. With these definitions and knowing that

$$\delta\gamma = \frac{\partial\gamma}{\partial\alpha}\delta\alpha + \frac{\partial\gamma}{\partial\Theta}\delta\theta \qquad \text{(B.VI.58)}$$

one has (note that $\partial\gamma/\partial\alpha = \gamma_\varphi(\alpha - \Theta)$ and $\partial\gamma/\partial\Theta = -\gamma_\varphi(\alpha - \Theta)$):

$$\delta F_s = \int \left(-\gamma_\varphi \frac{d}{ds}\delta h - \gamma_\varphi \frac{\partial \Theta}{\partial \xi}\delta h + \frac{\gamma}{R}\delta h\right) ds$$

$$= b.t. + \int \left[\frac{\gamma_{\varphi\varphi}}{R} - \frac{\gamma_{\varphi\varphi}}{R_{//}} - \frac{\gamma_\varphi}{R_\perp} + \frac{\gamma}{R}\right] \delta h \, ds \qquad \text{(B.VI.59)}$$

where γ_φ etc. are the partial derivatives of γ with respect to φ and "b.t." the boundary terms resulting from the integration by parts.

The bulk term, after an interface displacement $\delta h(s)$ at fixed texture, varies by:

$$\delta F_v = \int ds\, \delta h \left[\Delta f + \frac{1}{2} K (\nabla\theta)^2\right]$$

$$= \int ds \left[\Delta f + \frac{1}{2} K \left(\frac{1}{R_{//}^2} + \frac{1}{R_\perp^2}\right)\right] \delta h \qquad \text{(B.VI.60)}$$

At equilibrium $\delta F_s + \delta F_v = 0$ ($\forall\, \delta h$), which, taking into account eqs. B.VI.59 and B.VI.60, gives:

$$\frac{\gamma + \gamma_{\varphi\varphi}}{R} - \frac{\gamma_{\varphi\varphi}}{R_{//}} - \frac{\gamma_\varphi}{R_\perp} + \Delta f + \frac{1}{2} K \left(\frac{1}{R_{//}^2} + \frac{1}{R_\perp^2}\right) = 0 \qquad \text{(B.VI.61)}$$

In an ordinary crystal, $R_{//}$ and $R_\perp$ are infinite because the crystallographic axes do not turn, and the Gibbs-Thomson equation reduces to $(\gamma + \gamma_{\varphi\varphi})/R + \Delta f = 0$, only involving the surface stiffness $\gamma + \gamma_{\varphi\varphi}$. In nematics, the director can easily turn, giving rise to additional terms such as the elastic corrections in $1/R_{//}^2$ and $1/R_\perp^2$.

Finally, the texture must also be equilibrated. In this case, minimization is done at fixed interface. When varying θ by $\delta\theta$, the free energy changes by:

$$\delta F = \int dV\, K\, \nabla\Theta . \nabla\delta\Theta + \int ds\, (-\gamma_\varphi)\, \delta\Theta$$

$$= \int dV\, K\, \Delta\Theta\, (-\delta\Theta) + \int ds\, K \frac{\partial \Theta}{\partial \xi} \delta\Theta + \int ds\, (-\gamma_\varphi)\, \delta\Theta \qquad \text{(B.VI.62)}$$

yielding two equilibrium equations:

$$\Delta\theta = 0 \tag{B.VI.63a}$$

$$\frac{K}{R_\perp} = \gamma_\varphi \tag{B.VI.63b}$$

The first one states that, in the absence of an external torque, the bulk elastic torque is zero. The second expresses the equilibrium between elastic torque and interface anchoring torque.

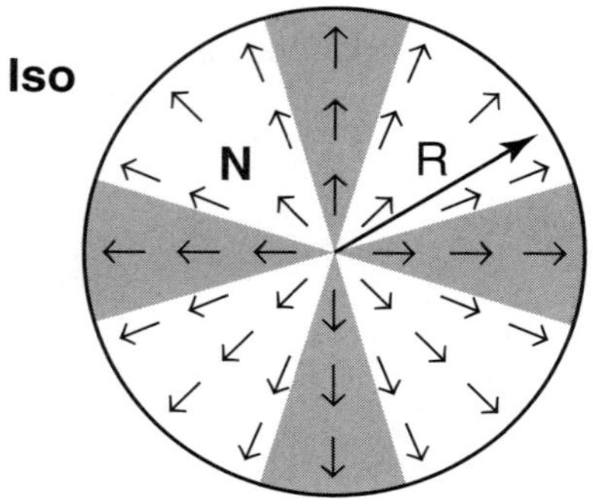

Fig. B.VI.13 Circular domain in radial configuration. Between crossed polarizers, it appears as a black cross.

Exactly solving the equation system B.VI.61 and B.VI.63 represents one of the challenges of recent years. Two exact solutions to this problem are known, for a surface energy of the type:

$$\gamma(\varphi) = \gamma_o - W_a \cos(n\varphi) \tag{B.VI.64}$$

with integer n. This anchoring term is the first of a Fourier series development of the surface energy. In ordinary nematics, the value $n = 2$ is imposed by the $\mathbf{n} \leftrightarrow -\mathbf{n}$ symmetry. The favored anchoring is then homeotropic for $W_a > 0$ and planar if $W_a < 0$ (the following $\cos(4\varphi)$ term must be preserved to describe oblique anchoring). In a Langmuir or a smectic C* film, the director is the unit vector tangent to the molecular projection in the film plane. Thus, **n** and −**n** are no longer equivalent (**n** is similar to a spin) and the first non-null term of the series has $n = 1$. In the following, we shall assume $n = 1$ and take $W_a > 0$ (which favors homeotropic anchoring).

The most trivial solution is shown in figure B.VI.13, depicting a circular domain (of radius R) containing a radial director field. In this configuration it is obvious that $\Delta\theta = 0$, as $\Theta = \theta$. On the other hand, θ being ξ independent, $1/R_\perp = \partial\theta/\partial\xi = 0$. The anchoring is homeotropic at the interface, $\gamma_\varphi = 0$ ensuring the equilibrium of surface torques. Now, the Gibbs-Thomson equation immediately gives the equilibrium temperature of the nematic domain because $R = R_{//}$:

$$T = T_c\left(1 - \frac{d_o}{R} - \frac{d_o l}{2R^2}\right) \qquad \text{(B.VI.65)}$$

In this formula, $d_o = \gamma_o/\Delta H$ is the "usual" capillary length (of the order of 0.1 Å in nematics) while $l = K/\gamma_o$ is a new length, of a few thousand Ångstroms, measuring the competition between surface tension and curvature elasticity. In concrete terms, when the domain radius is comparable to this length, the elastic correction to the transition temperature becomes of the same order of magnitude as the capillary correction. This solution is straightforwardly applied to the nematic case (corresponding to n = 2).

The second solution, much less obvious, is given by the so-called "boojum" configuration (the name was introduced by Mermin to describe the point vortex at the surface of a spherical volume of superfluid ^{3}He [15]), where the defect core is virtual, i.e., expelled from the nematic domain. For this specific configuration, the defect is an m = 2 disclination (compared to m = 1 in the previous example). Its director field is given by the law:

$$\Theta = 2\theta \qquad \text{(B.VI.66)}$$

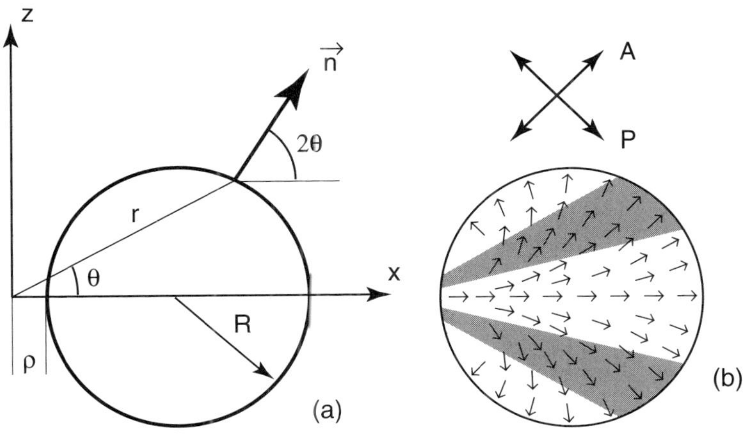

Fig. B.VI.14 "Boojum" and the associated director field. The m = 2 defect core is at the coordinate origin. Extinction branches between crossed polarizers are shown in gray.

Rudnick and Bruisma [16] recently showed that the equilibrium shape of the nematic domain remains circular. To prove it, assume the boundary is a circle of radius R and let ρ be the distance at which the defect is expelled (Fig. B.VI.14). Clearly, the chosen director field (eq. B.VI.66) satisfies the equation of bulk torque equilibrium ($\Delta\Theta = 0$). One can also show that:

$$\gamma_\varphi = W_a \sin(\alpha - 2\theta) \tag{B.VI.67a}$$

$$\gamma_{\varphi\varphi} = W_a \cos(\alpha - 2\theta) \tag{B.VI.67b}$$

$$\gamma + \gamma_{\varphi\varphi} = \gamma_o \tag{B.VI.67c}$$

$$\frac{1}{R_\perp} = \frac{2}{r}\sin(\alpha - \theta) \tag{B.VI.67d}$$

$$\frac{1}{R_{//}} = \frac{2}{r}\cos(\alpha - \theta) \tag{B.VI.67e}$$

with α the angle between the x axis and the interface normal.

With these relations, the surface torque equation B.VI.63b becomes:

$$W_a \sin(\alpha - 2\theta) = \frac{2K}{r}\sin(\alpha - \theta) \tag{B.VI.68a}$$

while the Gibbs-Thomson equation B.VI.61 yields:

$$\frac{\gamma_o}{R} - \frac{2W_a}{r}\cos\theta + \frac{2K}{r^2} + \Delta f = 0 \tag{B.VI.68b}$$

describing a circle, in agreement with our starting hypothesis. These two relations allow for explicitly calculating the domain radius R and the expelling distance ρ of the defect core. A first relation between these parameters is obtained by eliminating angles α and θ from equation B.VI.67:

$$\frac{\rho(\rho + 2R)}{\rho + R} = \frac{2K}{W_a} \tag{B.VI.69}$$

Moreover, comparing equation B.VI.68b to the polar coordinates formula of a circle with radius R:

$$1 + \frac{\rho^2 + 2\rho R}{r^2} - \frac{2(\rho + R)}{r}\cos\theta = 0 \tag{B.VI.70}$$

one gets two more relations:

$$\frac{2K}{\Delta F + \gamma_o/R} = \rho(\rho + 2R) \tag{B.VI.71a}$$

$$\frac{W_a}{\Delta F + \gamma_o/R} = \rho + R \tag{B.VI.71b}$$

Note that eq. B.VI.69 is found again by dividing these equations side by side.

Consequently, the circle is an **exact** solution to this problem. At a fixed supersaturation Δf, there is only one solution, whose parameters (ρ, R) are obtained by solving two out of the three eqs. B.VI.69 and B.VI.71a and b. The final result is particularly simple when the domain radius R is much larger than the anchoring penetration length $l_a = K/W_a$, as in this limit:

$$\rho \approx l_a \qquad \text{(B.VI.72a)}$$

$$R \approx \frac{\gamma_o - W_a}{\Delta H} \frac{T_c}{T_c - T} \qquad \text{(B.VI.72b)}$$

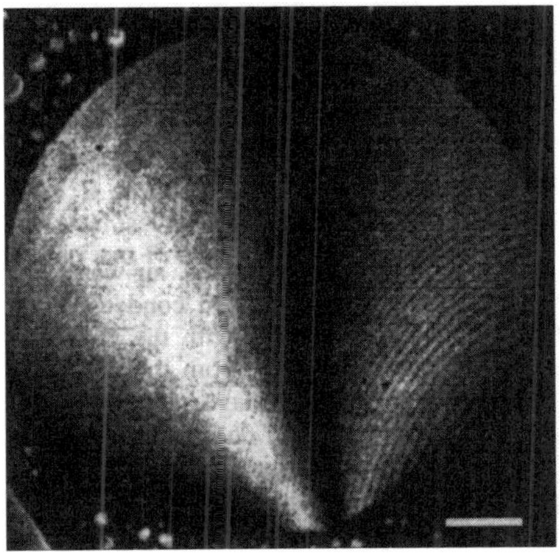

Fig. B.VI.15 Brewster angle microscopy observation in polarized light of a two-dimensional nematic drop (condensed liquid) in the isotropic liquid (gas phase). The system is a monolayer of palmitic acid spread on water at pH = 5.5. The drop radius is about 200 μm (from ref. [19]).

This boojum configuration was experimentally obtained in smectic C* free films [17] and in Langmuir films of fatty acids on water [18,19] (Fig. B.VI.15). In both cases, the defect core is relatively far from the domain boundary, leading to an anchoring penetration length of the order of several tens of micrometers. Thus, the anchoring energy in these systems is weaker than in ordinary nematics, where $l_a \approx 1$ μm. On the other hand, experiment shows that the domains are never perfectly circular, due to the fact that $K_1 \neq K_3$ in these systems. The director field $\Theta = 2\theta$ still satisfies the bulk torque equation [20] but the circle is no longer a solution of the Gibbs-Thomson equation, which now contains an additional term expressing elastic anisotropy. For a detailed discussion of this problem, the reader is referred to the paper by Galatola and Fournier [21].

VI.5 Directional growth of the nematic-isotropic liquid front

Directional growth is by now a well-known technique allowing growth of crystals at a controlled speed in a temperature gradient. In metallurgy, this technique is employed for material purification by zone melting and for obtaining monocrystals. Its main practical limitation is given by the growth front developing a cell instability above a critical velocity, sometimes as low as a few μm/s. This Mullins-Sekerka instability [22], named after the theorists who performed its first correct analysis, is due to the impurities present in the crystal being rejected and then diffusing into the liquid. The resulting impurity micro-segregation can strongly alter the mechanical, chemical (corrosion resistance), or electrical properties of the solid. This instability also appears in liquid crystals, in particular at the nematic-isotropic liquid interface, with some specificities we shall try to emphasize in this section. First observed in nematics by Armitage and Price in 1978 [23], its systematic study in directional growth only started ten years afterwards [24, 25]. To keep this chapter from becoming too cumbersome, the basics of growth theory and the corresponding calculations are concentrated in chapter B.IX, to which the reader is referred for proof of the equations employed in the following.

VI.5.a) Experimental observation of the Mullins and Sekerka instability

Let us start with a brief description of the growth cell (see ref. [26] for more details).

It consists of two ovens, i.e., two blocks of copper maintained at different temperatures, onto which slides the liquid crystal sample confined between two glass plates (Fig. B.VI.16). The two ovens are 4 mm apart, so the growth front and its instabilities can be observed under the polarizing microscope. A step-by-step engine, acting on a micrometric screw by a speed reducer, can move the sample both ways, with controlled speed. The pulling speed ranges between 0.01 μm/s and 300 μm/s, while the temperature gradient can be set between 1 and 100°C/cm.

The liquid crystal sample is usually prepared between two glass plates treated in homeotropic anchoring. The anchoring must be strong in order that the nematic phase completely wets the glass; otherwise, the instability is hidden by nematic phase nucleation ahead of the front.

The photos in Fig. B.VI.17 show the interface behavior with increasing speed. The 30 μm sample consists of impure 8CB, with $\Delta T = T_l - T_s \approx 0.15°C$, where T_l and T_s are the temperatures of the liquidus and the solidus, respectively. The imposed temperature gradient is 25°C/cm.

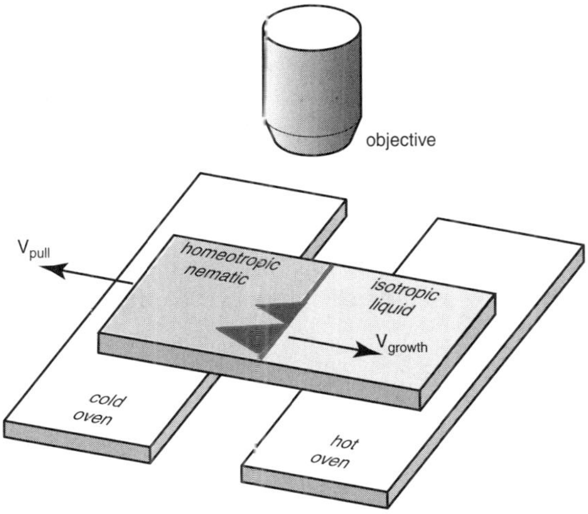

Fig. B.VI.16 Diagram of a directional growth experiment. The sample is pulled at constant speed in a fixed temperature gradient. The planar triangles will be discussed in the next section.

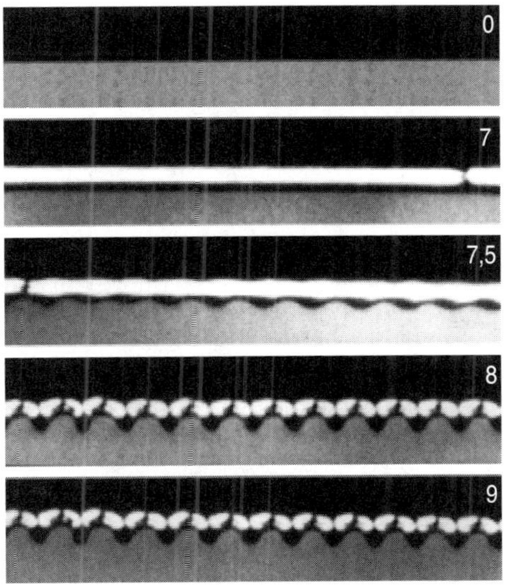

Fig. B.VI.17 Destabilization of the nematic-isotropic liquid front in an 8CB sample (d = 30 μm, G = 25°C/cm, and $\Delta T \approx 0.15°C$). The photos, taken between crossed polarizers, are labeled with their speed in μm/s. The nematic phase is at the bottom and grows into the isotropic phase, at the top of the image. In the first photo, the front is at rest (V = 0): the interface is very narrow, below the microscope resolution. With increasing speed, a diffuse stripe appears ahead of the front, its width saturating rapidly (second photo at V = 7 μm/s). This stripe is due to a meniscus forming in the thickness of the sample. The vertical part of the meniscus is destabilized, producing an undulating pattern above a critical speed close to V = 7.5 μm/s. When the speed is increased further, the modulation amplitude increases and the meniscus is rapidly "decorated" with umbilics. This bifurcation is supercritical (from ref. [24]).

At rest, the front is straight. Its apparent width is very small, lower than the instrumental resolution (a few micrometers at most).

At low speed, a diffuse stripe, strongly scintillating between crossed polarizers, appears ahead of the front. Its width increases very rapidly with the speed, then saturates at a value inversely proportional to the temperature gradient. This saturation width does not change with sample thickness and corresponds to a temperature difference very close to ΔT. It is now well established that this stripe corresponds to front destabilization in the thickness of the sample and to the formation of a large meniscus separating the two phases.

At a critical speed V_c, the front is destabilized in the horizontal plane. More precisely, the vertical part of the meniscus takes the shape of a sinusoid with a well-defined wavelength λ_c. This cell bifurcation is supercritical and exhibits no hysteresis.

Systematic measurements performed using cyanobiphenyls of the nCB series show that the critical speed V_c decreases with increasing ΔT or decreasing G [24, 25]. This behavior shows that the impurity diffusion field destabilizes the interface, while the temperature gradient has a stabilizing action: this is the Mullins-Sekerka instability.

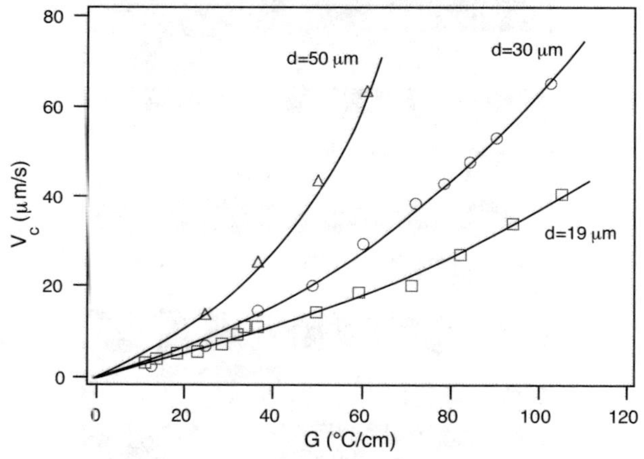

Fig. B.VI.18 Critical destabilizing speed measured for an 8CB sample for three different values of the thickness as a function of the temperature gradient (from ref. [24]).

At this point, let us recall the theoretical predictions for stability loss of a plane front. A linear stability analysis yields the critical speed and the wavelength at the threshold (see chapter B.IX):

$$V_c = \frac{D_I G}{\Delta T}\left(1 + K\frac{D_N}{D_I}\right) \qquad \text{(B.VI.73a)}$$

$$\lambda_c = 2\pi \left(\frac{4\,\gamma\, D_I T_c\, \Delta T}{K\, G^2\, (D_I + D_N)\, \Delta H} \right)^{1/3} \qquad \text{(B.VI.73b)}$$

In these formulae, D_I is the impurity diffusion constant in the isotropic liquid, D_N is the impurity diffusion constant in the nematic phase (here, in the direction normal to the director), and K is the impurity partition coefficient (close to 0.9). These relations are only valid if the diffusion lengths at the threshold, D_I/V_c and D_N/V_c, are much larger than the wavelength λ_c (in practice, a factor of two or three is enough for 10% accuracy).

Rigorously speaking, these formulae are not correct, as they do not take into account the existence of the meniscus or the solute convection sometimes appearing ahead of the front [25,27]. These two effects profoundly change front behavior, as shown in figure B.VI.18, where the critical speed is traced as a function of the temperature gradient for three different values of sample thickness. Besides the fact that the critical speed usually increases faster than G (in strict disagreement with formula B.VI.73a), V_c clearly increases with the sample thickness. Furthermore, the measured wavelength is markedly larger than predicted by the linear plane front theory. Consequently, the capillary length d_0 obtained from formula B.VI.73b is 10 to 1000 times larger than the static one, the discrepancy becoming stronger with increasing thickness [28,29]. Note that elastic effects, which could appear as a result of strong molecular anchoring at the interface (see the next section), cannot explain this difference [30,31]. On the other hand, a recent study showed that zero-thickness extrapolation of the critical speed and the threshold wavelength measured in different temperature gradients are in fairly good agreement with the predictions of the linear plane front theory. In this (experimentally hard to reach) limit, formulae B.VI.73 seem to apply [32].

Let us emphasize that the Mullins-Sekerka instability can also be observed during melting, provided the meniscus is "reversed," i.e., it is the isotropic phase that wets the substrate rather than the nematic one. Among several surface treatments leading to this result, a common one consists of evaporating a thin silica layer under oblique incidence, resulting in planar director orientation. In this case, the roughness of the silica layer favors the isotropic phase. Another equally effective treatment uses a short-chain silane, dichlorodimethylsilane, rinsing with toluene and then rubbing according to the Châtelain method. It also leads to planar molecular anchoring with dewetting of the nematic phase. The same silane, however, when employed as described by Cognard [32], leads to homeotropic anchoring with (sometimes complete) dewetting of the nematic phase. This orientational change of the liquid crystal and the reasons for nematic dewetting in homeotropic anchoring are still unexplained. The melting instability was observed both in planar anchoring [25, 33] and in homeotropic anchoring [34]. Systematic study showed that the results are not substantially different from the solidification case [33] (Fig. B.VI.19).

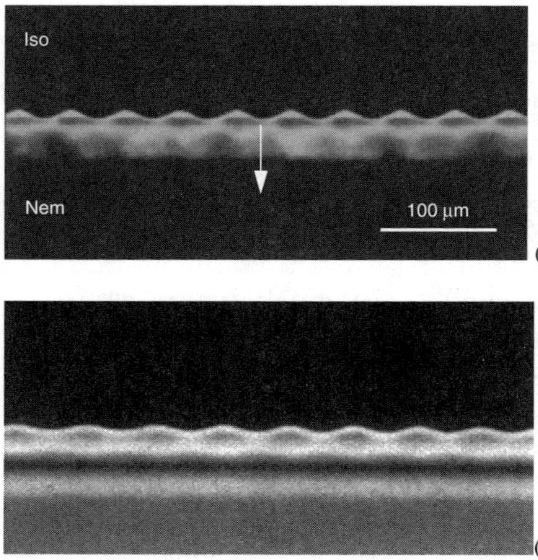

Fig. B.VI.19 Melting instability of the nematic-isotropic liquid front, observed between crossed polarizers. In (a), the nematic is homeotropic (photo by F. Picano). In (b), it is oriented in planar anchoring (from ref. [33]). The meniscus is clearly visible in both cases.

In conclusion, an accurate study of the meniscus and its linear stability is yet to be done in order to understand fully the dynamics and destabilization of the nematic-isotropic liquid front in directional solidification or melting.

In the following, we shall describe the front behavior beyond the instability threshold, where nonlinear effects become dominant.

V.5.b) Secondary instabilities in the rapid solidification regime and absolute restabilization

In this section, we will show that liquid crystals, nematics in particular, allow for the study of "rapid solidification" regimes, hardly accessible in metals or usual plastic crystals (such as succinonitrile), thanks to the particular values of their physical constants, which are very different from those in classical materials. Before going into the numerical values, let us first review the relevant parameters of the problem.

First come the two diffusion constants for the impurities, D_I and D_N. Combined with the imposed growth speed V, they give two diffusion lengths:

$$l_D^I = \frac{D_I}{V} \qquad l_D^N = \frac{D_N}{V} \tag{B.VI.74}$$

In nematics, these two lengths are close, as both phases are fluid ($D_I \approx D_N$ within a factor of 2). We shall then set $l_D = D/V$ irrespective of the phase.

Next are the surface tension γ and the latent heat per unit volume ΔH; their ratio provides a capillary length:

$$d_o = \frac{\gamma}{\Delta H} \tag{B.VI.75}$$

However, only the chemical capillary length is relevant:

$$d_o^c = \frac{\gamma}{\Delta H}\frac{T_c}{\Delta T} \tag{B.VI.76}$$

Finally, a thermal length can be defined in terms of ΔT and the temperature gradient G:

$$l_T = \frac{\Delta T}{G} \tag{B.VI.77}$$

These lengths quantify the destabilizing effect of the diffusion field and the stabilizing effect of the capillarity and the temperature gradient. It is their ratio that fixes front behavior. This discussion is summarized in figure B.VI.20, schematically showing the complete marginal stability curve of the front in the plane of the driving parameters (G,V). The front is unstable inside a closed domain, bounded by two curves joining at a certain value G_{max} of the temperature gradient.

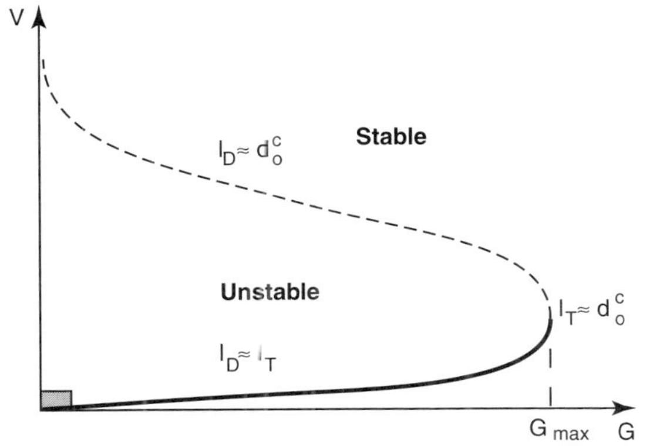

Fig. B.VI.20 Generic shape of the marginal stability curve in the plane of the driving parameters (G,V). The solid line = the destabilizing curve for the plane front; the dotted line = the absolute restabilization curve for the plane front. Only a very small region of the diagram is explored using a classical crystal (the shaded rectangle, typically). With nematics, the diagram is almost entirely accessible.

The lower curve fixes the destabilization threshold of the plane front. It can be shown that for a small gradient G capillarity has little effect. The destabilizing threshold is then given by:

$$l_D \approx l_T \qquad (G << G_{max}) \tag{B.VI.78}$$

which gives again the previously encountered formula B.VI.73a. Capillarity only plays a role to fix the wavelength, which follows the scaling law $\lambda_c \approx (d_0^c)^{1/3}(l_T)^{2/3}$ (see eq. B.VI.73b). This dependence is not intuitive and cannot be obtained by mere dimensional reasoning. Note that the $G^{-2/3}$ dependence of λ_c is well enough verified experimentally, in spite of the meniscus effects [24, 25, 33].

The upper curve defines the limit of absolute restabilization of the plane front. Along it, the diffusion length is small enough as to be comparable to the chemical capillary length:

$$l_D \approx d_0^c \tag{B.VI.79}$$

This relation determines the absolute restabilization speed V*:

$$V^* \approx \frac{\Delta T}{T_c}\frac{D}{d_0} \tag{B.VI.80a}$$

A more rigorous calculation shows that, for $G << G_{max}$, one has:

$$V^* \approx \frac{1}{K}\frac{\Delta T}{T_c}\frac{D}{d_0} \tag{B.VI.80b}$$

Finally, the two curves join for

$$l_D \approx d_0^c \approx l_T \tag{B.VI.81}$$

These relations give the value of the temperature gradient G_{max}, above which the front is always stable:

$$G_{max} \approx \frac{1}{d_0}\frac{(\Delta T)^2}{T_c} \tag{B.VI.82a}$$

A more rigorous estimate (see chapter B.IX), taking into account the stabilizing effect of diffusion in the nematic phase shows that:

$$G_{max} \approx \frac{1}{4K(1+K\beta)d_0}\frac{(\Delta T)^2}{T_c} \tag{B.VI.82b}$$

where $\beta = D_N/D_I$.

Let us now see how much of the bifurcation diagram can be explored experimentally, first for a classical material (metal or plastic crystal), and then for a nematic.

In a plastic crystal such as succinonitrile, the typical values are $D \approx 10^{-5}$ cm²/s, $\beta \approx 10^{-4}$, $K \approx 0.1$, $d_0 \approx 1$ Å, and $\Delta T \approx 1$°C, yielding $G_{max} \approx 10000$°C/cm. Such a gradient is not experimentally attainable. Let us now determine V_c and V* for G = 100°C/cm. We find $V_c \approx 10$ µm/s and $V^* \approx 30$ cm/s. Thus, if the interface is easy to destabilize, it is nonetheless impossible to reach absolute restabilization using a current setup. In conclusion, only a small part of the bifurcation diagram (the small rectangle close to the origin in figure B.VI.20 can be explored in classical materials.

In nematics, the situation differs radically. The first reason is that the diffusion constant for impurities is smaller than in the previously discussed materials by at least one order of magnitude: $D \approx 10^{-6}\ cm^2/s$. The destabilizing speed being proportional to D, interface instability appears sooner. Second, the impurity partition coefficient is close to 1 (0.9, typically), as both phases are fluid. This has a twofold outcome: it reduces V* and G_{max}, both varying as 1/K, and it allows for mixtures with very small ΔT. As a consequence, at equal impurity concentration, ΔT is ten times smaller in nematics than in plastic crystals or in metals. This strongly decreases both G_{max} (proportional to $(\Delta T)^2$) and the restabilization speed V* (linear in ΔT). For pure 8CB, used as received from the manufacturer, $\Delta T \approx 10^{-2}$ °C. With this value for ΔT and taking $d_o \approx 0.1$ Å, one has $G^* \approx 30$ °C/cm while $V_c \approx 20$ µm/s and $V^* \approx 300$ µm/s for G = 10°C/cm. These order of magnitude calculations show that the bifurcation diagram should be almost entirely accessible for a very pure nematic [35, 36].

This result is confirmed by the experimental bifurcation diagram in figure B.VI.21, obtained with a 9CB-10CB mixture [36].

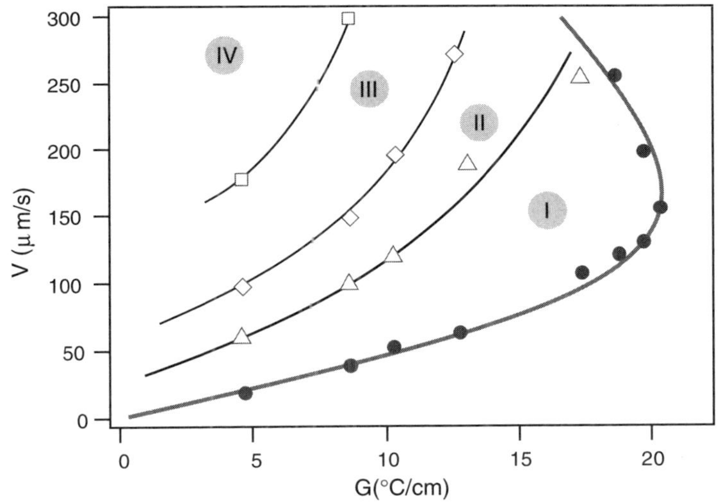

Fig. B.VI.21 Experimental bifurcation diagram for a 9CB-10CB mixture. The thick solid line shows the marginal stability curve. The dynamic behavior of the front in areas I to IV is discussed in the text (from ref. [36]).

Several areas (numbered from I to IV) can be distinguished inside the marginal stability curve, each one corresponding to a different behavior of the cellular front.

In area (I), the cellular front is stationary: the cells, of sinusoidal shape close to the threshold, become more and more "squared" and narrow with increasing speed. They do not drift and can locally readjust the wavelength by splitting or by the disappearance of isolated cells (Fig. B.VI.22).

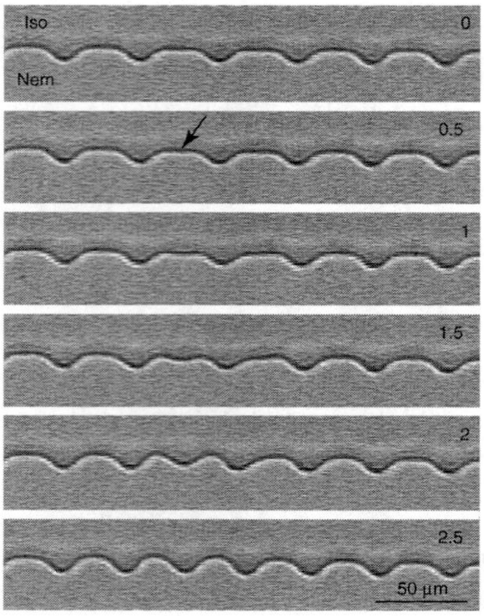

Fig. B.VI.22 Local cell division after a change in speed (from 70 to 80 µm/s) in area I of the bifurcation diagram. The time in seconds is shown in the upper right corner of each photo (9CB-10CB sample, d = 10 µm, G = 8.7°C/cm).

In area II, cell dynamics change. Asymmetric cells appear at the sample edges and drift along the front. These solitary modes, discovered by Simon et al. [37, 38], break the mirror symmetry of the interface and allow the system to select a new wavelength, for instance after a jump in pulling speed (Fig. B.VI.23).

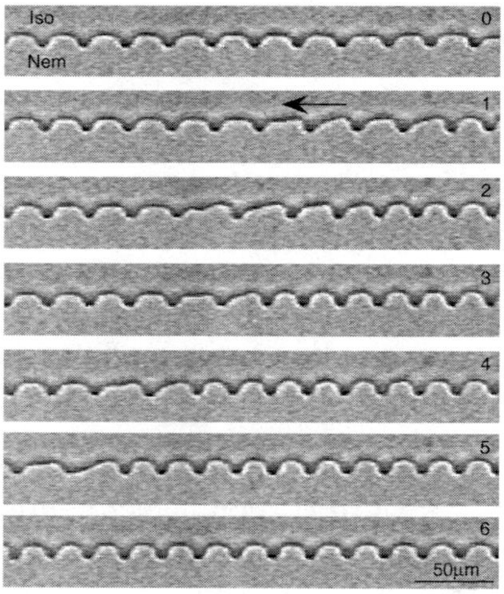

Fig. B.VI.23 Solitary mode propagating along the front (area II in the bifurcation diagram). The time in seconds is shown in the upper right corner of each photo. After the soliton passage, the wavelength decreases (9CB-10CB sample, d = 10 µm, G = 8.7°C/cm, V = 110 µm/s).

In area III, solitary modes disappear, being replaced by a new instability where the grooves between the cells pinch off, giving rise to isotropic liquid droplets which then rapidly diffuse in the nematic phase. This time-periodic instability leads to doubling of the spatial period, as the odd grooves and the even grooves alternately shed their droplets (Fig. B.VI.24).

Finally, droplet-shedding becomes chaotic in region IV.

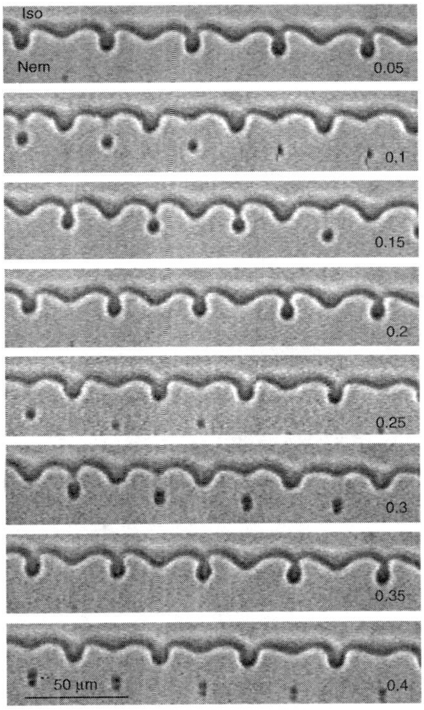

Fig. B.VI.24 Isotropic droplet formation by pinching off of the grooves between the cells (area III in the bifurcation diagram). The time in seconds is shown in the upper right corner of each photo (9CB-10CB sample, d = 10 µm, G = 8.7°C/cm, V = 160 µm/s).

A final point to be made is that these instabilities belong to the 10 generic instabilities of one-dimensional fronts as classified by Coullet and Iooss [39] on symmetry grounds. They have prompted considerable theoretical work, reaching far beyond the domain of liquid crystals. The interested reader is referred to the specialized papers [40] and to the recent book by K. Kassner [41] for a more detailed study.

We conclude this chapter by discussing another instability, typical of the nematic phase and involving a disclination line; this instability is often observed in thick samples.

VI.5.c) Depinning of a disclination line

If the speed of the solidification front exceeds a certain critical value (G and ΔT independent in this case), fairly rapidly, triangle-shaped planar regions appear on the nematic side (Fig. B.VI.25).

Triangle formation is triggered by the pinning on dust particles of the disclination line that forms close to the meniscus. The presence of the line is imposed by the molecular anchoring conditions on the front and on the glass plates (Fig. B.VI.26a). Examined under the polarizing microscope, the line is of purely wedge type (forming a planar configuration) in the 9CB-10CB mixture, as shown in figure B.VI.26a where we assumed homeotropic anchoring at the nematic-air interface. Thus, the line does not break the mirror symmetry of the interface and is not at the origin of the already described solitary modes. When it gets detached from the interface (after the passage of a dust particle), a planar region appears at the center of the sample.

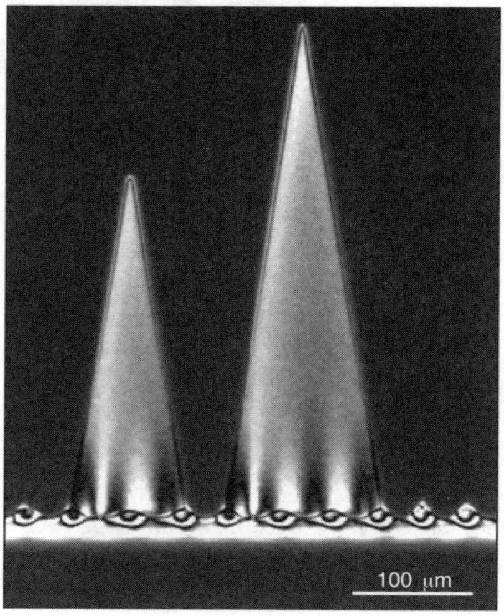

Fig. B.VI.25 Depinning of a disclination line at the nematic-isotropic liquid interface (d = 30 μm, V = 30 μm/s, G = 20°C/cm).

There are two possibilities: if the line moves slower than the front the size of the planar region increases; if it goes faster, it will rapidly catch up with the meniscus. Thus, in order to obtain a planar region, the solidification speed V must be higher than the speed V_l of the disclination line.

In fact, this description cannot successfully account for the observations; indeed, the line is strongly pinned to the interface and does not detach easily. This is because the favored molecular anchoring angle at the nematic-isotropic interface is closer to 45° than to 90° [6]. From an energy point of view, the line can be shown to become virtual, which amounts to displacing it to the isotropic side [42]. It is then easily understood that depinning always takes place locally, after "virtual line pinning" on a dust particle crossing the growth front. This dust particle usually becomes loose after giving rise to a planar orientation isosceles triangle with vertex angle 2θ.

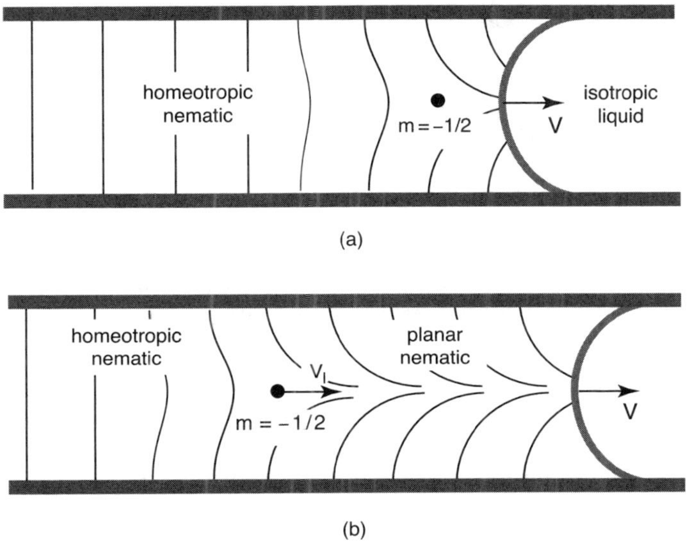

Fig. B.VI.26 a) Diagram demonstrating the presence of a −1/2 disclination line attached to the interface (i.e., to the meniscus); b) this line, imposed by the anchoring conditions on the plates and on the meniscus, can only move away from the interface if its speed V_l is lower than the growth speed V of the nematic phase. In this case, a planar region develops.

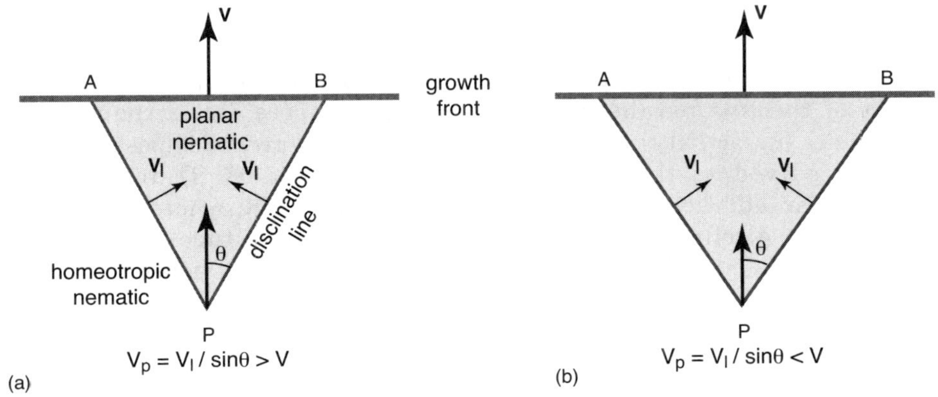

Fig. B.VI.27 Triangle evolution for stable N-I interface (high enough temperature gradient to avoid Mullins-Sekerka instability). a) The triangle disappears because its vertex angle is too small; b) the triangle is wider and grows homothetically.

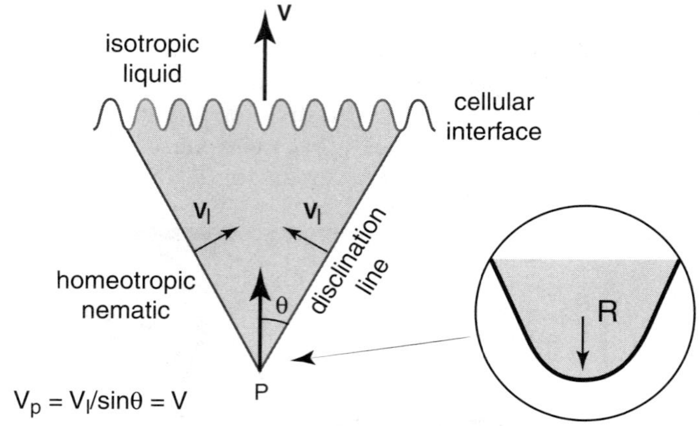

Fig. B.VI.28 Stationary triangle. Points A and B are locked by the cells and cannot drift freely along the macroscopic interface. The triangle then evolves towards a stationary equilibrium shape by adjusting the angle θ and the curvature radius R at the tip.

Triangle evolution then depends on the growth speed and the interface state:

– either the interface is stable and the pinning points A and B can drift freely (Fig. B.VI.27). The triangle then evolves homothetically, growing for $V > V_l/\sin\theta$, and shrinking to the point of disappearance for $V < V_l/\sin\theta$;

– or the interface is unstable and undulates as in the figure B.VI.25 photo (Fig. B.VI.28). The pinning points A and B are then strongly locked at the bottom of the grooves between cells. The triangle then evolves in a completely different manner, progressively changing its vertex angle θ and finally reaching a stationary shape. Obviously, this is only possible for $V > V_l$, when $\sin\theta = V_l/V$. This dependence is well verified experimentally [25] and allows for precise determination of V_l (Fig. BIV.29). The result is in good agreement with the law B.IV.42 in chapter IV, established for a wedge line, even though the line at the edge of a triangle is almost always of mixed type.

Finally, one can inquire about what happens at the tip of a triangle, as the part of the line parallel to the interface advances faster than the two triangle sides (in fact, at the interface speed V). This is rendered possible by the line being curved at this point (with curvature radius R) and therefore subjected to an additional inward force of magnitude T/R, where T is the line tension of the disclination (of the same order of magnitude as its energy, namely a few K). The balance of forces acting on this segment yields

$$V = V_l + \mathrm{Cst}\,\frac{K}{\gamma_1}\frac{1}{R} \tag{B.VI.83}$$

with Cst a factor of order unity. This law is equally verified experimentally (Fig. B.VI.30 [25]).

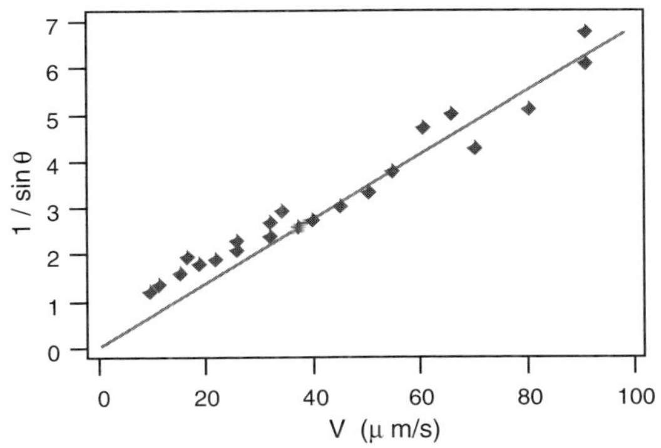

Fig. B.VI.29 Vertex angle (1/sinθ) for a stationary triangle as a function of the growth speed (8CB sample, d = 28 μm). The points were obtained for different values of the temperature gradient (between 36 and 112°C/cm) but fall on a unique curve (from ref. [25]).

Sometimes the disclination line does break the mirror symmetry of the interface, as in 8CB, where the director field lines tilt out of the plane normal to the interface (see chapter B.IV on the stability of wedge lines). This three-dimensional distortion of the director field induces cell drift close to the Mullins and Sekerka threshold [35, 36], the impurity diffusion field being anisotropic in the nematic phase in spite of the homeotropic alignment on the plates.

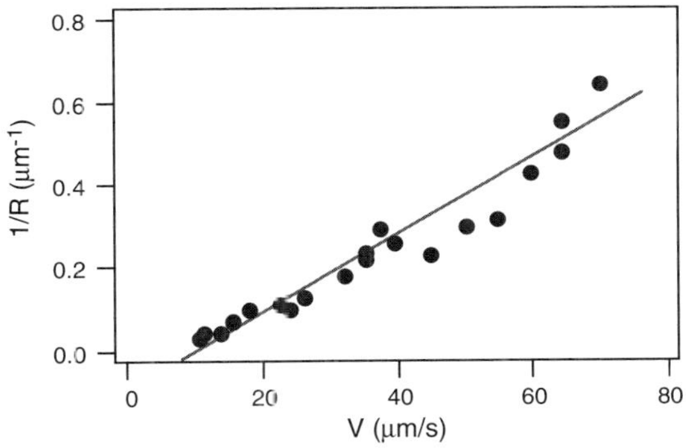

Fig. B.VI.30 Tip curvature of a stationary triangle as a function of the growth speed (8CB, d = 28 μm) (from ref. [25]).

This drift is not observed in the 9CB-10CB mixture, where the distortion is inhibited by the proximity of the smectic phase [36]. The cell drift close to the threshold therefore depends on the elastic properties of the material and is not due to nonlinear effects. Unlike solitary modes observed in all one-dimensional fronts, cell drift is system-specific.

In conclusion we shall briefly touch upon some recent experimental investigations of the cholesteric-isotropic liquid front. The situation is more complex because of the coupling between the Mullins-Sekerka wavelength and the cholesteric pitch, sometimes leading to the appearance of "breathing modes," where cell width oscillates in time [43]. Very disordered states, similar to orientational glasses, can appear at high speed [44]. Finally, the disclination line attached to the interface can also become depinned, leading to the formation of peculiar cholesteric fingers [45]. We shall return to these topics in the next chapter.

BIBLIOGRAPHY

[1] Vilanove R., Guyon E., Pieranski P., *J. Physique (France)*, **35** (1974) 153.

[2] Williams R., *Mol. Cryst. Liq. Cryst.*, **35** (1976) 349.

[3] Adamson A.W., *Physical Chemistry of Surfaces*, 2nd ed., Wiley-Interscience, New York, 1967, p. 28.

[4] Press M.J., Arrott A.S., *Phys. Rev. A*, **8** (1973) 1459.

[5] Faetti S., Palleschi V., *J. Chem. Phys.*, **81** (1984) 6254.

[6] a) Faetti S., Palleschi V., *Phys. Rev. A*, **30** (1984) 3241; b) ibid., *J. Physique Lettres (France)*, **45** (1984) L313.

[7] Meunier J., Langevin-Cruchon D., *J. Physique Lettres (France)*, **43** (1982) L185.

[8] Berremann D.W., *J. Opt. Soc. Am.*, **62** (1972) 502.

[9] Madhusudana N.V., Pratibha R., *Mol. Cryst. Liq. Cryst.*, **89** (1982) 249.

[10] De Gennes P.G., *Mol. Cryst. Liq. Cryst.*, **12** (1971) 193.

[11] Prost J., Marcerou J.P., *J. Physique (France)*, **38** (1977) 315.

[12] Barbero G., Dozov I., Palierne J.F., Durand G., *Phys. Rev. Lett.*, **56** (1986) 2056.

[13] McMullen W.E., *Phys. Rev. A*, **40** (1989) 2649.

[14] Fournier J.B., *Phys. Rev. Lett.*, **75** (1995) 854.

[15] Mermin N.D., *Quantum Fluids and Solids*, Eds. Trickey S.B., Adams E.D., and Dufty J.F., Plenum Press, New York, 1977, p. 3.

[16] Rudnick J., Bruisma R., *Phys. Rev. Lett.*, **74** (1995) 2491.

[17] Langer S., Sethna J.P., *Phys. Rev. A*, **34** (1986) 5035.

[18] Schartz D.S., Tsao M.-W., Knobler Ch., *J. Chem. Phys.*, **101** (1994) 8258.

[19] Rivière S., Meunier J., *Phys. Rev. Lett.*, **74** (1995) 2495.

[20] Note that in isotropic elasticity any vector field of the form $\Theta = m\theta$ (for any

m) is a solution of the bulk torque equation ($\Delta\Theta = 0$). In anisotropic elasticity ($K_1 \neq K_3$), m can only take the values m = 0 (uniform domain), m = 1 (radial or circular director field), and m = 2 (corresponding to the experimentally observed configuration).

[21] Galatola P., Fournier J.B., *Phys. Rev. Lett.*, **75** (1995) 3297.

[22] Mullins W.W., Sekerka R.F., *J. Appl. Phys.*, **35** (1964) 444.

[23] Armitage D., Price F.P., *Mol. Cryst. Liq. Cryst.*, **44** (1978) 33.

[24] Oswald P., Bechhoefer J., Libchaber A., *Phys. Rev. Lett.*, **58** (1987) 2318.

[25] Bechhoefer J., Ph.D. Thesis, University of Chicago, 1988.

[26] Oswald P., Moulin M., Metz P., Géminard J.C., Sotta P., Sallen L., *J. Phys. III (France)*, **3** (1993) 1891.

[27] Bechhoefer J., Simon A., Libchaber A., Oswald P., *Phys. Rev. A*, **40** (1989) 2042.

[28] Figueiredo J.M.A., Santos M.B.L., Ladeira L.O., Mesquita O.N., *Phys. Rev. Lett.*, **71** (1993) 4397.

[29] Figueiredo J.M.A., Mesquita O.N., *Phys. Rev. E*, **53** (1996) 2423.

[30] Bechhoefer J., Langer S.A., *Phys. Rev. E*, **51** (1995) 2356.

[31] Misbah C., Valance A., *Phys. Rev. E.*, **51** (1995) 1282.

[32] Cognard J., *Mol. Cryst. Liq. Cryst.*, **78** (1982) 1.

[33] Ignes-Mullol J., Oswald P., *Phys. Rev. E*, **61** (2000) 3969.

[34] Picano F., Graduate internship report, Ecole Normale Supérieure de Lyon, 1996.

[35] Simon A., Libchaber A., *Phys. Rev. A*, **41** (1990) 7090.

[36] Oswald P., *J. Phys. II (France)*, **1** (1991) 571.

[37] Bechhoefer J., Simon A., Libchaber A., Oswald P., "Directional solidification of liquid crystals," in *Random Fluctuations and Pattern Growth: Experiments and Models*, Eds. Stanley H.E. and Ostrowsky N., Kluwer Academic Publishers, Dordrecht, 1988.

[38] Simon A., Bechhoefer J., Libchaber A., *Phys. Rev. Lett.*, **61** (1988) 2574.

[39] Coullet P., Iooss G., *Phys. Rev. Lett.*, **64** (1990) 866.

[40] Kassner K., Misbah C., Müller-Krumbhaar H., Valance A., *Phys. Rev. E*, **49** (1994) 5477; ibid., *Phys. Rev. E*, **49** (1994) 5495.

[41] Kassner K., *Pattern Formation in Diffusion-Limited Growth*, World Scientific, Singapore, 1996.

[42] Ignés-Mullol J., Baudry J., Lejcek L., Oswald P., *Phys. Rev. E*, **59** (1999) 5562.

[43] Cladis P., Gleeson J.T., Finn P.L., Brand H.R., *Phys. Rev. Lett.*, **67** (1991) 3239.

[44] Brand H.R., Cladis P.E., *Phys. Rev. Lett.*, **72** (1994) 104.

[45] Baudry J., Pirkl S., Oswald P., *Phys. Rev. E*, **57** (1998) 3038.

Chapter B.VII

Cholesterics: the first example of a frustrated mesophase

The concept of **frustration** was introduced in physics by Gérard Toulouse and Philip Anderson to designate the absence in certain magnetic systems (spin glasses) of a **global configuration** of the spins allowing **all pairs** of neighboring spins to achieve a **local configuration** of minimal energy. An obvious example of magnetic frustration is provided by a triangular lattice of spins in antiferromagnetic interaction (i.e., two neighboring spins prefer a "head to tail" orientation.

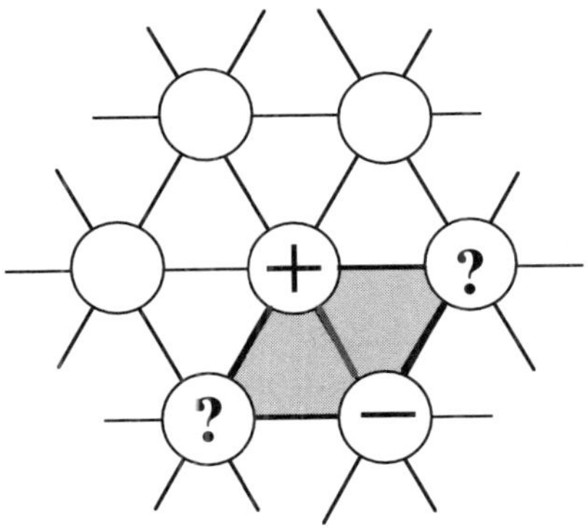

Fig. B.VII.1 Frustration in an antiferromagnetic spin system on a triangular lattice.

This kind of conflict between local tendencies and the impossibility of globally satisfying all of them also appears in other physical systems, such as glasses, amorphous metals, quasi-crystals, and certain mesophases.

Cholesterics represent the oldest and most studied case of a frustrated mesophase. Other examples are the Blue Phases (to be described in the next chapter) and certain smectic phases, such as the smectic A phase of chiral materials, the presence of the layers being incompatible with director field twist (see chapter C.IV of the second volume). One also encounters twisted smectic A phases (chapter C.V), where chirality is expressed by the appearance of twist walls composed of screw dislocations. Most of the lamellar phases in lyotropic systems are also frustrated: each lamella, flat at equilibrium, is made up of two monolayers, each of them having a finite spontaneous curvature.

Thus, whatever the frustration, physical systems manage to minimize it, giving rise to phases which are sometimes simple (smectics A and cholesterics), and sometimes odd and complicated, such as Blue Phases or twisted smectic A phases (also known as TGB_A).

This chapter, exclusively dedicated to the cholesteric phase, is structured as follows. In section VII.1 we return to the origin of cholesteric frustration and show that it can result in unexpected structural diversity. Section VII.2 provides an overview of the symmetry elements in the cholesteric phase and discusses the relevant order parameter describing the phase transition towards the isotropic liquid. We shall see that the transition can be second order if the cholesteric spontaneous twist is high enough. Section VII.3 explains some of the optical properties of this phase, such as its exceptionally high rotatory power or its sometimes vivid colors created by selective reflections on the cholesteric layers. The chapter continues with the description of the defect lines observed in the Cano wedge experiment (section VII.4), and with the study of the unwinding transition of the cholesteric helix, transition induced by a field or by geometric confinement between two homeotropically treated plates (section VII.5). Finally, we present some of the dynamic phenomena specific to this phase, such as permeation flow across the layers and the Lehmann rotation of the cholesteric helix in a temperature or electric potential gradient (section VII.6).

VII.1 Cholesteric frustration

Consider the two following mesogenic systems:

1. The nematic phase of a non-chiral material to which chiral impurities were added in a concentration c^*, and

2. An enantiomer mixture with an excess concentration $c^* = c_+ - c_-$, exhibiting a uniaxial nematic phase when racemic ($c^* = 0$).

At $c^* = 0$, the two systems are in the uniaxial nematic phase. Orientational molecular order is quadrupolar, spatial orientation being described by the director field n(r). In the absence of boundaries and applied external fields, $\mathbf{n}(\mathbf{r})$ = Cst at thermodynamic equilibrium.

For finite c^*, this homogeneous nematic order is perturbed by the local tendency of neighboring molecules to form a certain angle. As shown in figure

B.VII.2 below, this tendency resembles that of two springs fitting their spirals together.

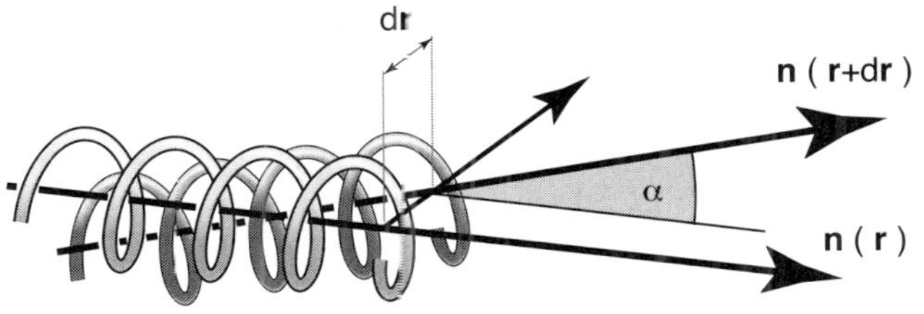

Fig. B.VII.2 Two neighboring chiral molecules tend to form a small angle α like these two springs.

Without going into the details of intermolecular interaction, let us point out that, in a pure enantiomer, the typical angle between two neighboring molecules does not exceed 1°.

We now ask the essential question [1a]:

Is there a global molecular configuration such that any two neighboring molecules achieve the twisted configuration in figure B.VII.2?

The answer is "no," as we shall prove by an *ad absurdum* argument. Suppose the answer were in the affirmative and consider two different paths leading from point 0, where the molecules are parallel to z, to the same point P (Fig. B.VII.3).

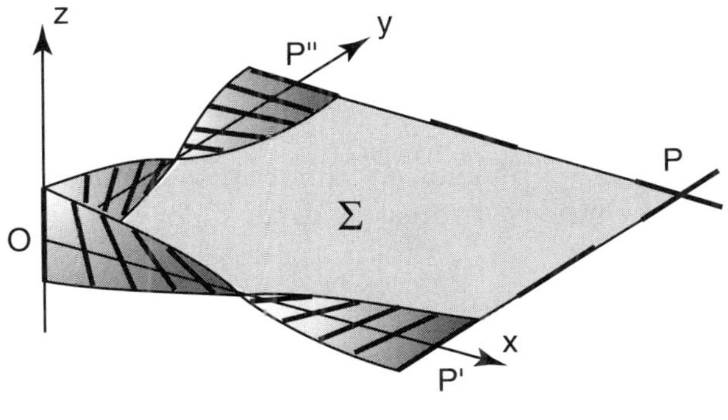

Fig. B.VII.3 Origin of cholesteric frustration: in order to satisfy the local tendencies on the paths OP'P and OP''P, different molecular orientations are obtained upon arrival at point P.

Let us now follow the path OP', satisfying the local tendencies as we go: the length of the path OP' is chosen such that at point P' the molecules will have turned by $\pi/2$, and are now parallel to y. Continue on the path from P' to

P. Molecules do not need to turn, so their orientation at P is the same as at P': parallel to y.

Now take the other path, from O towards P". If OP' = OP", the molecules are parallel to x at point P". From P" to P, the molecules do not turn, remaining parallel to x.

In conclusion, by following two different paths leading to the same point, two different orientations are obtained at the arrival.

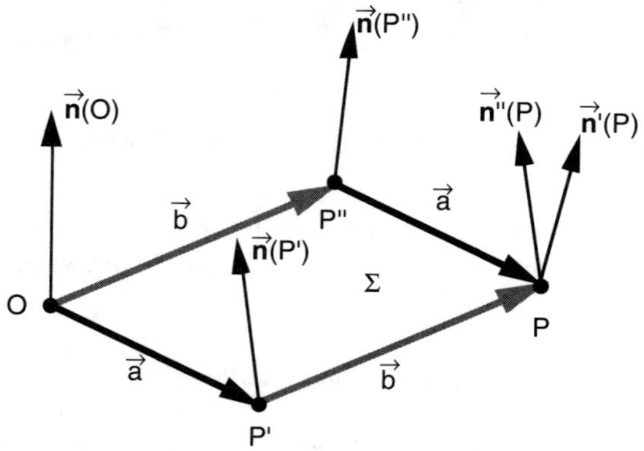

Fig. B.VII.4 Demonstration of the cholesteric frustration on infinitesimal paths.

The above reasoning can be generalized to any two points (O, P) infinitesimally close to one another and to all paths (a + b) and (b + a) from O to P going through P' and P", respectively [1b].

For determining the changes in molecular orientation along these two paths, let us first analytically express the tendency of molecules to make an angle with their neighbors. Supposing that local orientational order is uniaxial, molecular orientation can be described by a director field $\mathbf{n}(\mathbf{r})$ (Fig. B.VII.4).

In this uniaxial approximation, the local molecular tendency of turning one with respect to the other is expressed by the following differential equation:

$$\frac{\partial n_j}{\partial x_i} = -q\, e_{ijk}\, n_k, \qquad q = 2\pi/p \tag{B.VII.1}$$

where p is the distance on which molecules turn by 2π.
Follow now the path $\mathbf{a} + \mathbf{b}$. One has:

$$n_j(P') = n_j(O) - q\, a_i\, e_{ijk}\, n_k(O) \tag{B.VII.2}$$

and

$$n'_j(P) = n_j(P') - q\, b_i\, e_{ijk}\, n_k(P') \tag{B.VII.3}$$

whence, using eq. B.VII.1:

$$n'_j(P) = n_j(O) - q\, a_i\, e_{ijk}\, n_k(O) - q\, b_i\, e_{ijk}\, n_k(O) + q^2\, e_{ijk}\, e_{lkm}\, a_l\, b_i\, n_m(O) \qquad \text{(B.VII.4)}$$

Going along path **b** + **a**, the result is:

$$n''_j(P) = n_j(O) - q\, b_i\, e_{ijk}\, n_k(O) - q\, a_i\, e_{ijk}\, n_k(O) + q^2\, e_{ijk}\, e_{lkm}\, b_l\, a_i\, n_m(O) \qquad \text{(B.VII.5)}$$

The difference between **n'** and **n"** at point P is therefore:

$$n''_j\ (P) - n'_j(P) = q^2\, a_i\, b_l\, n_m(O)\, [\delta_{jl}\delta_{im} - \delta_{ji}\delta_{lm}] \qquad \text{(B.VII.6)}$$

or, in vector notation:

$$\mathbf{\delta n} = \mathbf{n''}(P) - \mathbf{n'}(P) = -\,q^2[\ \mathbf{n}(O) \times \mathbf{\Sigma}\ ] \qquad \text{with} \qquad \mathbf{\Sigma} = \mathbf{a} \times \mathbf{b} \qquad \text{(B.VII.7)}$$

In conclusion, trying to satisfy local tendencies on an infinitesimal closed circuit Σ brings about an orientational defect δn proportional, on the one hand, to the cross product n(O)×Σ and, on the other hand, to the square of the wave vector q.

This is an analytical formulation of cholesteric frustration.

Notice that the frustration is more or less "serious" according to the magnitude of the wave vector q. In other words, the larger the angle between two neighboring molecules, the stronger the cholesteric frustration.

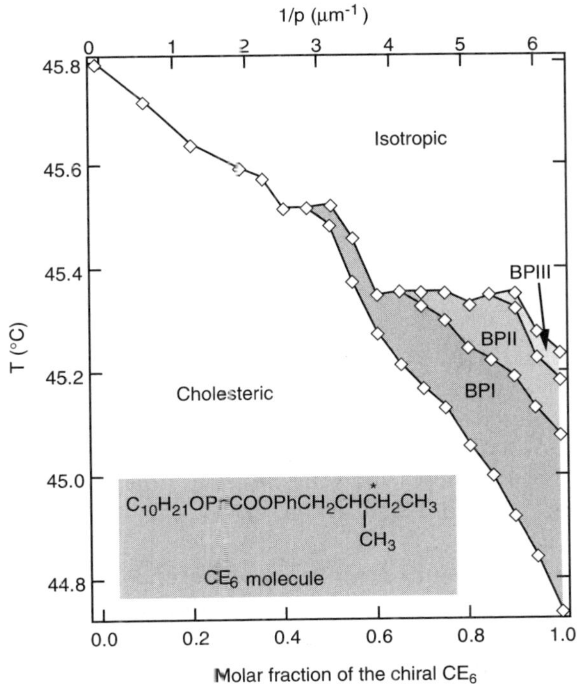

Fig. B.VII.5 Phase diagram of an enantiomeric mixture. Ph represents a 1,4 substituted phenyl ring (from D.K. Yang and P. P. Crooker [2]).

At this point, it is worth mentioning that cholesteric frustration disappears in a four-dimensional space, as shown by Sethna et al. [1c]. Inspired by this result, B. Pansu et al. [1d] constructed a perfect Blue Phase in S^3 space.

Let us come back to our three-dimensional euclidian space, where frustration is unavoidable and generates a wide structural diversity, as exemplified by the phase diagram of an enantiomeric mixture (CE6).

In the diagram, the abscissa represents the excess concentration $c^* = c_+ - c_-$. It also shows the degree of cholesteric frustration, increasing with c^*. It is apparent that, between 44.6°C and 46°C, the number of phases increases with the degree of frustration.

At $c^* = 0$, in the racemic mixture, only the nematic phase appears below the isotropic phase.

For $0 < c^* < 0.5$, there is still a unique phase below the isotropic one. This phase, a simply twisted version of the nematic phase, is the "cholesteric" phase (Ch).

For $c^* > 0.5$, two new phases appear between the isotropic and the cholesteric phases. They are termed "Blue Phases I and II" and will be presented in the next chapter. A third Blue Phase, denoted as BP III, will be discussed briefly in this book, as it is very seldom encountered.

Generally, as for the example in figure B.VII.5, Blue Phases only exist in a very narrow temperature range. This is why their existence has only been demonstrated 50 years after the era of Reinitzer, Lehmann, and Friedel. But let us now concentrate on the cholesteric phase.

VII.2 Cholesteric order parameter and the cholesteric-isotropic liquid phase transition

VII.2.a) Structure and symmetry

In figure B.VII.5, no transition line separates the cholesteric phase (Ch) from the nematic one (N). It follows that the nematic phase is only a particular case of the cholesteric one, defined by a strictly racemic composition of the enantiomeric mixture: $c^* = 0$.

When c^* is infinitesimal, so is the difference between the nematic and the cholesteric phases. To specify this difference, consider a finite but macroscopic volume V, of typical size L.

We know that, in the absence of external fields and boundary effects, the nematic phase exhibits **uniaxial quadrupolar** orientational long-range order. At thermodynamic equilibrium the orientation of the director **n** is homogeneous, so the symmetry of the nematic phase is $D_{\infty h}$.

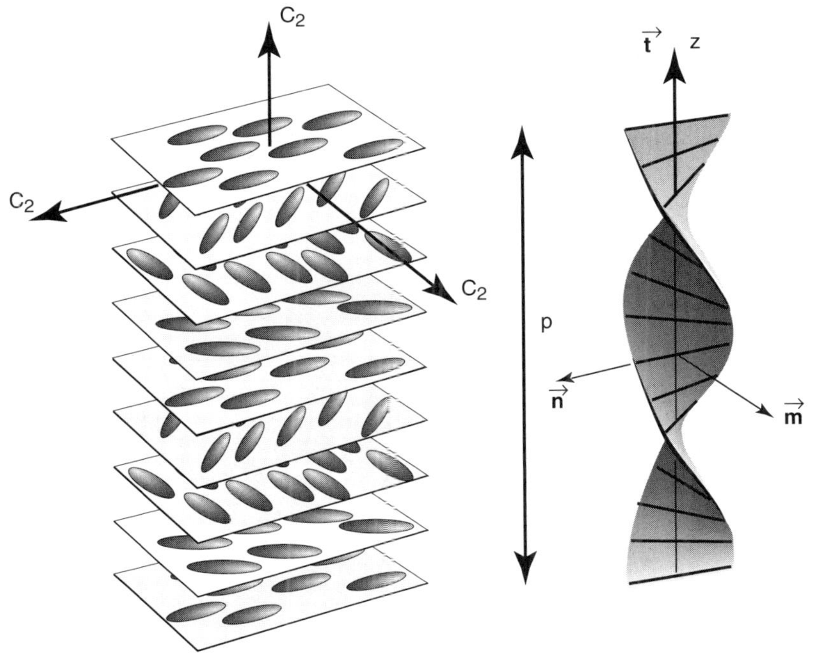

Fig. B.VII.6 Structure and symmetry of the cholesteric phase (left-handed helix, $p < 0$). The planes depicted here are fictitious, as there is no positional order.

The cholesteric phase differs from the nematic one in two essential points:

1. The average molecular orientation **n** turns about one (and only one) space direction **t** orthogonal to **n.** In a coordinate system with axis **z** parallel to **t**, the director components are:

$$n_x = \cos(qz)$$

$$n_y = \sin(qz) \qquad \text{(B.VII.8)}$$

$$n_z = 0$$

with $q = 2\pi/p$, where p is the helix pitch. For low chirality ($c^* << 1$), the wave vector q is proportional to c^*. Thus, in the limit $c^* \to 0$, the helix pitch diverges: $p \to \infty$.

2. The second difference is that local orientational order (defined in a volume of size much smaller than the pitch, but still macroscopic!) is no longer strictly uniaxial, but becomes **biaxial**.

Indeed, the $D_{\infty h}$ symmetry of the nematic phase is broken as soon as the helix appears, as the C_∞ axis parallel to **n** is replaced by a second-order C_2 axis. Therefore, only two among the C_2 axes orthogonal to C_∞ are conserved: the one parallel to the helix **t** and the one normal to both **n** and **t**.

In conclusion, the point symmetry of the cholesteric phase is D_2. The phase is therefore biaxial due to its helical configuration.

A last point to be made is that the cholesteric phase solves the frustration problem by only creating twist along one direction **t**. The downside to this compromise is that twist tendencies in all directions other than **t** are unsatisfied.

VII.2.b) Order parameter for the cholesteric phase

Because of this symmetry reduction with respect to the nematic phase, the order parameter of the cholesteric phase is biaxial quadrupolar, rather than uniaxial. Furthermore, one axis of the Q_{ij} quadrupole must be parallel to **z** (with **z** // **t**), and the other two must turn uniformly when moving along this axis.

The most general expression for a quadrupole with these properties is the following:

$$Q_{ij} = -\frac{Q_o}{\sqrt{6}}\begin{pmatrix} -1 & 0 & 0 \\ 0 & -1 & 0 \\ 0 & 0 & 2 \end{pmatrix} + \frac{Q_2}{\sqrt{2}}\begin{pmatrix} \cos(2qz) & \sin(2qz) & 0 \\ \sin(2qz) & -\cos(2qz) & 0 \\ 0 & 0 & 0 \end{pmatrix} \tag{B.VII.9}$$

Due to the invariance **n** ↔ −**n**, **t** ↔ −**t** and **m** ↔ −**m**, this tensor is of periodicity p/2. The first term of the expression represents the average orientational order of the cholesteric phase. Here, the "average" is taken over a much larger scale than the cholesteric pitch. This tensor is uniaxial, of symmetry axis parallel to **t** // **z**. The second term is the periodic part of the order parameter. It is a purely biaxial tensor, turning around **z**. Their sum is a new tensor, the biaxiality of which depends on the ratio

$$r = \frac{Q_2}{Q_o} \tag{B.VII.10}$$

In the z = 0 plane, the order parameter has a very simple expression:

$$Q_{ij}(z=0) = -\frac{Q_o}{\sqrt{6}}\begin{pmatrix} -\sqrt{3}r - 1 & 0 & 0 \\ 0 & \sqrt{3}r - 1 & 0 \\ 0 & 0 & 2 \end{pmatrix} \tag{B.VII.11}$$

This local order parameter becomes uniaxial when:

1) $r = +\sqrt{3}$ or $-\sqrt{3}$:

In these cases, the tensor axis in the z = 0 plane is oriented along **x** or **y**, respectively.

It is easily checked that, for these two values of r, the cholesteric order parameter is everywhere **locally uniaxial**. The tensor's revolution axis (i.e., the director **n**), parallel to the (x,y) plane, turns around z with the cholesteric helix. These two cases are distinguished by a p/4 helix translation along **z**.

2) r = 0

In this case, the Q_{ij} tensor, oriented along **z**, is z-independent: **this corresponds to the nematic uniaxial phase oriented in the z direction.**

For all other value of r, the local order parameter is biaxial.

VII.2.c) Grebel, Hornreich, and Shtrikman theory of the isotropic liquid → cholesteric transition

The usual method for the study of this transition [3] consists of developing the average free energy density in a power series of the quadrupolar order parameter Q_{ij}. It can be proved (as we shall admit in the following) that the most general development compatible with the system symmetries can be written, up to fourth-order terms, as:

$$F = V^{-1}\int\left\{\frac{1}{2}[aQ_{ij}Q_{ij} + c_1 Q_{ij,l}Q_{ij,l} + c_2 Q_{ij,i}Q_{lj,l} - 2de_{ijl}Q_{in}Q_{jn,l}] - \beta Q_{ij}Q_{jl}Q_{li} + \gamma\,(Q_{ij}Q_{ij})^2\right\} \tag{B.VII.12}$$

with $Q_{ij,l} = \partial Q_{ij}/\partial x_l$. In this development, "a" is proportional to a reduced temperature, going to zero and changing sign at a certain temperature T*; the other parameters, c_1, c_2, d, β, and γ are assumed to be temperature independent. The "d" term contains the chirality: it must therefore be zero in the racemic mixture and change sign with the chirality. Finally, for reasons of thermodynamic stability, coefficients c_1 and γ must be positive.

To describe the transition, let us replace Q_{ij} by its expression B.VII.9 in the general formula. After minimization with respect to q, this leads to

$$q = \frac{d}{2c_1} \tag{B.VII.13}$$

and to the following form for the average free energy density as a function of Q_o and Q_2:

$$F = [AQ_o^2 + (A-k^2)Q_2^2] - B[Q_o^3 - 3Q_oQ_2^2] + C[Q_o^2 + Q_2^2]^2 \tag{B.VII.14}$$

where

$$A = \frac{a}{2} = a_o\,(T-T^*)$$

$$k^2 = \frac{d^2}{2c_1} = 2q^2c_1, \tag{B.VII.15}$$

$$B = \frac{\beta}{\sqrt{6}} \qquad \text{and} \qquad C = \gamma$$

Note that k^2 goes to zero in the racemic mixture, where $c^* = 0$.

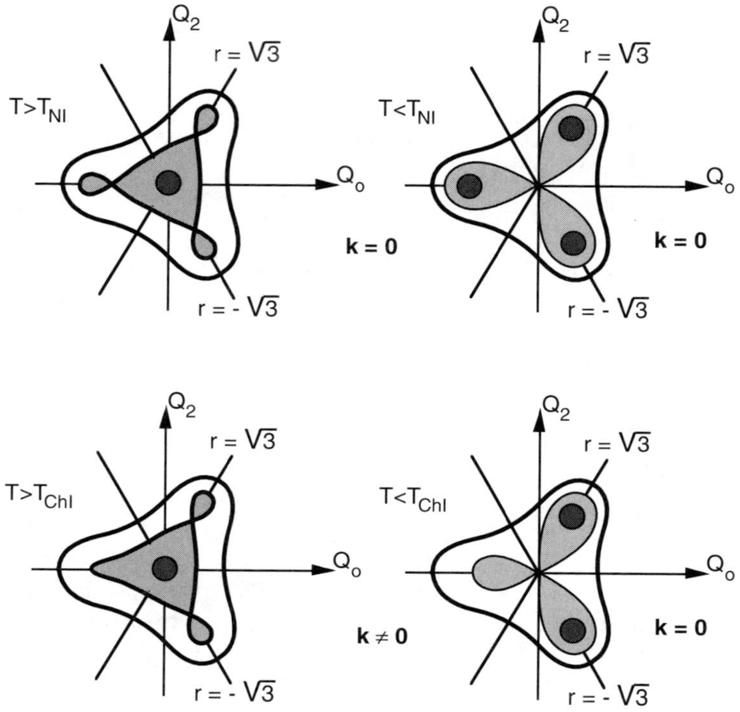

Fig. B.VII.7 Profile of the free energy F as a function of the amplitudes Q_o and Q_2 of the order parameter. For $k = 0$, F is invariant with respect to a $2\pi/3$ rotation in the (Q_o, Q_2) plane; this symmetry ensures the equivalence between the three orientations x, y, and z of the uniaxial nematic order parameter. For $k \neq 0$, the two $Q_2 \neq 0$ minima corresponding to the cholesteric phase are deeper than the third one, corresponding to the nematic phase.

Some comments are in order concerning eq. B.VII.14:

– first, it is invariant with respect to a change in the sign of Q_2, as this only amounts to shifting the helix by $p/4$;

– second, it is invariant under a $2\pi/3$ rotation in the parameter plane (Q_o, Q_2) when the mixture is racemic ($k = 0$). Actually, second- and fourth-order terms only depend upon the magnitude $[Q_o^2 + Q_2^2]$ of the (Q_o, Q_2) vector, thus being invariant under all rotations in the (Q_o, Q_2) plane. As to the third-order term, it is a very particular combination of Q_o^3 and $Q_oQ_2^2$, invariant under a $2\pi/3$

rotation in the (Q_0,Q_2) plane. This symmetry of the $F(Q_0,Q_2)$ function in the $k = 0$ limit shows that the three possible nematic states, corresponding to $r = \pm\sqrt{3}$ and $r = 0$ (see the previous section), with the director oriented along **x**, **y** and **z** respectively, are energetically equivalent.

Figure B.VII.7 shows the free energy profile as a function of Q_0 and Q_2. The gray levels represent the energy values, with the minima at the center of the darker areas.

In the racemic case $k = 0$, at high temperature ($T > T_{NI}$) there is only one minimum in $Q_0 = Q_2 = 0$ corresponding to the isotropic phase; at low temperature ($T < T_{NI}$), three minima with identical energies appear, placed on the axes given by $r = 0$ and $r = \pm\sqrt{3}$. These minima correspond to the uniaxial nematic phase oriented along x, y, or z.

For a cholesteric, $k \neq 0$ and the threefold symmetry is broken. At high temperature ($T > T_{ChI}$), the isotropic phase ($Q_0 = Q_2 = 0$) is still the energy minimum. However, when $T < T_{ChI}$ two real minima of equal depth appear on the $r = \pm\sqrt{3}$ axes, corresponding to the same cholesteric phase (in real space, the two states are only distinguished by a p/4 shift of the helix).

To determine these minima analytically and predict the transition order, the expression of $F(Q_0,Q_2)$ can be simplified introducing the following variables:

$$\mu_i = \frac{C}{B} Q_i, \qquad f = \frac{F}{B^4/C^3}, \qquad \frac{t}{4} = \frac{AC}{B^2}, \qquad \frac{\kappa^2}{4} = \frac{k^2 C}{B^2} \qquad \text{(B.VII.16)}$$

We are then left with:

$$f = \frac{1}{4} t\, \mu_0^2 + \frac{1}{4}(t - \kappa^2)\, \mu_2^2 + (\mu_0^3 - 3\mu_0\mu_2^2) + (\mu_0^2 + \mu_2^2)^2 \qquad \text{(B.VII.17)}$$

only containing two relevant parameters, t and κ, representing temperature shift with respect to T* and chirality, respectively.

Search for solutions is even simpler in the polar coordinates (μ, θ) defined by:

$$\mu_0 = \mu \sin\theta\ , \qquad \mu_2 = \mu \cos\theta \qquad \text{(B.VII.18)}$$

In this coordinate system, the conditions giving the reduced critical temperature t_{ChI} and the values of the order parameters m_0 and m_2 at the isotropic → cholesteric transition read:

$$\frac{f_{Ch} - f_{Iso}}{\mu^2} = \frac{t_{ChI}}{4} - \frac{\kappa^2}{4}\cos^2\theta + (\sin^3\theta - 3\sin\theta\cos^2\theta)\,\mu + \mu^2 = 0 \qquad \text{(B.VII.19a)}$$

$$\frac{1}{\mu}\frac{\partial f}{\partial \mu} = \frac{t_{ChI}}{2} - \frac{\kappa^2}{2}\cos^2\theta + 3(\sin^3\theta - 3\sin\theta\cos^2\theta)\,\mu + 4\mu^2 = 0 \qquad \text{(B.VII.19b)}$$

$$\frac{1}{\mu^2\cos\theta}\frac{\partial f}{\partial \theta} = \frac{\kappa^2}{2}\sin\theta + 3(4\sin^2\theta - 1)\,\mu = 0 \qquad \text{(B.VII.19c)}$$

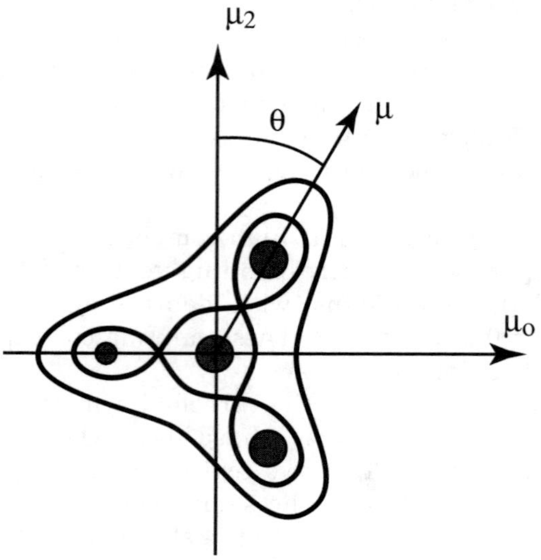

Fig. B.VII.8 Polar coordinates in the (μ_0, μ_2) plane.

Combining eqs. B.VII.19a and b we obtain a first relation between μ and θ at the I→Ch transition:

$$\mu = -\frac{1}{2}(\sin^3\theta - 3\sin\theta\cos^2\theta) \qquad \text{(B.VII.20)}$$

then substituting eq. B.VII.20 in eq. B.VII.19c, we obtain an expression relating θ to the chirality κ:

$$4\sin^2\theta = 2 - \sqrt{1 + \frac{\kappa^2}{3}} \qquad \text{(B.VII.21)}$$

It is clear from eq. B.VII.21 that for $\kappa \geq 3$ the only solution is $\theta = 0$ and $\mu = 0$, yielding $\mu_0 = \mu_2 = 0$. As there is no jump in order parameter at the transition, the latter is second order. The reduced transition temperature has then a very simple form, given by eq. B.VII.19a:

$$t_{ChI} = \kappa^2 \qquad (\kappa \geq 3) \qquad \text{(B.VII.22a)}$$

On the other hand, for $\kappa < 3$ the θ and μ equations B.VII.20 and 21 have two non-null solutions (figures B.VII.9 and 10). Note that the two symmetrical branches in figure B.VII.9 represent the same cholesteric phase. The order parameter has a discontinuity at the critical temperature and the transition is first order, with critical temperature:

$$t_{ChI} = \frac{1}{2}\left[1 + \kappa^2 + \left(1 + \frac{1}{3}\kappa^2\right)^{3/2}\right] \qquad (\kappa < 3) \qquad \text{(B.VII.22b)}$$

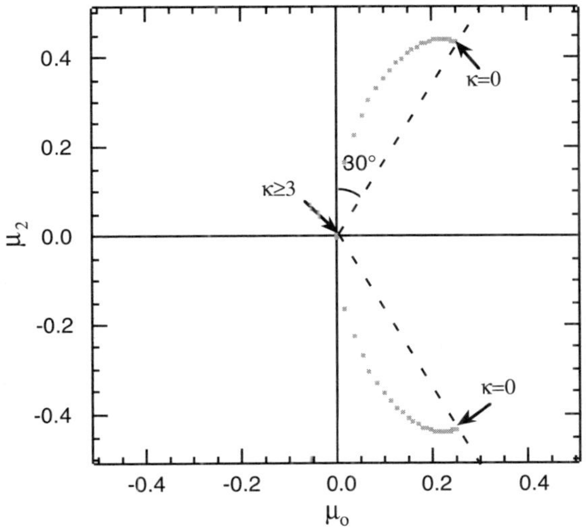

Fig. B.VII.9 Amplitudes μ_0 and μ_2 of the cholesteric order parameter as a function of the chirality κ at the critical temperature. For $\kappa \geq 3$, $\mu_0 = \mu_2 = 0$, the cholesteric-isotropic transition is second order. The following points correspond to values of κ decreasing in 0.1 steps. The transition is then first order. The last point corresponds to $\kappa = 0$ (racemic) and to the nematic-isotropic phase transition.

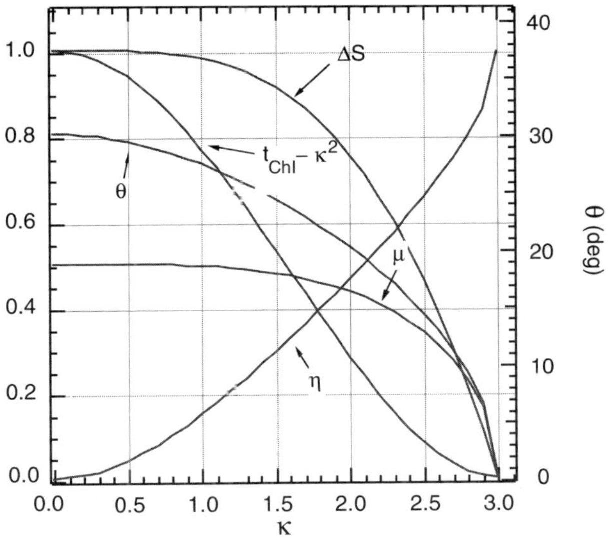

Fig. B.VII.10 Components θ and μ of the order parameter, transition entropy ΔS, reduced transition temperature t_{ChI}, and biaxiality degree η as a function of the chirality κ when the transition is first order.

One can equally determine the transition latent heat or, in different terms, the entropy jump $\Delta S \equiv \Delta S(\kappa)/\Delta S(0)$. From eq. B.VII.18:

$$\Delta S = 16\frac{df}{dt}(t = t_{ChI}) = 4\,\mu^2(t_{ChI})$$

$$= \frac{1}{4}\left[2 + 3\left(1 + \frac{1}{3}\kappa^2\right)^{1/2} - \left(1 + \frac{1}{3}\kappa^2\right)^{3/2}\right] \qquad (\kappa \leq 3) \qquad \text{(B.VII.23)}$$

This quantity goes to zero for $\kappa = 3$ as the transition becomes second order (Fig. B.VII.10).

The figure B.VII.9 graph also shows that the amplitudes μ_o and μ_2 of the cholesteric order parameter at the I → Ch transition vary with the chirality κ. In agreement with the threefold symmetry of the function $f(\mu_o, \mu_2)$ at zero chirality, the points corresponding à $\kappa = 0$ (racemic) are situated on the axes $\theta = 30°$ and $\theta = 150°$, giving the order parameter μ_{ij} the uniaxial character expected for the nematic phase. With increasing κ, angle θ decreases, leading to the biaxiality of the local order parameter $\mu_{ij}(z)$. The biaxiality degree of the order parameter μ_{ij} (or Q_{ij}) can be quantified by diagonalizing matrix $[Q_{ij}(\mathbf{r})]$. In the cholesteric phase, this diagonal form is **r**-independent and reads:

$$[Q_{ij}] \equiv Q \begin{bmatrix} \frac{1}{2}(-1-\eta) & 0 & 0 \\ 0 & \frac{1}{2}(-1+\eta) & 0 \\ 0 & 0 & 1 \end{bmatrix} \qquad \text{(B.VII.24)}$$

where parameter η characterizes the biaxiality degree of the quadrupolar order parameter at the transition temperature:

$$\eta = 1 - \frac{2\tan\theta}{\sqrt{3} - \tan\theta} \qquad \text{(B.VII.25)}$$

As shown in figure B.VII.10, η varies between 0, for $\kappa = 0$, and 1 for $\kappa \geq 3$.

The main result of the GHS model is that above $\kappa = 3$, i.e., in strongly chiral materials, the isotropic → cholesteric transition becomes second order.

In this case, the ensuing question is whether or not a first-order phase transition, at higher temperature than t_{ChI}, might lead to a new phase, of lower energy than the isotropic and cholesteric phases.

From the phase diagram in figure B.VII.5, we already know that the answer to this question is sometimes in the affirmative. Actually, experiment shows that, in strongly chiral products, the Blue Phases appear between the cholesteric phase and the isotropic liquid. We shall return to them in the following chapter.

Let us now study the optical properties of the cholesteric phase.

VII.3 Optical properties of the cholesteric phase

When the cholesteric pitch p is comparable to the wavelength of visible light, a well-oriented cholesteric sample between two plates treated in homeotropic anchoring exhibits very nice specular reflections.

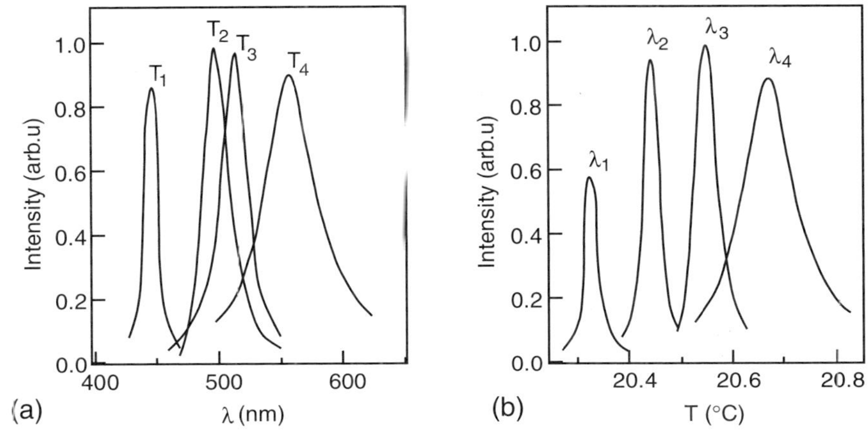

Fig. B.VII.11 Intensity of light selectively reflected by a cholesteric sample of cholesteryl oleyl carbonate: a) $T_1 = 20.68°C$, $T_2 = 20.55°C$, $T_3 = 20.50°C$; $T_4 = 20.44°C$; b) $\lambda_1 = 650$ nm; $\lambda_2 = 550$ nm; $\lambda_3 = 500$ nm; $\lambda_4 = 450$ nm. This material, extremely sensitive to temperature, has one of the highest temperature coefficients (from ref. [4]).

As plotted in figure B.VII.11, these reflections have very narrow bandwidth. Furthermore, they are (almost) circularly polarized. These two features show that some circularly polarized modes cannot propagate inside the liquid crystal so they are reflected. They can be considered as **Bragg reflections** of the light on the periodic helical structure of the cholesteric phase.

Light propagation in the cholesteric phase is described by the Maxwell equations:

$$\mathbf{rot\,E} = -\frac{\partial \mathbf{B}}{\partial t} \tag{B.VII.26a}$$

$$\mathbf{rot\,B} = \frac{1}{\varepsilon_0 c^2}\frac{\partial \mathbf{D}}{\partial t} \tag{B.VII.26b}$$

$$\mathbf{div\,B} = 0 \qquad \text{and} \qquad \mathbf{div\,D} = 0 \tag{B.VII.26c, d}$$

where

$$D_i = \varepsilon_0 \varepsilon_{ij} E_j \tag{B.VII.27}$$

The spatial periodicity of the dielectric susceptibility tensor ε_{ij} makes finding the eigenmodes a complicated task.

To simplify the analysis, assume that light propagates along the helix axis **t**//**z**. The electric field only has two non-null components:

$$E_x = E_x(z)\, e^{-i\omega t} \tag{B.VII.28a}$$

and

$$E_y = E_y(z)\, e^{-i\omega t} \tag{B.VII.28b}$$

The two first Maxwell equations yield:

$$\mathbf{e}_z \times \frac{\partial \mathbf{E}}{\partial z} = i\omega\, \mathbf{B} \tag{B.VII.29a}$$

and

$$\mathbf{e}_z \times \frac{\partial \mathbf{B}}{\partial z} = \frac{-i\omega}{\varepsilon_0 c^2}\, \mathbf{D} \tag{B.VII.29b}$$

with $\mathbf{e}_z$ the unit vector of the **z** axis.

Replacing eq. B.VII.29b in eq. B.VII.29a, one has:

$$\mathbf{D} = -\frac{\varepsilon_0 c^2}{\omega^2}\frac{\partial^2 \mathbf{E}}{\partial z^2} \tag{B.VII.30}$$

whence, using eq. B.VII.27:

$$\frac{d^2 E_i}{dz^2} = -\left(\frac{\omega}{c}\right)^2 \varepsilon_{ij}\, E_j, \qquad i, j = x, y \tag{B.VII.31}$$

In cartesian coordinates with axis **z** // **t**, the restriction to the (x,y) plane of the dielectric tensor ε_{ij} corresponding to a right-handed cholesteric helix reads, in the uniaxial approximation (Fig. B.VII.6):

$$\underline{\varepsilon}(z) = \frac{\varepsilon_{//} + \varepsilon_\perp}{2}\begin{bmatrix} 1 & 0 \\ 0 & 1 \end{bmatrix} + \frac{\varepsilon_{//} - \varepsilon_\perp}{2}\begin{bmatrix} \cos(2qz) & \sin(2qz) \\ \sin(2qz) & -\cos(2qz) \end{bmatrix} \tag{B.VII.32}$$

with $q = 2\pi/p$ and $p > 0$.

Defining:

$$k_0^2 = \left(\frac{\omega}{c}\right)^2 \frac{\varepsilon_{//} + \varepsilon_\perp}{2} \tag{B.VII.33a}$$

and

$$k_a^2 = \left(\frac{\omega}{c}\right)^2 \frac{\varepsilon_{//} - \varepsilon_\perp}{2} = k^2\,\frac{\varepsilon_a}{2} \tag{B.VII.33b}$$

one gets from equations B.VII.31-32:

$$-\frac{d^2E_x}{dz^2} = k_o^2E_x + k_a^2[\cos(2qz)\,E_x + \sin(2qz)\,E_y] \qquad \text{(B.VII.34a)}$$

$$-\frac{d^2E_y}{dz^2} = k_o^2E_y + k_a^2[\sin(2qz)\,E_x - \cos(2qz)\,E_y] \qquad \text{(B.VII.34b)}$$

With the new variables:

$$E_\pm = E_x \pm iE_y \qquad \text{(B.VII.35)}$$

these equations become:

$$-\frac{d^2E_+}{dz^2} = k_o^2E_+ + k_a^2 e^{2iqz}E_- \qquad \text{(B.VII.36a)}$$

$$-\frac{d^2E_-}{dz^2} = k_o^2E_- + k_a^2 e^{-2iqz}E_+ \qquad \text{(B.VII.36b)}$$

The solutions [$E_+(z)$, $E_-(z)$] to these two coupled differential equations are usually taken in the form

$$E_+(z) = a\,e^{i(l+q)z} \qquad \text{(B.VII.37a)}$$

$$E_-(z) = b\,e^{i(l-q)z} \qquad \text{(B.VII.37b)}$$

which has the advantage of the exponential terms disappearing upon substitution in eqs. B.VII.36. Before writing down the corresponding dispersion relation and the equation linking a and b, let us first analyze the wave polarization. According to definition B.VII.35, the complex amplitudes of the electric field are:

$$E_x(z) = \frac{E_+(z) + E_-(z)}{2}, \qquad E_y(z) = \frac{E_+(z) - E_-(z)}{2i} \qquad \text{(B.VII.38)}$$

yielding the (real) components of the electric field:

$$E_x = \mathrm{Re}[E_x(z)e^{-i\omega t}] = \frac{1}{2}a\cos[(l+q)z - \omega t] + \frac{1}{2}b\cos[(l-q)z - \omega t] \qquad \text{(B.VII.39a)}$$

$$E_y = \mathrm{Re}[E_y(z)e^{-i\omega t}] = \frac{1}{2}a\sin[(l+q)z - \omega t] - \frac{1}{2}b\sin[(l-q)z - \omega t] \qquad \text{(B.VII.39b)}$$

Figure B.VII.12 shows the wave polarization in the $z = 0$ plane for different values of the amplitudes a and b. Polarization is circular right-handed if $a \neq 0$ and $b = 0$ (E_+ wave), circular left-handed if $a = 0$ and $b \neq 0$ (E_- wave), linear if $a/b = \pm 1$, and elliptical otherwise.

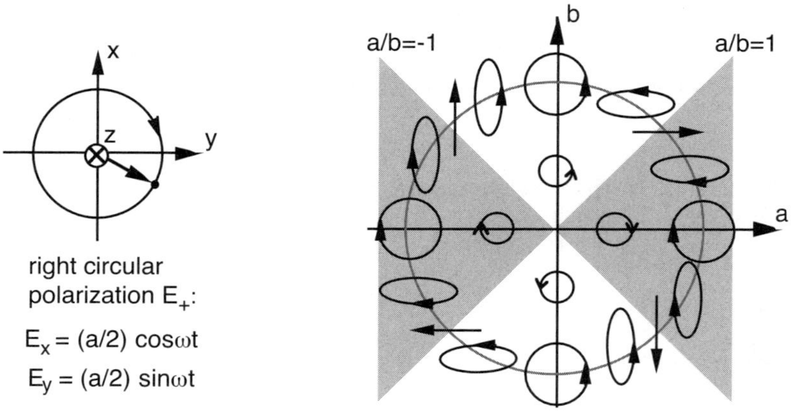

Fig. B.VII.12 Wave obtained by the superposition of two circularly polarized waves, right- and left-handed, respectively.

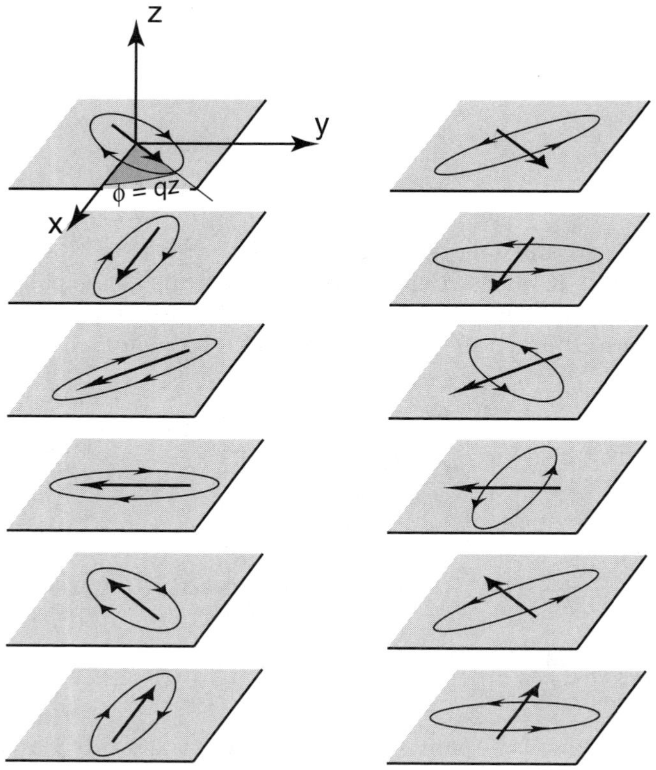

Fig. B.VII.13 The eigenmodes l_1 and l_2 of a cholesteric are right- and left-handed elliptically polarized; the ellipses' axes follow the director in its rotation along the helix (right-handed in the diagram).

For arbitrary z, the total wave is the superposition of the two circularly polarized waves (eq. B.VII.39), one right-handed (E_+ wave of amplitude a) and the other left-handed (E_- wave of amplitude b). Thus, the general result is an elliptically polarized wave. Moreover, as the wave vectors l + q and l – q of the two circular vibrations differ by 2q, **the axes of the polarization ellipsis of the total vibration follow the director in its rotation along the cholesteric helix** (Fig. B.VII.12). This is an essential feature of the eigenmodes.

Let us now return to their detailed calculation. Replacing eqs. B.VII.37a and b in eqs. B.VII.36a and b we obtain a system of two coupled linear equations for the amplitudes a and b:

$$[(l+q)^2 - k_o^2]\, a - k_a^2\, b = 0 \qquad \text{(B.VII.40a)}$$

$$[(l-q)^2 - k_o^2]\, b - k_a^2\, a = 0 \qquad \text{(B.VII.40b)}$$

Setting the determinant to zero yields the **dispersion relation** relating the wave vector l to the frequency ω depending on k_a and k_o:

$$[(l+q)^2 - k_o^2]\,[(l-q)^2 - k_o^2] - k_a^4 = 0 \qquad \text{(B.VII.41)}$$

For a given ω, this fourth-order equation in l has four solutions l(ω) grouped into two pairs of opposite sign as shown in the diagrams of figures B.VII.13a and b:

$$\pm l_2 \;, \qquad \pm l_1 = \pm\sqrt{(k_o^2 + q^2) \pm \sqrt{4k_o^2 q^2 + k_a^4}} \qquad \text{(B.VII.42)}$$

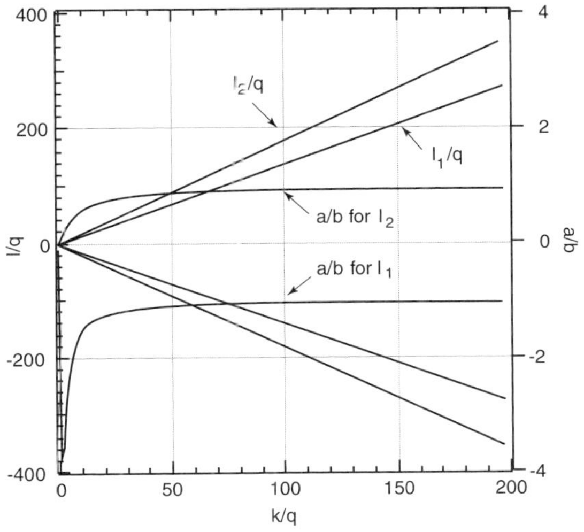

Fig. B.VII.13a Eigenmodes of a cholesteric in the limit k >> q, where the helix pitch is much larger than the light wavelength.

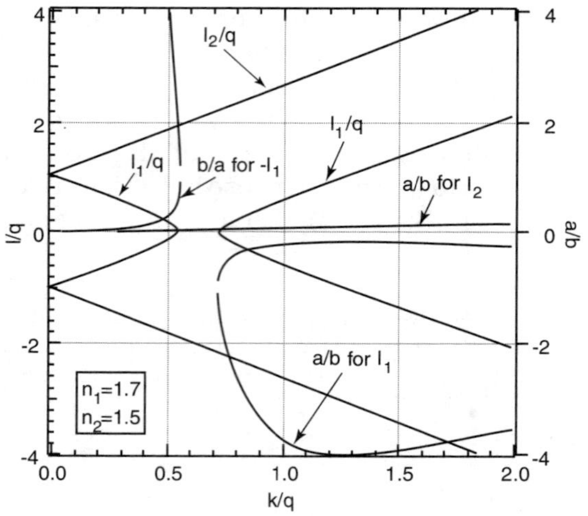

Fig. B.VII.13b Optical eigenmodes of a cholesteric in the limit of vacuum wave vector $k = \omega/c$ close to the wave vector q of the helix.

These diagrams also show the ratio a/b (or b/a) corresponding to each of the modes l_1 and l_2, obtained from the equations

$$a/b = k_a^2/[(l+q)^2 - k_0^2] \qquad \text{(B.VII.43a)}$$

or

$$b/a = k_a^2/[(l-q)^2 - k_0^2] \qquad \text{(B.VII.43b)}$$

In the limit k >> q, i.e., when the helix pitch is very large compared to the light wavelength, the two solutions of eq. B.VII.41 are:

$$l_2^2 \ , \quad l_1^2 \approx k_0^2 \pm k_a^2 \qquad \text{(B.VII.44a)}$$

with:

$$l_1 \approx \sqrt{\varepsilon_\perp}\, k = n_o k \qquad \text{and} \qquad l_2 \approx \sqrt{\varepsilon_{//}}\, k = n_e k \qquad \text{(B.VII.44b)}$$

Replacing l by their expressions B.VII.44 in formulae B.VII.43, one finds that the two modes l_1 and l_2 are linearly polarized at right angles to each other (see the diagram in figure B.VII.12).

The dispersion relations B.VII.44 further show that the refractive indices correspond to the extraordinary and ordinary indices of the twisted nematic. Thus, in the adiabatic limit, we recover the extraordinary and ordinary modes of twisted nematic cells (see section I.5.d in chapter B.I).

When the k/q ratio decreases, modes l_1 and l_2 become right- and left-handed elliptically polarized.

The ratio a/b of mode l_2 uniformly decreases with decreasing k, the mode becoming parfectly left-handed circular at k = 0.

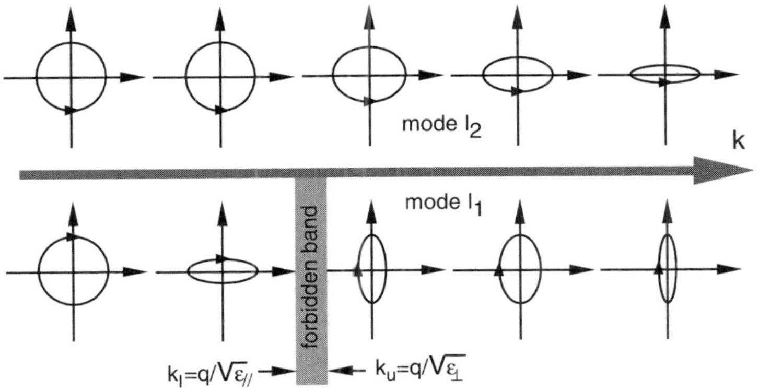

Fig. B.VII.14 Profile of the eigenmodes l_1 and l_2 as a function of the wave vector k.

As to the l_1 mode, its polarization is again linear when the wave vector k decreases to a certain value k_u (see Fig. B.VII.14). Then, **between k_u and k_l**, the solution l_1 is no longer real, meaning that the mode cannot propagate without attenuation. These two particular values of k are obtained by setting l_1 equal to zero. They are:

$$k_u = \frac{q}{\sqrt{\varepsilon_\perp}} \quad \text{and} \quad k_l = \frac{q}{\sqrt{\varepsilon_{//}}} \tag{B.VII.45}$$

(assuming $\varepsilon_{//} > \varepsilon_\perp$). In the absence of absorption, mode l_1 is reflected (Bragg reflection). The range of wave vectors corresponding to Bragg reflection has a finite width δk depending on the optical anisotropy ε_a. This spectral band where Bragg reflection takes place is named after Darwin, who first determined its width for X-ray diffraction. From B.VII.45, it is:

$$\frac{\delta k}{q} = \frac{1}{n_o} - \frac{1}{n_e} \approx 10^{-1} \tag{B.VII.46}$$

Note that, in comparison with X-rays, where refractive index modulation inside a crystal is of the order of 10^{-6}, the Darwin band in a cholesteric is a factor of 10^5 wider.

For $k = k_u$, mode l_1 is linearly polarized at 90° from the $k = k_l$ polarization.

Finally, for $k < k_l$ the polarization is again right-handed elliptical and tends to become circular in the limit $k \to 0$.

Thus, **in the limit $k << q$**, the two modes l_1 and l_2 are practically circularly polarized: left-handed for l_2 (as a/b goes to zero) and right-handed for l_1 (b/a also going to zero). The wave vectors of these modes

$$k_1 = -l_1 + q \qquad \text{and} \qquad k_2 = l_2 - q \tag{B.VII.47}$$

are slightly different, as seen from the plots in figures B.VII.15 and B.VII.16. This difference in wave vectors, and hence in phase velocity between the two modes gives the rotatory power

$$\rho = \frac{k_1 - k_2}{2} \tag{B.VII.48}$$

characteristic of the cholesteric phase in the wave vector domain where the eigenmodes are almost circularly polarized.

This rotatory power, numerically determined from eq. B.VII.42, is given as a function of the incident wave vector k/q in the figure B.VII.16 diagram below (solid line). In the approximation

$$k_1^4 << 4k_0^2q^2 \tag{B.VII.49}$$

solution B.VII.42 of the dispersion relation can be developed, yielding the following analytical expression:

$$\rho = \frac{\varepsilon_a^2\, k^4}{8q(q^2 - k_0^2)} \;, \tag{B.VII.50}$$

known as the **de Vries** equation [5], after the author of one of the first optical models for cholesterics. The de Vries expression is shown in the dotted line in the figure B.VII.16 graph.

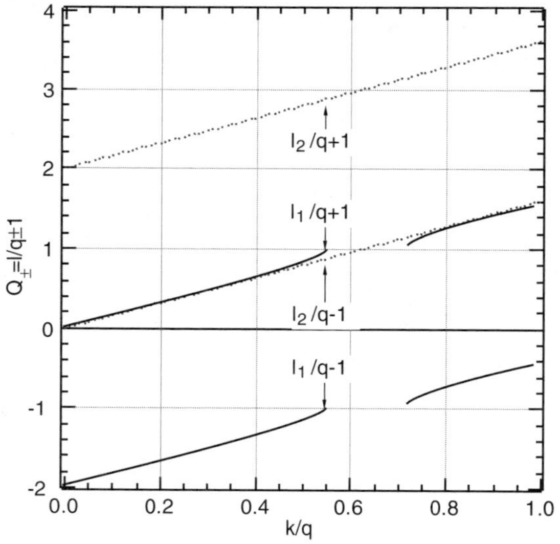

Fig. B.VII.15 Wave vectors $l \pm q$ of the right- and left-handed circular components of modes l_1 and l_2.

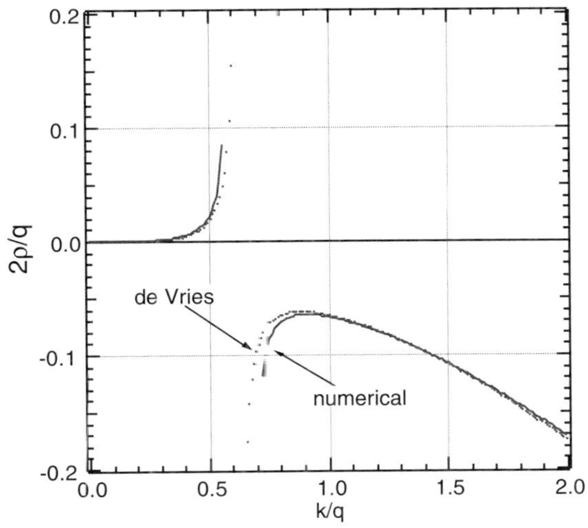

Fig. B.VII.16 Rotatory power of the cholesteric phase, calculated as the difference between the wave vectors k_1 and k_2 of the right- and left-handed elliptically polarized modes. Solid line: numerical calculation from eq. B.VII.42; dotted line: de Vries formula (eq. B.VII.50).

All these calculations predict that:

1. The rotatory power changes sign across the Darwin band [k_l, k_u].

2. The rotatory power increases when the wave vector k approaches the limits k_l and k_u of the Darwin band.

These predictions are in agreement with experimental determinations, such as those on cholesteryl cinnamate at 177°C (Fig. B.VII.17).

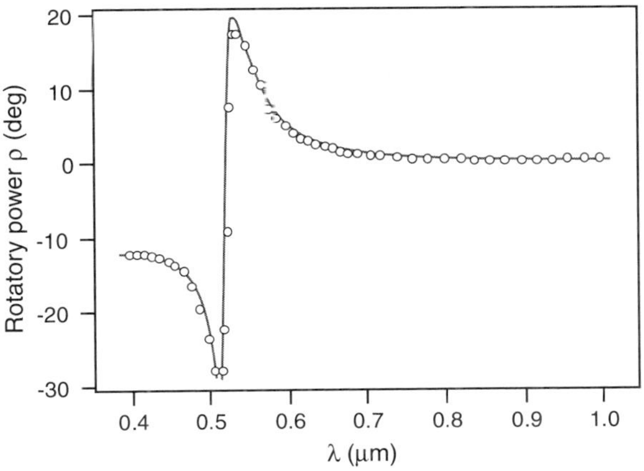

Fig. B.VII.17 Rotatory power of cholesteryl cinnamate (from ref. [6a]).

In conclusion, let us note that the optical properties of cholesterics and the enormous variation in helix pitch with composition, temperature, and pressure have found practical applications [6b]. Thus, the variation p(T) is used in thermography to reveal the temperature distribution for the surface of a body previously covered with a thin cholesteric layer. One can also find composite plastic sheets enclosing minute cholesteric droplets. These sheets change color with temperature and can therefore serve as thermometers. Finally, cholesterics are employed for painting (visual arts) and in displays.

VII.4 Defects and textures of the cholesteric phase

The cholesteric phase, like most mesophases, can be identified by the very characteristic textures it exhibits under the polarizing microscope, in transmission, or in reflection. The texture of a sample depends on several factors, such as sample shape and size, anchoring conditions at the boundaries, the helix pitch or the presence of external fields.

We start this description with the most classical texture, obtained when the helix axis **t** is normal to the sample surfaces. We shall show that its observation allows for accurate determination of the cholesteric pitch.

VII.4.a) "The Grandjean-Cano wedge"

Uniform orientation of the cholesteric helix is easily achieved by placing the liquid crystal between two surfaces treated in **planar anchoring.** The surfaces can be of freshly cleaved mica or glass plates covered with a rubbed polymer layer or, better yet, a SiO film evaporated under oblique incidence.

A particularly interesting geometry is one where the distance h(x, y) between the two surfaces slowly varies along one direction [7–10]. In the figure B.VII.18a example, the sample is sandwiched between a glass plate and a cylindrical lens. The thickness, almost zero at the contact point, increases from left to right.

The photo, taken in transmission between crossed polarizers, shows a row of clearly separated stripes. It is easily checked that, in this texture, **the number n (= 0,1,2,...) of helix half-pitches fitting in the space between the two glass plates changes discontinuously from one stripe to the next**. In the given example, this number varies by one unit between two neighboring stripes, until n = 10 at least. The most distant black stripes are fringes of equal rotatory power. Note that they are split in three, indicating a sawtooth variation of the lightwave rotation. This is because the local half-pitch of the helix $p_l(x)/2$ changes with x as follows:

$$\frac{p_l(x)}{2} = \frac{h(x)}{n} \tag{B.VII.51}$$

One expects the twist energy to be minimal if n is an integer such that the local pitch is as close as possible to the natural pitch p. One therefore expects the stripes to be centered on areas of normal twist, which shall be rigorously proved later. The left edge of each stripe (Fig. B.VII.18b) is then situated at abscissa x_m and thickness:

$$h(x_m) = (2m-1)\frac{p}{4}, \quad m = 1, 2, \ldots \tag{B.VII.52}$$

as confirmed in experiments.

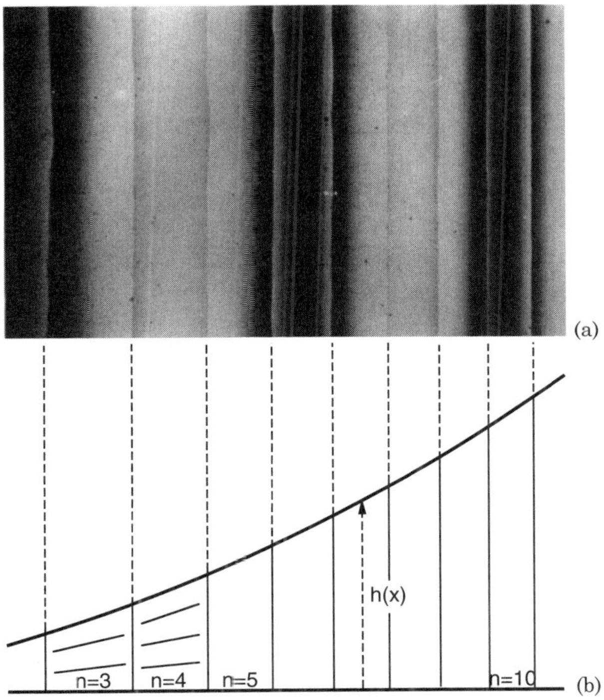

Fig. B.VII.18 a) Cholesteric texture in a "Cano wedge" between a glass plate and a cylindrical lens with radius R = 10 cm. Observation in transmitted light between crossed polarizers. The sample is a mixture of PAA and cholesterol benzoate (5%). b) Arrangement of the cholesteric layers (photo by M. Brunet-Germain [9]).

For convenience, one often employs the sphere-plane geometry, rather than cylinder-plane. In this case:

$$h(x) = \frac{x^2}{R} \tag{B.VII.53}$$

with R the sphere radius and:

$$x_m^2 = (2m - 1)R\frac{p}{4} \tag{B.VII.54}$$

This linear dependence ($x_m^2 \propto m$) was checked experimentally by many groups, such as Feldman et al. [10], who went to order m = 10 (Fig. B.VII.19) in a very small pitch cholesteric.

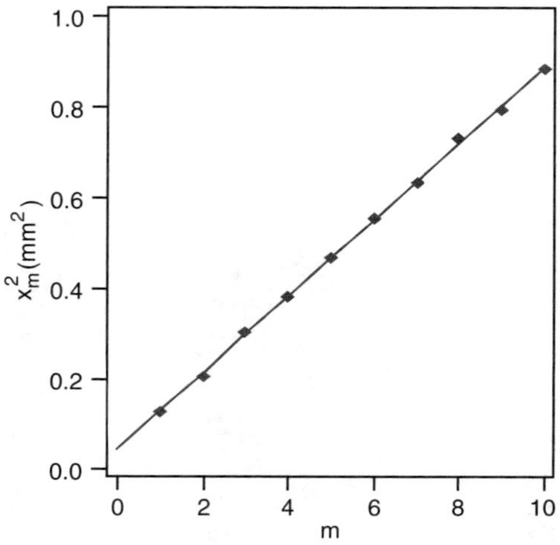

Fig. B.VII.19 Experimental verification of the law B.VII.53 in a cholesteric with small pitch $p \approx 0.65$ µm (from Feldman et al. [10]). The law is verified up to an additive constant, as the sphere and the plane are never in contact at the center.

A more rigorous proof of the result B.VII.52 can be obtained by calculating the elastic distortion energy of the cholesteric inside the nth stripe. For convenience, we assume that the director remains parallel to the glass plates and that the twist is constant in the transverse direction. In x_n, the thickness is np/2. On both sides of this point, the twist is given by **n.curln** – $n\pi/h(x)$, while the number of half-pitches remains unchanged. The free energy density f_n in the nth band at x is then, neglecting K_1 and K_3 contributions:

$$f_n(x) = \frac{1}{2} K_2 \left(q - \frac{n\pi}{h(x)}\right)^2 \tag{B.VII.55}$$

with $q = 2\pi/p$ the spontaneous cholesteric twist. In figure B.VII.20, f_n is shown for the different stripes when the surfaces limiting the sample are two planes making an angle α. Clearly, the passage from one stripe to the next occurs where the curves intersect. This condition ensures minimal total twist for the system. Thus, the position of the mth separation line, between stripes m – 1 and m, is given by:

$$f_{m-1}(x_m) = f_m(x_m) \tag{B.VII.56}$$

yielding

$$q - \frac{(m-1)\pi}{h(x_m)} = -\left[q - \frac{m\pi}{h(x_m)}\right] \tag{B.VII.57}$$

whence

$$h(x_m) = (2m-1)\frac{p}{4}, \quad m=1, 2, \ldots \tag{B.VII.58}$$

This equation is the same as eq. B.VII.52 which we stated without proof. Although well verified experimentally, this law is not yet rigorously demonstrated. However, it will prove to be very robust.

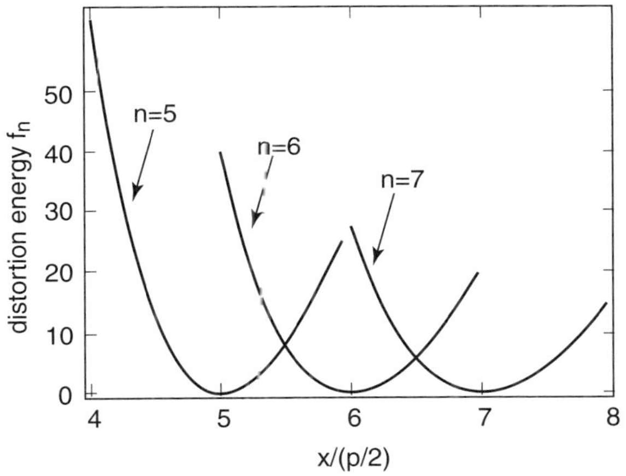

Fig. B.VII.20 Twist free energy stored in the nth stripe.

The equilibrium position x_m of the mth line is immediately obtained in the plane/plane geometry, where $h(x) = \alpha x$. One has

$$x_m = \left(m - \frac{1}{2}\right)\frac{p}{2\alpha}, \qquad m=1, 2, \ldots \tag{B.VII.59}$$

In order to determine the cholesteric pitch p, one only needs to measure the stripe width $p/2\alpha$. The method can be employed even for a sub-micron pitch p; for small enough α, the stripe width is easily determined.

Let us now tackle a more delicate problem, that of the narrow transition zone between two stripes. Does it consist of a real discontinuity plane, as sketched in Fig. B.VII.18a (highly unlikely) or, on the contrary, does a singular line appear in the bulk? To look for the answer, let us determine the director field in the planar approximation, but without the assumption of constant twist along z [11, 12]. More precisely, let us look for an equilibrium configuration, singular along a line (the y axis), such that the pitch number be m – 1 left of the line and m on the right. We assume molecular anchoring at the surfaces (in $z = \pm h/2$) to be along the x axis and take $K = K_1 = K_3$. Let $\varphi(x,z)$ be the angle between the director and the x axis and F the Frank free energy, expressed as:

$$F = \frac{1}{2} K_2 \iint \left[\frac{K}{K_2} \left(\frac{\partial \varphi}{\partial x} \right)^2 + \left(q - \frac{\partial \varphi}{\partial z} \right)^2 \right] dxdz \tag{B.VII.60}$$

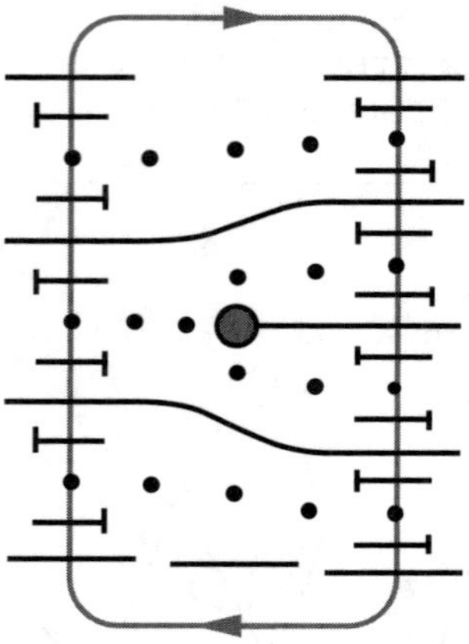

Fig. B.VII.21 χ disclination. Around the Burgers circuit, the director turns by an angle $\Omega = 2\pi m$ (with integer or half-integer m) around the cholesteric axis. Here, $\Omega = \pi$ so this χ disclination is of rank m = 1/2.

The corresponding Euler equation is:

$$\frac{\partial^2 \varphi}{\partial x^2} + \frac{K_2}{K} \frac{\partial^2 \varphi}{\partial z^2} = 0 \tag{B.VII.61}$$

or

$$\Delta \varphi = \frac{\partial^2 \varphi}{\partial \xi^2} + \frac{\partial^2 \varphi}{\partial z^2} = 0 \tag{B.VII.62}$$

with the new variable $\xi = (K_2/K)^{1/2} x$. **In a boundless medium**, the solution to this equation is:

$$\varphi = \frac{1}{2} \arctan \left(\frac{z}{\xi} \right) + qz + \varphi_0 \tag{B.VII.63}$$

describing a χ disclination of rank 1/2 (see next section for the nomenclature

[13]). Such a defect is sketched in figure B.VII.21. It resembles a wedge disclination in a lamellar material, as it amounts to inserting half a cholesteric layer on one side of the line. This is the defect to be introduced in the Cano wedge.

Let us prove it by searching for a solution of the same type, satisfying both equation B.VII.62 and the following boundary conditions:

$$\lim_{x \to -\infty} \frac{\partial \varphi}{\partial z} = (m-1)\frac{\pi}{h} \qquad \text{(B.VII.64a)}$$

$$\lim_{x \to +\infty} \frac{\partial \varphi}{\partial z} = m\frac{\pi}{h} \qquad \text{(B.VII.64b)}$$

$$\varphi\left(x, z = -\frac{h}{2}\right) = 0 \qquad \text{(B.VII.65a)}$$

$$\varphi\left(x, z = \frac{h}{2}\right) = \begin{cases} (m-1)\pi & (x<0) \\ m\pi & (x>0) \end{cases} \qquad \text{(B.VII.65b)}$$

The first two conditions express constant twist far from the line, while the last two ensure strong anchoring at the walls (Fig. B.VII.22).

This problem was solved by Malet [14] using the method of images. Because the calculations are tedious, we only give the essential results. The first step consists of superposing solutions of the type B.VII.63 in order to satisfy the anchoring conditions B.VII.65. The first χ disclination of rank 1/2 is placed in O: it is the experimentally observed one; the following ones (images) are virtual and placed symmetrically with respect to the plates. This yields:

$$\varphi(\xi, z) = qz + \frac{1}{2}\sum_{n=-\infty}^{n=+\infty}\left[\arctan\left(\frac{z-2nh}{\xi}\right) + \arctan\left(\frac{z-(2n-1)h}{\xi}\right)\right] \qquad \text{(B.VII.66)}$$

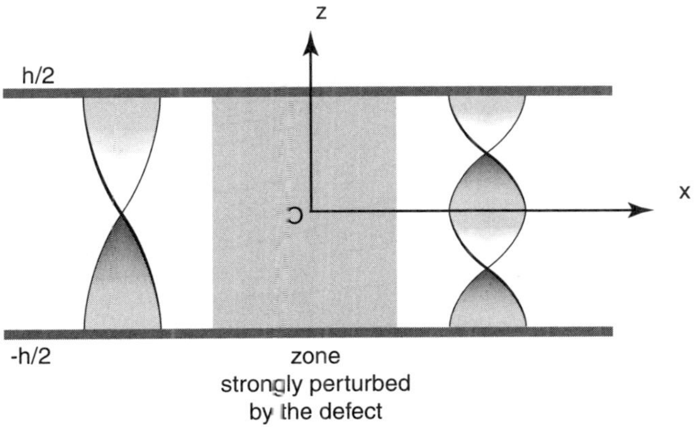

Fig. B.VII.22 χ disclination line in a constant thickness cholesteric slice.

The twist is:

$$\frac{\partial \varphi}{\partial z} = q + \frac{\pi}{4h}\,\mathrm{sh}\left(\frac{\pi\xi}{h}\right)\left[\frac{1}{\mathrm{ch}\left(\frac{\pi\xi}{h}\right) - \cos\left(\frac{\pi z}{h}\right)} + \frac{1}{\mathrm{ch}\left(\frac{\pi\xi}{h}\right) + \cos\left(\frac{\pi z}{h}\right)}\right] \qquad \text{(B.VII.67)}$$

yielding the sample thickness for which the boundary conditions B.VII.64 are satisfied:

$$q = (2m-1)\frac{\pi}{2h} \qquad \text{(B.VII.68)}$$

This equation is similar to eq. B.VII.58 giving the equilibrium position of the mth 1/2 χ line in a sample of varying thickness. By means of this relation, the infinite series in expression B.VII.66 can be summed, giving the final form for the solution:

$$\varphi = \left(m - \frac{1}{2}\right)\frac{\pi z}{h} + (m-1)\frac{\pi}{2} + \frac{\pi}{2}E(\xi) + \frac{1}{2}\left\{\arctan\left[\coth\left(\frac{\pi\xi}{2h}\right)\tan\left(\frac{\pi z}{2h}\right)\right]\right.$$

$$\left. + \arctan\left[\mathrm{th}\left(\frac{\pi\xi}{2h}\right)\tan\left(\frac{\pi z}{2h}\right)\right]\right\} \qquad \text{(B.VII.69)}$$

with $E(\xi)$ the Heaviside function ($E(\xi) = 0$ for $\xi < 0$ and 1 for $\xi > 0$). The director field corresponding to the m = 2 planar line is shown in figure B.VII.23.

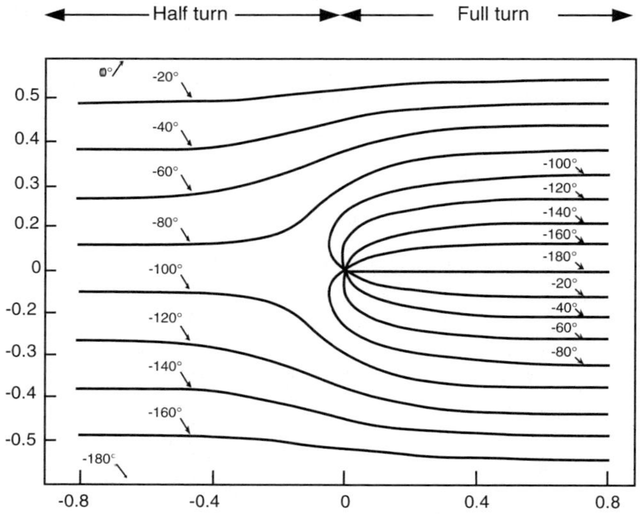

Fig. B.VII.23 Lines of equal director tilt around the second (m = 2) χ disclination line of rank 1/2. This model assumes planar configuration, i.e., that the director remains in the (x, y) plane (from ref. [12]).

Expression B.VII.69 also gives the excess energy E_d of the mth line. This calculation, performed by Malet, yields:

$$E_d \approx \frac{\pi}{4}\sqrt{K_2 K}\left[\ln\left(\frac{h}{4\pi r_c}\sqrt{\frac{K}{K_2}}\right)+\frac{2G}{\pi}\right]+E_c \qquad \text{(B.VII.70)}$$

for any m. In this expression r_c is a core radius (assumed to be much smaller than the thickness), G is the Catalan constant ($G \approx 0.916$), and E_c is a typical core energy.

This line energy adds to the twist energy of the stripes in a Cano wedge. Competition between these two energy contributions may lead to the presence of rank 1 χ lines in a Cano wedge (see section VII.3.c). Before describing these double lines (double, because this time the number of half-pitches varies by two across the line), let us first classify the other types of disclination usually encountered in cholesterics.

VII.4.b) λ, τ, and χ disclinations

In the previous chapter we saw that disclinations can be generated through the Volterra process.

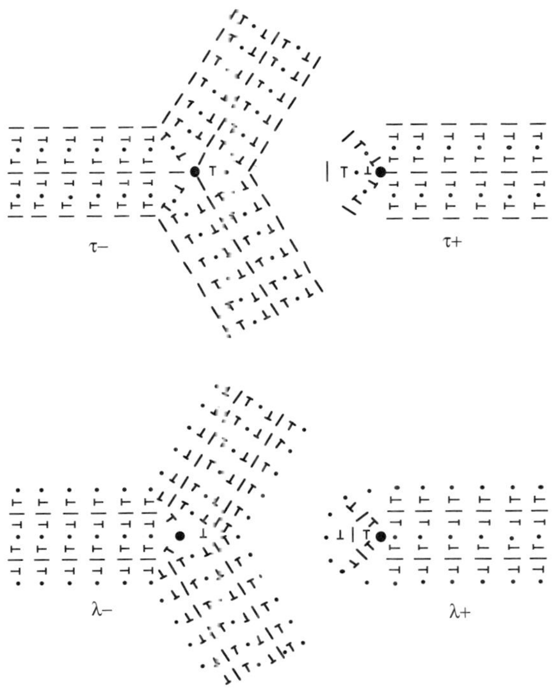

Fig. B.VII.24 τ^+, λ^+, τ^-, and λ^- disclination lines.

In a cholesteric there are essentially two types of disclinations [13]:

1. χ disclinations, already described in the previous section (Fig. B.VII.21), corresponding to $\Omega = 2\pi m$ rotations around an axis parallel to the cholesteric helix. These disclinations can have any shape and are assigned a topological rank m (integer or half-integer).

2. τ or λ disclinations, corresponding to $\Omega = 2\pi m$ rotations around an axis in the cholesteric plane. This axis can be parallel to the director (τ disclination) or perpendicular to it (λ disclination) in the plane of the cut. These disclinations are necessarily straight lines parallel to the rotation axis. The simplest configurations appear for the smallest values of m, i.e., $m = \pm 1/2$. A rotation of $+\pi$ generates a λ^- or τ^- line, while a $-\pi$ rotation leads to the appearance of a λ^+ or τ^+ line. The minus sign signifies that matter must be added between the two sides of the cut, rotated during the Volterra process; the plus sign means that matter has to be removed. These four types of line are shown in figure B.VII.24. Notice that the core of λ lines is continuous, while that of τ lines is singular. One can then expect λ disclinations to appear more often than τ lines.

Let us now return to the Cano wedge, in the case of large cholesteric pitch.

VII.4.c) "Double" line in a Cano wedge

The figure B.VII.25 photo shows a large-pitch ($p \approx 30$ μm) cholesteric sample between a plane surface and a cylindrical lens with radius R = 80 mm. The sample is observed in transmission and the light is linearly polarized parallel to the anchoring direction (perpendicular to the wedge). The thickness increases from left to right; one first notices four thin lines marked SL1 through SL4 in the figure, then two thicker lines marked DL1 and DL2.

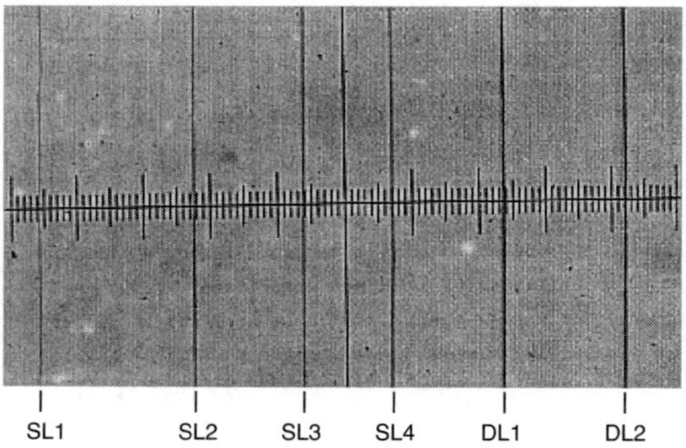

Fig. B.VII.25 Simple (SL) and double (DL) lines in a Cano wedge (p = 29.9 μm and R = 80 mm). The observation light is polarized along the anchoring direction (itself perpendicular to the edge of the wedge, to the left of the photo). One division of the eyepiece micrometer corresponds to 34 μm (photo by G. Malet).

The first lines, separating the stripes from 0 to 4 are closer together as the thickness increases, because of the cylindrical geometry; one can easily check that the number of half-steps increases by one across a simple line SL. These lines are similar to those described in § VII.3.a): they are χ disclinations of rank 1/2 (Fig. B.VII.21).

The situation changes starting with the fourth stripe, which is wider than the third. On the other hand, the contrast of DL lines is clearly different from that of SL lines. Finally, the number of half-steps changes by two across DL lines, which are therefore called "double lines." They are topologically equivalent to rank 1 χ lines.

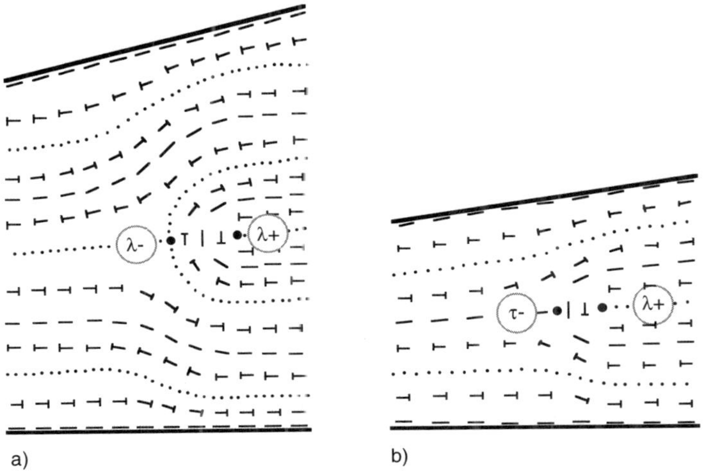

Fig. B.VII.26 Disclination pair model for double and simple χ lines. a) (λ^-, λ^+) pair: the line is DL1 after three simple lines; b) (τ^-, λ^+) pair: the line is SL2.

The internal structure of these lines prompted extensive theoretical work. In particular, the unusual width of double lines made Kléman and Bouligand conjecture that they were dissociated in a (λ^+,λ^-) line pair (Fig. B.VII.26a), which has the additional advantage of suppressing core singularity. This interpretation is backed up by the zigzag instability these lines undergo in the presence of a magnetic field (see section VII.5a). A similar model, proposed by Bouligand for simple lines of order m > 1, involves splitting in a (τ^-,λ^+) disclination pair (Fig. B.VII.26b) [15]. This time, however, the singular core persists. These models, confirmed by the optical measurements of Malet, explained some of the features of line kink observed in the samples. Indeed, the lines are not always at the center of the sample (even if this is the position of minimal energy); thus, each time they change level in the sample thickness, clearly visible kinks occur. Only the simple SL1 line seems always to have a planar structure.

Why do the lines change in nature with increasing sample thickness

(SL-DL passage)? To answer this question, consider a sample limited by two surfaces making an angle α and the liquid crystal slab between two vertical sections $[x_n, x_{n+2}]$ such that $h(x_n) = n(p/2)$ and $h(x_{n+2}) = (n+2)(p/2)$. The question is whether two simple lines SL (Fig. B.VII.27a) are more favorable than one double line DL (Fig. B.VII.27b). To this end, let us compare the energies E_S and E_D of these two configurations. They comprise the energies of the lines themselves (E_{SL} for a simple line and E_{DL} for a double line) and the distortion energy of the cholesteric layers, obtained by integrating the energy density B.VII.55 over the entire slab.

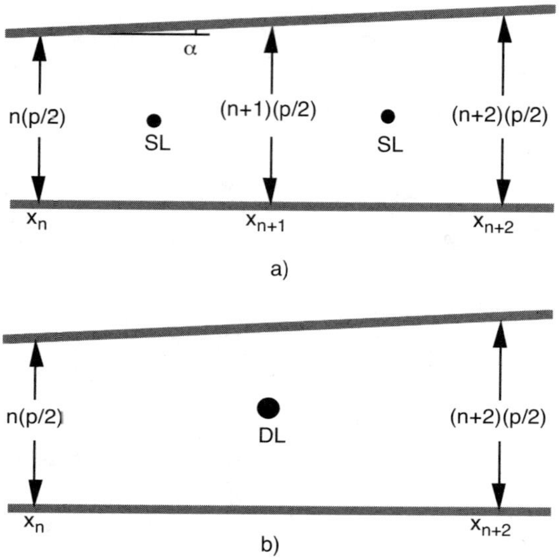

Fig. B.VII.27 Two possible configurations.

For two simple lines one has:

$$E_S = 2E_{LS} + \pi^2 \frac{K_2}{2\alpha} f_S(n) \qquad \text{(B.VII.71a)}$$

with

$$f_S(n) = n^2 \ln\left(\frac{2n+1}{2n}\right) + (n+1)^2 \ln\left(\frac{2n+3}{2n+1}\right) + (n+2)^2 \ln\left(\frac{2n+4}{2n+3}\right) - 2n - 2 \qquad \text{(B.VII.71b)}$$

while for a double line:

$$E_D = E_{LD} + \pi^2 \frac{K_2}{2\alpha} f_D(n) \qquad \text{(B.VII.72a)}$$

with

$$f_D(n) = n^2 \ln\left(\frac{n+1}{n}\right) + (n+1)^2 \ln\left(\frac{n+2}{n+1}\right) - 2n \qquad \text{(B.VII.72b)}$$

We have plotted $f_S(n)$ and $f_D(n)$ in figure B.VII.28. As expected, $f_D(n)$ is always larger than $f_S(n)$, whatever the value of n. Furthermore, the elastic energy outside the lines varies as $1/\alpha$. Consequently, simple lines will always be preferable to double lines in the limit of α going to 0 (whatever the energy of the two line types). **For double lines to appear, one must necessarily have**

$$2E_{LS} > E_{LD} \qquad \text{(B.VII.73)}$$

The reason this situation is experimentally achieved is certainly related to the fact that SL lines have a singular core, contrary to the double lines DL.

Let us now determine the thickness at which the first double line appears. This happens when both configurations have the same energy, namely for:

$$\pi^2 \frac{K_2}{2\alpha}[f_D(n) - f_S(n)] = 2E_{LS} - E_{LD} \qquad \text{(B.VII.74)}$$

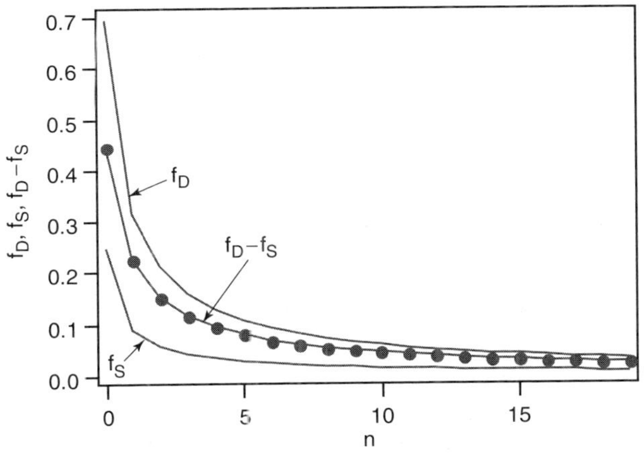

Fig. B.VII.28 Functions f_D, f_S, and their difference.

At fixed angle α, i.e., between two plane surfaces, this equation directly yields the number n* of simple lines that will appear.

In cylindrical geometry, the angle α changes with x. It is easily verified that, at point $x_n = np/2$, α is $(np/R)^{1/2}$. If we still denote by n* the number of simple lines, the first double line appears close to x_{n^*+1}, where the angle α has the value $[(n^*+1)p/R]^{1/2}$. n* is then given by the following equation:

$$\frac{1}{2} K_2 \pi^2 \sqrt{\frac{R}{(n+1)p}}[f_D(n) - f_S(n)] = 2E_{LS} - E_D \qquad \text{(B.VII.75)}$$

In Feldman et al.'s experiment [10], the cholesteric pitch is very small ($p \approx 0.65$ μm) and the lens must have a large radius of curvature ($R \approx 260$ mm) to produce clearly visible stripes. The previous formula then predicts a large n* (several tens). All lines experimentally observed in this experiment are indeed simple.

On the other hand, in the Malet experiment the pitch is large ($p \approx 30$ μm); hence the choice of a smaller radius of curvature for the lens ($R = 80$ mm), yielding $n^* = 4$. Eq. B.VII.75 provides an estimate of $2E_{LS} - E_{LD} \approx 10\,K_2$, which is of the expected order of magnitude; eq. B.VII.70 yields $E_{LS} \approx 8K_2$ for a planar simple line, so one would have $E_{LD} \approx 6\,K_2$.

VII.4.d) Fan-shaped textures and polygonal fields

These are other very usual textures observed in cholesterics. In fan-shaped textures, observed as early as 1922 by G. Friedel [16], the cholesteric layers are normal to the glass plates and free to turn in the sample plane. Very nice fan-shaped textures were obtained by Bouligand by dissolving Canada balsam in MBBA [17] (Fig. B.VII.29). His samples had a thickness of a few tens of μm for a cholesteric pitch of a few micrometers. Bouligand specifies that the fans are easier to obtain when the sample is close to its melting point and does not dismiss the possibility of a thin isotropic layer appearing between the cholesteric and the glass plates. This texture contains several patterns, themselves formed by the association of λ^+ or λ^- disclinations (rather than singular-core τ lines).

Fig. B.VII.29 Fan-shaped texture in a thin sample of MBBA and Canada balsam (photo by Y. Bouligand). The elementary patterns (a), (b) (c) and (d) are described in the next figure. (e) is an inclusion of isotropic liquid.

The most frequent patterns include [17]:

1. The "elementary pinch" (Fig. B.VII.30a) leaving the layer topology unchanged at infinity.

2. The "zigzag," forming a disorientation between two domains in which layer orientation is on the average well defined (Fig. B.VII.30b).

3. The "wedge dislocation" (in the cholesteric layer system) already encountered in the Cano wedge (Fig. B.VII.30c).

4. The "diamond" formed by associating two λ^+ and two λ^- lines (Fig. B.VII.30d).

Bouligand also described numerous variants of these patterns.

The "polygonal fields" are a third example of a characteristic texture, extensively studied by Bouligand (Fig. B.VII.31) [15]. This splendid texture is obtained by letting cholesteric bubbles grow in the isotropic phase of an MBBA-Canada balsam mixture. It is more complicated than the fan-shaped texture, the cholesteric layers being no longer vertical, but strongly inclined with respect to the glass plates. This is shown by successively focusing on the upper and on the lower plate, which also reveals the existence of two conjugated lattices of double spirals (Fig. B.VII.31a and b). The Bouligand model of figure B.VII.32 explains these observations.

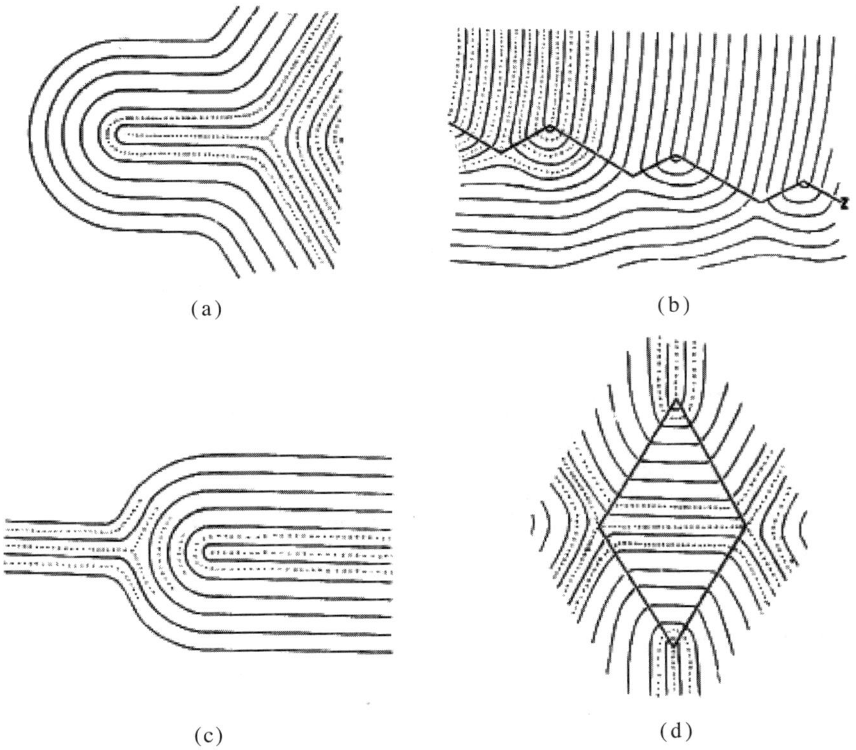

Fig. B.VII.30 The most frequent patterns met in the fan-shaped texture (from Y. Bouligand [17]).

The reader is referred to the very useful papers of Y. Bouligand and F. Livolant in the *Journal de Physique* for a detailed description of these textures and their relevance for biological systems [15, 17–20].

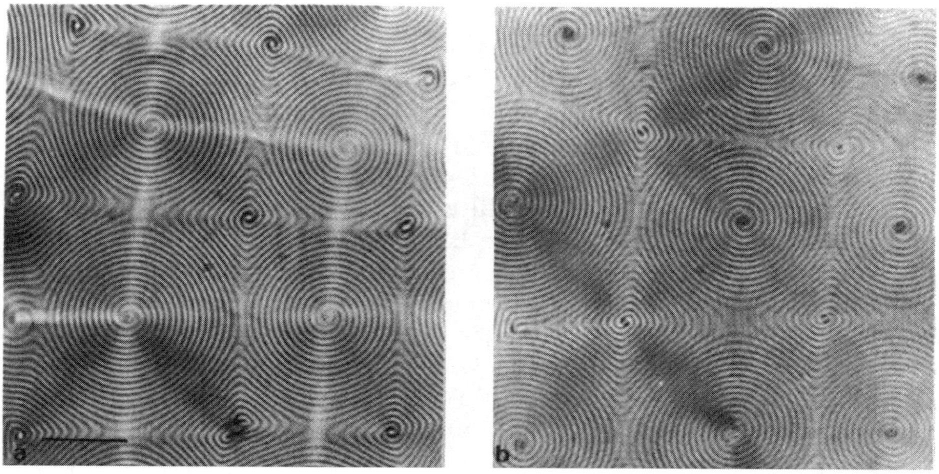

Fig. B.VII.31 Natural light observation of a polygonal field in an MBBA-cholesteryl benzoate mixture; a) the upper plate is in focus; b) the lower plate is in focus (photo by Y. Bouligand [15]).

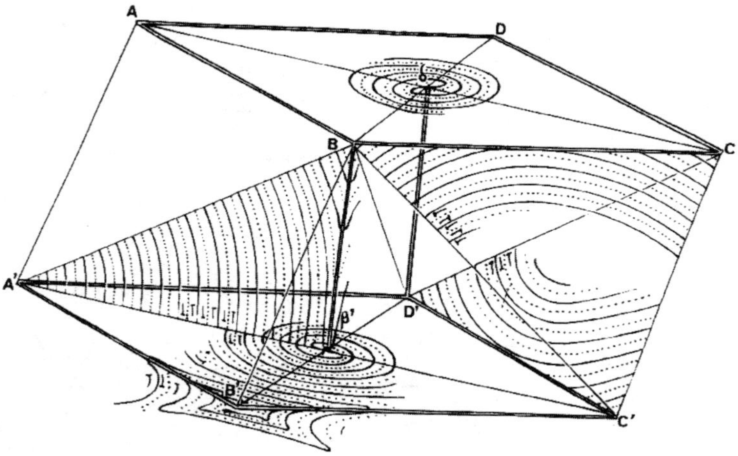

Fig. B.VII.32 Domain arrangement and arrangement of the cholesteric layers in a unit parallelepiped of the quadratic lattice (illustration by Y. Bouligand [15]).

In the next section we discuss the unwinding transition in cholesterics. We also describe another texture, observed in homeotropic anchoring and commonly known as the "fingerprint texture."

VII.5 Unwinding transition

The cholesteric-nematic transition can be induced in one of two ways. The first way is to apply a field (electric or magnetic) to the cholesteric layers. If the (electric or magnetic) anisotropy is positive, the field will tend to unwind the cholesteric helix. The second possibility is to impose to the sample boundary conditions that are topologically incompatible with the helicoidal structure of the director field but favor nematic alignment. These two methods can also be combined, diversifying the richness of the phenomena observed. Let us first analyze the action of a magnetic field

VII.5.a) Magnetic field unwinding

De Gennes [21] and Meyer [22] were the first to calculate the evolution of the cholesteric pitch when a magnetic field is applied in a direction normal to the helix axis (Fig. B.VII.33). Let us summarize this calculation employing the notations of de Gennes.

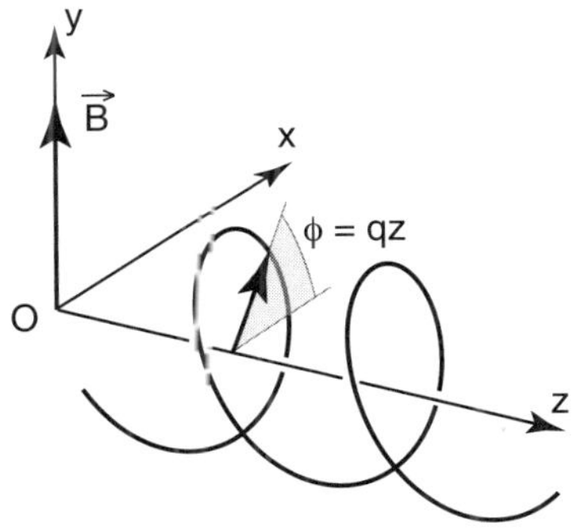

Fig. B.VII.33 Geometry of the unwinding transition under a magnetic field.

Under the action of the field **B** (parallel to the y axis), director **n** has the components:

$$n_x = \cos\varphi(z), \qquad n_y = \sin\varphi(z), \qquad n_z = 0 \tag{B.VII.76}$$

The final structure must minimize the Frank free energy:

$$F = \frac{1}{2}\int dz \left[K_2 \left(\frac{\partial \varphi}{\partial z} - q\right)^2 - \frac{\chi_a}{\mu_o} B^2 \sin^2\varphi \right] \tag{B.VII.77}$$

requiring that angle φ satisfy the Euler equation:

$$\xi^2 \frac{d^2\varphi}{dz^2} + \sin\varphi \cos\varphi = 0 \qquad \text{(B.VII.78)}$$

with the magnetic coherence length:

$$\xi = \frac{1}{B}\sqrt{\frac{\mu_0 K_2}{\chi_a}} \qquad \text{(B.VII.79)}$$

This equation has the first integral:

$$\xi^2 \left(\frac{d\varphi}{dz}\right)^2 + \sin^2\varphi = \frac{1}{k^2} = \text{Const} \qquad \text{(B.VII.80)}$$

where k is a constant to be determined, related to the repeat distance Λ of the deformed helix by the relation:

$$\Lambda = \int_0^{\pi} \frac{dz}{d\varphi} d\varphi = 2\xi\, k\, K(k) \qquad \text{(B.VII.81)}$$

In this expression, K(k) is the complete elliptical integral of the first kind:

$$K(k) = \int_0^{\pi/2} \frac{d\varphi}{\sqrt{1 - k^2 \sin^2\varphi}} \qquad \text{(B.VII.82)}$$

k is determined by minimizing F with respect to it. With the reduced parameters

$$B^* = (q\xi)^{-1}, \quad F^* = \frac{2F}{K_2 q^2 \int dz} \qquad \text{(B.VII.83)}$$

and defining

$$C = \frac{1}{k^2} - 1 \quad \text{and} \quad J = 2\int_0^{\pi} d\varphi \sqrt{C + \cos^2\varphi} \qquad \text{(B.VII.84)}$$

it is easily verified that

$$F^* = 1 - \frac{2\pi}{q\Lambda} + \frac{B^* J}{q\Lambda} - B^{*2}(1 + C) \qquad \text{(B.VII.85)}$$

Writing that dF/dC = 0 yields:

$$B^* = \frac{\pi k}{2E(k)} \qquad \text{(B.VII.86)}$$

where E(k) is the complete elliptical integral of the second kind:

$$E(k) = \int_0^{\pi/2} \sqrt{1 - k^2 \sin^2\varphi}\, d\varphi \qquad \text{(B.VII.87)}$$

Using relations B.VII.79, 81, 86, the repeat distance is obtained as a function of k:

$$\frac{\Lambda}{p/2} = \left(\frac{2}{\pi}\right)^2 K(k)\, E(k) \qquad \text{(B.VII.88)}$$

The repeat distance changes very little in weak fields, as

$$\frac{\Lambda}{p/2} = 1 + \frac{B^{*4}}{32} + \ldots \quad \text{for} \quad B << 1 \qquad \text{(B.VII.89)}$$

On the other hand, the repeat distance diverges with the elliptical integral K(k), i.e., for k = 1. One then has E(k) = 1 and $B = B_c = \pi/2$ or, returning to the physical parameters:

$$B = B_c = \frac{\pi}{2} q \sqrt{\frac{\mu_0 K_2}{\chi_a}} \qquad \text{(B.VII.90)}$$

Above this critical field, the cholesteric helix is completely unwound and the sample is in the nematic phase.

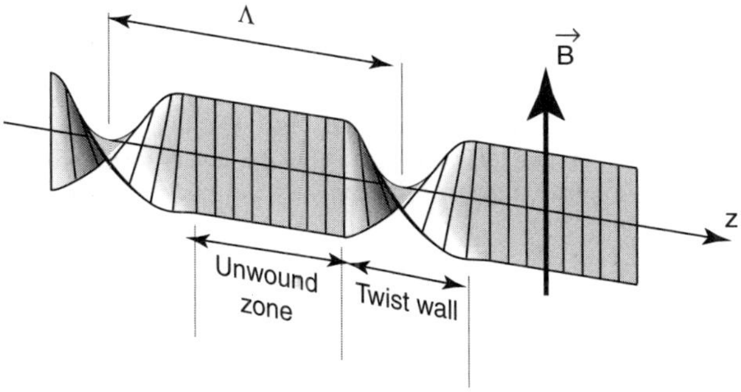

Fig. B.VII.34 Schematic representation of the cholesteric helix for $0 < B < B_c$ (projected in the (y,z) plane).

Angle φ(z) can be determined numerically for $B < B_c$ by integrating eq. B.VII.80. The cholesteric helix is represented in figure B.VII.34 for $0 < B < B_c$. The magnetic field generates large stripes where the director remains almost parallel to the field, separated by very narrow regions where the director turns by π (twist walls). The wall width remains finite as $B \to B_c$.

The pitch of the structure can be easily determined by placing the cholesteric in a Cano wedge; the field must be parallel to the anchoring direction, hence perpendicular to the axis. As the pitch increases, the

Grandjean lines move out of the wedge. Their new equilibrium positions give the pitch increase corresponding to the applied field. This method, simultaneously employed by Meyer [23] and by Durand et al. [24], allowed for verifying the theory. The experimental results of Durand et al. are shown in figure B.VII.35. These authors also verified the predicted 1/p dependence of the critical field.

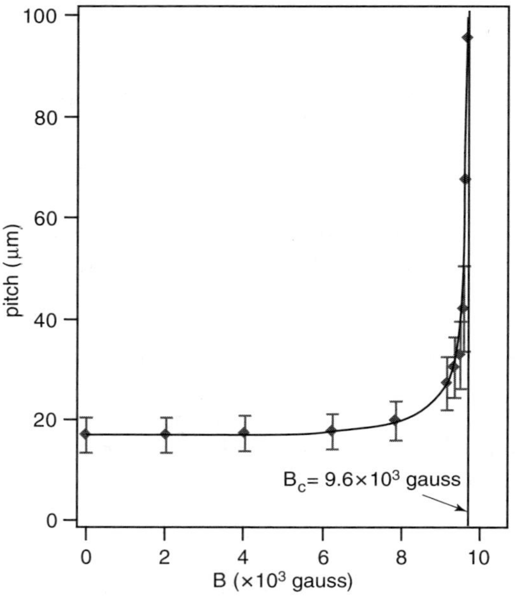

Fig. B.VII.35 Pitch increase with the magnetic field (from the experimental data of Durand et al. [24]). The cholesteric is a mixture of PAA and cholesteryl chloride (0.02 % by weight). The temperature is 129°C. The solid line indicates the theoretical prediction of de Gennes.

A final point is that displacement under a magnetic field of the Grandjean lines in a Cano wedge can induce a **zigzag instability of the double lines** (Fig. B.VII.36). This microstructure typically develops for $B > B_c/2$ and lasts even when the lines have reached their new equilibrium positions. It does not affect simple lines. Kléman and Friedel proposed that the instability originates in the (λ^+,λ^-) pair structure of the double lines. To prove it, they calculated the line tension of the pair as a function of the applied field and showed that it becomes negative for $B > B_c/2$. The line is then absolutely unstable. This instability is of the same nature as the Herring instability of the facets of a crystal with negative stiffness (see chapter C.VII on equilibrium shapes). The same calculation predicts a similar instability for simple lines composed of a (τ^-,λ^+) pair, but this time the critical field is of the order of B_c. As the lines disappear for this value of the field, the instability can no longer be observed. An additional reason given by Malet to explain the stability of simple lines is that their molecular configurations tend to become planar again under the applied field.

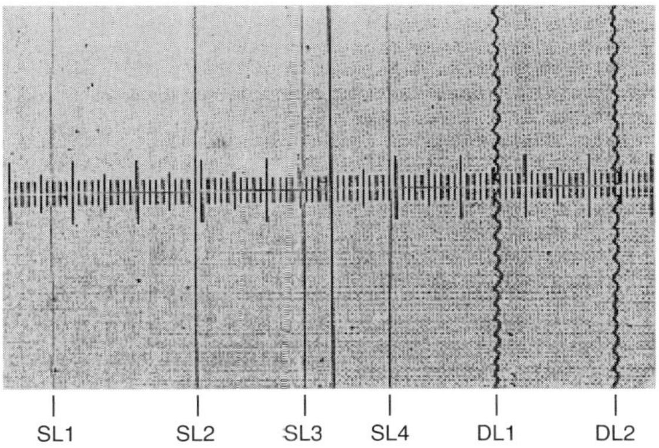

Fig. B.VII.36 Zigzag instability of double lines under a horizontal magnetic field normal to the Cano lines (B = 6480 Gauss). The sample is the same as in figure B.VII.25 (photo by G. Malet [14]).

Let us now discuss confinement-induced unwinding.

VII.5.b) Unwinding transition induced by geometrical frustration: experimental evidence

The textures of the cholesteric phase confined between two plates treated in homeotropic anchoring have been extensively studied [25]. While in planar anchoring the boundary conditions are satisfied by orienting the axis of the helix normal to the plates, no orientation is compatible with homeotropic anchoring. The liquid crystal is therefore **frustrated**, and the more so with decreasing sample thickness.

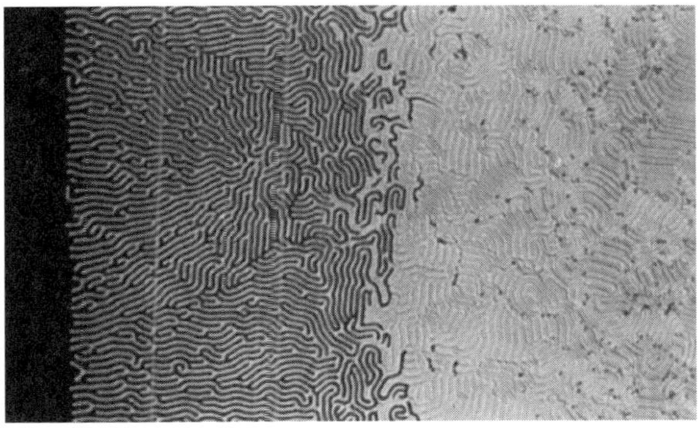

Fig. B.VII.37 Homeotropic anchoring texture in a wedge (between crossed polarizers). From left to right, one can distinguish a black zone (homeotropic nematic), then fingers separated by homeotropic nematic and finally adhering fingers. The natural cholesteric pitch is p = 6.7 μm and the angle α is 3 μm/mm (photo by S. Pirkl [26]).

Experiment confirms that the texture strongly depends on the ratio $C = h/p$ between the thickness h and the cholesteric pitch p. The substance employed in ref. [26], a mixture of 96.5 wt% ZLI1981 nematic (marketed by Merck) and 3.5 wt% cholesteryl nonanoate, has a 6.7 μm pitch at room temperature. The mixture is placed between two homeotropically treated plates forming a small angle α. Microscopic observation immediately shows the existence of a **critical confinement ratio**, $C_c \approx 1.25$, such that (Fig. B.VII.37):

1. For $C < C_c$, the cholesteric is completely unwound, and the director is everywhere parallel to the **z** axis, normal to the plates (homeotropic nematic phase).

2. For $C > C_c$, "fingerprint" textures appear, composed of cholesteric fingers, first separated from one another by homeotropic nematic for $C_c < C < C^*$, then adhering together for $C > C^*$ (in this mixture, $C^* \approx 1.65$).

It should be clear from the beginning that the cholesteric-nematic transformation is not a phase transition in the usual thermodynamic sense, but a bifurcation, having as control parameter the confinement ratio C.

What is the nature of this bifurcation?

Valuable information can be obtained by studying the nucleation and growth of cholesteric fingers in the homeotropic nematic as a function of the confinement ratio. This ratio can be modified by changing either the thickness or the pitch. The first method is not applicable, as the induced flow strongly perturbs the director field. The second, which is more convenient, has two variants. The first consists of varying the pitch by changing the temperature: its application is very limited, as the pitch cannot be varied very quickly due to the thermal inertia of the sample. The second possibility is to apply an electric field along z, which requires that the sample be prepared between two conducting electrodes. If the cholesteric has positive dielectric anisotropy, it can always be unwound by applying a high enough voltage (see the following section for the critical field calculation). By removing the field, nucleation and growth of the cholesteric phase inside the nematic phase are achieved. Subsequent evolution strongly depends on the confinement ratio.

The experiments to be described were performed with a mixture of the nematic ZLI 2452 and the chiral molecule ZLI 811 (about 0.6 wt%) [27, 28], both obtained from Merck. At room temperature, the pitch in this mixture is 16.7 μm. Experiment shows that if $C < C_c \approx 0.9$, the homeotropic nematic persists after turning off the field, and for $0.9 < C < 0.92$, isolated finger segments appear and slowly grow in length. Their nucleation is difficult and always heterogeneous. As shown in figure B.VII.38, these segments have two different tips, one pointed and the other one rounded. These fingers usually wiggle during growth and always end up by meeting one another: one notices that tips of the same type always repel each other, while tips of different kinds can merge together without a trace (Fig. B.VII.39). Furthermore, a rounded tip is always repelled by the edge of a finger, while a pointed tip can "cut in" forming a T-junction (Fig. B.VII.40);

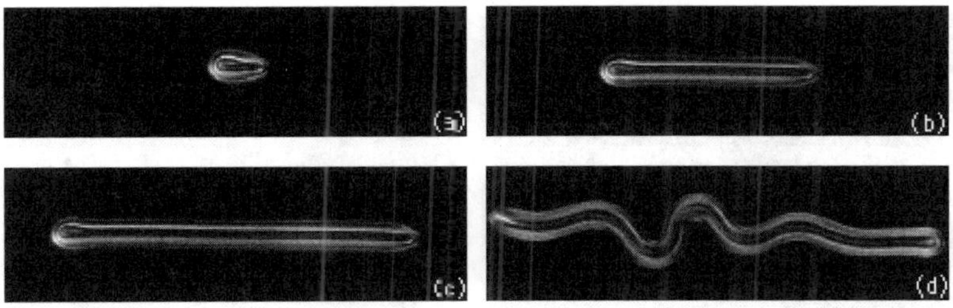

Fig. B.VII.38 Growth of an isolated finger. At exactly the critical thickness, the finger remains straight during the growth (a) (b) (c). Above it, the finger wiggles during growth (d).

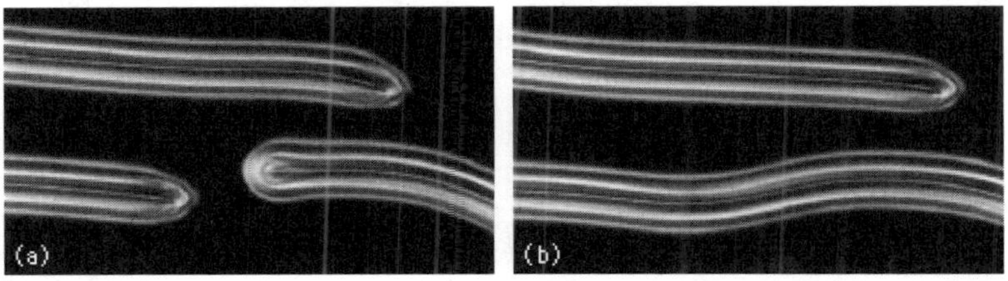

Fig. B.VII.39 Two different tips (rounded and pointed) merge together: a) just before touching; b) after the connection (from ref. [28]).

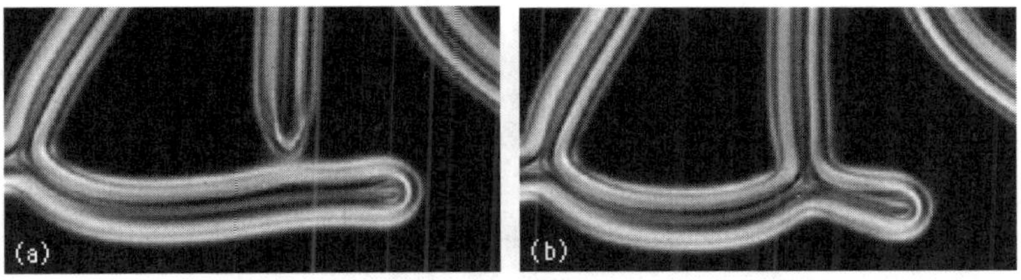

Fig. B.VII.40 A pointed tip reaches the edge of another finger: a) just before the connection; b) after the connection, a T-junction is formed (from ref. [28]).

For $0.92 < C < C^* \approx 1.2$, finger growth is totally different. This time, the rounded tip is always unstable and splits up, while the pointed end remains intact. This instability often leads to the formation of a circular domain (Fig. B.VII.41). When the growth process is completed the fingers, still separated by homeotropic nematic, form a periodic pattern containing numerous edge dislocations.

If $C > C^*$, the nematic phase becomes absolutely unstable; immediately after the electric field is removed, the bulk of the sample "lights up" (in a fraction of a second) and presents a transient, translationally invariant configuration, or TIC for short (Fig. B.VII.42). This texture then slowly relaxes (over a few minutes) towards a periodic structure composed of adhering fingers, no longer separated by homeotropic nematic.

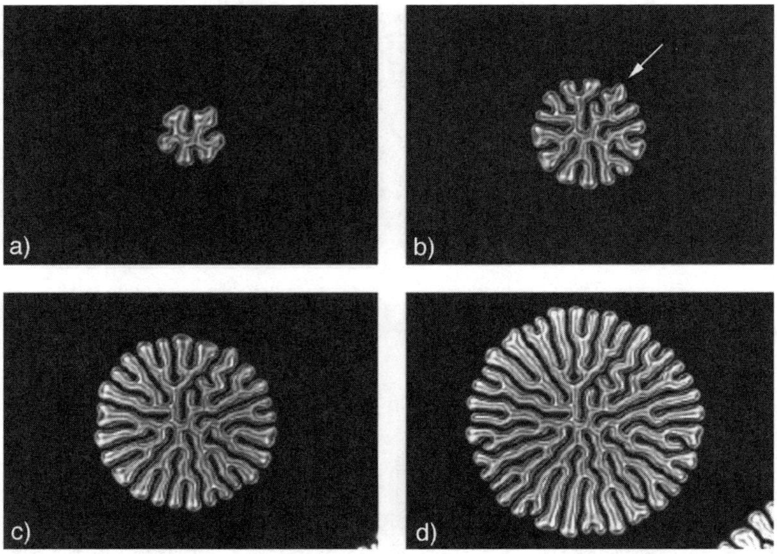

Fig. B.VII.41 "Flowerlike" growth of a periodic finger pattern. Only the rounded tip of the initial finger grows and then splits. This instability gives rise to two rounded-tip segments that split again and so on. The pointed tip, which never splits, is indicated in the photos by an arrow. This radial growth process creates a periodic finger pattern containing numerous edge dislocations. These defects maintain the average finger spacing constant (from ref. [28]).

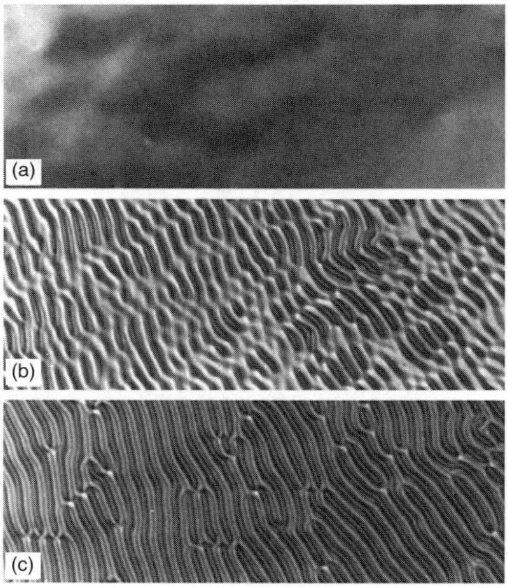

Fig. B.VII.42 a) Immediately after the field is removed, a translationally invariant configuration (TIC) appears (C = 1.92). b) and c) This configuration then slowly relaxes towards a fingerprint texture (photos taken 1 and 2 min, respectively, after removing the field) (from ref. [28]).

These observations clearly prove that the bifurcation is **subcritical**. In the cited example, $C_c \approx 0.9$ is the critical value of the confinement ratio for which the two phases have exactly the same energy. As to the value $C^* \approx 1.2$, it corresponds to the spinodal limit of absolute stability for the homeotropic nematic.

Let us see how this transition can be theoretically described. The method employed is not rigorous, but allows for simple interpretation of most of the experimental results. It is based on representing the textures on the surface of the unit S2 sphere.

VII.5.c) Representing the TIC and the fingers on the surface of the S2 sphere: application to the confinement-induced unwinding transition

In each space point **r**, the direction of the director is given by the two Euler angles $\theta(\mathbf{r})$ and $\varphi(\mathbf{r})$. In principle, they can be determined by solving the two nonlinear coupled differential equations describing minimization of the Frank free energy. These two complicated equations can only be solved numerically. This is what Press and Arrott did in 1976 [29] under the simplifying assumption of isotropic elasticity $K_1 = K_2 = K_3 = K$. These authors found a critical confinement ratio $C_c = 1/2$ below which the only stable solution is the homeotropic nematic, while above it appears a finger-like periodic solution, **topologically continuous everywhere**. Later, Stieb [30] and Ribière et al. [27, 28] showed that this continuous model explained the topological properties of the fingers (nature of the tips and merging rules) as well as their optical contrast under the microscope [31]. On the other hand, the simulations of Press and Arrott lead to a supercritical bifurcation, at variance with the experimental result. It will be shown later that elastic constant anisotropy must be taken into account to explain bifurcation subcriticality.

Another, more phenomenological, approach consists in looking for a relevant order parameter allowing for a "Landau-like" description of the bifurcation. This method is based on the geometrical representation of the observed textures (TIC and fingers) on the surface of the unit S2 sphere; first proposed by Thurston and Almgren [32] in nematics, this idea was then generalized to cholesterics by F. Lequeux et al. [33]. The principle is very simple: a point on the S2 sphere is associated with the director orientation. As the director changes its orientation in real space, the representative point describes a curve on the surface of the S2 sphere. In the simplest case, that of the homeotropic nematic, the director has the same orientation everywhere (parallel to the z axis) and the entire director field is represented by a unique point that we shall choose at the North pole (Fig. B.VII.43a). For a free cholesteric, the S2 image is a great circle: if the helix axis is along z, this circle is the equator; if it is normal to z, it will be a meridian. Note that the same coordinate system (x, y, z) is employed in real space and in the S2 sphere.

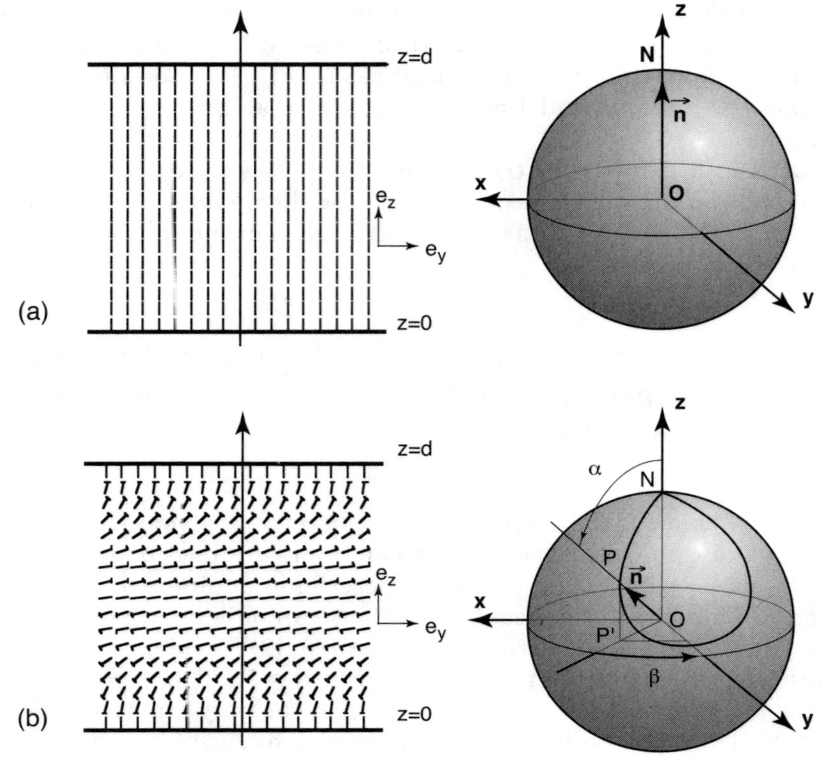

Fig. B.VII.43 Representation on the surface of the S2 sphere (right) and in real space (left) of the nematic and the TIC. a) Homeotropic nematic. The image of a straight line along the z axis reduces to the North pole. b) Invariant configuration in the (x, y) plane (TIC). The image of a vertical line is a closed curve passing through the N pole. Molecules are represented as nails pointing out of the image.

Let us now consider the TIC. This configuration being invariant under translation in the (x, y) plane, it is completely characterized by the S2 sphere image of a line parallel to the z axis. This image is a closed curve C going through the North pole in z = 0 and z = d in order to satisfy homeotropic anchoring on the plates. A point P on this curve is defined by the two angles α(z) and β(z) (Fig. B.VII.43b). At height z, the director has the components:

$$n_x = \sin\alpha(z)\sin\beta(z)$$

$$n_y = \sin\alpha(z)\cos\beta(z) \qquad \text{(B.VII.91a)}$$

$$n_z = \cos\alpha(z)$$

These formulae establish the correspondence between curve C in S2 and the director field in real space, represented in figure B.VII.43b. Before giving the analytical expression of functions α(z) and β(z), let us see how the finger is represented on the surface of the S2 sphere.

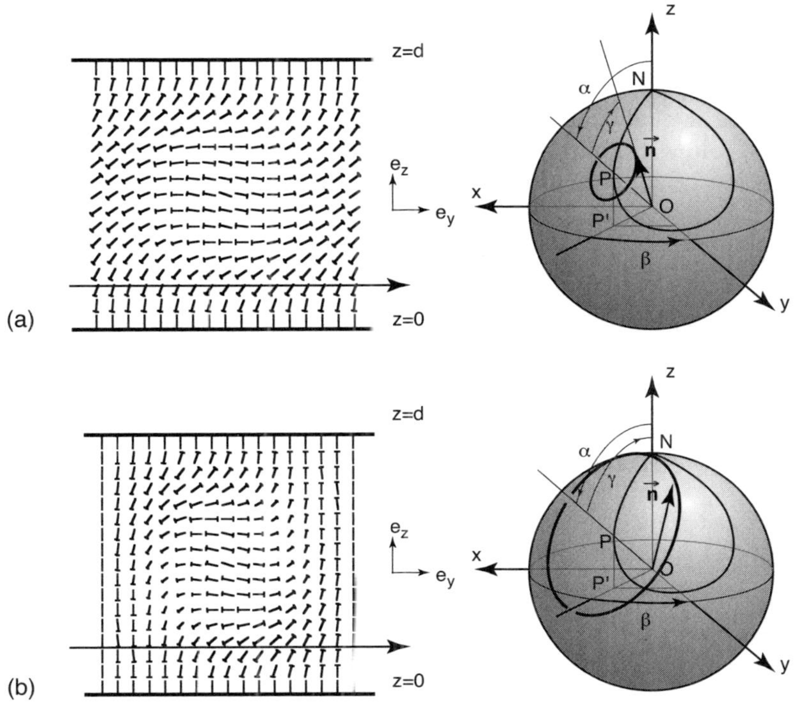

Fig. B.VII.44 S2 and real space representation of the fingers. a) Fingers parallel to the x axis. The fingers are built upon a periodic modulation of the TIC, obtained by varying the director position along y around the average position P corresponding to $n_{TIC}(z_o)$, so that $n_{finger}(z_o, y)$ describes a loop. The simplest choice is presumably a circle (thick line on the surface of the sphere, image of the horizontal line). O is the center of the S2 sphere; β is the angle between Oy and OP', with P' the projection of P onto the (x, y) plane; α is the angle between the vertical and OP; and γ is the half-opening angle of the circle describing TIC modulation, as seen from 0. b) Homeotropic edge finger, α(z) = γ(z). The thick-line circle, tangent to the N pole, is the image of the horizontal line. This finger is topologically identical to the one numerically obtained by Press and Arrott in isotropic elasticity [29].

Experiment clearly shows the topological continuity between the fingerprint texture and the TIC. As the fingers (assumed to be straight) are only invariant along one space direction (say x), a complete description requires the S2 images of all straight lines parallel to the y axis that cross the finger. If the fingers are edged with homeotropic nematic, a straight line at height z is represented by a closed curve going through the North pole. In the following, we shall assume that these curves are circles. One goes continuously from the TIC to the finger by "opening up" circles representing **n**(z, y) around each point P of the curve C giving **n**(z) in the TIC (Fig. B.VII.44a). Let γ(z) be the half-opening angle of the circles, assumed to be traced at constant velocity. The director field is then, with λ the finger width and k = 2π/λ:

$$n_x = \cos\beta \sin\gamma \sin ky - \cos\alpha \sin\beta \sin\gamma \cos ky + \sin\alpha \sin\beta \cos\gamma$$

$$n_y = -\sin\beta\,\sin\gamma\,\sin ky - \cos\alpha\,\cos\beta\,\sin\gamma\,\cos ky + \sin\alpha\,\cos\beta\,\cos\gamma \qquad \text{(B.VII.91b)}$$

$$n_z = \sin\alpha\,\sin\gamma\,\cos ky + \cos\alpha\,\cos\gamma$$

Obviously, these equations reduce to those of the TIC for $\gamma = 0$.

The following, more tedious, step is to calculate the Frank free energy per unit area in the horizontal plane. The result is:

$$\frac{F}{K_2 q} = C\left[\left(\frac{k}{q}\right)^2 J - 2\frac{k}{q}\left(I + \frac{H}{2C}\right)\right] + L + \frac{M}{C} \qquad \text{(B.VII.92)}$$

where quantities I, J, L, and M depend on the elastic constants, on α, β, γ, and their z derivatives. For instance,

$$I = \int_0^{\pi} \cos\beta\,\sin\alpha\,\sin^2\gamma\,dZ \qquad \text{(B.VII.93)}$$

where $Z = \pi z/d$. The other functions, similar in form, are given in refs. [27, 34]. The energy B.VII.92 has an obvious minimum for

$$k = q\,\frac{I + H/2C}{J} \qquad \text{(B.VII.94)}$$

This expression yields the thermodynamic equilibrium wavelength of the periodic finger pattern as a function of the angles α, β, and γ.

The absolute instability threshold of the nematic (spinodal limit C*) can be found by writing down the Euler equations for angles α, β, and γ in the limit $\alpha \ll 1$ and $\gamma \ll 1$. The behavior of integrals I, H, and J in this limit shows that $k \to 0$, since I and H vary like α^3 while J varies like α^2 and γ^2. This means that, immediately above the absolute instability threshold, the solution is of the TIC type (transiently, at least). The Euler equations are then written as:

$$\dot{\beta} = -\frac{2C}{K_{32}}$$

$$\ddot{\alpha}\,K_{32} = -\frac{4C^2}{K_{32}}\alpha \qquad \text{(B.VII.95)}$$

$$\ddot{\gamma}\,K_{32} = -\frac{4C^2}{K_{32}}\gamma$$

with $\dot{(\ )} = d/dZ$. This linear equation system has a non-trivial solution satisfying homeotropic boundary conditions for:

$$C = C^* = \frac{K_{32}}{2} \qquad \text{(B.VII.96)}$$

in which case:

$$\alpha = \alpha_o \sin Z, \qquad \gamma = \gamma_o \sin Z \qquad \text{and} \qquad \beta = -\frac{2C}{K_{32}}\left(Z - \frac{\pi}{2}\right) \tag{B.VII.97}$$

For reasons of symmetry, we took $\beta = 0$ in the center of the sample. This value of C^* defines the spinodal limit of the nematic phase.

What is the bifurcation order? Answering this question requires a nonlinear analysis close to C^*. One can reasonably assume that, in this limit, the solution is still of the type B.VII.97. The couple (α_o, γ_o) then provides a natural choice for the **order parameter**, as in the nematic phase $\alpha_o = \gamma_o = 0$ while the fingerprint texture corresponds to $\alpha_o\gamma_o \neq 0$. Taking symmetry into account, the most general development of the free energy around $C = C^*$ to fourth order in α_o and γ_o can be shown to be [27, 34]:

$$\frac{F}{K_2 q} = \nu\,\alpha_o^2 + \nu\,\gamma_o^2 + B\,\alpha_o^4 + B\gamma_o^4 + D\,\alpha_o^2\,\gamma_o^2 + \ldots \tag{B.VII.98}$$

where the coefficients can be explicitly calculated as:

$$\nu = \frac{\pi}{2K_{32}C}\left(\frac{K_{32}}{2} - C\right)\left(\frac{K_{32}}{2} + C\right) \tag{B.VII.99a}$$

$$B = \frac{\pi}{16K_{32}}(3 + K_{12} - 3K_{32}) \tag{B.VII.99b}$$

$$D = \frac{\pi(15 + 26\,K_{12} + 7K_{12}^2 - 18K_{32} - 22K_{12}K_{32} - K_{32}^2)}{32K_{32}(1 + K_{12})} \tag{B.VII.99c}$$

with $K_{12} = K_1/K_2$ and $K_{32} = K_3/K_2$. It is clear from B.VII.98 that the bifurcation is first order for $B < 0$ or $D + 2B < 0$ and second order for $B > 0$ and $D + 2B > 0$.

Let us examine the supercritical case. Above the critical threshold ($C > C^* = C_c$ or $\nu < 0$), the solutions belong to one of two types:

1. Two solutions (which are in fact identical)

$$\alpha_o = 0,\ \gamma_o = \pm\sqrt{\frac{-\nu}{2B}} \qquad \text{and} \qquad \gamma_o = 0,\ \alpha_o = \pm\sqrt{\frac{-\nu}{2B}} \quad \text{(for k=0)} \tag{B.VII.100a}$$

with the same energy

$$\frac{F}{K_2 q} = -\frac{\nu^2}{4B} \tag{B.VII.100b}$$

describing the same TIC, or:

2. One solution of the type "periodic fingers with homeotropic edges" for which

$$\alpha_o = \gamma_o = \pm\sqrt{\frac{-\nu}{D+2B}} \qquad \text{(B.VII.101a)}$$

and of energy

$$\frac{F}{K_2 q} = -\frac{\nu^2}{D+2B} \qquad \text{(B.VII.101b)}$$

These formulae also show that the fingers are more favorable than the TIC for D – 2B < 0. Conversely, the TIC is preferred for D – 2B > 0.

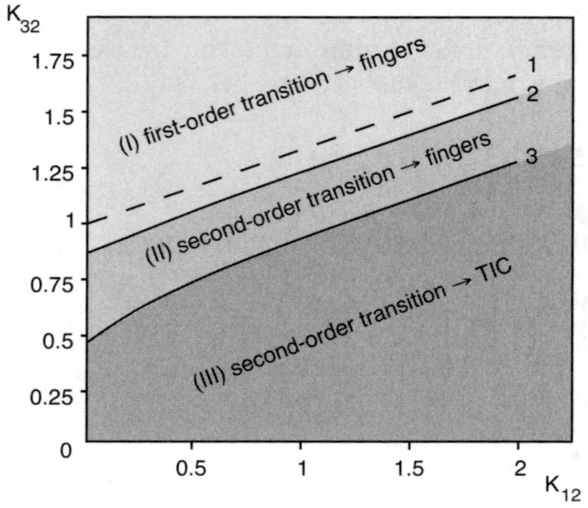

Fig. B.VII.45 Nature of the frustration transition under zero field. Depending on the values of K_{12} and K_{32}, the transition order and the nature of the solution change. Dashed line 1 corresponds to a change in order for the homeotropic nematic-TIC transition; it is not observed, since the fingers are then more stable than the TIC. Along line 2, the nematic-fingers transition changes order. Along line 3, fingers and TIC have the same energy at the transition (from ref. [34]).

This discussion can be graphically summarized in the plane of the elastic parameters (K_{12}, K_{32}) (Fig. B.VII.44). Three curves must be considered:

1. Curve 1, corresponding to B = 0 (in $C = K_{32}/2$); from B.VII.99b, its equation is:

$$K_{32} = \frac{1}{3} K_{12} + 1 \qquad \text{(B.VII.102)}$$

2. Curve 2, corresponding to D + 2B = 0 (in $C = K_{32}/2$), which reads according to B.VII.99b and c:

$$K_{32} = -15 - 17K_{12} + 2\sqrt{63 + 138K_{12} + 75K_{12}^2} \qquad \text{(B.VII.103)}$$

3. Curve 3 corresponding to D – 2B = 0 (in $C = K_{32}/2$):

$$K_{32} = -3 - 5K_{12} + 2\sqrt{3 + 10K_{12} + 7K_{12}^2} \qquad \text{(B.VII.104)}$$

These three curves divide space in several domains. Thus, below lines 1 and 2 (hence, below line 2), the transition is second-order. In this half-space a distinction should be made between region II, of fingers with homeotropic edges, and region III, where the TIC is more stable. Above line 2, in region I, the transition is first order; the solution, consisting of periodic fingers with homeotropic edges, must then be determined numerically. These predictions are in agreement with the numerical simulations of Press and Arrott, finding a supercritical bifurcation towards fingers with homeotropic edges for $K_{12} = K_{23} = 1$ (this point is indeed in region II).

Experimentally, most of the liquid crystals belong to region I. In agreement with theory, they exhibit a subcritical bifurcation and fingers with homeotropic edges. Note, in addition, that the theory also explains the existence of isolated fingers close to the critical thickness when the transition is subcritical. These fingers are replaced by a periodic pattern for a value of C between C_c and C^*, which can be numerically determined.

An alternative approach can be used for treating the frustration problem; it is entirely phenomenological because it is uniquely based on symmetry reasons and allows numerical determination of the dynamics of finger growth. We shall present it in the following section.

VII.5.d) Landau-Ginzburg model derived only on symmetry grounds

Before describing the experiment under an electric field, it is interesting to search for a different description, exclusively based on symmetry arguments. Following Frisch et al. [35], we take as the order parameter the complex number $A = n_z(n_x + in_y)$ where $\mathbf{n}(x, y)$ is the director in the middle plane $z = d/2$ (rather than the number $n_x + in_y$, not invariant under $\mathbf{n} \leftrightarrow -\mathbf{n}$).

Writing that the system is invariant under rotation around the z axis and under reversal (exchanging the upper and lower plate), these authors showed that the solutions must satisfy a dynamical equation whose most general form is:

$$\frac{\partial A}{\partial t} = -\frac{\partial F}{\partial \bar{A}} = \mu A + A_{\chi\bar{\chi}} + \delta \bar{A}_{\chi\chi} + i\eta(A A_{\bar{\chi}} + \bar{A} A_{\chi}) - |A|^2 A + O(A^4) \qquad \text{(B.VII.105)}$$

with $\partial\chi = \partial x + i\partial y$. The reader is referred to the original paper for the enumeration of the fourth-order saturation terms (nine of them in the general case [35]). On the right-hand side of expression B.VII.105, the linear term is proportional to $C - C^*$, where C^* is the confinement parameter at the stability limit of the nematic phase. The second term is a Laplacian, also present in isotropic elasticity. The following terms (third and fourth) express the anisotropy of the elastic constants but only the fourth, not invariant under mirror symmetry, contains the chirality of the medium; this term is proportional to $1/p$.

Studying this equation and its solutions, Frisch et al. assumed that $\delta = 0$ (stating that this coefficient only controls the shape of the fingers and the way they connect). It can then be shown that there are four kinds of solutions in the parameter plane (μ, η): homeotropic nematic (A = 0), TIC, periodic fingers (that these authors call modulated TIC), or isolated fingers. These results are presented in figure B.VII.46, formally equivalent to figure B.VII.45 as it shows that the nature of the transition and the observed solutions depend on the control parameter η (given by the elastic anisotropy). Thus:

1. When $0 < \eta < 0.72$, the transition is second order and leads to the TIC.

2. For $0.72 < \eta < 1.44$, the transition is still second order but leads to periodic fingers.

3. Finally, when $\eta > 1.44$, the transition becomes first-order; numerical simulations show that isolated fingers appear close to the coexistence line (gray line in the figure), while a periodic pattern is observed for $\mu > 0$.

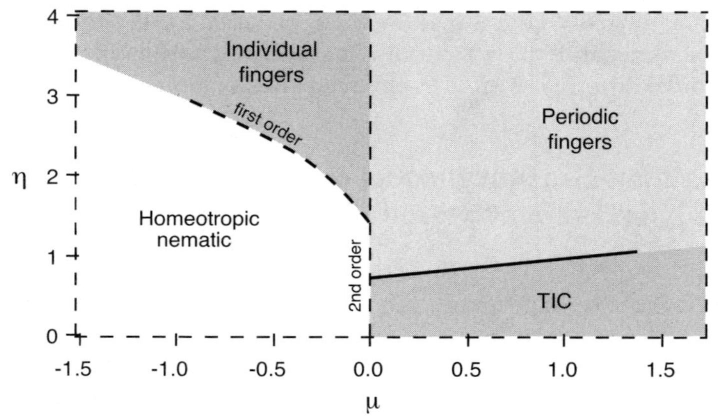

Fig. B.VII.46 Phase diagram of a cholesteric in homeotropic anchoring obtained by analyzing the Landau-Ginzburg normal form in the parameter space (μ, η). Solid lines were analytically calculated. The dotted line was determined numerically (from ref. [35]).

These results are in accordance with those of the previous section, confirming the essential role of elastic anisotropy in this problem.

VII.5.e) The combined action of frustration and an electric field

This experiment and the corresponding theory can be easily extended to the case where the sample is also under the influence of an electric (or magnetic) field normal to the glass plates. Its behavior strongly depends upon the sign of the dielectric anisotropy of the medium, as the field favors a cholesteric structure for $\varepsilon_a < 0$ and the nematic phase for $\varepsilon_a > 0$.

Two experimental phase diagrams are given in figures B.VII.47 and B.VII.48. In the abscissa is the confinement parameter C and in the ordinate is the r.m.s. voltage V between the two electrodes. The experiments are performed under an AC field (1 kHz) in order to avoid convective instabilities.

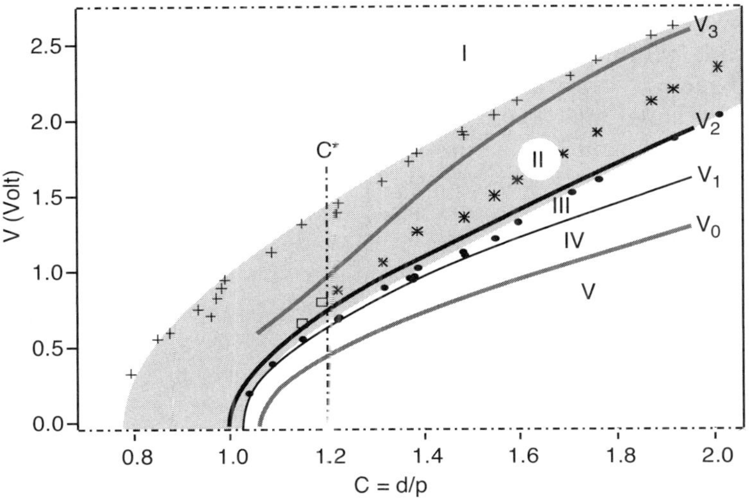

Fig. B.VII.47 Experimental phase diagram for the 8CB/ZLI811 mixture. Thick line: the critical line for coexistence of straight fingers and nematic ($V = V_2(C)$). The two outer gray lines correspond to the spinodal lines for fingers ($V = V_3(C)$) and nematic ($V = V_0(C)$). The thin line ($V = V_1(C)$) determines the limit between *isolated* fingers and a *periodic* finger pattern. Spherulites (see the next section) are stable for very long times (several days, at least) in the shaded region. The crosses define the limit above which spherulites are unstable and spontaneously disappear; dots (●) mark the limit below which spherulites destabilize and grow giving rise to *isolated* fingers with rounded tips (or *normal*). For an explanation of the stars and the squares, see the next section.

The first diagram (Fig. B.VII.47) shows that, for $\varepsilon_a > 0$, the bifurcation is always first order. The substance is a mixture of 8CB with the chiral molecule ZLI 811 from Merck (0.46 wt% and T = 39°C). The texture observed at equilibrium consists of fingers below the critical line V_2 and of homeotropic nematic above it. Lines V_0 and V_3 correspond to the spinodal limits of the nematic and the fingers, respectively. A linear stability analysis yields an analytical formula for V_0:

$$V_0(C) = 2\pi\sqrt{\frac{K_2}{\varepsilon_a\varepsilon_o}}\sqrt{\frac{C^2}{K_{32}} - \frac{K_{32}}{4}} \tag{B.VII.106a}$$

The thin line V_1 gives the limit between isolated fingers (region III and Fig. B.VII.38) and periodic fingers (regions IV and V). In region IV, the cholesteric phase appears after nucleation and growth of a circular germ (see Fig. B.VII.41) while below the spinodal line V_0 the first to appear is a transient TIC which then slowly develops homogeneous modulations (see Fig. B.VII.42).

The second diagram (Fig. B.VII.48) corresponds to a material of negative dielectric anisotropy, a mixture of the nematic Roche 2860 and the chiral molecule ZLI 811 (0.792 wt%). In this case, by applying an electric field to the sample in the nematic phase (thinner than the critical value $pK_{32}/2$ given by formula B.VII.99a) the cholesteric phase is induced, under the form of fingers or TIC for a high enough field. The phase diagram is now more complicated, exhibiting three different thickness ranges separated by two particular points, indicated by P and P' in the diagram. Left of P, the bifurcation is supercritical and leads to TIC. Between P and P', the transition is still supercritical, but the solution consists in periodic fingers. P is a triple point, as three phases coexist here. To the right of P', the transition is first order with the formation of fingers. As before, $V_2(C)$ is the coexistence line and $V_0(C)$ and $V_3(C)$ are the spinodal limits while $V_1(C)$ separates two types of finger growth (isolated below this line and in the shape of circular germs above it). The spinodal limit of the nematic phase is:

$$V_0(C) = 2\pi \sqrt{\frac{K_2}{-\varepsilon_a \varepsilon_o}} \sqrt{\frac{K_{32}}{4} - \frac{C^2}{K_{32}}} \qquad \text{(B.VII.106b)}$$

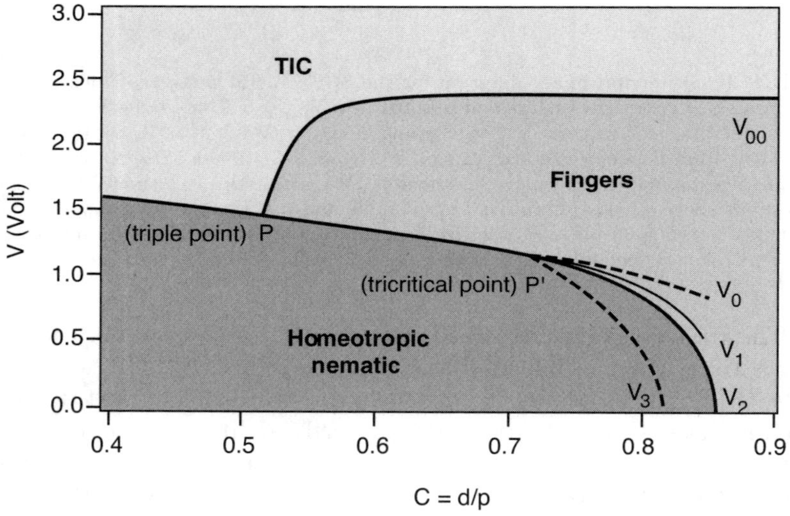

Fig. B.VII.48 Experimental phase diagram of mixture Roche 2860+S811. To the left of the triple point, the nematic-TIC transition is second-order. Between this point and the tricritical point, the nematic-fingers transition is second order. To the right of the tricritical point, the nematic-fingers transition is first order.

Point P', where the transition changes order, is a tricritical point. Note that left of P' the coexistence line and the two spinodals merge together, the bifurcation being supercritical.

It can be shown [27, 34] that the abscissae of the tricritical and triple points are given by D + 2B = 0 and D − 2B = 0, respectively, with:

$$D+2B=-\frac{64CK_{32}^6(-3-K_{12}+K_{32})^2\cos^2(\pi C/K_{32})}{\pi(1+K_{12})(2C-3K_{32})^2(2C+3K_{32})^2(2C-K_{32})^2(2C+K_{32})^2}+\frac{3\pi}{16CK_{32}^2}(12C^2-12C^2K_{32}+K_{12}K_{32}^2) \qquad \text{(B.VII.107a)}$$

$$D-2B=-\frac{64CK_{32}^6(-3-K_{12}+K_{32})^2\cos^2(\pi C/K_{32})}{\pi(1+K_{12})(2C-3K_{32})^2(2C+3K_{32})^2(2C-K_{32})^2(2C+K_{32})^2}+\frac{\pi}{16CK_{32}^2}(12C^2-12C^2K_{32}+K_{12}K_{32}^2) \qquad \text{(B.VII.107b)}$$

These two equations can be solved numerically and their results are in excellent agreement with the experimental data.

As to the line V_{00} separating the fingers from TIC, it must be numerically determined.

So far, we have only presented continuous structures close to equilibrium, but other structures also appear in homeotropic samples. Some of them, like spherulites (or cholesteric bubbles), are metastable and contain singularities. Finally, there are other types of fingers different from the ones already discussed; in the dynamical regime, they can lead to the appearance of spirals under a low-frequency electric field. These new phenomena are presented in the three following sections.

VII.5.f) Spherulites

In usual cholesteric fingers and the TIC, the director field is continuous. We describe here another frequently encountered configuration, commonly known as **spherulite** or **cholesteric bubble** [36].

The optical contrast of these bubbles depends on the observation conditions, as shown in figure B.VII.49. Numerous bubbles appear at the cholesteric-isotropic transition when the sample is rapidly cooled (a few degrees per minute). They appear where the last droplets of isotropic phase turn into cholesteric, certainly because of the higher concentration of chiral molecules. Their existence domain in the plane of the parameters (V,C) corresponds to the shaded area in the phase diagram of the 8CB-S811 mixture in figure B.VII.47. The spherulites are stable (metastable, in fact) in a wide area, ranging roughly from the coexistence line (more precisely, the line connecting the dots) to an upper boundary (corresponding to their absolute stability limit) well above the spinodal line $V_3(C)$ of the usual cholesteric fingers. Immediately below their lower boundary, spherulites are unstable and lengthen, forming a finger with two rounded tips.

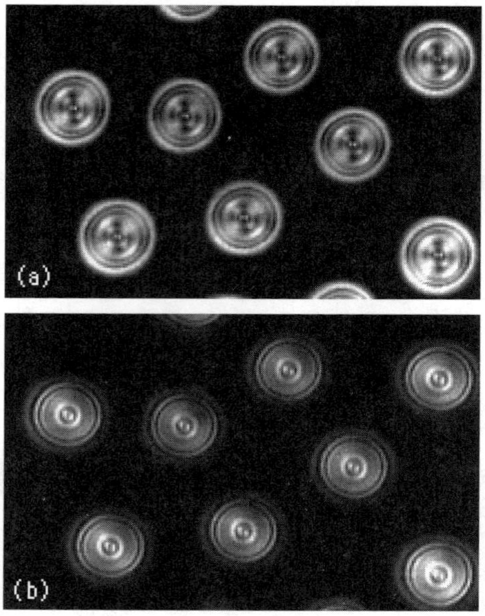

Fig. B.VII.49 Spherulites observed in the 8CB-S811 mixture (C = 1.36, V = 1.07 V); a) crossed polarizers; b) circularly polarized light (a quarter-wave plate at 45° from the analyzer produces extinction of the optically isotropic areas).

As the two tips are identical, a singularity necessarily persists inside the finger (Fig. B.VII.50).

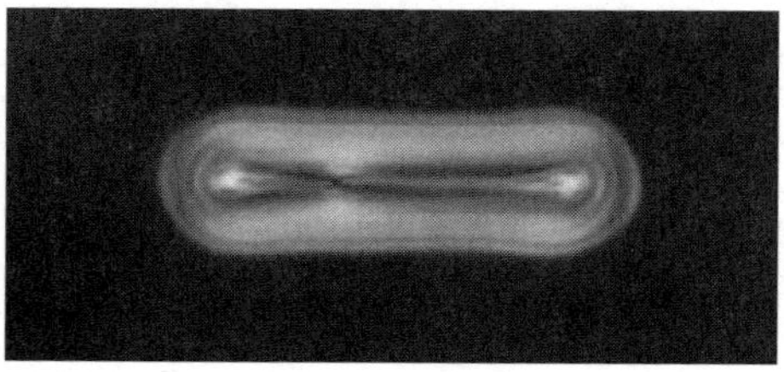

Fig. B.VII.50 Finger segment with two identical rounded tips, obtained after lengthening of a cholesteric bubble. It necessarily exhibits a central defect, clearly visible in the photo (from ref. [28]).

Another, more spectacular, way of obtaining a spherulite consists of applying an increasing electric field to a finger loop [37]. If the sample is thick enough (namely for $C > C^* \approx 1.2$), the loop turns suddenly and irreversibly into a spherulite as soon as $V > V^*(C)$ (limit indicated by stars in Fig. B.VII.47). In thin samples (when $C < C^*$), the loop disappears without forming a spherulite when the field is increased (this time, above the open squares in the phase diagram). A spherulite never yields back a finger loop upon reducing the field.

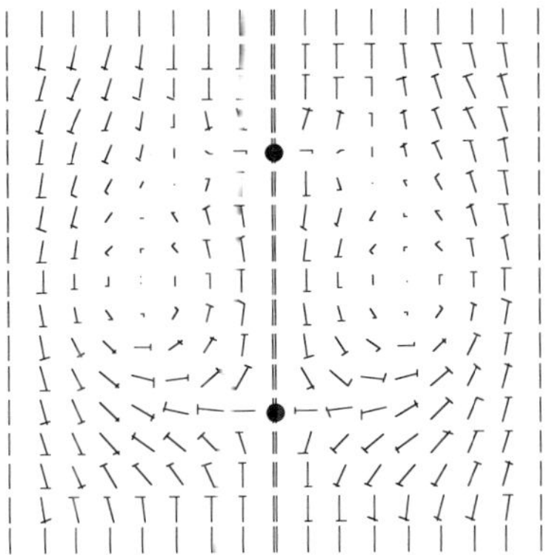

Fig. B.VII.51 Presumptive director field inside a spherulite. In this model, two point defects appear on the spherulite axis (from ref. [37]).

Several topological models for the spherulite structure have been put forward. Some authors suggested the existence of two disclination loops attached to the plates [38], which is hardly compatible with the strong anchoring conditions employed. We prefer the model given in figure B.VII.51, where the singularity (a line [30] or two point defects [37]) is along the spherulite axis.

Spherulites can also be obtained in samples with negative dielectric anisotropy, for instance by inducing turbulent flow in the sample for a few seconds using a low-frequency field (a few Hz) [39]. By subsequently increasing the frequency (above 500Hz, typically), spherulite domains of well-controlled density can be achieved. To our knowledge, the stability range for spherulites in the parameter plane (C,V) has not yet been accurately determined for materials with negative dielectric anisotropy.

VII.5.g) Fingers of the second and third kinds

Recent directional growth experiments at the cholesteric-isotropic liquid interface [40] showed that the fingers are not always of the previously described type. At least three different kinds of fingers can spontaneously nucleate at this interface. We shall then term cholesteric fingers of the first kind (or CF-1) those we have already presented, spontaneously nucleating in the samples (from the TIC, in particular).

These fingers are topologically continuous and their extremities are clearly distinct (Fig. B.VII.38).

The two other types are distinguished by the difference in static and dynamic properties.

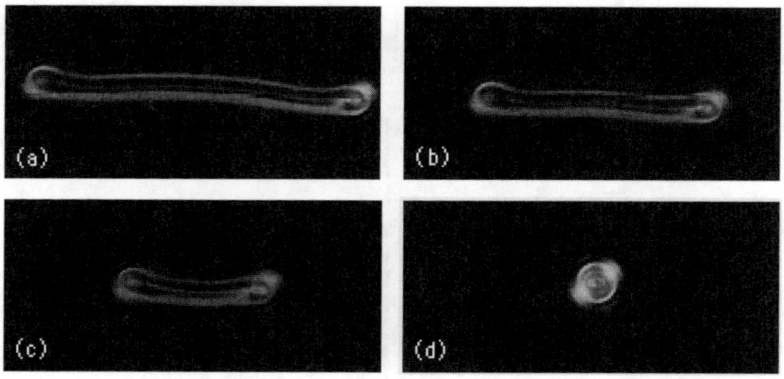

Fig. B.VII.52 Segment of a finger of the second kind (CF-2). Its two tips are almost identical. Contrary to the CF-1, which disappears completely, a CF-2 segment systematically yields a cholesteric bubble (from ref. [41]).

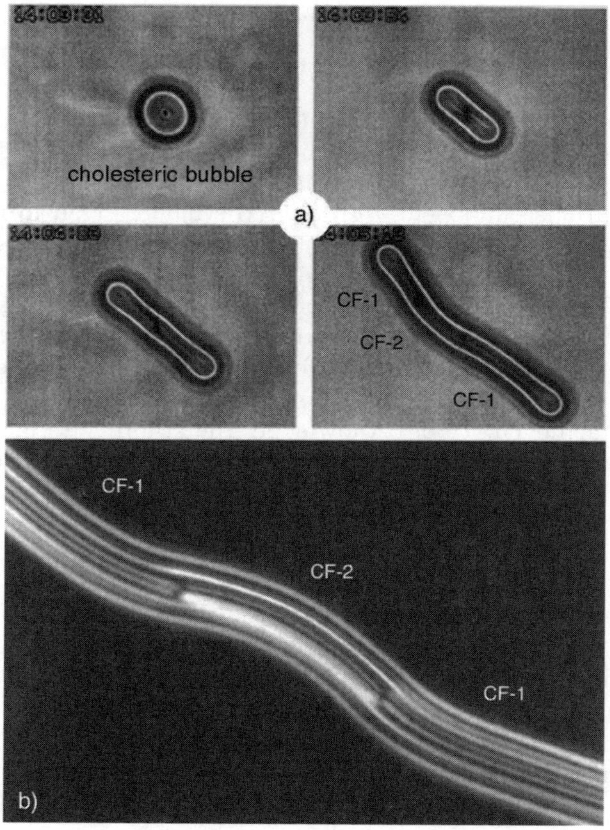

Fig. B.VII.53 a) Formation of a CF-2 segment from a cholesteric bubble in a cholesteric mixture of positive dielectric anisotropy. Natural light observation: b) the difference in optical contrast between a CF-1 and a CF-2 becomes obvious when they are observed between crossed polarizers in a tip-to-tip configuration. The CF-2 is much more asymmetric than the CF-1 (from ref. [42]).

One of them, called "finger of the second kind" (CF-2), strongly resembles the CF-1, explaining why the distinction has not been made for a long time. They are, however, very different topologically. Thus, a CF-2 segment has two almost identical extremities and always ends up as a cholesteric bubble when shortening under the action of a moderate electric field (Fig. B.VII.52) [41].

The reciprocal transformation is not impossible [42], but much more difficult, as shown in figure B.VII.53a. It can only be induced by lowering the electric field below the stability threshold of the bubble. At first, the bubble lengthens, forming two CF-1 segments; the central defect then splits, leading to the formation of a CF-2 segment. The terminal CF-1 segments can then be eliminated by temporarily increasing the electric field. A CF-2 segment like the one in figure B.VII.52 is thus obtained.

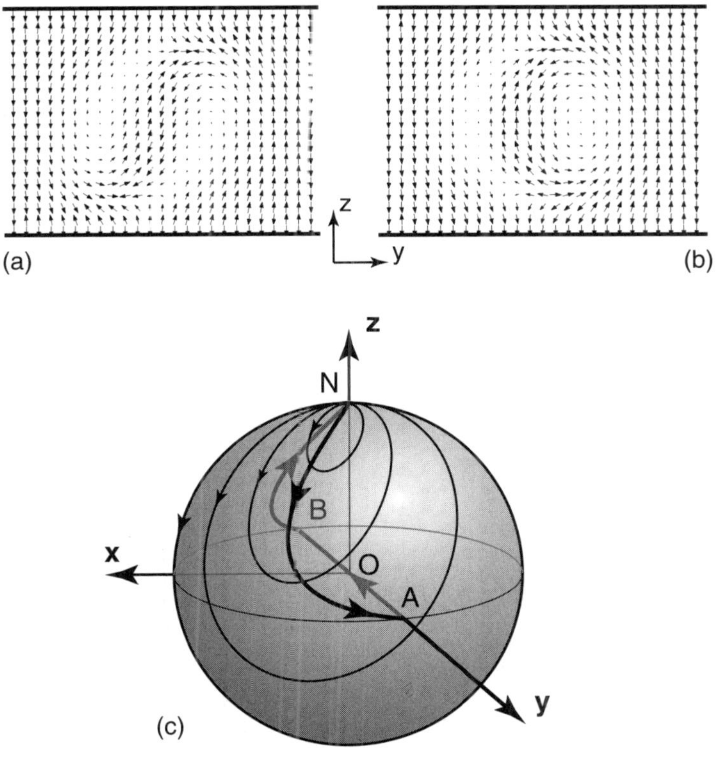

Fig. B.VII.54 a) Director field inside a finger of the first kind; b) director field inside a finger of the second kind. As opposed to the first one, this finger is not invariant under reversal (from ref. [43]) (simulations by J. Baudry); c) on the surface of the S2 sphere, the centers of director trajectories describe an "open" curve formed by two pieces (NA) and (BN) tangent to the equator plane and related by a rotation of π about the z axis. Points A and B are topologically equivalent. In these two points, corresponding to the middle sample plane, the director follows the great circle in the thick line on the surface of the S2 sphere. As before, smaller circles are the images of straight lines parallel to the y axis.

The director field inside a finger of the second kind is displayed in figure B.VII.54. This finger, numerically obtained by Gil and Gilli [43], is still topologically continuous over most of its length. It is no longer reversal-invariant, which explains its optical asymmetry (see Fig. B.VII.53b). It can be represented schematically on the surface of the S2 sphere by the images of straight lines D perpendicular to the finger axis in real space (approximated as a family of circles). In a CF-1, the centers of the circles describe a closed "pear-shaped" curve going through the North pole. By reducing this curve to a unique point (the North pole) one can go continuously from the CF-1 to the homeotropic nematic without creating a defect. This mechanism explains why there is no defect between the two extremities of a CF-1 segment. In a finger of the second kind, the curve described by the centers is no longer closed, but rather composed of two distinct pieces, related by a π rotation around the z axis and ending in the equator plane in two diametrally opposed (and topologically equivalent) points A and B. In these two points, the image of the director trajectory along a straight line in the middle sample plane and normal to the finger axis is a great circle on the surface of the S2 sphere. This model predicts a point-like defect at each tip of a CF-2 segment, as the finger can no longer be continuously reduced to a point on the S2 sphere. The same feature explains the exceptional electric field resistance of these fingers [41]. Indeed, as homogeneous elimination is topologically impossible, they can only disappear by shortening from the tips (which always leads to a cholesteric bubble).

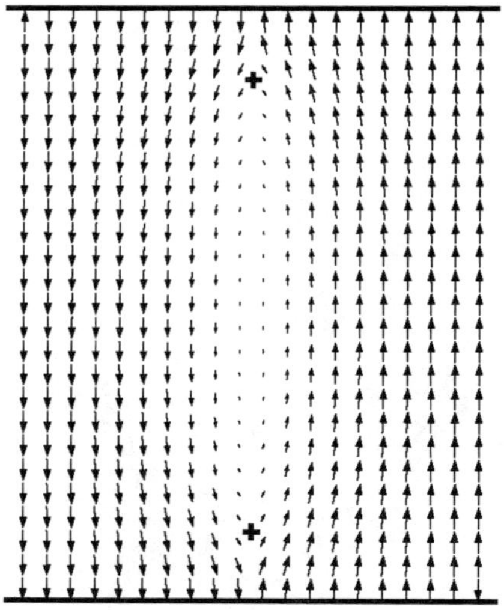

Fig. B.VII.55 Finger of the third kind. The director turns by π in the middle plane. Note the presence of two twist disclinations (marked by crosses) inside the finger, localized close to the plates.

The last of the thoroughly studied configurations is the finger of the third kind (CF-3). Experimentally, it is half as thick as the other two and corresponds to a director rotation of π in the middle plane (for the first two, the director turns by 2π). This finger possesses two disclination lines along its axis, a model first put forward by Cladis and Kléman [44], then numerically obtained by several authors [40, 45]. The corresponding director field is shown in figure B.VII.55. These fingers do not drift under an AC electric field.

VII.5.h) Dynamics of the fingers of the first and second kind: crawling and cholesteric spirals

The differences between fingers of the first and second kind are spectacularly demonstrated by their dynamic properties under the action of an AC electric field. Experiment shows that, at low enough frequency, the fingers can spontaneously move inside the samples, but in very different ways.

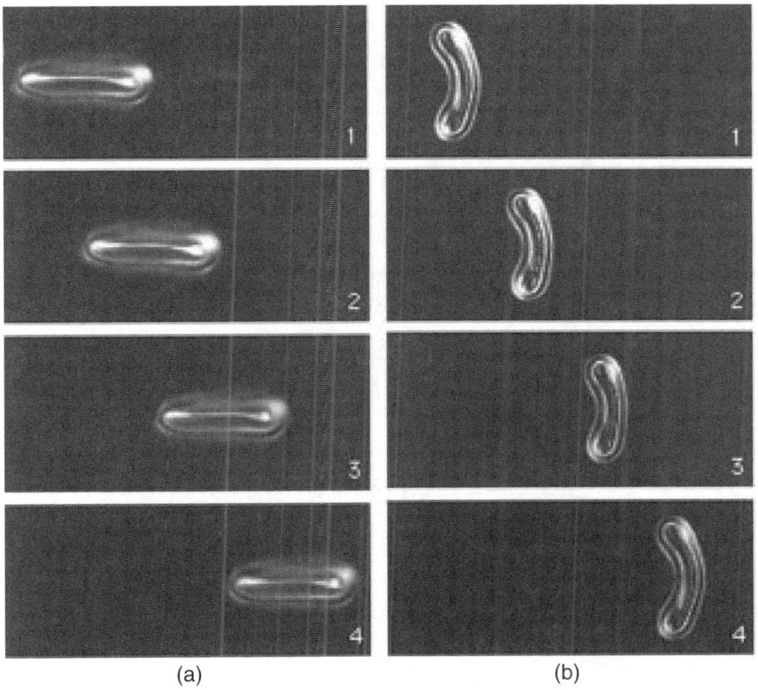

Fig. B.VII.56 Finger movement under a low frequency electric field. a) Crawling of a CF-1 segment; b) transverse drift of a CF-2 segment (from ref. [41]).

Fingers of the first kind have a crawling movement along their axis [41]. Figure B.VII.56a shows the example of a segment of constant length (a well-adapted voltage is required). This extremely slow movement has gone unnoticed for a long time.

On the other hand, fingers of the second kind drift in the transverse direction. Figure B.VI.56b shows the example of a segment of constant length (again requiring a well-adapted voltage). When the voltage is slightly decreased, these segments deform and grow, forming spirals (Fig. B.VII.57). The resulting spirals can be simple (Archimedes' spirals far from the center) or appear in pairs, depending on the initial conditions (Fig. B.VII.58). These spirals are produced by the transverse finger drift (of lower normal velocity when the curvature increases) combined with their lengthening. These spirals were simultaneously observed for the first time by Gilli and Kamayé [46] and by Mitov and Sixou [47] in a polymer mixture, and a little later by Ribière et al. [41,48,52] in mixtures of small molecules.

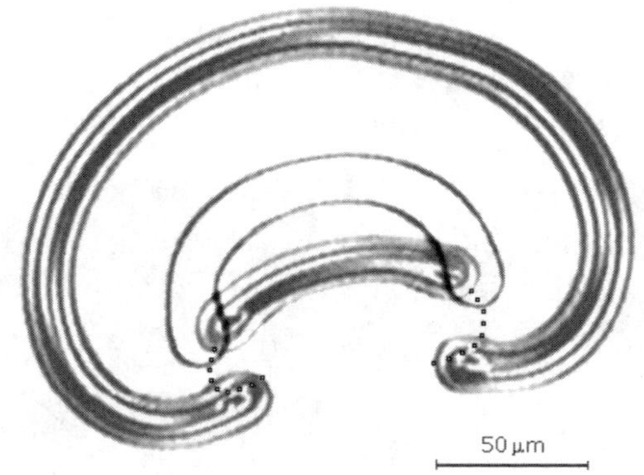

Fig. B.VII.57 Initial deformations of a CF-2 segment (from ref. [42]). The segment, represented in three successive positions, drifts normally to its axis and at the same time grows by the tips. The small squares describe the tip trajectories.

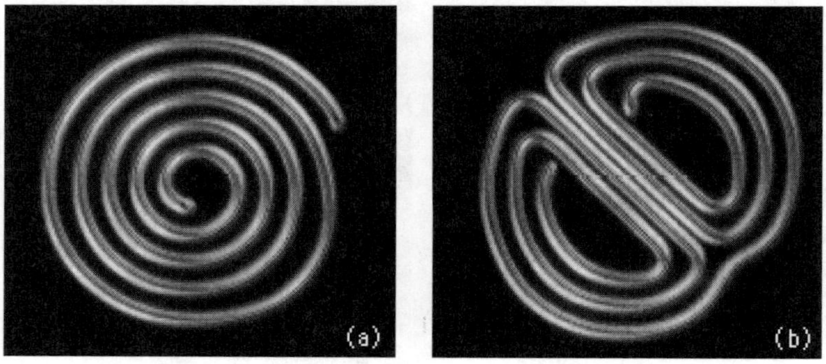

Fig. B.VII.58 Evolved spirals. a) Simple spiral. The inner tip describes a circle at constant angular frequency. The other tip describes a logarithmic spiral; b) twin spirals (from ref. [42]).

In large pitch cholesterics (10 μm or more), the spirals sometimes need several hours to form, which explains why they were only recently observed. The spirals form for a wide frequency range (from a few tens of Hz to a few kHz), close to the critical line $V_2(C)$, and always in materials with positive dielectric anisotropy (note that they are most easily obtained and observed slightly above this line).

The origin of drift in fingers of the second kind is clearly related to their asymmetry. Indeed, these fingers differ from the others by not being invariant under reversal. A CF-2 has no reason for reversing its velocity when the field is reversed, hence the possibility of drift under an AC field. Several drift mechanisms were proposed by Gil et al. [43,49]. One of them involves the Lehmann effect we shall describe further in this chapter. Another model is based on flexo-electricity. Precise measurements invalidate these models by showing that the fingers only drift in the conductive regime, when the frequency of the applied field is lower than the frequency of charge relaxation in the liquid crystal [50]. Recent numerical simulations confirm that, indeed, electrohydrodynamic effects may be responsible for the drift of the CF-2s [51].

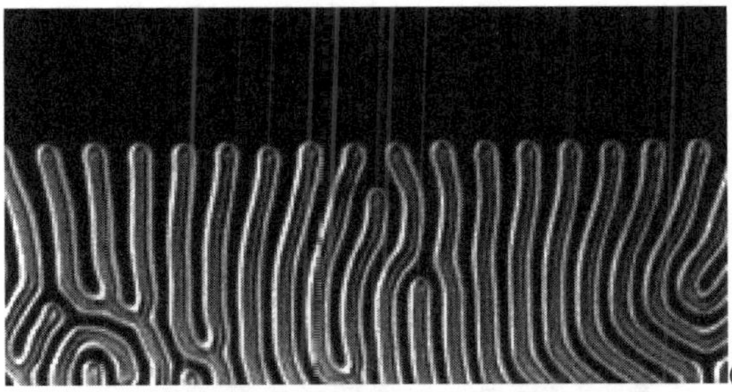
(a)

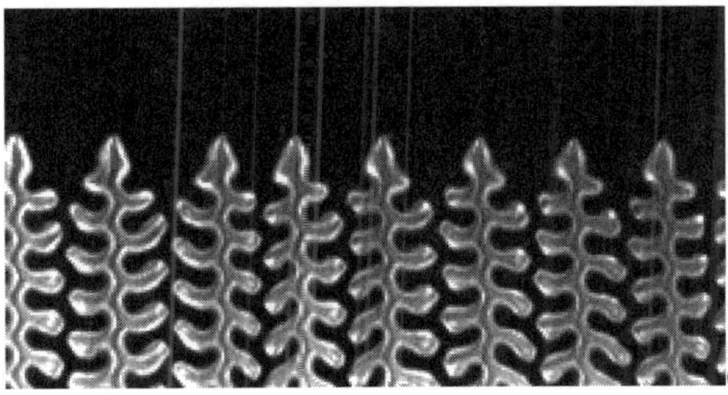
(b)

Fig. B.VII.59 Cellular (a) to dendritic (b) morphological transition at a moving cholesteric-nematic interface. The cholesteric unwinds because the pitch diverges in the temperature gradient (from ref. [52]).

Finally, we mention that several experimental results about the selection of cholesteric fingers in directional growth experiments remain unexplained [33a]. One of them (not mentioned in ref. [33a]) is shown in figure B.VII.59. In this experiment, the pitch diverges and the cholesteric unwinds because a smectic phase is present on the cold side. In this way, a nematic band forms, in the case of homeotropic anchoring, between the cholesteric phase and the smectic A phase. The cell-to-dendrite transition is due to a rotation by π of fingers of the first type, which successively expose at the front their rounded tips at small pulling velocity, and their pointed tips at large velocity [52a, b]. Note that this transition is only observed in thin enough samples [52c].

VII.5.i) When the sample is much thicker than the cholesteric pitch: Helfrich-Hurault instability

In the previous experiments, the samples were strongly confined, their thickness being of the order of the cholesteric pitch. In this section, we shall consider the opposite limit and discuss the behavior under field of samples much thicker than the cholesteric pitch [53–55].

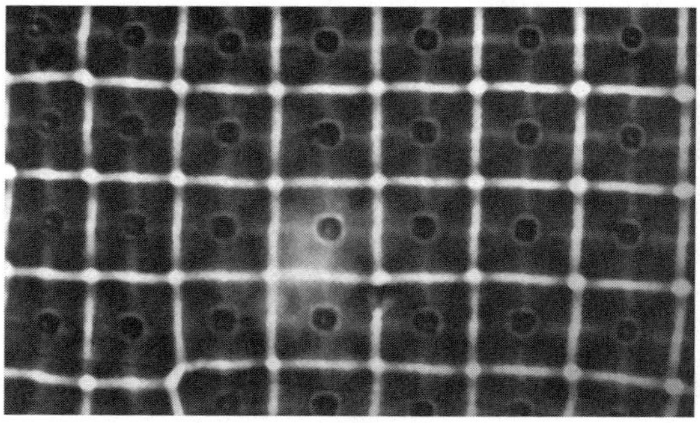

Fig. B.VII.60 Square lattice of undulations in a planar cholesteric under a vertical magnetic field (photo by F. Rondelez).

Consider the case of a planar sample, with the helix axis normal to the glass plates. If molecular anchoring on the plates is strong enough, the helix will maintain its orientation at contact under all circumstances. Let us now apply a magnetic field **B** normal to the glass plates. If the field is much weaker than the unwinding threshold, the helix will locally maintain its structure. However, its axis will try to turn in order to decrease its energy of interaction with the field (the diamagnetic susceptibility anisotropy is assumed positive). This coupling can lead to a two-dimensional undulation of the layers for a high enough field (Fig. B.VII.60) (but still much smaller than the unwinding field). Undulation amplitude will be a maximum in the middle of the sample and zero on the plates, where the layers are strongly anchored.

This buckling instability is in fact generic for lamellar fluid systems, also appearing in smectics A or C (and even smectics B). For this reason, it will be theoretically analyzed in chapter C.II dedicated to smectic elasticity and to the study of this instability. In particular, we shall see that a cholesteric can be described like a smectic, provided that the cholesteric helix is not too distorted (i.e., that the applied field is much weaker than the unwinding field).

VII.6 Cholesteric hydrodynamics

In this section we shall concentrate on two very important facts.

The first is that a cholesteric has a much larger apparent viscosity than the corresponding nematic (racemic). This surprising feature was qualitatively explained by Helfrich [56] who introduced the fundamental concept of **permeation** across the cholesteric layers.

The second important fact concerns the rotation of the cholesteric helix when placed in a temperature gradient [57]. This phenomenon, observed by Lehmann in 1900, shortly after the discovery of cholesterics, has no equivalent in nematics.

VII.6.a) Permeation

The Poiseuille flow experiments of Sakamoto et al. [58] showed that the apparent viscosity of cholesterics is much larger than that of ordinary liquids. On the other hand, recent experiments by Candau et al. showed that cholesterics are non-newtonian [59].

As a qualitative explanation of these results, W. Helfrich introduced the concept of **permeation** [56]. As the name indicates, this flow corresponds to the molecules crossing the cholesteric layers, assumed to be immobile (Fig. B.VII.61). To fix the layers, one can imagine (but this is almost impossible experimentally) that the molecules are strongly anchored at the sample boundaries but still keep their helix-like configuration. Molecule flow across the layers is achieved by imposing a pressure gradient along the axis of the helix. This flow can be theoretically analyzed using the equations of nematodynamics presented in chapter B.III. A simplifying assumption is that the velocity **v** of molecules is constant and parallel to the helix axis (parallel to z). Under these conditions, eq. B.III.69 gives the energy dissipation which is simply:

$$T\overset{\circ}{S} = \mathbf{h}.\frac{D\mathbf{n}}{Dt} \qquad \text{(B.VII.108)}$$

since $\partial v_j/\partial x_j = 0$ (uniform flow). As D/Dt is a total time derivative, we can write in the stationary regime:

$$\frac{D\mathbf{n}}{Dt} = v\frac{\partial \mathbf{n}}{\partial z} \qquad \text{(B.VII.109)}$$

The director turns with the cholesteric helix of wave vector q, so:

$$\frac{\partial \mathbf{n}}{\partial z} = q\,(\mathbf{z}\times\mathbf{n}) \qquad \text{(B.VII.111)}$$

with **z** the unit vector along the z axis.

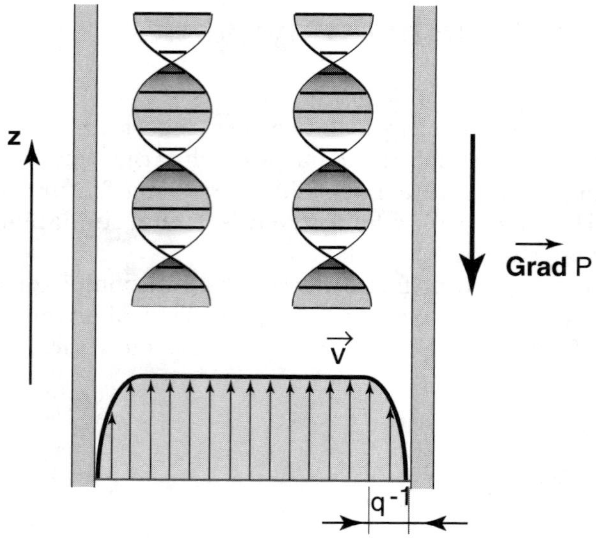

Fig. B.VII.61 Permeation flow in a cholesteric, as imagined by W. Helfrich. Two boundary layers appear close to the wall, where the velocity goes to zero. Beyond these layers, the velocity is constant and proportional to the imposed pressure gradient. One speaks of "plug flow," in analogy with flow in concentrated suspensions or porous media.

As to the molecular field **h**, it is given by the material constitutive law (see eq. 4.9b of appendix 4 to chapter B.III):

$$\mathbf{h} = \gamma_1 \mathbf{N} + \gamma_2 \mathbf{n}\underline{A} \qquad \text{(B.VII.112)}$$

reducing here to

$$\mathbf{h} = \gamma_1 \frac{D\mathbf{n}}{Dt} = \gamma_1 q v\, (\mathbf{z} \times \mathbf{n}) \qquad \text{(B.VII.113)}$$

because $A_{ij} = 0$. The dissipation is finally:

$$T\overset{\circ}{s} = \gamma_1 q^2 v^2 \qquad \text{(B.VII.114)}$$

This dissipated energy equals the work done by the pressure, so one has immediately:

$$p'v = \gamma_1 q^2 v^2 \qquad \text{(B.VII.115)}$$

so the flow velocity is

$$v = \frac{1}{\gamma_1 q^2} p' \qquad \text{(B.VII.116)}$$

This formula is outstanding; it indicates that the velocity of the flow is proportional to the imposed pressure gradient. This is similar to the Darcy law for porous media.

Thus, in this experiment the cholesteric behaves as a porous medium with permeability $1/\gamma_1 q^2$, with the original feature that the cholesteric plays both the role of the medium and that of the fluid traveling across it.

We still need to take care of the boundary conditions, requiring that the velocity go to zero at the walls limiting the sample. One can no longer assume that v = Cst, which complicates the equations. We shall only give the result and refer the reader to the book by S. Chandrasekhar (see ref. [2] p. 4) for the detailed calculations. The main result is that the velocity goes to zero close to the walls over two boundary layers of thickness 1/q (roughly, the cholesteric pitch) and is constant in the center of the sample. The configuration, represented in figure B.VII.61, is that of a plug flow.

In conclusion, let us compare the apparent viscosity of cholesterics to that of an ordinary fluid. For a definition of the apparent viscosity, assume that the material flows through a tube of radius R under a pressure gradient p'.

In an ordinary fluid, the velocity profile is parabolic and the flow rate is proportional to the pressure and inversely proportional to the dynamic viscosity η:

$$Q = \frac{\pi}{8}\frac{\rho R^4}{\eta}p' \qquad \text{(B.VII.117)}$$

In a cholesteric, the same dependence exists between the flow rate and the pressure gradient, so the apparent viscosity η_{app} of the cholesteric can be defined by the relation:

$$Q = \frac{\pi}{8}\frac{\rho R^4}{\eta_{app}}p' \qquad \text{(B.VII.118)}$$

Using eq. B.VII.116 and neglecting the presence of the boundary layer (which amounts to considering that $R >> q^{-1}$), one immediately has:

$$\frac{\eta_{app}}{\eta} = \frac{1}{8}\frac{\gamma_1}{\eta}(qR)^2 \approx 0{,}1\,(qR)^2 \qquad \text{(B.VII.119)}$$

This relation shows that the apparent viscosity depends on the radius of the tube and becomes noticeably larger than for an ordinary fluid as soon as $qR >> 1$. This difference can be considerable: for R = 1 mm and $q = 10^5\ \text{cm}^{-1}$, one has $\eta_{app} \approx 10^7\eta$. One should however keep in mind that this calculation is not realistic, as it assumes the helical structure is fixed. This constraint, difficult to enforce experimentally, will be discussed again in the chapter on smectic hydrodynamics, where a similar permeation process exists.

VII.6.b) Lehmann rotation

In the early 20th century, Lehmann made an astounding observation [57]. Having placed a cholesteric drop on a glass plate heated from below, microscope observation showed that the drop spun continuously, never stopping. Closer observation showed that it was not the drop itself which turned, but the texture, i.e., the director and the cholesteric helix.

This phenomenon, not encountered in nematics, is related to the absence of mirror symmetry in cholesterics. Indeed, there is no equivalence between applying the temperature gradient **grad**T along the wave vector **q** of the chiral structure or in the opposite direction. This is obvious in hydrodynamics, when a ship propeller is placed in a uniform flow. Reversing flow direction, the direction of rotation is also reversed.

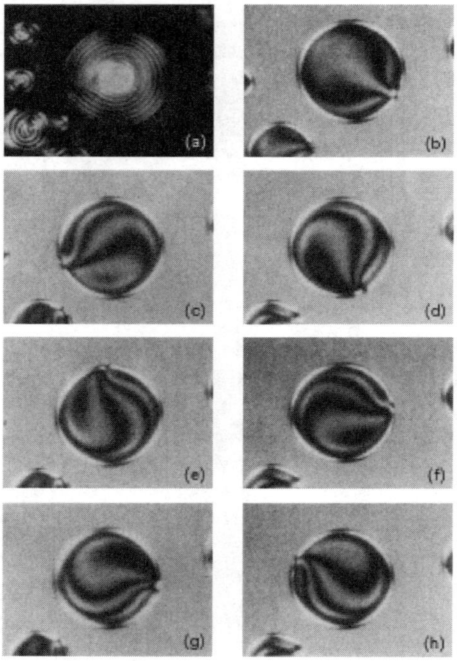

Fig. B.VII.61 Almost spherical cholesteric droplets in a sample of thickness 150 μm. a) Photo between crossed polarizers. A χ defect line of rank 2 appears on the curved side of the droplet. In the center, the structure is essentially planar; b) photo of a different drop in zero field, with the polarizers uncrossed by 20°; c, e, g) under a +2 V DC voltage, the same droplet starts turning clockwise; d, f, h) applying a −2 V voltage, the texture turns the opposite way (from ref. [60]).

This thermomechanical coupling between rotation of the cholesteric helix and a heat flow is not the only example of this kind. A diffusion current of a chemical species different from the liquid crystal, or an electric field must also induce rotation of the cholesteric helix. This last phenomenon was recently demonstrated by Madhusudana and Pratibha [60]. The experiment is as follows. A cholesteric (mixture of two carboxylic esters with 5% cholesteryl chloride) is doped with a small quantity of epoxy resin (Lixon). This mixture, exhibiting a wide cholesteric-isotropic coexistence range at room temperature, is placed between two electrodes. The experiment consists of monitoring the rotation under DC field of the internal structure of an isolated cholesteric droplet suspended in the isotropic liquid. The Lixon plays an essential part: because of its high affinity for glass, it favors the formation, between the glass and the drop being studied, of an isotropic liquid layer of major practical import, as it avoids rigid director anchoring on the plates (that would unavoidably block helix rotation). Figure B.VII.62 presents two series of images showing that the texture spins clockwise under an upwards pointing DC field, and counterclockwise when the field is reversed. Madhusudana and Pratibha equally showed that the rotation velocity does not depend on the size of the drop, changes sign when changing the enantiomer and is linear in the applied voltage (Fig. B.VII.63).

In order to explain these observations, the theory of nematodynamics must be generalized. Clearly, all conservation laws remain valid and only the material constitutive equations must be modified to take into account the absence of mirror symmetry. This problem, first mentioned by Oseen [61], was formalized by Leslie [62]. Because the theory is complicated, we present here only the essential results, before analyzing the particular case of the Lehmann rotation.

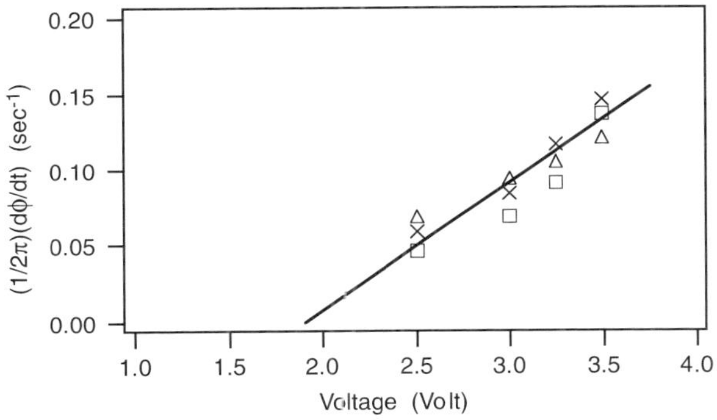

Fig. B.VII.62 Rotation speed of the texture as a function of the applied DC voltage. The linear fit does not go through the origin, the electric field being screened up to a voltage of about 1.9 V. This means that the redox potential of at least one of the components of the mixture is 1.9 V. Different symbols correspond to different droplets. Sample thickness is d ≈ 8 μm (from the experimental data of ref. [60]).

As in nematics, the general form of irreversible entropy is (see appendix 4 of chapter B.III):

$$T\overset{\circ}{s} = \sigma_{ij}^{s} A_{ij} + (\mathbf{h} + \mathbf{h}^{M}).\mathbf{N} + \mathbf{J}.\mathbf{E} \qquad \text{(B.VII.120)}$$

with **J** the current conjugated to the field **E**. In practice, **J** can be an electric current, a heat flux (so far denoted by **q**), or a diffusion current. The field **E** derives from a potential and, in the previous examples, it is given by:

$$\mathbf{E} = -\,\mathbf{grad}\,V \qquad (V = \text{electrical potential}) \qquad \text{(B.VII.121a)}$$

$$\mathbf{E} = -\frac{\mathbf{grad}T}{T} \qquad \text{(B.VII.121b)}$$

$$\mathbf{E} = -\,\mathbf{grad}\mu \qquad (\mu = \text{chemical potential}) \qquad \text{(B.VII.121c)}$$

Taking as fluxes the quantities **A**, **N**, and **E** and as conjugated forces $\underline{\sigma}^s$ (the symmetric part of the viscous stress tensor), $\mathbf{h} + \mathbf{h}^M$, and **J**, the phenomenological relations relating these quantities can be shown to be:

$$\sigma^s_{ij} = \sigma^s_{ij}\,(\text{Nem}) + \frac{1}{2}(\mu_1 + \mu_2)\,[\,(\mathbf{E}\times\mathbf{n})_i n_j + (\mathbf{E}\times\mathbf{n})_j n_i] \qquad \text{(B.VII.122)}$$

$$\mathbf{h} + \mathbf{h}^M = (\mathbf{h} + \mathbf{h}^M)\,(\text{Nem}) + \nu\,\mathbf{n}\times\mathbf{E} \qquad \text{(B.VII.123)}$$

$$\mathbf{J} = \sigma_\perp\mathbf{E} + (\sigma_{//} - \sigma_\perp)\,(\mathbf{n}.\mathbf{E})\mathbf{n} + \nu\,\mathbf{n}\times\mathbf{N} - (\mu_1+\mu_2)\mathbf{n}\times(\mathbf{n}\underline{A}) \qquad \text{(B.VII.124)}$$

In the last equation, $\sigma_\perp$ and $\sigma_{//}$ represent the usual conductivities (thermal, electrical, or chemical). Quantities marked (Nem) are the same as in ordinary nematics.

In order to return to the complete viscous stress tensor, one needs to write the balance of the viscous and elastic torques acting on the director (eq. B.III.51). Taking into account B.VII.123, one has:

$$\mathbf{\Gamma}^v = -\,\mathbf{n}\times(\mathbf{h} + \mathbf{h}^M) = -\,\mathbf{n}\times(\gamma_1\mathbf{N} + \gamma_2\mathbf{n}\underline{A}) + \nu\mathbf{E}_\perp \qquad \text{(B.VII.125)}$$

with $\mathbf{E}_\perp$ the component of **E** perpendicular to the director **n** ($\mathbf{E}_\perp = (\mathbf{n}\times\mathbf{E})\times\mathbf{n}$). On the other hand,

$$\Gamma^v_i = -\,e_{ijk}\sigma^a_{jk} \qquad \text{(B.VII.126)}$$

with $\underline{\sigma}^a$ the antisymmetric part of the viscous stress tensor $\underline{\sigma}^v$. These two relations yield $\underline{\sigma}^a$ and then $\underline{\sigma}^v$:

$$\sigma^v_{ij} = \sigma^v_{ij}(\text{Nem}) + \mu_1(\mathbf{E}\times\mathbf{n})_i n_j + \mu_2(\mathbf{E}\times\mathbf{n})_j n_i \qquad \text{(B.VII.127)}$$

as well as an additional relation between coefficients μ_i and ν:

$$\nu = \mu_1 - \mu_2 \qquad \text{(B.VII.128)}$$

A final point to this theoretical discussion is that, as shown by eq. B.VII.125, a field **E** can induce a torque in a cholesteric. This is only possible because the cholesteric is different from its mirror image. Note that this effect is **linear**, the torque being proportional to the applied field. Finally, it is easily shown that this "driving" torque does not contribute to the dissipation, since

$$\nu(\mathbf{n}\times\mathbf{E}).\mathbf{N} + \nu(\mathbf{n}\times\mathbf{N}).\mathbf{E} = 0 \qquad \text{(B.VII.129a)}$$

$$\frac{1}{2}(\mu_1 + \mu_2)\,[\,(\mathbf{E}\times\mathbf{n})_i n_j + (\mathbf{E}\times\mathbf{n})_j n_i]\,A_{ij} - (\mu_1 + \mu_2)\,[\mathbf{n}\times(\mathbf{n}\underline{A})\,].E = 0 \qquad \text{(B.VII.129b)}$$

Let us now consider a simple case in detail. Suppose the field **E** is applied along the helix axis (taken as the z axis). In this geometry, the hydrodynamic speed is zero ($\underline{A} = 0$). The balance of torques equation B.VII.125 becomes, letting φ be the angle made by the director with a fixed direction x:

$$\gamma_1 \frac{\partial \varphi}{\partial t} = K_2 \frac{\partial^2 \varphi}{\partial z^2} - \nu E \qquad \text{(B.VII.130)}$$

This equation must be supplemented by the two boundary conditions on the two plane surfaces limiting the sample (in z = 0 and z = d). Two cases are possible.

In the first case the surfaces are free, a situation close to the previously described experiments. The elastic torque then goes to zero on the two surfaces, yielding:

$$K_2 \left(\frac{\partial \varphi}{\partial z} - q \right) = 0 \quad \text{in} \quad z = 0 \text{ and } z = d \qquad \text{(B.VII.131)}$$

with q the spontaneous cholesteric twist. One can equally verify $\underline{\sigma}\mathbf{v} = 0$ on the two free surfaces, with $\underline{\sigma}$ the total stress tensor. These equations have an obvious solution in the stationary regime:

$$\varphi = qz - \frac{\nu E}{\gamma_1} t + \text{Cst} \qquad \text{(B.VII.132)}$$

This equation shows that the cholesteric helix continuously turns at the angular velocity $\nu E/\gamma_1$. The experimental data of Madhusudana and Pratibha (Fig. B.VII.55) give an order of magnitude $\nu \approx 0.32\ 10^{-3}\ \text{gs}^{-2}\text{V}^{-1}$ taking $\gamma_1 = 0.7$ poise.

In the second case the director is anchored on one surface (or both). The rotation is then blocked. Indeed, if

$$\varphi = 0 \quad \text{in} \quad z = 0 \quad \text{and} \quad K_2 \left(\frac{\partial \varphi}{\partial z} - q \right) = 0 \quad \text{in} \quad z = d \qquad \text{(B.VII.133)}$$

the only solution is static:

$$\varphi = \left(q - \frac{\nu E}{K_2} d \right) z + \frac{1}{2} \frac{\nu E}{K_2} z^2 \qquad \text{(B.VII.134)}$$

A final comment is in order. One might wonder where does chirality enter in eq. B.VII.132. As a matter of fact, the signs of μ_1 and μ_2 (and thus that of ν) depend on the chirality and must vanish for q = 0 (the coupling disappears when the nematic phase is composed either of non-chiral molecules or of a racemic mixture of two opposite enantiomers). Consequently the rotation must be reversed when going from a right-handed cholesteric to a left-handed one, thus predicting a relation of the form:

$$\nu = \nu_0 q + O(q^3) \qquad \text{(B.VII.135)}$$

as the twist goes to 0.

The experiments of Madhusudana et al. [60b] show that this relation is well verified in the case of electromechanical coupling.

Moreover, these authors show that ν changes sign at the compensation point (q = 0) of a mixture of two molecules with opposite chirality. This result, coherent with the hydrodynamic theory, proves that the presented electromechanical effect is of structural rather than molecular origin.

Note that Eber and Janossy reached an opposite conclusion in the case of a thermomechanical coupling [63]. Indeed, these authors observe a finite thermomechanical coefficient ν in a compensated mixture (q = 0) of 8CB and cholesteryl chloride. In this case, the Lehmann effect would be of purely molecular origin. This somewhat surprising result may be due to erroneous interpretation of the experimental results [64].

BIBLIOGRAPHY

[1] a) For a review of frustration in liquid crystals, see M. Kléman, "Effects of frustration in liquid crystals and polymers," CMD Meeting, Pisa, April 1987, *Physica Scripta*, **19** (1987) 565.
b) Our presentation of frustration is inspired by the work of Barbet-Massin R., "Structure and optical properties of the Blue Phase I of cholesteric liquid crystals," M.Sc. Thesis, University of Paris XI, Orsay, 1985.
c) Sethna J.P., Wright D.C., Mermin N.D., *Phys. Rev. Lett.*, **51** (1983) 467.
d) Pansu B., Dubois-Violette E., Dandeloff R., *J. Physique (France)*, **48** (1987) 305.

[2] Yang D.K., Crooker P.P., *Phys. Rev. A*, **35** (1987) 4419.

[3] Grebel H., Hornreich R.M., Shtrikman S., *Phys. Rev. A*, **28** (1983) 1114.

[4] Ennulat R.D., *Mol. Cryst. Liq. Cryst.*, **13** (1971) 337.

[5] De Vries H., *Acta Cryst.*, **4** (1951) 219.

[6] a) Chandrasekhar S., Shashidhara Prasad J., *Mol. Cryst. Liq. Cryst.*, **14** (1971) 115.
b) Stasiek J., *Heat and Mass Transfer*, **33** (1997) 27.

[7] Grandjean F., *C. R. Acad. Sci. (Paris)*, **172** (1921) 71.

[8] Cano R., *Bull. Soc. Fr. Minér. Crist.*, **91** (1968) 20.

[9] Brunet-Germain M., D.Sc. Thesis, University of Sciences and Techniques of Languedoc, Montpellier, CNRS order number: AO 6962, 1972. See also Brunet-Germain M., *C. R. Acad. Sci. (Paris)*, **274** Série B (1972) 1036.

[10] Feldman A.I., Crooker P.P., Goh L.M., *Phys. Rev. A*, **35** (1987) 842.

[11] de Gennes P.-G., *C. R. Acad. Sci. (Paris)*, **266** Série B (1968) 571.

[12] Scheffer T.J., *Phys. Rev. A*, **5** (1972) 1327.

[13] a) Kléman M., Friedel J., *J. Physique (France)*, **30** Coll. C4 (1969) 43.
b) Kléman M., Friedel J., "Application of dislocation theory to liquid crystals," Conference on Fundamental Aspects of the theory of dislocations, Gaithersburg, April 1969, *NBS Special Pub.*, **317** (1970) 607–636.

[14] Malet G., "Structure et dynamique des lignes de Grandjean dans les cristaux liquides cholestériques de grands pas," D.Sc. Thesis, University of Sciences and Techniques of Languedoc, Montpellier, 1981.

[15] Bouligand Y., *J. Physique (France)*, **33** (1972) 715.

[16] Friedel G., *Annales de Physique*, **18** (1922) 273.

[17] Bouligand Y., *J. Physique (France)*, **34** (1973) 603.

[18] Bouligand Y., Livolant F., *J. Physique (France)*, **45** (1984) 1899.

[19] Bouligand Y., *J. Physique (France)*, **30** (1969) C4-90.

[20] Bouligand Y., *J. Physique (France)*, **36** (1975) C1-331.

[21] de Gennes P.-G., *Sol. Stat. Com.*, **6** (1968) 163.

[22] Meyer B., *Appl. Phys. Lett.*, **12** (1968) 281.

[23] Meyer B., *Appl. Phys. Lett.*, **14** (1969) 208.

[24] Durand G., Leger L., Rondelez F., Veyssie M., *Phys. Rev. Lett.*, **22** (1969) 227.

[25] a) Brehm M., Finkelmann H., Stegemeyer H., *Ber. Bunsenges. Phys. Chem.*, **78** (1974) 883.
b) Harvey T., *Mol. Cryst. Liq. Cryst.*, **34** (1978) 224.

[26] Pirkl S., *Cryst. Res. Technol.*, **26** (1991) 5, K111.

[27] Ribière P., "Déroulage d'un cholestérique frustré en champ électrique," Ph.D. Thesis, Claude Bernard University-Lyon I, Order number 289.92, 1992.

[28] Ribière P., Oswald P., *J. Physique (France)*, **51** (1990) 1703.

[29] a) Press M.J., Arrott A.S., *J. Physique (France)*, **37** (1976) 387.
b) Press M.J., Arrott A.S., *Mol. Cryst. Liq. Cryst.*, **37** (1976) 81.

[30] Stieb A., *J. Physique (France)*, **41** (1980) 961.

[31] Ribière P., Pirkl S., Oswald P., *Liq. Cryst.*, **16** (1994) 203.

[32] Thurston R.N., Almgren F.J., *J. Physique (France)*, **42** (1981) 413.

[33] a) Oswald P., Bechhoefer J., Libchaber A., Lequeux F., *Phys. Rev. A*, **36** (1987) 5832.
b) Lequeux F., "Quelques instabilités élastiques locales dans les cholestériques," Ph.D. Thesis, University of Paris XI, Orsay, 1988.

c) Lequeux F., *J. Physique (France)*, **49** (1988) 254.
d) Lequeux F., Oswald P., Bechhoefer J., *Phys. Rev. A*, **40** (1989) 3974.

[34] Ribière P., Pirkl S., Oswald P., *Phys. Rev. A*, **44** (1991) 8198.

[35] Frisch T., Gil L., Gilli J.M., *Phys. Rev. E*, **48** (1993) R4199.

[36] Nawa N., Nakamura K., *Jap. J. Appl. Phys.*, **17** (1978) 219.

[37] Pirkl S., Ribière P., Oswald P., *Liq. Cryst.*, **13** (1993) 413.

[38] Hirata S., Akahane T., Tako T., *Mol. Cryst. Liq. Cryst.*, **38** (1981) 47.

[39] a) Kawachi M., Kogure O., Kato Y., *Jap. J. Appl. Phys.*, **13** (1974) 1457.
b) Haas W.E.L., Adams J.E., *Appl. Phys. Lett.*, **25** (1974) 263 and 535.

[40] Baudry J., Pirkl S., Oswald P., *Phys. Rev. E*, **57** (1998) 3038.

[41] Ribière P., Oswald P., Pirkl S., *J. Phys. II (France)*, **4** (1994) 127.

[42] Pirkl S., Oswald P., *J. Phys. II (France)*, **6** (1996) 355.

[43] Gil L., Gilli J.M., *Phys. Rev. Lett.*, **80** (1998) 5742.

[44] Cladis P., Kléman M., *Mol. Cryst. Liq. Cryst.*, **16** (1972) 1.

[45] Gil L., *J. Phys. II (France)*, **5** (1995) 1819.

[46] a) Gilli J.M., Kamayé M., *Liq. Cryst.*, **11** (1992) 791.
b) Gilli J.M., Kamayé M., *Liq. Cryst.*, **12** (1992) 545.

[47] a) Mitov M., Sixou P., *J. Phys. II (France)*, **2** (1992) 1659.
b) Mitov M., Sixou P., *Mol. Cryst. Liq. Cryst.*, **231** (1993) 11.

[48] Pirkl S., Oswald P., *Liq. Cryst.*, **28** (2001) 299.

[49] Gil L., Thiberge S., *J. Phys. II (France)*, **7** (1997) 1499.

[50] Baudry J., Pirkl S., Oswald P., *Phys. Rev. E*, **60** (1999) 2990.

[51] a) Tarasov O.S., Krekhov A.P., Kramer L., *Phys. Rev. E*, **68** (2003) 031708.
b) Tarasov O.S., Ph.D. Thesis, University of Bayreuth, 2003.

[52] a) Baudry J., "Statique et dynamique des doigts cholestériques sous champ électrique," Ph.D. Thesis, Ecole Normale Supérieure de Lyon, 1999.
b) Oswald P., Baudry J., Pirkl S., *Phys. Rep.*, **337** (2000) 67.

c) Oswald P, Baudry J., Rondepierre T., *Phys. Rev. E*, **70** (2004) 041702.

[53] Helfrich W., *J. Chem. Phys.*, **55** (1971) 839.

[54] Hurault J.P., *J. Chem. Phys.*, **59** (1973) 2068.

[55] Rondelez F., Hulin J.P., *Sol. State Commun.*, **10** (1972) 1009.

[56] Helfrich W., *Phys. Rev. Lett.*, **23** (1969) 372.

[57] Lehmann O., *Ann. Physik*, Leipzig, **2** (1900) 649.

[58] Sakamoto K., Porter R.S., *Mol. Cryst.*, **8** (1969) 443.

[59] Candau S., Martinoty P., Debeauvais F., *C. R. Acad. Sci. (Paris)*, **277** Série B (1973) 769.

[60] a) Madhusudana N.V., Pratibha R., *Mol. Cryst. Liq. Cryst. Lett.*, **5** (1987) 43.
b) Madhusudana N.V., Pratibha R., *Liq. Cryst.*, **5** (1989) 1827.
c) Madhusudana N.V., Pratibha R., Padmini H.P., *Mol. Cryst. Liq. Cryst.*, **202** (1991) 35.
d) Padmini H.P., Madhusudana N.V., *Liq. Cryst.*, **14** (1993) 497.

[61] Oseen C.W., *Trans. Faraday Soc.*, **29** (1933) 883.

[62] a) Leslie F.M., *Proc. R. Soc. London*, **307** Ser. A (1968) 359.
b) Leslie F.M., *Advances in Liquid Crystals*, Vol. 4, Ed. Brown G.H., Academic Press, New York, 1979, p. 1.

[63] a) Eber N., Janossy I., *Mol. Cryst. Liq. Cryst. Lett.*, **72** (1982) 233.
b) Eber N., Janossy I., *Mol. Cryst. Liq. Cryst. Lett.*, **102** (1984) 311.
c) Eber N., Janossy I., *Mol. Cryst. Liq. Cryst. Lett.*, **5** (1988) 81.

[64] a) Pleiner H., Brand H.R., *Mol. Cryst. Liq. Cryst. Lett.*, **5** (1987) 61.
b) Pleiner H., Brand H.R., *Mol. Cryst. Liq. Cryst. Lett.*, **5** (1988) 183.

Chapter B.VIII

Blue Phases: a second example of a frustrated mesophase

This is what Sir Charles Frank said about Blue Phases in 1983:

"They are totally useless, I think, except for one important intellectual use, that of providing tangible examples of topological oddities, and so helping to bring topology into the public domain of science, from being the private preserve of a few abstract mathematicians and particle theorists".

The main reason why Blue Phases have no practical application is related to their very limited existence range. **Indeed, they are always inserted between the liquid isotropic phase and the cholesteric phase in a very narrow temperature interval** (fig. B.VII.5). This marginal position in the phase diagrams is also a possible explanation for the little interest they aroused for more than half a century, until the English chemist George Gray (well known for synthesizing the cyanobiphenyls) studied them again (after Reinitzer, Friedel and Lehmann) about 1956 and noticed an anomalous texture very close to the clearing point of the cholesteric phase [1a]. Later on, Gray and his coworkers, Coates and Harrison, refined these initial observations and concluded that not one but **three different textures** were present between the isotropic and cholesteric phases [1b–d]. These observations aroused the interest of the scientific community and prompted new research. This chapter summarizes the results of this research, mainly on Blue Phases I and II.

To begin with, we point out some of the macroscopic properties of Blue Phases, showing that they have a cubic structure (section VIII.1). We then see how to describe them theoretically, first using the double-twist cylinder model (section VIII.2), which assumes that the medium is locally uniaxial, then using a more general, biaxial model (section VIII.3), on which a Landau theory of phase transitions can be built (section VIII.4). The theoretical considerations are followed by the presentation of a series of experiments ranging from the Kossel diagrams to faceting without neglecting the influence of the electric field (section VIII.5). Finally, we briefly discuss the problem of the Blue Phase III, whose structure remains a mystery.

VIII.1 Experimental evidence for the cubic symmetry of Blue Phases I and II

VIII.1.a) Saupe's crucial experiment and the O^5 model

The first unequivocal proof that Blue Phases do not correspond to some peculiar textures of the cholesteric phase, but have completely different symmetry, came from Alfred Saupe [2]. In a paper published in 1969, he reported the results of a very simple, yet crucial experiment (Fig. B.VIII.1), which consisted of observing a large quantity of cholesteryl *p*-nonylphenylcarbonate in a glass beaker maintained at a very homogeneous temperature.

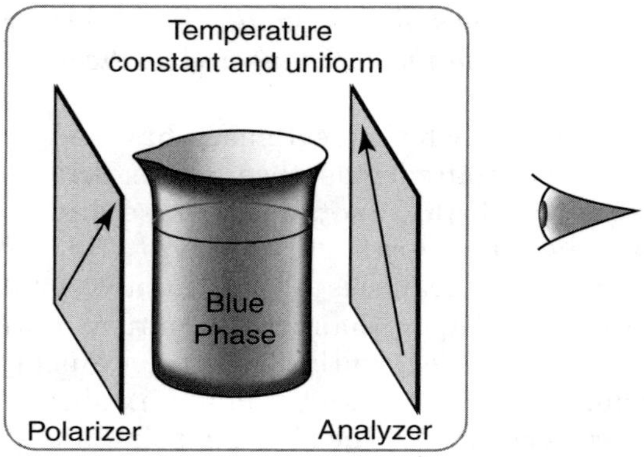

Fig. B.VIII.1 Experimental evidence for the optical isotropy of Blue Phases by A. Saupe [2].

Saupe noticed that, immediately below the isotropic liquid, this product remained optically isotropic, but exhibited enormous rotatory power in comparison to the isotropic phase. Furthermore, its consistency was not at all that of an isotropic liquid, but much more like a jelly that vibrates when gently shaken.

Given the optical isotropy of the medium, Saupe concluded that the mesophase could only have **cubic symmetry**. He also put forward a structure **both crystalline and liquid**, consisting of a lattice of **giant cubic cells** of size close to $a \approx 0.2\ \mu m$ (Fig. B.VIII.2). The volume of a cell V being $V = a^3 \approx 10^{-14}\ cm^3$, each of them would have had about $\mathbf{10^7}$ **molecules**, the molecular volume being of the order of $5 \times 5 \times 25\ Å^3 \approx 10^{-21}\ cm^3$. Note that, in contrast with solid crystals where the molecules have fixed positions (and orientations) inside the unit cell, in the Blue Phases I and II, the 10^7 molecules are free to diffuse from one place to another, provided they maintain on average the good orientation throughout the medium.

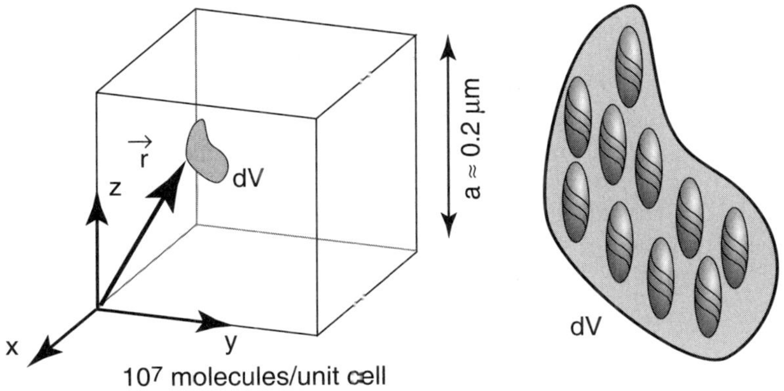

Fig. B.VIII.2 "Liquid crystalline" structure of the Blue Phases I and II: inside a giant cubic cell, the molecular order is purely orientational.

The distribution of molecular orientational order as a function of the position **r** inside the unit cell represents the pattern of the cell. The pattern proposed by Saupe assigned to the Blue Phase the O^5 (I432) symmetry.

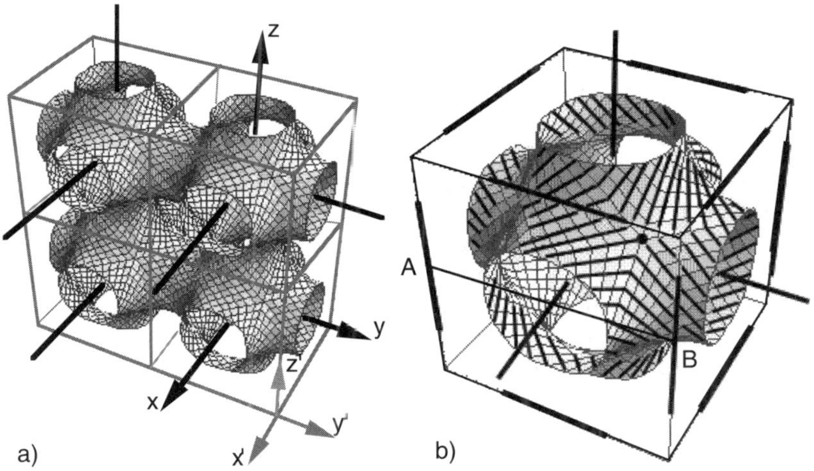

Fig. B.VIII.3 Saupe model for the Blue Phases build upon the minimal periodic surface P. a) Four cubic cells; b) orientational pattern of a cubic unit cell. The thick lines give the average molecular orientation.

In figure B.VIII.3b, we present the Saupe model, indicating the molecular orientation along the axes of two reference systems (x, y, z) and (x', y', z') belonging to two interpenetrating cubic lattices, as well as on a surface (close to the minimal periodic surface P) halfway between these two lattices. The orientation of the molecules in other points of the cell is easily obtained by extrapolation. In this way we find that molecular orientation is twisted almost everywhere, like, for instance, along the path AB. Nevertheless, one can equally see that in certain points of the unit cell the orientation of the director remains undetermined: this is the case for the nodes of the two lattices and, more generally, along the axes of order 3 along the diagonals of the cube. Examining the director configuration on the P surface close to its intersection points with the threefold axes shows the presence of $-\pi$ disclinations along these axes. Thus, Saupe found out that a **lattice of disclinations** was needed in order to build a twisted director field which is periodical in three dimensions.

The structure proposed by Saupe (of space group O^5) is on average (i.e., on scales much larger than the size of the unit cell) optically isotropic. However, it is not the only structure having this property, as all (very numerous!) space groups belonging to the cubic system are also optically isotropic, by symmetry.

To distinguish between these different structures, physicists studied the Bragg reflections they give. This is how they showed that the model with symmetry O^5 was in disagreement with the experiments, as we shall see further.

VIII.1.b) Bragg reflections and polycrystalline textures

The first measurements of the light spectrum $I(\lambda)$ reflected by cholesteryl esters were performed by Stegemeyer et al. [3]. They revealed, in agreement with conclusions of Gray et al. [1a–d], that cholesteryl nonanoate presents **not one, but at least two** Blue Phases. Stegemeyer et al. reached this conclusion by measuring the temperature dependence of the wavelength λ_B of a Bragg peak in the $I(\lambda)$ spectrum. Their results can be summarized as follows: in the cholesteric phase, λ_B is 360 nm and only slightly varies in temperature; at 91°C, the temperature where the substance loses its birefringence, λ_B undergoes a first discontinuity, going from 360 to 500 nm; but at 91.3°C they discovered a second discontinuity, λ_B falling to 400 nm (Fig. B.VIII.4). As their sample remains on the average optically isotropic at this second transition, they inferred the existence of two Blue Phases of cubic symmetry, that they called BPI and BPII.

This conclusion was later confirmed by Meiboom and Sammon [4], who measured the spectrum of light transmitted through polycrystalline samples, i.e., composed of crystallites smaller in size than the diameter of the light beam and usually oriented at random.

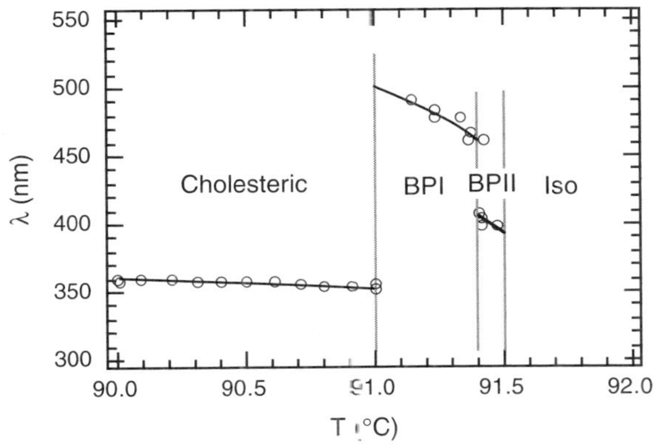

Fig. B.VIII.4 Wavelength variation for a Bragg reflection in cholesteryl nonanoate (from ref. [3]).

These polycrystalline samples were then observed by Marcus [5] under the reflecting microscope (Fig. B.VIII.5). He noticed that among the dozens of polygonal crystallites separated by grain boundaries that compose the texture, only a small number were shining, exhibiting monochromatic Bragg reflection at the wavelength λ_B. In Blue Phase II of cholesteryl nonanoate, $\lambda_B \approx 400$ nm, the bright crystallites have a blue-violet color. On the other hand, in Blue Phase I, the wavelength λ_B goes to 500 nm, such that the bright crystallites change color and become green! Marcus also noticed that by cooling the samples from the isotropic phase, the "mosaic" textures of the Blue Phases I and II are preceded by another texture he describes as "foggy" and which he interpreted as signaling a third, amorphous Blue Phase that he named BPIII (or "Blue Fog"). We emphasize that all these textures of Blue Phases I, II, and III were previously identified by Gray et al. [1a–d].

Fig. B.VIII.5 Typical polycrystalline texture of Blue Phase II observed under the reflecting microscope. The brighter crystallites show Bragg reflection of the light (from ref. [5]).

In conclusion, the spectroscopy experiments coupled with the direct observation of the textures of polycrystalline samples in transmitted or reflected light showed that the phase sequence of cholesteryl nonanoate actually contains **three Blue Phases, within a very narrow temperature interval of only 0.6°C**:

$$\text{Chol} \overset{91.35°\text{C}}{\rightarrow} \text{BPI} \overset{91.76°\text{C}}{\rightarrow} \text{BPII} \overset{91.84°\text{C}}{\rightarrow} \text{BPIII} \overset{91.95°\text{C}}{\rightarrow} \text{Iso}$$

The same phase sequence was subsequently found in other materials, such as, for instance, MMBC [6] (Fig. B.VIII.6), CE2, CE4, CE5, CE6 [7], or mixtures of CB15 and E9 [8].

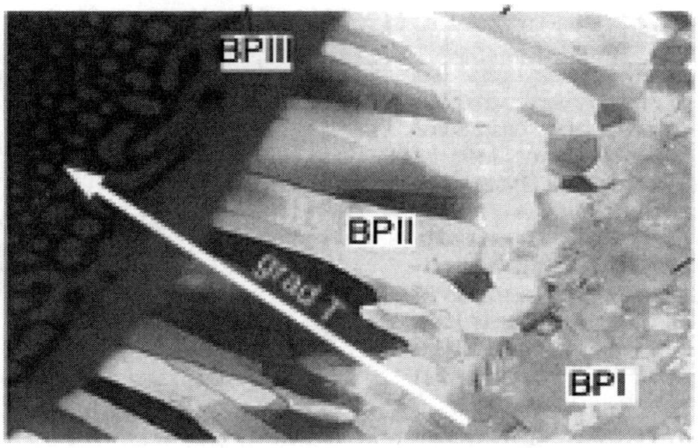

Fig. B.VIII.6 In a temperature gradient, the textures of chiral MMBC observed using transmission microscopy reveal the existence of the three Blue Phases (from ref. [6]).

If the outcome of these experiments clearly demonstrated that Blue Phases I and II have cubic symmetry, the problem of indexing the Bragg reflections and eventually determining the structures still remained.

VIII.1.c) Bragg reflections from monocrystals

The decisive progress in this direction was first made by Johnson, Flack, and Crooker [8], and later by R. Barbet-Massin et al. [9]. Their study considered the Bragg reflections of Blue Phase **monocrystals** in mixtures of the eutectic nematic E9 and the chiral product CB15. With respect to cholesteryl esters, using the CB15/E9 mixtures has several practical advantages:

1. The temperature intervals where the Blue Phases exist in these mixtures are close to room temperature, facilitating **precise** temperature regulation using the experimental setup in figure B.VIII.7. This setup comprises a reflecting microscope fitted with an immersion objective and

allows both spectroscopic measurements and direct observation of the monocrystal shape.

2. By conveniently choosing the composition of the mixture, **monocrystals** of Blue Phases I or II can be grown from the isotropic phase.

3. Finally, the size of the cubic unit cell can be tuned by changing the concentration so that several reflections (hkl) are in the visible range.

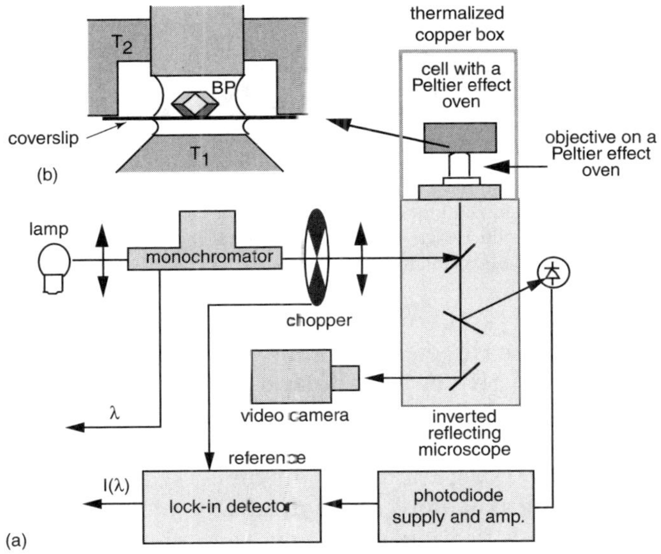

Fig. B.VIII.7 Setup used for growing Blue Phase monocrystals, studying them using reflecting microscopy and measuring the Bragg reflections. When the surfaces of the coverslip and the glass cylinder are fitted with electrodes, this setup is also convenient for studying the behavior of the Blue Phases under electric fields.

Let us give some practical details concerning the setup in figure B.VIII.7. In this setup, the sample is supported by capillarity in an adjustable gap (100 μm, typically) between a coverslip and a glass cylinder. The sample is in thermal contact with the objective (through the coverslip and the immersion oil) so the temperature of the objective T_1 must be regulated using Peltier effect elements. The upper surface of the sample is also in thermal contact with the glass cylinder, independently regulated at a temperature T_2. Using this double temperature regulation a vertical temperature gradient can be obtained and Blue Phase crystals can be nucleated on the coverslip. Using an immersion objective avoids the stray reflections on the glass plate.

The qualities of this setup are illustrated by the photos in figure B.VIII.8, showing monocrystals of Blue Phase I grown from the isotropic phase. When the wavelength of the incoming light, selected using the monochromator, corresponds to the Bragg reflection given by the (hkl) planes orthogonal to the optical axis of the microscope, the crystal reflects the light and appears bright against the black background of the isotropic liquid.

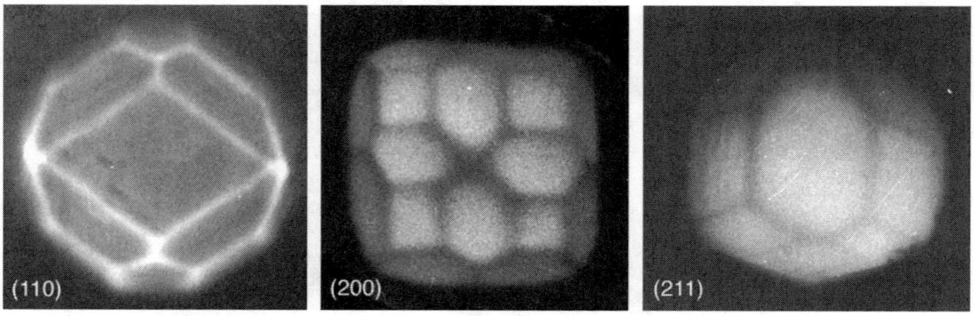

(a) (b) (c)

Fig. B.VIII.8 Images of Blue Phase I crystallites in E9/CB15 mixtures. Reflecting microscopy shows bright crystallites against the black background of the isotropic phase when the wavelength of the lighting corresponds to the Bragg reflection on the (110) (a), (200) (b), and (211) (c) planes. This indexing is required by the faceting symmetries.

The envelope of these crystals consist of facets of the type (110) and (211), so their orientations in these three images are easily determined. Thus:

1. In figure B.VIII.8a, the second-order symmetry identifies the diamond-shaped facet as being of the type (110). The reflection given by this crystal is therefore due to the (110) planes.

2. In figure B.VIII.8b, the fourfold symmetry shows that the crystal is oriented with the [100] axis along the optical axis of the microscope. Because the wavelength of the reflection is shorter than the one resulting from the (110) reflection, it can be assigned to the (200) planes.

3. In figure B.VIII.8c, the crystal turns its (211) facet towards the objective, so the reflection can be attributed to the (211) planes.

Finally, using a polarizer and a quarterwave plate one can experimentally demonstrate that these reflections are **circularly polarized**.

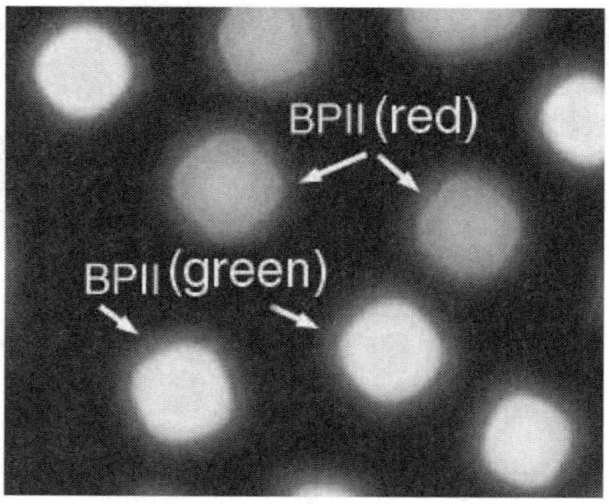

Fig. B.VIII.9 Coexistence of the Blue Phase I (green crystallites) and the Blue Phase II (red crystallites) in a sample of the CB15/E9 mixture. Due to the applied electric field, the crystallites turn their [100] axis towards the objective.

In conclusion, these images are direct proof that, in the BPI phase, circularly polarized Bragg reflections are given by the (110), (200), and (211) planes. Similar observations on monocrystals of Blue Phase II identified two circularly polarized Bragg reflections due to the (100) and (110) planes.

The image in figure B.VIII.9 shows that for certain concentrations the BPI, BPII, and isotropic phases can coexist and also that the (100) reflection of Blue Phase II is red, while the (200) reflection of Blue Phase I is green.

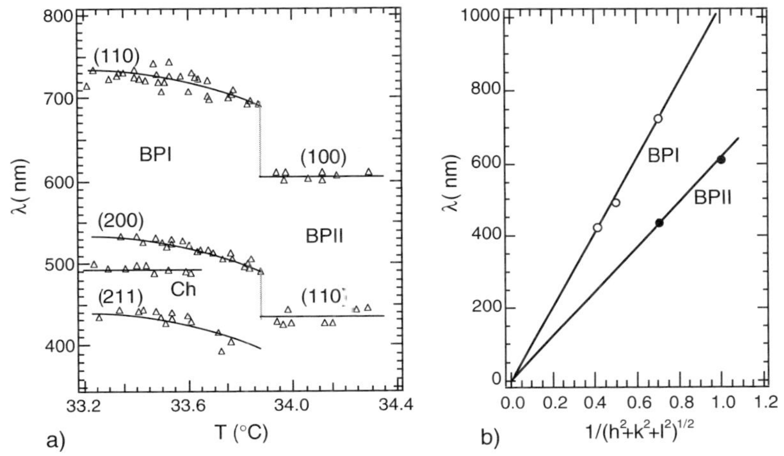

Fig. B.VIII.10 Bragg reflections in a mixture of 50/50% CB15 and E9 as a function of the temperature (from Johnson, Flack, and Crocker, ref. [8]).

The results of systematic measurements of the wavelengths associated with the three reflections in the BPI phase and the two reflections in the PBI phase are given in figure B.VIII.10a. Plotting the measured values for the wavelengths as a function of $1/(h^2+k^2+l^2)^{1/2}$ (Fig. B.VIII.10b), we can verify that in Blue Phase I at 33.6°C, the Bragg relation

$$\lambda = 2nd_{hkl} = \frac{2na}{\sqrt{h^2+k^2+l^2}} \qquad \text{(B.VIII.1)}$$

is fulfilled, with 2na = 1008 nm, where n is the average refractive index and a is the size of the cubic cell. Similarly, Blue Phase II yields 2na = 610 nm.

Indexing the Bragg reflections in the cubic Blue Phases represented another step forward in determining their symmetries.

Indeed, the series (110), (200), and (211), characteristic for **Blue Phase I**, is the signature of the **body-centered cubic** Bravais lattice, for which the selection rule is that the sum h+k+l must be even. This would seem to confirm the model of Saupe, as the subgroup of pure translations of the O^5 (I432) space group forms the cubic-centered lattice. In fact, this is not the case. Anticipating the rigorous proof to be developed in section B.VIII.3, we shall delineate here the reason why the model of symmetry O^5 must be discarded. Let us start by recalling that the reflection given by the (200) planes of Blue Phase I is **circularly polarized**. On the other hand, all models O^5 with symmetry have, by definition, a fourfold axis (ordinary, not helicoidal) in the [100] direction. It ensues that the average of the dielectric tensor ε_{ij} taken in a (y, z) plane for a generic position along x must be invariant with respect to a $\pi/2$ rotation, implying that $<\varepsilon>_{ij}$ is a uniaxial tensor of optical axis parallel to [100]. The two eigenvalues of this tensor, ε_{xx} and $\varepsilon_{yy} = \varepsilon_{zz}$, can be modulated along x, but this modulation has no chirality.

As to the (100), (110) series characterizing **Blue Phase II**, it is only compatible with the **simple cubic** Bravais lattice.

Having presented the experimental facts allowing identification of the three Blue Phases and assignment of the body-centered cubic and simple cubic Bravais lattices to BPI and BPII, respectively, let us return to a discussion of the theory.

VIII.2 Uniaxial models for the Blue Phases: disclination lattices

VIII.2.a) Double twist cylinders

The Saupe model, which we have just discarded, is a uniaxial model. In this type of model, the local orientational order parameter $Q_{ij}(\mathbf{r})$ is uniaxial [10]. The local isotropy axis of the tensor is thus given by the usual director $\mathbf{n}(\mathbf{r})$. In the previous chapter we saw that the cholesteric phase was frustrated at the molecular level, because each molecule wants to form a small angle with its nearest neighbors. An important (and less intuitive) point is that the molecule being chiral does not automatically ensure that this local configuration, termed **double twist**, minimizes the energy. This can be shown by comparing the local energies of a simple twist configuration (ordinary cholesteric) and of a double twist configuration similar to the one depicted in figure B.II.5. In the first case, using the notations of chapter B.II ($t_1 = t \neq 0$; $t_2 = 0$), one has:

$$f_{ST} = \frac{1}{2} K_2 (t + q)^2 \qquad \text{(B.VIII.2a)}$$

This energy has a minimum for $t = -q$, yielding:

$$f_{ST} = 0 \qquad \text{(B.VIII.2b)}$$

For a double twist configuration (Blue Phase), the Gauss term in K_4 must be considered, since $t_1 = t_2 = t$:

$$f_{DT} = \frac{1}{2} K_2 (2t + q)^2 + K_4 t^2 \qquad \text{(B.VIII.3a)}$$

Minimization with respect to t yields

$$t_{DT} = \frac{-K_2 q}{K_4 + 2K_2} \qquad \text{(B.VIII.3b)}$$

whence:

$$f_{DT} = \frac{K_2 K_4 q^2}{2(K_4 + 2K_2)} \qquad \text{(B.VIII.3c)}$$

It is immediately clear that the double twist configuration is locally more favorable than the first one if

$$-2K_2 < K_4 < 0 \qquad \text{(B.VIII.4)}$$

These are necessary conditions for the existence of a Blue Phase; are they also sufficient? We can already guess the response will be negative, having seen in the previous chapter that the double twist condition cannot be simultaneously satisfied everywhere in a euclidian three-dimensional space (frustration).

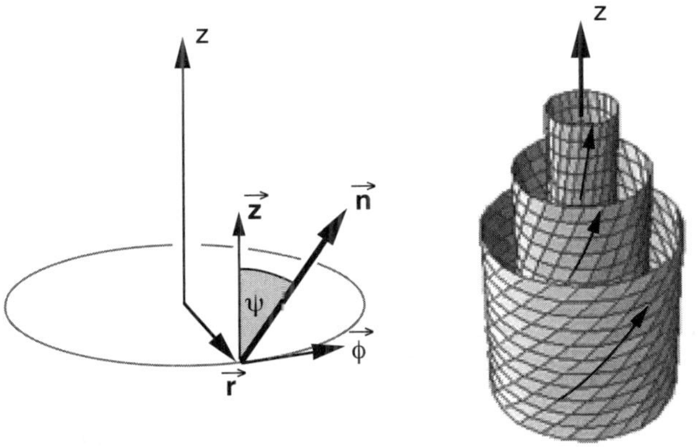

Fig. B.VIII.11 Structure of a double-twist cylinder.

Let us nevertheless try to satisfy it to the greatest extent possible and try to obtain a configuration that is **globally** better from energy point of view than the simple helix configuration of the cholesteric phase. One possible solution is the assembly of double twist cylinders we shall now describe.

We recall that in such a cylinder, parallel (say) to the z axis, the director turns in the following way (Fig. B.VIII.11):

$$\mathbf{n} = \mathbf{z}\cos qr + \boldsymbol{\phi}\sin qr \qquad \text{(B.VIII.5)}$$

with **z** and **ϕ** the unit vectors in cylindrical coordinates (r,ϕ,z).

The equation B.VIII.5 expresses the fact that the molecules prefer a configuration twisted in all directions perpendicular to the cylinder axis. When moving away from this axis, eq. B.VII.2, expressing the local tendency of the molecules to turn with respect to one another, is always perfectly satisfied along **r** but less and less fulfilled along **ϕ**. Furthermore, when the angle $\psi = qr$ of director tilt with respect to the z axis approaches $\pi/2$, an energetically unfavorable curvature distortion appears.

It can be shown that the distortion energy of such a cylinder with radius $r = p/8$ (this choice will be justified later) is lower than in a cholesteric phase of the same volume. In conclusion, the double twist cylinder (thus named because **n.curl n** = 2q along its axis) locally represents a better solution to the cholesteric frustration than the cholesteric phase, provided the inequalities B.VIII.4 are fulfilled.

VIII.2.b) Simple cubic model with O^2 symmetry

Now, the double twist cylinders must be assembled in a structure with cubic symmetry. Meiboom, Sethna, Anderson, and Brinkman [10] put forward a structure where the cylinders are placed along the three directions x, y, and z (Fig. B.VIII.12).

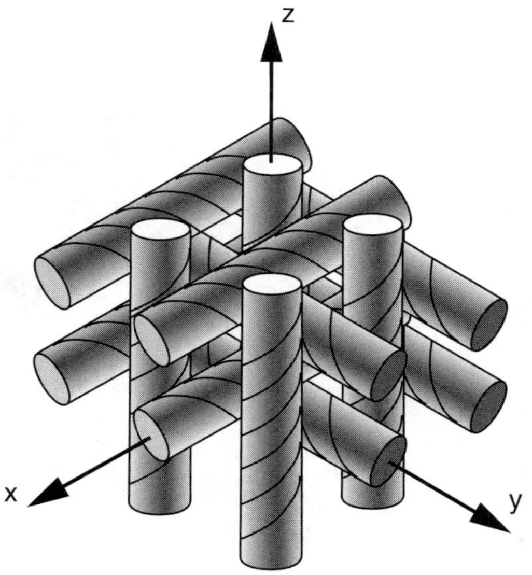

Fig. B.VIII.12 Uniaxial model for the Blue Phase II with symmetry O^2: a structure of double twist cylinders with radius p/8.

In this construction, the orientations of the director **n** at the contact points of each cylinder pair are identical (as $r = p/8$). Consequently, the director field **n**(**r**) is everywhere continuous in the space occupied by the cylinders.

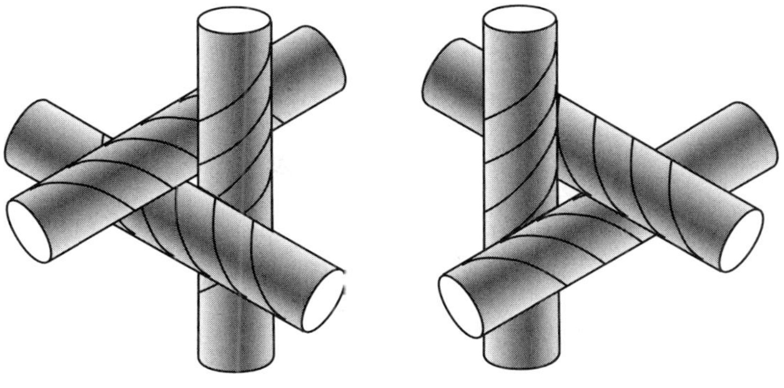

Fig. B.VIII.13 Trying to join the director field in the space between the cylinders, one finds that disclinations appear along some diagonals.

To complete this model, the gaps between the cylinders must be filled. When trying to join the director field in these spaces, one finds that this is impossible along some cube diagonals passing between three cylinders perpendicular to one another (Fig. B.VIII.5). Along these threefold axes must be introduced **wedge disclinations of rank −1/2**. These disclinations form two interpenetrating diamond-type lattices, as shown in figure B.VIII.14.

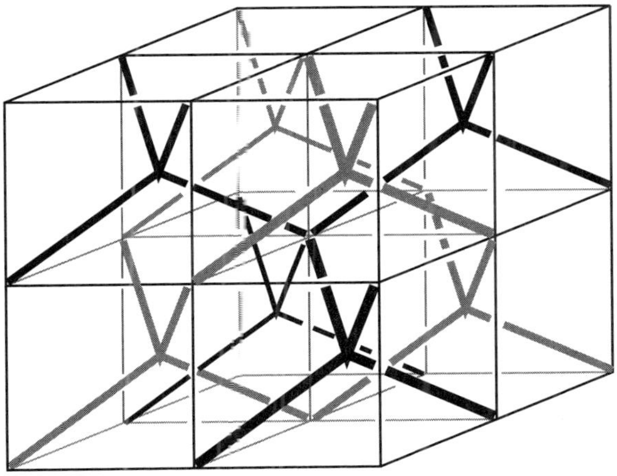

Fig. B.VIII.14 Disclination lattice corresponding to the assembly of double twist cylinders in figure B.VIII.4. Blue Phase II, symmetry C^2.

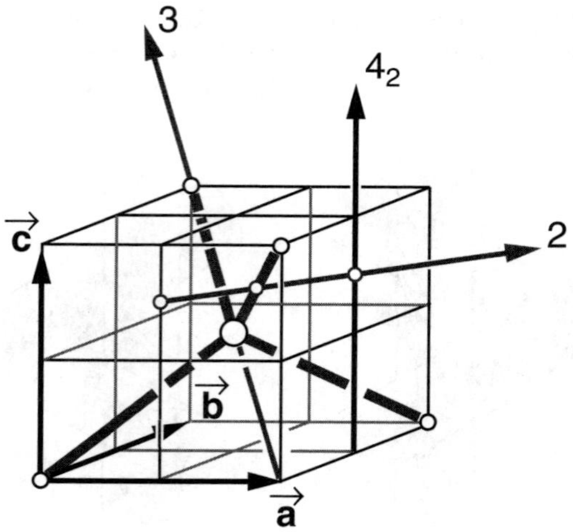

Fig. B.VIII.15 Position of the symmetry axes in a unit cell; only one axis of each type is represented in this diagram.

The symmetry of this model is O^2, or $P4_232$ in international notation. The symmetry elements of this space group are (Fig. B.VIII.15) for **translations**:

1. Those of a simple cubic Bravais lattice (P),

and for **rotations**:

2. The fourfold axes directed along the base vectors of the Bravais lattice. These axes are helicoidal, of the type 4_2, i.e., the $\pi/2$ rotation must be followed by a a/2 translation along the axis.
3. The threefold axes directed along the cube diagonals.
4. The twofold axes parallel to the **c** axis.

Comparison between this model and the one of the cholesteric phase shows that the Blue Phase is, on the one hand, preferable to the cholesteric phase because it contains double twist cylinders (instead of twist along one direction only) in a considerable volume fraction of the unit cell. The energy density for the distortion being of the order of Kq^2, the energy gain per unit cell can be estimated as

$$F_{dt} \approx -A'K(1/a)^2a^3 = -A'Ka \qquad \text{(B.VIII.6)}$$

with A' a numerical constant. On the other hand, the need for a disclination lattice leads to an energy increase comprising two terms:

1. The core energy of the disclinations (eq. B.IV.15):

$$F_c \approx B'Ka \qquad \text{(B.VIII.7)}$$

whose length per unit cell is proportional to a, and

2. the elastic energy of the disclinations:

$$F_e \approx C'aK \ln\left(\frac{a}{r_c}\right) \qquad \text{(B.VIII.8)}$$

depending logarithmically on the distance between disclinations (of the order a).

Summing these contributions and knowing that a = p/2 (Fig. B.VIII.12), the energy difference per unit cell between the Blue Phase and the cholesteric phase reads:

$$\Delta F = F_{bp} - F_{ch} = K(B - A)\,p + CpK \ln\left(\frac{p}{r_c}\right) \qquad \text{(B.VIII.9)}$$

with A, B, and C the three constants. If the core energy of the disclinations is low enough (as it happens close to the transition temperature towards the isotropic phase), then B−A is negative and for some values of the cholesteric pitch p $\Delta F < 0$. More specifically

$$\Delta F < 0 \qquad \text{if} \qquad p < p_c = r_c \exp\left(\frac{A - B}{C}\right) \qquad \text{(B.VIII.10)}$$

This calculation shows that **the pitch must be small enough for a Blue Phase to appear**.

These predictions (small pitch, proximity of the isotropic phase) are in qualitative agreement with the experimental results.

VIII.2.c) Body-centered cubic model with O^8 symmetry

Using an assembly of double twist cylinders one can also build models with different symmetries for the Blue Phases. For instance, starting from the previous model, with symmetry O^2, one can remove every other cylinder, as shown in figure B.VIII.16a.

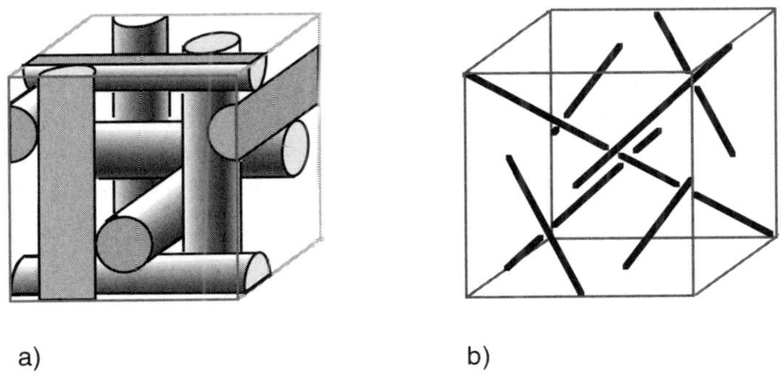

Fig.B.VIII.16 Uniaxial model with symmetry O^8 for Blue Phase I, obtained from the O^2 model by removing every other cylinder [11]: a) structure of the lattice of double twist cylinders; b) structure of the disclination lattice.

The final structure is body-centered cubic and its space group is O^8, or $I4_132$ in international notation. As for the simple cubic model, this new structure requires disclinations along the cube diagonals, forming the entanglement represented in figure B.VIII.16b.

Replacing the disclinations in this structure by double twist cylinders yields a new model with the same O^8 symmetry. In this case, the disclinations form two interlocking lattices of opposite chirality.

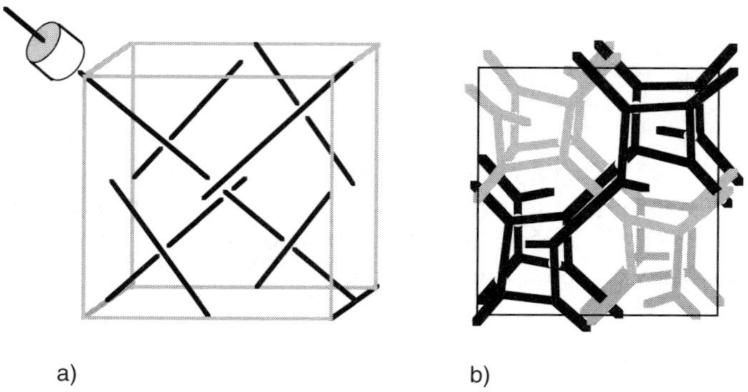

Fig. B.VIII.17 Uniaxial model of the Blue Phase I obtained from the O^8 model in figure B.VIII.16b by replacing the disclinations by double twist cylinders. This new model also has O^8 symmetry. a) Structure of the double twist cylinder lattice. For the sake of simplicity, only the axes of the cylinders are represented. b) Structure of the disclination lattice.

VIII.2.d) Blue Phases and minimal periodic surfaces

A very convenient way of building uniaxial models for the Blue Phases uses minimal periodic surfaces. It is based upon the fact that, by definition, the total curvature is zero everywhere on a minimal surface (see for instance ref. [11]).

We shall illustrate the importance of this property for the Blue Phases by first pointing out that two line systems start from each point of a minimal surface (Fig. B.VIII.18):

1. Extremal curvature lines C_1 and C_2 (also termed principal curvature lines). These lines intersect at straight angles on any surface (minimal or not).

2. Zero curvature lines, also called asymptotic lines A_1 and A_2. On a minimal surface, these lines are the bisectors of the principal lines and therefore also intersect at straight angles. They provide a surface parameterization.

On a minimal periodic surface, for instance the D surface in figure B.VIII.19a, the asymptotic lines form two distinct systems which, **taken separately, can be seen as two director fields of opposite chirality**. By choosing one of them, a double twist director field can be built.

This construction is illustrated in figure B.VIII.18, showing that the director field $\mathbf{n} // \mathbf{A}_2$ is twisted along the $\mathbf{A}_1$ direction. The asymptotic lines can also be defined on surfaces close to a minimal surface (shifted by a constant amount d along the normal $\mathbf{m}$ to the minimal surface). One can then show that the director field defined in this way is equally twisted along the $\mathbf{m}$ direction orthogonal to the surface and that the twist along $\mathbf{m}$ is of the same sign as the one along $\mathbf{A}_1$.

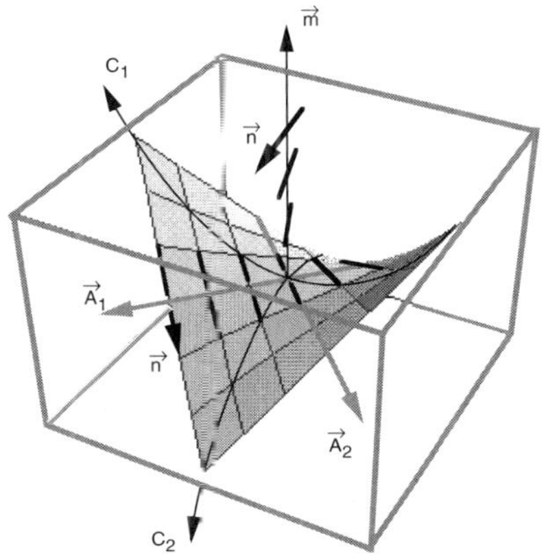

Fig. B.VIII.18 Double twist director field built upon the system of asymptotic lines on a minimal surface and the surfaces parallel to it.

This method of constructing a double twist director field is graphically illustrated for three examples of minimal surfaces. Thus, the P surface gives the field with symmetry O^5 shown in figure B.VIII.3. The D surface gives the field with symmetry O^2 in figure B.VIII.19, while the G surface gives the field with symmetry O^8 in figure B.VIII.20.

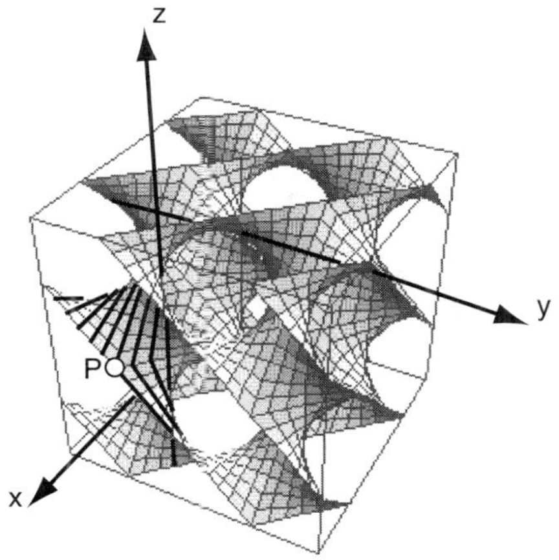

Fig. B.VIII.19 Uniaxial model with O^2 symmetry built upon the minimal periodic surface D; (assembly of eight unit cells). The axes of three double twist cylinders are shown. Point P is on the intersection of a disclination with the D surface.

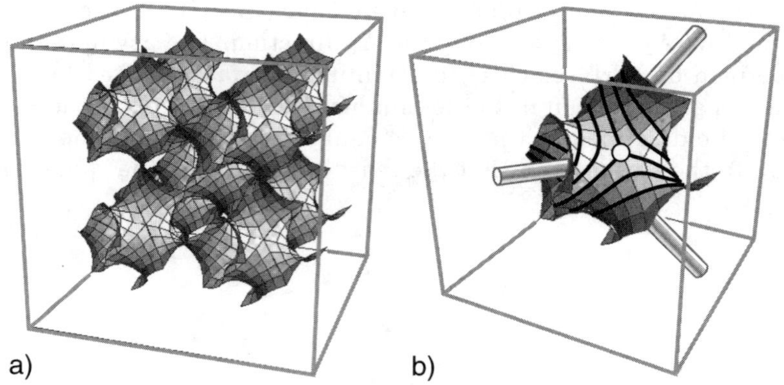

Fig. B.VIII.20 Uniaxial model with O^8 symmetry built upon the minimal periodic surface G; a) assembly of nine unit cells placed on the sites of a body-centered cubic lattice; b) unit cell. The axes of three double twist cylinders are shown. Point P is on the intersection of a disclination with the G surface.

All these fields contain disclination systems, because on every minimal periodic surface there are so-called flat points, where the main curvatures are zero (umbilic points). Clearly, the direction of the asymptotic lines cannot be defined in these points; as a result, the asymptotic line fields exhibit singularities at the umbilic points.

Because the structure of the director field in the body-centered uniaxial model with O^8 symmetry is complex, we shall not discuss it in detail. However, we shall prove that this model predicts a circularly polarized Bragg reflection in normal incidence to the (100) planes, as opposed to the O^5 model in which the fourfold symmetry axes forbid it (see section B.VIII.1c).

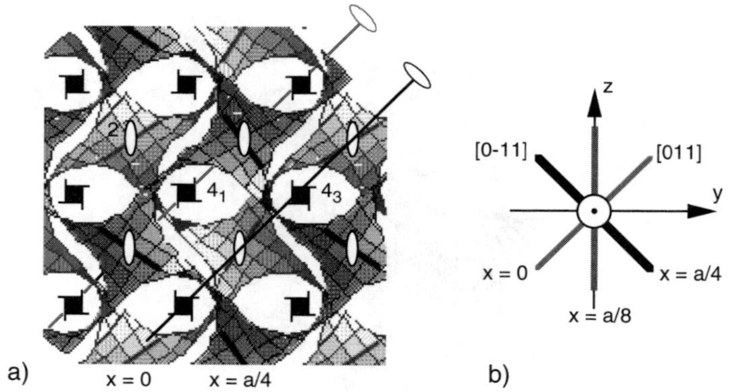

Fig. B.VIII.21 Chiral structure of the director field in the model with 0^8 symmetry built upon the minimal periodic surface G. a) View of two slabs of thickness a/16 normal to the [001] axis and placed at x = 0 and x = a/4. b) Average director orientation in the slabs of coordinates x = 0, a/8, and a/4.

As a starting point, note that, in the O^8 space group, the fourth-order revolution symmetry is not ordinary, consisting of the presence of the helicoidal axes 4_1 and 4_3. For this reason, the average of the dielectric tensor $<\varepsilon_{ij}>_{\Delta x}$ along y and z in a slab of thickness Δx orthogonal to the [100] axis need not be invariant with respect to a $2\pi/4$ rotation and can thus be anisotropic. This property of the O^8 model is made clear by the diagram in figure B.VIII.21 where the director field is shown in two slabs of thickness $\Delta x = a/16$.

Indeed, in the second plane slab, centered at $x = 0$, the director field is on the average oriented along the [011] direction. This local anisotropy is also visible in the slab in the first plane, centered at $x = a/4$, where the director field is identical to the one in the previous slab, only now the average director orientation is along the [0–11] direction.

The rotation of the director field through $2\pi/4$ between the $x = 0$ slab to the $x = a/4$ one is, of course, due to the action of the 4_1 helicoidal symmetry axes, also shown in the diagram of figure B.VIII.21. Owing to the $\mathbf{n} = -\mathbf{n}$ identity, the director field in figure B.VIII.21 is invariant with respect to the binary axes C_2 parallel to [100] and, consequently, also invariant with respect to the helicoidal axes 4_3. The symmetry with respect to these binary axes also entails that one of the eigenvectors of the average tensor $<\varepsilon_{ij}>_{\Delta x}$ must be parallel to the [100] direction, whatever the x coordinate of the (y,z) slab over which the average is taken: $\mathbf{e}_1$// [100]. Nevertheless, the orientation of the other two eigenvectors, $\mathbf{e}_2$ and $\mathbf{e}_3$, depends on the position x of the slab. For instance, in the slab $x = 0$, the symmetry with respect to the binary axes C_2 parallel to [011] requires that the second eigenvector of the $<\varepsilon_{ij}>_{\Delta x}$ tensor must be directed along this [011] direction: $\mathbf{e}_2$// [011]. Similarly, due to the helicoidal 4_1 (or 4_3) symmetry, in the slab shifted by $a/4$ this second axis of the $<\varepsilon_{ij}>_{\Delta x}$ tensor must turn through $2\pi/4$ (or $-2\pi/4$): $\mathbf{e}_2$//[0–11]. Applying once more this symmetry operation it can be shown that, in the $x = a/2$ slab, the $<\varepsilon_{ij}>_{\Delta x}$ tensor is identical to the one at $x = 0$: $\mathbf{e}_2$// [011]. One still needs to determine the position of the average tensor $<\varepsilon_{ij}>_{\Delta x}$ for intermediate values of x, where the O^8 space group imposes no a priori constraint on the orientations of the eigenvectors $\mathbf{e}_2$ and $\mathbf{e}_3$ or on the eigenvalues. In the figure B.VIII.20 model, built upon the asymptotic lines of the minimal surface G, the average director orientation in the slab at $x = a/8$ is parallel to the [001] direction, whence $\mathbf{e}_2$// [001]. Due to the helicoidal symmetry 4_1, $\mathbf{e}_2$// [010] in the slab at $x = 3a/8$, too.

From these data it can be inferred that the average tensor $<\varepsilon_{ij}>_{\Delta x}$ taken along the x direction approximately has a helix-like configuration similar to the one of the cholesteric phase with pitch $p = a$ (Fig. B.VIII.21b). Due to this resemblance, the Bragg reflection for incidence normal to the (200) planes can be circularly polarized, in agreement with the experimental data in section B.VIII.1c.

VIII.3 Biaxial model of the Blue Phases by Grebel, Hornreich, and Shtrikman

The uniaxial models form a subset of a much larger group of models where **the local orientational order** is of the **biaxial** quadrupolar type [12].

This observation is already relevant in the case of the cholesteric phase, where the local order parameter is intrinsically biaxial owing to its D_2 local symmetry. In order to obtain the most general expression B.VII.9 for the biaxial order parameter of the cholesteric phase:

$$Q_{ij}(\sigma) = -\frac{1}{\sqrt{6}} Q_0 \begin{bmatrix} -1 & 0 & 0 \\ 0 & -1 & 0 \\ 0 & 0 & 2 \end{bmatrix} + \frac{1}{\sqrt{6}} Q_2 \begin{bmatrix} \cos(2qz) & \sin(2qz) & 0 \\ \sin(2qz) & -\cos(2qz) & 0 \\ 0 & 0 & 0 \end{bmatrix} \qquad \text{(B.VIII.11)}$$

we used two of its physical properties (see the previous chapter), namely:

1. Over scales much larger than the cholesteric pitch, the average orientational order is negative uniaxial, the molecules being perpendicular to the twist axis **t**. The first term in expression B.VIII.11 accounts for this property.

2. The cholesteric phase gives a circularly polarized Bragg reflection, as it possesses a helicoidal symmetry. This led us to the second term in the expression B.VIII.11. We highlighted the fact that, for certain values of the ratio r between the amplitudes of these two terms, the order parameter becomes uniaxial and, in the $p \rightarrow \infty$ limit, the symmetry of the nematic phase is restored.

Let us now return to the Blue Phases. As Grebel, Hornreich, and Shtrikman [12] noticed by analogy with the cholesteric phase, there is no particular reason for the local orientational order in these phases to be uniaxial. Consequently, the most general way of describing their structure is to build their order parameter from a second-rank traceless tensor field such as, for instance, the quadrupolar moment density $Q_{ij}(\mathbf{r})$, or the anisotropy of the local dielectric permittivity $\varepsilon_{ij}(\mathbf{r})$.

However, this tensor field must be invariant under the symmetry operations of the space group of the Blue Phase. To ensure invariance with respect to the pure discrete translations engendering the cubic Bravais lattices (simple for the Blue Phase II and body-centered for the Blue Phase I), Grebel, Hornreich, and Shtrikman suggested that the tensor field $Q_{ij}(\mathbf{r})$ be developed over a three-dimensional Fourier series

$$Q_{ij}(\mathbf{r}) = \sum_{hkl} Q_{ij}(h,k,l) \exp\left[iq(hx+ky+lz)\right] + \text{c.c.} \qquad \text{(B.VIII.12)}$$

In this formula, the sum is taken over the wave vectors $\mathbf{q}_{hkl} = q(h,k,l)$ of the cubic reciprocal lattices defined by $q = 2\pi/a$ (with a the size of the cubic lattice) and by the Miller indices h,k,l.

For the Blue Phase II, where the reciprocal lattice is simple cubic, the Miller indices take on all integer values:

$$h,k,l = 0, \pm 1, \pm 2, \ldots \tag{B.VIII.13a}$$

On the other hand, for the Blue Phase I, where the reciprocal lattice is face-centered, the sum of the Miller indices must be even:

$$h+k+l = 2n \tag{B.VIII.13b}$$

When the sum h+k+l is odd, the $Q_{ij}(h,k,l)$ coefficient is zero by symmetry.

Moreover,

$$\sigma = h^2 + k^2 + l^2 \tag{B.VIII.13c}$$

and

$$S_{hkl} = \sqrt{\frac{n_1!}{(3!)2^{3-n_0}}} \tag{B.VIII.13d}$$

where n_0 and n_1 represent, for each combination (h,k,l), the number of $|h|$, $|k|$, $|l|$ which are zero and which are equal, respectively. Note that S_{hkl} represents the number of wave vectors [h, k, l] with the same length σ. As to the $Q_{ij}(\sigma)$, they are complex tensor amplitudes which can be decomposed over a basis of five mutually orthogonal matrices, yielding:

$$Q_{ij}(\sigma) = \sum_{m=-2}^{2} Q_m(\sigma) \exp[i\Psi_m(h,k,l)]\, M_m(h,k,l)$$

$$= \frac{1}{2} Q_2 \exp[i\Psi_2] \begin{bmatrix} 1 & i & 0 \\ i & -1 & 0 \\ 0 & 0 & 0 \end{bmatrix} + \frac{1}{2} Q_{-2} \exp[i\Psi_{-2}] \begin{bmatrix} 1 & -i & 0 \\ -i & -1 & 0 \\ 0 & 0 & 0 \end{bmatrix} +$$

$$+ \frac{1}{2} Q_1 \exp[i\Psi_1] \begin{bmatrix} 0 & 0 & 1 \\ 0 & 0 & i \\ 1 & i & 0 \end{bmatrix} + \frac{1}{2} Q_{-1} \exp[i\Psi_{-1}] \begin{bmatrix} 0 & 0 & -1 \\ 0 & 0 & i \\ -1 & i & 0 \end{bmatrix} +$$

$$+ \frac{1}{\sqrt{6}} Q_0 \exp[i\Psi_0] \begin{bmatrix} -1 & 0 & 0 \\ 0 & -1 & 0 \\ 0 & 0 & 2 \end{bmatrix} \tag{B.VIII.14a}$$

with

$$\Psi_m(h,k,l) = -\Psi_m(-h,-k,-l) \tag{B.VIII.14b}$$

In this development of the complex amplitude $Q_{ij}(\sigma)$, the basis matrices

M_m(h,k,l) are given, for each vector **q**(h,k,l), in a local coordinate system with the 3-axis parallel to **q**.

In terms of this Fourier expansion, the structure of a Blue Phase is determined by:

1. The set of wave vectors **q**(h,k,l) forming the **reciprocal** lattice. This lattice is face-centered cubic when the Bravais lattice of the Blue Phase is body-centered cubic (O^8 model) or simple cubic, when the Bravais lattice of the Blue Phase is simple cubic (O^2 model).

2. The amplitudes $Q_m(\sigma)$ and the phases $\Psi_m(\sigma)$ that define the space group, for a given kind of reciprocal lattice. According to the space group, some amplitudes $Q_m(\sigma)$ must be zero, while others need not be.

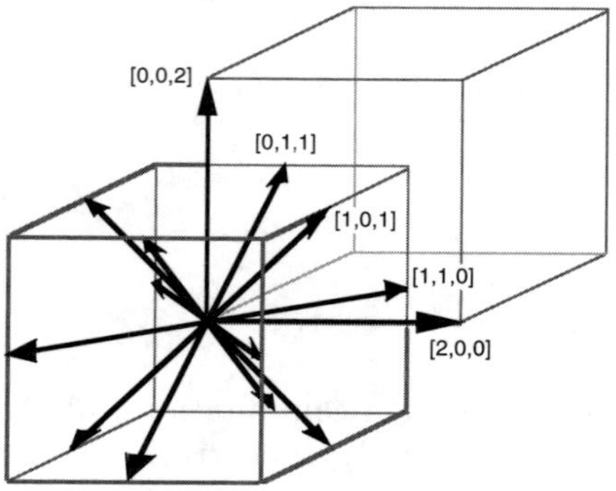

Fig. B.VIII.22 The wave vectors **q**(h,k,l) of a body-centered cubic Blue Phase form a face-centered cubic reciprocal lattice. In this kind of lattice, the sum h+k+l must be even. In this diagram, the 12 vectors of the (1,1,0) family are shown.

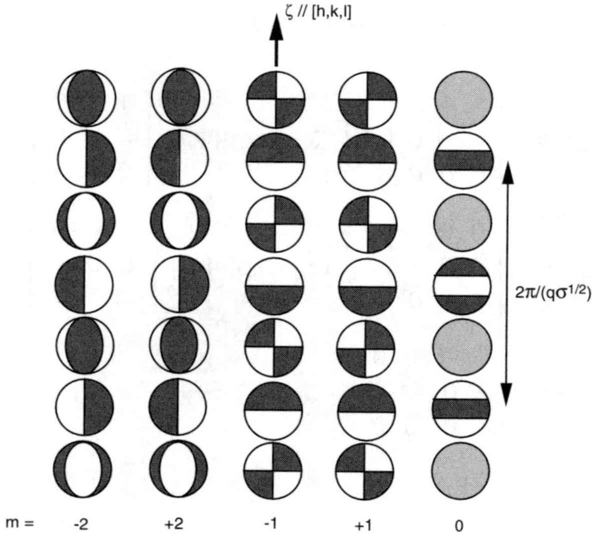

Fig. B.VIII.23 Schematic representation of the five Fourier components of the quadrupolar order parameter to be examined for each wave vector **q**(h,k,l).

Example:

Consider a model for the Blue Phase with symmetry O^8, or $I4_132$ whose Bravais lattice is body-centered cubic:

1. The cubic symmetry requires that the $Q_{ij}(0)$ Fourier component be zero, as it represents the average order parameter of the structure (the only second-rank tensor compatible with the cubic symmetry is δ_{ij}, but its traceless part is zero).

2. The fact that the cubic lattice is body centered imposes a selection rule for the Miller indices: their sum must be even. Thus, the wave vector [1,0,0] is forbidden, while [1,1,0], [2,0,0], [2,1,1] and [2,2,0] are allowed. In the simplest model, the choice of the wave vectors could be limited to the 12 wave vectors of the [1,1,0] family (Fig. B.VIII.22).

For each wave vector $\mathbf{q}$(h,k,l) in the reciprocal lattice, the amplitudes $Q_m(\sigma)$ and the phases $\Psi_m(\sigma)$ must then be examined. This is when the symmetry operations other than translations must be taken into account. To find and apply the selection rules, it is more convenient to use a real representation of the five components $m = -2,-1,0,1,2$ corresponding to a given wave vector $\mathbf{q}$(h,k,l) (Fig. B.VIII.23). In this representation, each of these components is a sort of "tensorial wave." To understand this outline, consider one of the five components $Q_m M_m \exp[iq(hx+ky+lz)+i\Psi_m]+\text{c.c.}$ of the tensor order parameter corresponding to a wave vector [h,k,l] of the reciprocal lattice. In the coordinate system (1,2,3), or (ξ,η,ζ), of axis 3 parallel to the wave vector [h,k,l], this component reads $Q_m M_m \exp[iq\sigma^{1/2}\zeta+i\psi_m]+\text{c.c.}$ and corresponds, for each ζ, to a certain real tensor $M_{ij}(\zeta)$. With this tensor, one can associate a function of the spherical coordinates $f(\theta,\varphi) = n_i M_{ij}(\zeta) n_j$, where $\mathbf{n} = (\sin\theta\cos\varphi, \sin\theta\sin\varphi, \cos\theta)$. This function is nothing other than one of the five spherical harmonics $Y_{lm}(\theta,\varphi)$ of order $m = 2$. A sketchy representation of this function is shown in each of the spheres in figure B.VIII.23. In this representation, each sphere is divided by one or several nodal lines where the function f changes sign. By convention, f is positive in the white areas, negative in the black areas, and zero in the gray areas.

After these explanations, let us consider the Fourier components of the type $\mathbf{q}(1,1,0)$ in the model with symmetry $I4_132$. As shown in figure B.VIII.8, the wave vectors of this type are directed along the twofold symmetry axes. It is readily apparent from figure B.VIII.23 that the $m = \pm 1$ components must be zero, as they are not invariant, changing sign after a rotation through π.

As to the $\mathbf{q}(2,0,0)$ vector, it is directed along a fourfold helicoidal axis of type 4_1, so we need to check what happens when a $\pi/2$ rotation is followed by a translation through a quarter of the lattice parameter. The $m = 0$ component is invariant with respect to all rotations, but its space period for the $\mathbf{q}(2,0,0)$ wave vector is a/2, so it changes sign after an a/4 translation. Consequently, this component must be zero. The $m = \pm 2$ components change sign when rotated through $\pi/2$, and the following a/4 translation changes their sign yet again. In conclusion, the $m = \pm 2$ components are compatible with the O^8 symmetry. Similarly, it can be proved that the $m = \pm 1$ components must be zero.

The results of applying these selection rules to the O^2, O^5, and O^8 symmetries are summarized in figure B.VIII.25.

Finally, one needs to determine the numerical values of the amplitudes $Q_m(\sigma)$ and the phases $\psi_m(\sigma)$ allowed by the symmetry. This task can be completed either experimentally or theoretically.

In conclusion, the description of the Blue Phases in terms of the Fourier expansion of the quadrupolar order parameter is universal and useful, both from the theoretical and the experimental points of view.

VIII.4 Landau theory of the Blue Phases by Grebel, Hornreich, and Shtrikman

Grebel, Hornreich, and Shtrikman [12] put forward a Landau theory of the Blue Phases starting, on the one hand, from the Fourier transform of the tensor order parameter $Q_{ij}(\mathbf{r})$ and, on the other hand, from the following expression for the average free energy density:

$$F = \frac{1}{V}\int d\mathbf{r}\left\{\frac{1}{2}\left[aQ_{ij}^2 + c_1 Q_{ij,l}^2 + c_2 Q_{ij,i}Q_{lj,l} - 2de_{ijl}Q_{in}Q_{jn,l}\right]\right.$$

$$\left. - \beta Q_{ij}Q_{jl}Q_{li} + \gamma(Q_{ij}^2)^2\right\} \qquad \text{(B.VIII.15)}$$

In this development, a is a coefficient (reduced temperature) which changes sign at $T = T^*$ and c_1, c_2, d, β, and γ are parameters that do not depend on the temperature. Furthermore, $Q_{ij,l} = \partial Q_{ij}/\partial x_l$,... As usual, this development contains the terms F_2, F_3, and F_4 of second, third, and fourth order in Q_{ij}. We shall discuss each of them separately.

Among all the second-order terms, only one, $2de_{ijl}Q_{in}Q_{jn,l}$, changes sign under inversion and, consequently, makes the distinction between the left-handed and right-handed chiralities. This coefficient must therefore be zero for a racemic mixture and change sign between the two enantiomers. This can be illustrated by determining the average F_2 for a Fourier component $Q_{ij}(\sigma)$, corresponding to a generic wave vector $\mathbf{q}(h,k,l)$:

$$F_2 = \frac{1}{2S_{hkl}}\sum_{m=-2}^{2}\left\{a - dm\sqrt{q^2\sigma} + \left[c_1 + \frac{c_2}{6}(4 - m^2)\right]q^2\sigma\right\} Q_m^2(\sigma)$$

$$= \frac{1}{2S_{hkl}}\sum_{m=-2}^{2} F_2(m,\sqrt{q^2\sigma})\, Q_m^2(\sigma) \qquad \text{(B.VIII.16)}$$

For thermodynamic stability reasons, $c_1 > 0$ and $c_1 + (2/3)c_2 > 0$.

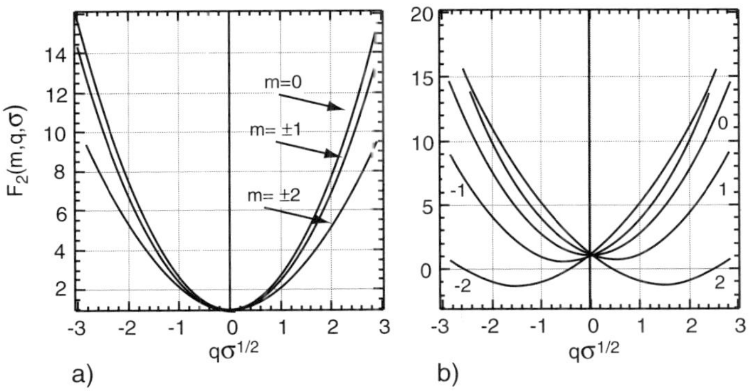

Fig. B.VIII.24 a) Variation of the F_2 term as a function of the wave vector $qs^{1/2}$ for a racemic mixture with d = 0 (we took $c_1 = c_2 = 1$); b) the same curve for a chiral product with d = 1.5 ($c_1 = c_2 = 1$).

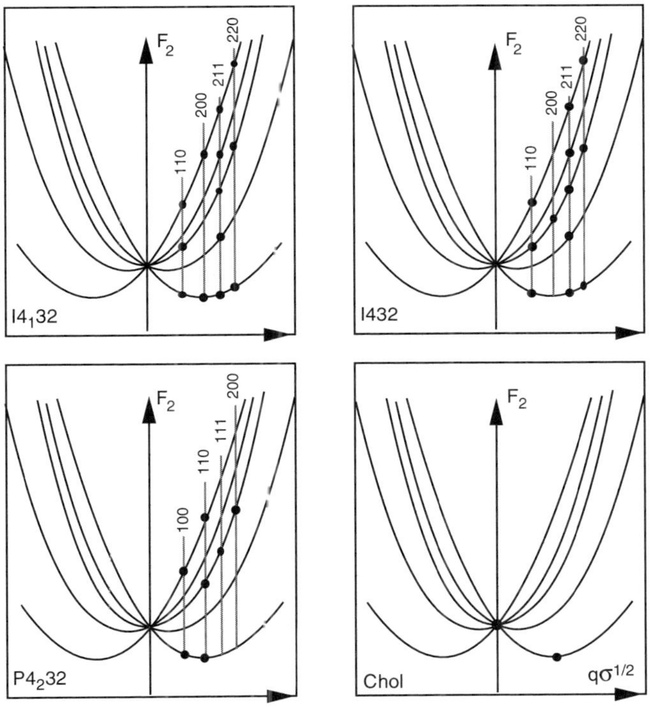

Fig. B.VIII.25 Selection rules for the $Q_m(\sigma)$ components allowed by the symmetries O^8, O^5, and O^2 of the Blue Phases and the cholesteric phase.

The curves in figure B.VIII.24 show how the quadratic term F_2 varies as a function of q, for the five values of m and for two values of the chiral coefficient d. In the case of figure B.VIII.24a, we set d = 0, corresponding to the case of a racemic mixture. The difference between the modes with different $|m|$ is a result of the elastic anisotropy accounted for by the c_2 term. In the case of figure B.VIII.24b, we set d = 1.5, corresponding to a chiral material. As an effect of chirality, the modes m with opposite signs are separated. Clearly, for q > 0, the chiral modes m = +2 and m = +1 are energetically more favorable than modes m=0, −1, and −2.

The figure B.VIII.24 diagrams provide a very convenient framework for a schematic presentation of the selection rules, based upon the symmetry requirements mentioned above. These rules are summarized in figure B.VIII.25 for the cholesteric phase and the three Blue Phases with symmetry O^2, O^5, and O^8. One can also see that, by only keeping the quadratic term F_2, among all amplitudes $Q_m(\sigma)$ allowed by the symmetries only those corresponding to the lower branch m = 2 will have a significant amplitude. Moreover, if the Fourier components of a single (h,k,l) family are preserved, then the absolute value $q\sigma^{1/2}$ of their wave vectors will correspond to the minimum of the function $F_2(m = 2,q,\sigma)$:

$$[q\sigma^{1/2}]_{min} = q_c = \frac{d}{c_1} \qquad \text{(B.VIII.17)}$$

In this case, each Fourier component Q_2(h,k,l) gives the following contribution to the second-order term:

$$\frac{1}{2}\left(a - \frac{d^2}{c_1}\right) Q_2^2(\sigma) \qquad \text{(B.VIII.18)}$$

We shall now consider the third-order term in the expansion B.VIII.15, which gives the isotropic liquid → Blue Phase transition its first-order character. The larger this term, the higher the transition temperature. The expression of this term contains integrals of the type

$$\frac{1}{V}\int d^3\mathbf{r}\, \exp[i(\mathbf{q}_1 + \mathbf{q}_2 + \mathbf{q}_3)\,.\,\mathbf{r} \qquad \text{(B.VIII.19)}$$

involving the sum of three wave vectors $\mathbf{q}_1$, $\mathbf{q}_2$, and $\mathbf{q}_3$, corresponding to the Fourier components appearing in the product $Q_{ij}Q_{jl}Q_{li}$. These integrals are only zero if the sum of the three vectors is zero:

$$\mathbf{q}_1 + \mathbf{q}_2 + \mathbf{q}_3 = 0 \qquad \text{(B.VIII.20)}$$

In the cholesteric phase, where there is a Fourier component $Q_o(0)$ of null wave vector and another one, $Q_2(2)$ of wave vector q_c, applying this rule yields two third-order terms. One of them corresponds to $q_1 = q_2 = q_3 = 0$, the other to $q_1 = -q_2 = q_c$; $q_3 = 0$. In this way we obtain again the expression B.VII.17 of the Landau energy given in the previous chapter.

The third-order term is even simpler in the case of a planar hexagonal Blue Phase where the order parameter

$$Q_{ij}(\mathbf{r}) = \frac{1}{\sqrt{6}} \sum_{n=1}^{3} Q_2(2) \left(M_2(2n) \exp\{i[\mathbf{q}_{2n}.\mathbf{r} + \Psi_2(2n)]\} + c.c. \right) \qquad \text{(B.VIII.21)}$$

would be the superposition of three Fourier components of the type m = 2, with the corresponding wave vectors (of absolute value q_c) forming an equilateral triangle (Fig. B.VIII.26).

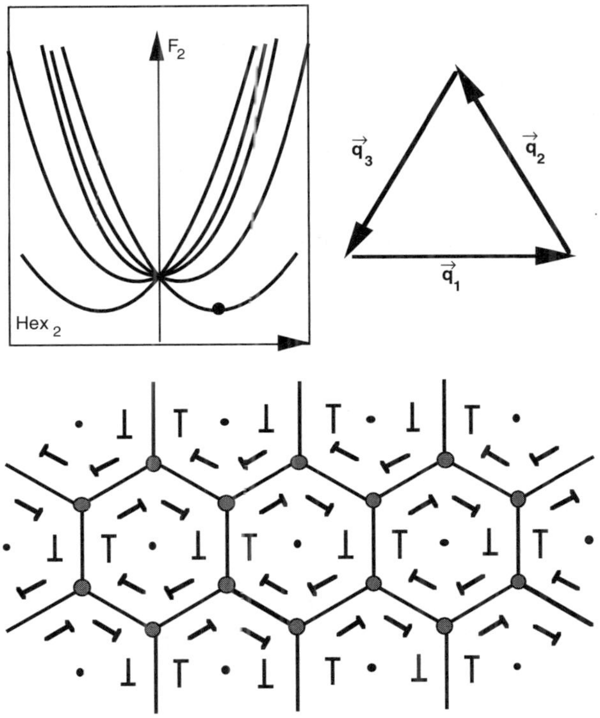

Fig. B.VIII.26 Model of a planar hexagonal Blue Phase proposed by Brazovskii and Dmitriev [13] and built upon three Fourier components Q_2 with wave vectors $\mathbf{q}_1$, $\mathbf{q}_2$, and $\mathbf{q}_3$ forming an equilateral triangle.

Such a structure was first proposed as early as 1976 by Brazovskii and Dmitriev [13]. Obviously, in this case there is only one third-order term. Using the same dimensionless variables as for the cholesteric phase (eq. B.VII.17), the free energy for this example reads:

$$f_{hex} = f_{iso} + \frac{t - \kappa^2}{4} \mu_2^2 - \frac{27}{32} \mu_2^3 + \frac{233}{192} \mu_2^4 \qquad \text{(B.VIII.22)}$$

The transition between the isotropic phase and the hexagonal Blue Phase occurs when

$$f_{hex} - f_{iso} = 0 \qquad \text{and} \qquad \frac{\partial f_{hex}}{\partial \mu_2} = 0 \tag{B.VIII.23}$$

These equations lead to the following expression for the transition temperature as a function of the chirality κ:

$$t_{IH} = \frac{3^7}{2^4 \times 233} + \kappa^2 = 0.587 + \kappa^2 \tag{B.VIII.24}$$

In body-centered cubic Blue Phases, the condition B.VIII.20 can be fulfilled by forming triangles, using for instance wave vectors of the type [1,1,0], as shown in figure B.VIII.27. In the limit of strong chirality, when the Fourier components of type m = 2 of wave vector q_c dominate all the other possible components $Q_m(\sigma)$, one only need consider the wave vectors of the type [1,1,0], forming equilateral triangles.

In this case, the expression of the free energy is identical to that for the hexagonal phase, up to the numerical constants:

$$f_{O^5} = f_{iso} - \frac{t - \kappa^2}{4}\mu_2^2 - \frac{23\sqrt{2}}{32}\mu_2^3 + \frac{499}{384}\mu_2^4 \tag{B.VIII.25}$$

The transition temperature varies as a function of κ in the following manner:

$$t_{O^5} = \frac{1587}{1996} + \kappa^2 = 0.795 + \kappa^2 \tag{B.VIII.26}$$

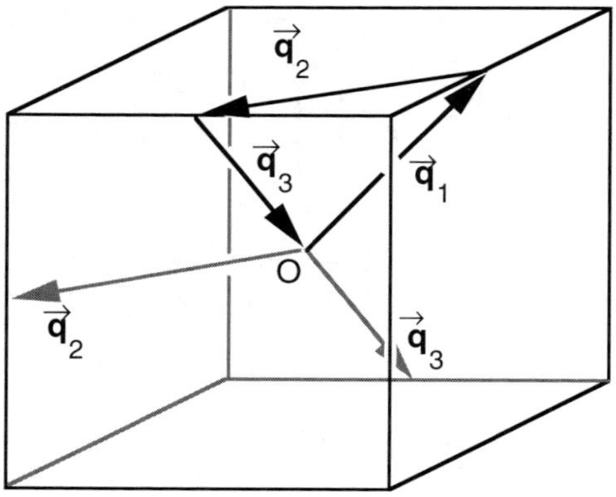

Fig. B.VIII.27 Equilateral triangle formed by three wave vectors of the type [1,1,0].

The results of the GHR theory for these two simple models of Blue Phases, planar hexagonal and O^5 cubic, are summarized in figure B.VIII.28, where the transition temperatures $t_{IChol} - \kappa^2$, $t_{IHex} - \kappa^2$, and $t_{IO^5} - \kappa^2$ are also shown as a function of the chirality κ.

This diagram shows that, for high enough chirality, the O^5 Blue Phase and the hexagonal Blue Phase are more stable than the cholesteric phase.

As we shall see, however, the experimentally encountered Blue Phases are not of the O^5 symmetry (as already mentioned in section B.VIII), but O^2 and O^8. In the framework of the GHS theory, other families [h,k,l] of wave vectors must be taken into account to distinguish between these symmetries.

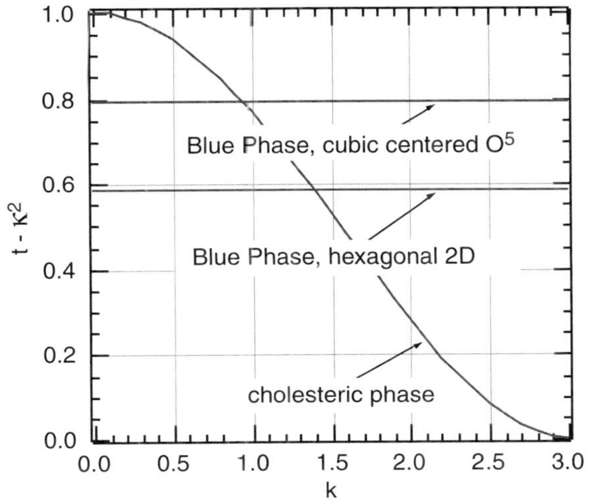

Fig. B.VIII.28 Transition temperatures as a function of the chirality: the cubic and hexagonal Blue Phases precede the cholesteric phase for high enough chirality.

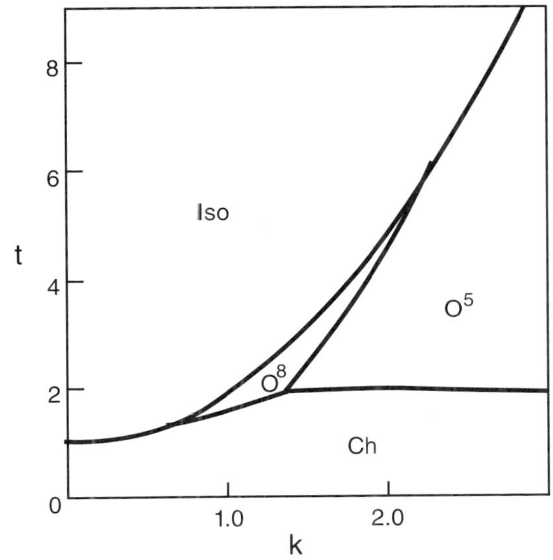

Fig. B.VIII.29 One of the phase diagrams obtained in the GHS theory when only the I, Ch, O^5, and O^8 phases are allowed (from ref. [12])

In this case, calculating the third-order term becomes complicated, since the number of possible combinations for grouping the wave vectors into triangles increases when other wave vectors [h,k,l] are considered. At the same time, the expression of the free energy becomes more complicated, no longer depending uniquely on the amplitude $\mu_2(2)$, but on all considered amplitudes $\mu_m(\sigma)$. For more details, the reader is referred to the original paper by Grebel, Hornreich, and Shtrikman [12], from which we present only one of the phase diagrams they obtained (Fig. B.VIII.29).

VIII.5 Experiments

The most important features of the theoretical models for the Blue Phases we have just discussed are their symmetries. But what symmetries really exist? This is the question we shall now try to answer.

To determine experimentally the symmetry of a Blue Phase, light diffraction is the most direct method (for the theory see [9], [14], [15], and [16]). Indeed, in terms of the Fourier expansion (eqs. B.VIII.12–14), the structure of a Blue Phase is determined by the amplitudes $Q_{ij}(\mathbf{q})$ of the quadrupolar order parameter, itself proportional to the dielectric susceptibility tensor $\varepsilon_{ij}(\mathbf{q})$ governing light propagation. This way, the selection rules for the amplitudes $Q_m(\mathbf{q})$ established as a function of the symmetry (the figure B.VIII.25 diagrams) can be directly transposed to the Fourier components of the dielectric tensor $\varepsilon_m(\mathbf{q})$. As for the ε_2 component of the cholesteric phase (eq. B.VII.9), each non-zero component $\varepsilon_m(\mathbf{q})$ of the dielectric tensor can, in principle, give a specific Bragg reflection.

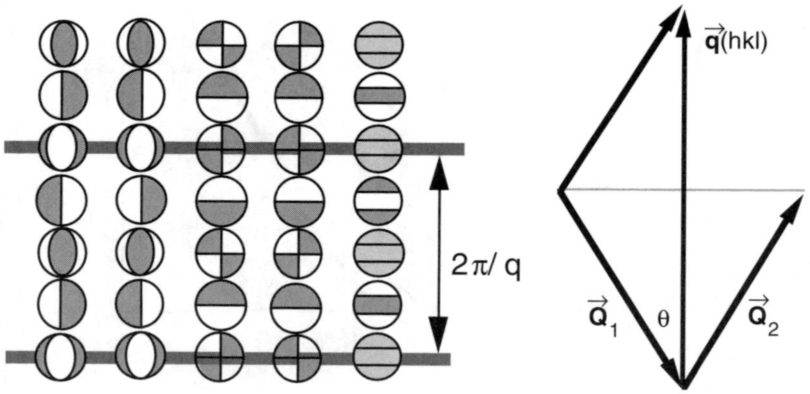

Fig. B.VIII.30 Bragg relation in the Blue Phases.

Strictly speaking, searching for the Bragg reflections of the Blue Phases involves determining the eigenmodes of the Maxwell equations taking into account the periodical 3D structure of the dielectric tensor ε_{ij} [9a, b, c]. This is a formidable task because, in contrast to X-rays, the amplitudes $\varepsilon_m(\mathbf{q})$ are strong in Blue Phases, and one cannot rule out the possibility of multiple reflections in the same direction as a simple reflection. For instance, in incidence parallel to the [100] axis, the double reflection on the (110) and (1–10) planes is superposed on the simple reflection on the (200) planes. Neglecting all but the simple reflections, a light wave with wave vector $\mathbf{Q}$ will give a Bragg reflection if the two following conditions are satisfied. First, its direction θ in the crystal must be such that:

$$2Qn\cos\theta = q(h,k,l) \qquad \text{(B.VIII.27)}$$

where n is an effective refractive index. Second, the polarization must be correct. For instance, in the case of Blue Phase I, of structure $I4_132$, the light wave incident along direction [110] can undergo reflection on the three non-null Fourier components: $\varepsilon_{-2}(110)$, $\varepsilon_0(110)$, $\varepsilon_2(110)$ (Fig. B.VIII.25). Components $\varepsilon_2(110)$ and $\varepsilon_{-2}(110)$ reflect light if its polarization is circular right- and left-handed, respectively. The component $\varepsilon_0(110)$ can give a linearly polarized reflection. However, in a right-handed chiral Blue Phase ($q > 0$) one has, according to the figure B.VIII.25 diagram, $\varepsilon_{-2}(110)<\varepsilon_0(110)<\varepsilon_2(110)$ and only the right-handed circular reflection is strong. For a direction of incidence along [211], the $\varepsilon_{-1}(211)$ and $\varepsilon_1(211)$ components can also contribute. One can show that the corresponding reflections are circularly polarized, but their intensity is zero in normal incidence (parallel to [211]).

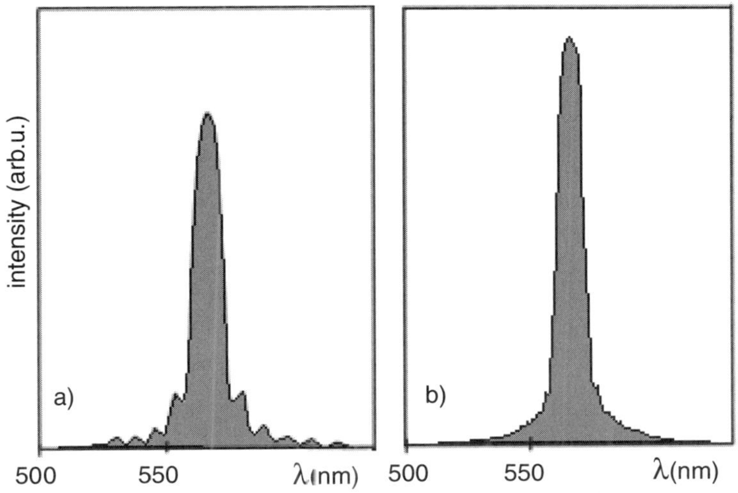

Fig. B.VIII.31 Spectrum of the Bragg reflection in normal incidence to the (110) planes. Crystal thickness: a) 12900 nm; b) 28100 nm (from ref. [9c]).

An example of Bragg reflections under normal incidence on the (110) planes is given in figure B.VIII.31. The experiment shows that:

1. This reflection is circularly polarized, and therefore due to the $\varepsilon_2(110)$ component.

2. Its width is finite, of the order of 20 nm for a wavelength λ of 570 nm. According to the analysis in section 3 of chapter B.VII, this width indicates a strong amplitude of the $\varepsilon_2(211)$ component.

3. The spectrum exhibits oscillations at the edge of the peaks, more apparent in the thinner sample. This phenomenon, known in the theory of X-ray diffraction as "Pendellösung," is due to the finite thickness of the crystal and is only present if the crystal is perfect, without dislocations. We shall see later that perfect crystals of Blue Phases I and II can indeed be grown.

VIII.5.a) Kossel diagrams of Blue Phases I and II

One of the most successful methods for establishing the structure of Blue Phases is the Kossel diagrams [17, 18]. This method simultaneously shows a large number of Fourier components with wave vectors **q**(h,k,l) and allows one to establish the angular relations between them. Its principle is as follows (Fig. B.VIII.32):

Let O be a point source of monochromatic radiation inside a Blue Phase monocrystal. This source generates waves with wave vectors **Q** pointing in all possible directions.

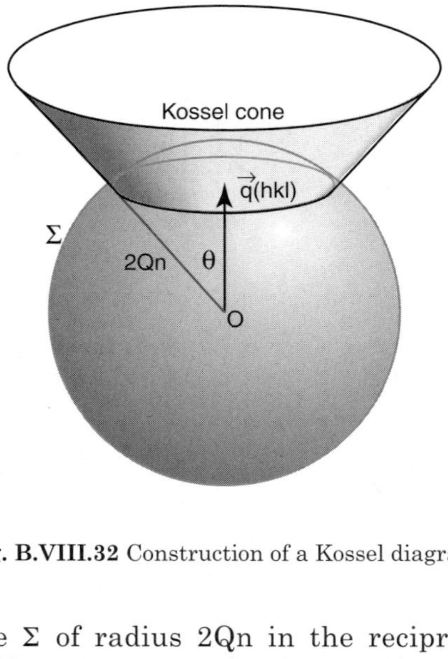

Fig. B.VIII.32 Construction of a Kossel diagram.

Let us draw a sphere Σ of radius 2Qn in the reciprocal space, where the structure of the Blue Phase is represented by the set of vectors **q**(h,k,l) of the reciprocal lattice (we assume the index n is constant). We recall that for each vector **q**(h,k,l) there are five independent Fourier amplitudes ε_m. For a wave vector **q**(h,k,l) inside the sphere Σ, the Bragg relation B.VIII.27 will be fulfilled by all wave vectors $\mathbf{Q}_{hkl}$ on a cone of axis $\mathbf{q}$(h,k,l) and with opening angle θ. Obviously, light propagation along these directions will be influenced by the Bragg reflections inside the crystal. On a screen around the crystal, each Kossel cone will leave an annular trace. The diameters of the rings and their relative positions will give us information on the lengths of the wave vectors in the reciprocal lattice and on the crystal symmetry.

Experimentally implementing the Kossel method is fairly straightforward. For instance, one can use a reflecting microscope with monochromatic lighting and equipped with an objective of high numerical aperture (an immersion objective, for instance). The trajectories of the light rays are shown in the figure B.VIII.33 diagram. The incident light rays, making an angle θ with the (h,k,l) planes, are reflected towards the objective. These rays form a ring in the focal plane of the objective.

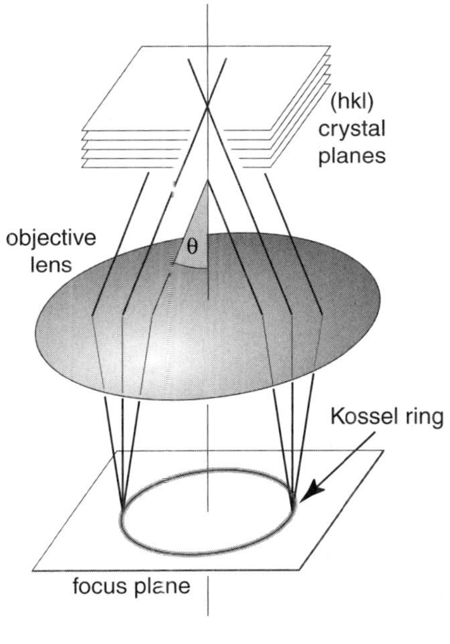

Fig. B.VIII.33 Formation of the Kossel rings in the focal plane of the objective.

The Kossel diagrams obtained in this manner, which are typical for the Blue Phases I and II, are shown in the figures B.VIII.34 and 35 photos. These diagrams allow easy identification of the type of the Bravais lattice, face-centered cubic for the Blue Phase I and simple cubic for the Blue Phase II. Without going into a detailed discussion of these diagrams, we shall highlight a few important features.

On the Kossel diagrams of Blue Phase I one can identify several types of rings (hkl) forming, in the order of decreasing diameter, the following series: (110), (200), then (211). This last ring is not visible in figure B.VIII.34: a shorter wavelength should be used to see it.

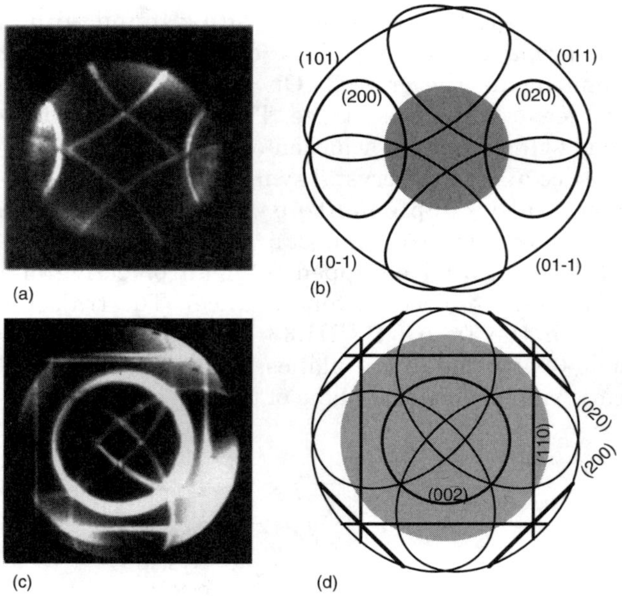

Fig. B.VIII.34 Typical Kossel diagrams of Blue Phase I in a mixture of 42.5% CB15 with E9 [17]: a) a view along the twofold symmetry axis; b) ring indexing; c) view along the fourfold symmetry axis [18]; d) ring indexing. Due to the finite aperture of the objective, only the central part of the diagram (inside the gray disk) is visible in the experiment.

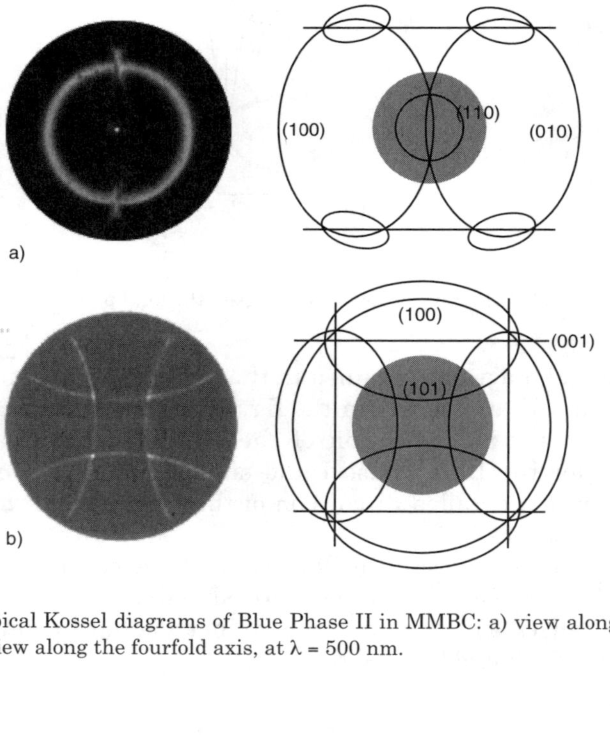

Fig. B.VIII.35 Typical Kossel diagrams of Blue Phase II in MMBC: a) view along the twofold axis, at $\lambda = 602$ nm; b) view along the fourfold axis, at $\lambda = 500$ nm.

The identification of the (hkl) indices is made easier by the position of the rings, which follows the symmetry rules. For instance, the figure B.VIII.34 diagram is clearly symmetric with respect to a twofold axis going through the center. Let us call this axis [110]. The Miller indices of the four rings forming a diamond-shaped figure are (101), (10–1), (01–1) and (011). The two other rings (smaller diameters) go through the intersection points of the four larger rings. Using the construction principle of the Kossel diagrams, it can be shown that in cubic symmetry the (200) ring goes through the intersection points of the (101) and (10–1) rings (Fig. B.VIII.34b). The absence of the (100) ring shows that the structure cannot be simple cubic (for instance, of symmetry $P4_232$, Fig. B.VIII.25). On the other hand, the fact that the intensities are comparable shows that the structure must be $I4_132$, rather than I432.

In one of the Kossel diagrams of Blue Phase II, one can find a ring structure with twofold symmetry (Fig. B.VIII.35a). The Miller indices of the central ring in this diagram are (110). This ring appears at the intersection of two larger rings, with indices (100) and (010). This shows that the Bravais lattice of the Blue Phase II is simple cubic. The fact that the intensities of these two types of rings are similar indicates that the space group is $P4_232$ (Fig. B.VIII.25).

We shall see later that the Kossel diagram technique is a very effective way of following the deformations of Blue Phases under the action of an electric field or a mechanical stress.

VIII.5.b) Faceting of cubic Blue Phases

We saw that the nucleation and growth of Phase Blue crystallites can be followed at the transition from the isotropic phase [19–21]. These crystallites are bright against the dark background of the isotropic phase owing to their Bragg reflections (the color of which depends on the orientation).

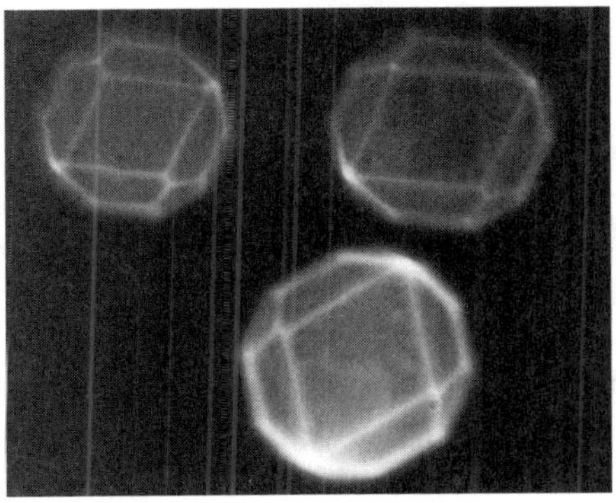

Fig. B.VIII.36 Crystallites of Blue Phase I in coexistence with the isotropic phase observed using the setup in figure B.VIII.7. The wavelength of the lighting is slightly shifted with respect to the maximum of the Bragg reflection peak.

The size of the crystallites increases when the temperature is slowly decreased, the two phases coexisting in a certain temperature range, the span of which depends on the composition of the mixture studied (for a discussion on coexistence, see chapter B.IX). This behavior is typical for a first-order transition. Furthermore, the boundary between the two phases is very sharp, indicating a certain interface energy. When the size of the crystallites reaches a few tens of micrometers, it becomes apparent that they are faceted. The figure B.VIII.36 photo shows typical Blue Phase I crystallites oriented with their twofold axis perpendicular to the plane of the photo. All these crystallites are limited by two types of facets, (110) and (211). On other crystallites of the BPI phase (Fig. B.VIII.37a), (222), (310), and (321) facets can also be seen. On the complete crystal habit of the Blue Phase I, presented in figure B.VIII.37b, one can easily identify the two-, three-, and fourfold symmetry axes of the O symmetry group, once more confirming the cubic symmetry of the phase. Observing the crystal habit gives even further information on the structure: indeed, the lack of any (100) facets identifies the space group as O^8, according to the rule of Donnay and Harker according to which: "the prominence order of the facets of a crystal is the same as the decreasing order of the interplanar distances, taking into account an eventual reduction by a factor of two, three or four as a result of the group space symmetries." This empirical rule has the following theoretical justification.

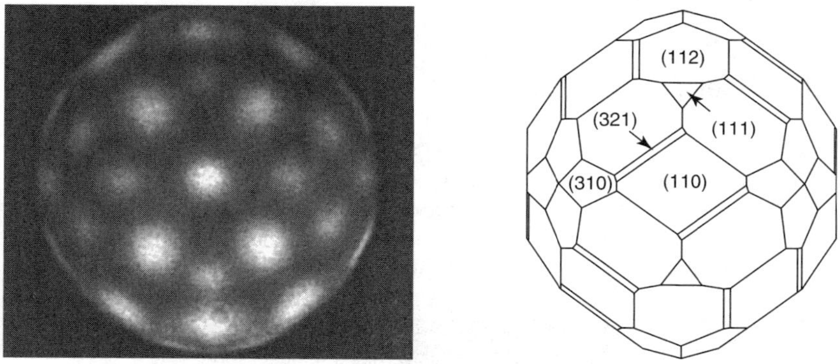

(a) (b)

Fig. B.VIII.37 Faceting of the Blue Phase I. Photo of a Blue Phase I crystallite in the E9/CB15 mixture. The central facet is of the type type (110). The others are of type (211), (310), and (222); b) diagram of the crystal habit of the Blue Phase I. The two-, three- and fourfold symmetry axes are easily identified. Note the absence of the (100) facets (from ref. [20, 21]).

Crystal faceting at thermodynamic equilibrium is a consequence of minimizing the surface energy at constant volume. Consequently, the equilibrium shape of a crystal depends on the anisotropy of its surface energy. Let $\gamma(\mathbf{m})$ be the energy (per unit surface) of a surface element oriented along $\mathbf{m}$. When the crystal shape is not only anisotropic, but also exhibits an (hkl) facet, the function $\gamma(\mathbf{m})$ has a cusp singularity in the direction $\mathbf{m}$ orthogonal to the (hkl) facet. As shown in figure B.VIII.38, this singularity is due to the presence of surface steps on the surface tilted with respect to the (hkl) planes. This surface exhibits steps, with a density (number per length unit) proportional to the absolute value of the angle θ between **m** and the [hkl] direction. Each step adds an additional energy β with respect to a surface strictly parallel to the (hkl) planes, the energy of the vicinal surface being given by:

$$\gamma(\theta) = \gamma_{hkl} + \beta|\theta| \qquad \text{(B.VIII.28)}$$

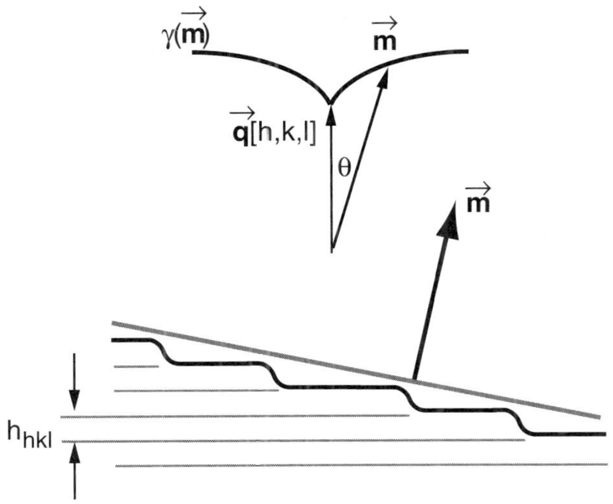

Fig. B.VIII.38 Relation between the singularity of the function γ(**m**) and the presence of steps on the crystal surface.

Having related the existence of facets to the energy β of the steps appearing on a vicinal surface, one must now consider the origin of the steps on a surface. In other words, we are looking for the reason why the surface, instead of being curved and tilted with respect to the (hkl) planes, becomes decomposed into terraces separated by steps. The physical reason is that, at the microscopic level, the interface energy between the isotropic liquid and a periodic medium (here, the Blue Phase) depends on the height z of the interface with respect to the structure inside the crystal. To put it differently, the level z where the system goes from the Blue Phase to the isotropic phase counts. In figure B.VIII.38, the surface can be seen to "stick" to certain discrete levels, avoiding as much as possible the intermediate positions. **The higher the distance h_{hkl} between the equivalent levels, the stronger this preference**. This distance depends not only on the Miller indices (hkl), but also on the presence of helicoidal axes, orthogonal to the (hkl) planes.

The Miller indices (hkl) define the periodicity of the crystal along the [hkl] direction; the distance between the equivalent planes inside the crystal is:

$$d_{hkl} = \frac{a}{\sqrt{h^2+k^2+l^2}} \qquad \text{(B.VIII.29)}$$

For a body-centered cubic crystal, where the sum of the Miller indices must be even, the series of reticular planes (hkl) in decreasing order of the d_{hkl} distance would be:

(110), (200), (211), (310), (222), and (321) (B.VIII.30a)

In a crystal without helicoidal axes, $d_{hkl} = h_{hkl}$, so the (200) facets should be present on the crystal habit of the Blue Phase I if it has O^5 symmetry.

However, by arranging in decreasing prominence order the facets actually observed (see Fig. B.VIII.37), we get the following series:

(110), (211), (310), (222), and (321) (B.VIII.30b)

Comparing these two series, one notices the lack of the (200) facet on the real crystal. Thus, a helicoidal axis along the [100] direction must modify the order of the facets. But, in the case of the O^8 symmetry, the fourfold axis is of the 4_1 type. Due to this axis, the distance h_{hkl} between the equidistant levels (from the point of view of the interface), is reduced by a factor of 4 with respect to d_{hkl}:

$$h_{hkl} = \frac{d_{hkl}}{4} \qquad \text{(B.VIII.31)}$$

Indeed, under the action of the helicoidal axis 4_1, the pattern existing at level z is repeated at level $z+h_{hkl}$ though rotated by $\pi/2$. Now, as far as the energy of the interface with an isotropic phase is concerned, the orientation of the pattern is irrelevant.

In conclusion, due to the helicoidal axis 4_1, the distance h_{200} goes from second to last position in the series B.VIII.30a.

VIII.5.c) Growth and steps

The observation of Blue Phases under the reflecting microscope in monochromatic lighting (using the setup in figure B.VIII.7), allows detection of the steps on the crystal surfaces. When the crystal nucleates on the coverslip, the appearance of the steps depends on the type of interface on which they develop.

For instance, the figure B.VIII.39 photo shows the typical appearance of the steps at the BPI/glass interface. These steps form at the intersection of the crystalline planes (110) with the glass surface. They can only be explained by admitting the BPI crystal is deformed because, at this scale, the glass surface is flat.

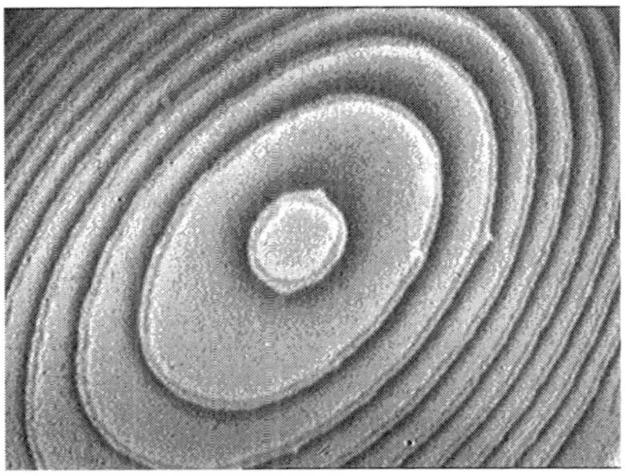

Fig. B.VIII.39 Steps at the BPI/glass interface. Their arrangement suggests the crystal is deformed.

One can also encounter steps on the BPI/isotropic liquid interface. By studying them, the growth and melting of the crystal can be followed.

Thus, in the example of figure B.VIII.40, the first photo (a) shows a partially melted crystal. On its main facet (110), one can notice the presence of a step system forming a kind of a terrace pyramid. By subsequently growing the crystal again, one notices that each terrace grows at the edges until it reaches the neighboring facets, of type (211) (Fig. B.VIII.40b). This process continues until the steps are completely eliminated (Fig. B.VIII.40c).

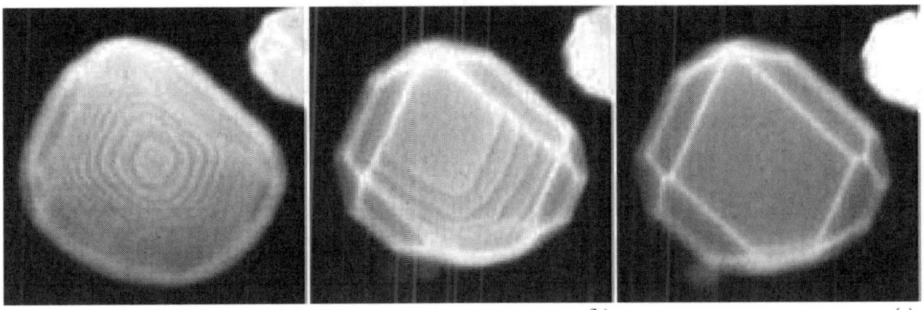

(a) (b) (c)

Fig. B.VIII.40 The behavior of steps at the BPI / isotropic liquid interface illustrates the growth of a crystal free from screw dislocations. a) Partially melted crystal; b) intermediate state; c) crystal when all the steps have disappeared.

From this point on, new terraces must be nucleated for the growth process to continue. This growth mode is typical for a crystal without screw dislocations.

On the surface of some crystals one can sometimes notice steps that abruptly stop in the middle of a facet: this is the sign of a screw dislocation piercing the crystal at that point (Fig. B.VIII.41).

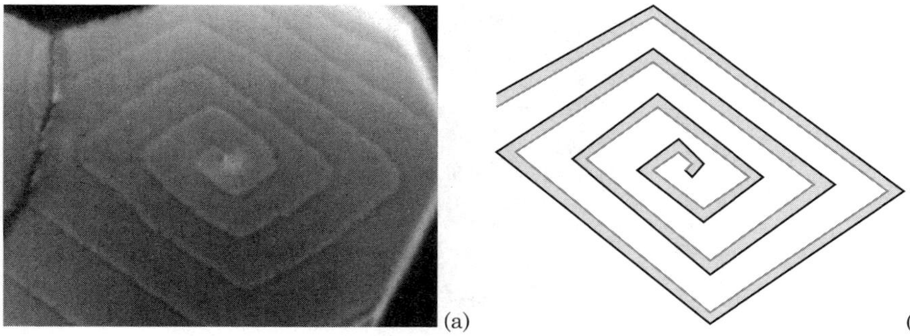

Fig. B.VIII.41 a) Spiral step on the surface of a BPI crystal; b) Frank-Read process of crystal growth: the movement of the step generates crystalline growth. The diagram shows two successive positions of the step, shown by gray and black lines, respectively. Multiplying the surface swept by the step (shown in light gray) by the width h_{110} of the step yields the crystal volume created by growth.

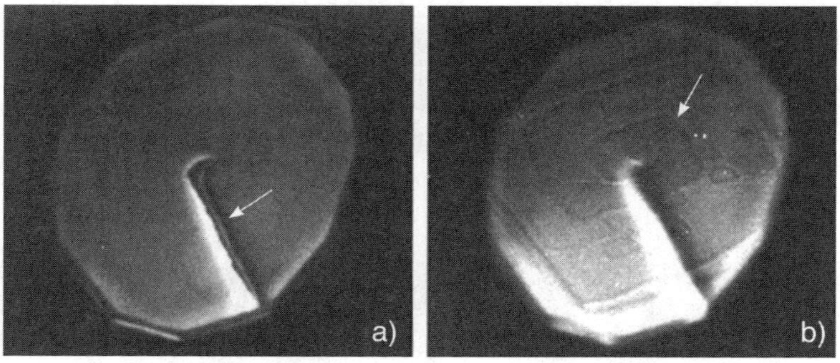

Fig. B.VIII.42 Monocrystal of the BPI phase pierced by a screw dislocation; a) focus on the lower glass plate. The arrow indicates a straight step going from the core of the dislocation towards the edge of the crystallite; b) focus on the BPI-isotropic liquid interface: in this case, the arrow shows a spiral step originating at the emergence point of the screw dislocation. The nucleation of a few terraces is also visible (from ref. [21]).

Frank and Read showed that this step assumes the shape of a spiral which rotates during crystal growth, favoring the progress of the facet. This phenomenon is also shown in the figure B.VIII.42 photo, at the interface BPI/isotropic liquid. When focusing on the BPI/glass interface, one only sees a straight step (no growth on this side).

VIII.5.d) Effect of an electric field on the Blue Phases

It was shown experimentally that the crystallites of Blue Phases I and II nucleate most frequently on surfaces of a cell (Fig. B.VIII.43). Thus, in Blue

Phase I, the (110) or (211) planes are oriented along the solid wall, while for Blue Phase II the preferred orientations are (111) and (110).

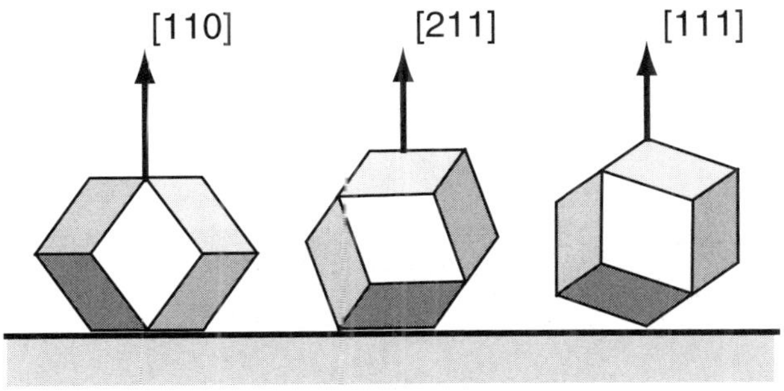

Fig. B.VIII.43 Preferred orientations of the Blue Phases at a wall.

When the crystals are faceted, it is very easy to identify their orientation from their morphology. Sometimes the Blue Phase crystals float in the isotropic phase; in this case, their orientation is random.

When an electric field is applied using two transparent electrodes (ITO), its effects depend on several parameters, the most important being:

1. The dielectric anisotropy ε_a of the material.
2. The field intensity E.
3. The position of the crystallites with respect to the field.
4. Their freedom of rotation.

Two kinds of effects (orientation and deformation) are observed, and we shall successively describe both of them. For more details, the reader is referred to a very thorough review of electric field effects on Blue Phases, published by H.-S. Kitzerow [18].

Crystal orientation by the field

Let us consider a material with positive dielectric anisotropy forming crystals in suspension or weakly anchored at the walls. It was shown experimentally that the electric field orients the fourfold axis [100] of these crystals along its direction. This effect is easily demonstrated using reflecting microscopy because, in the presence of the field, all crystals have the same color (due to the same Bragg reflection) and the same square shape [22] (Fig. B.VIII.44).

In order to explain the rotation of the crystals under the electric field [22,23], consider the torque

$$\boldsymbol{\Gamma} = \mathbf{P} \times \mathbf{E} \qquad \text{(B.VIII.32)}$$

exerted by the electric field **E** on the polarization **P**.

Because the Blue Phases have no spontaneous polarization because of their cubic symmetry, this can only be induced and, as such, it can be expanded in a power series of the electric field:

$$P_i = V\,\varepsilon_o\,(\chi_{ij}E_j + \chi_{ijk}E_jE_k + \chi_{ijkl}E_jE_kE_l + \ldots) \qquad \text{(B.VIII.33)}$$

with V the volume of the crystal. The tensors χ appearing in this expansion must be symmetrical with respect to the exchange of all index pairs. Moreover, these tensors must be invariant under all the symmetry elements of the Blue Phases.

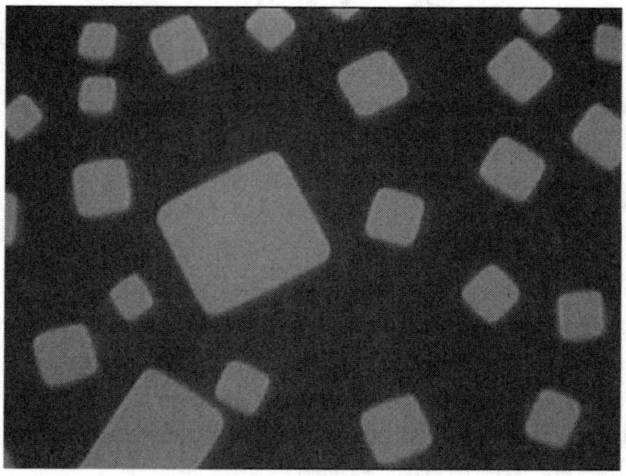

Fig. B.VIII.44 Crystals of Blue Phase II oriented by the electric field: all crystals have the same color.

Owing to their cubic symmetry, the linear dielectric susceptibility χ_{ij} of the Blue Phases must be isotropic:

$$\chi_{ij} = \chi\,\delta_{ij} \qquad \text{(B.VIII.34)}$$

and cannot explain the above-mentioned effect.

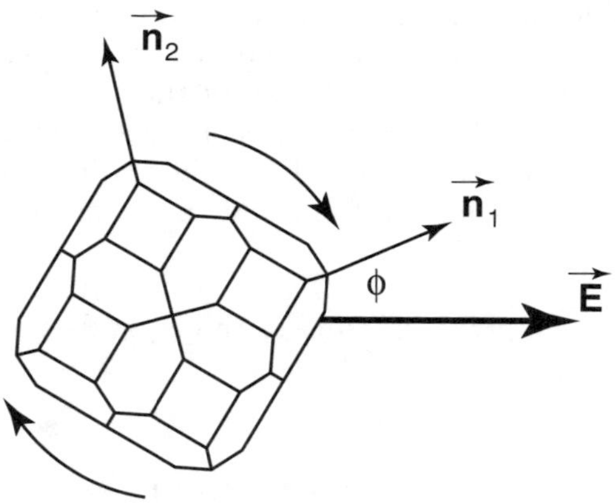

Fig. B.VIII.45 Torque exerted by the electric field on a Blue Phase crystal.

The nonlinear susceptibility χ_{ijk} must be zero for the same symmetry reasons.

We are left with the susceptibility χ_{ijkl}. This is a fourth-rank tensor which can be anisotropic even in cubic symmetry. It can be built upon three unit vectors n_i^α parallel to the fourfold axes of the crystal (Fig. B.VIII.45).

Using these vectors and the Kronecker symbol, two independent tensors can be built:

$$\chi_{ijkl}^{iso} = \delta_{ij}\delta_{kl} + \delta_{ik}\delta_{jl} + \delta_{il}\delta_{jk} \tag{B.VIII.35}$$

and

$$\chi_{ijkl}^{aniso} = \sum_{\alpha=1}^{3} n_i^\alpha n_j^\alpha n_k^\alpha n_l^\alpha \tag{B.VIII.36}$$

The fourth-rank susceptibility tensor can thus be written as a linear combination of these two tensors:

$$\chi_{ijkl} = A\,\chi_{ijkl}^{aniso} + B\,\chi_{ijkl}^{iso} \tag{B.VIII.37}$$

Plugging the expressions B.VIII.33–37 in eq. B.VIII.32, we have for the torque:

$$\mathbf{\Gamma} = VA\varepsilon_o \sum_{\alpha=1}^{3} (\mathbf{n}^\alpha.\mathbf{E})^3(\mathbf{n}^\alpha \times \mathbf{E}) \tag{B.VIII.38}$$

The effect of this bulk torque is most conveniently discussed in the reference frame of the vectors $\mathbf{n}^\alpha$, where it reads:

$$\mathbf{\Gamma} = VA\varepsilon_o\, \mathbf{h} \times \mathbf{E} \tag{B.VIII.39}$$

with

$$\mathbf{h} = (E_1^3, E_2^3, E_3^3) \tag{B.VIII.40}$$

yielding explicitly

$$\Gamma_1 = VA\varepsilon_o\, E_2E_3(E_2^2 - E_3^2) \tag{B.VIII.41a}$$

$$\Gamma_2 = VA\varepsilon_o\, E_3E_1(E_3^2 - E_1^2) \tag{B.VIII.41b}$$

$$\Gamma_3 = VA\varepsilon_o\, E_1E_2(E_1^2 - E_2^2) \tag{B.VIII.41c}$$

Thus, the torque $\mathbf{\Gamma}$ is zero when one of the following conditions is fulfilled:

a) $|E_1| = |E_2| = |E_3|$

b) $E_i = 0$ and $|E_j| = |E_k|$, $i = 1, 2, 3$, $j \neq k \neq i$

c) $E_i \neq 0$, and $E_j = E_k = 0$, $i = 1,2,3$, $j \neq k \neq i$

These three cases correspond, respectively, to the following geometries:

a) **E** parallel to the threefold axis [111]

b) **E** parallel to the twofold axis [110]

c) **E** parallel to the fourfold axis [100]

For all intermediate orientations, the field exerts a finite torque on the crystal, making it turn until it reaches the nearest stable orientation.

The stereogram in figure B.VIII.46 displays a diagram of the stable and unstable orientations for a positive coefficient A.

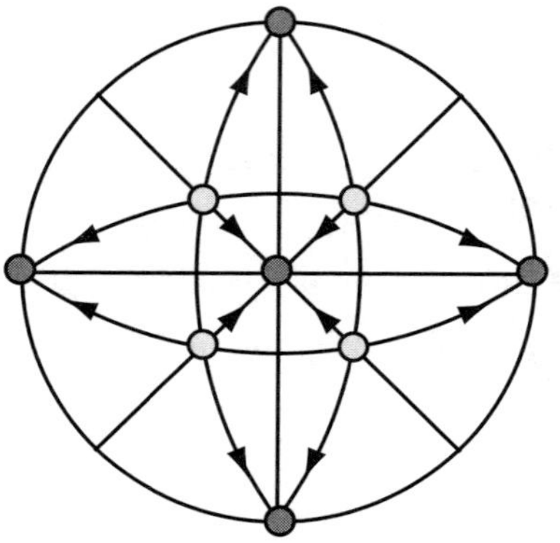

Fig. B.VIII.46 Stereographic representation of the stable and unstable orientations for A > 0.

The order of magnitude of the nonlinear anisotropy A can be estimated from the characteristic time needed by a crystallite with volume V to reorient in a field **E**. Let ϕ be the angle between the $\mathbf{n}_1$ axis and the field **E** (Fig. B.VIII.45). During rotation, the crystallite is subjected to the dielectric torque

$$\Gamma_{el} \approx VA\varepsilon_o E^4 \phi \qquad \text{(B.VIII.42a)}$$

and the viscous torque

$$\Gamma_{visc} \approx 6V\eta \frac{d\phi}{dt} \qquad \text{(B.VIII.42b)}$$

The reorienting time is then close to

$$\tau \approx \frac{6\eta}{A\varepsilon_o E^4} \qquad \text{(B.VIII.43)}$$

The typical observed time is of the order of 1 s for a field of 30 V/100 μm = 3×10^5 V/m. The susceptibility can thus be estimated at:

$$A \approx 10^{-10}\ \mathrm{m^2\ V^{-2}}$$

for $\eta = 1$ poise. Expressing it as the squared reciprocal of a characteristic field:

$$A = \frac{1}{E_c^2} \qquad \text{(B.VIII.44)}$$

We have

$$E_c \approx 10^3\ \mathrm{V/cm}$$

This field is much lower than the molecular field involved in the nonlinear susceptibility mechanisms of ordinary materials: $E_{mol}=10^8$ V/cm, meaning that its origin in the Blue Phases must be completely different.

Let us now discuss the second effect of the electric field.

Deformation of the Blue Phases by the field, electrostriction

The explanation of the giant nonlinear susceptibility of the Blue Phases is based upon their being easily deformed in an electric field. This deformation was demonstrated using the Kossel diagrams [18, 24]. Figure B.VIII.47 shows a series of Kossel diagrams of a Blue Phase I crystallite in a E9/CB15 mixture with positive dielectric anisotropy, initially oriented with the binary axis [110] along the field (the field is perpendicular to the image plane).

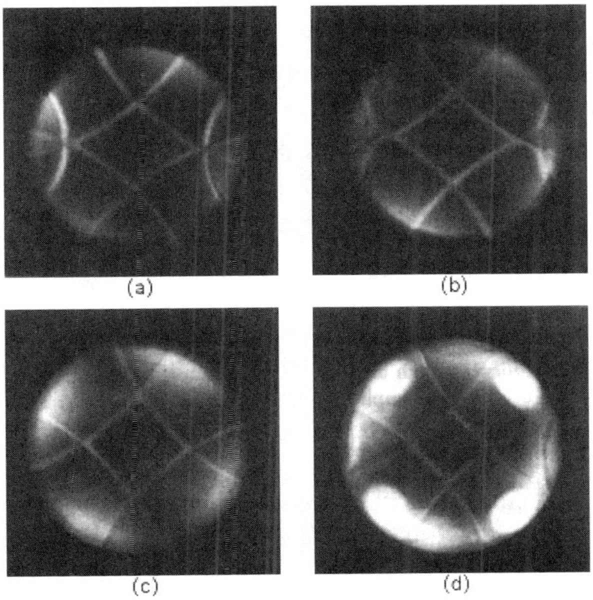

Fig. B.VIII.47 Kossel diagrams for a Blue Phase I under an increasing electric field (from a to d), in a mixture of CB15 (42.5%) and E9 (from ref. [24]).

Diagram (a) shows an undistorted Blue Phase with the binary axis [110] perpendicular to the plane of the page. In this diagram, one can identify four (110) rings, forming a diamond-shaped pattern, and two (200) rings, going through the intersections of the (110) rings (we already know this to be a feature of cubic crystals). In the presence of the electric field, this diagram is deformed (Fig. B.VIII.47b): the (200) and (110) rings no longer intersect at only one point. This is a clear sign that the crystal is deformed, losing its cubic structure. At high enough fields, the binary symmetry of the Kossel diagram changes. Figure B.VIII.47d shows that the Kossel diagram has a fourfold symmetry, so a phase transition must have occurred.

The deformation of the Kossel diagram reveals the deformation of the Blue Phase I under the action of the electric field. More specifically, the field deforms both the unit cell and the orientational pattern it contains. The deformation of the orientational pattern can be very complex, and the deformation of the cubic lattice corresponding to the Kossel diagrams in figure B.VIII.47 can be easily visualized as shown in figure B.VIII.48 (from H.-S. Kitzerow [18]).

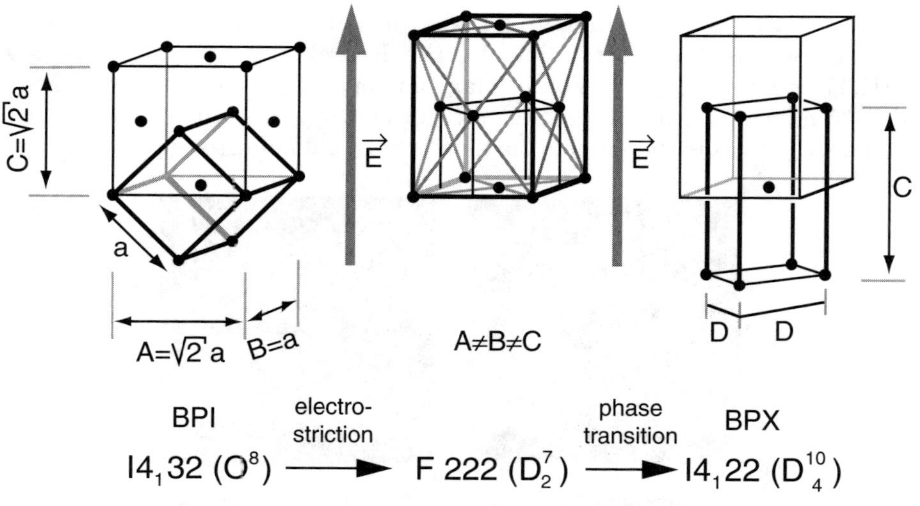

Fig. B.VIII.48 Deformation of the cubic unit cell of Blue Phase I to a tetragonal cell under the action of an electric field (from H.-S. Kitzerow [18]).

The first drawing shows that the body-centered cubic lattice built from cubic cells of size a can be seen as a particular case of an orthorhombic lattice with cell dimensions B = a and A = C = √2a. Under the action of an electric field applied along the direction C of the orthorhombic lattice (corresponding to the [110] direction of the cubic lattice), this one becomes elongated along the field, so the equality A = C is lost. As a result, the $I4_132$ symmetry of the Blue Phase I is broken, being replaced by the F 222 symmetry, as shown in the center diagram of figure B.VIII.48. The elongation of the orthorhombic cell is accompanied by a reduction in the A/B ratio. In the Kossel diagrams, this effect is translated by a change in the angles of the diamond formed by the (110) rings. Finally, at high enough fields the rectangular basis AB of the orthorhombic unit cell becomes square, with A/B = 1. This particular case corresponds to an increase in system symmetry, which goes from the space group F 222 to the $I4_122$ group. This phase transition will be discussed later. The electric field-induced deformation of the Blue Phases was quantitatively studied by Kitzerow et al. [25, 26]. More specifically, these authors determined the λ_{hkl} of the Bragg reflections as a function of the intensity E of the applied electric field. The diagram in figure B.VIII.49a shows that, in agreement with the previous discussion, the wavelength of the (110) reflection of the Blue Phase I increases with the field. On the contrary, the wavelength λ_{200} slightly decreases with the field when it is applied along the [100] direction of Blue Phase I.

One can also apply an electric field to crystals of the Blue Phase II, using more CB15 in the mixtures of E9 and CB15. In this case, figure B.VIII.49a shows that the two wavelengths λ_{100} and λ_{110} both increase with the field.

H.-S. Kitzerow also studied the action of an electric field on the Blue Phases in mixtures with negative dielectric anisotropy (for instance, the chiral product MW190 and the nematic ZLI 2585). Figure B.VIII.49b shows that the variation of the wavelengths λ_{hkl} with E changes sign with respect to the previous case of positive dielectric anisotropy.

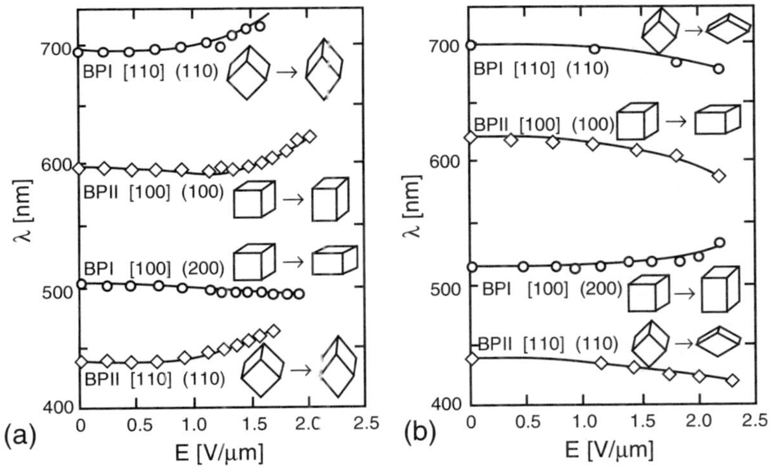

Fig. B.VIII.49 Measurements of the wavelength of the Bragg reflections as a function of the electric field intensity: a) in a mixture of 49.2% CB15 with E9, with positive dielectric anisotropy [25a]; b) in a mixture of 30.7% MW190 with ZLI 2585, with negative dielectric anisotropy [25b].

From a theoretical point of view, the electrostriction phenomenon is governed by the equation

$$\varepsilon_{ij} = \gamma_{ijkl} E_k E_l \tag{B.VIII.45}$$

relating the deformation tensor and the deformation of the Bravais lattice

$$\varepsilon_{ij} = \frac{1}{2}\left(\frac{\partial u_i}{\partial x_j} + \frac{\partial u_j}{\partial x_i}\right) \tag{B.VIII.46}$$

to the electric field **E**. In the case of cubic symmetry, the fourth-rank tensor γ_{ijkl} has three independent coefficients. In the reference frame (1,2,3) of the cubic unit cell, eq. B.VIII.45 can be written as follows [27]:

$$\begin{bmatrix} \varepsilon_{11} \\ \varepsilon_{22} \\ \varepsilon_{33} \\ 2\varepsilon_{23} \\ 2\varepsilon_{13} \\ 2\varepsilon_{12} \end{bmatrix} = \begin{bmatrix} \gamma_{1111} & \gamma_{1122} & \gamma_{1122} & 0 & 0 & 0 \\ \gamma_{1122} & \gamma_{1111} & \gamma_{1122} & 0 & 0 & 0 \\ \gamma_{1122} & \gamma_{1122} & \gamma_{1111} & 0 & 0 & 0 \\ 0 & 0 & 0 & 4\gamma_{2323} & 0 & 0 \\ 0 & 0 & 0 & 0 & 4\gamma_{2323} & 0 \\ 0 & 0 & 0 & 0 & 0 & 4\gamma_{2323} \end{bmatrix} \begin{bmatrix} E_1E_1 \\ E_2E_2 \\ E_3E_3 \\ E_2E_3 \\ E_1E_3 \\ E_1E_2 \end{bmatrix} \tag{B.VIII.47}$$

When the electric field is applied along the [110] direction, one has

$$\begin{aligned} &\varepsilon_{33} = \gamma_{1122} E^2 \\ &\varepsilon_{11} = \varepsilon_{22} = (1/2)(\gamma_{1111} + \gamma_{1122}) E^2 \\ &\varepsilon_{12} = \gamma_{2323} E^2 \end{aligned} \tag{B.VIII.48}$$

In the reference frame (1',2',3) such that the 2' axis is along the [1–10] direction perpendicular to the electric field, one has:

$$\begin{aligned} &\varepsilon_{1'1'} = \varepsilon_{11} + \varepsilon_{12} = (\gamma_{1111}/2 + \gamma_{1122}/2 + \gamma_{2323}) E^2 \\ &\varepsilon_{2'2'} = \varepsilon_{11} - \varepsilon_{12} = (\gamma_{1111}/2 = \gamma_{1122}/2 - \gamma_{2323}) E^2 \\ &\varepsilon_{33} = \gamma_{1122} E^2 \end{aligned} \tag{B.VIII.49}$$

As the three coefficients appearing in these equations are independent, the deformations of the dimensions A and B of the orthorhombic cell defined in the figure B.VIII.48 are in general not identical, meaning that the ratio A/B depends on the field.

Tetragonal and hexagonal Blue Phases induced by the field

The Kossel diagrams a) and b) in figure B.VIII.47 show that the Blue Phase I crystal is distorted, but the twofold symmetry of its Kossel diagram is preserved while the field is not too high. However, when the field exceeds a critical value $E_{O/T}$, the Kossel diagram suddenly acquires a fourfold symmetry (Fig. B.VIII.47c and d). This change cannot be interpreted simply as the fourfold axis reorienting parallel to the field, since the (200) rings are absent from these diagrams. The emerging conclusion is that the electric field induced a phase transition towards a Blue Phase of tetragonal symmetry, denoted by BPX [22]. We recall that this phase transition corresponds to the precise moment when the rectangular basis AB of the orthorhombic cubic cell becomes square (A / B = 1, see figure B.VIII.48c). This particular case corresponds to an increase in system symmetry, which goes from an F 222 space group to the $I4_122$ group (see figure B.VIII.54).

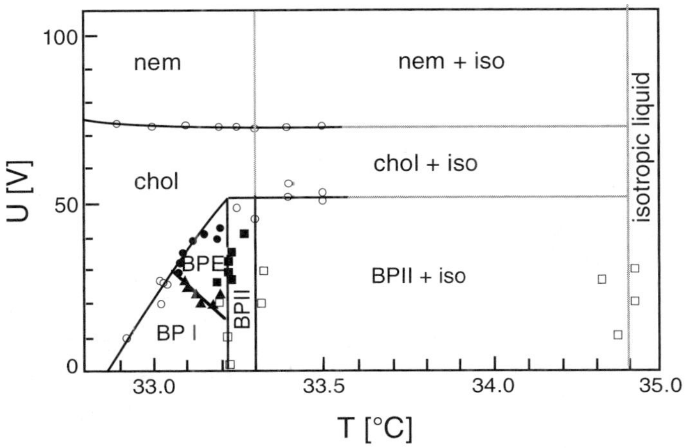

Fig. B.VIII.50 Phase diagram of the CB15/E9 mixture under the action of an electric field (F. Porsch and H. Stegemeyer [28]). The Blue Phase E induced by the electric field corresponds to the tetragonal phase BPX. Under even stronger fields, the Blue Phase is replaced by the cholesteric, then by the nematic phase (completely unwound cholesteric).

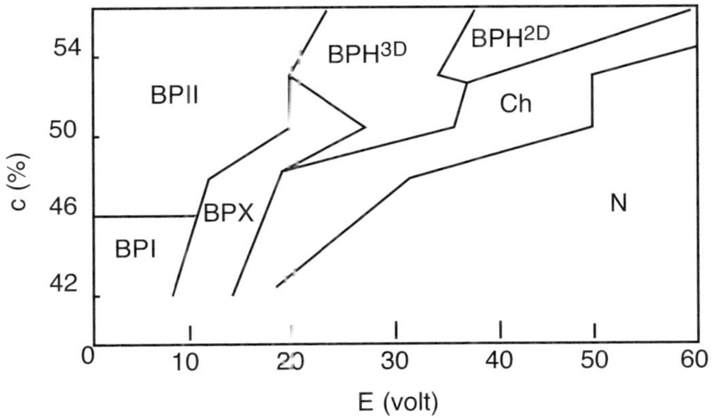

Fig. B.VIII.51 Phase diagram of the CB15/E9 mixture under an electric field. The ordinate gives the CB15 weight concentration (from ref. [29]).

The phase transition BPI → BPX can be experimentally determined without using the Kossel diagrams, simply by direct optical observation under the microscope, as F. Porsch and H. Stegemeyer did [28]. In their "voltage vs.

temperature" phase diagram of the E9/CB15 mixture, the Blue Phase induced by the electric field starting with the Blue Phase I was labeled BPE. In fact, it corresponds to the tetragonal symmetry Blue Phase X, denoted BPX, that we have already discussed. The BPI → BPX transition is also represented on the complete phase diagram of the E9/CB15 mixture under an electric field [29]. This diagram, represented in figure B.VIII.51, shows indeed that the BPII phase, obtained at higher CB15 concentration, successively changes with increasing voltage to a three-dimensional hexagonal Blue Phase (BPH^{3D}) (Fig. B.VIII.52), then to a two-dimensional hexagonal Blue Phase (BPH^{2D}), identical to the one theoretically predicted by Brazovskii et al. [13].

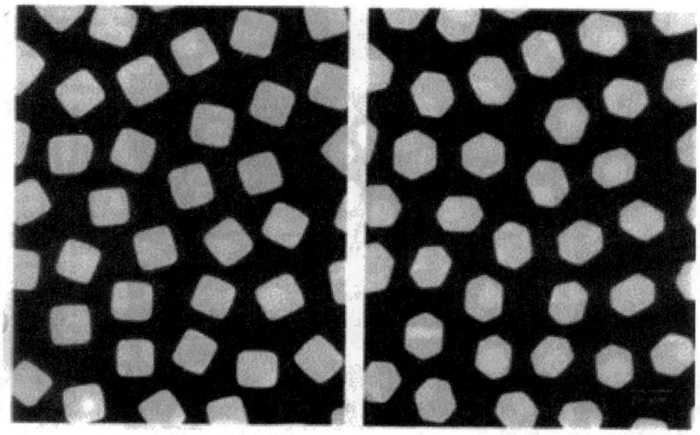

Fig. B.VIII.52 The crystallites of Blue Phase II first orient with their fourfold axis along the electric field (a), then change shape when the field exceeds a certain threshold (b); the $PBII \rightarrow BPH^{3D}$ transition has just taken place. The BPH^{3D} Blue Phase is three-dimensional, as there is still a reflection (the crystallites are bright against the black background of the isotropic phase) under incidence parallel to the six-fold axis. This reflection disappears in a stronger field, when the phase becomes BPH^{2D}.

The transition between the Blue Phase II and the Hexagonal Blue Phase is illustrated in figure B.VIII.52.

The possibility that a Blue Phase with hexagonal symmetry might appear under an electric field was theoretically predicted by Hornreich, Kugler, and Shtrikman [30]. Without going into the details of their calculation, let us give some qualitative, but nevertheless very convincing arguments. Contrary to the Blue Phases with cubic symmetry, the tetragonal and hexagonal symmetry phases can have a finite average dielectric anisotropy, just like the cholesteric phase. Then, in the framework of a Landau theory, the $\varepsilon_{ij}(0)=\varepsilon_a(e_ie_j-\delta_{ij}/3)$ component of the order parameter, where **e** is the unit vector parallel to the electric field, is present and there is an additional term in the free energy B.VIII.15, representing the coupling with the field. Thus:

$$f(E) = f(0) - \frac{1}{2}\varepsilon_0\varepsilon_a E^2 \qquad \text{(B.VIII.50)}$$

where f(0) is calculated along the same lines as in section VIII.4. Hornreich, Kugler, and Shtrikman showed that, in the presence of a high enough field, the hexagonal phases are energetically favored over the cubic phases.

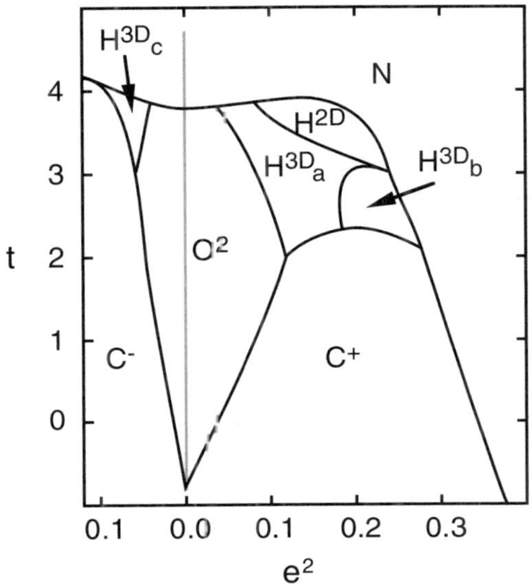

Fig. B.VIII.53 Theoretical phase diagram, obtained by Hornreich and Shtrikman [31]. To the right and to the left of the vertical gray line the dielectric anisotropy is positive and negative, respectively. t and e^2 are dimensionless variables (temperature and the square of the field). (The chirality, expressed by the parameter k, is 1.7.)

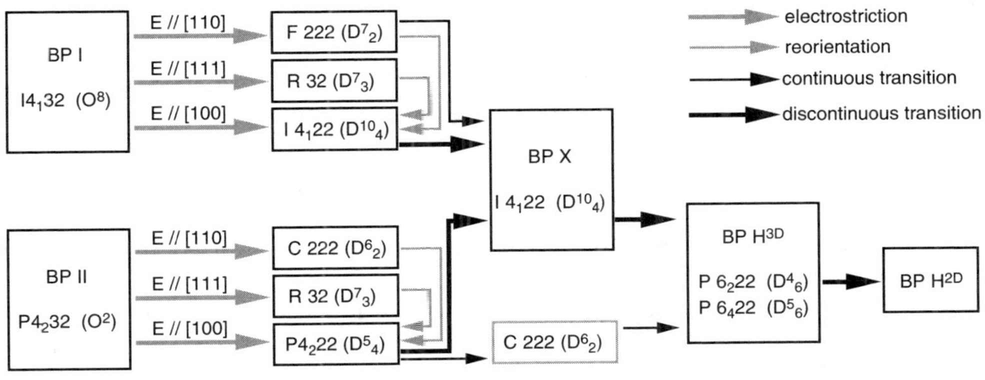

Fig. B.VIII.54 Symmetry of the Blue Phases I and II, deformed by the electric field and the new field-induced phases BP X and BP H (from ref. [18]).

This feature is illustrated in the theoretical phase diagram in figure B.VIII.53, where the dielectric anisotropy of the molecules is taken into account: positive is to the right and negative to the left. This diagram shows that:

1. When the dielectric anisotropy is positive, two three-dimensional and one two-dimensional hexagonal phases must exist, which partly corresponds to the observations, since a unique three-dimensional hexagonal phase was identified.

2. When the anisotropy is negative, a three-dimensional hexagonal phase appears, that was experimentally found by Kitzerow et al. [32] in a mixture of S811 (Merck) and EN18 (Chisso Corp.), with negative dielectric anisotropy.

We conclude this section with figure B.VIII.54, which summarizes the symmetry changes induced by the electric field in Blue Phases [18].

VIII.5e) Elasticity of the Blue Phases

As Saupe [2] first noticed, the Blue Phase kept in a beaker does not flow like the cholesteric phase, behaving instead rather like a gel. This observation clearly shows that the Blue Phases are endowed with elasticity. In a crystal with cubic symmetry, the elasticity tensor λ_{ijkl} has the same structure as the electrostriction tensor γ_{ijkl} (eq. B.VIII.47) and has three independent coefficients λ_{ijkl}, for instance, λ_{1111}, λ_{1122}, and λ_{2323}. The first two determine the isotropic compressibility

$$B = \frac{1}{3}(\lambda_{1111} + 2\lambda_{1122}) \qquad \text{(B.VIII.51)}$$

and the shear modulus

$$G_1 = \frac{1}{2}(\lambda_{1111} - \lambda_{1122}) \qquad \text{(B.VIII.52)}$$

measured when the deformation **u** is along the [−1,10] direction and varies along direction [110]. As to the λ_{2323} coefficient, it corresponds to the shear modulus

$$G_2 = \lambda_{2323} \qquad \text{(B.VIII.53)}$$

measured when the deformation **u** is along the [001] direction and varies along direction [110].

The isotropic compressibility of the Blue Phases is comparable to that of the isotropic liquid: $B \approx 10^{11}$ dyn/cm^2. On the other hand, the shear moduli must be very low, as the Blue Phases have the consistency of a gel.

To date, no one has determined the two shear moduli separately because the Blue Phase monocrystals are too small to be individually deformed. It is however easy to work with samples in the centimeter range, which are necessarily polycrystals. Using such samples, the average shear modulus G was determined by two methods represented in figure B.VIII.55.

In the first method, used by Kleinman et al. [33], the sample is in a small cylindrical container of volume $\approx 1\ cm^3$. This container, as well as the rod and the massive stand it rests upon, were lathed from a monocrystalline block of copper. A system of electrodes induces oscillations of the container about the axis defined by the rod. This torsion oscillator, which has a very high quality factor when empty, is sensitive to the viscoelastic properties of the liquid crystal.

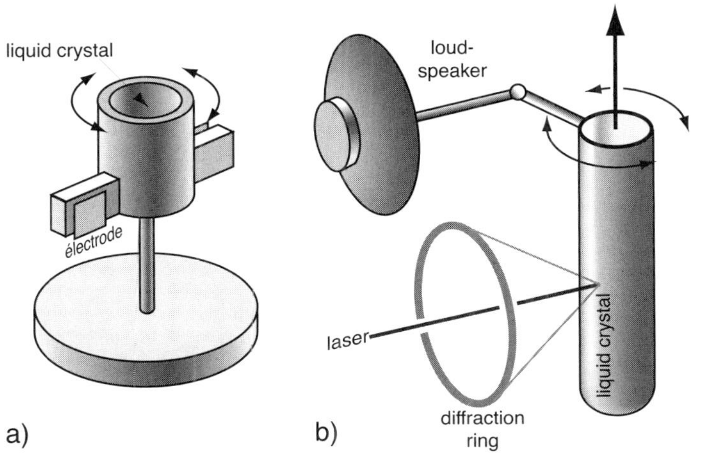

Fig. B.VIII.55 Experimental setups for measuring the shear modulus of the Blue Phase I: a) the one used by Kleinman et al. [33]; b) the one used by Cladis et al. [34].

Using this setup, Kleinman et al. [33] were able to determine the real part G' and the imaginary part G'' of the shear modulus G as a function of the temperature for a sample of cholesteryl nonanoate, at several excitation frequencies. We recall that the complex shear modulus

$$G^* = G' + iG'' \tag{B.VIII.54}$$

is defined by the relation between the oscillating shear deformation of frequency ω :

$$\varepsilon_{12} = \varepsilon e^{i\omega t} \tag{B.VIII.55}$$

and the stress generated by this deformation:

$$\sigma_{12} = G^* \varepsilon_{12} \tag{B.VIII.56}$$

In diagram (a) of figure B.VIII.56, the measurement of the elastic modulus G' (at $\omega = 52$ Hz) is correlated with a calorimetric record. In this way, the curve G'(T) is more easily interpreted. It is thus clear that the elastic modulus G' is only zero in the isotropic phase, while the Blue Fog Phase, i.e., the BPIII phase, already has an elastic response. At the BPIII→BPII transition, appearing on the calorimetric record as a very modest peak (very low latent heat), the elastic modulus jumps up by an order of magnitude. The BPII→PBI transition is also visible on the record of G'(T), although the elastic modulus stays at about the same value. Finally, the elastic modulus G' drops by about a factor of three at the transition to the cholesteric phase. Figure B.VIII.56b shows that the elastic modulus G' decreases with the shear frequency.

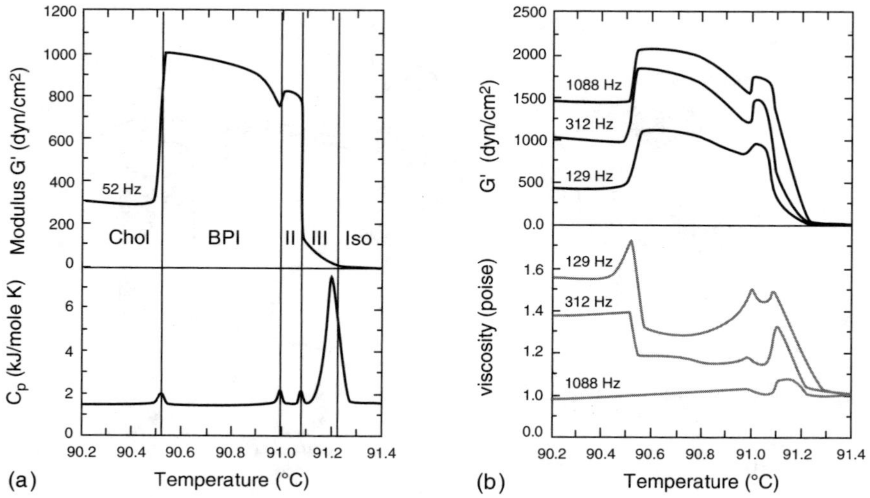

Fig. B.VIII.56 Measurements of the shear elastic modulus and viscosity of Blue Phases in cholesteryl nonanoate. a) Shear modulus (at 52 Hz) and heat capacity as a function of temperature; b) shear modulus and viscosity for different frequencies over the same temperature range as in a) (from ref. [33]).

We shall now describe the experiment of Cladis et al. [34] (Fig. B.VIII.55b). They used a mixture of CB15 (58% by weight) and the nematic ZLI1840 (Merck), which has the advantage of exhibiting a Blue Phase I over a large temperature range, between 21.8 and 27°C. This mixture is held in test tubes of radius 0.6 cm, giving a polycrystalline sample of macroscopic size. The experiment consists of making the tube oscillate about its axis with a small amplitude (Fig. B.VIII.55b); at the same time, the deformation of the Blue Phase is measured using the reflection of a He-Ne laser on the (110) planes. At rest, due to the polycrystalline structure of the sample, the diffraction diagram is a ring (powder diagram). Under shear, this ring becomes distorted, the deformation being detected by a photodiode. The record of the deformation amplitude of the ring as a function of the excitation frequency f reveals a resonance peak corresponding to the first cylindrically symmetric vibration mode of the crystal. Its frequency is given by the formula:

$$f_o = \frac{\mu}{2\pi}\frac{1}{R}\sqrt{\frac{G'}{\rho}} \qquad \text{(B.VIII.57)}$$

where μ is the first root of the Bessel function J_0, G' the shear modulus, and ρ the density of the Blue Phase. The frequency f_0 varies from 15 to 2 Hz when the temperature goes from 21.8 to 27°C. With $\rho \approx 1$ g/cm^3, R = 0.6 cm, and $\mu = 2.4$, one has $100 < G' < 500$ dyn/cm^2. This value, very weak compared to that of an ordinary crystal, is typical for a colloidal crystal with a lattice parameter in the micrometer range.

VIII.6 BPIII or Blue Fog

To conclude this chapter, we shall now take a closer look at the most elusive of Blue Phases – Blue Phase III. In the photo in figure B.VIII.6, it appears as milky **droplets** of a bluish hue, **surrounded by the transparent isotropic phase** (it is due to this appearance that Blue Phase III is also known as Blue Fog). From this photo we can draw a first conclusion: in MMBC, the transition between the liquid isotropic phase and BPIII is first order. This conclusion remains true for most of the materials studied, as confirmed by microcalorimetry measurements. For instance, in figure B.VIII.57, we have the record obtained by Thoen et al. [35]. Using a very sensitive and precise calorimeter shows that all phase transitions, Iso → BPIII → BPII → BPI → Chol, are first order in cholesteryl nonanoate. Among them, the Iso → BPIII transition has the largest latent heat, showing that the loss of orientational entropy of the molecules is almost entirely concentrated at the transition from the isotropic phase to Blue Phase III.

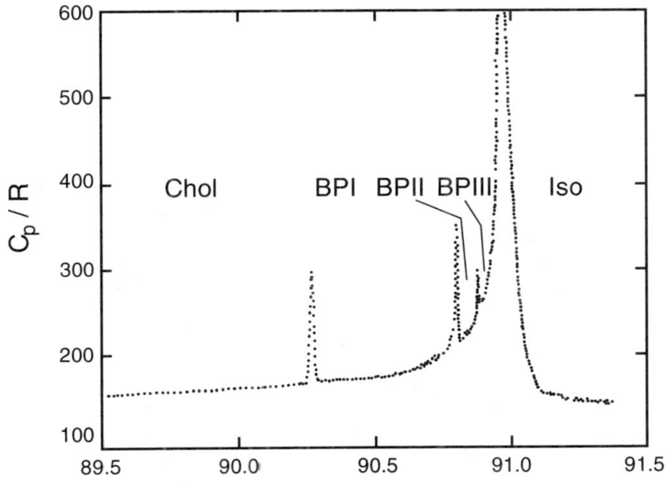

Fig. B.VIII.57 Calorimetric scan in cholesteryl nonanoate, by Thoen et al. (from ref. [35]).

On the other hand, optical reflectivity determinations show that, contrary to Blue Phases I and II, Blue Phase III does not exhibit Bragg reflections although its spectrum $I_r(\lambda)$ is not flat, possessing a very broad maximum (Fig. B.VIII.58) around a wavelength close to the (100) reflection of Blue Phase II.

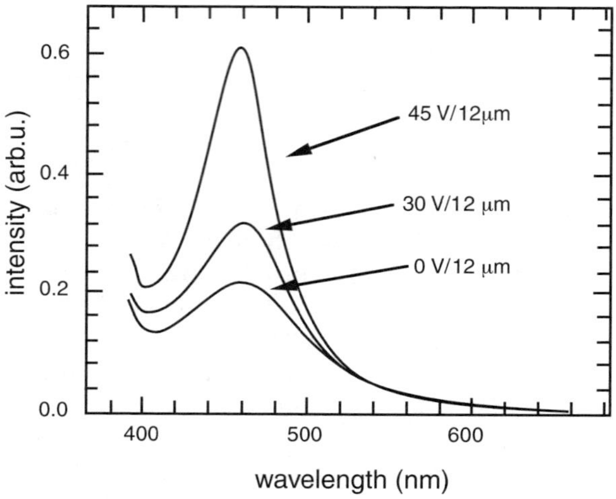

Fig. B.VIII.58 Reflectivity spectra of Blue Phase III under an electric field. Mixture of 30% S811 in EN18 (negative dielectric anisotropy) (from ref. [36]).

One must then admit that Blue Phase III has no three-dimensional orientational long-range order. However, the orientation of the molecules is certainly correlated over distances of the order of the lattice size in Blue Phase II.

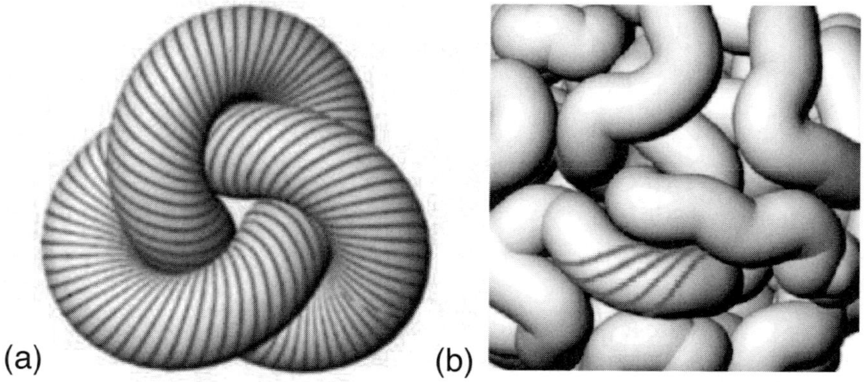

Fig. B.VIII.59 Models for the Blue Phase III: a) emulsion of droplets consisting in double twist cylinders; b) random entanglement of double twist cylinders (drawing by Piotr Pieranski).

Within the framework of uniaxial models, Hornreich suggested that Blue Phase III might look like a random entanglement of double twist cylinders. Another version of this model describes Blue Phase III as some sort of emulsion consisting of knotted double twist cylinders (see refs. [37] and [38]).

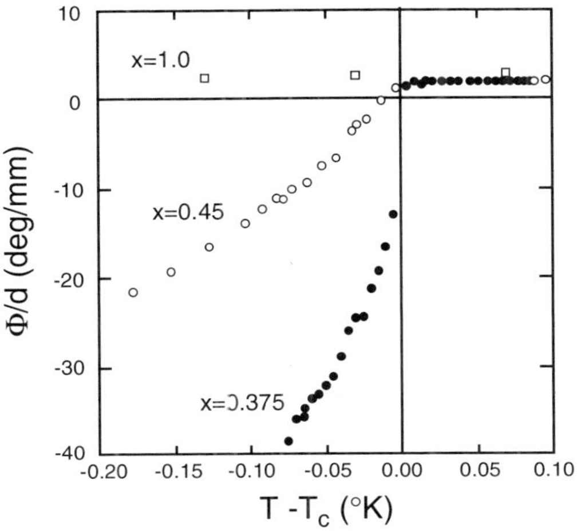

Fig B.VIII.60 Measurements of the rotatory power as a function of temperature in mixtures of R and S enantiomers of the product CE2. $x = (c_R - c_S)/(c_R + c_S)$ is the chiral fraction (from ref. [39]).

Since there is no symmetry breaking at the isotropic → BPIII transition, this first-order transition could end in the phase diagram by a critical point. Indeed, measurements of rotatory power performed by Kutnjak et al. [39] on mixtures of CE2 enantiomers show that, for chiral fractions x above 0.45, the discontinuity in rotatory power characteristic of the isotropic → BPIII transition vanishes.

To conclude this chapter on Blue Phases, we would like to recommend the review paper by P.P. Crooker cited as ref. [38], which contains an exhaustive list of references.

BIBLIOGRAPHY

[1] a) Gray G.W., *J. Chem. Soc.*, (1956) 3733.
b) Coates D., Harrison K.J., Gray G.W., *Mol. Cryst. Liq. Cryst.*, **22** (1973) 99.
c) Coates D., Gray G.W., *Phys. Lett.*, **45A** (1973) 115.
d) Coates D., Gray G.W., *Phys. Lett.*, **51A** (1975) 335.

[2] Saupe A., *Mol. Cryst. Liq. Cryst.*, **7** (1969) 59.

[3] Stegemeyer H., Bergmann K., "Experimental results and problems concerning Blue Phases," in *Liquid Crystals of one- and two-dimensional order*, Eds. Helfrich W. and Heppke G., Springer Series in Chemical Physics, vol. 11, Springer-Verlag, Heidelberg, 1980, p.161.

[4] Meiboom S., Sammon M., *Phys. Rev. Lett.*, **44** (1980) 882.

[5] Marcus M., *J. Physique (France)*, **42** (1981) 61.

[6] Grelet E., "Structural study of Smectic Blue Phases," Ph.D. Thesis, University of Paris XI, Orsay, 2001.

[7] Yang D.K., Crooker P.P., *Phys. Rev. Lett.*, **35** (1987) 4419.

[8] Johnson D.L., Flack J.H., Crooker P.P., *Phys. Rev. Lett.*, **45** (1980) 641.

[9] a) Barbet-Massin R., "Structure and optical properties of the Blue Phase I in cholesteric liquid crystals," M.Sc. Thesis, University of Paris XI, Orsay, 1985.
b) Barbet-Massin R., Pieranski P., *J. Physique (France)*, **46** Coll. C3 (1985) 61.
c) Barbet-Massin R., Pieranski P., *J. Physique Lettres (France)*, **45** (1984) L-799.

[10] Meiboom S., Sethna J.P., Anderson P.W., Brinkman W.F., *Phys. Rev. Lett.*, **46** (1981) 1216.

[11] a) Pansu B., Dubois-Violette E., *J. Physique (France)*, **51** Coll. C7 (1990) 281.
b) Pansu B., "Cubic symmetry and fluidity: from colloidal crystals to the Blue Phases," Habilitation Thesis, University of Paris XI, Orsay, 1994.

[12] Grebel H., Hornreich R.M., Shtrikman S., *Phys. Rev. A*, **28** (1983) 1114.

[13] Brazovskii S.A., Dmitriev S.G., *Sov. Phys. JETP.*, **42** (1976) 497.

[14] Belyakov V.A., Dmitrenko V.E., Osadchii S.M., *Sov. Phys. J.E.T.P.*, **56** (1982) 322.

[15] Hornreich R.M., Shtrikman S., *Phys. Rev. A*, **28** (1983) 1791.

[16] Crooker P., *Phys. Rev. A*, **31** (1985) 1010.

[17] a) Jérôme B., Pieranski P., *Liq. Cryst.*, **5** (1989) 799.
b) Jérôme B., Pieranski P., Godec V., Haran G., Germain C., *J. Physique (France)*, **49** (1988) 837.

[18] Kitzerow H.-S., *Mol. Cryst. Liq. Cryst.*, **202** (1991) 51.

[19] a) Blümel Th., Stegemeyer H., *J. Cryst. Growth*, **66** (1984) 163.
b) Stegemeyer H., Blümel Th., Hiltrop K., Onusseit H., Porsch F., *Liq. Cryst.*, **1** (1986) 3.

[20] Barbet-Massin R., Cladis P.E., Pieranski P., *Phys. Rev. A*, **30** (1984) 1161.

[21] Pieranski P., Barbet-Massin R., Cladis P.E., *Phys. Rev. A*, **31** (1985) 3912.

[22] Pieranski P., Cladis P.E., Garel T., Barbet-Massin R., *J. Physique (France)*, **47** (1986) 139.

[23] Saidachmetov P., *J. Physique (France)*, **45** (1984) 761.

[24] Pieranski P., Cladis P.E., *Phys. Rev. A*, **35** (1987) 355.

[25] a) Heppke G., Jérôme B., Kitzerow H.-S., Pieranski P., *J. Physique (France)*, **50** (1989) 549.
b) Heppke G., Jérôme B., Kitzerow H.-S., Pieranski P., *J. Physique (France)*, **50** (1989) 2991.

[26] Kitzerow H.-S., Crooker P.P., Rand J., Xu J., Heppke G., *J. Phys. II (France)*, **2** (1992) 279.

[27] Nye J.F., *Physical Properties of Crystals*, Clarendon Press, Oxford, 1957.

[28] Porsch F., Stegemeyer H., *Liq. Cryst.*, **2** (1987) 395.

[29] Jérôme B., Pieranski P., *Liq. Cryst.*, **5** (1889) 799.

[30] Hornreich R.M., Kugler M., Shtrikman S., *Phys. Rev. Lett.*, **54** (1985) 2099.

[31] Hornreich R.M., Shtrikman S., *Phys. Rev. A*, **41** (1990) 1978.

[32] Heppke G. Jérôme B., Kitzerow H.-S., Pieranski P., *Liq. Cryst.*, **5** (1989) 813.

[33] Kleinman R.N., Bishop D.J., Pindak R., Taborek P., *Phys. Rev. Lett.*, **53** (1984) 2137.

[34] Cladis P.E., Pieranski P., Joanicot M., *Phys. Rev. Lett.*, **52** (1984) 542.

[35] Thoen J., *Phys. Rev. A*, **37** (1987) 1754.

[36] Kitzerow H.-S., Crooker P.P., Kwok S.L., Heppke G., *J. Physique (France)*, **51** (1990) 1303.

[37] Collins P.J., in *Handbook of Liquid Crystal Research*, Eds. Collins P.J. and Patel J.S., Oxford University Press, Oxford, 1997, p. 99.

[38] Crooker P.P., "Blue Phases," in *Chirality in Liquid Crystals*, Partially Ordered Systems Series, Springer-Verlag, New York, 2001.

[39] Kutnjak Z., Garland C.W., Schatz C.G., Collings P.J., Booth C.J., Goodby J.W., *Phys. Rev. E*, **53** (1996) 4955.

Chapter B.IX

Overview of growth phenomena and the Mullins-Sekerka instability

In this chapter we shall briefly review crystalline growth and the Mullins-Sekerka instability. As a simplification, we shall assume that the interface is **rough** at the atomic scale and that there are no forbidden orientations (rendering the surface stiffness $\gamma + \gamma''$ negative). We do not consider the case of faceted crystals, a few aspects of which will be discussed in chapter C.VI, dedicated to smectic B plastic crystals. Nevertheless, this chapter is very general in nature and applicable to all kinds of fronts and materials. For instance, it introduces the concepts necessary for understanding most of the solid-liquid interfaces in metals or plastic crystals, and also many of the liquid-crystal interfaces when the associated phase transition is first order. We shall generically term "solid" the most ordered phase and "liquid" the disordered one. In a pure substance, the rate of growth for the solid is given by the kinetics of molecular attachment to the interface and by diffusion of the latent heat rejected during crystallization. When the material is impure (which is generally difficult to avoid), impurity rejection from one phase to the other and its diffusion in the two phases must also be taken into account. The resulting inhomogeneous distribution of the impurity concentration field can destabilize the front in free or directional growth, just as the temperature field does. This is the Mullins and Sekerka instability [1, 2], whose nonlinear evolution can lead to very complex morphologies. The most studied is the dendritic morphology because of its importance in metallurgy. This microstructure, spontaneously appearing during solidification of metals and alloys, influences many of their properties, including mechanical, plastic, thermal, electrical, or chemical, such as corrosion resistance.

This introductory chapter to growth phenomena will discuss the physical mechanisms determining the growth rate of an interface and then explain how it becomes unstable. We will not discuss the nonlinear regime and dendrite selection, because they are difficult problems whose theoretical analysis is beyond the scope of this book. This topic is extensively treated in more specialized books, such as those of P. Pelcé [3] or K. Kassner [4]. The reader is also referred to several review articles on growth phenomena, such as the one by J. Langer [2] (excellent in spite of its age) as well as the more recent ones by Brener and Mel'nikov [5] on dendrite selection and by B. Billia and R. Trivedi [6] on directional growth.

The chapter is structured as follows. We start with a brief overview of the phase diagram in binary mixtures. We then deduce the Gibbs-Thomson relation yielding the equilibrium temperature of a front as a function of its local curvature and of the impurity concentration (section IX.1). Next we discuss the three fundamental mechanisms determining the growth rate of an interface. This will provide us with a "minimal model" containing all equations to be solved (section IX.2). Finally, we try to determine elementary solutions to this problem. The front plane will be considered, in the kinetic and the diffusive regimes (sections IX.3 and IX.4, respectively), the circular germ in sections IX.5 and IX.6, and the dendrite in section IX.7. Whenever possible, the solution considered will be analyzed from the point of view of stability with respect to the Mullins-Sekerka instability.

IX.1 Gibbs-Thomson relation and the phase diagram of a diluted binary mixture

Synthesizing a completely pure organic substance is practically impossible. It will always contain impurities in solution (solute) in a weight concentration which is seldom less than 10^{-3}. Consider the simple case of a **diluted binary** mixture containing N solvent molecules and n solute molecules. This mixture, with a fixed number of particles, is in equilibrium with a reservoir at pressure P_L and temperature T. Under these conditions, the quantity to be minimized at equilibrium is the Gibbs free energy $G = F + P_L V$, the differential of which is, taking into account the surface tension between the two phases:

$$dG = (P_L - P_S)\,\delta V_S + (\mu_S - \mu_L)\delta N_S + (\nu_S - \nu_L)\delta n_S + \delta \int \gamma(\mathbf{n})\, dS \qquad \text{(B.IX.1)}$$

In this expression, μ is the chemical potential of the solvent and ν that of the solute. Indices "S" and "L" denote the solid and the liquid, respectively. Since at equilibrium dG = 0, algebraic calculation yields:

$$P_S - P_L = \frac{\tilde{\gamma}_1}{R_1} + \frac{\tilde{\gamma}_2}{R_2} \qquad \text{(B.IX.2a)}$$

$$\mu_L(P_L,T,C_L) = \mu_S(P_S,T,C_S) \qquad \text{(B.IX.2b)}$$

$$\nu_L(P_L,T,C_L) = \nu_S(P_S,T,C_S) \qquad \text{(B.IX.2c)}$$

with C_S and C_L the solute concentrations in the solid and the liquid, respectively ($C_S = n_S/N_S$ and $C_L = n_L/N_L$), R_1 and R_2 the principal radii of curvature of the surface, and $\tilde{\gamma}_1$ and $\tilde{\gamma}_2$ the values of the surface stiffness in the two principal planes:

$$\tilde{\gamma}_i = \gamma + \frac{d^2\gamma}{d\theta_i^2} \qquad (i = 1, 2) \qquad \text{(B.IX.3)}$$

Angles θ_1 and θ_2 give the orientation of the normal **n** to the interface with respect to two fixed directions of the crystal chosen in the two principal planes.

In a diluted system, the chemical potentials for the solvent and the solute are [7]:

$$\mu = \mu(P,T) - k_B T C \qquad \text{(B.IX.4a)}$$

$$\nu = \Psi(P,T) + k_B T \ln C \qquad \text{(B.IX.4b)}$$

with $\mu(P,T)$ the chemical potential of the pure solvent and Ψ a function only of P and T proportional to the free energy of a solute molecule.

Let P_a be the reference pressure (atmospheric pressure), T_c the melting temperature of the pure substance, v_L and v_S the specific volumes of the liquid and the solid, respectively, and L the transition latent heat per solid unit volume. Eq. B.IX.2b expressing the equality of solvent chemical potential can be developed about the reference point (P_a, T_c, $C = 0$) and, employing eq. B.IX.2a, one obtains the Gibbs-Thomson relation for a dilute solution:

$$\delta P \left(\frac{v_L}{v_S} - 1 \right) + L \frac{\delta T}{T_c} = \frac{\tilde{\gamma}_1}{R_1} + \frac{\tilde{\gamma}_2}{R_2} + \frac{k_B T_c}{v_S}(C_L - C_S) \qquad \text{(B.IX.5)}$$

with $\delta P = P_L - P_a$ and $\delta T = T_c - T$. In a similar way, equality of the solute chemical potential across the interface (eq. B.IX.2c) yields:

$$\frac{C_S}{C_L} = \exp\left(\frac{\Psi_L - \Psi_S}{k_B T} \right) = K \qquad \text{(B.IX.6)}$$

The coefficient K is the **partition coefficient** of the solute. It is practically constant (or at least we shall make this assumption), because Ψ varies very little with pressure in a solid or a liquid and because T remains close to T_c in the diluted regime.

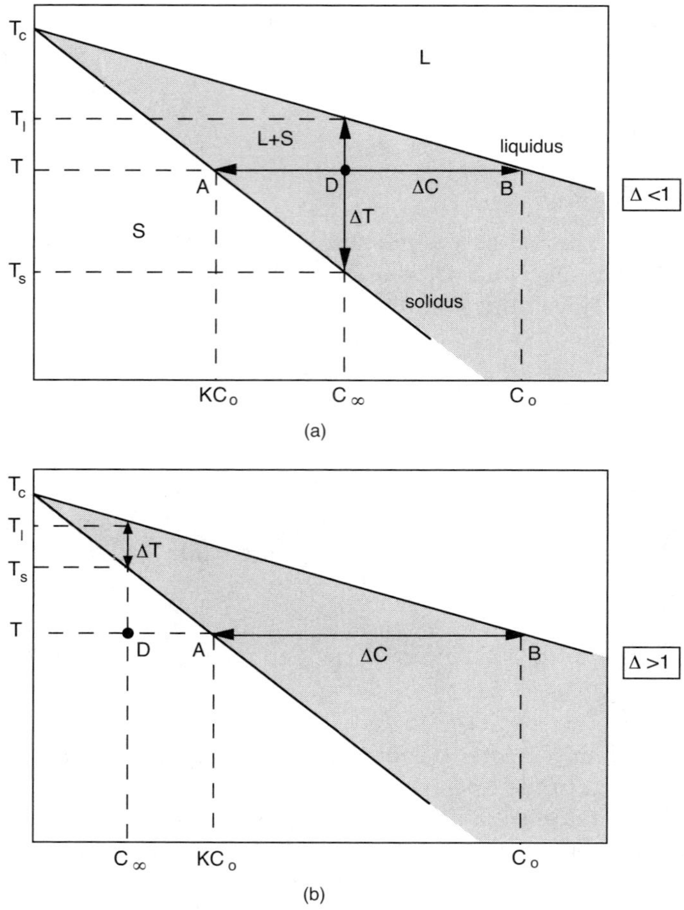

Fig. B.IX.1 Phase diagram of a binary mixture. Two situations can appear during growth, for supersaturation lower (a) or higher (b) than 1.

The preceding equations also allow us to obtain the phase diagram of a dilute solution at atmospheric pressure in the case where the interface separating the two phases is planar ($P_L = P_S = P_a$). In this limit one has, with C_{Lo} and C_{So} the concentrations in the liquid and the solid (the van't Hoff law):

$$C_{Lo} - C_{So} = \frac{L}{k_B T_c^2} v_S (T_c - T) \qquad \text{(B.IX.7a)}$$

$$C_{So} = K\, C_{Lo} \qquad \text{(B.IX.7b)}$$

or:

$$T - T_c = m\,C_{Lo} \quad \text{with} \quad m = \frac{(K-1)k_B T_c^2}{L v_S} \quad \text{(liquidus)} \tag{B.IX.8a}$$

$$T - T_c = \frac{m}{K} C_{So} \quad \text{(solidus)} \tag{B.IX.8b}$$

These relations define in the (T, C) plane two lines, the **liquidus** and the **solidus** (Fig. B.IX.1). The slope m of the liquidus is usually negative, with the partition coefficient K in general less than 1. The slope of the solidus is given by m/K. To simplify the notations, let $C_{Lo} = C_o$, $C_{So} = KC_o$ and C_∞ be the average impurity concentration in the sample.

At thermodynamic equilibrium and for $T_s < T < T_l$, the system separates into two phases, solid and liquid ("demixing" occurs). In the liquid, the concentration is:

$$C_L = C_o \tag{B.IX.9a}$$

while in the solid:

$$C_S = K\,C_o \tag{B.IX.9b}$$

The concentration C_o is given by the phase diagram:

$$T = T_c + mC_o \tag{B.IX.9c}$$

Writing total impurity conservation, the volume fraction of solid f_S at equilibrium is:

$$f_S = \frac{C_o - C_\infty}{C_o(1-K)} \tag{B.IX.10}$$

Note that a **concentration jump $\Delta C = C_o(1 - K)$ appears at the interface.** In the following, we shall use $\Delta T = T_l - T_s$ and take as the control parameter the **chemical supesaturation**:

$$\Delta = \frac{C_o - C_\infty}{C_o(1-K)} \tag{B.IX.11}$$

This dimensionless quantity varies between 0 and 1 when $T_s < T < T_l$ (Fig. B.IX.1a). In this case, it is identical to the volume fraction of solid at equilibrium.

On the contrary, $\Delta > 1$ when $T < T_s$ (Fig. B.IX.1b). In this regime, another definition of the supersaturation, called thermal supersaturation, is sometimes employed as it is more convenient experimentally:

$$\Delta = \frac{T_s - T}{\Delta T} \tag{B.IX.12}$$

To avoid confusion, we shall not employ this definition in the following.

IX.2 The minimal model

When the liquid is cooled below the liquidus temperature T_l, solid germs appear (nucleate) and then grow with time. During growth, the solute concentration in the two phases is no longer homogeneous. Let C_{Li} be its value close to the interface in the liquid. C_{Li} is presumably different from C_o and from C_∞. If the interface does not move too fast (i.e., if its velocity v is small compared to D_L/a, where D_L is the solute diffusion coefficient in the liquid and "a" a molecular distance), the partition coefficient of the impurities is almost unchanged (no trapping). A concentration jump $C_{Li}(1 - K)$ at the interface ensues, due to the fact that **the growing solid is rejecting solute in the liquid**. This solute rejection produces a local increase in the solute concentration in the liquid close to the interface, tending to melt the solid. The solute must then diffuse in the two phases for the growth to continue. The latent heat has a similar effect, as it locally heats and melts the solid. Again, latent heat must diffuse for the growth to be pursued. The effect of latent heat is generally negligible compared to that of the solute, as this latter diffuses much slower than the heat. On the other hand, heat can easily "escape" in the plates limiting the sample, unlike the solute. As this is the case in most laboratory experiments, we shall neglect latent heat rejection and the local heating it induces.

These considerations lead us to the "**minimal model**" comprising four equations.

First are the two concentration diffusion equations in the solid and the liquid:

$$D_S \nabla^2 C = \frac{\partial C}{\partial t} \quad \text{et} \quad D_L \quad \nabla^2 C = \frac{\partial C}{\partial t} \tag{B.IX.13}$$

with D_S and D_L the solute diffusion constants in the solid and the liquid, respectively (they vary between 10^{-7} and 10^{-6} cm^2/s in liquid crystals).

Next we have the law of impurity conservation at the interface:

$$C_{Li}(1 - K)\mathbf{v}.\mathbf{n} = - D_L (\mathbf{grad}C)_L.\mathbf{n} + D_S (\mathbf{grad}C)_S.\mathbf{n} \tag{B.IX.14}$$

with **n** the unit vector normal to the interface and directed from the solid towards the liquid. This equation shows that the quantity of solute rejected at the interface per unit time and surface $C_{Li}(1-K)\mathbf{v}.\mathbf{n}$ is balanced by the diffusion currents in the two phases.

Finally, in the dynamic regime, the Gibbs-Thomson equation must be generalized to take into account the **kinetics of molecular attachment at the interface**. Indeed, the interface must shift a little from its equilibrium temperature B.IX.5 in order to move with velocity v. Let T be its temperature during growth and T_e its equilibrium temperature. If the interface is rough and the velocity not too high, the kinetic law is linear in the first approximation:

$$v = \mu (T - T_e) \tag{B.IX.15}$$

with $v = \mathbf{v}.\mathbf{n}$ the normal interface velocity. Coefficient μ is the **kinetic coefficient**, giving the distance from thermodynamic equilibrium and describing order parameter dynamics inside the interface. A Landau-Ginzburg model relates it to the energy dissipated in the interface by unit time and surface:

$$\Phi_{surf} = \frac{\mu L (T - T_e)^2}{T_e} \qquad \text{(B.IX.16)}$$

Using eq. B.IX.5, equation B.IX.15 can be rewritten as (assuming $P_L = P_a$):

$$T = T_c + mC_{Li} - \frac{\gamma T_c}{L} C - \frac{v}{\mu} \qquad \text{(B.IX.17)}$$

with C_{Li} the local solute concentration and C the average interface curvature (unless stated otherwise, we take in the following γ = cst; then $C = 1/R_1 + 1/R_2$). The equation can be rewritten in the more convenient form:

$$C_{Li} = C_o + \frac{\gamma T_c}{mL} C + \frac{v}{m\mu} \qquad \text{(B.IX.18)}$$

where C_o is the equilibrium concentration in the liquid at temperature T, as given by the phase diagram. In the following, we assume that the kinetic coefficient μ does not depend on the solute concentration, which is a valid approximation in a dilute mixture. Equations B.IX.13, B.IX.14, and B.IX.18 represent the ***"minimal model"*** [9]. To these equations one must add the very important condition of **global impurity conservation**. This model explains most of the observed phenomena. Its complexity resides not only in the nonlinearity of the boundary conditions B.IX.14 and B.IX.18, but also from the need to impose these conditions on a boundary, the interface, whose equation is not known and can only be obtained by solving the equations. This is a **free boundary problem**. Note that similar problems appear in hydrodynamics (Saffman-Taylor fingering [8]), combustion (flames [3]), and even biology (growth of algae, neurons, etc. [3]).

In the next section, we shall discuss solutions of the stationary plane front type. We shall then see how they can be experimentally "forced" in directional growth before studying their stability.

IX.3 Stationary plane front

IX.3.a) Existence condition

Consider a plane front propagating at a constant velocity v. Far in the liquid the concentration is C_∞. In the reference frame moving with the front, the concentration profile (Fig. B.IX.2):

$$C_S = C_\infty \qquad \text{for} \qquad Z<0 \qquad \text{(B.IX.19a)}$$

$$C_L = C_\infty + \left(\frac{C_\infty}{K} - C_\infty\right)\exp\left(-\frac{Z}{l_D}\right) \qquad \text{for} \qquad Z>0 \qquad \text{(B.IX.19b)}$$

with $l_D = D_L/v$ (**diffusion length**), is a solution of the diffusion equations B.IX.13 and 14. Moreover, it satisfies the law of global impurity conservation in the system, as:

$$\frac{d}{dt}\int_{-\infty}^{+\infty} C\,dx = 0 \qquad \text{(B.IX.20)}$$

This condition (just like the condition of local impurity conservation) shows that, in the stationary regime, concentration in the solid must be C_∞.

Finally, the Gibbs-Thomson relation B.IX.18 yields:

$$C_{Li} = \frac{C_\infty}{K} = C_0 + \frac{v}{m\mu} \qquad \text{(B.IX.21)}$$

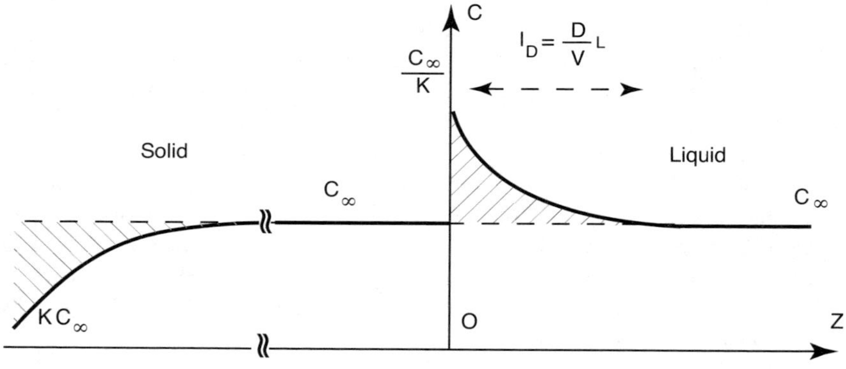

Fig. B.IX.2 Solute concentration profile ahead of a stationary plane front. The solute mound carried by the front is formed in the transient regime. Global solute conservation imposes equality of the two hatched areas.

This relation fixes the front velocity as a function of the solidus temperature or the chemical supersaturation (Fig. B.IX.1):

$$v = \mu(T_s - T) = \frac{\Delta T}{K}\mu(\Delta - 1) \qquad \text{(B.IX.22)}$$

A very important point is that **v must be positive**, otherwise the concentration diverges in the liquid, according to eq. B.IX.19b; this requires

$\Delta > 1$ or, equivalently, $T < T_S$. **The stationary plane front only exists for a supersaturation higher than 1.** One then speaks of a **kinetic regime**, the velocity depending only on the kinetic coefficient μ and the temperature.

Note that, in the kinetic regime, solute diffusion does not hinder front propagation. However, the interface carries with it a "mound" of solute formed during the transient regime preceding the stationary one (Fig. B.IX.2). During this transient regime (lasting for $t = l_D^2/KD_L$ [10]), the solute rejected from the solid increases the concentration of the liquid ahead of the front. Thus, the solid is purified during the transient regime; then, in the stationary regime, the solute accumulated ahead of the front is carried towards the other end of the sample. This effect, allowing for purification of the first diffusion length in the material, is known in metallurgy under the name of "zone melting." After the first passage, solute concentration is KC_∞ in the first diffusion length. After n passages, solute concentration becomes K^nC_∞. Purification is therefore more effective for small K and increases with the number of passages.

Let us now see how a plane front can be experimentally observed.

IX.3.b) Plane front in directional growth

A convenient experimental method for imposing a plane front consists of placing the sample in a temperature gradient G. In the laboratory reference system, the temperature inside the sample varies linearly with z (taken normal to the interface):

$$T = T_S + Gz \qquad \text{(B.IX.23)}$$

The origin is taken at the temperature $T = T_s$ (solidus temperature). We implicitly assume that the temperature gradient imposed by the setup is identical in the sample and in the glass plates containing it. This is an excellent approximation if the sample is much thinner than the glass plates and if the drawing velocity is not too high; otherwise, the difference in conductivity between the solid and the liquid must be taken into account, as well as the heat exchange with the glass plates and the latent heat rejection, changing the temperature locally and rendering the problem much more complicated.

In practice, the sample is moved at a constant velocity v (chapter B.VI describes the experiment). This drawing velocity is taken as positive when the solid grows (solidification) and negative when it melts. The kinetic law B.IX.22 gives the front temperature in the stationary regime:

$$T_i = T_S - \frac{V}{\mu} \qquad \text{(B.IX.24a)}$$

determining its position in the gradient (Fig. B.IX.3a):

$$z_i = -\frac{v}{\mu G} \tag{B.IX.24b}$$

Note that when the kinetics are fast (large μ), the front shifts very little from the solidus temperature. The concentration profile in the stationary regime is given by eqs B.IX.19 (Fig. B.IX.3b). From eqs. B.IX.24, it can also be expressed as a function of the temperature, yielding:

$$C_L = C_\infty + C_\infty \frac{1-K}{K} \exp\left[-\frac{v}{D_L}\left(\frac{T-T_i}{G}\right)\right] \qquad T>T_i \tag{B.IX.25a}$$

$$C_S = C_\infty \qquad T<T_i \tag{B.IX.25b}$$

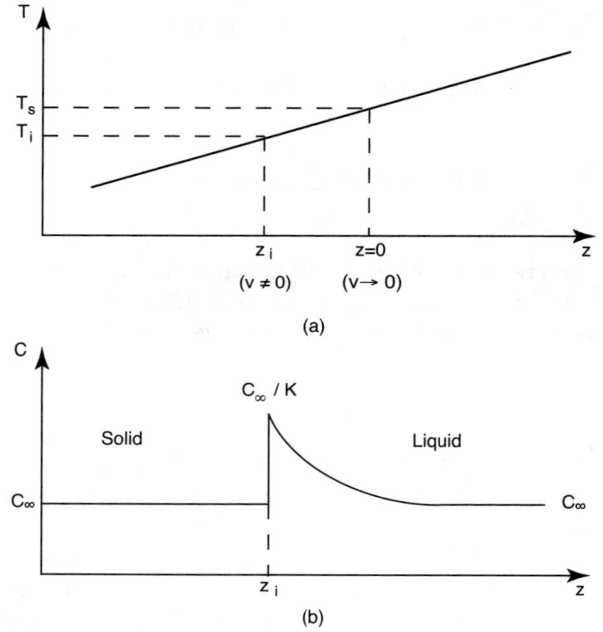

Fig. B.IX.3 Temperature and concentration profiles inside a sample in the stationary growth regime.

For the melting case, analogous calculations yield:

$$C_L = C_\infty \qquad T>T_i \tag{B.IX.26a}$$

$$C_S = C_\infty - C_\infty (1-K) \exp\left[\frac{v}{D_S}\left(\frac{T_i-T}{G}\right)\right] \qquad T<T_i \tag{B.IX.26b}$$

This time, there is a solute "hole" behind the front in the solid phase.

IX.3.c) Nucleation thresholds: the "constitutional undercooling" and "constitutional superheating" criteria

Experiment shows that nucleation ahead of the front sometimes appears above a certain drawing velocity. This phenomenon, hampering the observation of the Mullins-Sekerka instability, can occur during growth, but is more frequent during melting.

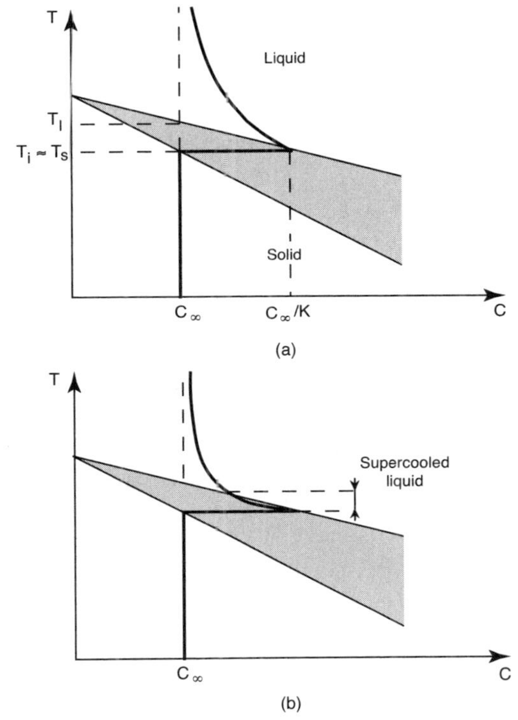

Fig. B.IX.4 Graphical method for determining the nucleation velocity during growth. In the absence of the nucleation delay it is given by the criterion of constitutional undercooling. In (a), the concentration profile does not penetrate in the freezing range. The system is stable. In (b), the profile penetrates in the freezing range. Ahead of the front appears a band of undercooled liquid, where solid germs can nucleate.

In the first case, solid germs nucleate in the liquid, while in the second liquid germs nucleate in the solid phase. In order to determine the threshold velocity, assume, as a first approximation, that there is no nucleation delay. Otherwise stated, nucleation occurs as soon as the liquid is undercooled (i.e., when its temperature is below the liquidus) or the solid is overheated (at a temperature above the solidus). Let us also neglect kinetic effects ($\mu = \infty$ and $T_i = T_s$). The simplest way of determining the nucleation velocity is to represent the theoretical concentration profile $C_{L,S}(T)$ in the phase diagram, as exemplified in figure C.VI.4 for the growth case. Obviously, there is a particular velocity v_n above which the concentration profile enters the freezing range of the two phases. A strip of undercooled liquid then appears ahead of the front, and solid germs can nucleate. Note that the nucleation rate reaches a maximum at a certain distance from the interface, where the local supersaturation is the highest (Δ_{max}). Thus, nucleation becomes possible as soon as the liquidus (negative) slope is larger than that of the concentration profile:

$$-\frac{K}{1-K}\frac{D_L G}{C_\infty v} > m \qquad \text{(B.IX.27)}$$

which fixes the "theoretical" nucleation velocity:

$$v_{cu} = \frac{D_L G K}{mC_\infty(K-1)} = \frac{D_L G}{T_l - T_s} \qquad \text{(B.IX.28)}$$

This **constitutional undercooling** criterion yields the minimum velocity v_{cu} above which solid germs can (in theory) nucleate ahead of the growth front.

This nucleation velocity is often underestimated because, in most systems, the nucleation rate of solid germs is only noticeable above a certain finite value Δ_{nucl} of the chemical supersaturation (Fig. B.IX.5). Consequently, the measured nucleation velocity v_{nucl}(exp.) satisfies the equation $\Delta_{max}(v) = \Delta_{nucl}$, with Δ_{max} the highest supersaturation in the liquid ahead of the front. One then has:

$$v_{nucl}(\text{exp.}) = \frac{D_L G}{T_l - T_s}\,\text{Cor}(\Delta_{nucl}) \qquad \text{(B.IX.29)}$$

where $\text{Cor}(\Delta_{nucl}) > 1$ is a correction factor that can be quite significant (a factor 2 or more [11]) and that can be easily obtained graphically, provided Δ_{nucl} and the phase diagram are known.

The same reasoning can be applied in the case of melting. Similarly, there is a velocity $\bar{v}_{cs}$ (negative by convention) below which liquid germs can nucleate in the solid phase, and which can be obtained by superposing the theoretical concentration profile B.IX.26 onto the phase diagram and writing that its slope equals that of the solidus (Fig. B.IX.6). This calculation yields:

$$\bar{v}_{cs} = \frac{D_S G K}{mC_\infty(1-K)} = -\frac{D_S G}{\Delta T} \qquad \text{(B.IX.30)}$$

This velocity is often very close to the experimentally determined nucleation velocity. Indeed, superheating a solid is very difficult: the presence of grain boundaries (defects in general) favoring nucleation of the liquid phase. One speaks of the **constitutional superheating** criterion.

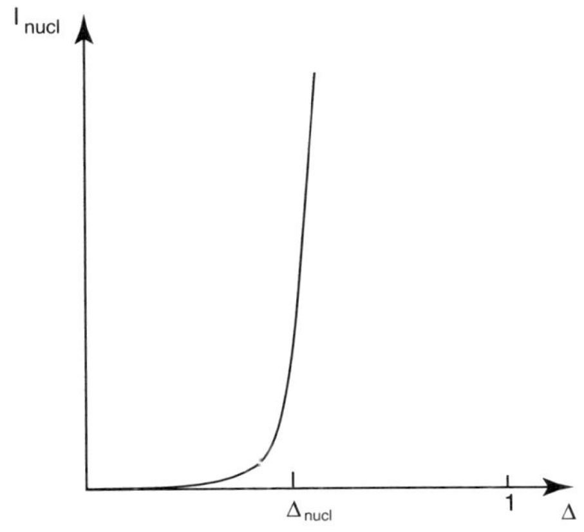

Fig. B.IX.5 Nucleation rate as a function of the supersaturation in a typical experiment. It is almost zero below Δ_{nucl} and grows very quickly above this value.

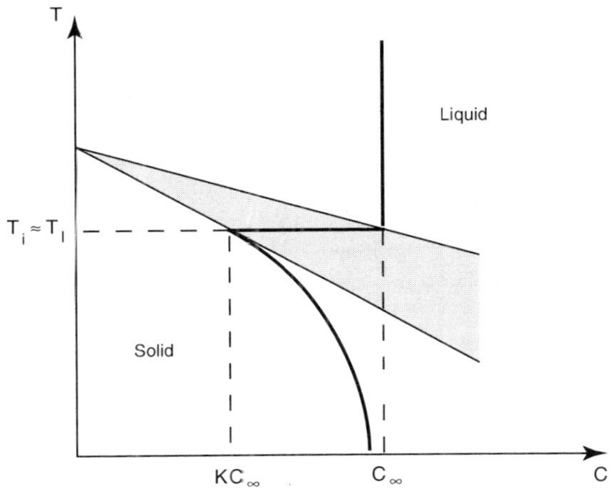

Fig. B.IX.6 Concentration profile during melting. The theoretical nucleation velocity is attained when the slope of the profile equals that of the solidus.

Let us now discuss the origin of the Mullins-Sekerka instability. It appears at a critical velocity whose absolute value is always greater than the constitutional undercooling velocity in solidification and the constitutional superheating velocity in melting. Experimental observation of the instability requires that the nucleation ahead of the front be moderate. Since undercooling a liquid without nucleation is easier than superheating a solid, the instability is more often observed in solidification than in melting. This is however not the only reason, as we shall see later.

IX.3.d) The mechanisms of the Mullins-Sekerka instability

To understand the origin of this instability in solidification, we start by assuming that G = O and neglect capillarity and the attachment kinetics. The interface is then a line of equal concentration, as $C_{Li} = C_o$ and $C_{Si} = KC_o$. Imagine now that a hump appears on the interface (Fig. B.IX.7a). Close to this hump, the lines of equal concentration get closer together, increasing the concentration gradient in the liquid. But **local front velocity is proportional to the local concentration gradient** since local impurity conservation yields:

$$\Delta C\,\mathbf{v.n} = -\,D_L\,\mathbf{grad}\,C_L\mathbf{.n} \qquad \text{(B.IX.31)}$$

Thus, the velocity of the solidification front increases at the tip of the hump. Conversely, the velocity decreases in a hollow, as the lines of equal concentration get farther apart. This very simple argument shows that the **diffusion field destabilizes the interface at any wavelength**.

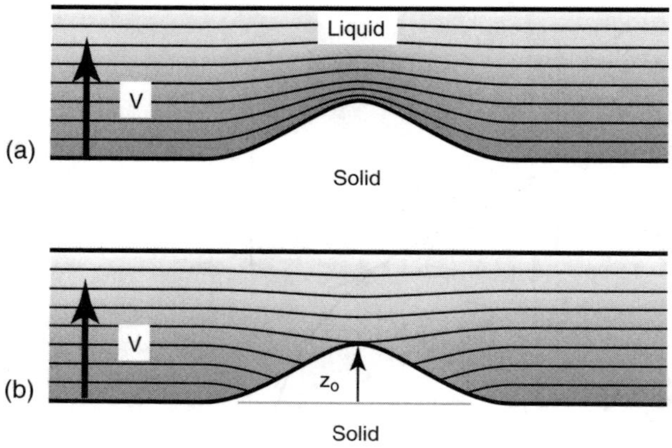

Fig. B.IX.7 a) Destabilizing effect of the concentration field. The lines of equal concentration get closer together and the velocity of the hump increases; b) stabilizing effect of the capillarity, of the attachment kinetics and of the temperature, locally "loosening" the lines of equal concentration. In the drawing, the hump goes slower than the plane front.

This argument applies whenever growth is controlled by a diffusion mechanism. Thus, the diffusion field of the temperature (related to latent heat rejection) can induce the Mullins-Sekerka instability in a pure substance. Diffusive instabilities also appear in hydrodynamics, as for instance, in the case of the Saffman-Taylor instability in porous media appearing at the interface between two immiscible fluids: here, the hydrodynamic pressure field plays the same role as the temperature or concentration fields in solidification, the velocity being proportional to the pressure gradient (Darcy's law).

This said, let us now look for the mechanisms stabilizing the interface.

In free growth (G = 0) there are at least two of them: the surface tension and the kinetics of molecular attachment. This is made apparent by the Gibbs-Thomson relation, giving the concentration in the liquid close to the interface:

$$C_{Li} = C_o + \frac{\gamma T_c}{mL} C + \frac{v}{m\mu} \qquad \text{with} \qquad m < 0 \tag{B.IX.32}$$

This relation shows that during solidification (v > 0), the concentration decreases in the neighborhood of a hump ($C > 0$). Under these conditions, the lines of equal concentration intersect the interface, leading to a decrease in the concentration gradient ahead of the hump (Fig. B.IX.7b) and thus to a reduction of its growth velocity. Conversely, the concentration gradient increases in a hollow and locally increases the growth velocity. These two effects are both stabilizing, although their actions are different. Indeed, the kinetic correction, proportional to the front velocity, does not depend on the wave vector of the perturbation and will only reduce its growth rate. The capillary correction, on the other hand, stabilizes the small wavelength perturbations ($\delta C_{cap.} \propto k^2$), as it acts against the increase in interface area.

In directional growth, the term Gz_0/m, originating in the deformation of the $z_0(x, t)$ front in the temperature gradient must be added to the Gibbs-Thomson equation B.IX.32. This term, proportional to the amplitude of the deformation with respect to the plane front (Fig. B.IX.7b), is again negative close to a hump ($z_0 > 0$), reducing its growth rate. The opposite occurs in a hollow. The temperature gradient is thus strongly stabilizing, whatever the perturbation wavelength. This result is easily understood, as the temperature increases close to a hump and decreases in a hollow. The hump will then tend to melt, and the hollow to crystallize.

To quantify these results, let us perform the linear stability analysis of the plane front in the $\Delta > 1$ stationary regime.

IX.3.e) Linear stability of the plane front in the kinetic regime

In order to analyze the linear stability of the plane front in directional growth, consider in the laboratory reference frame (which is also that of the interface) a sinusoidal perturbation with wave vector k and amplitude ε. The origin of the z axis is taken on the stationary plane front, whose position in the temperature gradient is fixed by the equation $T_i = T_s - v/\mu$, where v is the imposed drawing velocity. In this frame, the equation of the deformed front is (Fig. B.IX.7):

$$z_0(t) = \varepsilon \sin(kx) \exp\left(\frac{t}{\tau}\right) \tag{B.IX.33}$$

with $1/\tau$ the growth rate of the perturbation. The instability threshold ($1/\tau = 0$) is given by the dispersion relation relating the growth rate to the wave vector.

i) General dispersion relation

To fix the ideas, we perform the calculations in the solidification case. The first step consists of writing the diffusion equations in the laboratory frame:

$$\frac{\partial^2 C_L}{\partial x^2}+\frac{\partial^2 C_L}{\partial z^2}+\frac{v}{D_L}\frac{\partial C_L}{\partial z}=\frac{1}{D_L}\frac{\partial C_L}{\partial t} \qquad \text{(B.IX.34a)}$$

$$\frac{\partial^2 C_S}{\partial x^2}+\frac{\partial^2 C_S}{\partial z^2}+\frac{v}{D_S}\frac{\partial C_S}{\partial z}=\frac{1}{D_S}\frac{\partial C_S}{\partial t} \qquad \text{(B.IX.34b)}$$

To these equations must be added the boundary conditions:

$$C_L(z=+\infty)=C_\infty \qquad \text{(B.IX.35)}$$

$$v_n\,(1-K)\,C_{Li}=-D_L\,(\mathbf{grad}\,C_L)_i\mathbf{n}+D_S\,(\mathbf{grad}\,C_S)_i\,.\,\mathbf{n} \qquad \text{(B.IX.36)}$$

$$C_{Li}=C_o+\frac{Gz_o}{m}+\frac{d_oT_c}{m}z_o''+\frac{v_n}{m\mu} \qquad \text{(B.IX.37)}$$

$$C_{Si}=K\,C_{Li} \qquad \text{(B.IX.38)}$$

Here, C_o is the concentration in the liquid at temperature T_i if the system were at thermodynamic equilibrium ($T_i = T_c + mC_o$). We also note $d_o = \gamma/L$ (capillary length) and $C = z_o'' = \partial^2 z_o/\partial x^2$. Finally, v_n is the velocity normal to the front with respect to the solid phase. To first order in ε, it reads:

$$v_n = v + \frac{z_o}{\tau} \qquad \text{(B.IX.39)}$$

We shall look for solutions of the form:

$$C_L = C_\infty + C_\infty\frac{1-K}{K}\exp\left(-\frac{vz}{D_L}\right)+A_L\,z_o\exp(-b_L\,z) \qquad (z>0) \qquad \text{(B.IX.40a)}$$

$$C_S = C_\infty + A_S\,z_o\exp(-b_S\,z) \qquad (z<0) \qquad \text{(B.IX.40b)}$$

Substituting in the two diffusion equations B.IX.34a,b, we have:

$$b_L=\frac{v}{2D_L}+\sqrt{\left(\frac{v}{2D_L}\right)^2+k^2+\frac{1}{\tau D_L}}>0 \qquad \text{(B.IX.41a)}$$

$$b_S = \frac{v}{2D_S} - \sqrt{\left(\frac{v}{2D_S}\right)^2 + k^2 + \frac{1}{\tau D_S}} < 0 \tag{B.IX.41b}$$

Impurity partition at the interface (eq. B.IX.38) yields:

$$A_S = -\frac{v\,C_\infty}{D_L}(1-K) + KA_L \tag{B.IX.42}$$

while the impurity conservation law B.IX.36 allows us to obtain a first relation between the growth rate and the quantity A_L:

$$A_L = \frac{G_c K v \left(1 - \frac{D_S b_S}{v} + \frac{D_L}{Kv^2\tau}\right)}{v(1-K) - D_L(b_L - K\beta b_S)} \tag{B.IX.43}$$

where

$$\beta = \frac{D_S}{D_L} \tag{B.IX.44a}$$

and

$$G_c = -\frac{v}{D_L}C_\infty\frac{1-K}{K} = \frac{v\Delta T}{mD_L} \tag{B.IX.44b}$$

This quantity is the concentration gradient in the liquid ahead of the plane unperturbed front. Finally, the Gibbs-Thomson relation (eq. B.IX.37) yields:

$$\frac{d_o T_c}{m}k^2 + \frac{1}{\tau\mu m} + \frac{G}{m} = G_c + A_L \tag{B.IX.45}$$

The dispersion relation is obtained by eliminating A_L between the two equations B.IX.43 and B.IX.45:

$$\left[\frac{1}{\mu}\left(-\frac{v}{D_L}(1-K) + b_L - K\beta b_S\right) + \frac{mG_c}{v}\right]\frac{1}{\tau} = mG_c\left(-\frac{v}{D_L} - b_L\right) +$$

$$+ (G + d_o T_c k^2)\left[(1-K)\frac{v}{D_L} - b_L + K\beta b_S\right] \tag{B.IX.46}$$

For every drawing velocity v, this relation yields the growth rate $1/\tau$ of the perturbation as a function of the wave vector. The kinetic constant only appears in the coefficient of the growth rate, which is easily shown to be always positive. Thus, its only effect is to reduce the growth rate of the instability when this appears, an effect becoming more marked with slower kinetics or smaller μ.

This fairly complicated relation allows for determining the interface instability threshold. Although it is valid both in free (G = 0) and in directional growth ($G \neq 0$), these cases will be treated separately in the following. The results will then be extended to directional melting.

ii) Stability of the stationary plane front in free growth (G = 0)

In free growth, the system temperature T is experimentally imposed (with $T < T_s$ in the kinetic regime). Under these conditions, $T_i = T$ and the velocity of the plane front is $v = \mu(T_s - T)$.

Let us now return to the general dispersion relation B.IX.46. When G = 0, $\tau^{-1} = 0$ for k = 0, irrespective of the front velocity. As expected, the translation mode is neutral, since the position of the plane front is not determined when G = 0.

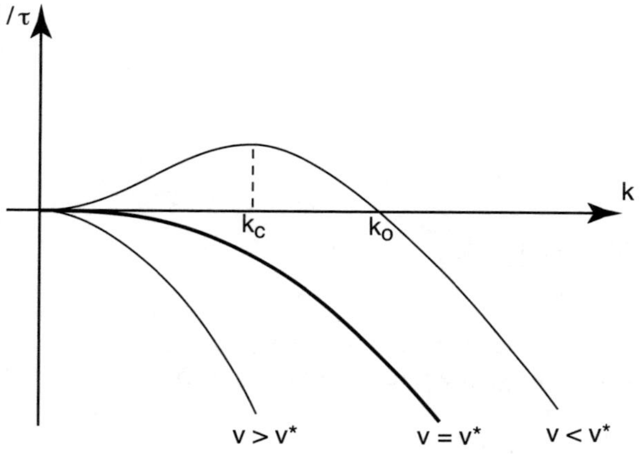

Fig. B.IX.8 Dispersion relation for the plane front when G = 0. The velocity v* is the absolute restabilization velocity.

In figure B.IX.8, we represented schematically the dispersion relation $\tau^{-1} = f(k, v)$ for different velocities. Note the presence of a particular value v* (termed **"absolute restabilization velocity"**) above which the front is stable and below which it is unstable for all wave vectors $k \in [0,k_o]$. In this case, a particular wavelength $\lambda_c = 2\pi/k_c$ appears, for which the growth rate is maximal. The wavelength λ_c is fundamental in free growth, as it comes up in all stability problems. We shall first determine it, before we determine v*.

k_c can be obtained analytically in the limit where the solidification velocity is small enough so that the perturbation wavelength is much smaller than the two diffusion lengths (which are comparable in liquid crystals):

$$\lambda << l_{DS} = 2D_S / v \quad \text{and} \quad \lambda << l_{DL} = 2D_L / v \tag{B.IX.47}$$

then:

$$b_L \approx -b_S \approx k \tag{B.IX.48}$$

and the dispersion relation B.IX.46 reduces to:

$$\frac{1}{\tau} = \mathrm{Cst}\,[kv - D_L d_o^c(1 + K\beta)k^3] \tag{B.IX.49}$$

In this expression, the multiplying constant is always positive. Let us define the **chemical capillary length** d_o^c as:

$$d_o^c = d_o \frac{T_c}{mC_{Li}^o(K-1)} \tag{B.IX.50a}$$

where C_{Li}^o designates the concentration in the liquid close to the interface assumed to be planar (hence the superscript ^{o}). For $\Delta > 1$, $C_{Li}^o = C_\infty/K$ in the stationary regime, yielding:

$$d_o^c = d_o \frac{KT_c}{mC_\infty(K-1)} = d_o \frac{T_c}{T_l - T_s} \qquad (\Delta > 1) \tag{B.IX.50b}$$

Relation B.IX.49 gives the wave vectors k_o and k_c corresponding to the neutral and the most unstable mode, respectively (Fig. B.IX.8):

$$k_o = \sqrt{\frac{1}{1+K\beta}\frac{1}{d_o^c\,(D_L/v)}} \tag{B.IX.51a}$$

$$k_c = \frac{k_o}{\sqrt{3}} \tag{B.IX.51b}$$

The most unstable wavelength of the plane front can be written as:

$$\lambda_c = 2\pi\sqrt{3(1+K\beta)\,d_o^c\,\frac{D_L}{v}} \tag{B.IX.52}$$

Typically, λ_c is about ten times the geometric average between the chemical capillary length and the diffusion length.

Let us now determine the absolute restabilization velocity. For an analytical estimate of the velocity above which the front is stable again, one must determine the curvature at the origin of the dispersion curve $1/\tau = f(k, v)$. As the growth rate goes to zero in this point, one only needs to develop this function to second order in the wave vector, yielding:

$$\frac{1}{\tau}(k\to 0) = \mathrm{Cst}\left[mG_c\frac{D_L}{v} - d_oT_cK\frac{v}{D_L}\right]k^2 + o(k^4) \tag{B.IX.53}$$

Thus, the curvature at the origin goes to zero for a velocity:

$$v^* = \frac{K-1}{K^2}\frac{mC_\infty}{T_c}\frac{D_L}{d_o} = \frac{D_L}{Kd_o^c} \qquad \text{(B.IX.54)}$$

This **absolute restabilization velocity** is attained for a growth temperature T* such that:

$$\frac{T_s - T^*}{T_l - T_s} = \frac{D_L / d_o}{K\mu T_c} \qquad \text{(B.IX.55)}$$

The corresponding chemical supersaturation is worth:

$$\Delta^* = 1 + \frac{D_L}{d_o^c \, \mu \, \Delta T} \qquad \text{(B.IX.56)}$$

Below T* or, equivalently, for a supersaturation higher than Δ^*, the front is stable. Also note that, close to the absolute restabilization point, the instability wavelength diverges.

Let us now see what these results become in the case of directional growth. The situation is more complicated due to the presence of the gradient and of a new length scale it introduces.

iii) Stability of the plane front in directional growth (G ≠ 0)

Let us now assume that the sample is placed in a temperature gradient $G > 0$. One can easily see that the growth rate is always negative in the limit $k \to 0$. The translation mode is then damped, an expected result, since the front position in the gradient is fixed. In the other limit, $\tau^{-1} \to -\infty$ for $k \to +\infty$, the capillarity stabilizing the small wave length perturbations.

The interface is thus unstable if there is at least one value of $k > 0$ for which the instability growth rate is zero. As we shall see, this depends on the experimental parameters (drawing velocity, gradient, etc.).

We shall start by analyzing the case of "small" drawing velocities (a few μm/s, typically). Figure B.IX.9 represents the typical shape of the growth rate $1/\tau$ versus k close to the destabilization threshold. This figure shows that a critical velocity v_c appears, below which the front is linearly stable and above which appears a range of unstable wave vectors.

For an analytical estimate of v_c and of the most unstable wave vector k_c of the plane front, let us assume (as we shall subsequently verify) that the wave vector k is much larger than v/D_L and v/D_S and let $1/\tau = 0$. In this limit:

$$b_L \approx k + \frac{v}{2D_L} \qquad \text{(B.IX.57a)}$$

$$b_S \approx -k + \frac{v}{2D_S} \qquad \text{(B.IX.57b)}$$

Substituting in the dispersion relation B.IX.46 and using eq. B.IX.44b, one obtains a first relation between v_c and k_c:

$$-2d_oT_c(D_L + KD_S)k_c^3 + d_oT_cv_c(1-K)k_c^2 + 2[(v_c\Delta T - G(D_L + KD_S)]k_c + $$
$$+ Gv_c(1-K) - \frac{v_c^2}{D_L}\Delta T = 0 \qquad \text{(B.IX.58)}$$

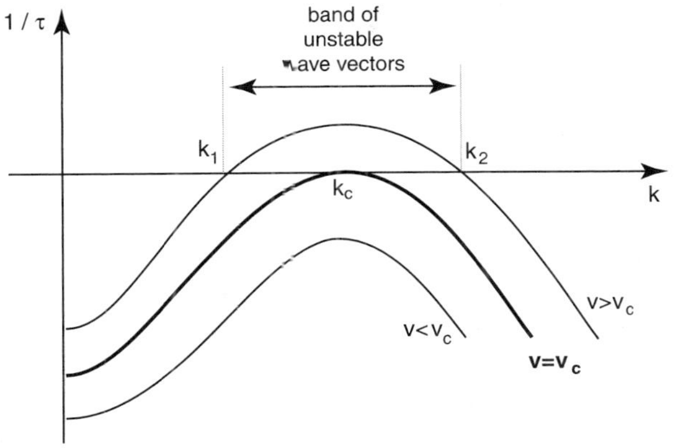

Fig. B.IX.9 Dispersion relation in the case of directional growth.

At the destabilization threshold, $d(1/\tau)/dk = 0$ (Fig. B.IX.9), yielding an additional relation:

$$-3d_oT_c(D_L + KD_S)k_c^2 + d_oT_cv_c(1-K)k_c + [(v_c\Delta T - G(D_L + KD_S)] = 0 \qquad \text{(B.IX.59a)}$$

If our approximation is valid, the second term of this expression can be neglected, so:

$$G = \frac{v_c\,\Delta T}{D_L + KD_S} - 3d_oT_c\,k_c^2 \qquad \text{(B.IX.59b)}$$

Substituting for G its expression in eq. B.IX.58, we have:

$$k_c = \left[\frac{K\Delta T(D_L + D_S)v_c^2}{D_L(D_L + KD_S)^2 4d_oT_c}\right]^{1/3} \qquad \text{(B.IX.60)}$$

Finally, the two previous equations show that the critical velocity is a solution of the equation:

$$G = \frac{v_c \Delta T}{D_L + KD_S}\left[1 - 3\left(\frac{K}{4}\right)^{2/3}\left(\frac{d_o T_c}{\Delta T}\right)^{1/3}\left(\frac{D_L + D_S}{D_L}\right)^{2/3}\frac{1}{(D_L + KD_S)^{1/3}} v_c^{1/3}\right] \quad \text{(B.IX.61)}$$

Neglecting the capillary term in this expression, which is justified if:

$$3\left(\frac{K}{4}\right)^{2/3}\left(\frac{d_o T_c}{\Delta T}\right)^{1/3}\left(\frac{D_L + D_S}{D_L}\right)^{2/3}\frac{1}{(D_L + KD_S)^{1/3}} v_c^{1/3} << 1 \quad \text{(B.IX.62)}$$

the critical velocity has the simple expression:

$$v_c = \frac{(D_L + KD_S)G}{\Delta T} = \frac{D_L G}{\Delta T}(1 + K\beta) = v_{cu}(1 + K\beta) \quad \text{(B.IX.63)}$$

and the critical wave vector:

$$k_c = \frac{2\pi}{\lambda_c} = \left[\frac{K(D_L + D_S)G^2}{4\Delta T D_L d_o T_c}\right]^{1/3} \quad \text{(B.IX.64)}$$

Formula B.IX.63 shows that the Mullins-Sekerka critical velocity is larger than the constitutional undercooling velocity by a factor of $(1 + K\beta)$. This multiplying factor expresses the **stabilizing effect** of diffusion in the solid. In typical solids, the difference between v_c and v_{cu} is insignificant, as $\beta \approx 10^{-3}$ and $K \approx 0.1$. In a liquid crystal, K and β are often of order unity, leading to $v_c \approx 2v_{cu}$. Nucleation ahead of the front can thus hamper the observation of the Mullins-Sekerka instability. This depends on the material employed, on its purity, the content of dust particles, etc.

Formulae B.IX.63 and B.IX.64 are more telling when written as a function of the characteristic lengths of the problem. In directional growth, these are **the diffusion length**:

$$l_D = \frac{2D_L}{v} \quad \text{(B.IX.65a)}$$

the chemical capillary length:

$$d_o^c = d_o \frac{T_c}{\Delta T} \quad \text{(B.IX.65b)}$$

and **the thermal length**:

$$l_T = \frac{\Delta T}{G} \quad \text{(B.IX.65c)}$$

With these definitions, an order of magnitude estimate of the critical threshold is:

$$l_T \approx l_D \quad \text{(condition of front destabilization)} \quad \text{(B.IX.66)}$$

As to the threshold wavelength, it is given by the relation:

$$\lambda_c \approx 2\pi \left[\frac{4}{K(1+\beta)}\right]^{1/3} d_0^{c\,1/3}\, l_T^{2/3} \approx 7\, d_0^{c\,1/3}\, l_T^{2/3} \quad \text{(with } K \approx \beta \approx 1\text{)} \qquad \text{(B.IX.67)}$$

Note that λ_c varies as $C_\infty^{1/3} G^{-2/3}$.

Finally, we must establish the condition under which our working hypotheses are valid. First of all, we assumed that $l_D >> \lambda_c$, yielding:

$$l_T >> 350 d_0^c \qquad \text{(B.IX.68)}$$

As to the condition B.IX.62, it is less restrictive, only imposing that $l_T >> d_0^c$. In most of the experiments presented in this book, condition B.IX.68 is largely satisfied. The preceding simplified formulae are thus applicable (provided that the bifurcation is supercritical).

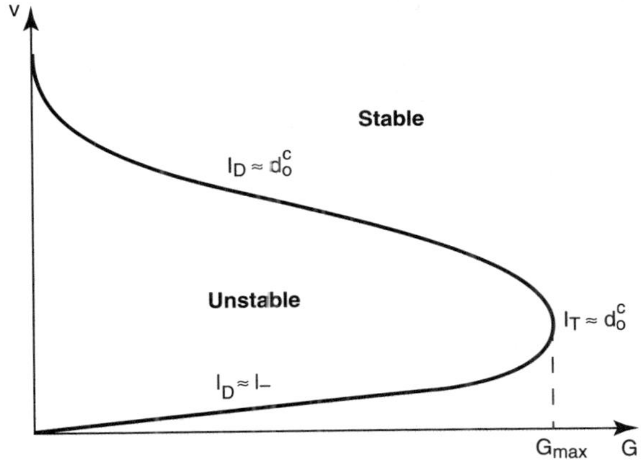

Fig. B.IX.10 Generic shape of the marginal stability curve in directional growth.

Let us now consider the "high velocity" case. We saw that, in free growth, the front is stable again above a certain velocity. The same phenomenon must occur in the presence of a thermal gradient, and for a lower velocity, as the gradient is stabilizing. Let us determine the absolute restabilization velocity, assuming that $k << v/D_L$ and $k << v/D_S$, so that the dispersion relation can be written as a power series of k. To fourth order, one has:

$$\text{Cst.}\ \tau^{-1} = -K + [l_D l_T - l_D^2(1 + K\beta^2) - K d_0^c l_T)]k^2$$
$$+ [(l_D^4(1 + K\beta^4) - l_D^3 l_T - d_0^c l_D^2 l_T (1 + K\beta^2)]\, k^4 \qquad \text{(B.IX.69)}$$

where Cst has a positive value. At the restabilization threshold, one must have $\tau^{-1} = 0$ and $\partial\tau^{-1}/\partial k^2 = 0$. These two equations yield the restabilization velocity v* as a function of G and then the corresponding wave vector k*. In the general case, this calculation cannot be performed analytically. However, the behavior of v* in the limit $G \rightarrow 0$ is easily obtained:

$$v^* = \frac{D_L}{Kd_o^c}\left[1 - 2\sqrt{(1+K)\,K\frac{d_o^c}{\Delta T}\,G}\,\right] \qquad \text{(B.IX.70)}$$

For G = 0, this formula gives again eq. B.IX.54.

One can also show the existence of a maximum gradient G_{max} above which the front is stable for all drawing velocities. An upper bound of G_{max} is easily obtained from the value of G (or l_T) for which the k^2 term in expression B.IX.69 becomes negative for all drawing velocities (or diffusion lengths l_D). This criterion yields:

$$G_{max} < \frac{\Delta T}{4Kd_o^c(1+K\beta^2)} = \frac{(mC_\infty)^2}{d_oT_c}\,\frac{(1-K)^2}{4K^3(1+K\beta^2)} \qquad \text{(B.IX.71)}$$

In order of magnitude, the maximum gradient is attained when $l_T \approx d_o^c$. Also, the value of G_{max} crucially depends on the partition coefficient. In metals or in ordinary plastic crystals, K is small (≈ 0.1) and $G_{max} \approx 10^6$ K/cm, an experimentally unattainable value. On the contrary, in a very pure nematic $K \approx 0.9$ and $G_{max} \approx 20$ K/cm, in good agreement with the experimental results (see chapter B.VI).

To summarize this discussion, we show in figure B.IX.10 the generic shape of the marginal stability curve in the parameter plane (G,v). Inside this contour, the front is unstable; outside, it is stable.

iv) Nature of the cell bifurcation in directional growth

As we have just seen, the front becomes unstable above a critical velocity v_c, meaning that every infinitesimal deformation with wave vector $k \in [k_1,k_2]$ is amplified by the diffusion field.

In the linear approximation, the time evolution of amplitude A_k corresponding to the mode of wave vector k is given by the equation:

$$\frac{dA_k}{dt} = \Omega(k)\;A_k \qquad \text{(B.IX.72a)}$$

where the growth rate $\Omega = \tau^{-1}$ is given by equation B.IX.46. If the mode is unstable, its amplitude increases exponentially. Consequently, this equation only describes the initial evolution of the perturbation. The nonlinear terms are needed to account for the amplitude saturation. The front being invariant under translation along the x axis (parallel to the front), this equation must remain invariant when exchanging A_k and $-A_k$ (which amounts to an x-translation of the sinusoid by half a wavelength). Thus, the first nonlinear term of the amplitude equation is the cubic one, such that:

$$\frac{dA_k}{dt} = \Omega(k)\ A_k - \alpha_1 A_k^3 + o(A_k^5) \qquad \text{(B.IX.72b)}$$

This "Hopf-Landau" equation is only valid close to the bifurcation threshold, when $(v - v_c)/v_c << 1$, i.e., when the linearly unstable modes satisfy the condition $|k - k_c|/k_c << 1$. As $\Omega(k)$ goes to zero on the marginal stability curve, its exact form must be kept in equation B.IX.72. On the other hand, the α_1 coefficient of the cubic term is generally finite at the bifurcation threshold, so in a first approximation $\alpha_1 = \alpha_1(v_c, k_c)$.

Consider the case $k = k_c$. Eq. B.IX.72 has two stationary solutions:

$$A_{k_c} = 0 \qquad \text{(B.IX.73a)}$$

which corresponds to the plane front, stable for $\Omega(k_c) < 0$, and

$$A_{k_c} = A_o = \sqrt{\frac{\Omega(k_c)}{\alpha_1}} \qquad \text{(B.IX.73b)}$$

The latter solution only exists for $\Omega(k_c)/\alpha_1 > 0$ and is stable only when $\Omega(k_c) < 0$ (as can be shown by noting that the quantity $\varepsilon = A_{k_c} - A_o$ satisfies the linearized equation $d\varepsilon/dt = -2\Omega\varepsilon$).

Two cases must then be distinguished, according to the sign of α_1:

1st case: $\alpha_1 > 0$

If $v < v_c$, $\Omega(k_c) < 0$ and the only stationary solution is the plane front.

If $v > v_c$, $\Omega(k_c) > 0$ and the plane front is unstable. On the other hand, solution "A_o" exists and is linearly stable. Starting from equation B.IX.72, it can be shown that:

$$\Omega(k_c) = k_c(v - v_c) + O[(v - v_c)^2] \qquad \text{(B.IX.74)}$$

up to a positive multiplicative constant. Hence, the amplitude of the mode with wave vector $k = k_c$ increases with the drawing velocity as $(v - v_c)^{1/2}$ (Fig. B.IX.11a). This variation is the distinctive mark of a **normal or supercritical bifurcation**.

2nd case: $\alpha_1 < 0$

The plane front is linearly unstable for $v > v_c$ and stable otherwise. The other solution B.IX.73b exists for $v < v_c$, but it is unstable because $\Omega(k_c) < 0$. One then expects a diagram as in figure B.IX.11b, describing a **reversed or subcritical bifurcation**. As the plane front is unstable below v_c, there must be another solution, of finite amplitude at $v = v_c$, not described by our amplitude equation. This solution corresponds to the thick line branch in figure B.IX.11b. Such a bifurcation must exhibit hysteresis: indeed, suppose that the front velocity slowly increases. As long as $v < v_c$, the plane front remains stable. As soon as $v > v_c$, however, a deformation of finite amplitude appears.

Upon slowly reducing the drawing velocity, this structure persists down to a velocity v_- below which the front is again planar. Note that, between v_- and v_c, the unstable solution "A_o" fixes a threshold that the deformation amplitude must exceed in order that the front develop a cellular structure.

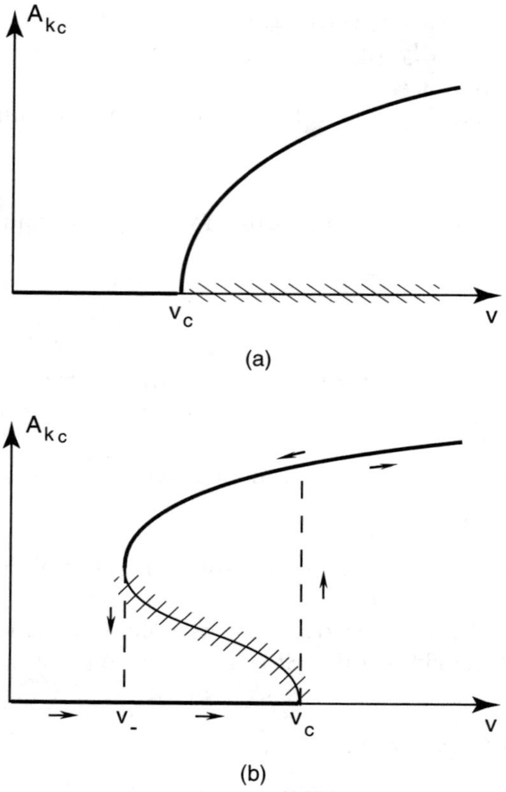

Fig. B.IX.11 Critical mode amplitude as a function of the drawing speed. a) "Normal" bifurcation; b) "reversed" bifurcation. The hatched branches correspond to unstable solutions.

The previous discussion demonstrates the importance of calculating α_1. This calculation, whose principle is presented in ref. [9], and with the assumption $D_S = 0$ (asymmetric model), yields for a thin sample:

$$\alpha_1 = \frac{D_L}{d_o^c\, l_c^3}\left(\frac{K^2 d_o^c}{2l_c}\right)^{1/3}\frac{K^2+4K-2}{2K} \qquad \text{(B.IX.75)}$$

with $l_c = D_L/v_c$ the diffusion length at the instability threshold. Obviously, the sign of α_1 is only determined by the value of K. For $K > 0.45$, α_1 is positive and the bifurcation is normal. Otherwise, the bifurcation is reversed (this is the case for many metallic alloys and for typical plastic crystals).

In liquid crystals, diffusion in the "solid" phase is sometimes as fast as in the "liquid" (the case of the nematic-isotropic liquid transition). On the other hand, K can be close to 1. The symmetric model ($K = 1$ and $D_S = D_L$) is thus better adapted for describing these situations. Langer and Turski [12] showed that in this case the bifurcation is always normal.

We conclude this section on the stationary plane front with some results on directional melting.

v) Mullins-Sekerka instability in directional melting

Let us first recall that, by definition, the velocity is negative during melting. In the stationary regime, the temperature of the plane front is equal to that of the liquidus (up to the kinetic effects): $T_i = T_l$ (Fig. B.IX.6). Under these conditions, the solution to the small perturbation problem can be written as:

$$C_S = C_\infty - C_\infty \frac{1-K}{K} \exp\left(-\frac{vz}{D_S}\right) + A_S z_0 \exp(-b_S z) \qquad (z<0) \qquad \text{(B.IX.76a)}$$

$$C_L = C_\infty + A_L z_0 \exp(-b_L z) \qquad (z>0) \qquad \text{(B.IX.76b)}$$

The calculation is exactly analogous to the solidification case and leads to the dispersion relation:

$$\text{Cst}\,\tau^{-1} = (G + d_o T_c k^2)\left(v - \sqrt{v^2 + 4D_L^2 k^2 + 4\,D_L \tau^{-1}}\right)$$

$$- K\left(G + d_o T_c k^2 + \frac{\Delta T v}{D_S}\right)\left(v + \sqrt{v^2 + 4D_S^2 k^2 + 4D_S \tau^{-1}}\right) \qquad \text{(B.IX.77)}$$

Assuming that $k >> v/D_L$ and v/D_S at the critical threshold (which is often the case in liquid crystals, where $D_L \approx D_S$), the (negative) melting velocity $\bar{v}_c$ below which the front is unstable with respect to the Mullins-Sekerka instability is:

$$\bar{v}_c = -\frac{(D_L + KD_S)G}{K\Delta T} \qquad \text{(B.IX.78)}$$

The corresponding critical wavelength is given by the relation:

$$\bar{\lambda}_c = 2\pi \left(\frac{4K\beta}{1+\beta}\right)^{1/3} l_T^{2/3} d_o^{c\,1/3} \qquad \text{(B.IX.79)}$$

Let us compare the critical Mullins-Sekerka velocity $\bar{v}_c$ to the nucleation velocity $\bar{v}_{cs}$ given by the constitutional superheating criterion. From eqs. B.IX.30 and B.IX.78, one gets:

$$\frac{\bar{v}_c}{\bar{v}_{cs}} = \frac{1 + K\beta}{K\beta} > 2 \qquad \text{(for } K \text{ and } \beta < 1\text{)} \tag{B.IX.80}$$

This relation shows that the Mullins-Sekerka instability during melting is almost always hidden by the nucleation ahead of the front. Nevertheless, it was observed during melting in nematic liquid crystals (see chapter B.VI) and in hexagonal lyotropic phases [11]. Finally, note that the ratio of the critical velocities during solidification and melting only depends on the partition coefficient, since:

$$\frac{\bar{v}_c}{v_c} = -\frac{1}{K} \tag{B.IX.81}$$

Thus, $|\bar{v}_c| > v_c$ because $K < 1$ generally.

Let us now study free growth for a supersaturation less than 1, starting with the simpler case of the plane front.

IX.4 Plane front in the diffusive regime ($\Delta < 1$)

During solidification at a chemical supersaturation less than 1, the front gradually slows down as the solute accumulates ahead of the front in the liquid.

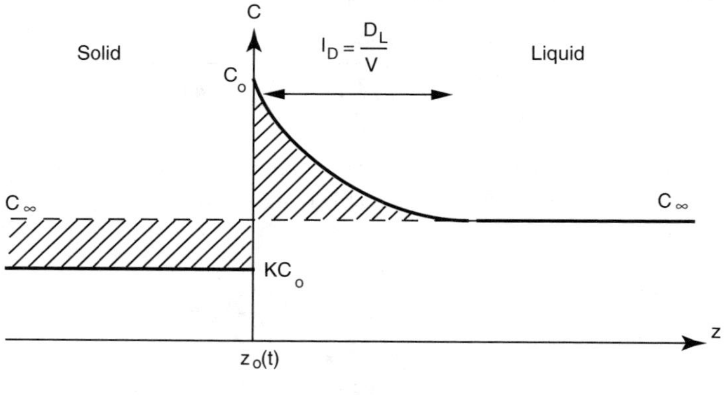

Fig. B.IX.12 Concentration profile in the diffusive regime. Global solute conservation imposes that the hatched areas on both sides of the interface remain equal at all times.

Since a stationary solution no longer exists in this case, we shall perform the calculations in the reference frame of the solid and designate by $z_0(t)$ the front position at time t. After a long enough time, front velocity is low enough so that the kinetic effects are negligible. In this limit, the front reaches thermodynamic equilibrium and the concentration in the liquid at the interface is worth

$C_{Li} = C_o$, while in the solid it is $C_S = KC_o$. The concentration profile at time t has the shape depicted in figure B.IX.12. In the liquid, the concentration decreases over a distance $l_D = D_L/v$ (diffusion length at the time t). The solute being globally conserved, the hatched areas on the two sides of the front in figure B.IX.12 must remain equal at all times. This leads to a progressive increase of l_D and hence to a slowing down of the front. We start by giving an approximate solution to this problem.

IX.4.a) Zener approximation

Front behavior in the asymptotic regime can be obtained by first accounting for global solute conservation. Equality of the two hatched areas in figure B.IX.12 gives:

$$z_o(t)\,(C_\infty - KC_o) \approx \frac{1}{2} l_D\,(C_o - C_\infty) \qquad \text{(B.IX.82)}$$

This first relation shows that the concentration field can be put in the form of a function $f(z/z_o)$, as l_D is proportional to z_o, an important observation to be exploited when looking for the exact solution.

The second relation is obtained by expressing local impurity conservation (eq. B.IX.14):

$$\dot{z}_o(t) C_o (1 - K) \approx D_L \frac{C_o - C_\infty}{l_D} \qquad \text{(B.IX.83)}$$

where $^\bullet$ designates the time derivative. Term-by-term multiplication of the two equations allows for eliminating the unknown diffusion length, yielding:

$$z_o \dot{z}_o(t) \approx \frac{1}{2} D_L \frac{\Delta^2}{1 - \Delta} \qquad \text{(B.IX.84)}$$

whence:

$$z_o(t) \approx \frac{\Delta\sqrt{D_L t}}{\sqrt{1 - \Delta}} \qquad \text{(B.IX.85)}$$

This $\sqrt{t}$ dependence is the hallmark of a diffusive regime we shall now rigorously describe.

IX.4.b) Exact calculation in the $\mu = 0$ case

Defining the dimensionless concentration field

$$u = \frac{C - C_o}{C_o(1 - K)} \tag{B.IX.86}$$

the diffusion equation in the liquid becomes:

$$D_L \Delta u_L = \frac{\partial u_L}{\partial t} \tag{B.IX.87}$$

At the interface, in $z = z_o(t)$:

$$D_L \frac{\partial u_L}{\partial t} = -\frac{dz_o}{dt} \qquad \text{and} \qquad u_L = 0 \tag{B.IX.88}$$

while far away from the front:

$$u_L = -\Delta \qquad \text{at} \qquad z = +\infty \tag{B.IX.89}$$

Finally, in the solid $u_S = -1$. As shown in the previous section, global solute conservation shows that the solution must be determined using the form $u = f(z/z_o(t))$. With the variable $Z = z/z_o(t)$, equation B.IX.87 becomes:

$$f'' + Zf'2P = 0 \tag{B.IX.90}$$

where P is a dimensionless number to be determined and is known as the Peclet number of the problem:

$$P = \frac{z_o \dot{z}_o}{2D_L} \tag{B.IX.91}$$

The boundary conditions are:

$$f(+\infty) = -\Delta, \qquad f(1) = 1, \qquad f'(1) = -2P \tag{B.IX.92}$$

Integrating eq. B.IX.90 taking into account the boundary conditions successively leads to:

$$f' = -2P\exp(P)\exp(-PZ^2) \tag{B.IX.93}$$

and then to:

$$f(Z) = -\Delta + 2P\exp(P)\int_Z^{+\infty} \exp(-P\xi^2)\, d\xi \tag{B.IX.94}$$

Finally, the condition f(1) = 0 fixes the constant P:

$$\Delta = \sqrt{\pi P}\exp(P)\,\mathrm{erfc}(\sqrt{P}) \tag{B.IX.95}$$

where

$$\operatorname{erfc}(t) = \frac{2}{\sqrt{\pi}} \int_z^{+\infty} \exp(-y^2)\,dy \tag{B.IX.96}$$

From eq. B.IX.91:

$$z_0(t) = \sqrt{4PD_L t} \tag{B.IX.97}$$

We shall not analyze the linear stability of this solution, as it is never observed in free growth. Instead, we shall generalize these results to the case of a circular germ and study its stability. This solution is observed in thin layers of liquid crystals or transparent plastic crystals (here, "thin" means "of thickness much smaller than the solute diffusion length").

IX.5 Free growth of a circular germ in the diffusive regime ($0 < \Delta < 1$)

Let us first establish the time evolution of the germ and then analyze its linear stability.

IX.5.a) Radius evolution law as a function of the supersaturation

This section is dedicated to analyzing the growth of an **isolated** circular germ. There is no analytical solution to the complete problem, including capillarity and kinetic effects, so we shall simplify it by neglecting the nonstationary term in the diffusion equation (**quasi-stationary approximation**). In this case, the concentration field verifies the Laplace equation everywhere:

$$\Delta C = 0 \tag{B.IX.98}$$

The quasi-stationary approximation assumes that the concentration field at time t only depends on the boundary conditions at the germ surface at time t or, equivalently, that the growth velocity of the germ dR/dt (R(t) being the germ radius at time t is much lower than the "diffusion velocity" D_L/R. This condition can be simply written as:

$$P = \frac{R\dot{R}}{2D_L} << 1 \tag{B.IX.99}$$

where P is by definition the Peclet number of the germ. As we shall see later, this condition is fulfilled if Δ is small. Let us now write the equations of the problem as a function of the dimensionless field u(r) given by eq. B.IX.86. The bulk equations for the two phases are simply:

$$\Delta u_{L,S} = \frac{1}{r}\frac{\partial}{\partial r}\left(r\frac{\partial u_{L,S}}{\partial r}\right) = 0 \qquad \text{(B.IX.100)}$$

while at the germ surface:

$$u_L\,(r=R) = -\frac{d_o^c}{R} - \frac{\Delta}{\mu\delta T}\frac{dR}{dt} \qquad \text{(B.IX.101)}$$

$$D_L\frac{du_L}{dr}(r=R) - D_S\frac{du_S}{dr}(r=R) = -\frac{dR}{dt}\left[1-(1-K)\left(\frac{d_o^c}{R}+\frac{\Delta}{\mu\delta T}\frac{dR}{dt}\right)\right] \qquad \text{(B.IX.102)}$$

$$u_S(r=R) = Ku_L(r=R) - 1 \qquad \text{(B.IX.103)}$$

where $\delta T = T_l - T$ is the **imposed undercooling** and $d_o^c = d_oT_c/[mC_o(K-1)]$ the **chemical capillary length** (depending on the chosen growth temperature). Finally, far from the germ:

$$u_L(+\infty) = -\Delta \qquad \text{(B.IX.104)}$$

Let us look for a solution of the form:

$$u_L = A\ln(r) + B \qquad \text{(B.IX.105a)}$$

$$u_S = E \qquad \text{(B.IX.105b)}$$

where the time-dependent parameters A, B, and E are to be determined. To avoid the logarithmic divergence, we shall introduce a cut-off radius R* to be specified later (for a spherical germ this difficulty does not appear, as u(r) varies as 1/r). Eqs. B.IX.104, B.IX.101, and B.IX.103 yield:

$$A\ln(R^*) + B = -\Delta \qquad \text{(B.IX.106)}$$

$$A\ln(R) + B = -\frac{d_o^c}{R} - \frac{\Delta}{\mu\delta T}\frac{dR}{dt} \qquad \text{(B.IX.107)}$$

$$E = K[A\ln(R) + B] - 1 \qquad \text{(B.IX.108)}$$

Eliminating B between the first two equations gives:

$$A\ln\left(\frac{R^*}{R}\right) = \frac{d_o^c}{R} - \Delta + \frac{\Delta}{\mu\delta T}\frac{dR}{dt} \qquad \text{(B.IX.109)}$$

Finally, from the local impurity conservation law B.IX.102 one has:

$$D_L \frac{A}{R} = -\frac{dR}{dt}\left[1 - (1-K)\left(\frac{d_o^c}{R} + \frac{\Delta}{\mu\delta T}\frac{dR}{dT}\right)\right] \tag{B.IX.110}$$

Note that the concentration gradient is zero in the solid, as a result of the quasi-stationary hypothesis entailing that diffusion in the solid is rapid at the scale of the germ. In liquid crystals, this hypothesis is well verified. Eliminating A between eqs. B.IX.109 and B.IX.110, one gets the evolution equation for the germ:

$$\ln\left(\frac{R^*}{R}\right) R\frac{dR}{dt} = D_L \frac{\Delta - \dfrac{d_o^c}{R} - \dfrac{\Delta}{\mu\delta T}\dfrac{dR}{dt}}{1 - (1-K)\left(\dfrac{d_o^c}{R} + \dfrac{\Delta}{\mu\delta T}\dfrac{dR}{dt}\right)} \tag{B.IX.111}$$

This equation, very complicated at first glance, shows that:

$$\frac{dR}{dt} = 0 \quad \text{if} \quad R = \frac{d_o^c}{\Delta} = R_c \tag{B.IX.112}$$

Radius R_c is the **critical nucleation radius** above which the germ grows and below which it shrinks out. In practice, $d_o^c \approx 1000$ Å (or less) and R_c is very small (a few thousand Å). The germs observed under the microscope always have $R >> R_c$, which we shall assume in the following. In this case, capillarity is negligible. On the other hand, we can verify a posteriori that $\Delta << 1$ (the quasi-stationary hypothesis is only valid in this limit) implies:

$$(1-K)\frac{\Delta}{\mu\delta T}\frac{dR}{dt} << 1 \tag{B.IX.113}$$

The evolution equation B.IX.111 can then be written with good accuracy:

$$\ln\left(\frac{R^*}{R}\right) R\frac{dR}{dt} = D_L \Delta\left(1 - \frac{1}{\mu\delta T}\frac{dR}{dt}\right) \tag{B.IX.114}$$

The only unknown left in this equation is R*. It can be determined by writing a supplementary condition expressing global impurity conservation in the system:

$$\pi R^2 (C_\infty - KCo) = \int_R^{R^*} (C_L - C_\infty) 2\pi r dr \tag{B.IX.115}$$

Let

$$\lambda^2 = \frac{\Delta}{2\ln\dfrac{R^*}{R}} \tag{B.IX.116}$$

It can be easily shown that, for small Δ, the condition B.IX.115 becomes:

$$\Delta_{eff} + \lambda^2 \ln\lambda^2 = 0 \qquad \text{(B.IX.117)}$$

where $\Delta_{eff.}$ is a time-dependent effective supersaturation:

$$\Delta_{eff} = \Delta \left(1 - \frac{1}{\mu\delta T}\frac{dR}{dt}\right) \qquad \text{(B.IX.118)}$$

Expression B.IX.117 is in fact the quasi-stationary approximation of the exact equation:

$$\lambda^2 \exp(\lambda^2)\ \mathrm{Ei}(-\lambda^2) + \Delta_{eff} = 0 \qquad \text{(B.IX.119)}$$

where the function Ei is the Exponential Integral defined by:

$$\mathrm{Ei}(t) = -\int_t^{+\infty} \frac{e^{-u}}{u}\, du\ .$$

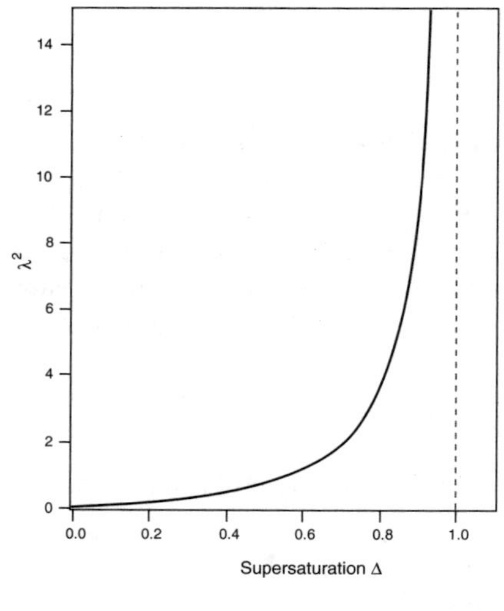

Fig. B.IX.13 Parameter λ^2 as a function of the chemical supersaturation Δ for a circular germ.

The function $\lambda^2(\Delta)$ is plotted in Fig. B.IX.13. It goes to 0 faster than Δ for $\Delta \rightarrow 0$ and to $+\infty$ for $\Delta \rightarrow 1$. Solving the two coupled equations B.IX.114 and B.IX.117 yields the evolution law for a small supersaturation germ.

Writing eq. B.IX.114 in the form:

$$R\frac{dR}{dt} = 2D_L\lambda^2\left(1 - \frac{1}{\mu\delta T}\frac{dR}{dt}\right) \qquad \text{(B.IX.120)}$$

allows us to introduce the characteristic radius:

$$\bar{R} = \frac{2D_L\lambda^2}{\mu\delta T} \tag{B.IX.121}$$

determining the crossover from the initial kinetic regime to the diffusive regime at long times (see figure B.IX.14). Thus:

$$\text{– for } R << \bar{R}, \quad R = \mu\,\delta T\,t = \mu\,(T_I - T)t \tag{B.IX.122a}$$

$$\text{– for } R >> \bar{R}, \quad R = 2\lambda\sqrt{D_L t} \quad \text{with} \quad \lambda^2 \ln(\lambda^2) + \Delta = 0 \tag{B.IX.122b}$$

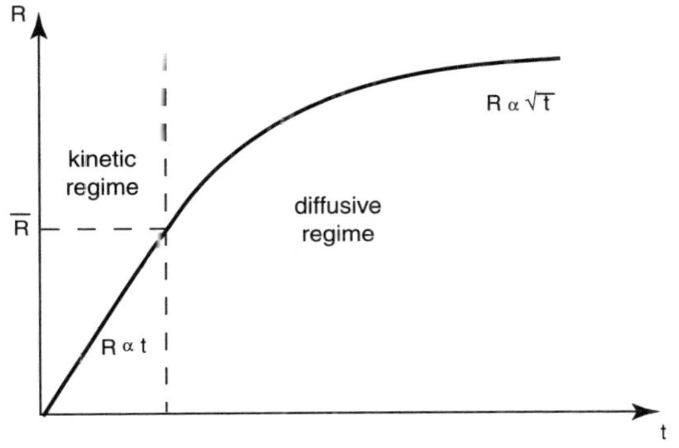

Fig. B.IX.14 Growth of a circular germ: kinetic and diffusive regimes.

Note that, in the diffusive regime, $\lambda^2 = R\dot{R}/2D_L$ is none other than the Peclet number of the problem.

Let us now analyze the stability of this solution.

IX.5.b) Linear stability analysis

The linear stability analysis of a circular germ can be performed similarly to that of the plane front, but the calculations are somewhat more complicated. On the other hand, the main results can be obtained by writing that the instability mode j, corresponding to a radius modulation of the form

$$R(\theta,t) = R_o(t) + \delta_j(t)\cos(j\theta) \tag{B.IX.123}$$

develops as soon as the germ perimeter is larger than $j\lambda_c$, where λ_c is the most unstable wavelength of the plane front given by formula B.IX.52. This criterion immediately yields the destabilization radius R_j of mode j:

$$R_j \approx j\sqrt{3(1+K\beta)\, d_o^c \frac{D_L}{v}} \qquad \text{(B.IX.124)}$$

where d_o^c is the capillary length, whose rigorous expression is given by eq. B.IX.50a. In order to determine the capillary length and the front velocity v at the time of the destabilization, two cases must be considered: either the germ is still in the initial kinetic regime, or it entered the asymptotic $t^{1/2}$ regime.

Let us start by considering the second case, which applies for small enough supersaturation. The chemical capillary length is then worth $d_o^c = d_o T_c/[mC_o(K-1)]$ and the evolution law of the germ is given by the equation:

$$R\frac{dR}{dt} = 2D_L\lambda^2 \qquad \text{with} \qquad \Delta + \lambda^2 \exp(\lambda^2)\, \mathrm{Ei}(-\lambda^2) = 0 \qquad \text{(B.IX.125)}$$

Substituting for the velocity its expression in eq. B.IX.124, one has:

$$R_j \approx \frac{3}{2}\frac{j^2}{\lambda^2}(1+K\beta)\, d_o^c \qquad \text{(B.IX.126)}$$

A complete stability calculation for mode j leads to:

$$R_j \approx \left[1 + j(j+1)(1+K\beta)\frac{\Delta}{2\lambda^2}\right]\frac{d_o^c}{\Delta} \approx \frac{1}{2}\frac{j(j+1)}{\lambda^2}(1+K\beta)\, d_o^c \qquad \text{(B.IX.127)}$$

and the growth rate:

$$\tau_j^{-1} = 2(j-1)D_L\lambda^2\frac{R_o - R_j}{R_o^3} \qquad \text{(B.IX.128)}$$

Obviously, the destabilization radius diverges for $\Delta \to 0$. On the other hand, the first unstable mode is $j = 2$, as the $j = 0$ mode, corresponding to a fluctuation in germ radius, is stable and the translation mode $j = 1$ is neutral. In practice, a cubic crystal favors mode 4 and a hexagonal crystal mode 6. This effect is due to the surface tension anisotropy of the crystal creating an initial germ deformation, subsequently amplified by the solute diffusion field.

When the supersaturation approaches 1, λ^2 diverges (Fig. B.IX.15). Formula B.IX.127 suggests that the destabilization radius tends to 0, but this is not observed experimentally. The discrepancy can be explained by taking into account the kinetics of molecular attachment. Let us first determine in which limit, i.e., for what supersaturation values, formula B.IX.127 remains valid. It implies that the destabilization takes place in the diffusive regime, meaning that R_j must be larger than the cut-off radius $\bar{R}$ (see eq. B.IX.121) between the diffusive and the kinetic regimes. For a hexagonal crystal ($j = 6$), after some elementary algebra this condition can be written as:

$$\lambda^2 < 3\sqrt{\frac{(1+K\beta)\Delta\mu T_c d_o}{D_L}} \tag{B.IX.129}$$

This inequality defines a value $\bar{\Delta}$ below which the destabilization radius is given by formula B.IX.127. In liquid crystals, $\bar{\Delta} \approx 0.7$ typically.

If the imposed supersaturation is higher than $\bar{\Delta}$ the germ gets destabilized while still in the kinetic regime. Its destabilization radius is more difficult to obtain. However, it can be estimated from eq. B.IX.124. The concentration at the interface in the liquid is close to C_∞, while the growth velocity is worth approximately $v = \mu(T_l - T)$. In this limit, the destabilization radius becomes:

$$R_j \approx j\sqrt{3(1+K\beta)\frac{D_L d_o^c}{\mu(T_l - T)}} \tag{B.IX.130}$$

where this time $d_o^c = d_o T_c/\Delta T$. This formula, valid for $\Delta > \bar{\Delta}$, shows that the destabilization radius does not go to zero for $\Delta \rightarrow 1$, but rather to a finite value proportional to $1/(\Delta T)^{1/2}$.

IX.6 Free growth of a circular germ in the kinetic regime ($\Delta > 1$)

In this regime, the interface behaves like a plane front as soon as $R >> R_c$ (negligible capillarity). Two growth regimes can be discerned: an initial regime, very brief, during which the stationary diffusion profile shown in figure B.IX.2 appears, and an asymptotic regime of stationary growth during which the radius grows linearly with time. Thus, the radius initially grows like:

$$R = \mu(T_l - T)t \qquad \text{(short times)} \tag{B.IX.131a}$$

Then the germ enters the asymptotic growth regime during which:

$$R = Cst + \mu(T_s - T)t \qquad \text{(long times)} \tag{B.IX.131b}$$

Note that the crossover between the two regimes occurs after a time $t \approx l_D^2/KD_L \approx D_L/K\mu^2\delta T^2$ corresponding to the $\bar{R} \approx D_L/\mu K\delta T$ (Fig. B.IX.15). This radius is usually very small (less than one micrometer).

Concerning germ stability, two cases can be distinguished according to the supersaturation value:

1. Either $\Delta > \Delta^*$, with Δ^* the absolute restabilization threshold given by expression B.IX.56. In this high-velocity regime, the front is always stable.

2. Or $\Delta < \Delta^*$. In this case, the front becomes unstable with respect to the Mullins-Sekerka instability. The instability wavelength is then roughly the

same as for the stationary plane front. Note that the destabilization radius diverges for $\Delta \rightarrow \Delta^*$ but remains finite for $\Delta = 1$.

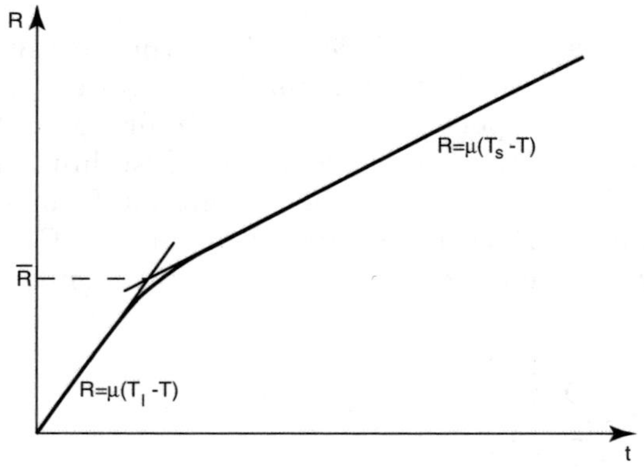

Fig. B.IX.15 Radius as a function of time in the kinetic regime ($\Delta > 1$).

Finally, in the case $\Delta = 1$, the germ radius grows asymptotically like $t^{2/3}$. This result can be qualitatively proved using the Zener approximation and taking into account the kinetic correction δC to C_0. Eq. B.IX.84 then becomes:

$$R\dot{R} \approx \frac{1}{2} D_L \frac{(C_0 - C_\infty + \delta C)^2}{(C_0 + \delta C)(1 - K)(C_\infty - KC_0 - K\delta C)} \tag{B.IX.132}$$

where $\delta C = \dot{R}/(m\mu) \ll C_0$. Since for $\Delta = 1$, $C_0 = C_\infty / K$, this equation reduces to:

$$R\dot{R} \approx \frac{1}{2} D_L \frac{(1 - K)C_0}{K(-\delta C)} \tag{B.IX.133}$$

Taking into account the definition of δC, the solution is:

$$R \approx \left[\frac{\mu D_L m C_\infty (K - 1)}{K^2}\right]^{1/3} t^{2/3}$$

This concludes the proof of the postulated scaling law. The 2/3 exponent was not experimentally observed because other possible solutions exist for $\Delta = 1$, and among them the dendrite we shall now describe.

IX.7 The Ivantsov dendrite ($0 < \Delta < 1$)

We saw that, for a small supersaturation, the velocity growth of a plane front or of a circular germ decreases with time, rapidly tending towards a diffusive solution in $1/t^{1/2}$. However, experiment proves the existence of a different solution, called **dendrite** by the metallurgists, parabolic in shape and propagating at constant velocity for supersaturations less than 1.

The existence of this solution has been demonstrated theoretically for the first time by Ivantsov in 1947 [13], under two strong assumptions (the problem being analytically unsolvable otherwise):

1. The solidification velocity of the curved front is slow enough so that the attachment kinetics are negligible ($\mu = 0$).

2. The radius of curvature of the interface is everywhere much larger than the critical nucleation radius, hence the surface tension effects can also be neglected ($\gamma = 0$).

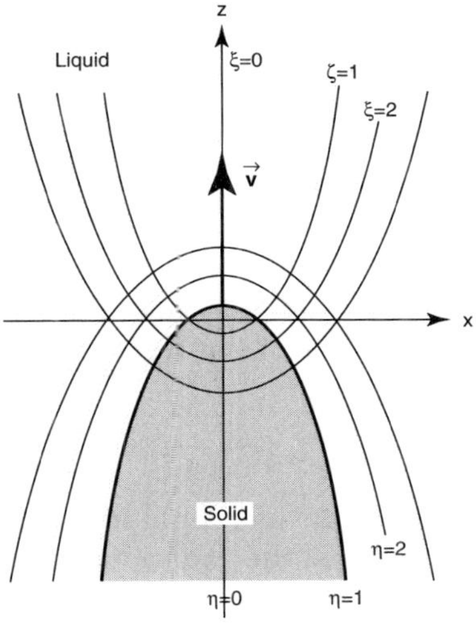

Fig. B.IX.16 Parabolic coordinates.

To prove the existence of a parabolic solution of the nonlinear equations B.IX.13, 14, and 18, it is more convenient to introduce parabolic coordinates in the plane (z, x) of the sample, which we assume to be two dimensional:

$$\xi = \frac{r - z}{\rho} \qquad \text{(B.IX.135a)}$$

$$\eta = \frac{r + z}{\rho} \qquad \text{(B.IX.135b)}$$

The length ρ is arbitrary for the time being and $r = \sqrt{x^2+z^2}$. In this coordinate system, the curve $\eta = 1$ represents a parabola with the equation:

$$z = \frac{1}{2\rho}\left(\rho^2 - x^2\right) \qquad \text{(B.IX.136)}$$

its radius of curvature at the tip being ρ (Fig. B.IX.16).

Suppose now that the chosen reference frame moves at a velocity U (parallel to z) with respect to the quiescent fluid (U is thus the solidification velocity at the dendrite tip). The stationary diffusion equation for the concentration field in the liquid reads, in parabolic coordinates:

$$-P\left(\eta\frac{\partial C_L}{\partial\eta}-\xi\frac{\partial C_L}{\partial\xi}\right)=\left[\sqrt{\eta}\frac{\partial}{\partial\eta}\left(\sqrt{\eta}\frac{\partial C_L}{\partial\eta}\right)+\sqrt{\xi}\frac{\partial}{\partial\xi}\left(\sqrt{\xi}\frac{\partial C_L}{\partial\xi}\right)\right] \qquad \text{(B.IX.137)}$$

where P is the Peclet number defined by:

$$P=\frac{\rho U}{2D_L} \qquad \text{(B.IX.138)}$$

On the interface, given by the equation $\eta=\eta(\xi)$, the solute conservation condition yields the equation:

$$\Delta C\left(\eta+\xi\frac{\partial\eta}{\partial\xi}\right)=-\frac{1}{P}\left[\eta\frac{\partial C_L}{\partial\eta}-\xi\frac{\partial\eta}{\partial\xi}\frac{\partial C_L}{\partial\xi}\right] \quad \text{in} \quad \eta=\eta(\xi) \qquad \text{(B.IX.139)}$$

assuming that the liquid is on the $\eta>\eta(\xi)$ side. Furthermore, $\Delta C=C_0(1-K)$, as the concentration in the solid is uniform and equal to KC_0 in the stationary regime. Finally,

$$C_L=C_\infty \quad \text{in} \quad \eta=+\infty \quad \text{and} \quad C_L=C_0 \quad \text{in} \quad \eta=\eta(\xi) \qquad \text{(B.IX.140)}$$

The equation system B.IX.137-140 has a solution of the type $C_L=C_L(\eta)$ and $\eta(\xi)=1$. With this particular choice, the equations become:

$$-P\sqrt{\eta}\frac{\partial C_L}{\partial\eta}=\frac{\partial}{\partial\eta}\left(\sqrt{\eta}\frac{\partial C_L}{\partial\eta}\right) \qquad \text{(B.IX.141)}$$

$$\Delta C=-\frac{1}{P}\frac{\partial C_L}{\partial\eta}(\eta=1) \qquad \text{(B.IX.142)}$$

$$C_L(\infty)=C_\infty \qquad C_L(1)=C_0 \qquad \text{(B.IX.143)}$$

After integration, the first two equations immediately yield:

$$\sqrt{\eta}\frac{\partial C_L}{\partial\eta}=-\Delta C\,P\exp(P)\exp(-P\eta) \qquad \text{(B.IX.144)}$$

whence, using the first boundary condition B.IX.143:

$$C_L=C_\infty+\Delta C\sqrt{P}\exp(P)\int_P^{+\infty}\frac{\exp(-u)}{\sqrt{u}}du \qquad \text{(B.IX.145)}$$

The second boundary condition B.IX.143 fixes the Peclet number as a function of the supersaturation:

$$\Delta = \sqrt{P}\exp(P)\int_{P}^{+\infty}\frac{\exp(-u)}{\sqrt{u}}du = 2\sqrt{P}\exp(P)\int_{P^{1/2}}^{+\infty}\exp(-u^2)\,du \qquad \text{(B.IX.146)}$$

This solution describes a parabolic dendrite, of curvature radius at the tip ρ, propagating along the z axis at a velocity U. Relation B.IX.146 only fixes the Peclet number, i.e., the ρU product, as a function of the imposed supersaturation Δ. This relation, first established by Ivantsov [13], shows that the smaller the curvature radius at the tip, the faster the dendrite propagates.

It turns out that we obtained a ***continuum of solutions***, which is explained by the fact that neglecting surface tension effects and the kinetics of molecular attachment left out all the characteristic lengths in the problem. It is therefore not surprising that the curvature radius at the tip is not defined. In order to find out which dendrite is "selected," one must take into account one length scale, e.g., the chemical capillary length d_o^c (still neglecting the kinetics).

This greatly complicates the mathematical problem, as the capillary term, although much smaller than the other terms in the Gibbs-Thomson equation ($\delta T_{cap.} \approx 10^{-3}$ K), behaves like a ***singular perturbation*** (second-order derivative term in the equations), so that classical perturbation methods based on Taylor series cannot be used. its nonlinearity further complicates the situation.

Nevertheless, the fact that the capillary correction is very small explains why the dendrite shape is always close to a parabola at the tip and why the Ivantsov relation remains quantitatively adequate.

Several theories have tried to solve the selection problem. One of them, from Langer, consists of assuming that the "chosen" dendrite is **marginally stable.** This criterion yields the value σ^* of the dimensionless quantity $\sigma = 2D_L d_o^c/\rho^2 U$. This quantity, proportional to $(\lambda_c/\rho)^2$ where λ_c is the most unstable wavelength of the plane front, is the stability constant. The value σ^* is termed the ***marginal stability constant***, all dendrites with $\sigma > \sigma^*$ being linearly stable, while those with $\sigma < \sigma^*$ are absolutely unstable. Knowledge of this parameter ($\sigma^* \approx 0.02$ for a crystal of cubic symmetry) combined with the Ivantsov relation (giving the ρU product) allow for separately calculating ρ and U.

Unfortunately, the Langer criterion is not rigorous and has no solid theoretical justification, although it is often in good agreement with the experimental data. It also gives the correct scaling law, the radius of curvature being indeed proportional to the most unstable wavelength for the plane front λ_c. One still needs to determine the value of σ^*.

After forty years of efforts, the current theory allows determination of this value [3-5]. The mathematics being quite involved, we shall only give its main results.

The first is that the stationary dendrites only exist for an **anisotropic** surface free energy $\gamma(\theta)$, where θ is the angle between the interface and a chosen crystallographic direction.

The second result is that, at low velocity, the stationary dendrites grow along the crystallographic axes of maximum surface tension or, equivalently, of minimum surface stiffness, in agreement with experiment.

Finally, the theory proves that the previously defined quantity σ^* is given by a law of the form:

$$\sigma^* = \frac{2D_L d_o^c}{\rho^2 U} = f(\varepsilon_m) \qquad \text{(B.IX.147)}$$

where $f(\varepsilon_m)$ is (in the limit of a small Peclet number) a universal function of the surface free energy anisotropy ε_m, defined by:

$$\gamma(\theta) = \gamma_o\,[\,1 + \varepsilon_m \cos(m\theta)\,] \qquad \text{(B.IX.148)}$$

In practice, m = 2, 4, or 6 according to the crystal symmetry. For small anisotropy, an asymptotic calculation yields:

$$f(\varepsilon_m) \propto \varepsilon_m^{7/m} \quad (\varepsilon \rightarrow 0) \qquad \text{(B.IX.149)}$$

It is now clear that the Ivantsov relation B.IX.146 and formula B.IX.147 fix both ρ and U. In particular, it can be shown that, when both the supersaturation Δ and the anisotropy ε_m are small, one has:

$$U \approx \frac{D_L}{d_o^c}\,\varepsilon_m^{7/m}\,\Delta^4 \qquad \text{(B.IX.150a)}$$

$$\rho \approx \frac{d_o^c}{\varepsilon_m^{7/m}\,\Delta^2} \qquad \text{(B.IX.150b)}$$

Finally, at high velocity (i.e., high supersaturation), the stability constant of a dendrite depends on the kinetic coefficient anisotropy [5]. In this case, the dendrites grow along the crystallographic axes of highest kinetic coefficient. The easy growth axes at low and high velocity may differ, so for a certain supersaturation dendrites tend to grow simultaneously along all directions. This competition can lead to a dense side-branching regime (see chapter C.X). Thickness effects can also be experimentally relevant, even if the sample is only a few μm thick. Much work remains to be done, both in theory and numerical simulation, to explain the entire body of experimental results.

Let us conclude by noting that the equivalent of the Ivantsov dendrite can be obtained in the kinetic regime ($\Delta > 1$). In this case, however, the dendrite is no longer parabolic, but rather triangular [14].

BIBLIOGRAPHY

[1] Mullins W.W., Sekerka R.F., *J. Appl. Phys.*, **35** (1964) 444.

[2] Langer J.S., *Rev. Mod. Phys.*, **52** (1980) 1.

[3] Pelcé P., *Théorie des Formes de Croissance: Digitations, Dendrites et Flammes*, EDP Science/CNRS Editions, Les Ulis, 2000.

[4] Kassner K., *Pattern Formation in Diffusion Limited Crystal Growth*, World Scientific, Singapore, 1996.

[5] Brener E.A., Mel'nikov V.I., *Adv. Phys.*, **40** (1991) 53.

[6] Billia B., Trivedi R., "Pattern Formation in Crystal Growth," Chapter 14 in *Handbook of Crystal Growth*, Vol. 1, Ed. Hurle D.T.J., Elsevier Science Publishers, Amsterdam, 1993.

[7] Landau L., Lifchitz E.L. *Statistical Physics*, Mir, Moscow, 1967.

[8] Sonin A.A., *The Surface Physics of Liquid Crystals*, Gordon and Breach, Luxembourg, 1995.

[9] Caroli B., Caroli C., Roulet B., "Instabilities of planar solidification fronts," in *Solids Far from Equilibrium*, Ed. Godrèche C., Cambridge University Press (Alea Saclay collection), Cambridge, 1991, p. 155.

[10] Caroli B., Caroli C., Ramirez-Piscina L., *J. Cryst. Growth*, **132** (1993) 377.

[11] Sallen L., Oswald P., Sotta P., *J. Phys. II (France)*, **7** (1997) 107.

[12] Langer J.S., Turski L.A., *Acta Metall.*, **25** (1977) 1113.

[13] Ivantsov G.P., *Dokl. Akad. Nauk. SSSR*, **58** (1947) 467.

[14] Temkin D., Géminard J.-C., Oswald P., *J. Phys. I (France)*, **4** (1994) 403.

INDEX

SUBJECT INDEX

When several page numbers are listed for an entry, the most important pages are boldface.

E

F

G

H

M

T

U

V

W

AUTHOR INDEX

Bibliography pages are marked in boldface.

R

S

X

Y

Z